High-Dimensional Knotting

An Illustrated Guide

SERIES ON KNOTS AND EVERYTHING

ISSN: 0219-9769

Editor-in-charge: Louis H. Kauffman *(Univ. of Illinois, Chicago)*

The Series on Knots and Everything: is a book series polarized around the theory of knots. Volume 1 in the series is Louis H Kauffman's Knots and Physics.

One purpose of this series is to continue the exploration of many of the themes indicated in Volume 1. These themes reach out beyond knot theory into physics, mathematics, logic, linguistics, philosophy, biology and practical experience. All of these outreaches have relations with knot theory when knot theory is regarded as a pivot or meeting place for apparently separate ideas. Knots act as such a pivotal place. We do not fully understand why this is so. The series represents stages in the exploration of this nexus.

Details of the titles in this series to date give a picture of the enterprise.

Published:

Vol. 79: *High-Dimensional Knotting: An Illustrated Guide*
by D. Roseman

Vol. 78: *Four-Dimensional Paper Constructions After Möbius, Klein and Boy*
by E. Ogasa

Vol. 77: *Quipu: Decorated Permutation Representations of Finite Groups*
by Y. Bae, J. S. Carter & B. Kim

Vol. 76: *Combinatorial Knot Theory*
by R. A. Fenn

Vol. 75: *Scientific Legacy of Professor Zbigniew Oziewicz: Selected Papers from the International Conference "Applied Category Theory Graph-Operad-Logic"*
edited by H. M. C. García, José de Jesús Cruz Guzmán,
L. H. Kauffman & H. Makaruk

Vol. 74: *Seeing Four-Dimensional Space and Beyond: Using Knots!*
by E. Ogasa

Vol. 73: *One-Cocycles and Knot Invariants*
by T. Fiedler

Vol. 72: *Laws of Form: A Fiftieth Anniversary*
edited by L. H. Kauffman, F. Cummins, R. Dible, L. Conrad, G. Elisbury,
A. Crompton & F. Grote

Vol. 71: *The Geometry of the Universe*
by C. Rourke

More information on this series can also be found at http://www.worldscientific.com/series/skae

K&E Series on Knots and Everything — Vol. 79

High-Dimensional Knotting

An Illustrated Guide

Dennis Roseman
University of Iowa, USA

World Scientific

NEW JERSEY • LONDON • SINGAPORE • BEIJING • SHANGHAI • HONG KONG • TAIPEI • CHENNAI • TOKYO

Published by

World Scientific Publishing Co. Pte. Ltd.
5 Toh Tuck Link, Singapore 596224
USA office: 27 Warren Street, Suite 401-402, Hackensack, NJ 07601
UK office: 57 Shelton Street, Covent Garden, London WC2H 9HE

Library of Congress Cataloging-in-Publication Data
Names: Roseman, Dennis, 1939– author
Title: High-dimensional knotting : an illustrated guide / Dennis Roseman, University of Iowa, USA.
Description: New Jersey : World Scientific, [2026] | Series: Series on knots and everything, 0219-9769 ; vol. 79 | Includes bibliographical references and index.
Identifiers: LCCN 2025032358 | ISBN 9789813237391 hardcover | ISBN 9789813237407 ebook for institutions | ISBN 9789813237414 ebook for individuals
Subjects: LCSH: Knot theory | Four-manifolds (Topology)
Classification: LCC QA612.2 .R67 2026
LC record available at https://lccn.loc.gov/2025032358

British Library Cataloguing-in-Publication Data
A catalogue record for this book is available from the British Library.

About the cover image: The cover graphic shows twelve stages of an isotopy produced by an energy flow (Section 11.3). Each stage is visualized as a color-gradient rendering (Section 6.4) of the projection (Chapter 6) of the stage split in half.

For any available supplementary material, please visit
https://www.worldscientific.com/worldscibooks/10.1142/10915#t=suppl

Desk Editors: Eshak Nabi Akbar Ali/Lai Fun Kwong

Typeset by Stallion Press
Email: enquiries@stallionpress.com

To Robin, for her loving support especially during the writing of this book, and to my mother Ruth Demsey Cohen, who, by her example, taught me to marvel and want to understand all that lay on life's pathway.

Preface

The study of knotting, purposeful or accidental, utilitarian or decorative, dates from prehistoric times. Sewing needles 50,000 years old have been found. Knotting was both a problem and solution to manipulation of long, thin, flexible objects. The variety and complexity of knotting prompts fascination, both aesthetically and mathematically. A circular path serves as a metaphor for spiritual binding.

Knot *theory* began in the 3rd century AD, perhaps earlier, with decorative drawings and carvings of complex tangled *circular* paths, [Jablan *et al.* (2012)]. With skill and patience, tangles in any sufficiently flexible rope can be unraveled by teasing the cord ends and slipping the "self-tangling" off the end. A circle, however, has no ends, so tangles can be trapped. Similarly, a collection of circles can be linked so there may be no way to separate.

The mathematical study of knotting began with the geometrical classification of circular paths in space. Since knotting was modeled on flexible objects, it became natural to view them as part of topology. The concept of dimension is a seemingly simple featureless abstraction. Although n-dimensional space, R^n, is now routinely introduced to scientists, engineers and mathematicians early in their education, spaces of dimension greater than three seem, for many, remote and mysterious.

Interest in higher dimensions is not entirely driven by a mathematical desire to generalize. Often, basic mathematical structure is most clearly revealed in higher dimensions. The circle, the ellipse, the parabola and the hyperbola are subsets of the plane (2-dimensional space). From antiquity, mathematicians have understood these are related as conic sections—intersections of a plane and a cone in 3-dimensional space.

We understand sets by subsets they can contain. Clearly, R^3 (space) and R^2 (plane) are different: a circular path can be knotted in R^3 while a circular path in R^2 cannot be knotted. Similarly, higher dimensional knots provide one way to understand the difference between spaces in higher dimensions.

To non-mathematicians, the mathematical use of the familiar word "dimension", frequently causes puzzlement. For example, let E denote the edges of a cube. What is the dimension of E? To a mathematician the answer is: E is composed of line segments, each line segment has only one dimension, therefore E is one-dimensional. A non-mathematician might say that a cube is three dimensional, you couldn't have the edges without the cube and so E must be three-dimensional. Someone else could bring up the fact that, using perspective drawing, the E can be drawn on a piece of paper and so, perhaps, E is two-dimensional. Mathematically, "dimension" is taken as an intrinsic property of a set, independent of how it may be represented as a subset.

As we will see circles do not knot in higher dimensional spaces as they do in three dimensions. It is a surface in four-dimensional space provides a natural generalization of "knot" in higher dimensions.

The first publication on mathematics of knots was [Vandermonde (1771)] followed a hundred years later by [Gauss (1863–1933)]. The first paper on higher dimensional knots was a paper on knotted 2-spheres in 4-dimensional space [Artin (1925)], followed by a one-page paper [Van Kampen (1928)] on linked 2-spheres.

In the 1960's, as geometric topology matured, knottings in dimensions higher than four were discovered and studied. Given the existence of higher dimensional knots, questions immediately arise. How does the nature of knotting change as we consider higher dimensions? What are some basic examples? How can we visualize such examples? This book will give some answers. Much is known, and we make no attempt to cover all aspects, but offer an introduction by concentrating on specific examples and concrete constructions, using graphical means as much as possible.

Many express the view that it is easy and natural to see in three dimensions—while four is unnatural and difficult, perhaps impossible, to understand. As illustrated by many examples in this book, three-dimensional visualization of any complex object, or group of objects can be challenging. However, the key to higher dimensional visualization is skill in three-dimensional visualization.

Dennis Roseman

About the Author

Photo shows author illustrating construction of a spun trefoil knotting.

Dennis Roseman is a Professor Emeritus at the University of Iowa. He received his bachelor's and master's degrees in mathematics from the University of Wisconsin (after a first two years at Case Institute of Technology). His Ph.D. in mathematics was from the University of Michigan (Frank Raymond, advisor). He was a mathematics faculty member at Wayne State University for one year, Indiana University for two years and the University of Iowa for 41 years.

In high school, Manning's book, *Geometry of Four Dimensions*, sparked an interest in higher dimensions. A year-long engineering drawing class at Case increased graphics and visualization skills. A class in Wisconsin with R. H. Bing, focused on concrete examples, steered that interest in a topological direction, leading to a research career in the geometric topology of four and five-dimensional space, resulting in this book.

Contents

Part 1
Introduction

The series "Knots and Everything" in which this book appears, is focused on mathematical knot theory with applications to physics, biology and chemistry. In the spirit of this series, our goal is to make this book, a sketchbook in spirit, accessible to a wide readership—not just geometric topologists and knot theorists Our primary goal is to explore some interesting examples of higher dimensional subsets. Our minimal background requirements are modest: standard courses in calculus (especially curves and surfaces in three-dimensional space), linear algebra and some introduction to topology (just basic concepts of continuity, topological spaces, and homeomorphism). So, we avoid certain important but mathematically advanced topics (function spaces, quotient spaces and abstract algebra), well covered in other texts. The introductory material may be familiar, but we need to establish some basic notations. A reader familiar with standard mathematical notations and terminology (or one eager to explore higher dimensions) is encouraged to scan the first sections, returning later to fill in details, if needed. To that end, we have provided cross-references and an extensive index to aid in such browsing. Computer graphics was done with Mathematica. Graphics derived from four-dimensional knottings were obtained from four-dimensional computational models. So, besides serving as illustrations, they are results of mathematical calculations.

Chapter 1

Basics

1.1 n-Space, Sets, Functions and Cartesian Products

We create n-dimensional spaces using two basic mathematical tools—the real numbers and the cartesian product. We explore n-dimensional spaces, especially 4-space, looking at interesting subsets called knottings.

Definition 1.1. **Real n-dimensional space**, often just called **n-dimensional space**, is denoted R^n and consists of all ordered n-tuples of real numbers $(x_1, \ldots, x_n)$. The set of all real numbers, also called **1-dimensional space**, will be denoted by R^1. An n-dimensional vector is also an n-tuple of real numbers (but viewed algebraically). If $x \in R^n$, we will switch to vector notation $\overrightarrow{x}$ if we intend to use vector addition, etc. We delete the arrow when we no longer need such algebraic interpretation.

It is technically useful to define the set consisting of the single number zero as **0-dimensional space** R^0, and the empty set as **(-1)-dimensional space**, denoted R^{-1}. We use $\mathbb{O}$ for the **origin of** R^n, so $\mathbb{O} = (0, \ldots, 0)$.

Distance in R^n is given by the generalized Pythagorean formula. For $x = (x_1, \ldots, x_n) \in R^n$ and $y = (y_1, \ldots, y_n) \in R^n$ the **distance between** x **and** y is $\|x - y\| = \sqrt{(x_1 - y_1)^2 + \cdots + (x_n - y_n)^2}$. We use notation $\|x\| = \|\overrightarrow{x} - \mathbb{O}\| = \sqrt{\sum_{i=1}^{i=n} x_i^2}$ for the **size of** x—that is, the length of vector $\overrightarrow{x}$ corresponding to the point x. ∎

For general sets $X \times Y$ is the **cartesian product** denoted $X \times Y$—the set of all ordered pairs (x, y) where $x \in X$ and $y \in Y$. For any n and m we identify $R^n \times R^m$ and R^{n+m} by identifying the pair $((x_1, \ldots x_n), (y_1, \ldots, y_m))$ with the $(n + m)$-tuple $(x_1, \ldots, x_n, y_1, \ldots, y_m)$. We write this as $R^{n+m} \approx R^n \times R^m$. Here $\approx$ means "the same" topologically.

Also see Notations 1.6 and 2.2). We refer to the first factor of $R^k \times R^{n-k}$ as **the standard** $R^k \subseteq R^n$. We write $R^n \approx R^1 \times \cdots \times R^1$.

Notation 1.1. Another common use of pair notation is for *subsets.* In general, if X is a set with $A \subseteq X$, a common notation for this is **pair notation**: (X, A). In pair notation, the standard $R^k \subseteq R^n$ is (R^n, R^k). In pair notation, if we have set Y and $B \subseteq Y$, a function $f\colon X \to Y$ with $f(A) \subseteq B$ we write this as a map of pairs $f : (X, A) \to (Y, B)$. ▌

Notation 1.2. Points $p \in R^n$ are written with subscript notation as $p = (x_1, \ldots, x_n)$. However, in low dimensions, to avoid subscripts, we often alternatively write $p \in R^2$ as $p = (x, y)$, and $p \in R^3$ as $p = (x, y, z)$.

For $p \in R^4$ we often use $p = (x, y, z, w)$ or $p = (x, y, z, t)$. When we refer to the fourth coordinate t as "time" we mean metaphorically, not literally. *In particular, all coordinates of* R^n *are equivalent, interchangeable and without a priori fixed interpretation.*

The line of points in R^n whose coordinates are all zero except for the i-th coordinate is the **i-th coordinate axis of R^n**, written $axis(x_i)$. Writing $p = (x, y, z, w) \in R^4$, we have $axis(x)$, $axis(y)$, $axis(z)$ or $axis(w)$. ▌

In ordinary discourse, a "disk" is 2-dimensional, a "ball" is 3-dimensional. However, in topology, we consider dimensions other than 2 or 3 and the terms "disk" or "ball" are commonly used interchangeably and applied to all dimensions. Thus, one sometimes refers to a "disk" as a 2-dimensional ball and a "ball" as a 3-dimensional disk. The reader should focus on the dimension involved not the disk/ball terminology choice.

Definition 1.2. If $p \in R^n$, the **open neighborhood of p in R^n** of radius ϵ is $N^n_\epsilon(p) = \{x \in R^n\colon \|x - p\| < \epsilon\}$. Other names for $N^n_\epsilon(p)$ are **open ball with center p and radius ϵ** or **open disk with center p and radius ϵ.** The **closed n-disk (or closed n-ball) with center p and radius ϵ** is $D^n_\epsilon(p) = \{x \in R^n : \|x - p\| \leq \epsilon\}$. ▌

Definition 1.3. An **open subset of R^n** is any union of open balls of R^n. If $X \subseteq R^n$ then $V \subseteq X$ is an **open subset of X** if $V = X \cap U$ for some open subset U of R^n. A subset of R^n whose complement is open is a **closed subset of R^n** and $C \subseteq X$ **is a closed subset of X** if $X - C$ is an open subset of X. ▌

Note that "U is open" and "U is closed" needs context of the containing set. For $U = \{(x, y) \in R^2 : 0 \leq x \leq 1 \text{ and } y = 0\}$, we have $U \subseteq R^1 \subseteq R^2$. Here, U is an open subset of R^1 but not an open subset of R^2.

Topology is often defined abstractly as "a set together with certain subsets". We will not do this. **In this book, all subsets, unless explicitly stated otherwise, will be subsets of some real n-dimensional space R^n and thus have the topology inherited from the notion of distance between points R^n.** This will keep mathematical requirements minimal and examples concrete.

The notions of continuity and differentiability are familiar from multivariable calculus: a function $F\colon R^n \to R^m$ can be expressed as m real-valued functions each of n-variables:

$$F(x_1, \ldots, x_n) = (f_1(x_1, \ldots x_n), \ldots, f_m(x_1, \ldots x_n))$$

A common example of such a function is a three-dimensional vector field for R^3 which can be expressed as $F(x, y, z) = (f_1(x, y, z), f_2(x, y, z), f_3(x, y, z))$.

Definition 1.4. F **is continuous** if each f_i is continuous. A function $f\colon X \to Y$ is **continuous** from $X \subseteq R^n$ to $Y \subseteq R^m$ if it is a restriction of a continuous function. ∎

Definition 1.5. F **is differentiable** if all partial $\frac{\partial f_i}{\partial x_j}$ derivatives exist and are continuous. ∎

Geometrically $F : R^1 \to R^3$ has continuous first derivatives if it has a continuously varying tangent line. In the case $F : R^2 \to R^3$ we have a continuously varying tangent plane. An oval in the plane made of two semicircles joined by two line segments has a tangent line (first derivative) at each point, but fails to have a second derivative at the join points since there is no well-defined radius of curvature.

Definition 1.6. ? F **is smooth** if all partial derivatives of all orders (including all mixed partials of all orders) exist and are continuous. A function $f\colon X \to Y$ is **smooth** from $X \subseteq R^n$ to $Y \subseteq R^m$ if it is a restriction of a smooth function. ∎

Often, a smooth function is called a C^∞**-function** since it is differentiable infinitely many times.

Definition 1.7. A subset $A \subseteq R^n$ is **bounded** if it is contained in some open n-ball. If A is a bounded closed subset, we say A **is compact**. ∎

Closed n-disks $D^n_\epsilon(p)$ are compact, any finite subset is compact, but R^n and open disks $N^n_\epsilon(p)$ are not compact. Compact subsets have important properties. The most familiar is the maximum/minimum theorem of multivariable calculus:

Prop 1.1. If $f\colon A \to R^1$ is continuous where $A \subseteq R^n$ is compact, then there exists a **maximum point**—that is, a point $a \in A$ with $f(x) \leq f(a)$ for all $x \in A$. Also, there exists a **minimum point** $b \in A$ with $f(b) \leq f(x)$ for all $x \in A$. ▮

We also will use these standard objects:

Definition 1.8. The subset $D^n = \{x \in R^n : |x| \leq 1\}$ is the **standard closed n-disk** in R^n or the **standard closed n-ball in R^n**. The subset $\overset{\circ}{D}{}^n = \{x \in R^n : |x| < 1\}$ is the **standard open n-disk** in R^n or the **standard open n-ball in R^n**. Additionally, $R^n - \overset{\circ}{D}{}^n$ is the **standard punctured R^n**. ▮

The 1-dimensional disk $D^1 \subseteq R^1$ is the closed interval: $D^1 = \{x \in R^1\colon -1 \leq x \leq 1\}$. The standard punctured R^1 is $\{x \in R^1 : x \leq -1 \text{ or } 1 \leq x\}$. We will also use the following:

Definition 1.9. The standard **n-sphere in R^{n+1}** is defined:

$$S^n = \{x \in R^{n+1}\colon \|x\| = 1\}.$$

For $n \neq 0$ the point $(0, \ldots, 0, 1) \in S^n$ will be denoted by N_0, and sometimes called the **north pole of S^n**. Similarly $(0, \ldots, 0, -1) \in S^n$ denoted S_0 is called the **south pole of S^n**.

The **standard n-dimensional upper hemisphere** H^n_+ is the subset of S^n with $0 \leq x_{n+1}$, and **standard n-dimensional lower hemisphere** H^n_- is the subset of S^n with $x_{n+1} \leq 0$

A subset $S^n = \{x \in R^{n+1}\colon \|x - a\| = r\}$ is **geometric n-sphere** (with center a and radius r). Suppose S is a geometric n-sphere and H_n is an n-dimensional hyperplane such that $H \cap S$ is an $(n-1)$-sphere. Then the closure of either component of $S - H$ will be called a **spherical cap.** So, a hemisphere is a special case of a spherical cap. ▮

If $n + 1 \leq k$, we write $R^k = R^{n+1} \times R^{k-n-1}$ Then the **standard n-sphere in R^k** is defined to be $S^n \times \{\mathbb{O}\}$. Just as we have defined the standard subsets $R^n \subseteq R^q$ using the first n coordinates we define:

Definition 1.10. If $n \leq q$, the **standard** $D^n \subseteq D^q$ is $D^n \times \{\mathbb{O}\} \subseteq D^n \times D^{q-n} \approx D^q$. The **standard** $S^n \subseteq S^q$ is $S^n \times \{\mathbb{O}\} \subseteq D^{n+1} \times D^{q-n} \approx D^{q+1}$. In pair notation, these are (D^q, D^n) and (S^q, S^n) and called the **standard disk pair (standard ball pair)**, or **standard sphere pair.** ▮

In non-subscript notations, the **standard 1-sphere in** R^2 is the unit circle in R^2, $x^2+y^2=1$. The **standard** S^1 **in** R^3 is $\{(x,y,z)\in R^3\colon x^2+y^2=1 \text{ and } z=0\}$. The **standard** S^2 **in** S^3 $\{(x,y,z)\in R^3\colon x^2+y^2+z^2=1\}$. The **standard** S^2 **in** S^4 $\{(x,y,z,w)\in R^4\colon x^2+y^2+z^2=1 \text{ and } w=0\}$. The standard **3-sphere in** R^4 is $\{(x,y,z,w)\in R^4\colon x^2+y^2+z^2+w^2=1\}$.

For the reader not used to higher dimensions, here are the details for the standard circle S^1 in the 3-sphere S^3. The 3-sphere is the unit sphere in R^4—so $S^3=\{(x,y,z,w)\in R^4\colon x^2+y^2+z^2+w^2=1\}$. Then $R^2\subseteq R^4$ and $S^1\subseteq R^2$ is the unit circle. The standard circle in S^3, $S^1\times\{\mathbb{O}\}$, corresponds to $\{(x,y,z,w)\in R^4\colon x^2+y^2=1, z=0 \text{ and } w=0\}$.

Remark 1.1. We define $D^0=R^0$ so that D^0 is a one-point set consisting of the number 0. Also, S^0 is the two-point subset of R^1 which consists of the numbers +1 and -1. ▮

A torus is an important set. We introduce it in two ways. First, we use parametric equations. Then, in Example 1.2, we view the torus as a subset of R^4.

Example 1.1. A torus is often described as "a tube of radius r about a standard circle of radius R in R^3". In parametric equations:

$$\pi(u,v)=((R+r\cos(u))\cos(v),(R+r\cos(u))\sin(v),r\sin(v))$$

where $0\le u\le 2\pi$ and $0\le v\le 2\pi$. Points on the u-axis and v-axis are sent to the two intersecting circles shown in Figure 1.1. Other useful subsets of the torus can be described using these equations. If $v=mu+b$ is the equation of a line $L\subseteq R^2$, then the image $L'=\pi(L)$ is a curve with parametric equations

$$\pi(u,v)=((R+r\cos(u))\cos(mu+b),(R+r\cos(u))\sin(mu+b),\\ r\sin(mu+b)).$$

A torus described with the above parametric equations can be useful, however, there is no *standard* parametric equation; we get different subsets depending on our choice of $r>0$ and $R>0$. We will choose other subsets to be "standard torii" in R^3, see Definition 1.42.

If L is the line $u=v$ then L' will correspond to the closed path on $S^1\times S^1$, Figure 1.7. If L consists of the two parallel lines $u=v$ and $u=v+1/2$ then L' will be two disjoint circles (called a **Hopf linking**) as in Figure 1.17. For L with equation $u=3/2v$, L' is a knotting of a circle called a **trefoil knotting**. For L with equation $u=-3/2v$, L' is a mirror image of this knotting Figures 1.2 and 1.3. Lines with rational slope will always be sent to a circle. The image of a line with irrational slope (for example, $v = u/\sqrt{2}$) will be sent to a curve which winds infinitely many times, as indicated in Figures 1.4 and 1.5. Such a curve is called an **irrational flow line**. In this case, although $\pi|_L$ is continuous and one-to-one, it is not an embedding. ▮

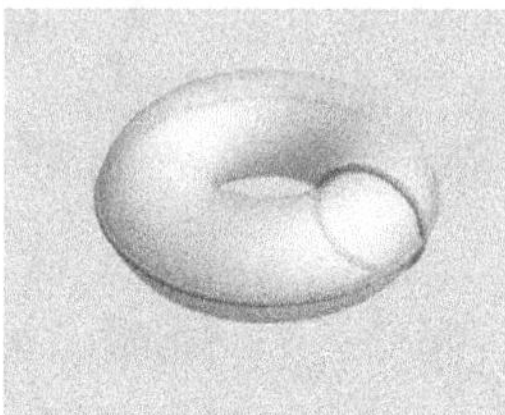

Fig. 1.1 Image under π of u and v axes.

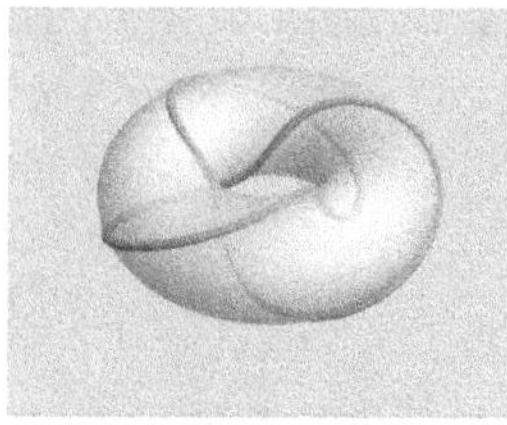

Fig. 1.2 L' for line $v = 3/2u$.

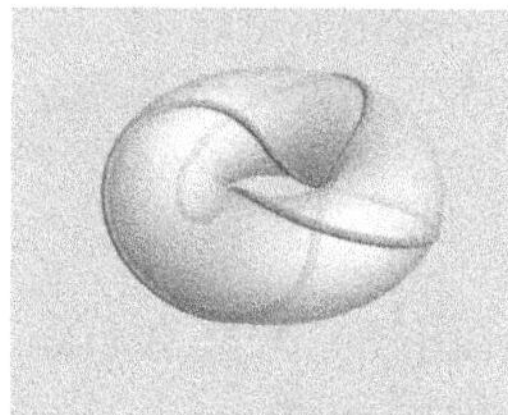

Fig. 1.3 L' for line $v = -3/2u$.

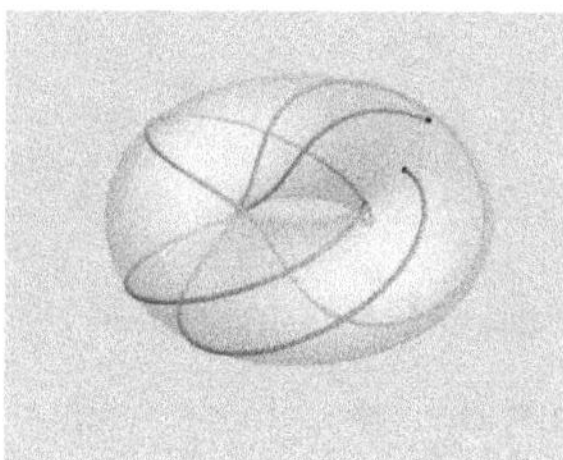

Fig. 1.4 Portion of L' for line $v = u/\sqrt{2}$.

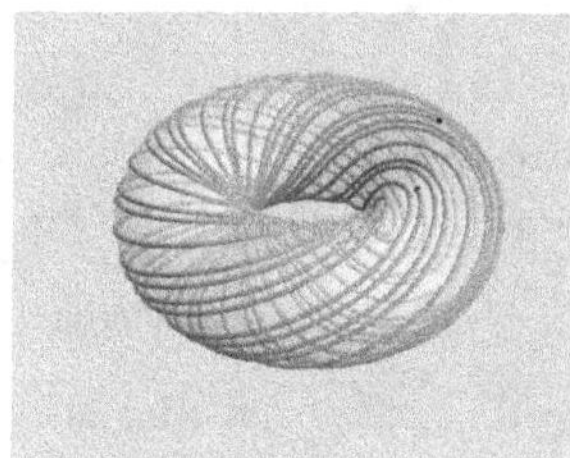

Fig. 1.5 larger portion of L' for line $v = u/\sqrt{2}$.

Basic to all mathematics we have:

Notation 1.3. A **set** is a collection of objects called **elements**. If S is a set, we use the notation $a \in S$ to mean a is an **element of** S.

If S is set and $a_1, \ldots, a_n$ is a finite collection of points of S then the **subset** of S of all these points is denoted $\{a_1, \ldots, a_n\}$. In particular, if $a \in S$ the subset consisting of the single point a is denoted $\{a\}$. So, $\{a\} \subseteq S$ and $a \in S$ are notationally correct, and $a \subseteq S$ incorrect.

The notation $F: X \to Y$ denotes F is a **function** from set X to set Y. There is no general expectation that F is an onto map. ▮

Notation 1.4. If X, Y and Z are sets, $A \subseteq X$, $B \subseteq Y$, $a \in X$, and $f: X \to Y$, $g: Y \to Z$ are functions:

(1) $f|_A$ is a **restriction of f to A.**
(2) $f(A)$ is the **image of A under f.**
(3) $id_X\colon X \to X$, the **identity map of X** is defined as $id_X(x) = x$.
(4) If there is a point $y_0 \in Y$ and $f(x) = y_0$ for all $x \in X$ we say **f is a constant map.**
(5) $f \circ g$, denotes **composition of functions** $f \circ g = f(g(x))$.
(6) The map $\pi_X\colon X \times Y \to X$ defined $\pi_X(x, y) = x$ is called **cartesian projection on the first factor.** The map $\pi_Y\colon X \times Y \to Y$ defined $\pi_Y(x, y) = y$ is called **cartesian projection on the second factor.**
(7) For $y_0 \in Y$ the map $i_{y_0}\colon X \to X \times Y$ defined $i_X(x) = (x, y_0)$ is called **a cartesian injection of the first factor.** For $x_0 \in Y$ the map $i_{x_0}\colon Y \to X \times Y$ defined $i_X(x) = (x_0, y)$ is called **a cartesian injection of the second factor.**
(8) If $A \subseteq X$ this will sometimes be notated as an ordered pair of spaces: (X, A). This is called **space pair notation.**
(9) Using the space pair notation we write $(X, A) \times Y = (X \times Y, A \times Y)$.
(10) More generally, write $(X, A) \times (Y, B) = (X \times Y, A \times B)$.
(11) If X is a set the **diagonal of** $X \times X$ is $\triangle(X) = \{(x, x) \in X \times X\}$. ∎

Example 1.2. For the two-point set $S^0 = \{-1, +1\} \in R^1$, the product $S^0 \times S^0 \subseteq R^2$ is the four-point set $\{(-1, -1), (-1, +1), (+1, -1), (+1, +1)\}$ and we can write: $(R^1, S^0) \times (R^1, S^0) = (R^1 \times R^1, S^0 \times S^0)$.

These four points are corners of a unit square in R^2 and so lie on a circle with radius $\sqrt{2}$. There are mathematical reasons to sometimes prefer these points to be contained in $S^1 \subseteq R^2$. We can think of the two-point set $S = \{-\frac{\sqrt{2}}{2}, \frac{\sqrt{2}}{2}\}$ as a 0-sphere of radius $\sqrt{2}/2$ and now $S \times S = \{(-\frac{\sqrt{2}}{2}, -\frac{\sqrt{2}}{2}), (-\frac{\sqrt{2}}{2}, \frac{\sqrt{2}}{2}), (\frac{\sqrt{2}}{2}, -\frac{\sqrt{2}}{2}), (\frac{\sqrt{2}}{2}, \frac{\sqrt{2}}{2})\}$ will be four points on the unit circle. The set $S \times S$ is a lower-dimensional version of the Clifford torus, Example 1.3.

If (R^2, S^1) is the standard circle in the plane then $(R^2, S^1) \times R^1 = (R^2 \times R^1, S^1 \times R^1)$ is an infinite cylinder in $R^3 \approx R^2 \times R^1$. ∎

Example 1.3. The torus, Example 1.1, can naturally be viewed as a subset of R^4. Start with the standard circle $S^1 \subseteq R^2$. The torus can be defined as $S^1 \times S^1$ naturally viewed as a subset of $R^4 \approx R^2 \times R^2$ since $(R^2, S^1) \times (R^2, S^1) = (R^2 \times R^2, S^1 \times S^1)$.

Note that if $(x, y, z, w) \in S^1 \times S^1$ then we must have $x^2 + y^2 = 1$ and $z^2 + w^2 = 1$, so $x^2 + y^2 + z^2 + w^2 = 2$.

It follows that this subset is a subset of a 3-sphere of R^4 of radius $\sqrt{2}$ and center $\mathbb{O}$. However we can tweak the definition slightly and get a subset of the standard $S^3 \subseteq R^4$. Let S be a circle in R^2 with radius $\sqrt{2}$ and center $\mathbb{O}$—that is $x^2 + y^2 = 1/2$. Then all points of $S \times S$ will have $x^2 + y^2 + z^2 + w^2 = 1$ and will be points of $S^3 \subseteq R^4$. This subset is called the Clifford torus and has interesting features that are not within the scope of this text. However, we will encounter this subset again as a "middle slice of S^3" in Section 1.8. ▮

Definition 1.11. The **unit interval** I is defined: $I = \{x \in R^1 : 0 \leq x \leq 1\}$. For $t \in I$ and any set X the subset $(X \times \{t\}) \subset (X \times I)$ is called the **level** t **set of** $X \times I$. ▮

Example 1.4. If I is the unit interval, $I \times I$ is a square and the diagonal $\triangle(I)$ is literally a diagonal of the square, Figure 1.6. More interesting is the diagonal of S^1. Here, $S^1 \times S^1$ is a torus. If we view it as a subset of R^3, $\triangle(S^1)$ corresponds to a circle shown in Figure 1.7, see Example 1.1. For some significance of the diagonal see Remark 2.4. ▮

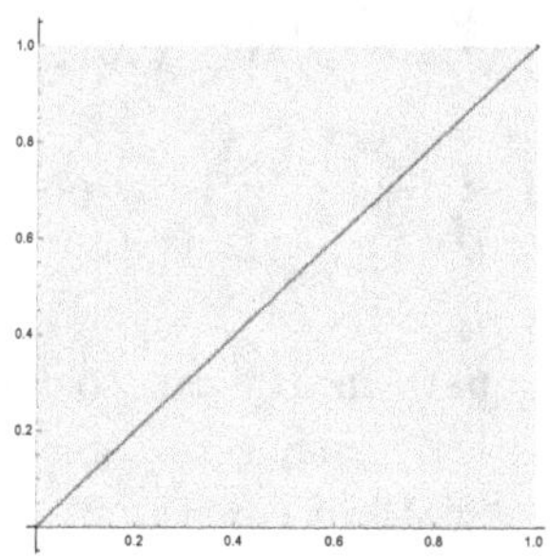

Fig. 1.6 The diagonal of $I \times I$.

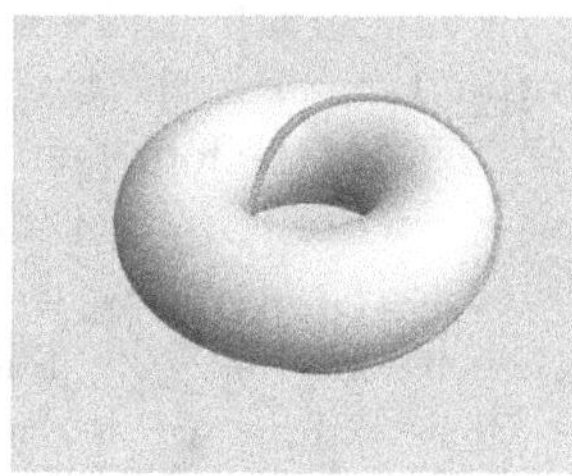

Fig. 1.7 The diagonal of $S^1 \times S^1$.

Cartesian injection in one factor is an inverse of projection that factor in the sense that for $y_0 \in Y$, $\pi_X \circ i_{y_0} = id_X$. In one way or another, Cartesian product and Cartesian projection, sometimes used locally, is the basis of most higher dimensional constructions and analysis in knot theory.

Definition 1.12. If $F \colon X \to Y$ the set

$$\Sigma(F) = \{y \in Y \colon F^{-1}(y) \text{ consists of more than 1 point}\}$$

is called the **self-intersection set of** F. ▮

Definition 1.13. If I is the unit interval and $F \colon X \times I \to Y \times I$ is a function, F is **level preserving** if, for all $t \in I$, $F(X \times \{t\}) \subseteq Y \times \{t\}$. ▮

Notation 1.5. If $F\colon X\times I\to Y\times I$ is level preserving then for $t\in I$ we let $F_t = F|_{X\times\{t\}}$—F_t is sometimes called the **level map of** F **at level** t . We often use F_t as a notation in place of F. Also, we identify X with $X\times\{t\}$ and Y with $Y\times\{t\}$ and thus view F_t as a map $F_t : X\to Y$. ▮

Definition 1.14. For a level preserving map $F\colon X\times I\to Y\times I$, the **reverse of** F_t is the level preserving map $F(x,1-t)$, which in other notation is F_{1-t}. If $G\colon X\times I\to Y\times I$ is another level preserving map with $f_1 = g_0$ the **concatenation** of F and G denoted $F*G$ is the level preserving map defined $F*G_t = F_{2t}$ if $0\le t\le 1/2$ and $(F*G)_t = G_{2t-1}$ if $1/2\le t\le 1$. ▮

If we view t as "time" then the reverse reverses time. Concatenation does F_t twice as fast, followed by G_t, also twice as fast.

The graph of a function $f\colon X\to Y$ is more than just a visualization aid. We use it to define certain subsets of $X\times Y$.

Definition 1.15. The **graph of** $f\colon X\to Y$ is the subset $\Gamma(f) = \{(x,y)\in X\times Y\colon y = f(x)\}$. ▮

For example, for any set X, the diagonal of X is the graph of the identity function of X.

1.2 Additional Topology Background

In some ways, the cube, Figure 1.8, seems easier to visualize than the ball because the cube's vertices and edges provide distinctive features that can be used for reference, while a ball is relatively featureless. In popular books on four dimensions, the **hypercube** Q^4, also called a **tesseract** is almost always used as an example, while the 4-ball is rarely mentioned.

Definition 1.16. The **standard** n**-cube** Q^n **in** R^n is $Q^n = \{(x_1,\ldots,x_n)\colon -1\le x_i\le 1 \text{ for } i=1,\ldots,n\}$. Similarly, the **unit** n**-cube** is $I^n = \{(x_1,\ldots,x_n)\colon 0\le x_i\le 1 \text{ for } i=1,\ldots,n\}$. ▮

Also, topologically an n-ball, is a standard n-simplex, Figure 1.9:

Definition 1.17. The **standard** n**-simplex** is $\Delta^n = \{(x_1,\ldots,x_{n+1})\in R^{n+1}\colon x_1+\cdots+x_{n+1} = 1 \text{ and } 0\le x_i \text{ for all } 1\le i\le (n+1)\}$.

The j**-th face** of the standard n-simplex is the subset where $x_j = 0$. ▮

Some useful constructions in R^n are the cone and the suspension.

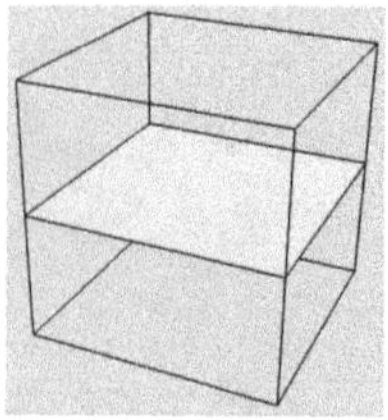

Fig. 1.8 The standard 2-cube as a subset of standard 3-cube.

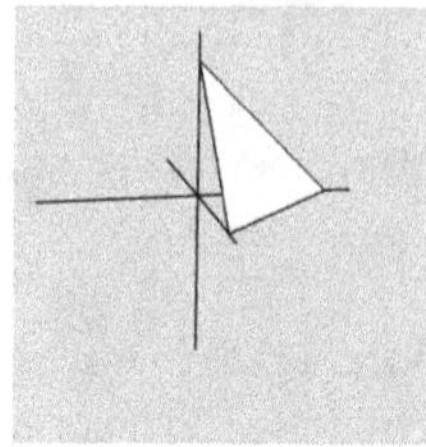

Fig. 1.9 The standard 2-simplex in R^3.

Definition 1.18. Suppose $A \subseteq R^{n-1}$:

(1) The **cone on A from P** $= (0, \ldots 0, p) \in R^n$, where $p \neq 0$, denoted $P * A$ is the union of all line segments N_a from P to $\{a\} \times \{0\}$ where $a \in A$.
(2) The **suspension of A from P and Q** denoted ΣA: $(P * A) \cup (Q * A)$ where $P = (0, \ldots 0, p)$ and $Q = (0, \ldots 0, -p)$, where $p \neq 0$. ▮

For the unit circle $S^1 \subseteq R^2$, the cone on S^1 is a geometric cone, Figure 1.10. The suspension of S^1 is topologically a 2-sphere, Figure 1.11. Also the standard simplex is a cone—the standard 2-simplex in R^3 is the cone from $(0, 0, 1)$ on the standard 1-simplex in R^2.

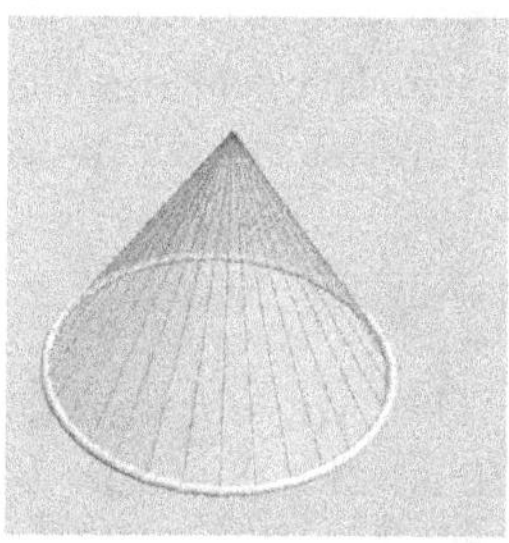

Fig. 1.10 The cone on $S^1 \subseteq R^2 \subseteq R^3$ from $(0, 0, 1)$.

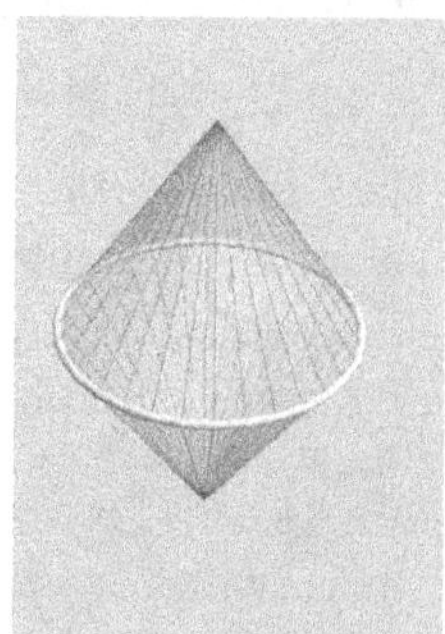

Fig. 1.11 The suspension of $S^1 \subseteq R^2 \subseteq R^3$ from $(0, 0, 1)$ and $(0, 0, -1)$.

Connectedness is a fundamental topological concept, however, for our examples, we mostly use *path connectedness*:

Definition 1.19. If $X \subseteq R^n$, X **is connected** if it cannot be expressed as a union of two disjoint non-empty open subsets. A **component, C, of**

X is a maximal connected subset of X. That is, $C \subseteq X$ is connected and no larger subset of X is connected.

A **path in** X is a continuous function $f\colon I \to X$. We say that f is a **path from** $f(0)$ **to** $f(1)$. We say X is **path connected** if, for any two points a and b of X, there is a path from a to b. A subset $C \subseteq X$ is a **path component of** X is a maximal path connected subset. That is, X is path connected and no larger subset of X is path connected. ▮

Many topology textbooks give examples of connected sets which are not path connected. In this book we do not encounter these, Proposition 2.3.

The number of components of a set is important. Homeomorphic sets (Definition 1.21) must have the same number of components.

The following definition is fundamental for geometric topology.

Definition 1.20. A subset $X \subseteq R^n$ is **simply connected** if for any continuous $f\colon S^1 \to X$ there is a continuous $F\colon D^2 \to X$ with $F|_{S^1} = f$. ▮

The unit circle S^1 is not simply connected. The proof, standard in many introductory topology texts, is elementary but non-trivial.

Any R^n is simply connected. Using polar coordinates, $S^1 \subset R^2$ is set of points $(1, \theta)$ and D^2 is set of points (r, θ) with $0 \leq r \leq 1$. Express points of R^n as n-dimensional vectors. Then, given any continuous $f\colon S^1 \to R^n$ we define $F\colon D^2 \to R^n$ by $F(r, \theta) = r\overrightarrow{f(1, \theta)}$. In other words, F maps the radius in D^2 between $(0, 0)$ and $(1, \theta)$ to the line segment in R^n between $\mathbb{O}$ and $f(1, \theta)$. Most uses of simple connectedness involve topics of algebraic topology. We will not go into this important concept further, only mention that in algebraic topology simple connectedness (for path connected spaces) is identical to a concept of " a trivial fundamental group".

1.3 Homeomorphisms, Embeddings, Knottings, Linkings, Immersions and Coverings

To avoid undue repetition we remind the reader that, in this book all sets we consider are subsets of some R^n.

Definition 1.21. A continuous function $f\colon X \to Y$ which is a 1-1 correspondence with continuous inverse is a **homeomorphism**. If there exists such a homeomorphism, we say X **and** Y **are homeomorphic.** ▮

For example, $f(x) = x^3$ is homeomorphism of R^1 (with inverse $x^{1/3}$).

If J is the half-open interval $[0, 2\pi)$, the map $g\colon J \to S^1$ defined $g(\theta) = (1, \theta)$ is an example of function that is a 1-1 correspondence but the inverse g^{-1} is not continuous. Roughly the reason is that $g^{-1}((1,0)) = 0$. However a point $p = (1, 2\pi - 1/100)$ is close to $(1,0)$. If g^{-1} were continuous $g^{-1}(p)$ would be close to 0. But $g^{-1}(p) = 2\pi - 1/100$, which is not close to 0.

Homeomorphism is the basic equivalence relation of topology.

Notation 1.6. If X and Y are homeomorphic, we write $X \approx Y$. In Notation 2.2 will extend this notation to *smooth* homeomorphisms (diffeomorphisms) where X and Y are smooth manifolds. ∎

For the standard 2-square we have $Q^2 = D^1 \times D^1 \subset R^1 \times R^1 \approx R^2$. Here, $D^1 \times D^1$ is *homeomorphic* to (but not equal to) the 2-disk D^2. The solid cylinder $D^2 \times D^1$ is homeomorphic to the 3-ball D^3. Also, for any n and k, $D^n \times D^k \approx D^{n+k}$ and $D^n \approx Q^n$.

Basic to the study of knotting is the related concept of an embedding—a homeomorphism from a set to its image $f(X)$.

Definition 1.22. Suppose $f\colon X \to Y$. We say f **is an embedding** if f is continuous, 1-1 and the inverse map $f^{-1}\colon f(X) \to X$ is also continuous. A subset $K \subseteq Y$ is also called a knotting: we say K **is a knotting in** Y. If there is an embedding $f\colon X \to Y$ with $K = f(X)$, we say K is a **knotting of** X **in** Y. ∎

Often in the literature, one considers a knotting to be an embedding (as opposed to the image of an embedding). This is good if one plans to present certain proofs which rely on results in an area known as function spaces. However, to keep our presentation simple, we will not go into such details.

The last part of Definition 1.22 is useful when we compare knottings and speak of two "knottings *of*" a particular set X. If X is a knotting of Y we often only need that an embedding $f : X \to Y$ exists and have no further need of f. Suppose we have two knottings $f\colon X \to Y$ and $g\colon X \to Y$ where $f(X) = g(X)$ and we want to investigate the relation between these two functions. We consider $h = g^{-1} \circ f\colon X \to X$. Now h is a homeomorphism of X to itself. In other words, if two knottings X have the same image $K \subseteq Y$, then they differ by a self-homeomorphism of X.

We will focus in this book on n-knottings, to be defined, Definition 3.4. But for our first two chapters we only need 1-knottings. A **1-knotting** K in R^3 (or S^3) is the image of a knotting of a compact closed 1-manifold (a circle or disjoint union of circles).

In this terminology, Figures 1.12 and 1.13 are "two 1-knottings of the standard circle S^1 in R^3". Details on knotting and the related term "knot" are found in Section 3.2. We might describe an embedding of S^1 for Figure 1.12 as $f(x,y) = (x,y,0)$. But we might also use $g(x,y) = (x,-y,0)$. In this case $h(x,y) = (x,-y)$ is our homeomorphism of S^1.

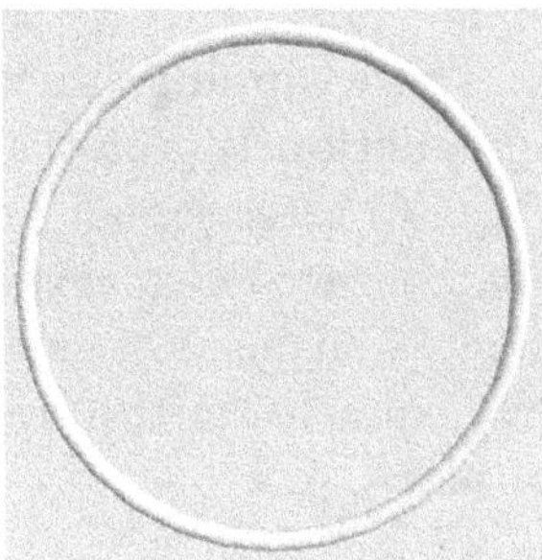

Fig. 1.12 A geometric circle $S^1 \subseteq R^3$.

Fig. 1.13 A knotting of a S^1 in R^3.

An embedding is an *into* homeomorphism that is *not* required to be an *onto* map. Technically speaking, every homeomorphism is an embedding, but, in general usage, an embedding that is a homeomorphism tends to be an exceptional situation.

The function]7]$g(x) = \arctan x$ *is* an embedding of R^1 whose image is the open interval $(-\pi/2, \pi/2)$. So, $\arctan x$ is continuous and has a continuous inverse on $(-\pi/2, \pi/2)$—namely $\tan(x)$. Here, $g(x)$ is *not* a homeomorphism of R^1 since it is not an onto map.

The unit interval is so central to topology we use a special term:

Definition 1.23. An **arc** is a set homeomorphic to the unit interval $I = [0,1] \subseteq R^1$. ▮

In other words, an arc is an image of an embedding of I (or equivalently of D^1). An arc of a circle is an *example* of an arc, but so is a straight-line segment. The graph of any continuous function $y = f(x)$ for $a \leq x \leq b$ is a subset of R^2 is also an arc. When we refer to an arc of a circle, we will generally use the term "**circular arc**".

A map that is a *locally* embedding is called an immersion. The theory of immersions is extensive and of interest in itself.

Definition 1.24. An **immersion** (or **topological immersion**) $f\colon X \to Y$ is a continuous function which is a local embedding. That is, for all

$x \in X$, there is an open set with $x \in U \subseteq X$ such that $f|_U : U \to Y$ is an embedding. We use the notation $f : X \looparrowright Y$ ∎

Example 1.5. The function $f(t) = (t(t^2 - 1), t^2)$ is an immersion $R^1 \looparrowright R^2$, Figure 1.14. This curve is 1-1 except that $f(-1) = f(1)$.

The wrapping function is a standard map which is an immersion of R^1 in $S^1 \subseteq R^2$. There are two common versions, w and W. Write points of $S^1 \subset R^2$ as $(1, \theta)$ using polar coordinates. The **wrapping function** w is defined $w(s) = (1, s)$ while the **wrapping function** $W(s)$ is defined $W(s) = (1, (2\pi)s)$. The distinction is that $w^{-1}((1, 0))$ is the set of all even multiples of 2π while $W^{-1}((1, 0))$ is the set of all integers. If we view R^2 as complex numbers, then S^1 corresponds to complex numbers of modulus 1. So, we can write $w(s) = \exp(2\pi sI)$ and $W = \exp(sI)$. For this reason, some call the wrapping functions **exponential maps.** ∎

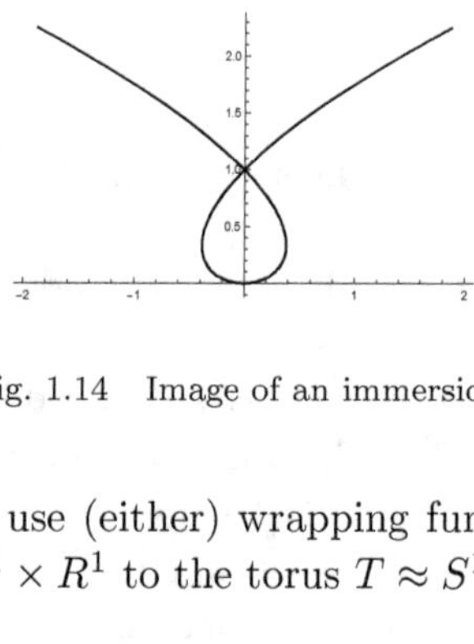

Fig. 1.14 Image of an immersion.

Example 1.6. We can use (either) wrapping function to obtain a useful immersion π of $R^2 \approx R^1 \times R^1$ to the torus $T \approx S^1 \times S^1$, namely $\pi(x, y) = (W(z), W(y))$. ∎

Example 1.6 is an example of an important special type of immersion with a uniformity condition called a **covering.**

Definition 1.25. Suppose E and B are path connected and $p\colon E \to B$ is an onto continuous map so that for every $x \in B$ there is an open subset $x \in U \subseteq B$ such that if $\widetilde{U} = p^{-1}(U)$ then, for every component C of $\widetilde{U}$, the restriction $p|_C$ is an onto homeomorphism of C onto U. Then we say that p **is a covering map** and that E is a **total space** for **base space** B. ∎

Example 1.7. For any positive integer n we define a covering $p_n : S^1 \to S^1$ by $p_n((1, \theta) = (1, n\theta)$. Here p_n is a covering map where for any $x \in S^1$, $p_n^{-1}(x)$ consists of n points. If we view S^1 as the unit modulus complex numbers, then p_n corresponds to the complex function z^n. ∎

Example 1.8. Projection on the y-coordinate maps a circle in R^2 to a line segment in R^1. This fails to be an immersion at two points p and q where the tangent line is vertical. However, it *is* an immersion at other points of the circle. For example, at $r = (1,0)$ or $r' = (-1,0)$ which are both sent to $r_* = 0$, Figure 1.15. Similarly, Figure 1.16 shows that the planar projection a sphere in R^3 to disk fails to be an immersion exactly at points of the equator. ∎

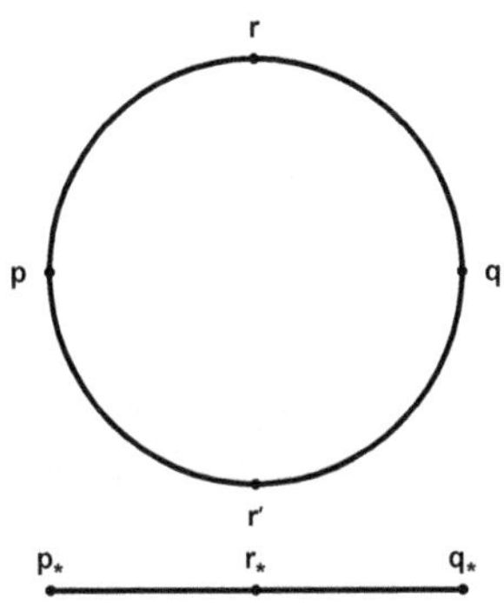

Fig. 1.15 Projection of circle to line is not an immersion at p and q.

Fig. 1.16 Projection of sphere to a disk is not an immersion at equatorial points.

In this book, when we speak of a knotting of A there is no requirement that the set A is connected. When we want to emphasize that A is not connected, we say A is "linking".

Definition 1.26. If $A = B \cup C$ is a set where B and C are components (or unions of components) of A and $f\colon A \to X$ is an embedding of A we say **the knotting $K = F(A)$ is a linking of $F(A)$ and $F(B)$.** ∎

Example 1.9. So, knotting shown in Figure 1.17, called a **Hopf linking**, is a linking of the two circles. The knotting of Figure 1.18 can be viewed as a linking of four circles. ∎

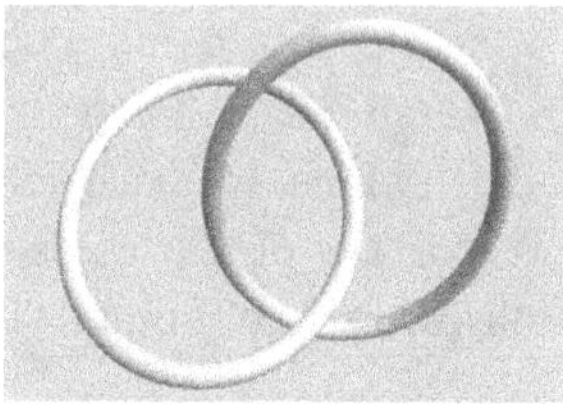

Fig. 1.17 The Hopf linking.

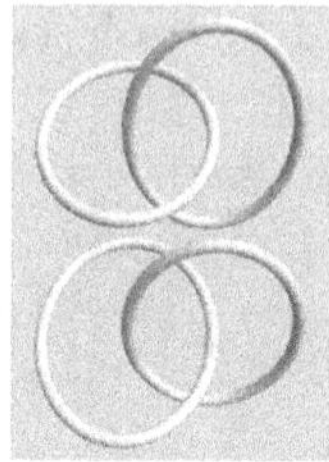

Fig. 1.18 A linking of four circles.

1.4 Stereographic Mapping

Stereographic mapping (also called stereographic projection) is a fundamental homeomorphism from the standard n-sphere S^n with a point removed (a subset of R^{n+1}) and R^n (as the standard $R^n \subseteq R^{n+1}$). Often it is easier to visualize subsets of R^n compared to subsets of S^n, even though often corresponding subsets of S^n may be mathematically preferred. Stereographic mapping relates these subsets. See Figure 1.19 for the case of a 2-sphere minus a point.

Prop 1.2. Let $S^n \subseteq R^{n+1}$ be the standard n-sphere and let $S_0 \in S^n$ be the south pole $(0, 0, \ldots, 0, -1)$. For $P \in S^n - \{S_0\}$ let l_P be the line in $\mathbf{R^{n+1}}$ containing P and S_0. Considering $R^n \subseteq R^{n+1}$ the intersection $Q = l_P \cap R^n$ will be a single point. This defines a function notated as $\mathcal{P}(P) = Q$ called the **stereographic mapping** which explicitly is given:

$$\mathcal{P}\big((x_1, x_2, \ldots, x_n, x_{n+1})\big) = \left(\frac{x_1}{x_{n+1}+1}, \frac{x_2}{x_{n+1}+1}, \ldots, \frac{x_n}{x_{n+1}+1}, \ 0\right). \blacksquare$$

In this book, projection is fundamental and refers to a map that involves going from a higher-dimensional object to a lower-dimensional object. The term stereographic "projection" is in common use, but we prefer stereographic "mapping" since it does not involve any dimension change. The map $\mathcal{P}\colon (S^n - \{S_0\}) \to \mathbf{R^n}$ is a smooth homeomorphism. Points of the lower hemisphere are sent by $\mathcal{P}$ to points *outside* the standard $S^{n-1} \subseteq R^n \subseteq R^{n+1}$, points of the upper hemisphere sent *inside*. The standard $S^{n-1} \subseteq S^n$ is sent to $S^{n-1} \subseteq R^n$, Figures 1.19 and 1.20.

The **inverse of stereographic mapping** is a smooth embedding $\mathcal{P}^{-1} : R^n \to S^n$. If $x \in R^n$ we write $\widehat{x} = \mathcal{P}^{-1}(x)$. Similarly, if $A \subseteq R^n$ we write $\widehat{A} = \mathcal{P}^{-1}(A)$. One can verify that

$$\mathcal{P}^{-1}((x_1, x_2, \ldots, x_n)) = \left(\frac{2x_1}{1+\Sigma x_i^2}, \frac{2x_2}{1+\Sigma x_i^2}, \ldots, \frac{2x_n}{1+\Sigma x_i^2}, \frac{1-\Sigma x_i^2}{1+\Sigma x_i^2} \right).$$

Stereographic mapping is **conformal** (i. e. angles between intersecting curves are preserved). For $\mathcal{P}\colon (S^2 - \{S_0\}) \to \mathbf{R^2}$, circles on S^2 correspond to circles correspond to circles in R^2, with the exception that circles on S^2 containing the south pole, S_0 correspond to lines in the plane. If circles C_1 and C_2 in $S^2 - \{S_0\}$ meet at a point p forming angle α, then circles $\mathcal{P}(C_1)$ and $\mathcal{P}(C_2)$ will meet at $\mathcal{P}(p)$ forming an angle of α. In three dimensions 2-spheres in S^3 correspond to 2-spheres in R^3, with the exception that spheres containing the south pole correspond to planes in R^3.

Remark 1.2. Let $C \subseteq R^2$ be a geometric circle in the plane bounding geometric disk D and let $c_0 \in \partial C$, Figure 1.22 with D opaque dark-grey. As an exercise in using stereographic mapping (and showing how subsets of R^2 can be related using R^3) we show that $(D - \{c_0\}) \approx R^2_+$.

We translate D in R^2, so that it is a disk D_1, centered at $\mathbb{O} \in R^2$ with image of c_0 called c_1, then scale D_1 so it becomes a *unit* disk D_2 with the image of d_1 labeled d_2. Next, rotate D_2 about $\mathbb{O} \in R^2$ so that c_2 is sent to $c_3 = (1, 0, 0)$. Call this rotated disk D_3, Figure 1.23. The inverse of the stereographic mapping $D_4 = \mathcal{P}^{-1}(D_3)$ is the northern hemisphere and c_3 is sent to itself, Figure 1.24. Apply be a rotation ρ by angle $\pi/2$ in R^3 about the y-axis the northern hemisphere to the hemisphere shown in Figure 1.25 so that $(1, 0, 0)$ is sent to S_0, call this rotated subset D_5. Finally, using stereographic mapping, the image of $D_5 - S_0$ is R^2_+, Figure 1.26. ∎

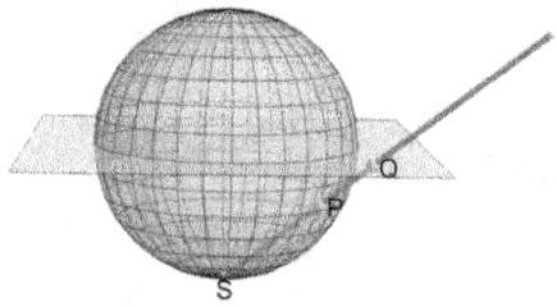

Fig. 1.19 Stereographic mapping sends P in lower hemisphere to Q outside unit circle of R^2.

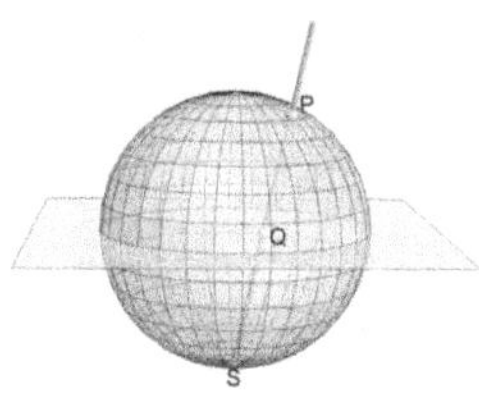

Fig. 1.20 Stereographic mapping sends P in upper hemisphere to Q inside unit circle of R^2.

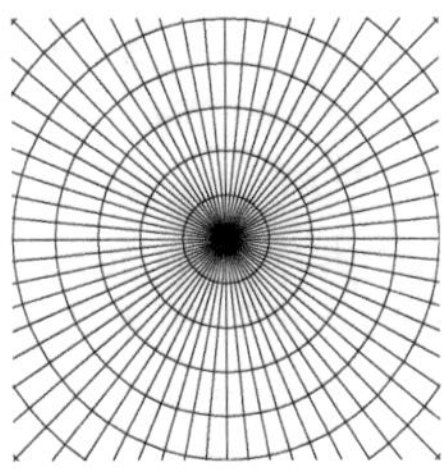

Fig. 1.21 Projection of coordinates near pole on sphere look like polar coordinates.

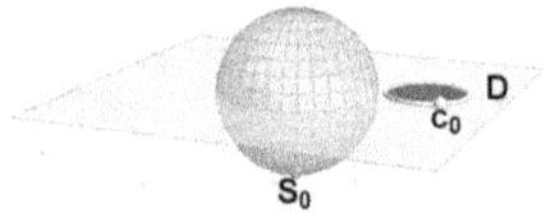

Fig. 1.22 D and c_0 in $R^2 \subseteq R^3$.

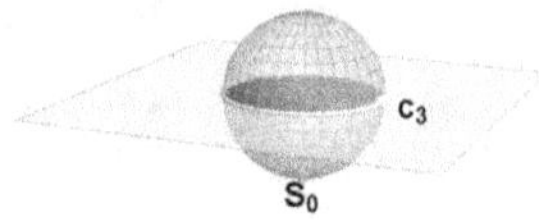

Fig. 1.23 $D_2 = D_3 = D^2 \subseteq R^2 \subseteq R^3$.

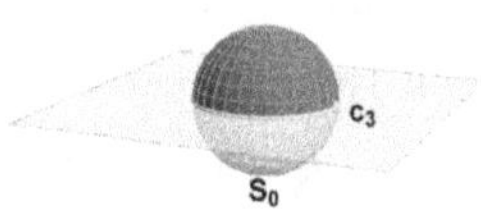

Fig. 1.24 D_4.

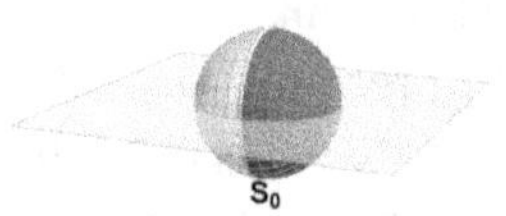

Fig. 1.25 D_5.

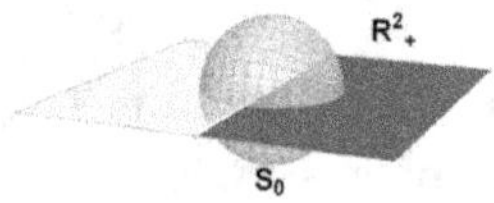

Fig. 1.26 $\mathcal{P}(D_5 - S_0) = R^2_+$.

Adopting geography terminology to S^2 we define quarter circles of great circles with one endpoint at a pole the other on the equator to be **longitudinal quarter-circles**. These are natural for our discussion of join-coordinates, Section 1.8. Also of interest to us is a semi-circle that goes from pole to pole—we will refer to this as a **longitudinal semi-circle** and is related to page-disks to be discussed also in Section 1.8. Geographic terminology for this semicircle is **longitudinal arc** or **meridian.**

Stereographic mapping of these latitudinal circles and longitudes yields a familiar grid of rays and circles of polar coordinates, Figure 1.21. Although a ray is not a 1-disk it does correspond to a 1-disk in S^2—a longitudinal semi-circle. To describe this without reference to R^3 we say a ray corresponds to a 1-disk with one endpoint at $\mathbb{O} \in R^2$ and the other a "point at infinity".

We get a grid associated with bipolar coordinates for R^2 from the geographic grid. Rotate S^2 in R^3 by $\pi/2$ so that the poles now lie in $R^2 \subseteq R^3$ then apply the stereographic mapping, Figure 1.27. The grid of latitudinal circles and longitudinal semi-circles maps to a grid for R^2, Figure 1.28.

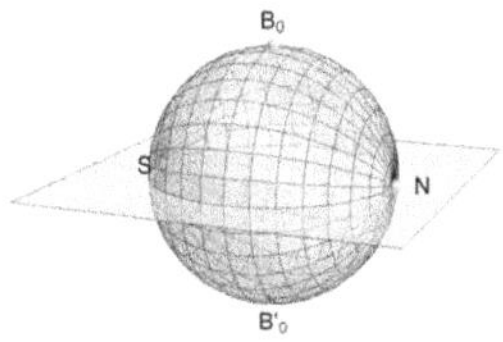

Fig. 1.27 Projecting the geographic grid of rotated S^2.

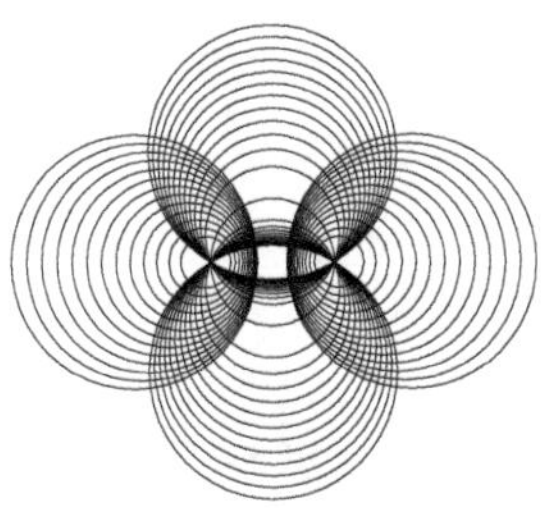

Fig. 1.28 The stereographic mapping of the grid of Figure 1.27 resembles bipolar coordinates.

1.5 Inversion in n-Space and Reflection in the n-Sphere

For $a \in R^{n+1}$, geometric inversion is a self-homeomorphism of $(R^{n+1} - \{a\})$ fixed on an n-sphere with center a, that sends each ray from a to itself and interchanges points inside the sphere with points outside. In S^n, it corresponds to reflection.

Definition 1.27. Standard inversion in R^{n+1} is the homeomorphism $\iota \colon (R^{n+1} - \{\mathbb{O}\}) \to (R^{n+1} - \{\mathbb{O}\})$ given by $\iota(\overrightarrow{x}) = \frac{1}{|\overrightarrow{x}|}\overrightarrow{x}$.

More generally, suppose $S \subseteq R^{n+1}$ is an n-sphere with center $\overrightarrow{a}$ and radius r. Let T be a map which is a translation and scaling of R^{n+1} so that $T(S) = S^n$. **Inversion in** S is the homeomorphism of $(R^{n+1} - \{\overrightarrow{a}\})$ fixed on S defined as $\iota_a = T^{-1} \circ \iota \circ T$. ▌

Consider the reflection map $\rho \colon S^n \to S^n$ given by $\rho(x_1, x_2, \ldots, x_{n+1}) = (x_1, x_2, \ldots, -x_{n+1})$.

Stereographic mapping transforms ρ to the standard inversion. That is, $\mathcal{P} \circ \rho = \iota \circ \mathcal{P}$. This is illustrated in Figure 1.29 for S^1. Inversion in R^1 is fixed on the 0-sphere $S^0 \subseteq R^1$. It sends $a = \frac{1}{2}$ to $b = \iota(\frac{1}{2}) = 2$. On the circle we have points $A = (\frac{4}{5}, \frac{3}{5})$ and $B = (\frac{4}{5}, -\frac{3}{5})$ where $\mathcal{P}(A) = a, \mathcal{P}(B) = b$ and that $\rho(A) = B$.

Remark 1.3. For any $X \subseteq R^{-}\{a\}$, $X' = \rho \circ \iota_a(X)$ is spherically equivalent to X, Definition 2.2. The ρ is used since inversion reverses orientation, see Section 3.5. So, if X and Y are two knottings in R^n they are spherically equivalent, if and only if there is an $a \in (R^n - X)$ and $\rho \circ \iota_a(X)$ is isotopic to Y. ▌

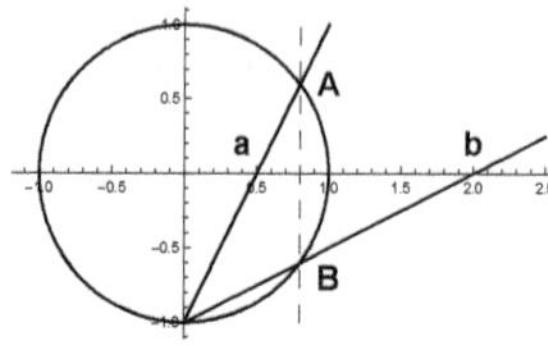

Fig. 1.29 By stereographic mapping $\mathcal{P}(A) = a, \mathcal{P}(B) = b$, by reflection $\rho(A) = B$. By inversion $\iota(a) = b$.

Stereographic mapping provides correspondence of other basic geometric notions in R^n and in S^n,

Definition 1.28. A **Möbius transformation** is a map $f\colon R^n \to R^n$ which is a composition of translations, rotations, inversions and scalings. ▮

For R^2, Möbius transformations can be described in terms of complex numbers as functions of the form $f(z) = (az+b)/(cz+d)$ for given complex numbers a, b, c, and d however we will not use this point of view. Möbius transformations appear in discussion of knot-energy, Section 4.8.

1.6 Coordinate Systems and Subsets of n-Space

Definition 1.29. A **hyperplane** H **in** R^n is a subset $H = \{(x_1, \ldots, x_n) \in R^n : a_1x_1 + \cdots + a_nx_n = b\}$ for fixed real numbers $a_1, \ldots, a_n, b$ and where $a_i \neq 0$ for at least one i. For $1 \leq j \leq n$, $H_j(b)$ denotes the hyperplane with equation $x_j = b$. If $b = 0$ these are **coordinate hyperplanes**. If we use letters rather than subscripts for coordinates, then we use notations such as $H_x(b), H_y(b), H_z(b)$, or $H_w(b)$. ▮

Remark 1.4. In multivariable calculus, one considers a tangent plane T to a surface $S \subseteq R^3$ at a point p and speaks of a point $q \in T$ as **"tangent vector based at** p**"**. The tangent plane is a 2-dimensional hyperplane but technically not a vector space unless it happens that $p = \mathbb{O} \in R^3$. However we can correctly write $\overrightarrow{q} = \overrightarrow{p} + \overrightarrow{(q-p)}$ and it is this $\overrightarrow{(q-p)}$ that corresponds to "tangent vector based at p". The vector $\overrightarrow{(q-p)}$ lies, not in T, but in the vector subspace parallel to T that contains $\mathbb{O}$. Such "point-based vector" terminology has widespread use in graphic depictions of vector fields. Technically, a point-based vector in R^n is a $(2n)$-dimensional object. We need $2n$ numbers since the point and the vector are *each* n-dimensional. We will see the utility of this view in Proposition 2.10. More generally a collection of linearly independent vectors based at a point p is called a **frame at** p.

A **line in** R^n is a 1-dimensional hyperplane; a **directed line in** R^n is a line with a preferred direction given by a vector. ▮

A hyperplane is a solution space of a single (non-trivial) linear equation. A generalization of this is an affine space:

Definition 1.30. An **affine space** is a non-empty set $A \subseteq R^n$ of solutions of a set of linear equations with n variables. ▮

A familiar example of an affine subspace is the tangent line L at x to a differentiable curve in R^3. Such a line could be described as an intersection of two hyperplanes with each such hyperplane corresponding to a linear equation. Points in L are called tangent *vectors* based at x. The solution space of a set V of *homogeneous* linear equations *is* a vector subspace of R^n- since $\mathbb{O} \in V$. However, for an *affine* space, the equations might not be homogeneous, so an affine space is not generally a vector space.

Definition 1.31. The **standard projection** $\pi\colon R^n \to R^{n-1}$ is the map $\pi(x_1, \ldots, x_{n-1}, x_n) = (x_1, \ldots, x_{n-1})$.

The point $x_* = \pi(x) \in R^{n-1}$ is called the **standard projection of** x, or sometimes just the **projection of** x. Also, x_n is called the **height** of the point x. If $M \subseteq R^n$, then $\pi(M) = M^*$ is the **standard projection of the knotting** M and $h\colon M \to R^1$ defined $h(x_1, \ldots, x_{n-1}, x_n) = x_n$ is the **height function for** M.

On occasion, we will consider other projections. The j**-th projection** π_j **of** R^n is the map $\pi_j(x_1, \ldots, x_j, \ldots x_n) = x_j$. ▮

Notation 1.7. If $X \subseteq R^n$ we generally use notation $X_* = \pi(X) \subseteq R^{n-1}$. However, we sometimes put the "$*$" as a *superscript* for readability. For example, if $A_0 \subseteq R^n$ we write $A_0^* = \pi(A_0)$, rather than $(A_0)_*$. ▮

Definition 1.32. The subset $R^n_+ = \{(x_1, x_2, \ldots x_n)$ with $0 \leq x_1\}$ is called n**-dimensional half-space**; R^2_+ is called the **half-plane**. We will have occasion to use associated sets $R^n_- = \{(x_1, x_2, \ldots x_n)$ with $x_1 \leq 0\}$ and $\partial R^n_+ = \{(x_1, x_2, \ldots x_n)$ with $x_1 = 0\}$ where the symbol ∂ is used in the sense of boundary, see Section 2.2. Finally $\overset{\circ}{R^n_+} = \{(x_1, x_2, \ldots x_n)$ with $0 < x_1\}$, is **open** n**-dimensional half-space**. That is, $\overset{\circ}{R^n_+} = (R^n_+ - \partial R^n_+)$. ▮

Generalized cylindrical coordinates are used often in higher dimensional knotting constructions, see Sections 10.2, 7.3 and 10.9. Most often generalized cylindrical coordinates are a hybrid where we use polar coordinates for the first two coordinates of R^n and cartesian coordinates for the rest. Occasionally we use spherical coordinates (of some dimension) followed by some cartesian coordinates. Context will make usage clear.

Definition 1.33. For $3 \leq n$ write $(x_1, x_2, \ldots, x_n) \in R^n$ as $(\rho, \theta, \ldots, x_n)$ where $\rho = \sqrt{x_1^2 + x_2^2}$ and $\tan(\theta) = \frac{x_2}{x_1}$ if $x_1 \neq 0$. If $n = 3$ this is called **cylindrical coordinates**; for $3 < n$ **generalized cylindrical coordinates.** ∎

Cartesian coordinates can express R^2 as a union of infinitely many disjoint coordinate lines, say those parallel to an axis. In polar coordinates, we express R^2 as a union of infinitely many rays. These rays are not disjoint— all met at the origin. With generalized cylindrical coordinates R^n can be viewed as a union of $n-1$-dimensional half-spaces called pages which meet in a common set called the binding:

Definition 1.34. For an angle, α, the subset of R^2 defined using polar coordinates $H^1_\alpha = \{(r, \theta) : \theta = \alpha, 0 \leq r\}$ is called **a ray**. For R^n, using generalized cylindrical coordinates, (see Notation 1.33) define $H^{n-1}_\alpha = \{(r, \theta, x_3, \ldots, x_n) \in R^n : \theta = \alpha\}$. Such a subset is called an $(n-1)$-**dimensional page of $\mathbf{R^n}$**. The subset of a page where $r \neq 0$ is called an **open page** denoted $\overset{\circ}{H}{}^{n-1}_\alpha$ and H^{n-1}_α called a **closed page.** The subset $\mathcal{Z} \subseteq R^n$, which in polar/cylindrical coordinates is all points $(r, \theta, x_3, \ldots x_n)$ with $r = 0$, is an $(n-2)$-dimensional subspace. For R^2, $\mathcal{Z}$ is the origin, for R^3 the z-axis. For dimensions 3 and larger we call $\mathcal{Z}$ the **binding set.**

The homeomorphism $i_\alpha \colon R^{n-1}_+ \to H^{n-1}_\alpha$ defined by $i_\alpha(x_1, x_2, x_3, \ldots, x_{n-1}) = (\sqrt{x_1^2 + x_2^2}, \alpha, x_3, \ldots, x_n)$ is called the **half-space identification of page α.**

If we have a subset $M_0 \subseteq H^{n-1}_0$ we can get a copy of it in each page: $M_\theta \subseteq H^{n-1}_\theta$ where: $M_\theta = j_\theta(M_0)$ where $j_\theta(r, 0, x_1, \ldots, x_{n-1}) = (r, \theta, x_1, \ldots, x_{n-1})$. The union of all these copies $\Sigma(M_0) = \bigcup_{0 \leq \theta \leq 2\pi} M_\theta$ is called the **spinning of M_0.** ∎

For example, in R^3, if M_0 is a semicircle, Figure 10.22, the spinning will be a sphere, as in Figure 10.26. If it is a circle as in Figure 6.124, the spinning will be a torus, as shown in Figure 6.128. The spinning construction $\Sigma(M_0)$ will be used for higher-dimensional knotting in Chapter 10.

Remark 1.5. Page identification relates the two subsets of R^3: R^2_+ and H^2_0. The distinction is $R^2_+ = \{(x, y, z)\colon 0 \leq x \text{ and } z = 0\}$ while $H^2_0 = \{(x, y, z)\colon 0 \leq x \text{ and } y = 0\}$. ∎

Remark 1.6. Standard projection for R^n, Definition 1.31, is projection on the last coordinate. **The standard projection for** R^n_+ is its restriction to R^n_+. It has image R^{n-1}_+. If $X \subseteq R^n_+$, we write the image under projection as $X_* \subseteq R^{n-1}_+$. So $X^*_0 = i_0(X_*) \subseteq H^{n-1}_0$ is a natural copy of X_* in the page. Then $\Sigma(X^*_0)$ is called the **spinning of the projection** X_*.

A property of spinning we will often use is: *the spinning of the projection is the projection of the spinning.* That is $\Sigma(X)_* = \Sigma(X_*)$. The first of many uses of this for higher-dimensional knotting is Example 10.4. ∎

The standard projection $\pi : R^n \to R^{n-1}$ maps pages to pages—$\pi(H^{n-1}_\theta) = H^{n-2}_\theta$ and the binding of R^n is sent to the binding of R^{n-1}.

An important example in this book is viewing R^4 as a union of 3-dimensional pages with binding the zw-coordinate plane. The standard projection $\pi : R^4 \to R^3$ helps visualize this. The projection of the binding is the z-axis of R^3 and each half-plane page of R^3 is the projection of a corresponding half-space page in R^4.

Spherical coordinates for R^3 use two angles and a distance from the origin: r, θ, ϕ. In **generalized spherical coordinates for** R^n points are written $(r, \theta, \phi, x_4, \ldots, x_n)$ where the first three coordinates are the usual spherical coordinates, the rest cartesian.

Remark 1.7. In mathematics, the third dimension is most often viewed as "height". So, graphically the third dimension is often represented as upwards on a page. However, in computer graphics, the third dimension is often represented as towards the viewer rising out of the page/screen. In either case, for visual clarity, one often offsets the viewpoint slightly. In this book, we use either method, depending on which best illustrates a particular mathematical point. Consider two graphics for the circle in R^3, given by $\{x^2 + y^2 = 1 \text{ and } z = 0\}$, together with axes. Figure 1.30 has z vertical on the page. In Figure 1.31, the z-axis is more or less in the direction of the viewer. This is termed an "above view". Figure 1.30 may seem more familiar, but note that the circle appears on the page as an ellipse, whereas in Figure 1.31 it is more clearly circular. In most cases, we do not show the axes graphically since it tends to clutter the graphics.

When we graphically show R^3_+ we will generally intend the positive x-axis to be horizontal on the page, directed to the right and the z-axis, the direction of projection, rising up from the page usually at a slight angle from the perpendicular. ∎

Fig. 1.30 A traditional view of R^3.

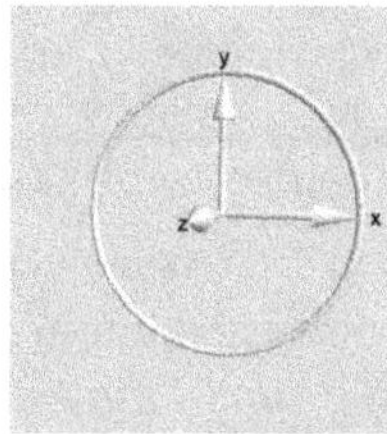

Fig. 1.31 An "above view" of R^3.

1.7 Isotopy

Informally, a knotting of a circle is a tangle of stretchable rope with two ends spliced together positioned in space. Two knottings are "the same" if we can physically move from one position to the other. All six knottings in Figure 3.1 are isotopic. To express this "motion" we could use an additional parameter t besides, our x, y and z, to mathematically account for "time". This gives four coordinates (x, y, z, t). We will express this motion in the language of subsets rather than processes. Because of the additional t coordinate, our motion will correspond to a four-dimensional subset.

Example 1.10. We express a simple isotopy of a circle in the plane R^2 as a subset of R^3. The square path $K \subseteq R^2$ and the unit circular path $K' \subseteq R^2$, shown in Figure 1.32 are isotopic subsets of R^2. One can move points radially starting with the square at $t = 0$ and ending with circle at $t = 1$. Several of the steps K_t are shown in Figure 1.33. We assemble these into a subset R^3 called the **trace of the isotopy** as shown in Figure 1.34. ∎

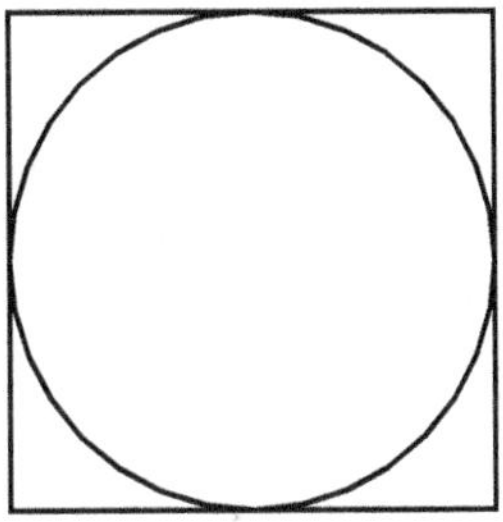

Fig. 1.32 Two isotopic subsets of R^2.

Fig. 1.33 Some intermediate steps—a discrete track.

Fig. 1.34 The trace of this isotopy—a subset of R^3.

Recall the definition of level preserving map, Definition 1.13.

Definition 1.35. For $t \in I$, the **level set of** $X \times I$ **at level** t is $X \times \{t\}$. If $F\colon X \times I \to Y \times I$ is a level preserving map such that F_t is a *knotting* of $X \times \{t\}$ in $Y \times \{t\}$, we say F **is an isotopy of** X **in** Y. The image $T = F(X \times I) \subseteq Y \times I$ is called the **trace of the isotopy** F.

If K and K' are knottings in Y they are **isotopic knottings of** X **in** Y if there is subset $T \subseteq Y \times I$ which is a trace of an isotopy F of where $F(X \times \{0\}) = K \times \{0\}$ and $F(X \times \{1\}) = K' \times \{0\}$. We say F **is an isotopy from** K **to** K'. The map $\lambda\colon T \to I$ defined $\lambda(X_t) = t$ is called the **level map of the trace.** ∎

Isotopy of a circle in $C \subseteq R^3$ as a flexible motion of C in R^3 is easier for most to understand than the trace of this isotopy, which is a knotting of a cylinder $C \times I$ in $R^3 \times I \subseteq R^4$. However, we will see that the trace concept, Definition 1.35, is central to our exploration of higher dimensions.

A common alternate definition of isotopy F is as a continuous 1-parameter family of embeddings. If we identify $X \times \{t\}$ with X and $Y \times \{t\}$ with Y, then for $t \in I$ we can view a level map of an isotopy as $F_t : X \to Y$. For each $t \in I$, write $X_t = F_t(X)$. Then our isotopy is a continuous 1-parameter *family of subsets* $X_t \subseteq Y$. So, two knottings K and K' of X are isotopic in Y if there is an isotopy X_t in Y such that $X_0 = K$ and $X_1 = K'$.

We allow "isotopy" to refer to a map $F\colon M \times J \to N$ where J is some other interval of real numbers (most frequently $0 \leq t \leq 2\pi$) if we explicitly mention our choice of J.

For $t \in I$ we will refer to X_t as a **step of the isotopy** or if needed more specificity, **step** t **of the isotopy**. The projection in Y of a trace, $p_Y(T) = \bigcup_{t=0}^{t=1} X_t$ is called the **track of the isotopy**.

The **track of an isotopy** is not a standard term and has use in this book only for visualization of simple isotopies. The term **step** is also not standard. Viewing an isotopy as a motion of X in Y, the track of the isotopy is all points in Y "visited" by the isotopy.

We often visualize an isotopy X_t as a finite sequence of steps $\{X_{t_0}, X_{t_1} \dots X_{t_p}\}$ where $t_0 < t_1 \cdots < t_p$ and $t_0 = 0, t_p = 1$, where the transition from each X_{t_i} to $X_{t_{i+1}}$ is apparent. A sequence of such steps is called a **discrete track for the isotopy** X_t. In simple cases we visualize a discrete track in a single graphic as union of steps $\bigcup_{t=0}^{t=p} X_t$. So, Figure 1.33 is a discrete track of the isotopy of Example 1.10. The (entire) track of this isotopy consists of all points of R^2 on the square, on the circle, or between the circle and square.

A vertical isotopy is a special simple isotopy.

Definition 1.36. An isotopy of F_t of X in R^n, where $\pi \circ F_t = F_t^*$, is a **vertical isotopy** if $F_t^* = F_0^*$ for all t. ▮

So, an isotopy is a **vertical isotopy** if, in terms of coordinate functions, if the only changes in X_t are in the last coordinate. For R^3 we can write $F_t(x) = (x_t(x), y_t(x), z_t(x))$ and a vertical isotopy can be written $F_t(x) = (x_0(x), y_0(x), z_t(x))$. Viewing an isotopy as a motion of a subset, a vertical isotopy is a motion in R^n whose shadow in R^{n-1} does not change.

More generally we will need to speak of isotopy of $K \subseteq M$ where our flexible motion is constrained within M. Figures 1.35, 1.36 and 1.37 show three knottings of a circle K, K' and K'' in torus M. It can be shown that K and K' are isotopic subsets of M but K and K'' are not.

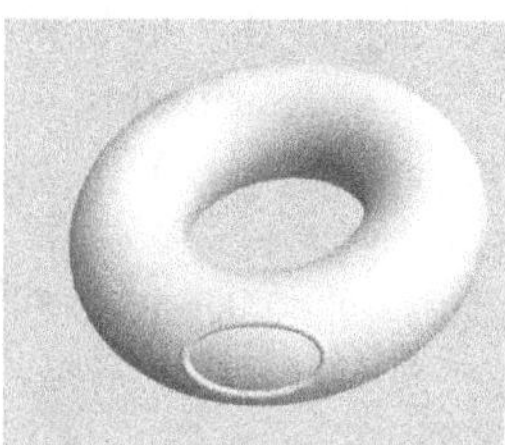

Fig. 1.35 A circle knotting K in torus M.

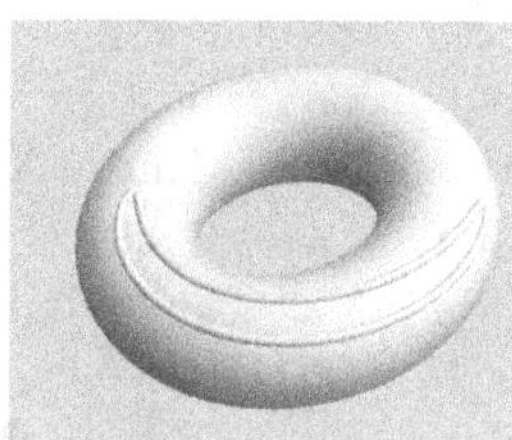

Fig. 1.36 An isotopic circle knotting $K' \subseteq M$.

Fig. 1.37 A non-isotopic circle knotting $K'' \subseteq M$.

Given an isotopy K_t of M we can obtain a second isotopy $F_t = K_{(1-t)}$. We think of F_t as doing the motion of K_t backwards in time starting with K_1 and ending with k_0.

If F_t and G_t are two isotopies $F_1 = G_0$ the **concatenation of F and G**, Definition 1.14 will also be an isotopy. Viewing an isotopy as a motion the concatenation is "the motion K_t done at double speed followed by the motion F_t at double speed".

The two knottings of an annulus in R^3, Figures 1.38 and 1.39 are not isotopic subsets. One annulus has a right-hand twist, the other a left-hand twist. More details on the topic of twist are in Section 4.3. No amount of flexing/stretching/bending of these subsets in R^3 of these sets will change one into the other. Similarly, the usual Möbius band with a right half-twist is not isotopic to the one with a left half-twist. ▮

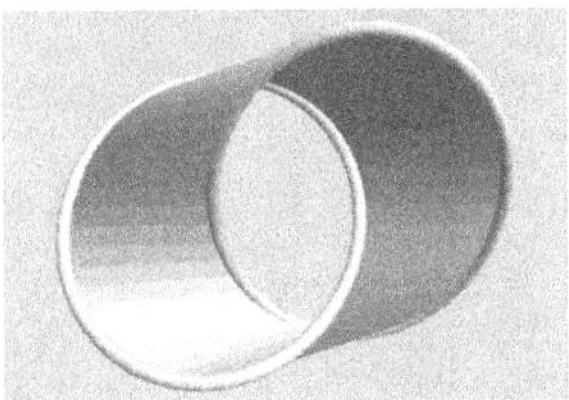

Fig. 1.38 An annulus with a left twist.

Fig. 1.39 An annulus with a right twist.

There are two distinct common meanings of "rotation of R^n"— sometimes it is a map, sometimes an isotopy (i.e. motion). As a map, a **rotation** can be described as a linear map r given by $r(\overrightarrow{x}) = M.\overrightarrow{x}$ where $\overrightarrow{x} \in R^n$ and M is an $n \times n$ orthogonal matrix of determinant ± 1. Other times rotation means an isotopy r_t of R^n such that for each t, r_t is a rotation. If $X \subseteq R^n$, then **rotation for X** is the restriction of a rotation of R^n, perhaps a map, perhaps an isotopy.

1.8 The Join Structure of the n-Sphere

The cartesian product allows us to express $R^{p+q} \approx R^p \times R^q$ and $D^{p+q} \approx D^p \times D^q$. Spheres are not cartesian products but there is a fundamental way of expressing high-dimensional spheres in terms of lower dimensional spheres called the **join structure**. Although this is a basic structure for high-dimensional spheres, we will not need reference until much later in this book. A broader notion of "join" is used elsewhere in topology. See [Hatcher (2002)] for topological definition, [Lefschetz (1942)] for complexes.

A point x on the globe is described by **longitude** and **latitude**. This can viewed as a triple (e, p, m), where e is a point on the equator and p is a choice of poles, north or south. Given e and p there is a unique quarter-circular longitudinal arc A on the globe joining them. We parameterize A to locate our point x on A by a number m— the latitude. Geographically, Chicago is at the triple $(87°39'W, N, 41°51')$, conventionally written as a pair $(41°51'N, 87°39'W)$. Join coordinates are a higher-dimensional versions of (e, p, m).

The join structure of S^n expresses S^n as a union of sub-arcs determined by the standard p-sphere and the co-standard $(n-p-1)$-sphere:

Definition 1.37. Write $R^n = R^{n-k-1} \times R^{k+1}$. The set

$$S^k = \{x \in R^n \colon x_1 = \dots x_{n-k-1} = 0 \text{ and } x_{n-k}^2 + \dots + x_n^2 = 1\}$$

is the **co-standard k-dimensional sphere in R^n.** ▮.

Remark 1.8. For any $\vec{x} \in S^n$ we define triple (a, b, t) for x called **join-coordinates**—here $a \in S^p, b \in S^q$, and $0 \leq t \leq 1$.

Using vector notation write $\vec{x} = (x_1, \dots x_{q+1}, x_{q+2}, \dots x_{n+1})$. Then let $\vec{x} = \vec{u} + \vec{v}$ where $\vec{u} = (x_1, \dots x_{p+1}, 0, \dots 0)$ and $\vec{v} = (0, \dots, 0, x_{p+2}, \dots, x_{n+1})$.

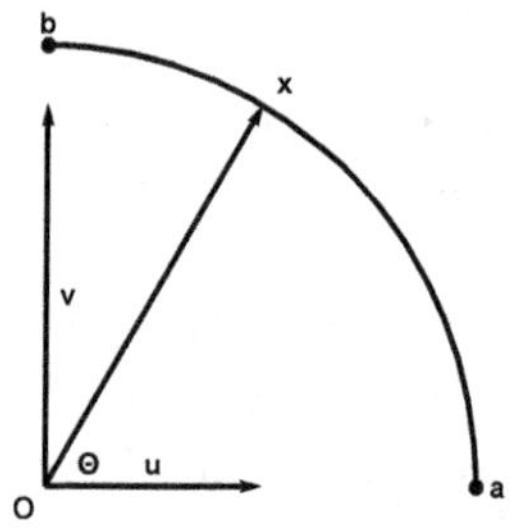

Fig. 1.40 Join-coordinates are calculated based on a 2-plane in R^n, see Remark 1.8.

Suppose, as is most often the case, we have $\vec{u} \neq \vec{0}$ and $\vec{v} \neq \vec{0}$. Let $\vec{a} = \vec{u}/\|u\|$, then $\vec{a} \in S^p$, see Figure 1.40. Let $\vec{b} = \vec{v}/\|v\|$, then $\vec{b} \in \vec{S}^q$. Now $\mathbb{O}$, $\vec{x}$, $\vec{a}$ and $\vec{b}$ all lie in a two-dimensional plane P in R^{n+1} and $P \cap S^n$ is a great circle C of S^n. The points $\vec{x}$, $\vec{a}$ and $\vec{b}$ all lie on C. Moreover, $\vec{a}$ and $\vec{b}$ are endpoints of a quarter-circle arc which contains $\vec{x}$. We call this the **arc joining a to b** or **join-arc of a and b** and denote it by α_a^b. We parameterize this circle by $[0, 1]$ so that 0 corresponds to $\vec{a}$ and 1 corresponds to $\vec{b}$ proportional to the angle θ between a point on the circle and $\vec{a}$, see Figure 1.40. We have $\theta = \arctan \frac{\|v\|}{\|u\|}$, so our third join coordinate of $\vec{x}$ is $t = \frac{2\theta}{\pi}$.

For the special case where $\vec{v} = \vec{0}$ then $\vec{u} \in S^p$, in which case $\vec{a} = \vec{v}$. The join-coordinates are not unique for $\vec{x}$ but are *all* triples $(\vec{a}, \vec{b}, 0)$ where $\vec{b}$ is any point in S^q. Similarly, if $\vec{u} = \vec{0}$ then $\vec{v} \in S^q$, in which case we let $\vec{b} = \vec{u}$. In this case the join-coordinates $\vec{x}$ consists of all triples $(\vec{a}, \vec{b}, 1)$ where $\vec{a}$ is any point in S^p. ∎

Notation 1.8. Write $S^p * S^q = S^{p+q+1}$ to indicate viewing S^{p+q+1} with join-coordinates for the standard p-sphere and co-standard q-sphere. ∎

Definition 1.38. We say that S^{p+q+1} is the **join** of S^q and S^q and that $S^p \subseteq S^{p+q+1}$ is the **first factor of** $S^p * S^q$ and that $S^q \subseteq S^{p+q+1}$ is the **second factor**. The knotting, which is the union of the first and second factors, is called the **standard knotting of** S^p **and** S^q **in** S^{p+q+1}. The standard knotting of S^n and S^n in S^{2n+1} is called the n**-dimensional Hopf Linking in** S^{2n+1}. ∎

We will be especially interested in the join structures $S^3 = S^0 * S^2 = S^1 * S^1 = S^2 * S^0$ and $S^4 = S^0 * S^3 = S^1 * S^2 = S^2 * S^1 = S^3 * S^0$.

The join structure can be used to describe other basic subsets of S^n.

Definition 1.39. If $A \subseteq S^p$ and $B \subseteq S^q$ we define the **join of** A **and** B to be $A * B = \{(a, b, t) \in S^p * S^q : a \in A \text{ and } b \in B\}$. ∎

We can use the join function to describe families of disks in spheres. We describe a natural onto map $S^p \times D^{q+1} \to S^p * S^q$, which is an embedding for $S^p \times int(D^{q+1})$. Also, we describe a similar map $D^{p+1} \times S^q \to S^p * S^q$.

Consider the case of $S^2 = S^1 * S^0 \subseteq R^3$ and $A = S^1 \times D^1 \subseteq R^3$, Figure 1.41. Note $S^2 \cap A = S^1$. For each $p \in S^1$, $l_p = \{p\} * S^0 \subseteq S^1 * S^0$ is a longitudinal arc of S^2. We get our map from A to S^2 by "stretching and bending each $\{p\} \times D^1 \subseteq A$ to fit onto l_p of S^2 ", Figure 1.42. Similarly, we have $D^2 \times S^0 \subseteq R^3$ is a disjoint union of two 2-disks, Figure 1.43. We stretch and bend the radii of these discs and fit them onto the sphere, forming the two hemispheres, Figure 1.44.

We can define a more general bending function.

Definition 1.40. Using generalized spherical coordinates write points $x \in D^q$ as $x = (z, t)$ where $z \in S^{q-1}$ and $0 \le t \le 1$ where points with $t = 0$ correspond to $\mathbb{O} \in D^q$. The **bending function** $\beta\colon S^p \times D^{q+1} \to S^p * S^q$ maps $\{x\} \times D^{q+1} \to \{x\} * S^q$ by $\beta(x, (y, t)) = (x, y, t)$ where we are using join-coordinates for $S^p * S^q$. ∎

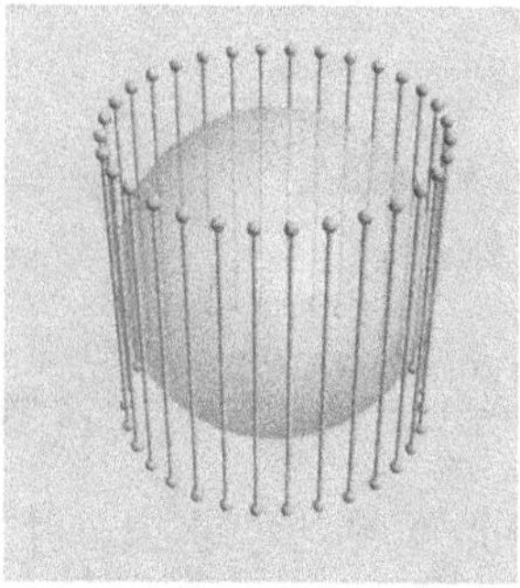

Fig. 1.41 $S^1 \times D^1 \subseteq R^2 \times R^1 \approx R^3$.

Fig. 1.42 Bending $S^1 \times D^1$ onto S^2.

Fig. 1.43 $D^2 \times S^0 \subseteq R^2 \times R^1 \approx R^3$.

Fig. 1.44 Bending $D^2 \times S^0$ onto S^2.

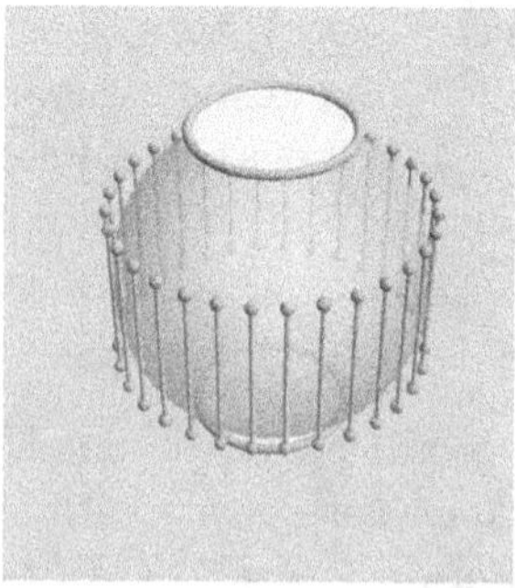

Fig. 1.45 Using 1-disks and 2-disks radius $\frac{1}{2}$.

Fig. 1.46 Bending disks Figure 1.45 onto sphere.

Additionally, we can view S^n as a union of two subsets each a sphere-disk product. For example if $D^p_{\frac{1}{2}} \subseteq D^p$ is the sub-disk of radius $\frac{1}{2}$, we apply the bending functions to each of $D^{p+1}_{\frac{1}{2}} \times S^q$ and $S^p \times D^{q+1}_{\frac{1}{2}}$. Figures 1.45 and 1.46 illustrate this for $S^2 = S^1 * S^0$.

Remark 1.9. View $S^p * S^q = S^{p+q+1}$. For $x \in S^p$, $\{x\} * S^q$, using notation of Definition 1.39, is a $(q+1)$-disk—a cone on S^q with vertex p. Similarly, for $y \in S^q$, $S^p * \{y\}$ is a homeomorphic to a $(p+1)$-disk. Thus, we have two interesting ways of expressing a sphere in terms of such sub-disks:

$$S^{p+q+1} = S^p * S^q = \bigcup_{x \in S^p} (\{x\} * S^q) = \bigcup_{y \in S^q} (S^p * \{y\}) .$$

In the first expression any two distinct $(q+1)$-disks intersect only in a common boundary which corresponds to the second factor of $S^p * S^q$. In the second expression any two distinct $(p+1)$-disks intersect only in a common boundary which corresponds to the first factor of $S^p * S^q$. ▌

Remark 1.10. For $S^3 = S^1 * S^1$ we call the 2-disks$\{x\} * S^1$ **page-disks of** S^3 and the 2-disks **cap-disks** $S^1 * \{y\}$, named after their images under stereographic mapping. The image of a page disk (minus south pole) is a page in R^3 and the image of a cap disk is a spherical cap of a sphere containing $S^1 \subseteq R^3$.

To see this, denote the first factor of $= S^1 * S^1$ by Z and the second factor by S. If $S_0 \in Z$ is the south pole, $Z - S_0$ is mapped to the z-axis of R^3 and S to the unit circle $S^1 \subseteq R^3$. Stereographic mapping takes each $(D_p - S_0)$ homeomorphically to the page of R^3 which contains $p \in S^1 \subseteq R^3$. (Recall, Remark 1.2, that a 2-disk with a single boundary point removed is homeomorphic to R^2_+).

In general, stereographic mapping sends any geometric 2-sphere in $(S^3 - S_0)$ to a geometric 2-sphere in R^3. So generally any of our cap-disks in S^3 (whose boundaries are all S) are sent to spherical caps with boundary $S^1 \subseteq R^3$. There are two exceptional cases. The disk D'_{N_0} through the north pole is sent to the standard $D^2 \subset R^2 \subseteq R^3$ so it is flat—we can think of this as "a hemisphere of a sphere whose center is the point at infinity". The other exceptional disk is D'_{S_0} which contains S_0—the image X under stereographic mapping of $D'_{S_0} - \{S_0\}$ is the standard punctured R^2. We can view this as "a hemisphere (which contains the point at infinity on its boundary) of a sphere whose center is the point at infinity". So, we have three families of surfaces in R^3 which correspond to families of surfaces in

S^4: open pages of $R^3 - axis(z)$, caps of $R^3 - S^1$, and finally concentric torii corresponding to $S^1 \times S^1 \times R^1$ for $R^3 - (S^1 \cup axis(z))$. ∎

Some of the next definitions will not be used until the proof of Proposition 10.7.

Definition 1.41. If $A \subseteq S^p$ and $B \subseteq S^q$, and $0 \leq t_0 \leq 1$, the subset of $S^p * S^q$ described in join-coordinates as $(S^p * S^q)|_{t_0} = \{(a, b, t) \in S^p * S^p : a \in A, b \in B \text{ and } t = t_0\}$ is called the **slice of $A * B$ at t_0**. Also, write $A \circledast B = \{(a, b, t) \in S^p * S^p : a \in A, b \in B, \text{ and } t = \frac{1}{2}\}$, the **middle slice** of $A * B$. If $J = [p, q]$ is a subinterval of $[0, 1]$ we write $(S^p * S^q)|_p^q = \{(a, b, t) \in S^p * S^p : a \in A, b \in B, \text{ and } p \leq t \leq q\}$. called the **$[p, q]$-section** of $A * B$.

The **first half of** $S^p * S^q$ is the subset $(S^p * S^q)|_0^{\frac{1}{2}}$. The **second half of** $S^p * S^q$ is the subset $(S^p * S^q)|_{\frac{1}{2}}^1$. ∎

Since all pairs of values a and b are possible for a slice $(S^p * S^p)|_{t_0}$, with $0 < t_0 < 1$, it follows that $(S^p * S^p)|_{t_0}$ is homeomorphic to $S^p \times S^q$.

Prop 1.3. The first half of $S^p * S^q$ is homeomorphic to $S^p \times D^{n-p+1}$; $S^p * S^q$ with the second factor removed is homeomorphic to $S^p \times R^{n-p+1}$.

The second half of $S^p * S^q$ is homeomorphic to $D^{n-q+1} \times S^q$; $S^p * S^q$ with the first factor removed is homeomorphic to $R^{n-q+1} \times S^q$. The middle slice is homeomorphic to $S^p \times S^q$. ∎

Another way to see this is to use the formula of Proposition 2.12.

It is sometimes helpful to remember, especially when discussing $S^n * S^n$, that the first half contains the first factor and the second half contains the second factor.

The second half of a join of spheres for $S^n = S^p * S^q$ contains the south pole of S^n so when we take stereographic mapping to R^n, the image of the first half of the join will be homeomorphic to $S^p \times D^{q+1}$. So, this will be a solid torus and the image of the second half (minus the south pole) will be the closure of the complement of this solid torus. In particular for $S^3 = S^1 * S^1$, the image of the first half is a solid torus homeomorphic to $S^1 \times D^2$. We will take these images to be our "standard" objects in R^3:

Definition 1.42. The **standard** $(S^p \times S^q)$**-torus in** R^{p+q+1} is the stereographic mapping of the middle slice of the join. The standard solid $(S^p \times D^{q+1})$-torus in R^{p+q+1} is the stereographic mapping of the first half of the join. ∎

Example 1.11. Consider $S^1 = S^0 * S^0 \subseteq R^2$. The standard S^0 consists of the two points $(1,0)$ and $(-1,0)$, the co-standard S^0 is the two-point set of $(0,1)$ and $(0,-1)$. The middle slice is the four-point set (a torus $S^0 \times S^0$), which written in polar coordinates, is $\{(1,\pi/4),(1,3\pi/4),(1,5\pi/4),(1,7\pi/4)\}$.

The standard solid torus corresponds to a pair of disjoint sub-arcs— one between $(1,3\pi/4)$ and $(1,5\pi/4)$ and one between $(1,7\pi/4)$ and $(1,3\pi/4)$ shown thickened in Figure 1.47. The stereographic mapping will be two disjoint sub-intervals of R^1, also shown thickened. A short calculation reveals that these two intervals are $[\sqrt{2}-1,\sqrt{2}+1]$ and $[-(\sqrt{2}+1),-(\sqrt{2}-1)]$. Note that the midpoints of these intervals are $\sqrt{2}$ and $-\sqrt{2}$ and we can view this two-point set as a "0-sphere of radius $\sqrt{2}$".

The above analysis, applied to the standard solid $S^1 \times D^2$ torus of $S^3 \subseteq R^4$, shows that the standard solid $S^1 \times D^2$ torus can be described as a solid tube of radius one, with center a planar circle of radius $\sqrt{2}$, Figure 1.48. This middle slice of $S^3 \subseteq R^4$ is known in the literature as the Clifford Torus. It has interesting mathematical properties beyond the scope of this book. ∎

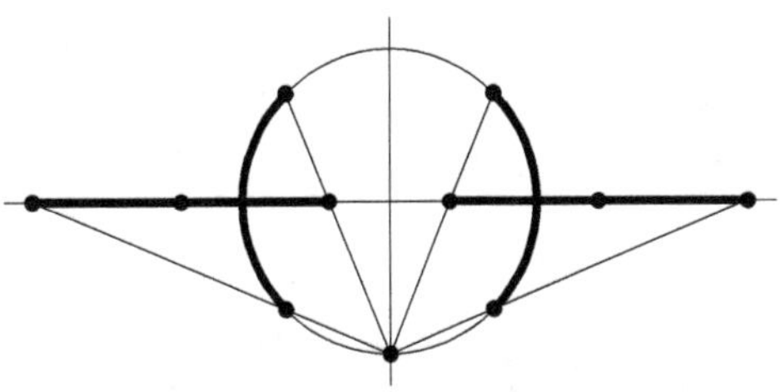

Fig. 1.47 Standard $S^0 \times D^1$ in S^1 and R^1.

Fig. 1.48 Standard solid $S^1 \times D^2$ torus in R^3.

Viewing S^2 as $S^1 * S^0$, the middle slice is homeomorphic to $S^1 \times S^0$— that is a disjoint union of two circles that correspond to latitudes 45°N and 45°S. The first half of $S^1 \times S^0$ is an annulus between these circles and the second half consists of two polar disks.

Much more interesting is $S^3 = S^1 * S^1$. The first and second factor are two disjoint circles in $S^3 \subseteq R^4$—the first $S = \{(x,y,z,w) : x^2+y^2 = 1, z = 0, w = 0\}$ and the second $Z = \{(x,y,z,w) : z^2+w^2 = 1, x = 0, y = 0\}$.

Each half of $S^1 * S^1$ is homeomorphic to a solid torus $S^1 \times D^2$. The south pole S_0 is a point of Z. Stereographic mapping $\mathcal{P} : (S^3 - \{S_0\}) \to R^3$ sends $(Z - S_0)$ to the z-axis and S to the unit circle in the xy-plane of R^3. If we rotate S^3 so that the S_0 is not a point of either factor circle, stereographic mapping will then send the circles S and Z to two disjoint circles—a subset of R^3 like that shown in Figure 4.63 called a **Hopf linking**.

In a geometric sense, $S^3 = S^1 * S^1$ gives rise to toroidal coordinates in R^3 as will be explained in the next remarks. Toroidal coordinates in R^3 are often described as obtained by "spinning a plane expressed as bipolar coordinates" about the z-axis. The family of 1-disks of bipolar coordinates will generate a family of topological 2-disks, spherical caps in R^3 whose boundary is $S^1 \subseteq R^2 \subseteq R^3$ with the exception of the 1-disk that contains the point at infinity—it generates a 2-disk which contains the point at infinity of R^3.

In toroidal coordinates each point not in the unit circle and not on the z-axis is the intersection of three surfaces: a torus which "contains" the standard $S^1 \subseteq R^3$, a 2-sphere which contains $S^1 \subseteq R^3$, a page that contains the z-axis.

We often use standard projection $\pi : R^n \to R^{n-1}$ to study a subset of K^n. There is a similarly useful tool for spheres we call **join projection**.

Definition 1.43. Let $X \subset S^p * S^q$ be the subset corresponding to the second factor of the join. The **join projection** Π: $(S^p * S^q - X) \to S^p$ is defined in join-coordinates by $\Pi(a, b, t) = a$. In the case of $S^n = S^{n-1} * S^0$, the join projection is called the **standard projection for** S^n. ▌

For S^2, view the first factor of the join $S^1 * S^0$ as the equator. Then, the standard projection of S^2 is simply the map that sends every non-polar point to the corresponding longitude point on the equator. Using stereographic mapping, S^3 minus the polar points corresponds to $R^3 - \{\mathbb{O}\}$, and the standard join projection of S^3 corresponds to radial projection in spherical coordinates to the unit sphere in R^3.

Also note that if $Z \subset S^p$ then $\Pi^{-1}(Z) \approx Z \times R^{q+1}$, in particular for the standard join projection $\Pi^{-1}(Z) \approx Z \times R^1$.

Viewing $S^4 = (S^1 * S^1) * S^0$ and using page/cap structures S^3, Remark 1.10 we can express S^4 in two ways. We could consider the family of 3-disks $D_p * S^0 \subseteq S^3 * S^0 \approx S^4$ where D_p is a page. Or we could consider the family of 3-disks $D'_p * S^0 \subseteq S^3 * S^0 \approx S^4$ where D'_p is a cap.

In each case, we express S^4 as a union of 3-disks, which are disjoint except for a common circle. This is the analog of expressing S^2 as a union of 1-disks (meridians), which have in common only a 0-sphere (two poles).

Page structure is used in our basic spinning constructions called Artin spinnings. The cap structure can then be used to view these knottings as spinnings in an alternate way as in Section 10.4.

Chapter 2

Elements of Geometric Topology

2.1 Manifolds—Topological and Smooth

Curves are very different from surfaces, but to a topologist, these are unified as special cases of a unifying concept—**manifold**. Informally, an **n-dimensional manifold** is a subset M of some R^m which "locally looks like R^n or R^n_+".

The 1-dimensional manifolds, informally called curves, are illustrated in many of our Figures such as Figures 3.19, 3.28 and 3.1. The 2-dimensional manifolds, informally called surfaces, are found in many of our figures such as Figures 2.42, 5.11, 5.13 and 5.14. For much of the study of circles in R^3 we will only need to consider manifolds of dimension three or less.

We will discuss *two* definitions: a topological manifold Definition 2.1 and a smooth manifold Definition 2.9. The distinction is due to two interpretations of "locally looks like". A third common notion, PL-manifold, will be discussed briefly later. Our focus will be on smooth manifolds.

Commonly "manifold" is defined to be an abstract topological space with certain properties, a point of view that has technical advantages for many proofs. We take a mathematically equivalent but more concrete approach and describe all manifolds as certain subsets of some R^n. This point of view has advantages when considering examples. It is called the **extrinsic point of view**; the more abstract approach is called the **intrinsic point of view**. The equivalence of these is via a mathematically non-trivial theorem called the **Whitney Embedding Theorem**.

Definition 2.1. A subset $M \subseteq R^m$ for some m is called a **topological n-manifold** (or **n-dimensional manifold**) if for every $x \in M$ there is map ϕ such that either:

(1) $\phi\colon R^n \to M$ is an embedding with $\phi(\mathbb{O}) = x$
(2) $\phi\colon R^n_+ \to M$ is an embedding with $\phi(\mathbb{O}) = x$

If the maps ϕ are all smooth, we say M is a **smooth n-manifold**, and n is the **dimension of the manifold**. We say that ϕ **is a local coordinate map for** M and its image is a **local coordinate patch of** M. ∎

For more complete treatments of the subject of smooth manifolds, see [Milnor (1997)], [Hirsch (1976)], [Guillemin and Pollack (2010)], [Lee (2013)].

In three-dimensional computer graphics terminology for surfaces, these local images as "coordinate surface patches" and the associated functions, ϕ, are local parameterizations.

Notation 2.1. Often, we use an *optional* superscript such as M^n as a helpful reminder of the dimension of an n-manifold. Other times, if it clutters the notation, we simply write M. ∎

Remark 2.1. For a number of reasons it is standard to say that the empty set is an n-manifold for any n. ∎

Definition 2.2. The subset of M where we need to use condition (2) of Definition 2.1 is called the **boundary of the manifold**, denoted $\partial(M)$. A point of a manifold that is not a boundary point is called an **interior point** of the manifold. The set of all interior points is called the **interior of the manifold** and is denoted $int(M)$. ∎

Our focus is primarily on smooth manifolds. Most basic propositions in theory of topological manifolds correspond to propositions for smooth manifolds. For example Proposition 2.1 is true if all manifolds are considered to be topological manifolds or if all manifolds are considered to be smooth. Rather than to point this out at each instance, in our introductory materials, we will simply use "manifold" to refer to either "smooth manifold" or "topological manifold", interpreted consistently. In those infrequent occasions where it does make a mathematical difference, we will revert to explicit topological/smooth usage.

Prop 2.1. The interior of n-manifold M is an n-manifold. If $\partial M \neq \emptyset$ then ∂M is an $(n-1)$-dimensional manifold. Also $\partial(\partial M) = \emptyset$. ∎

Example 2.1. For any $1 \leq n$, D^n, S^n, R^n and R^n_+ are n-manifolds. In terms of boundaries, $\partial R^n = \emptyset$, $\partial S^n = \emptyset$, $\partial D^n = S^{n-1}$ and $\partial R^n_+ = R^{n-1}$.

The inverse stereographic mapping $\mathcal{P}$ (Definition 1.2) is a coordinate patch, $\mathcal{P}\colon R^n \to (S^n - S_0)$ where S_0 is the south pole. Symmetrically, we can get a coordinate patch $R^n \to (S^n - N_0)$, where N_0 is the north pole. Using these two coordinate patches, we can show that S^n is a smooth n-manifold. ▮

Remark 2.2. Two knottings in R^n can be different, yet equivalent if we view these as knottings in S^n. Suppose knottings X and Y are not isotopic in R^n and let $\widetilde{X} = \mathcal{P}(X)$, let $\widetilde{Y} = \mathcal{P}(Y)$ be corresponding subsets in S^n. If $\widetilde{X}$ is isotopic to $\widetilde{Y}$, we say that knottings X and Y are **spherically equivalent**.

For example the three (dark) planar circles X in Figure 2.1 is not isotopic in R^2 three (dark) planar circles Y in Figure 2.2. They have non-homeomorphic complements: $R^2 - X$ consists of an open disk, two open annuli and a punctured R^2, while $R^2 - Y$ consists of two open disks, an annulus and a twice-punctured R^2. In each figure the stereographic image of $\widetilde{X}$ are the three white circles on S^2, one of which is in the lower hemisphere. In Figure 2.2, we can isotopically slide the white circle in the lower hemisphere of S^2 so encircles the south pole and corresponds to the white circle in the lower hemisphere of Figure 2.1. ▮

Fig. 2.1 Three circles, X, in R^2.

Fig. 2.2 Three circles, Y, in R^2.

Remark 2.3. In R^1, any interval, open, closed or half-open is a 1-manifold and any finite union of disjoint intervals is a 1-manifold.

The standard circle $S^1 \subseteq R^2$ is a 1-manifold. It is often said that "circle is two-dimensional since you need two dimensions to draw it". Indeed, in our definition, we have defined S^1 as a subset of R^2. However, in our mathematical terminology, the dimension refers to *what* an object is (it looks locally like an open interval of the line R^1) and not *where* it is (a subset of the plane). A circle in R^3 is still called a 1-manifold. Similarly, the 2-sphere S^2 is a 2-manifold even though one uses three dimensions to visualize it. Additionally, the torus $S^1 \times S^1$ is a 2-manifold. We use three dimensions to visualize it and, in fact, it is most naturally a subset of R^4: $S^1 \times S^1 \subseteq R^2 \times R^2 \approx R^4$. ▮

Example 2.2. The word "interior" is used in two common ways in topology that are often quite different. Note carefully the difference between "an interior point *of the subset* A" and "an interior point *of the manifold* A". In point set topology one usually defines $x \in A \subseteq X$ is an **interior point of** A if there is an open subset of X completely contained in A. Sometimes, the two uses coincide. Other times they differ.

For an example where the two notions coincide consider $A = D^2 \subseteq R^2$ and $x = (0,0)$. The interior of D^2 in the manifold sense is the standard open 2-disk $\overset{\circ}{D}{}^2 \subseteq R^2$. This is an open subset of R^2 and so is, also, the interior of D^2 in the subset sense.

For an example where the two notions differ consider $A = D^2 \subseteq R^3$. Now $\overset{\circ}{D}{}^2 \subseteq R^2 \subseteq R^3$ is still the interior of the manifold $D^2 \subset R^3$. However $\overset{\circ}{D}{}^2 \subseteq R^2 \subseteq R^3$ is not an open subset of R^3. An open subset of R^3 is a union of open 3-balls and $\overset{\circ}{D}{}^2$ contains no open 3-balls at all. ▮

Prop 2.2. Conditions (1) and (2) in Definition 2.1 are mutually exclusive. ▮

Definition 2.3. Suppose M^m is a smooth m-manifold, N^n is a smooth n-manifold and $f\colon M^n \to N^n$ is a function. We say f **is a smooth map** if, for every coordinate map ϕ of M^m and ψ of N^n, the maps $F = \psi^{-1} \circ f \circ \phi$ are smooth functions—from $R^m \to R^n$ in the case of interior points or from $R^m_+ \to R^n_+$ in the case of boundary points. Such a map F is called a **local parametrization for** f. A homeomorphism f which is smooth and has a smooth inverse is called a **diffeomorphism**. We say M^m **and** N^n **are diffeomorphic** and write $M^m \approx N^n$. ▮

The next critical point definitions, formally needed, and may make it seem like we will need extensive multivariable calculus calculations. But for our topological considerations we will later find rely on general theorems (such as Section 8.1) which will allow us to avoid such detail.

Definition 2.4. Suppose M^m is a smooth m-manifold, N^n is a smooth n-manifold and $f\colon M^n \to N^n$ is a smooth map. Consider a local parametrization F. Let $J(x) = \{\frac{\partial F_i}{\partial x_j}\}$ be the matrix of partial derivatives of $F = F_1(x_1, \ldots, x_m), \ldots, F_n(x_1, \ldots, x_m)$ called the **Jacobian matrix**, and we call the integer $r(x_1, \ldots, x_m)$ the **rank of f at** $(x_1, \ldots, x_m)$. A **critical point** $(x_1, \ldots, x_m)$ is one where the Jacobian does not have maximal rank. The image of a critical point is a **critical value.** ▮

Local parametrization, (and therefore the Jacobian matrix) is not unique. However, the *rank* of the Jacobian does not depend on choice of F.

The Jacobian is the linear transformation that best approximates F. This matrix has a geometric interpretation. We view a matrix as a linear function $R^m \to R^n$. The image of a linear transformation is a vector subspace of R^n. The rank of such a linear transformation is the dimension of its image. We will often consider the case that $m \geq n$ in which case we will have a critical point for this linear transformation if the dimension of its image is less than n.

For the special case, $F\colon R^m \to R^1$, the Jacobian is called the **gradient of F** and is written $\nabla(F)$. This $m \times 1$ matrix can be considered as a vector field. Here, $\nabla(F) = \{\frac{\partial F}{\partial x_j}\}$ is just the sequence of partial derivatives of F. So a critical point of F is a point where all partial derivatives vanish.

Definition 2.5. The **Hessian matrix of** $F\colon R^m \to R^1$, denoted $H(f)$ is the matrix of all second partial derivatives of F, so $H(F) = \{\frac{\partial^2 F}{\partial x_i \partial x_j}\}$. If p is a critical point of F we say it is a **non-degenerate critical point** if the Hessian of F at that point has maximal rank m (or equivalently has non-zero determinant), otherwise it is a **degenerate critical point.** ▮

In the case of two variables, $z = F(x, y)$, a **critical point** of a smooth function F occurs at (x_0, y_0) when $\frac{\partial h}{\partial x}(x_0, y_0) = 0$ and $\frac{\partial h}{\partial y}(x_0, y_0) = 0$. The **Hessian** $H(F)$ is the 2×2 matrix of all second partial derivatives at (x_0, y_0):

$$H(h, a) = \begin{vmatrix} \frac{\partial^2 F}{\partial x^2}(x_0, y_0) & \frac{\partial^2 F}{\partial x \partial y}(x_0, y_0) \\ \frac{\partial^2 F}{\partial y \partial x}(x_0, y_0) & \frac{\partial^2 F}{\partial y^2}(x_0, y_0) \end{vmatrix}.$$

In particular, if $F(x, y) = x^2 - y^2$, $\nabla(F) = (2x, -2y)$, $H(f)$ is $\begin{pmatrix} 2 & 0 \\ 0 & -2 \end{pmatrix}$ and the critical point at $(0, 0)$ is non-degenerate with critical value 0. The tangent plane at $(0, 0)$ is horizontal and its image is the single (0-dimensional) vector subspace $\{0\} \subseteq R^1$. Local models of non-degenerate critical points are discussed in Proposition 8.1.

Recall the second derivative test for critical point x_0 of $y = f(x)$. View the second derivative $f'')(x)$ as a 1×1 matrix. A critical point x_0 is non-degenerate means $f'')(x_0) = 0$.

Stereographic mapping is a diffeomorphism of an n-sphere minus a point (the south pole) and R^n.

It is not hard to see that if M is a smooth manifold, so is $M \times I$, see Section 2.3. Now we can define:

Definition 2.6. An isotopy $F\colon M \times I \to N \times I$ is a **smooth isotopy** if M, N and F are smooth. ∎

It can be shown that if M^m and N^n are homeomorphic then $m = n$. The special case where R^m and R^n are homeomorphic then $m = n$ is often stated as a corollary of a well-known non-trivial theorem known as "invariance of domain".

Suppose $f : R^m \to R^k$ is a smooth embedding. At each x_0 we have an $n \times k$ matrix $D(f)((x_0)$ of the partial derivatives $\frac{\partial f_i}{\partial x_j}$. For each x_0 $D(f)(x_0)$ will be of rank m (i.e. maximal rank). If $m = k = 1$, $D(f)$ is just the derivative of beginning calculus. If f is a diffeomorphism this derivative must be non-zero for all x_0. In the case $f : R^m \to R^m$, the embedding will be a diffeomorphism, the matrix will be square and the determinant of $D(f)((x_0)$ will be non-zero for each x_0.

For example, consider $f(x) = x^3$. This is a homeomorphism of the manifold R^1, but it is *not* a diffeomorphism since $\frac{df}{dx}(0) = 0$. Remark 2.8 contains a two-dimensional example.

As we will see, functions which are differentiable and with continuous second derivatives have strong geometric properties. For 2-manifolds in R^3, differentiable means that a manifold can be locally approximated by plane. Additionally, an intersection of such manifolds generally looks locally like an intersection of planes, and so we generally expect two smooth manifolds to intersect in curves. This is the basic idea behind general position such as in Definition 12.2. Having continuous second derivatives means that a manifold can be locally approximated by quadratic surfaces where we have three models—local maxima, local minima and saddle points. This is the basic idea behind Morse Theory as in Section 8.1.

The derivatives of higher order will not be directly addressed in our considerations but are used in the (omitted) proofs of basic theorems we will quote.

In terms of manifolds and connectedness:

Prop 2.3. Any connected manifold $M^k \subseteq R^n$ is path connected. ∎

2.2 Tangent Vectors, Tangent Spaces, Tangent Bundles

The familiar notions from multivariable calculus in R^3 of a tangent vector to a curve or a tangent plane to a surface extend naturally to a notion of a tangent space to a smooth n-manifold $M^n \subseteq R^k$. A **tangent vector** at $x \in M^n$ is the k-dimensional vector tangent to any smooth curve in M^n passing through x. At each $x \in M^n$ the **tangent space to** M^n **at** x is the n-dimensional affine space T_x that contains x and best approximates M^n near x. It is the union, T_x, of all tangent vectors of M^n at x. As pointed out in Remark 1.4 these are *based* vectors. A **smooth tangent vector field on a manifold** M^n is smooth function that assigns to $p \in M^n$ a vector based at p; a smooth function that assigns to p, a frame of n vectors based at p, is called a **smooth tangent framing of** M^n. ∎

There is a map called the **exponential map** $exp : T_x \to M^n$ defined as follows. Consider $\overrightarrow{v} \in T_x$. There is a unique geodesic $\gamma(t) \in M^n$ that begins at $\gamma(0) = x$ with tangent to this curve $d\gamma(0) = \overrightarrow{v}$. This γ is the path that a particle in M^n would take if it were to travel "always straight ahead, beginning at x in direction $\overrightarrow{v} \in T_v$". In the tangent plane, a geodesic is a straight line L_x containing x with direction v. Near x, the tangent plane is close to M^n. So γ is the curve in M^n close to L_x. We define $exp(\overrightarrow{v})$ to be the endpoint of γ which is a distance $|\overrightarrow{v}|$ from x as measured along γ. There is a number ϵ_x called a **radius of injectivity at** x, so that if $B = \{\overrightarrow{v} \in T_x \colon \|\overrightarrow{v}\| < \epsilon_x\}$, then $exp|_B$ will be an embedding. For example on S^2, a geodesic follows a great circle. Near x sub-arcs of great circles through x will be disjoint, except for x. However they will eventually (all) intersect at the point of S^2 antipodal to x. .

In cartesian coordinates for R^2 the line tangent to S^1 at $(1,0)$ is given by $T = \{(1,s) \text{ with } y \in R^1\}$. We can wrap T about S^1 in the way we wrapped R^1 about S^1 in Example 1.5. That is, viewing S^1 as unit complex numbers, define $E\colon T \to S^1$, by $E\left((1,s)\right) = exp(si)$. In this case, E is an exponential map in the sense of the previous paragraph as well the complex exponential function.

A smooth map $f : M^n \to N^k$ gives an affine map from T_x to $T_{f(x)}$ as follows. If $\overrightarrow{v} \in T_x(M)$ is represented by a curve $\gamma\colon R^1 \to M$ with $\gamma(0) = x$ then *df* is represented by the curve $f \circ \gamma$. If *df* is non-singular at $x \in M^n$

then the image of T_x is an affine n-space. If f is a diffeomorphism this map will be a 1-1 correspondence for each x.

A fundamental concept of smooth topology is the **tangent bundle** $T(M^n)$ for a smooth n-manifold M^n which corresponds to a union of all the tangent planes T_x for $x \in M^n$. In the literature this is often called the **total space of the tangent bundle**. We will see $T(M^n)$ is a $2n$-dimensional manifold. The first step is to show $T(M^n) \subset R^k$ for some k.

Example 2.3. Consider the 1-manifold R^1. For any $x \in R^1$, it is tempting to say that the tangent line T_x is the set R^1 itself. But if $x \neq y$, we must have T_x and T_y *disjoint* because a vector cannot be based at two different points. The tangent bundle $T(R^1) \approx R^2$ clears up this situation. Here the x-axis corresponds to R^1 and, for $x_0 \in R^1$, T_{x_0} corresponds to the vertical line of the points $\{(x_0, y)\}$. So, if $x \neq y$ then $T_x \cap T_y = \emptyset$. More generally $T(R^n) \approx R^n \times R^n$. ∎

More interesting is S^1. The tangent bundle $T(S^1) \approx S^1 \times R^1$.

Example 2.4. First, identify S^1 with the circle $C \subseteq R^2_+$ of radius 1, center $(2, 0)$ and write points of C as θ with $0 \leq \theta \leq 2\pi$. For each θ, represent tangent line T_θ to C at θ as a short open line segment, T'_θ, tangent at θ. However view $0 \leq \theta \leq 2\pi$, not as a subset of R^2_+ but rather as the corresponding subset T''_θ of the page H^2_θ. Figure 2.3 shows eight such T''_θ. (We also see eight copies of C, but these are only there to a visual aids in this graphic.) If we increase the number of segments we begin to see how these line segments T''_θ fit together, Figure 2.4. Now the $T(S^1)$ corresponds to the "union of all of T''_θ", which is clearly an open annulus Figure 2.5 and diffeomorphic to $S^1 \times R^1$. ∎

Fig. 2.3 Tangent line segments to circle in eight pages.

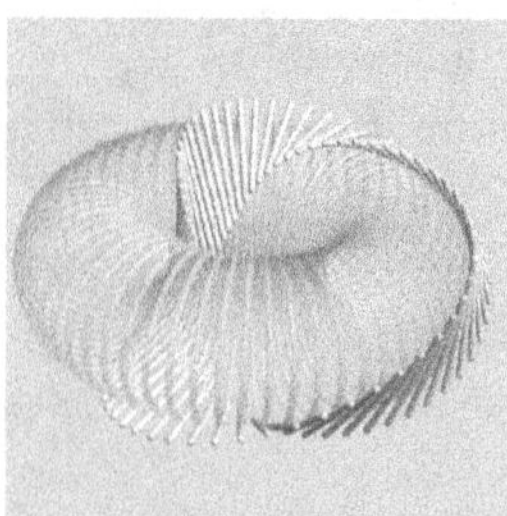

Fig. 2.4 More tangent line segments.

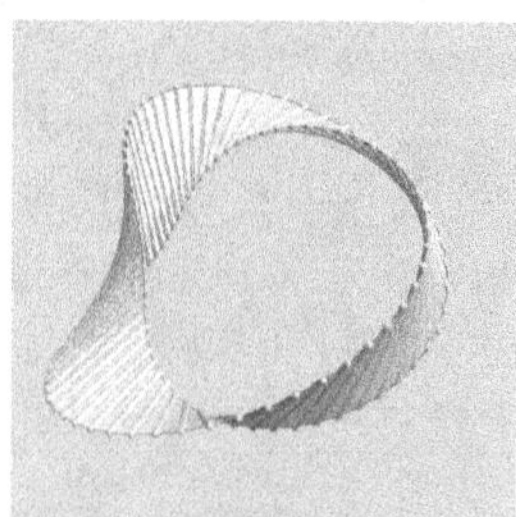

Fig. 2.5 The union of all is an annulus.

The argument of Example 2.4 relied on shrinking the tangent line and cylindrical coordinates. This was done to visualize $T(S^1)$ in R^3. There is another mathematically natural way to describe $T(S^1)$ in R^4. This method needs higher dimensions, but it is better in that it can be applied to the tangent bundle of any smooth manifold. We view $S^1 \subset R^2 = R^2 \times \{0\} \subset R^2 \times R^2 \approx R^4$. Write $p = (a, b) \in S^1$. The equation of the tangent line L_p will be of the form $A(x-a) + B(y-b) = 0$. Consider the line in R^4 with equations $x = a, y = b$, and $A(z-a) + B(w-b) = 0$. We have $L_p \cap L_y = \emptyset$ if $p \neq q$. The union of all these will be the 2-manifold $T(S^1) \subseteq R^4$.

It can be more generally shown that if $M^n \subset R^k$ is a smooth manifold, then $T(M^n)$ is a subset of R^{2k} and, in fact, a smooth $2n$-dimensional manifold. This manifold will be non-compact. However, if we consider *unit* tangent vectors for a compact manifold M^n then $T(M^n)$ will be a compact smooth manifold (of dimension $2n-1$).

Later we will give examples of the following:

Prop 2.4. If $M^n \subset R^k$ is a smooth manifold, the set of all unit vectors $T_1(M^n)$ is called the **unit tangent bundle of** M^n or the **unit sphere tangent bundle**; $T_1(M^n)$ is an $(2n-1)$-manifold. The set of all vectors of length 1 or less in $T_1(M^n)$ is called the **unit disk tangent bundle**; it is a $(2n)$-manifold whose boundary is $T_1(M^n)$. ∎

Remark 2.4. That $T(S^1)$ is a cartesian product $S^1 \times R^1$ is special to S^1. For higher dimensions $T(S^n) \approx S^n \times R^n$ only if $n = 1, 3$ or 7. An explanation of this important result is far beyond the scope of our discussion here. Any smooth manifold M^n such that $T(M^n) \approx M^n \times R^n$ is called a **parallelizable manifold.** ∎

2.3 Cartesian Products of Manifolds

In one way or another, cartesian product is part of almost every construction we consider.

Prop 2.5. If N^n is an n-manifold and M^m a m-manifold then $N^n \times M^m$ is a $(n+m)$-manifold. ∎

Proposition 2.5 is straightforward except in the smooth case where both manifolds have non-empty boundaries, see Remark 2.5. For example, an n-dimensional manifold N^n is locally (homeomorphic) diffeomorphic to R^n, an m-manifold M^m is locally (homeomorphic) diffeomorphic to R^m, it follows that $N^n \times M^m$ is locally (homeomorphic) diffeomorphic $R^n \times R^m = R^{n+m}$.

Remark 2.5. For smooth manifolds such that $\partial N^n \neq \emptyset$ *and* $\partial M^m \neq \emptyset$ there is a minor technical problem for the subset $\partial N^n \times \partial M^m$ called the **corner set.** There are many ways to define smooth coordinate maps (and therefore tangent vectors) for these corner points, but no simple natural way. The problem and solution are evident in the example of the square $I \times I$ where the corner set is literally the four corner points Figure 2.6, and the cube $I \times I \times I$ where the corner set is the 12 cube edges, Figure 2.8. There is a general method called "rounding the corners" that makes the topological manifold $N^n \times M^m$ into a smooth manifold such that if we have two different rounding of corners, the resulting smooth manifolds are diffeomorphic by a diffeomorphism which is the identity map outside a small neighborhood of the corner set. Figures 2.7 and 2.9 show results of such corner smoothing, but other shapes and sizes of smoothings are also possible. ∎

Fig. 2.6 square $I \times I \subseteq R^2$.

Fig. 2.7 Square with corners rounded.

Fig. 2.8 Cube $I \times I \times I \subseteq R^3$.

Fig. 2.9 Cube with corners rounded.

The important formula for the boundary of a product is discussed in Section 2.7. Other aspects of rounding corners are considered in Section 2.6.

The cartesian product of the standard spheres $S^p \subseteq R^{p+1}$ and $S^q \subseteq R^{q+1}$ is a smooth $(p+q)$-dimensional manifold $S^p \times S^q \subseteq R^{p+1} \times R^{q+1} \approx R^{p+q+2}$. called a **standard product of spheres.**

Definition 2.7. A manifold diffeomorphic to n copies of S^1 is called an n-**torus**. The 2-**torus** is mentioned so often in this book that unless made clear otherwise, "torus" will mean 2-torus. The four-point set $S^0 \times S^0$ 1s a **0-torus**—for more on this see Section 2.7. ∎

In a similar way, $D^p \times S^q$ or $S^p \times D^q$ are a smooth $(p+q)$-manifolds, both called a **standard solid torus.** These appear naturally as subsets of S^{p+q} in Section 1.8. Unless stated otherwise, **a solid torus** is a manifold diffeomorphic to $S^1 \times D^2$ and an **annulus** diffeomorphic to $S^1 \times D^1$.

Remark 2.6. In a 3-dimensional solid torus $M \approx S^1 \times D^2$, a circle corresponding to $\{x\} \times \partial D^2$ is called a **meridian of** ∂M. A circle corresponding to $S^1 \times \{x\}$ is called a **longitude of** ∂M. One sometimes finds the terms meridian and longitude applied to a 2-torus $T \approx S^1 \times S^1$ where it is understood from context that $T = \partial M$ for some solid torus. This standard terminology clashes with geographical nomenclature for a 2-sphere. ∎

Example 2.5. Any interval or finite union of intervals (open, closed or half-open) in R^1 is a 1-manifold. If D_i is a collection of disjoint closed 2-disks in $Int(D^2) \subset R^2$ the set $P = D^2 - \bigcup int(D_i)$ is called **a disk with** p **punctures**. Then P is a manifold, and its boundary is a disjoint union of $p+1$ circles. A disk with one puncture is an **annulus.** ∎

Remark 2.7. There is no mathematically compelling reason to allow the empty set to be considered as a manifold, smooth or otherwise. However, if we do *not*, then statements of many general theorems become awkward due to a need to explicitly address empty set cases. So, most authors *do* allow the empty set to be a (smooth) n-manifold and furthermore declare it to have dimension -1. We will see this used in Remark 2.15. In fact, there are circumstances when one wants to declare that the empty set has dimension n for *any* negative number. ∎

2.4 The Definition of Submanifold

It might seem natural to simply define a "submanifold" to be "a subset of a manifold which itself homeomorphic to a manifold" and then say that a knotting is a submanifold. However, unless we impose some restrictions, there are examples called "**wild knottings**" or "**wild submanifolds**", see Remark 3.1, Figure 3.2 and Section 5.5. Topologists have devised three definitions of submanifold that exclude wild knottings: **locally flat submanifold, smooth submanifold** and **PL-submanifold**. These definitions are equivalent for n-manifolds with $n = 1, 2, 3$. However, in high-dimensional settings, important and interesting differences appear. Our preference is for "smooth" manifolds and submanifolds and we will deal with these other types only in passing.

The basic approach to all three definitions is to say that a submanifold looks locally like some standard pair such as $R^k \subseteq R^n$ where "looks like" means "diffeomorphic to". For manifolds with non-empty boundary, there are two choices of "standard pair". Each is useful. Figure 2.10 and Figure 2.11 show a manifold $M^2 = D^2$ and a subset N^1 diffeomorphic to D^1. The endpoints of the interval in Figure 2.10 are modeled on a standard pair $R^1_+ \subseteq R^2$, see Figure 2.12 while the endpoints in Figure 2.11 are modeled on a standard pair $R^1_+ \subseteq R^2_+$ as in Figure 2.13.

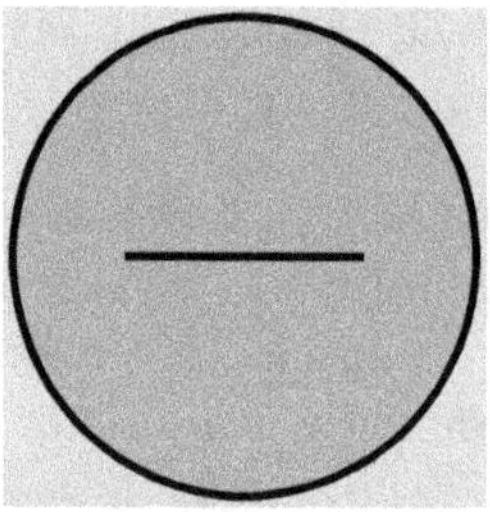

Fig. 2.10 A smooth 1-disk in the interior of D^2.

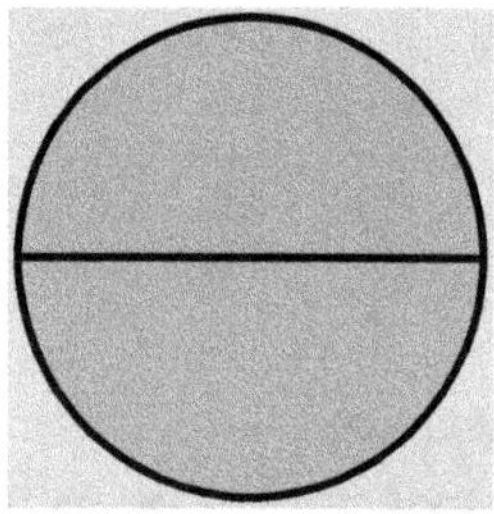

Fig. 2.11 A smooth *proper* 1-disk in D^2.

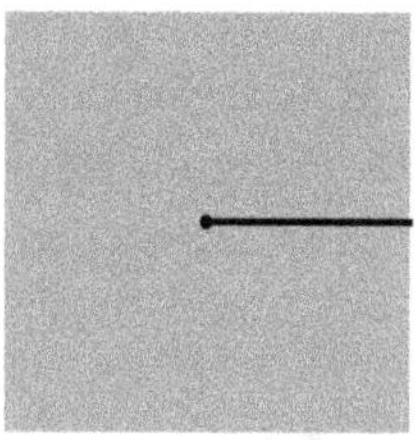

Fig. 2.12 The standard pair $R^1_+ \subseteq R^2$.

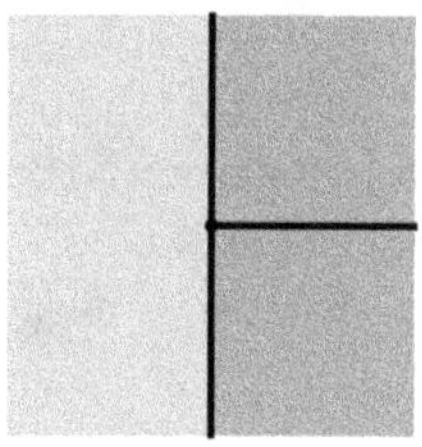

Fig. 2.13 The standard pair $R^1_+ \subseteq R^2_+$.

Piecewise-linear topology (PL-topology) views an n-manifold as a union of n-simplexes (of a certain "nice" kind. Much of the foundational work in geometric topology was done from this point of view, [Hudson (1969)],[Bryant (2002)], [Rourke and Sanderson (1982)].

To discuss the intersection of smooth manifolds (perhaps of different dimensions) we introduce:

Definition 2.8. Suppose N_0 and N_1 are smooth manifolds of R^m. We say N_0 **is transverse to** N_1 if for any $p \in N_0 \cap N_1$ every non-zero tangent vector to N_0 at p and non-zero tangent vector to N_1 at p form a non-zero angle. We say N_0 **is orthogonal to** $\cap N_1$ if every non-zero tangent vector to N_0 at p is orthogonal to any non-zero tangent vector to N_1 at p. ∎

Definition 2.9. If $N^k \subseteq M^n \subseteq R^m$ for some m, we say N^k **is a smooth** k**-submanifold (in the interior of** M^n**)** if for every $x \in N^k$ there is a smooth embedding $h\colon R^m \to R^m$ such that either:

(1) **(Condition for non-boundary points)** $h(\mathbb{O}) = x$, $h(R^n) \subseteq M$ and $h(R^k) \subseteq N$
(2) **(Condition for boundary points)** $h(\mathbb{O}) = x$, $h(R^n) \subseteq M$ and $h(R^k_+) = h(R^m) \cap N$.

We say N^k **is a proper smooth** k**-submanifold of** n**-manifold** M^n if for every $x \in N^k$ there is a smooth embedding $h\colon R^m \to R^m$ such that either:

(1) **(Condition for non-boundary points).** $h(\mathbb{O}) = x$, $h(R^n) \subseteq M^n$ and $h(R^k) \subseteq N^k$
(2) **(Condition for boundary points) (Local** R_+**-pair modeling).** $h(\mathbb{O}) = x$, $h(R^n_+) = h(R^m) \cap M^n$ and $h(R^k_+) = h(R^m) \cap N^k$.
(3) **(Orthogonality at the boundary).** Additionally, N^k is orthogonal to ∂M^n for all points of $\partial N^k \cap \partial M^n = \partial N^k$.

We will call any of the maps h, above, **local submanifold charts**. In each of the above cases, if the manifolds M^n and N^k are smooth and all maps h are smooth, we say that N^k is a **smooth submanifold** of M^n (or, in the proper case) a **proper smooth submanifold**. The **codimension of** N^k **in** M^n is $(n-k)$. ∎

The standard disk pair (D^n, D^k) is an example of a smooth proper submanifold; see Figure 2.25 for (D^3, D^2). More generally a pair (D^n, X) where K is a proper submanifold with $X \approx D^k$ is called a (n,k)-ball pair or (n,k)-disk pair.

It can be shown that any smooth embedding $\phi : R^n \to R^m$ can be extended to a smooth embedding $\Phi : R^n \to R^m$. Of course, R^m is a smooth manifold, so every smooth manifold $M^k \subseteq R^m$ is a smooth submanifold of R^m.

The definition of **locally flat submanifold** can be obtained from Definition 2.9 by replacing "smooth" by "topological" and omitting the third condition for proper embeddings. Submanifolds which are locally flat but not smooth are called "wild" and are mentioned only in passing in this book, Remark 3.1, Figure 3.2 and Section 5.5.

Remark 2.8. One implication of local R_+-pair modeling for the *smooth* case is that the submanifold N^k is required to meet the boundary ∂M^n transversely, instead of tangentially. For example, consider $f : R^2_+ \to R^2_+$ defined by $f(x, y) = (x, y + x^{1/3})$ is a homeomorphism but fails to be a diffeomorphism on the $axis(y)$, which is (∂R^2_+). It sends R^1_+ to a curve tangential to ∂R^2_+, see Figure 2.14.

A common definition of smooth proper manifold omits the orthogonality at the boundary condition. We have adopted the extrinsic view and will be interested in constructing submanifolds using unions. The orthogonality property helps with this. Figure 2.15 shows a proper arc E in a hemisphere, Figure 2.16 shows a proper arc F in another hemisphere. Taking the union. we get a smooth 2-sphere with a smooth 1-dimensional submanifold, Figure 2.17. We have smoothness at point p indicated by the tangent line, due to our orthogonality condition. Consider the sub-arc G shown in Figure 2.18. This satisfies the first two conditions of proper submanifold but not the orthogonality—G intersects the boundary *transversely* but *not* orthogonally. When we take the union, we have a locally flat submanifold but not a smooth one, Figure 2.19. Although we can "fix this" if we "straighten up" G slightly near p, as in Figure 2.20, we prefer to avoid this by using proper submanifolds. A more general discussion of unions of manifolds is in Section 2.6. ∎

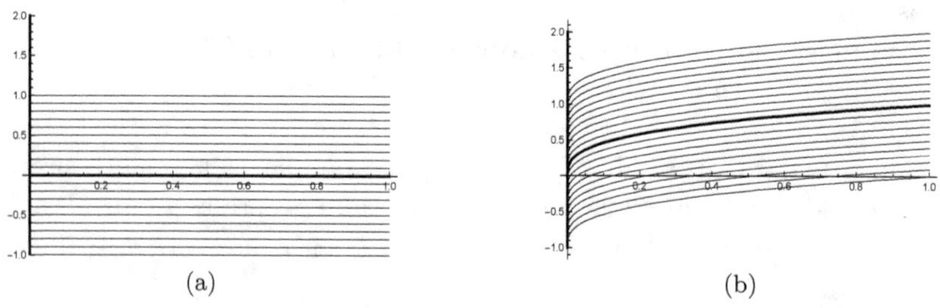

Fig. 2.14 The homeomorphism $f(x, y) = (x, y + x^{1/3})$ left to right is not a diffeomorphism at points of y-axis. It takes y-axis to itself and lines orthogonal to y-axis sent to curves tangent to y-axis.

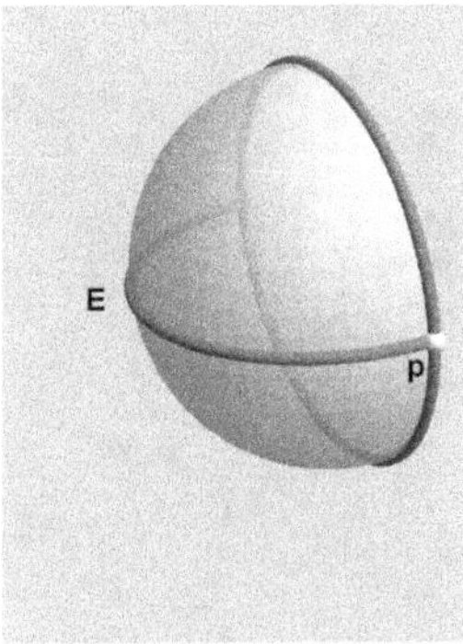

Fig. 2.15 Proper arc E in hemisphere.

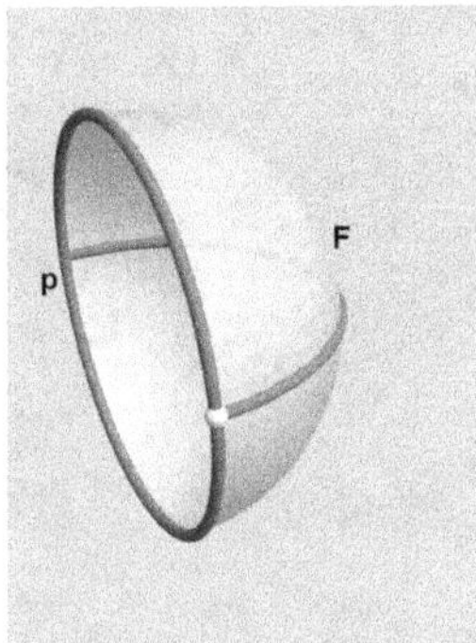

Fig. 2.16 Proper arc F in another hemisphere.

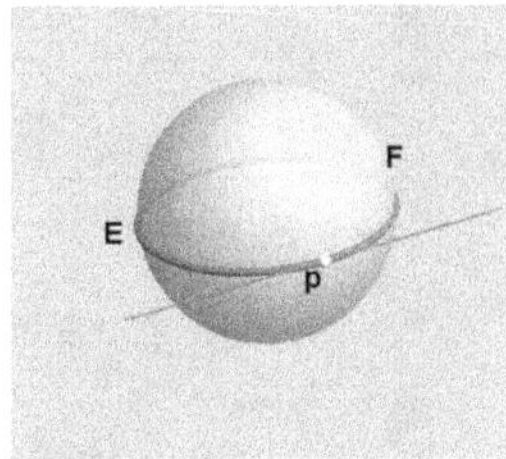

Fig. 2.17 Union of Figures 2.15 and 2.15.

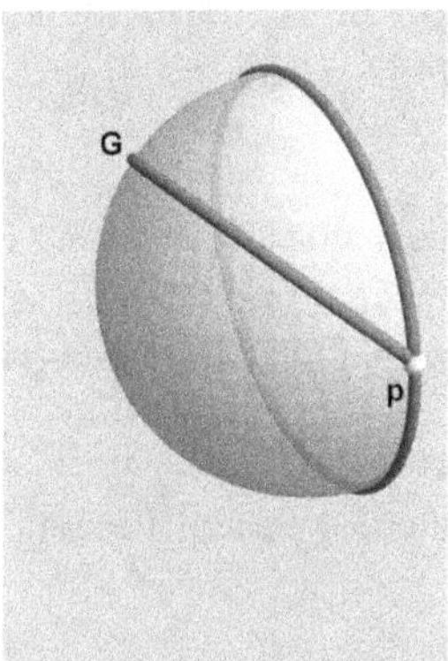

Fig. 2.18 An arc G transverse to boundary but not orthogonal.

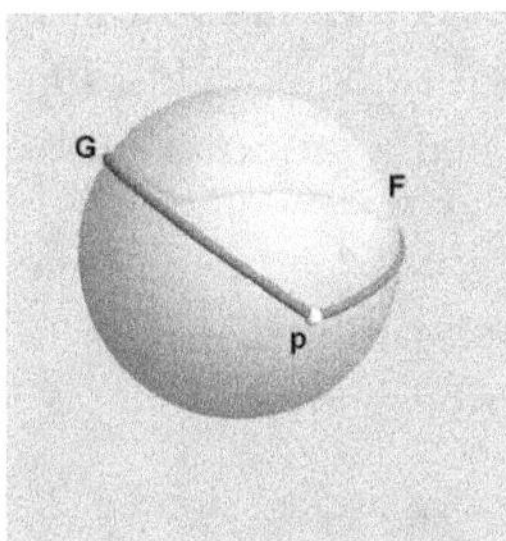

Fig. 2.19 Union of G and F not smooth at p.

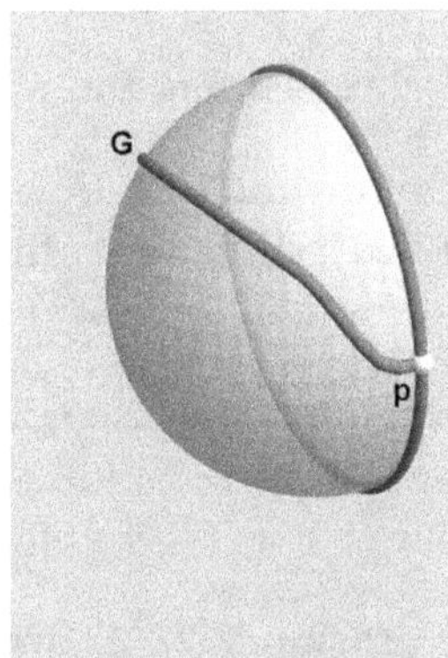

Fig. 2.20 Bending G slightly to make it proper.

In some settings as submanifold contained in a sub-ball is considered "uninteresting" or inessential.

Definition 2.10. If M^n is a smooth submanifold of smooth manifold N^k, B^k is a smooth k-ball with $M^n \subseteq B^k \subseteq N^n$ then we say M^n **is an inessential submanifold of** N^k. Otherwise, M^n is **an essential submanifold of** N^k or M^n **is an essential knotting in** N^k. ∎

Figure 2.21 shows two circles in a solid torus. Circle A is an essential submanifold, and B is not.

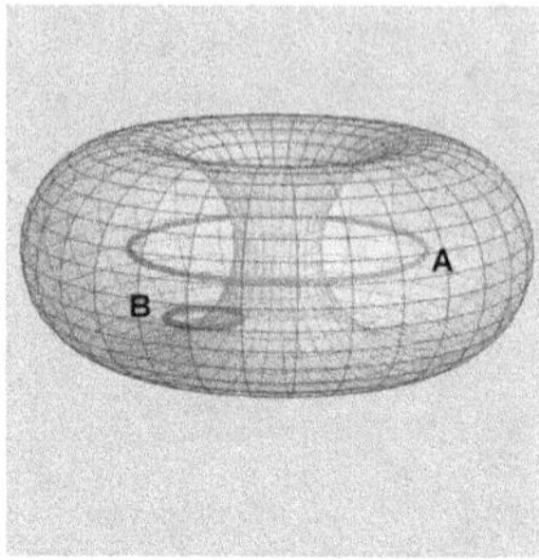

Fig. 2.21 Two circles in a solid torus.

The graph of a continuous (or smooth) function $f\colon R^k \to R^1$ (see Definition 1.15 for general definition) is a topological (or smooth) manifold in $R^k \times R^1 \approx R^{k+1}$. We illustrate some definitions using the graph of a real-valued function.

Definition 2.11. A smooth homeomorphism, F, with a smooth inverse, is called a **diffeomorphism**. A smooth embedding, F, with a smooth inverse (defined on its image) is called a **smooth embedding.** ▮

Example 2.6. Note that $y = x^3$ is not a diffeomorphism of R^1—it *is* a smooth homeomorphism but the inverse $y = x^{1/3}$ has a vertical tangent line at $x = 0$ and so fails to be smooth. ▮

Notation 2.2. We will use the notation $M \approx N$ to mean M and N are diffeomorphic. This is the same notation as homeomorphic, Notation 1.6. There are examples of smooth homeomorphic manifolds that are not diffeomorphic in high-dimensions mentioned briefly in Section 12.1. ▮

2.5 Normal Vectors, Normal Disks, Tubular Neighborhoods, Orientability

Tubular neighborhoods are fundamental to geometric topology. For a knotting of a circle in R^3, a tubular neighborhood is a small tube about the knotting. The usual definition of tubular neighborhood as in [Hirsch (1976)] or [Guillemin and Pollack (2010)], for example, has more structural detail than we use. Additionally, we consider tubular neighborhoods of images of immersions as well. We focus on a special case tubular neighborhood: a *product* tubular neighborhood.

Definition 2.12. Suppose N^k is a smooth submanifold of M^n with $x \in N^k$. We have a local submanifold chart $h\colon R^n \to M^n$ with $h(R^k) \subseteq N^k$ and $h(\mathbb{O}) = x$. At each $p \in R^k \subseteq R^n$ we have an $(n-k)$-disk $D_p^{n-k} = \{p\} \times D^{n-k} \subseteq R^k \times R^{n-k} \approx R^n$ whose image, $N_x^{n-k} = h(D_p^{n-k})$, is a smooth $(n-k)$-disk in M^n, containing x.

Each D_p^{n-k} meets $R^k \subseteq R^n$ orthogonally in a single point p and so each N_x^{n-k} meets N^k transversely in a single point $x = h(p)$. We say N_x^{n-k} **is a normal disk to** N^k **in** M^n **at** x. Using techniques based on the ideas of Example 3.3, we can, if we wish, improve this construction so that the disks N_x^{n-k} will, in fact, be orthogonal to N^k. However, in most applications transversality will be sufficient.

Figure 2.22 shows some normal disks, N_x^2, for a knotting of a circle in R^3. In this Figure the disks are not disjoint. However, we can choose smaller disks if necessary so they are disjoint as in Figure 2.23. A **tubular neighborhood** is a union of all such disks. It is a non-trivial fact that they fit together to form a smooth manifold, Figure 2.24. ▮

The **tubular neighborhood theorem** is basic in geometric topology. A treatment is found in all basic references. A tubular neighborhood is not a unique subset but any two are equivalent as submanifolds:

Prop 2.6. Tubular Neighborhood Theorem Any compact smooth submanifold $N^k \subseteq M^n$ has a tubular neighborhood T. If T' is another tubular neighborhood, there is a smooth isotopy F_t of M^n such that $F_1(T) = T'$ which is the identity on N^k. ▮

The structure of a tubular neighborhood can be complicated. To keep terminology simple, we will mostly consider product tubular neighborhoods where, as in Figure 2.24, these disks fit together to form a manifold diffeomorphic to $N^k \times D^{n-k}$. For such a case, we use the next definition.

Definition 2.13. For a proper submanifold $N^k \subseteq M^n$, a **product tubular neighborhood** of N^k is a smooth submanifold $T \subseteq M^n$ with $N^k \subseteq T$ where $T \approx N^k \times D^{n-k}$ and N^k corresponds to $N^k \times \{0\}$. If $\partial M^n \neq \emptyset$ then we also require that $T \cap \partial M^n$ is a (product) tubular neighborhood of $\partial N^{n-k} \subseteq \partial M^n$.

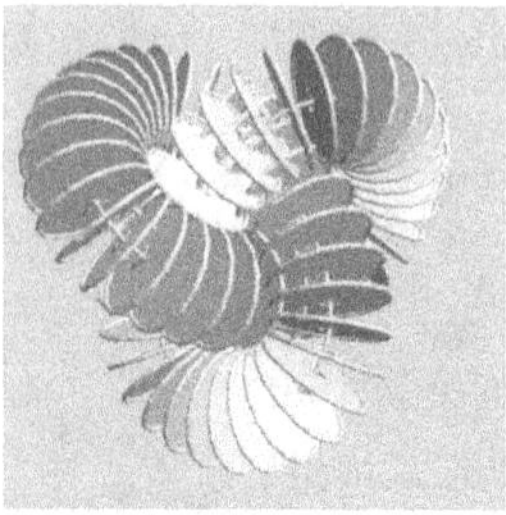

Fig. 2.22 Normal disks with large radii can intersect.

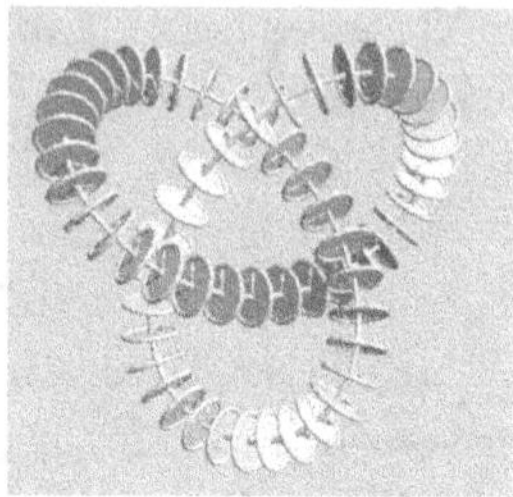

Fig. 2.23 For small enough radii normal disks will be disjoint.

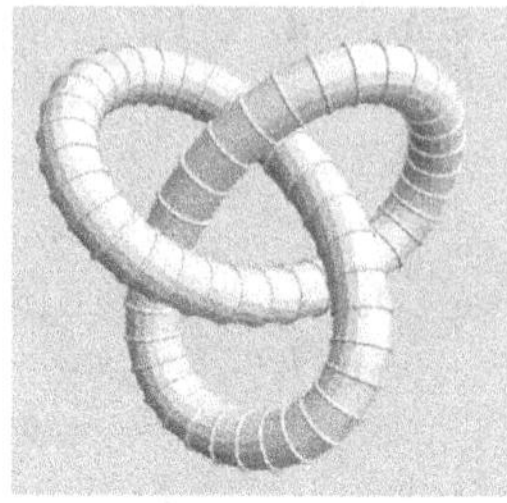

Fig. 2.24 The union of all such normal disks is a solid tube.

A choice of diffeomorphism $\phi\colon N^k \times D^{n-k} \to T$ (equivalently an embedding choice of $N^k \times D^{n-k}$ into M^n) is called a **trivialization of** T.

Additionally, if we have a proper smooth immersion $f \looparrowright N^k \to M^n$ a **product tubular neighborhood of** f is a proper smooth immersion $F \looparrowright N^k \times D^{n-k} \to M^n$ such that $F|_{N^k \times \{O\}} = f$. ▮

Remark 2.9. If C is a circle embedded in a 2-sphere or torus (or any orientable 2-manifold) then it will have a trivial tubular neighborhood—diffeomorphic to an annulus. However, the center line C of a Möbius band has a tubular neighborhood which is also a Möbius band which is an example of a non-trivial tubular neighborhood.

A smooth knotting of a circle in $N_1 \subseteq R^3$ always has a trivial tubular neighborhood T^3—each has infinitely many trivializations which correspond to twisting of the disk along the knotting. If $\phi\colon S^1 \times D^2 \to T$ is a smooth trivialization, then $A = \phi\colon S^1 \times D^1$ is a smooth annulus. The **twist number of this tubular neighborhood** is defined to be the twisting of this annulus.

The standard disk pair (D^3, D^2) is proper; Figure 2.25 shows a (product) tubular neighborhood $T \approx D^2 \times D^1$. Note that $T \cap \partial D^3$ (an annulus) is tubular neighborhood of ∂D^1 in ∂D^3. ▮

The homogeneity of manifolds basically says that for a connected n-manifold "all points basically look the same". This means that if p and q are any two points of M^n there is a homeomorphism $h\colon M^n \to M^n$ with $H(p) = q$. It is similarly true for a *smooth* manifold M^n: there is a *diffeomorphism* $h\colon M^n \to M^n$ with $H(p) = q$. One way to see this is to take a smooth arc A between p and q. It can be shown that A has a product tubular neighborhood T. This T is a smooth n-disk containing p

Fig. 2.25 Standard ball pair (D^3, D^2).

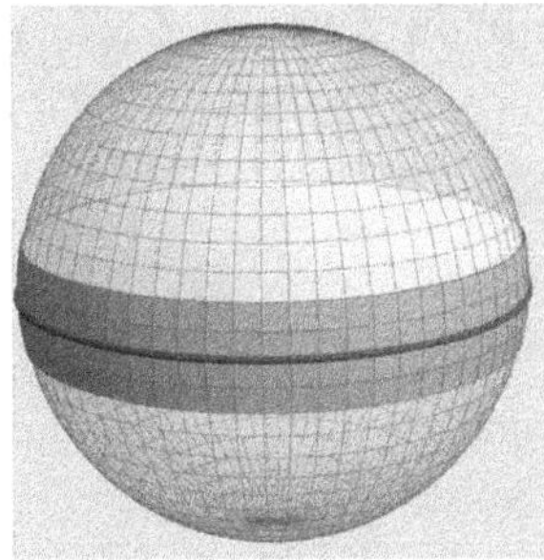

Fig. 2.26 Tubular neighborhood of (D^3, D^2).

and q. One can slide p along A to q inside T by a smooth isotopy that does not move points outside of T.

If we view a point as a 0-disk, homogeneity says that all 0-disks in a connected manifold are equivalent. Actually, this is true for k-disks. All k disks in a connected manifold are isotopic.

Prop 2.7. If M^n is closed connected smooth n-manifold and $P, Q \in M^n$ are any two subsets diffeomorphic to D^k then there is a smooth isotopy H_t of M^n with $H_1(P) = Q$. ∎

Suppose we have a smooth manifold M^n with two smooth n-disks, B^n and C^n with $b \in int(B^n)$ and $c \in int(C^n)$. We can consider $\{b\}$ and $\{c\}$ to be 0-dimensional submanifolds of M^n with product tubular neighborhoods B^n and C^n. Then Proposition 2.6 is the smooth homogeneity theorem Proposition 2.7.

It is also true that any two *embeddings* of a k-disk in a connected manifold are smoothly isotopic, *except for the case that* $k = n$. If $k = n$, one needs to consider orientation and orientability.

Not all submanifolds have *product* tubular neighborhoods. The center circle of a Möbius band does *not* have a product tubular neighborhood.

There are two basic kinds of manifold—orientable and non-orientable. All 1-manifolds are orientable. A 2-manifold is **non-orientable** if it contains a Möbius band—otherwise, it is **orientable**. A smooth knotting of a circle k in an n-manifold, $n \geq 2$, has a smooth tubular neighborhood T. There are two possibilities. Either k has a product tubular neighborhood $T \approx S^1 \times D^{n-1}$, or it does not have a product tubular neighborhood, in which case $T \approx M^2 \times D^{n-2}$ where M^2 is a Möbius band.

Definition 2.14. If M is an n-manifold with $2 \leq n$, M **is non-orientable** if there is a knotting of a circle in M that does *not* have a trivial tubular neighborhood—otherwise, it is an **orientable manifold.** ▮

Let $r : D^n \to D^n$ be reflection in the first coordinate, namely $r(x_1, \ldots, x_n) = r(x_1, \ldots, x_n)$—$r$ is orientation reversing. Then f and $f \circ r$ are smoothly isotopic if and only if M^n is non-orientable. For a connected orientable M^n, there are exactly two different embeddings of an n-disk up to smooth isotopy.

In R^3 we have two orientations—left-handed and right-handed. This is often described as follows. Suppose $B = \{b_1, b_2, b_3\}$ is an ordered basis for R^3. We define the **determinant of** B, $|B|$ to be the determinant of the 3×3 matrix whose i-th row is the coordinates of b_i. Since vectors B are linearly independent, the determinant is non-zero. If $|B| > 0$, we say $|B|$ is **left-handed**, otherwise **right-handed**. If B is a collection of three linearly independent vectors based at some point $p \in R^3$ (called a **frame**), we define orientation by translating B to $\mathbb{O}$ and using the previous definition for $|B|$.

If a manifold is orientable there is a related notion of two possible choices for orientation. Suppose $M^n \subseteq R^k$ is a smooth *connected orientable* manifold, $p \in M^n$. An **orientation of** M^n is an ordered choice B of n vectors based at p that gives an ordered basis for $T(M^n)_p$. Now consider $q \in M^n$ and a similar choice of basis vectors B' of $T(M^n)_q$. Using the orientability of M^n, we can compare B and B' as follows. We find a smooth circle in M^n that contains p and q. Since M^n is orientable this has a product tubular neighborhood $T \approx S^1 \times D^{n-1}$. If we take the standard $S^1 \subseteq R^n$, it also has a product tubular neighborhood T'. Choose a smooth embedding $\phi : T \to R^n$ such that $\phi(T) = T'$. Then, we have based n-dimensional vectors in R^n, $\phi(B)$ and $\phi(B')$ and we can define $|\phi(B)|$ and $|\phi(B')|$. We say B **has the same orientation as** B' if $|\phi(B)|$ and $|\phi(B')|$ have the same sign.

If we describe a smooth 1-manifold in terms of parametric curves, orientation choice corresponds to direction choice of parametrization. Suppose C is a component of ∂M^2 of an oriented 2-manifold M^2. Suppose the orientation of M^2 is given by a local basis B. This determines an orientation (direction) of C, as follows. Orient C so that at $c \in C$, a directed tangent vector to C followed by a vector pointing inwards to M^2 from C gives a basis for the tangent plane at c that agrees with B. For example, for $D^2 \subset R^2$, if we orient D^2 with the standard orientation of R^2, then the compatible orientation of $S^1 = \partial D^2$ will correspond to a counterclockwise parametrization as found in statements of Stoke's Theorem.

The next proposition is fundamental to geometric topology. The usual proofs involve algebraic topology and so are outside the scope of our presentation.

Prop 2.8. If $M^n \subseteq R^{n+1}$ is a closed smooth submanifold, then M^n is orientable. Also, if M_n is connected, then R^{n+1} has two components, A and B, and $\overline{A} \cap \overline{B} = M^n$. ▮

Remark 2.10. One can use the above result to obtain some non-embedding of some simple punctured 3-manifolds. These technical points will be useful much later in this text (such as in Section 10.11) and will not be needed until then. (These are most often considered as applications of algebraic topology.)

The manifold $(S^2 \overset{\circ}{\times} S^1)$ cannot be smoothly embedded in S^3. If there were an embedding $F\colon (S^2 \overset{\circ}{\times} S^1) \to S^3$, take $p = (a, b) \in (S^2 \overset{\circ}{\times} S^1)$ and consider the 2-sphere $B = S^2 \times \{b\}$ and the 1-sphere $A = \{a\} \times S^1$ then $A \cap B = \{p\}$. The set $A' = A - \{a\}$ is diffeomorphic to an open interval and so, is a connected set.

Now $F(B)$ is an embedded 2-sphere in S^3 and, by Proposition 2.8, we can write $S^3 - F(B) = E \cup F$ where E and F are disjoint non-empty open subsets of S^3. But if we let $E' = F^{-1}(E)$ and $F' = F^{-1}(F)$, then $E' \cap A'$ and $F' \cap A'$ are non-empty open subsets of A' whose union is A', contradicting the fact that A' must be connected. Having arrived at a contradiction, we conclude that, in fact. the embedding F cannot exist. Similarly, we can argue that $(S^1 \overset{\circ}{\times} S^1) \times S^1$ can be smoothly embedded in S^3. ▮

For *orientable* n-knottings we have (for a proof sketch, see Remark 12.9):

Proposition 2.1. *If $N^n \subseteq R^{n+2}$ is a smooth closed orientable submanifold then N^n has a product tubular neighborhood.* ▮

For an immersed circle in a R^2, a product tubular neighborhood is an immersed annulus as in Figure 2.34. For many 2-manifolds, any smooth proper immersion of a 1-manifold in M^2 has a product tubular neighborhood. These manifolds are called **orientable**, see Section 2.5.

Definition 2.15. A based vector at p that is orthogonal to every tangent vector at p is called a **normal vector** at p. A smooth function that assigns to each p a normal vector at p is called a **normal vector field.** If N^k is a smooth submanifold of M^n with a product tubular neighborhood a **normal product framing of** N^k is a smooth function that assigns a normal $(n-k)$-frame at each point of N^k. ∎

Our use of "normal" is not exactly the same as used in calculus, where one uses Frenet frames. In geometric topology all smooth curves have normal vector field; but a Frenet frame is only defined if the curvature is non-zero. A common alternative is a Bishop frame, see [Bishop (1975)].

A trivialization $\phi\colon N^k \times D^{n-k} \to T$ of a product tubular neighborhood gives rise to a normal $(n-k)$-framing of N_k. For each $p \in N_k$ take the standard $(n-k)$ unit frame $s(p)$ based at $\{p\} \times \mathbb{O}$ of $\{p\} \times D^{n-k}$. The normal frame is the image, $d\phi(s(p))$ of this under the differential of the trivialization.

Graphically a product tubular neighborhood is more easily visualized as an image of embedding $S^1 \times Q^2$ rather than $S^1 \times D^2$ Using a Q^2, we see boundary of the tubular neighborhood as a union of four annuli and the twisting of the framing that is apparent in the twisting of these annuli. The simplest with zero twists is shown in Figure 2.28. Figure 2.29 illustrates a single twist, Figure 2.30 illustrates two twists. In Figure 2.27 the vectors are shown from the center of Q to a vertex. In either Figures 2.31, 2.32, and 2.33, the frame consists of a first vector in black, the second in grey.

For any circle knotting in R^3—even for the standard $S^1 \subseteq R^3$—there are infinitely many distinct normal framings as measured by "twisting".

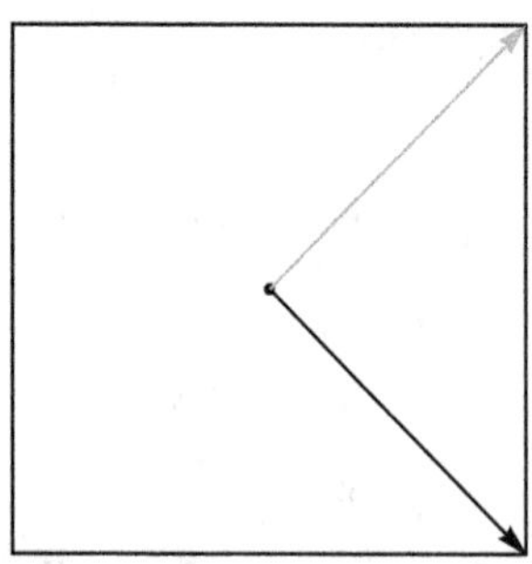

Fig. 2.27 A frame for Q^2.

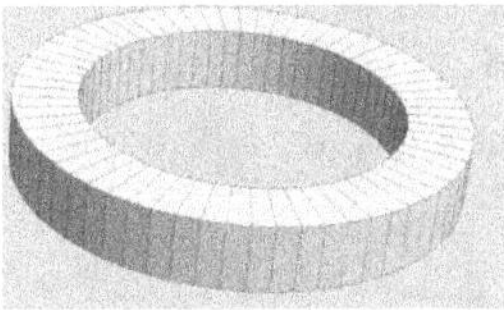

Fig. 2.28 Zero twists.

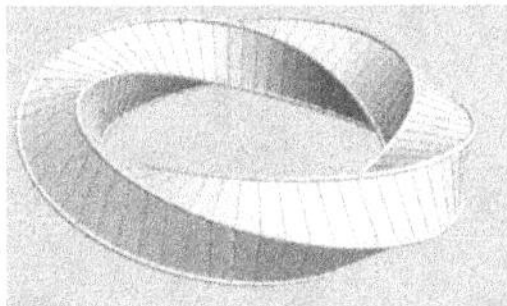

Fig. 2.29 One twist.

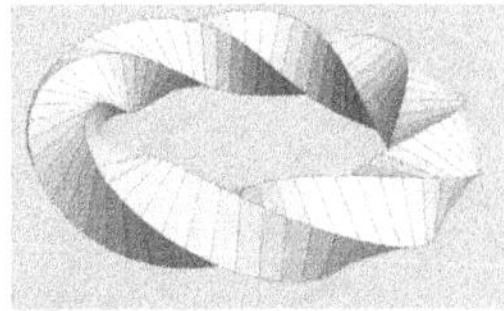

Fig. 2.30 Two twists.

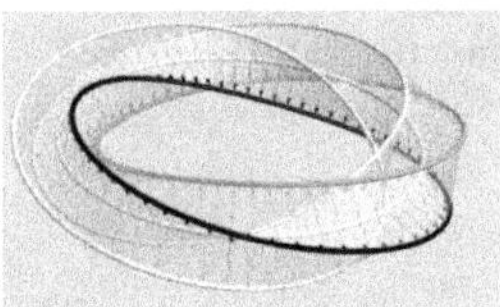

Fig. 2.31 Figure 2.29 showing normal frames.

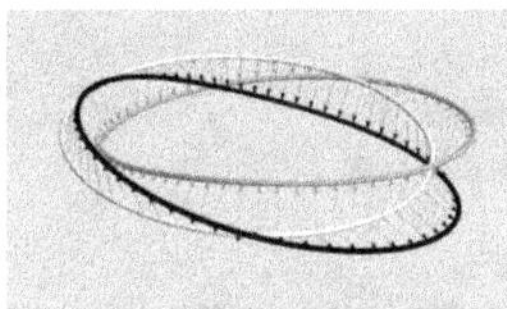

Fig. 2.32 The frames and vector endpoints.

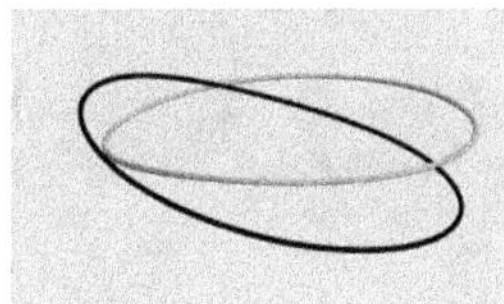

Fig. 2.33 Twisting can be measured by linking of these circles.

The untwisted normal framing of a knotting of a circle K has special mathematical significance but another framing called blackboard framing is sometimes easier to work with especially if a knotting K is given by a projection.

Example 2.7. Assume that we have a knotting of a circle $K \in R^3$ which is smooth and has a generic projection K_* which is a parameterized plane curve. Let T correspond to a product tubular neighborhood of K_* given by immersion $\tau \colon S^1 \times D^1 \to R^2$. For each $x \in S^1$ we let $D_x = \tau(\{x\} \times D^1)$ There is an annulus $A \subseteq R^3$ containing K so that $A_* = T$ namely A is a union $\bigcup_{x \in S^1} D_x \times \{h(x)\}$. Figure 2.35 shows this construction for the projection shown in Figure 2.34. The framing corresponding to annulus A is called the **blackboard framing.** ∎

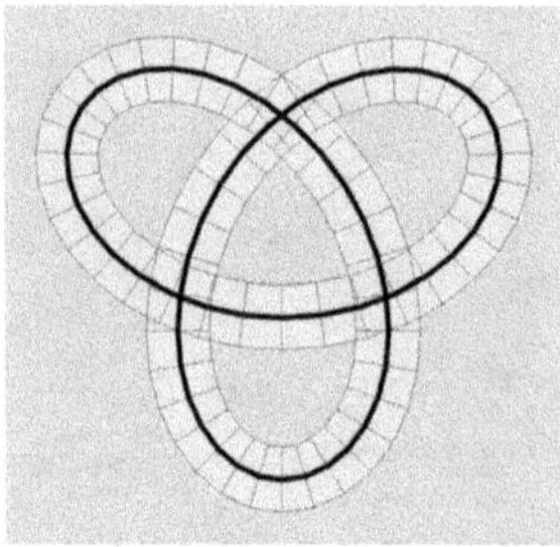

Fig. 2.34 A tubular neighborhood of an immersed circle is an immersed annulus.

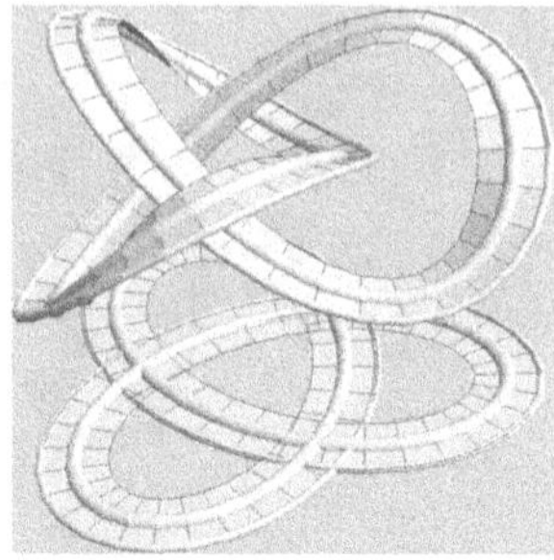

Fig. 2.35 Annulus knotting corresponding to immersed annulus..

A trivialization is an embedding so two trivializations are equivalent means they are ambiently isotopic embeddings.

A smooth annulus $A \subset R^3$ with center circle a 1-knotting $K \subset R^3$ has a product tubular neighborhood. A trivialization can be written $\phi\colon (S^1 \times D^1) \times D^1 \to R^3$ where $\phi((S^1 \times D^1) \times \{0\}) = A$ and $\phi((S^1 \times \{0\}) \times \{0\}) = K$. This can be considered as a trivialization of a product tubular neighborhood T of $K \subset R^3$. If we identify $(S^1 \times D^1) \times D^1 \approx (S^1 \times (D^1 \times D^1) \approx S^1 \times D^2$.

If A has n-twists, then the corresponding trivialization of T will have n-twists.

A trivialization is a knotting of $T \subset M^n$ and we use the following equivalence:

Definition 2.16. Two trivializations τ and τ' are **equivalent trivializations** of a tubular neighborhood if there is a smooth isotopy H_t of M^m such that $\tau' = H_0 \circ \tau$. ▌

For all orientable $K^n \subseteq R^{n+1}$ or orientable $K^n \subseteq R^{n+2}$ we will see that the normal bundle is trivial.

Example 2.8. If $C \subseteq R^2$ is a circle in the plane, it has a product tubular neighborhood A which is an annulus that is $A \approx S^1 \times D^1$. If we view C as a circle in R^3 via $C \subseteq R^2 \subseteq R^3$ it has a product tubular neighborhood, a solid torus T homeomorphic to $S^1 \times D^2$. It is clear we could view $T \approx A \times D^1 \approx (S^1 \times D^1) \times D^1 \approx S^1 \times D^2$. More interesting for us is that C has a product tubular neighborhood N in R^4 homeomorphic to $S^1 \times D^3$ since $N \approx T \times D^1 \approx (S^1 \times D^2) \times D^1 \approx S^1 \times D^3$. ▌

Definition 2.17. If $N^k \subseteq M^n$ is a proper submanifold with tubular neighborhood T then $Ext(M^n, N^k) = (M^n - int(T)) = \overline{M^n - T}$ is called the **exterior, in M^n, of N^k**. If we refer to the subset N^k as a knotting, we say E is the **exterior of the knotting N^k in M^n**. If M^n is understood, we simply write $Ext(N^k)$ and call it the **exterior of N^k**. Since it is commonly done, we will also refer to $Ext(N^k)$ as a **knot exterior**, rather than knotting exterior. ∎

We note that our *geometric topology* notion of exterior of a submanifold is quite different from the *general topology* concept of "exterior of a subset".

One focus of mathematical knot theory for a knotting $K \subseteq R^n$ is the complement $R^n - K$. This is sometimes put: "We study a knot by what it is not". We touch only briefly on this since, the tools used involve advanced topics in algebraic topology or involve difficult visualizations. One special case we *will* examine is for fibered knottings, see Section 10.11.

A submanifold N^k of M^n is a closed subset of M^n, so $M^n - N^k$ is an open subset of M^n. It is often technically preferable to use closed subsets—especially if they are also compact. The **exterior** of a submanifold is a closed subset that reflects many topological properties of $M^n - N^k$.

Example 2.9. For the standard (D^3, D^2) the exterior is a disjoint union of 3-disks whereas $D^3 - D^2$ has two components each diffeomorphic to R^3_+. For (D^3, D^1), $Ext(D^3, D^1)$ is diffeomorphic to a solid torus $S^1 \times D^2$. If we view $S^3 = S^1 * S^1$, then the first factor corresponds to the standard S^1, the first half is a tubular neighborhood of it. The second half corresponds to the exterior. So, for the standard $S^1 \subseteq S^3$, $Ext(S^3, S^1)$ is diffeomorphic to a solid torus. ∎

An important aspect of the boundary of a smooth manifold is that it is part of a smooth product submanifold called a **collar**. A fundamental property of smooth manifolds is the Collar Neighborhood Theorem:

Prop 2.9. If M^n is a smooth manifold with non-empty boundary ∂M^n, there is a smooth submanifold $C \subseteq M^n$ called **a collar** or a **collar neighborhood of the boundary** $C \approx \partial M^n \times I$ where ∂M^n corresponds to $\partial M^n \times \{0\}$. Furthermore, if $M_0 = \overline{M^n - C}$ then $M_0 \approx M^n$ with ∂M_0 corresponding to $\partial M^n \times \{1\}$, so we can write $M^n = M_0 + C \approx M + C$. ∎

Proposition 2.9 is easy if $M^n = D^n$. Writing points of D^n as n-vectors, let $C = \{\overrightarrow{x} \in D^n : 1/2 \leq ||\overrightarrow{x}|| \leq 1\}$. The map $f(x) = (\frac{\overrightarrow{x}}{||\overrightarrow{x}||}, 2||\overrightarrow{x}|| - 1)$is a diffeomorphism from C to $\partial D^n \times I$. Let $D_0 = D^n_{1/2} \subseteq D^n$ be the sub-disk of radius 1/2. The map $F(\overrightarrow{x}) = 2\overrightarrow{x}$ is a diffeomorphism from D_0 to D^n.

Figure 2.4, shows "small" tangent vectors to a circle fitting together to form an annulus. Topologically, we can describe "small" as "tangent vectors of length one or less". Using the exponential map, we can press this annulus down to the torus, where it becomes a tubular neighborhood of the diagonal circle $\triangle \subseteq S^1 \times S^1$. In general, for any smooth manifold M^n we define the unit **tangent disk manifold** $T_1(M^n)$ to be the tangent vectors of $T(M^n)$ of length 1 or less. It is a smooth manifold. Another common name for $T_1(M^n)$ is **the total space of the tangent disk bundle of** M^n. A connection between the tangent bundle and tubular neighborhoods is:

Prop 2.10. For a smooth manifold M^n, $T_1(M^n)$ is diffeomorphic to a tubular neighborhood of the diagonal $\triangle \subseteq (M^n \times M^n)$. ▌

Remark 2.11. In this book, we will have occasion to consider a certain uncommon version of product tubular neighborhood for *improper* submanifolds. We discuss here only the special case we need. Suppose $\mathbb{N}^n \subseteq int(M^{n+1})$ is a smooth manifold with $\partial\mathbb{N}^n \neq \emptyset$ and suppose $\mathbb{N}^n \approx N + C$ where C is a collar neighborhood of $\partial\mathbb{N}^n$. An **improper product tubular neighborhood** of N^n is a smooth submanifold $T \approx \mathbb{N}^n \times D^1 \subseteq int(M^{n+1})$ where $\mathbb{N}^n \times \{0\} = \mathbb{N}^n$. Note that $N^n \subseteq int(T)$. A simple example in cylindrical coordinates would be a disk $N^2 = \{(r,\theta,z) : r \leq 1/2, z = 0\}$, a larger disk $\mathbb{N} = \{(r,\theta,z) : r \leq 1, z = 0\}$ and a solid cylinder $\mathbb{T} = \{(r,\theta,z) : r \leq 1, -1 \leq z \leq 1\}$.

One can show that an orientable n-submanifold which is a subset of the interior of an orientable $(n+1)$-manifold has an improper product tubular neighborhood. ▌

2.6 Cutting and Gluing of Manifolds

Many constructions in topology are described as "gluing simple pieces together to make a more complex object" or a "reverse" process: "taking a complex object and cutting it into simple pieces". However, these are simple intuitive ideas that have subtleties that make mathematical formulation non-simple. This formulation is found in the (rather lengthy) Definition 2.19. The examples we offer should serve as a guide to the notation. Once the reader is convinced that the only problems with cutting/gluing are minor and technical, they will not have need of such detail in later Sections.

To manipulate manifolds, we want to be able to assemble big manifolds by gluing together small manifolds or cut big manifolds into smaller ones.

Also, we want "assemble" and "cut" to be reverse processes. The northern hemisphere of S^2 has the equator E as an edge; the same for the southern hemisphere. We can assemble S^2 by gluing these together along E. At first, the reverse of this process would seem to be: cut S^2 along the equator E, means "consider $S^2 - E$". The result is homeomorphic the disjoint union of two *open* 2-disks. These no longer have an edge that we can use for a re-gluing instruction.

We resolve this situation by using regular neighborhoods and collar neighborhoods. We first focus on cutting. In a more general setting, we describe cutting

Definition 2.18. Let $M^n \subseteq R^k$ be an n-manifold and $A^{n-1} \subseteq M^n$ a proper $(n-1)$-manifold with a product tubular neighborhood $N^n \approx A^{n-1} \times D^1$. Then M^n **cut along** A^{n-1} will be the exterior manifold, of A^{n-1} in M^n, $E^n = M^n - \overset{\circ}{N^n}$, or equivalently $E^n = \overline{M^n - N^n}$. ∎

To reverse this process we begin with the following notations. Let $\widetilde{A'} = A \times \{1\} \subseteq \partial E$ and $\widetilde{A} = A \times \{-1\} \subseteq \partial E$. For any $a \in A$ let $\widetilde{a'} = \{a\} \times \{+1\}$, $\widetilde{a} = \{a\} \times \{-1\}$. So now we want to "glue A_+ and A_- together so that, for all $a \in A$, $\widetilde{a}$ is glued to $\widetilde{a'}$".

A very general topological construction which captures a gluing process is called a **quotient space**. The quotient space construction has several names — **decomposition space, adjunction space, orbit space, identification space**—and can be technically challenging. A quotient space is a way of taking any topological space X together with an equivalence relation in that space and constructing a new topological space. However, in this book, we focus on subsets of R^n that are smooth manifolds, and so can use a simple special version we call **gluing**. A difficulty with more general quotients is that even if $X \subseteq R^p$, it is not always the case that a resulting quotient space is a subset of some R^n. Even if it *is* a subset of some R^n it may not be easy to show this.

Example 2.10. Figure 2.36 shows a proper arc A in a Möbius band M. Figure 2.37 shows a product tubular neighborhood $N \approx A \times D^1$. Topologically the exterior $\widetilde{M} = Ext(M, A)$, Figure 2.38. is just a closed rectangle with $\widetilde{A}$ and $\widetilde{A'}$ opposite sides.

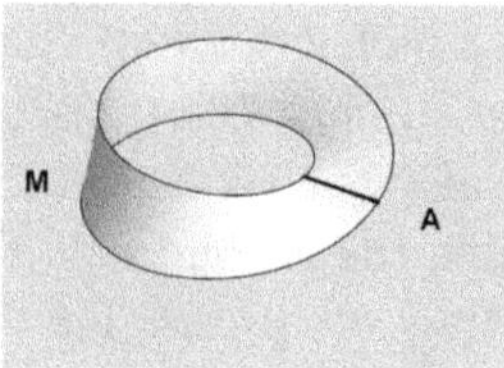

Fig. 2.36 A is proper manifold of M.

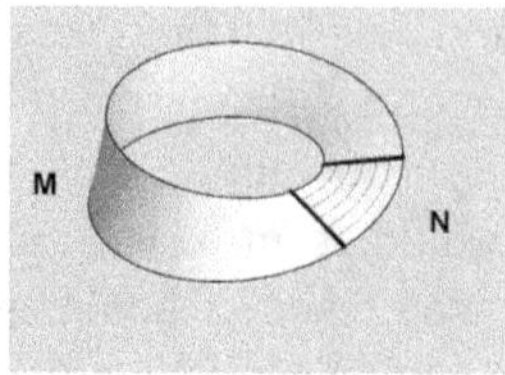

Fig. 2.37 N is a product tubular neighborhood of A.

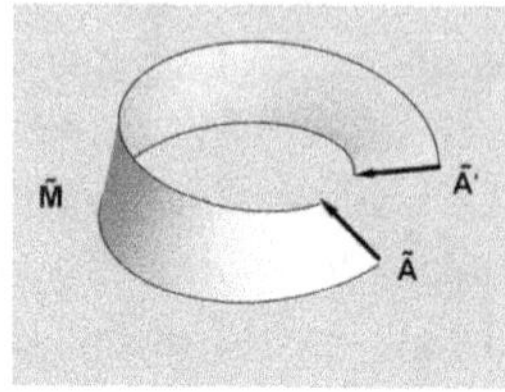

Fig. 2.38 removing N from M.

To reconstruct the Möbius band, we will glue together opposite sides, $\widetilde{A}$ and $\widetilde{A'}$, of rectangle $\widetilde{M}$. However, there is more than one way to unite these edges and get a smooth manifold. We can construct a smooth manifold M which is either an annulus or a Möbius band from a rectangle $\widetilde{M}$ by "gluing two opposite sides together" depending on the "gluing instructions" as indicated in Figures 2.39 and 2.40. The gluing is captured in either case by a smooth map $\phi\colon \widetilde{M} \to M$.

We detail this process to help navigate notations of the next definition. We have a line segment $A \subset M$ and line segments $\widetilde{A}$ and $\widetilde{A'}$ which are to be "glued together" to form A. This process is captured by a diffeomorphism $g\colon \widetilde{A} \to \widetilde{A'}$ so that the two points $a \in \widetilde{A}$ and $g(a) \in \widetilde{A'}$ correspond to a single point of A. Essentially this is a requirement that $\phi|_{\widetilde{A}} = \phi|_{\widetilde{A'}} \circ g$. In Figures 2.39 and 2.40 g is indicated graphically by direction arrows along $\widetilde{A}$ and $\widetilde{A'}$ which indicate the gluing alignment. ∎

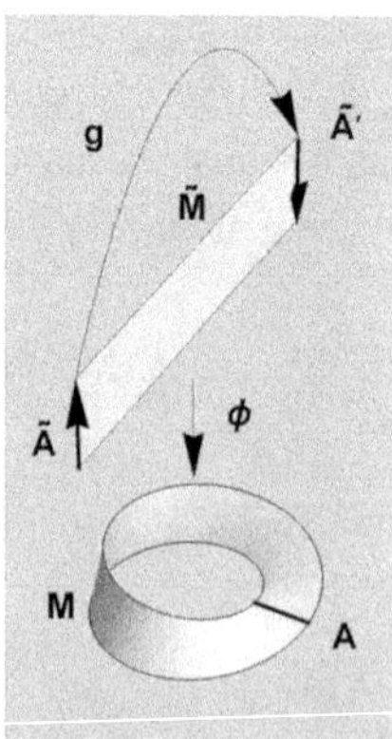

Fig. 2.39 Constructing a Möbius band from a rectangle.

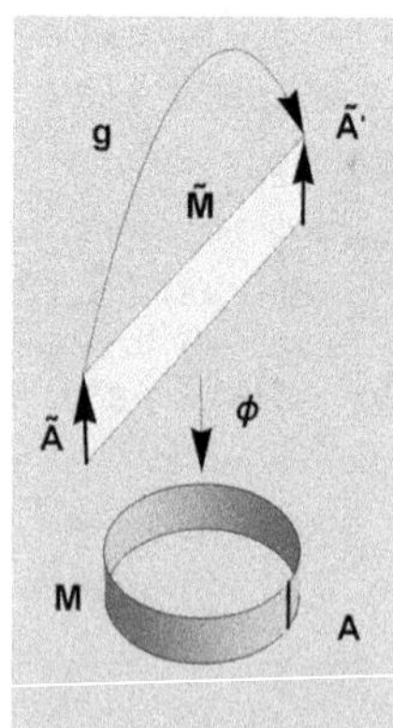

Fig. 2.40 Constructing an annulus from a rectangle.

Our next notations are a bit non-standard since we are avoiding a more general discussion of quotient spaces and are considering manifolds extrinsically.

Definition 2.19. We say M **is** $\widetilde{M}$ **with** $\widetilde{A}$ **glued to** $\widetilde{A'}$ **via** g if:

(1) (M, A) a smooth proper manifold pair dimension where dimension of M is n, dimension of A is $(n-1)$;
(2) smooth $(n-1)$-manifolds $\widetilde{A}$ and $\widetilde{A'}$ where $\widetilde{A} \subseteq \partial\widetilde{M}$ and $\widetilde{A'} \subseteq \partial\widetilde{M}$;
(3) the set $\widetilde{C} = \partial\widetilde{A} \cup \partial\widetilde{A'}$ is called the **corner set** in $\widetilde{M}$. Often we will have $\widetilde{A}$ and $\widetilde{A'}$ disjoint, but if not, we require $\widetilde{A} \cap \widetilde{A'} \subseteq C$;
(4) diffeomorphism $g : \widetilde{A} \to \widetilde{A'}$;
(5) a homeomorphism $\phi\colon \widetilde{M^n} \to M$ which is smooth on $\widetilde{M^n} - C$;
(6) For all points of $\widetilde{A}$ we have $\phi = \phi \circ g$.

We will write $M = \widetilde{M} \circlearrowleft g$ and refer to it as M **glued to itself via** g. For common situations we use a simpler notation, Notation 2.3. ∎

Example 2.11. An important special case is when F is a smooth $(n-1)$-manifold and $\widetilde{M} = F \times I$, $\widetilde{A} = F \times \{0\}$, $\widetilde{A'} = F \times \{1\}$ and $g\colon \widetilde{A} \to \widetilde{A'}$ a diffeomorphism.

There are two common names for $(F \times I) \circlearrowleft g$. The first is **mapping torus**. One reason for this terminology is that if $F = S^1$ and g is the map $g((x,0)) = (x,1)$ then $\left(S^1 \times I\right) \circlearrowleft g \approx S^1 \times S^1$.

A second name is a **fibration over a circle**. For $x \in I$, **the fiber at** x is $F_x = \phi(F \times \{x\})$, The map g is called the **monodromy map**. The wrapping function W maps I onto S^1. We define a smooth map $p\colon \left(\widetilde{M} \circlearrowleft g\right) \to S^1$ called the **projection of the fibration** by sending each F_x to $W(x)$. ∎

Example 2.12. If the gluing map is the identity map of the fiber F then the mapping torus will be diffeomorphic to the product $F \times S^1$.

For $0 < n$, the mapping torus of the reflection map of D^n is diffeomorphic to $M^2 \times D^{n-1}$ where M^2 is a Möbius band and is called a **thickened Möbius band**. Also, for $n = 2$, the name **solid Klein bottle** is used. ∎

There is a pesky minor technical problem with gluing that goes by the name **"rounding the corners"**. The problem is well illustrated even in the annulus M of Example 2.10. We would like to say that we have a smooth map g from the rectangle $\widetilde{M}$ to the annulus M that maps $\widetilde{A}$ and $\widetilde{A'}$ to A.

If we want to consider rectangle $\widetilde{A}$ as a smooth manifold, we need to round the corners so that corners look like half-planes. However, if we then glue two such half-planes together we get a plane, not a half-plane. Annoying as this is, it is not a serious mathematical obstacle to our construction. It is rarely mentioned explicitly. Our method of dealing with this is to not to require our gluing to be smooth at the corner points, condition (5) in Definition 2.19.

Often when gluing is considered, the manifold $\widetilde{M}$ is not connected and separate components of $\widetilde{M}$ are joined by this process, allowing a less awkward notation. Examples are connected sums, Definition 2.20, and handle addition, Section 5.3.

Notation 2.3. Suppose $\widetilde{M}$ is a smooth n-manifold which is a union of *disjoint* (non-empty) n-manifolds $\widetilde{M} = X \cup Y$. Also, suppose we have $(n-1)$-manifolds $\widetilde{A} \subset \partial X$, $\widetilde{A'} \subset \partial Y$ and a diffeomorphism $g\colon \widetilde{A} \to \widetilde{A'}$. We write $\widetilde{M} \urcorner g$ as $X \underset{g}{+} Y$ or alternatively $X \underset{\widetilde{A}=\widetilde{A'}}{+} Y$ or even $X \underset{\widetilde{A}}{+} Y$.

We call $X \underset{g}{+} Y$ a **union of X and Y with identifications via** g or a **union with identifications on** $\widetilde{A}$,

Also, to make notation easier to read, we will often write $\left(\left(X \underset{g}{+} Y\right) \underset{h}{+} Z\right) = X \underset{g}{+} Y \underset{h}{+} Z.$

Additionally, we will often consider a special situation where X is an n-manifold the second space Y is an n-disk. In this case, we will sometimes write $X \underset{h}{+} Y$ more simply as $X + h$.

Our conditions imply that $\widetilde{A} \cap \widetilde{A'} = \emptyset$. Thus, we do not glue points of X together nor points of Y together. As a consequence, the map $\phi\colon \widetilde{M} \to X \underset{g}{+} Y$ is an embedding when restricted to either X , Y or $\widetilde{A}$. Often $\phi(X)$ is identified with X and we write $X \subset (X \underset{\widetilde{A}}{+} Y)$ and $Y \subset (X \underset{\widetilde{A}}{+} Y)$. ∎

Other important and useful examples of gluing can be found in Section 5.3 and can be read without needing materials in much of the rest of this chapter or Chapters 3.1 or 4.

We think of S^2 as two hemispheres glued together along the equator and would like to write $S^2 \approx D^2 \underset{id}{+} D^2 \approx D^2 \underset{S^1}{+} D^2$ where $id : \partial D^2 \to \partial D^2$ is the identity map. However, to use $+$ as we have defined requires the sets on either side to be disjoint—in our case, they are the same.

We resolve this by allowing the replacement of Y by Y' diffeomorphic to Y. If $h\colon Y \to Y'$ is a diffeomorphism we extend our definition and say

$X \underset{h\circ g}{+} Y' \approx X \underset{g}{+} Y$ (even though we might not have $X \cap Y' = \emptyset$). We similarly allow $X' \underset{h^{-1}\circ g}{+} Y$ if $h\colon X \to X'$ is a diffeomorphism.

In this way, $X \underset{A}{+} Y$ means Z is a union of X and Y where $X \cap Y = A$ where the gluing map is understood to be the identity map of A.

Listed below for future reference are some useful gluing propositions.

Prop 2.11. Suppose M is a smooth n-manifold, A smooth $(n-1)$-manifold with $A \subseteq \partial M$ and for any manifold N, $N_0 = N \times \{0\} \subseteq N \times I$.

(1) $N \times I \approx (N \times I) \underset{N_1 = N_0}{+} (N \times I)$;
(2) more generally, $M \approx M \underset{\partial M = A_0}{+} A \times I$;
(3) If g and g' are isotopic gluing maps from A to $\partial M'$, then $M \underset{g}{+} M' \approx M \underset{g'}{+} M'$. ∎

Item (2) of Proposition 2.11 can be thought of as a topological vanishing. Suppose $A_0 = \partial M$. Then $M \approx M \underset{\partial M = \partial M_0}{+} \partial M \times I$ roughly says "glue a copy of the $\partial M \times I$ onto ∂M—it gets absorbed into M and seems to vanish". If we call A_0 a **partial-collar**, we can similarly say that "even for partial-collar A_0, if you glue on a copy, it gets absorbed and seems to vanish".

Item (3) of Proposition 2.11 tells us that isotopic gluing maps give the same resulting manifold. The map $g'((x,-1)) = (x/2, 1)$ is isotopic to the map g used in Remark 2.10 for a Möbius band. The result, Figure 2.41, is different from Figure 2.36 but diffeomorphic once we round the corners.

Remark 2.12. A circle can be formed by gluing two semicircles at the endpoints—in our notation this can be expressed $S^1 \approx D^1 \underset{id|_{\partial D^1}}{+} D^1$. Similarly, a torus can be obtained by gluing together two annuli $S^1 \times S^1 \approx S^1 \times D^1 \underset{id|_{S^1 \times \partial D^1}}{+} S^1 \times D^1$. Here the circles of the second factor of $S^1 \times S^1$ are as in previous example. The set $S^1 \times \partial D^1$ corresponds to two disjoint circles and so our gluing map is a diffeomorphism from a pair of disjoint circles to itself. Using (3) of Proposition 2.11, it can be shown that there are only two different manifolds we can get by gluing together two annuli—the torus and the Klein bottle, see Section 7.3.

Gluing considerations become more interesting when we look at $S^1 \times S^2 \approx S^1 \times D^2 \underset{id|_{S^1 \times \partial D^2}}{+} S^1 \times D^2$. Here the 2-spheres of the second factor of $S^1 \times S^2$ are two copies of D^2 glued along the edge. In this construction, we glue two solid torii together along the set $S^1 \times \partial D^2$, which is a 2-torus. There are many possible gluing maps—for example, S^3, Remark 2.13. ▌

Of special interest is the case where one of these manifolds (say N^n) is well understood—in the next example, N^n is an n-disk.

Example 2.13. Suppose $M^n \subseteq R^m$ is an n-manifold and $h\colon R^m \to M$ is a local coordinate map, with $h(R^n) \subseteq M^n$ see Definition 2.1 and let $B = h(D^n)$. If we write $\overset{\circledcirc}{M}{}^n = M^n - h(intD^n)$, we can express M^n as a gluing of two submanifolds: $M^n = \overset{\circledcirc}{M}{}^n \underset{\partial B}{+} B$.

We call $\overset{\circledcirc}{M}{}^n$ **the punctured** M^n**.** Figure 2.42 shows a torus with a single puncture. By homogeneity of disks in a manifold, Proposition 2.7, and two n-disks are equivalent subsets in M^n, and it follows that the diffeomorphism type of $\overset{\circledcirc}{M}{}^n$ is well defined. ▌

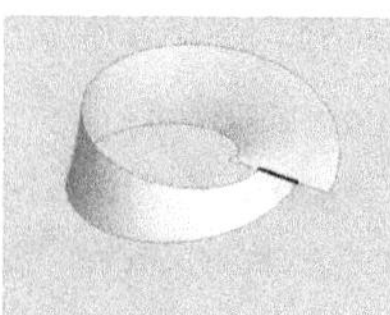

Fig. 2.41 Using gluing g'.

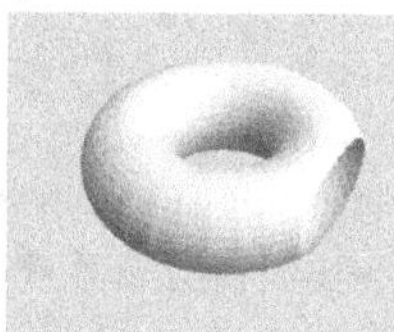

Fig. 2.42 A punctured torus.

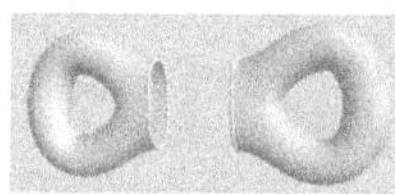

Fig. 2.43 Two punctured torii.

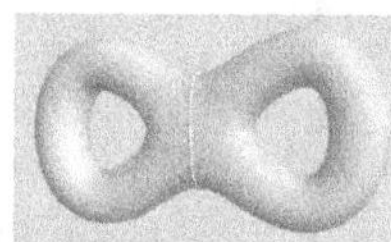

Fig. 2.44 Surface of genus 2 as connected sum.

If $\overset{\circledcirc}{M}{}^n$ and $\overset{\circledcirc}{N}{}^n$ are two smooth punctured connected n-manifolds, then $\partial \overset{\circledcirc}{M}{}^n$ and $\partial \overset{\circledcirc}{N}{}^n$ are diffeomorphic to S^{n-1} so we have a diffeomorphism $h\colon \partial \overset{\circledcirc}{M}{}^n \to \partial \overset{\circledcirc}{N}{}^n$. The smooth manifold $\overset{\circledcirc}{M}{}^n \underset{h}{+} \overset{\circledcirc}{N}{}^n$ is called the **connected sum of** M^n **and** N^n. Formally we define:

Definition 2.20. The n-**knotting** W^n **is the connected sum of the connected** n-**knottings** M^n **and** N^n if $W^n = \overset{\circledcirc}{M}{}^n \cup \overset{\circledcirc}{N}{}^n$. Notationally we write $W^n = M^n \# N^n$. We also say that the **manifold** W^n **is the connected sum of the manifolds** M^n **and** N^n. ▌

Figure 2.47 shows a connected sum of 2-manifolds, Figures 2.46 and 2.45. (Smoothing, as in Figure 2.44 has not yet been done for Figure 2.47.)

Note: Definition 2.20 requires manifolds to be connected (that is why it is called a *connected* sum). Because of homogeneity of manifolds, $[M^n \# N^n]$

is well defined 2-*knot* if both M^n and N^n are connected. A surface of genus 2 can be expressed as a connected sum of two punctured torii, Figures 2.43 and Figures 2.44.

If the manifolds are not connected, we can still do the gluing and obtain useful description of a *knotting*, but the result will depend on which components are punctured. Suppose K is a torus and if $L = S \cup T$ is a 2-component manifold consisting of sphere S and torus T. If we puncture T and glue it to a punctured K we will get a genus 2 manifold $K\#T$ and the sphere S. But if we puncture the sphere of S and glue to punctured K we get two torii: $K\#S$ and T. See Section 6.10 for other knotting examples.

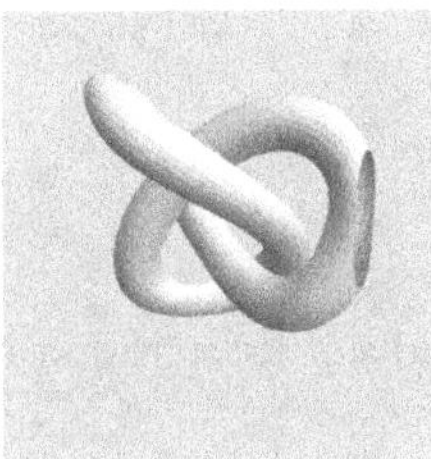

Fig. 2.45 $\overset{\circledcirc}{M}{}^n$.

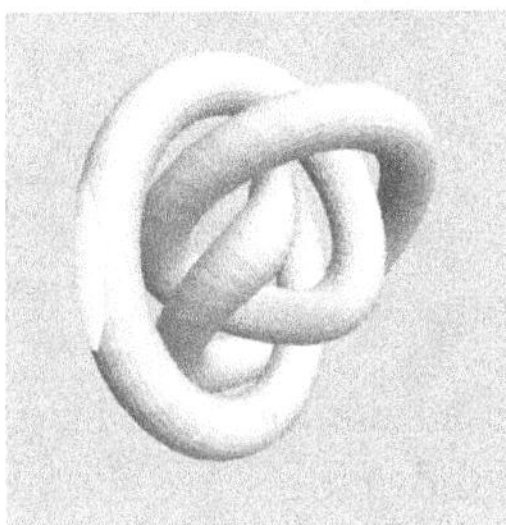

Fig. 2.46 $\overset{\circledcirc}{N}{}^n$.

Fig. 2.47 $W^n = M^n \# N^n$.

A connected sum $K_1 \# K_2$ where each factor is equivalent to a standard torus knotting will be taken to be the **standard genus 2 knotting**. A connected sum $K_1 \# \cdots K_n$ where each factor is equivalent to a standard torus knotting will be taken to be the **standard genus n knotting**. A split knotting where each component is a standard genus n knotting will be called a **standard orientable 2-knotting**.

2.7 Product Constructions

Prop 2.12. If M^m and N^n are two n-dimensional manifolds, then the cartesian product $M^m \times N^n$ is an $(m+n)$-dimensional manifold, and $\partial(M^m \times N^n)$ is an $(m+n) - 1$-dimensional manifold. With $\partial M^m \subseteq M^m$ and $\partial N^n \subseteq N^n$ we can naturally view $\partial M^m \times \partial N^n$ as subsets of either $M^m \times \partial N^n$ or $\partial M^m \times N^n$. The following formula holds:

$$\partial(M^m \times N^n) \approx (M^m \times \partial N^n) \underset{\partial M^m \times \partial N^n}{+} (\partial M^m \times N^n). \blacksquare$$

Remark 2.13. In Remark 1.10, it was noted that S^3 can be expressed as a union of two solid tori that intersect in their common boundary. An alternate way to express this is by using a formula for the boundary of the product of manifolds and our gluing notation:

Write $D^n \approx D^k \times D^{n-k}$. Also, consider

$$S^{n-1} \approx \partial D^n \approx \partial(D^k \times D^{n-k}) \approx (\partial D^k \times D^{n-k}) \underset{\partial D^k \times \partial D^{n-k}}{+} (D^k \times \partial D^{n-k})$$

And, in particular, we have

$$S^3 \approx \partial(D^2 \times D^2) \approx (S^1 \times D^2) \underset{S^1 \times S^1}{+} (D^2 \times S^1).$$

Switching the order of the factors, we have $D^2 \times S^1 \approx S^1 \times D^2$. It follows that we can write

$$S^3 \approx (S^1 \times D^2) \underset{g}{+} (S^1 \times D^2)$$

where g is the factor switching map of $S^1 \times S^1$. ▮

The next product example is important for the study of knotting later in this book. More details and figures are found in Section 10.1. Suppose (D^3, α) is a proper knotting of an arc. Then $(D^3, \alpha) \times D^1) \approx (D^3 \times D^1, D^1 \times \alpha) \approx (D^4, D^1 \times \alpha)$. Now $\alpha \times D^1$ is proper knotting of a 2-disk in D^4.

Definition 2.21. If $M \subseteq R^k$ is a manifold, the **double of** M**, denoted** $D(M)$ is defined $D(M) = \partial(M \times D^1) \subseteq R^k \times R^1 \approx R^{k+1}$. ▮

Remark 2.14. If $\partial M = \emptyset$ then $D(M)$ is two disjoint copies of M. Most frequently, this construction is used when $\partial M \neq \emptyset$ in which case $D(M)$ is frequently described as "two copies of M, M_- and M_+ glued together by the identity map of the boundaries". So, $D(M) \approx M \underset{\partial M}{+} M$.

For example, $D(D^n) \approx S^n = D^n \underset{\partial D^n}{+} D^n$ expresses the n-sphere as two n-disk hemispheres intersecting along an equatorial $(n-1)$ sphere. If A is an annulus, $D(A) \approx S^1 \times S^1$, see Figures 2.48, 2.49 and 2.50. In Figure 2.49, we have a copy of A with a copy of $\partial A \times I$ glued on. It is diffeomorphic to A itself. If M is a Möebius band then $D(M)$ will be a Klein bottle expressed as a union of two Möebius bands joined on their common circular edge, see Example 6.25. Closed orientable 2-manifolds are doubles. This is far from the case in higher dimensions. A basic knotting construction called ribbon knotting is restricted to doubles, see Chapter 9 and Section 12.4. ▮

Fig. 2.48 The double of an annulus $D(A)$.

Fig. 2.49 The bottom half of $\partial(D(A))$.

Fig. 2.50 The top half of $\partial(D(A))$.

Cobordism of manifolds is a major topic in algebraic and geometric topology. Most aspects are beyond the scope of this book. The basic concept is:

Definition 2.22. Two n-manifolds M_0^n **and** M_1^n **are cobordant** if there is manifold W^{n+1} with $\partial W^{n+1} \approx M_0^n \cup M_1^n$. Such a manifold W^{n+1} is called a **cobordism between** M_0^n **and** M_1^n. ∎

Remark 2.15. A closed n-manifold with non-empty boundary is called **null-cobordant**. If one considers the empty set as a manifold, Remark 2.7, one can view a null-cobordant manifold as one that is cobordant to the empty set. ∎

2.8 Branched Coverings

Covering spaces and branched covering spaces are important constructions in geometric topology. Branch covering spaces are covered in detail in [Fox (1957)], where the branch points are called "singularities". Covering spaces (with empty branch sets) is a topic in almost all standard topology textbooks.

In this book, we focus on finite cyclic covers (branched and non-branched) and infinite cyclic covers.

Example 2.14. Consider the map $f\colon R^2 \to R^2$ defined, using polar coordinates, as $f(r,\theta) = (r, m\theta)$. Alternatively, viewing R^2 as complex numbers we can write f as $f(z) = z^m$. The map f is an m-to-1 map, except it is 1-1 at the origin $\mathbb{O} \in R^2$. We say that $\mathbb{O}$ is a **branch point of** f and that f is an m-fold cover outside of this branch point.

Let X be a simple closed curve $X \subseteq R^2 - \{\mathbb{O}\}$ that winds twice around a point A, Figure 2.51. Take a ray, R, (half-line) from A (Figure 2.52) and cut the plane along A—this will cut X into two arcs. If we open up this cut (like a partial closing of a fan), this cut plane becomes an interior of an angle of $\frac{2\pi}{3}$ with the points of the curve sent to two arcs, Figure 2.53. The 3-fold branched cyclic cover is formed by putting three copies of this together, as shown in Figure 2.53. Then $\widetilde{X}$ becomes a curve with a 3-fold symmetry about A, Figures 2.54 and 2.55. ∎

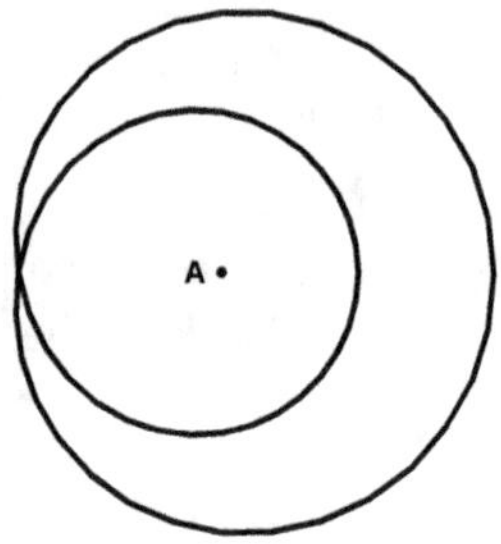

Fig. 2.51 A curve in R^2 that wraps twice about a point A.

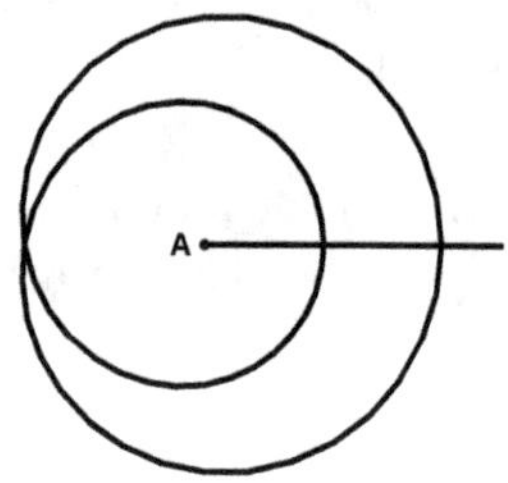

Fig. 2.52 Figure 2.51 with a "slicing ray" from A.

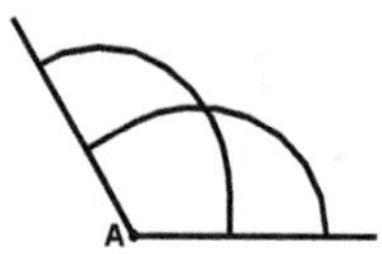

Fig. 2.53 After cutting and using 1/3 of polar angle we get two proper arcs in an angle interior.

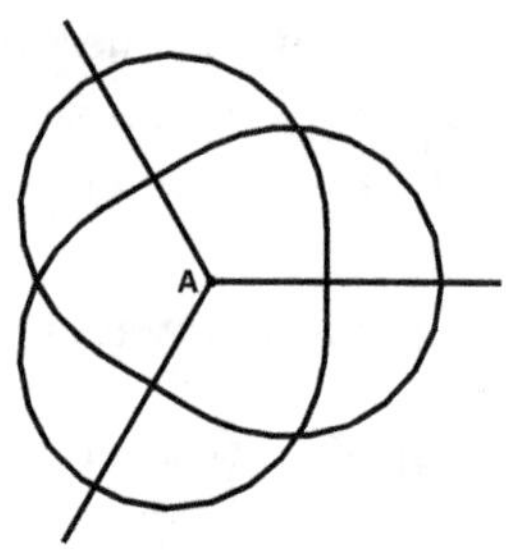

Fig. 2.54 Putting together three copies.

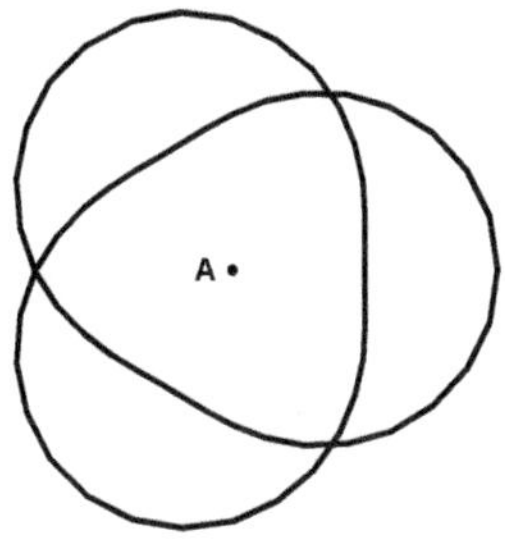

Fig. 2.55 The three fold cover with cutting rays removed showing curve $\widetilde{X}$.

More generally, we can use generalized cylindrical coordinates to get m-fold cyclic branched covers F of $R^n = R^2 \times R^{n-2}$ with branch set $\mathbb{O} \times R^{n-2}$, which in cartesian coordinates is $azis(z)$. If $f\colon R^2 \to R^2$ is our m-fold cover, define $F = f \times Id|_{R^{n-2}}$.

Example 2.15. For a more interesting construction in $R^3 = R^2 \times R^1$, we take a 1-knotting $X \subseteq R^3 - A$ where $A = -axis(z)$.

Here, our branch set is the A shown as a line segment in Figure 2.56 and our X is a curve that wraps twice about A. We cut by a half-plane with edge A, and spread. We obtain two proper arcs in a region of R^3 bounded by the union of two half-planes shown in Figure 2.57. We assemble three copies as shown in Figure 2.58, and see that $\widetilde{X}$ is a trefoil knotting, and we have $F(\widetilde{X}) = X$.

Using the same curve X and axis A of Figure 2.56, the 2-fold branched cyclic cover $\widetilde{X}$ is a knotting for the Hopf link, Figure 4.63. In our terminology, the trefoil knotting has 3-fold rotational symmetry, with quotient a trivial knotting X and the Hopf linking has 2-fold rotational symmetry with the same quotient knotting X. ∎

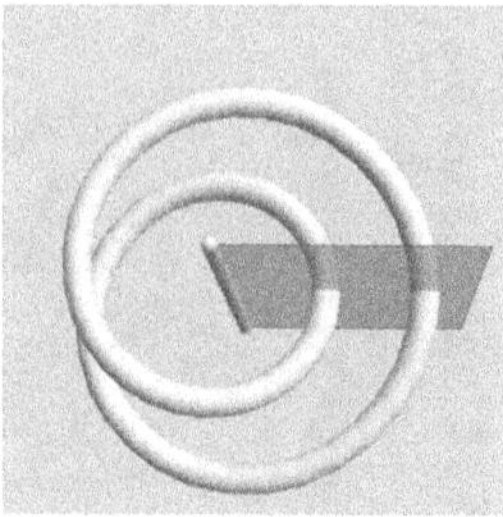

Fig. 2.56 In a 3-dimensional version of Figure 2.52 we cut with half-plane instead of ray.

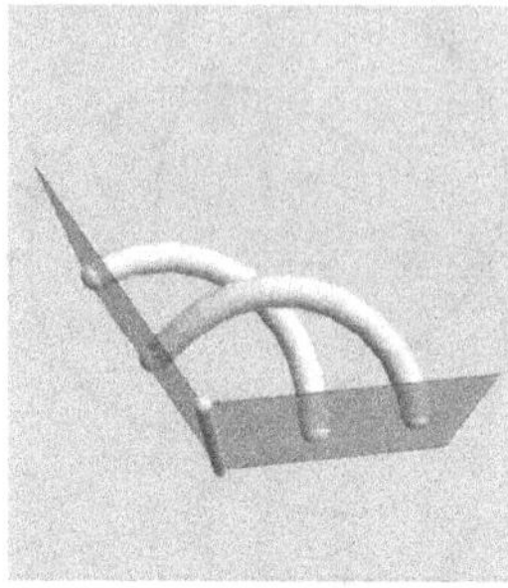

Fig. 2.57 Cutting along the half-plane then spreading out.

Fig. 2.58 Putting three copies together we obtain trefoil knotting in $R3$.

Example 2.16. Suppose X is a trefoil knotting with a branch set A as shown in Figure 2.59. Cut R^3 along the half-plane, open it up and the trefoil knotting will correspond to a proper arc in R^3_+, Figure 2.60. The 2-fold branched cyclic cover, $\widetilde{X}$ will be knotting, Figure 2.61, called a **granny knotting.** ▮

Fig. 2.59 A trefoil knotting and half-plane in R^3.

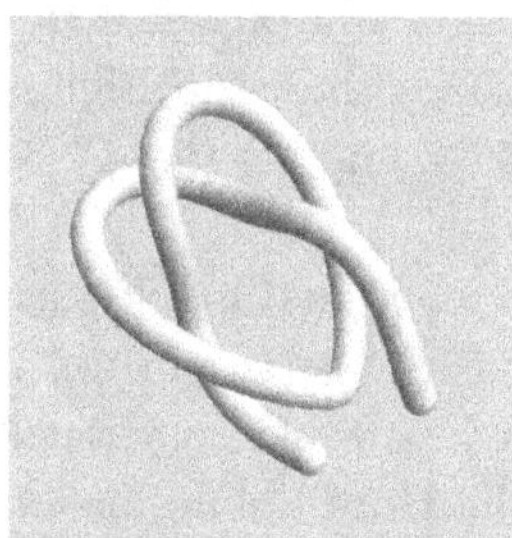

Fig. 2.60 Cut and divide angle coordinate by 2.

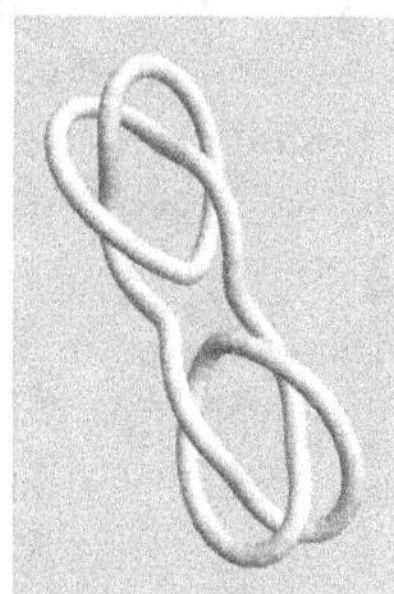

Fig. 2.61 Result is a granny knotting.

Example 2.17. Figure 2.62 shows a knotting of a 2-sphere in R^3. The z-axis, which will be the branch set, meets the sphere in four points. The two-fold cyclic cover of the sphere branched along these four points is a torus. Cut along the half-plane and open it up and the points of our sphere will correspond to a proper annulus in ∂R^3_+, Figure 2.63. Adding a similar copy of this annulus in R^3_-, we get a torus in R^3 as a double cover branched along the z-axis shown rotated to a familiar position in Figure 2.64. ▮

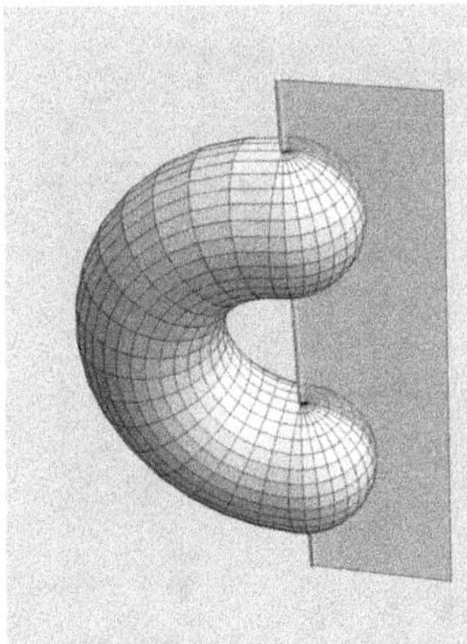

Fig. 2.62 Half-plane, boundary z-axis meets sphere in four points.

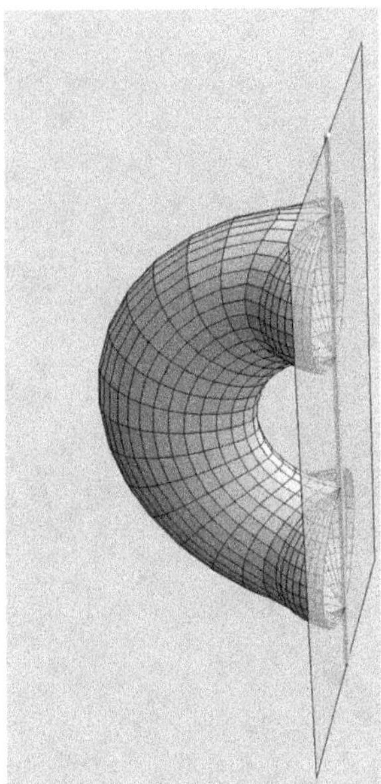

Fig. 2.63 The sphere split, then opened to make a full plane.

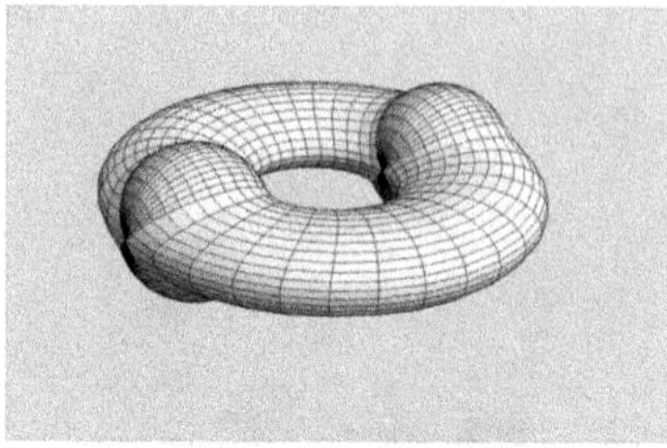

Fig. 2.64 The torus constructed as two-fold cover.

Chapter 3

Overview of Knottings

In its most general form, knot theory studies the "ways to place a set A into another set X"—when are two placements the " same" and when are they "different"? Various knot theories arise from choices of A, X and meanings of "place" and "same/different". Consider Figure 3.1—how can we express the idea that these six placements of a simple closed path are "the same"?

Historically, mathematical knot theory is "the mathematical study circles and disjoint unions of circles in space"—so A is a 1-manifold (with empty boundary) Definition 2.1, and X is R^3. The focus of this book is to give examples of some extensions of this knot theory to higher dimensional spaces. But we first discuss some knotting topics for circle. Readers familiar with this knot theory, or those who are adventurous, may wish to advance to the higher dimensions part of this book, Part 2, and rely on references and the index to fill in the two background if needed.

Remark 3.1. There are some interesting examples called "**wild knots**" as shown in Figure 3.2. Such wild knottings have infinitely many smaller and smaller "self-tanglings" which converge to one or more points. Due to this book's space and time constraints, we avoid detailed treatment of this interesting topic. However, one should be aware of some basic examples since they motivate our focus on "smooth" knottings. There are many more such strange wild subsets, see [Fox (1962a)] and [Rushing (1973)]. A good article on generating diagrams for wild knots is [Browne (2006)].

The knotting of Figure 3.2 is not a locally flat submanifold, Definition 2.9. The local flatness fails at one point—the point where "all the shrinking sub-knots converge". It is not elementary to prove, but it can be shown that this subset fails to look like $R^1 \subseteq R^3$ at that point. ▮

Two different styles of drawing a knotting, one angular (called piecewise-linear), one smooth, correspond to mathematical ways of avoiding wildness. The common way to describe such subsets is as images of parametric equations. Parametric equations that are defined as linear on a finite number of sub-intervals are called **piecewise-linear**. For technical reasons, we will insist a "smooth" parametrization will have derivatives of all order, not just the first derivative.

The knotting indicated in Figure 3.2 is smooth, except at one point x_0, where the small knots seem to converge. Intuitively, smoothness at x_0 fails because the tangent lines swing about wildly as we get close to x_0, so the derivative will fail to be continuous at x_0, and a second derivative not defined. Intuitively, the knotting is not piecewise linear since it would take infinitely many linear segments to represent this knotting.

An ellipse is a smooth curve—we can use familiar parametric equations $(a\cos(2\pi t), b\sin(2\pi t))$. However, an oval (two semi-circles joined by two line segments) is not. The radius of curvature is defined using the second derivative. A line segment has an infinite radius of curvature, the radius of the circle is its radius of curvature. At a point x of an oval, where a semicircle joins a line segment, there is a well-defined tangent line, but the second derivative at x is not well defined.

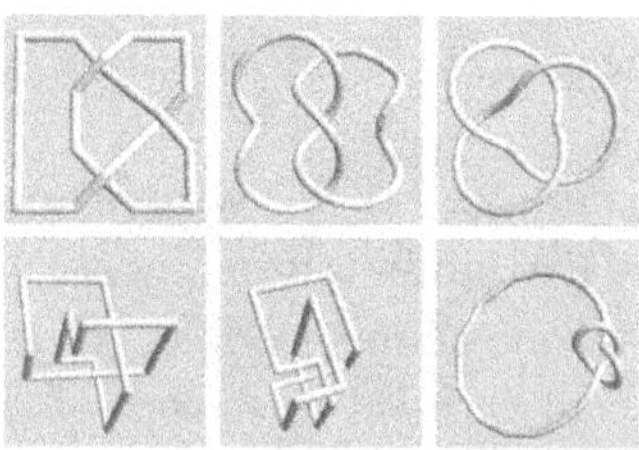

Fig. 3.1 Six knottings all representing the same knot.

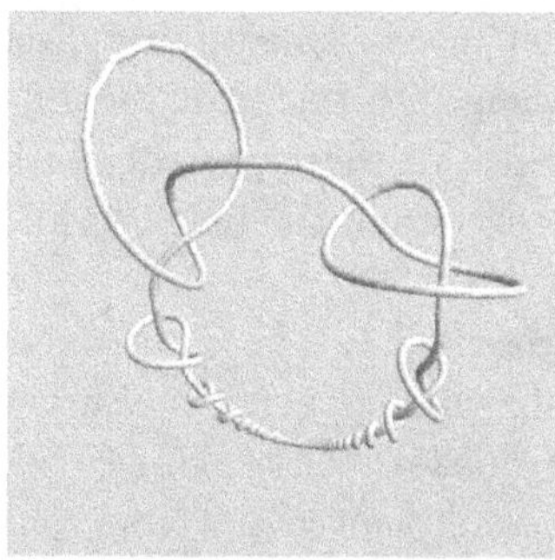

Fig. 3.2 A wild knotting—infinitely many smaller and smaller trefoils, most obscured by tube thickness.

3.1 Vision, Visualization and Higher Dimensions

To see something clearly, we open our eyes wide, but to visualize something, we often close our eyes. Outstanding blind mathematicians have done great work in geometry and topology such as Leonhard Euler, Louis Antoine, Bernard Morin and Lev Pontryagin—a testament to the distinction between biological sight and mathematical vision [Jackson (2002)].

When the term **visualization** is used in this book, it should be considered synonymous with "mathematical understanding".

A standard picture of a line is shown in Figure 3.3. Really what we see in this picture is a very thin rectangle R. The terms "line thickness" or "line width" are so common we rarely reflect on the fact that these terms, taken literally, are oxymorons. Also, mathematically a line is infinite in extent, yet any drawing of the line must terminate at the edge of the Figure. We cannot *see* something that is infinitely thin, and we cannot *see* something that is infinite in extent. Yet we (correctly) unhesitatingly feel that we can *visualize* a line in a plane.

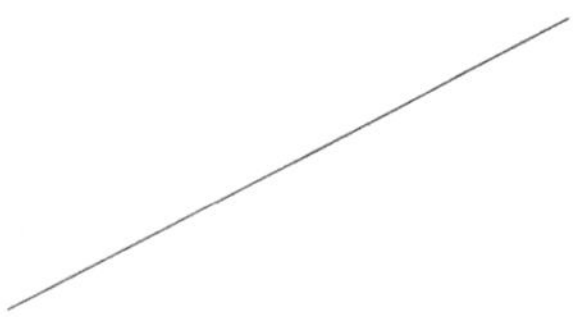

Fig. 3.3 Traditional representation of a line in the plane is actually a very thin rectangle.

Fig. 3.4 A line in space represented as a thin tube.

Additional problems occur in considering graphics for a line in space. It is natural to graphically "thicken" the line in two directions and represent a line in space as a thin tube. As with line thickness we are expected to mentally dismiss this thickening. So it is that our graphics for circles in space will be as thin tubes such as in Figure 3.4.

What is the connection, if any, between vision and dimension?

It is commonly held that, since we have two eyes, and thus binocular depth perception, that our vision is clearly three-dimensional. The number of eyes is not important but, having multiple images is. In binocular vision, distance information is obtained by mentally comparing two similar related images—each image being essentially two-dimensional. I once had a one-eyed cat who had all the usual feline agility and skills. She compensated by combining images as she moved, sometimes only moving her head.

Even accepting that three-dimensional space is a mathematical abstraction, we might still try to argue that the world we see is three-dimensional. By this, we mean that it is the best mathematical model of what we see—that three independent numbers are all we need.

However, a reader who knows modern physics might argue that "the world around us is not static"—the world is composed of "events". An event is something that happens at a certain place and time. So, even viewing location as having three dimensions, we need to add a fourth dimension for time. This view is so common that many believe that, if there is a fourth dimension, it must be time. Knot theorists often are attracted to puns (e.g. "What I study is not theory") . In this vein, one could say of high-dimensional mathematicians, "Some of us have time for the fourth dimension, and some of us don't".

To be clear, in this book, we take a purely mathematical view that each dimension corresponds to a real number, *where no dimension has a preassigned role.*

Returning to the question of how many dimensions we need to describe the world we see, an object with absolutely no color would not be visible. Human vision of color is a complex topic. A standard computer graphics notion of color is as a triple of independent numbers—red, green and blue in one common scheme. In part, this has a basis in the tri-chromatic physiology of human vision based on three types of proteins called opsins. So now a visible object needs *at least seven* dimensions: three for space, one for time, three for color. (Dragon flies—even apart from the extreme compound nature of their eyes–have as many as 30 opsins so, in a sense, dragonfly fly vision may need an enormous dimension for a model.)

Remark 3.2. There are two basic methods to visualize a large multistory building. Each involves using multiple two-dimensional images. These two approaches generalize to two visualization techniques for higher dimensions.

In mathematical terminology, these methods are called **projection** and **slicing**, but in terms of our building visualization, they might be termed "photographic" and "floor-plan" methods.

In the photographic method, we walk outside and inside the building photographing. Each photograph is two-dimensional, but depth can be deduced by comparing photographs taken with nearby views. These images can then be sequenced, in many different ways, to give virtual walk-throughs that allow us to visualize the entire building. A second method of visualization is a sequence of detailed floor-plans such as were likely used in the construction of the building. Each floor-plan is a drawing that shows walls as lines, with other specialized graphics to indicate features such as stairs, doors, furniture and appliances.

An advantage of the photographic method is that it seems more natural. Also, we have additional rich information from color, texture, light and shadows. In contrast, the floor-plans, usually monochromatic (in fact, traditionally called blueprints), take considerable instruction for interpretation. However, the floor-plans have detailed precision on measurements that allow actual construction of the building. It is unlikely that one would construct a building based on a virtual walk-through. Using the floor-plans, one could understand the detailed structure of the plumbing and electrical systems of the building, most of which is hidden from view and difficult to piece together from a virtual walk-through. ▮

In summary, we have two broad visualization methods, each with merits. Using *both* will give the best understanding.

3.2 Knottings vs Knots; Functions vs Images

In this book, a **knotting** is a subset, and a **knot** is an *equivalence class* of knottings. Thus, a *knotting is an element of a knot*, as will be detailed in Section 3.2.

Definition 3.1. Suppose $M \subseteq R^p$ for some p. We say K **is a knotting of** M **in** X (or sometimes more briefly simply a **knotting in** X) if $K \subseteq X$ and K is homeomorphic to M.

If $\phi\colon M$ is an embedding with $\phi(M) = K$, then we say **the knotting** K **is a** ϕ**-image of** M. ▮

Remark 3.3. For example, we say $S^1 \subseteq R^2$ is the standard circle, and each of the six figures of Figure 3.1 is a knotting of S^1 in R^3.

A knotting is an image of many embeddings—not just one. If K is a ϕ-image of M, and also a ψ-image of M, then $\psi^{-1} \circ \phi$ is a self-homeomorphism of M. Also, we can combine an embedding $\psi\colon M \to K$ with a self-homeomorphism $h\colon M \to M$ and obtain an embedding $\psi \circ h\colon M \to K$.

In the literature a technical reason to view knottings as embeddings arises when considering the proofs of some basic theorems. An embedding is a function, and so belongs to a function space. Powerful function space theorems can then be applied to provide proofs of many knot theory fundamentals. However we do not go into these details in this book and so most of the time will view a knotting as a subset. ▮

3.3 Definition of Smooth Knotting and Smooth Knot

We focus is on knottings that are locally flat manifolds. (In fact, soon we will further restrict attention to smooth manifolds.) For a manifold

$K \subseteq R^q$, $K \times I$ is also a submanifold of $R^q \times I$, and the trace T of an isotopy of K is a submanifold. But T might not be locally flat even if each K_t is locally flat. This is not just a technical objection. We will no longer have an interesting theory of knotted circles since all knottings of a circle in R^3 will be isotopically the same, using a **"pull-tight" isotopy**:

Example 3.1. This is our first example that involves visualization in four dimensions, so we proceed in detail. We describe an isotopy K_t of a circle in R^3 of type we will want to avoid.

A knotting K_0 of S^1, such as in Figure 3.6, can be viewed as a union of two smooth arcs, A and B, which share two points: $A \cap B = \{a, b\}$, Figure 3.5. Grab B with one hand near a, with the other hand near b, then pull, causing this self-tangled portion B of the knotting to get smaller and smaller for $0 \leq t \leq 3/4$, until "vanishes" to a point c at $t = 3/4$. Graphically, we show curves in R^3 as thin tubes. But mathematically, this knotting has no thickness, and we can cause B to shrink to a point c at $t = 3/4$, resulting in a knotting K_1 see Figures 3.7— 3.10. One detail: K_t will be unchanging (that is, $K_t = K_1$) for $3/4 \leq t \leq 1$. This last step is not essential, but it will keep our point of vanishing away from the boundary of our manifold. Sometimes, this is called the **pull-tight isotopy**.

The trace $\mathbb{K}$ of K_t is a knotting of $S^1 \times I$ in $R^3 \times I \subseteq R^3 \times R^1 \approx R^4$. This manifold is *not* locally flat at $k = (c, 3/4)$.

To visualize the trace we project it into R^3. In Figures 3.7— 3.10 the z-axis is assumed to be "out of the page, directed at the viewer".

There are two natural projections—the standard projection of R^3, $(x, y, z, t) \to (x, y, z)$ and a non-standard projection $(x, y, z, t) \to (x, y, t)$. This *non-standard* projection is important for the study of isotopies; it is called the **trace projection**. We show the trace projection of $\mathbb{K}$ in Figure 3.13. (To view details more clearly, we have rotated this in R^3 so that the projection of the trefoil at the top rather than the bottom.)

For $3/4 \leq t \leq 1$, the projection of the trace is cylindrical, Figure 3.11. The trefoil knotting of Figure 3.5 has three crossing points in the projection. These three crossing points will occur in the projections of all of the knottings K_t with $0 \leq t < 3/4$, giving rise to three line segments of crossings in the trace of the projection, Figure 3.12. The entire trace projection is shown in Figure 3.13. ▌

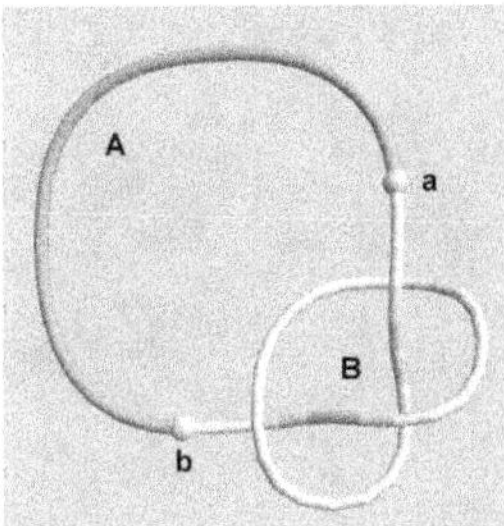

Fig. 3.5 A trefoil knotting K as union of arcs A and B.

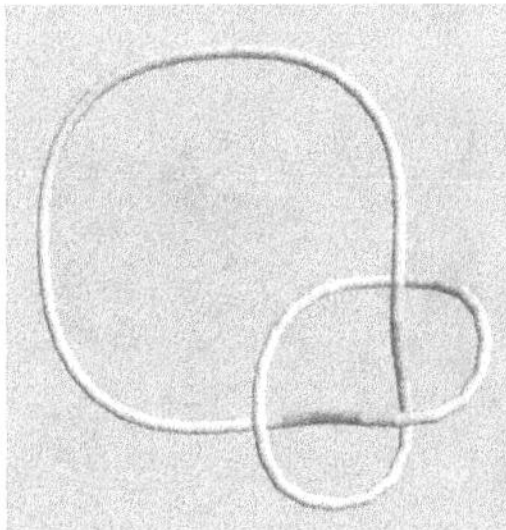

Fig. 3.6 Beginning of isotopy with K_0.

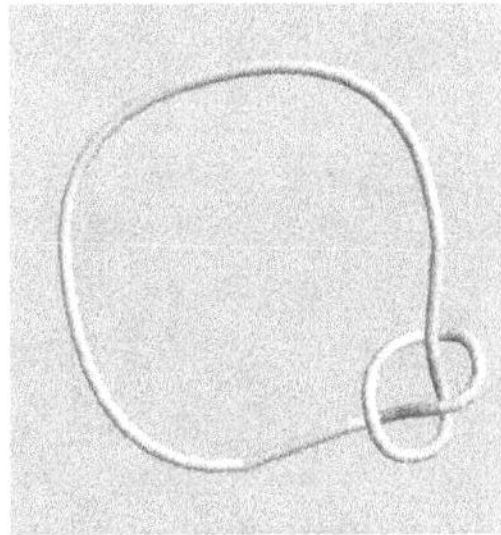

Fig. 3.7 The pull-tight "shrinks" the trefoil portion—here is $K_{1/4}$.

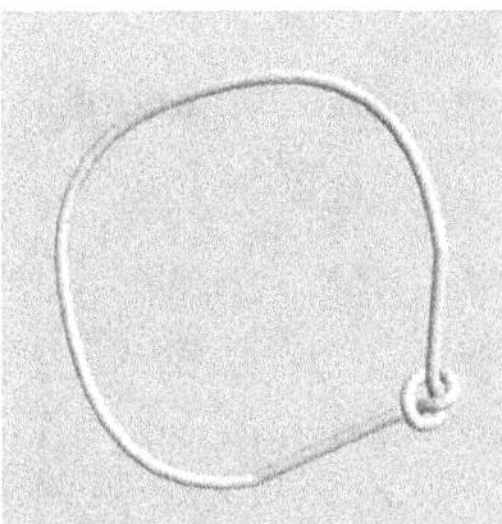

Fig. 3.8 More shrinking at $K_{1/2}$.

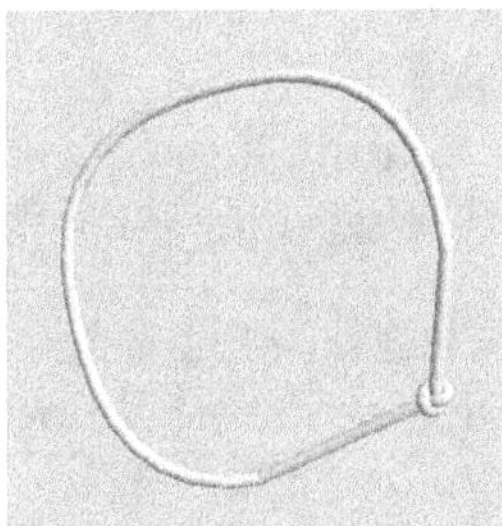

Fig. 3.9 More shrinking close to $t = 3/4$—here is $K_{0.7}$.

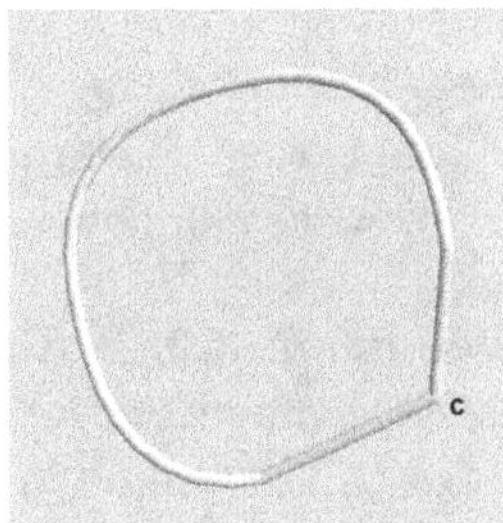

Fig. 3.10 At $K_{3/4}$ trefoil tangle shrunk to single point c.

In Figure 3.13 everything seems ordinary except at the one point of the trace corresponding to $k \in R^3 \times I$. The trace fails to be locally flat here, although it is non-trivial to prove this. Visually it seems that our trace Figure 3.13 is irretrievably crinkled at c_0 and thus not locally flat. We will see this behavior in a different context in Section 5.5.

Fig. 3.11 Projection of the trace for $3/4 \leq t \leq 1$.

Fig. 3.12 Projection of the trace for $0 \leq t \leq 3/4$.

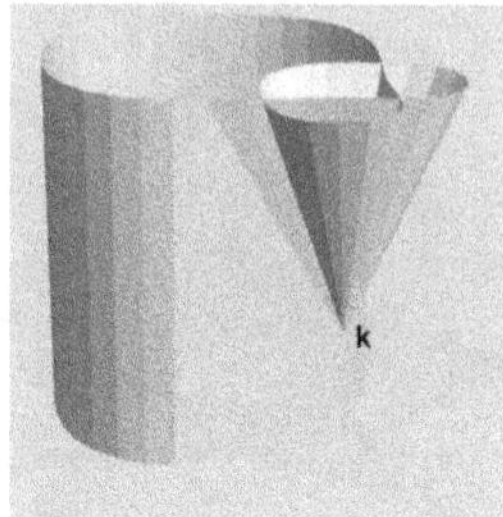

Fig. 3.13 At $k = (c, 3/4)$ the trace is *not* locally flat.

Our method of avoiding such non-locally flat points is to require smoothness, Definition 2.9.

An isotopy is a **smooth isotopy** if the trace is a smooth submanifold.

Now we can define what we mean by a knot:

Definition 3.2. If $K \subseteq R^q$ the set of all subsets of R^q smoothly isotopic to K, denoted by $[K]$ is called a **smooth knot** in R^q. If $K \subseteq M \subseteq R^q$ the set of all subsets of M smoothly isotopic in M to K, denoted by $[K]$ called a **smooth knot in** M. ▮

More generally we can extend this idea to include smooth proper pairs of manifolds such as proper knottings of arcs in D^3.

Definition 3.3. If $K \subseteq M$ and $K' \subseteq M$ are smooth proper manifold knottings, **K and K' are smoothly isotopic** if there is a smooth map $f\colon M \times I \to M$ such that if we let $K_t = f(M \times \{t\})$ then each K_t is a smooth proper knotting of M in X and $K_0 = K$, $K_1 = K'$. ▮

3.4 n-Knots and n-Knottings

For most of this book we focus on smooth knottings of closed n-dimensional submanifolds $M^n \subseteq R^q$ or $M^n \subseteq S^q \subseteq R^{q+1}$ (rather than a more general $M^n \subseteq N^m \subseteq R^p$). Also, we will see that the codimension 2 case (i. e. $q = n + 2$) is mathematically special. For most considerations, knottings in S^q are very similar to corresponding knottings in R^q. When there is a difference usually knottings in R^q are easier to visualize, but the knottings in S^q are slightly simpler, mathematically, as in Section 4.7.

Definition 3.4. A knotting of a closed smooth n-manifold—either $K^n \subseteq R^{n+2}$ or $K^n \subseteq S^{n+2}$ is called an n**-knotting**. The smooth isotopy equivalence class of an n-knotting is called an n**-knot**. ∎

In our terminology, an n-knotting can have more than one component. In the literature, often "knot" refers to a class of a knotting of a *single* component while "link" to a knotting with *several* components.

The word **surface** is used in several different ways in mathematics. In this book, a surface is a 2-manifold. A 2-knotting is also called **knotting of a surface**.

On rare, but significant, occasions (for example, Definition 1.38 or Example 12.11), we will need to consider subsets that are unions manifolds of *different* dimensions or have codimension other than two:

Definition 3.5. Suppose $X \subseteq M^n$ is a finite union of closed smooth submanifolds of M^n, possibly different dimensions. We say X is a **manifold-knotting in** M^n. The smooth isotopy equivalence class X in M^n is called a **manifold-knot**. ∎

Remark 3.4. One can usefully study knottings that are not manifolds, such as **spacial graphs**. We do not pursue this in this book. A **spacial graph** G is a finite union of arcs e_i called **edges** in R^3, that are either smooth or piecewise linear. These arcs are disjoint except (perhaps) if they share a common endpoint. The endpoints of the arcs are called **vertices**. In contrast to other uses of the term "graph", two distinct edges are allowed to share the same two vertices, and loops are allowed consisting of an edge from one vertex to itself. With this definition, a knotting of a circle can be viewed as a spatial graph with one edge and one vertex.

A basic spacial graph is a **Θ-curve**. [Kinoshita (1972)]. Topologically, this is a union of a circle with a diameter—thus it is homeomorphic to the letter Θ. One of the main topics for spacial graphs is the study of the "subknots" one obtains by deleting an edge from a knotting of a graph. For example, there are three circle knottings obtained by removing one of the three edges of a Θ-curve knotting. The paper [Conway and Gordon (1983)] discusses knottings of other simple well-known graphs. Good surveys are [Ramírez Alfonsín (2005)], [Foisy and Ludwig (2009)] and [Flapan *et al.* (2017)]. ∎

The simplest way to combine two disjoint knottings in M, P and Q is to take the union. The **split sum** is a special case where they are separated by an $(n-1)$-sphere $\Sigma \subseteq M$. The idea is that P and Q may be self-tangled but not tangled with each other since Σ that acts as a barrier. The knotting in Figure 1.17 is not split, but the knotting of Figure 1.18 is split. Using Proposition 2.7, the next definition is well-defined:

Definition 3.6. If $K^n \subseteq M^q$ is a knotting of an n-manifold K^n in a q-manifold M^q we say K^n **is split** if there is a q-ball $B^q \subseteq M^q$ with $B^q \cap K^n \neq \emptyset$, $K^n \cap \partial B^q = \emptyset$ and $K^n - B^q \neq \emptyset$. We also say that K^n is **split by the sphere** ∂B^q. If K^n is a knotting of r n-spheres which bounds a knotting of r disjoint $(n+1)$-balls,that we say K^n is a **trivial** n**-knotting of spheres** and that the equivalence class is a **trivial split** n**-link.** ∎

Historically, for 1-knottings, split knottings are most often ignored since one can usually consider the split components separately. However, for higher-dimensional knottings this may not be desirable. One example is slice knottings, Sections 6 and 8, as well as some notions of "random knottings" where split links can be common.

3.5 Knotting and Self-Diffeomorphisms

Given a self-diffeomorphism of smooth manifold $F : N \to N$ and a subset $K \subseteq N$, then $K' = F(K)$ is another smooth knotting. As we will discuss, K may or may not be isotopic to K'. Additionally, if we are viewing a knotting as an *image of an embedding* $K = \phi(M)$, and we have a self-diffeomorphism $G\colon M \to M$, then $K' = \phi \circ G(M)$ is an image of another smooth embedding, where the embedding ϕ may or may not be isotopic to the embedding $\phi \circ G$ (even though they have the same image).

A diffeotopy is a special case of smooth isotopy, Definition 2.6:

Definition 3.7. A **diffeotopy** of a smooth manifold M is a smooth isotopy $F_t\colon M \to M$ where each F_t is a diffeomorphism. We say F_0 and F_1 are **diffeotopic maps.** If $K \subseteq N$ is a knotting, then $F_0(K)$ and $F_1(K)$ are **diffeotopic knottings.** ∎

Example 3.2. Using polar coordinates, the map $F_t = (1, \theta + t)$ is a diffeotopy of R^2 where F_t rotates the circle upon itself by an angle of t radians, $0 \leq t \leq 1$. Here, F_0 is the identity map of S^1, and F_1 is a rotation by 1 radian. ∎.

A basic theorem of smooth topology is the **isotopy extension theorem**:

Prop 3.1. If ϕ and ϕ' are smoothly isotopic embeddings of a smooth manifold M into N, then there is a diffeotopy F_t of N such that $F_0 = \phi$ and $F_1 = \phi'$. ▮

For a knotting $K \subseteq N$, the manifold N is sometimes called an **ambient manifold** for the knotting, and a diffeotopy called an **ambient isotopy**.

If we have an ambient isotopy $F_t \colon N \to N$ and $M \subseteq N$ is a smooth submanifold, then $F_t|_M$ is a smooth isotopy of M. Thus, the isotopy extension theorem shows that knottings are isotopic if and only if they are ambiently isotopic. So, we could use ambient isotopy as our basic equivalence of knottings, as is commonly done in the literature. In some mathematical situations, this approach has advantages, but for looking at examples of isotopies—our particular interest—it is rarely used. To directly show that two knottings K and K' of a circle in R^3 are *ambiently* isotopic in R^3, we are required to find a diffeotopy of *all of* R^3 that moves K to K'.

Example 3.3. A square $n \times n$ matrix M of real numbers corresponds a linear map $f \colon R^n \to R^n$ via matrix multiplication: $f(\overrightarrow{v}) = M\overrightarrow{v}$. If M has non-zero determinant, it is called **non-singular**, and f is a diffeomorphism. If the determinant of M is ± 1, we say this **map is orthogonal.**

A 2×2 matrix can be visualized as an ordered pair of 2-dimensional vectors that correspond to the column vectors of M. If the matrix is non-singular, the vectors will be independent. Figure 3.14 is such a visualization for $M_0 = \begin{pmatrix} 3\ 3 \\ 3\ 7 \end{pmatrix}$.

We will show that a non-singular linear map is diffeotopic to an orthogonal map. A standard process in linear algebra, **Gram-Schmidt orthogonalization**, will convert any non-singular matrix M_0 to an orthogonal matrix M_1. For any $n \times n$ matrix this is usually done using discrete manipulations on the column vectors. For a 2×2 matrix M, orthogonalization replaces the first vector with a corresponding unit vector, replaces the second vector by projection orthogonal to the first, and finally replaces that vector with a corresponding unit vector. Doing each step smoothly we can obtain a smoothly changing matrix M_t with M_1 orthogonal. Each step replaces one vector $\overrightarrow{v_0}$ by another $\overrightarrow{v_1}$. Such a discrete change can also be done by a smooth change $\overrightarrow{v_t}$ where $1 \leq t \leq 1$ where $\overrightarrow{v_t} = (1-t)\overrightarrow{v_0} + t\overrightarrow{v_1}$.

Figures 3.14 through 3.18 give a visualization of an ambient isotopy of R^2 showing a smooth orthonormalization of M_0. ▮

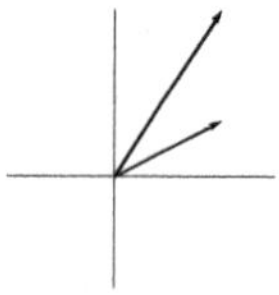
Fig. 3.14 M_0.

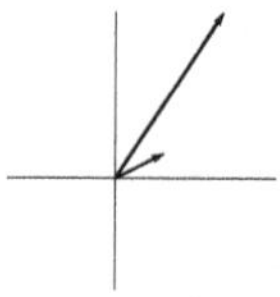
Fig. 3.15 $M_{0.25}$.

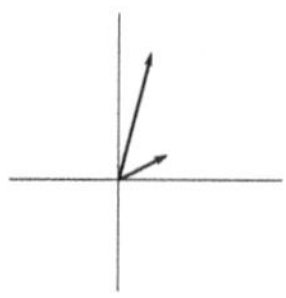
Fig. 3.16 $M_{0.5}$.

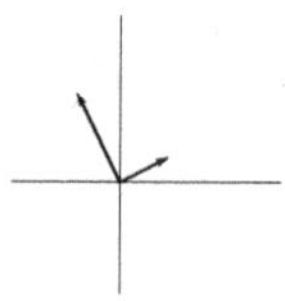
Fig. 3.17 $M_{0.75}$.

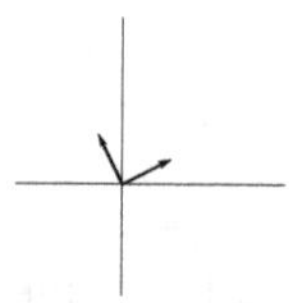
Fig. 3.18 M_1.

There are many self-homeomorphisms of a smooth manifold but many are diffeotopic. The standard S^n, D^n and R^n have special properties in this regard.

Definition 3.8. The **standard reflection of** R^n is $r\colon R^n \to R^n$ defined $r(x_1, x_2, \dots, x_n) = (-x_1, x_2, \dots, x_n)$. The **standard reflection of** $S^n \subseteq R^{n+1}$ is $r|_{S^n}$ is $r|S^n$ where r is the standard reflection in R^{n+1}. ▮

In general (except for 0-dimensional manifolds) a manifold will have infinitely many self-diffeomorphisms but for S^n, D^n or R^n there are only two that are different (up to diffeotopy):

Prop 3.2. Any self-diffeomorphism of the standard manifold S^n, D^n or R^n is either diffeotopic to the identity or to a reflection. ▮

The identity map of S^n (in fact for any orientable manifold) is **orientation preserving**. The reflection is **orientation reversing**. Thus Proposition 3.2 could be restated: there are exactly two diffeotopy classes of self-diffeomorphisms of S^n, D^n or R^n—orientation preserving and orientation reversing.

Even for the case $n = 2$ a proof of Proposition 3.2 is beyond the scope of this brief introduction. However, the proof for D^1 is easy. A smooth embedding $f\colon D^1 \to D^1$ is either monotone increasing or monotone decreasing. The identity diffeomorphism $id\colon D^1 \to D^1$ is defined $f(x) = x$ the reflection diffeomorphism is $r(x) = -x$. If f is monotone increasing, we define $F(x,t)\colon D^1 \times I$ by $F(x,t) = (1-t)f(x) + tx$ then $F_0 = f, F_1 = id$, and each F_t is an embedding since it is smooth and monotone increasing. In the case that f is monotone decreasing, we would define $F(x,t) = (1-t)f(x) - tx$.

Proposition 3.2 is special for S^n, D^n and R^n. The set of all diffeotopy classes of self-diffeomorphisms of a manifold is a well-studied subject called the **mapping class group**. For 2-manifolds, this topic is a large area of geometric topology, well covered in books such as [Casson and Bleiler (1988)], [Farb and Margalit (2012)] [Birman (1974)]. The torus $S^1 \times S^1$ has infinitely diffeotopy classes of self-diffeomorphisms that can be roughly described as twisting of a torus. The projective plane $\mathbb{P}^2$, which we will later consider, has only one diffeotopic class—that is, any self-diffeomorphism is diffeotopic to the identity diffeomorphism. The Klein bottle has only four diffeotopic classes. There are many examples known of smooth orientable manifolds that have no orientation reversing self-diffeomorphism, such as the 3-dimensional Poincaré sphere, and the 4-dimensional complex projective 2-space. A good summary is in [Müllner (2009)].

We noted in Remark 3.3 that the distinction between a knotting of M as a subset and a knotting as an image of an embedding of M is a self-homeomorphism of M. Similarly, the distinction between an isotopy of M and isotopy of an *embedding* of M is a self-isotopy of M.

So if $H_t\colon M \to M$ is a self-isotopy of M and F_t is an isotopy of embeddings of M in N, then $F_t \circ H_t$ is an isotopy of M in N.

In particular, if we have an embedding $\psi\colon M \to N$ and self-homeomorphism $h : M \to M$, which is isotopic to the identity map of M, then ψ and $\psi \circ h$ are isotopic embeddings. But what happens if h is not isotopic to the identity?

Although smooth isotopy is a natural way to consider equivalence of knottings, there is a commonly used slightly different alternative.

Definition 3.9. Given two knottings $K \subseteq M$ and $K' \subseteq M$, we say these knottings are **diffeomorphic as pairs** to if there is a diffeomorphism $H\colon M \to M$ with $H(K) = K'$. ∎

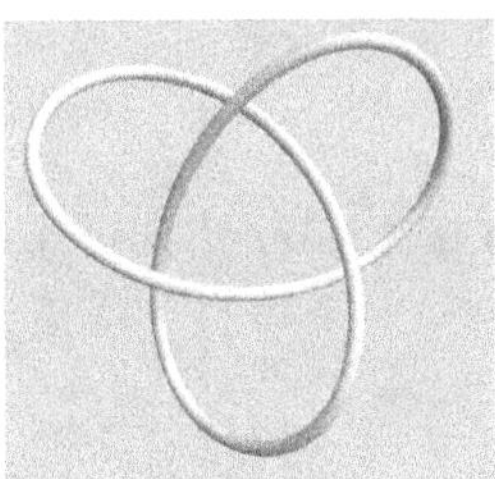

Fig. 3.19 A left-handed trefoil.

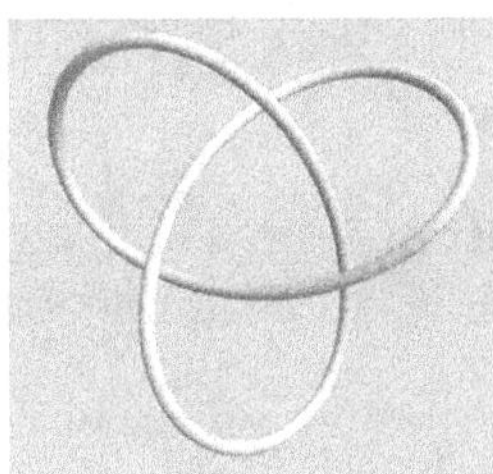

Fig. 3.20 A right-handed trefoil.

Remark 3.5. If two knottings are smoothly isotopic, then they are diffeomorphic as pairs, since given an ambient isotopy F we have diffeomorphism F_1 with $F_1(K) = K'$. However, there are knottings diffeomorphic as pairs that are not smoothly isotopic, such as the left-hand trefoil Figure 3.19 and the right-hand trefoil 3.20, originally shown in [Dehn (1914)] with a modern proof in [Kauffman (1990)]. ▮

If we have a knotting $K \subseteq M$ and a homeomorphism $H\colon M \to M$, we could let $K' = H(K)$ and study if K and K' are equivalent knottings or not. If we have embedding $\phi\colon N \to M$ and a self-homeomorphism $h\colon N \to N$ then $\phi \circ H$ is another embedding of N we also have another embedding if we combine both: $H \circ \phi \circ h$. We now can consider which (if any) of the following embeddings of N into M are isotopic: $\phi, H\circ\phi, \phi\circ h$ and $H\circ\phi\circ h$.

Example 3.4. This topic is familiar for $S^1 \subseteq R^2$, where (as in Stoke's Theorem) we distinguish the circle oriented clockwise, or oriented is counter-clockwise. If $\phi\colon S^1 \to R^2$ is the standard (identity) map, and $h\colon S^1 \to S^1$ the standard reflection of S^1, the images of ϕ and $\phi \circ h$ are the same set. The standard way to visualize h is to draw a curved arrow on the unit circle, Figures 3.21 and 3.22. The effect of this map is to reverse the direction of the circle. These two embeddings of the circle into the plane are not isotopic. ▮

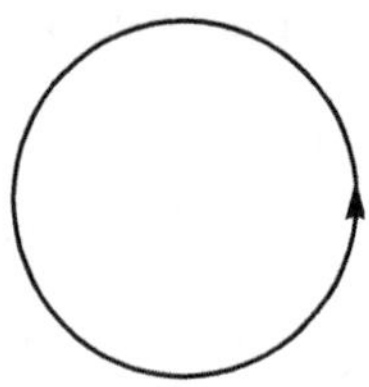

Fig. 3.21 Counter-clockwise embedding.

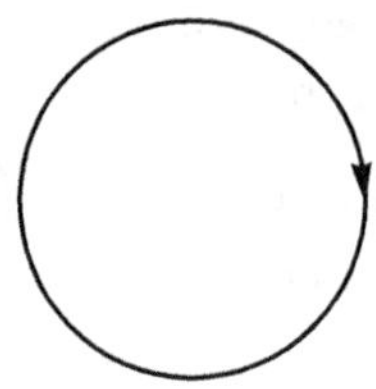

Fig. 3.22 Clockwise embedding.

Definition 3.10. For knotting K in R^n (or S^n), the **mirror image of K**, denoted ꓘ, is the image of K under standard reflection in R^n (S^n). ▮

Orientation is discussed in Section 2.5.

Definition 3.11. An oriented knotting is a knotting K together with some choice of direction we use the notation $\overrightarrow{K}$. The knotting with the opposite orientation of will be denoted $\overleftarrow{K}$ and called the **reverse of $\overrightarrow{K}$**. ▮

Given an oriented knotting $\overrightarrow{\mathrm{K}}$ there are three other knottings we can obtain if we consider orientation: $\overleftarrow{\mathrm{K}}$, $\overrightarrow{ꓘ}$, $\overleftarrow{ꓘ}$.

Definition 3.12. If $[\overrightarrow{K}] = [\overleftarrow{K}]$ we say that [K] is **invertible**. ▮

Not all knottings of a circle give an invertible knot. The knotting, which is 8_{17} in the standard knot table, is the first non-invertible knot [Trotter (1963)]. For more complex knottings, it appears that most of them are non-invertible. A standard reference on non-invertible knots is [Hartley (1983)].

A knot that is invertible via an involution of S^3 is called **strongly invertible**.

For the remainder of this section we will only consider n-knottings of spheres.

Definition 3.13. An n-knotting of a sphere K is **amphicheiral** (sometimes spelled amphichiral, or more briefly achiral) if $[\mathrm{K}] = [\reflectbox{K}]$, otherwise it is called **chiral**. If a knotting is oriented, it is called **positively amphicheiral** or **(+)-amphicheiral** if, in realizing this equivalence, the orientation is preserved, or **negatively amphicheiral** or **(−)-amphicheiral** if the orientation is reversed. ▮

The Figure-8 knotting is positively amphicheiral. Negative amphicheiral means an oriented knotting is equivalent to the reverse of its mirror image. This is not often found in relatively simple knottings.

The table in Figure 3.23 summarizes five types:

Fully Chiral Four knots: $[\overrightarrow{\mathrm{K}}], [\overleftarrow{\mathrm{K}}], [\overrightarrow{\reflectbox{K}}], [\overleftarrow{\reflectbox{K}}]$.
Fully Amphichiral One knot $[\overrightarrow{\mathrm{K}}] = [\overleftarrow{\mathrm{K}}] = [\overrightarrow{\reflectbox{K}}] = [\overleftarrow{\reflectbox{K}}]$.
Reversible Two knots: $[\overrightarrow{\mathrm{K}}] = [\overleftarrow{\mathrm{K}}], [\overrightarrow{\reflectbox{K}}] = [\overleftarrow{\reflectbox{K}}]$
Positive Amphicheiral Three knots: $[\overrightarrow{\mathrm{K}}] = [\overrightarrow{\reflectbox{K}}], [\overleftarrow{\mathrm{K}}], [\overleftarrow{\reflectbox{K}}]$
Negative Amphicheiral Three knots: $[\overrightarrow{\mathrm{K}}] = [\overleftarrow{\reflectbox{K}}], [\overleftarrow{\mathrm{K}}], [\overrightarrow{\reflectbox{K}}]$

<table>
<tr><th></th><th>Chrial</th><th>Amphicheiral</th></tr>
<tr><td rowspan="2">Non-invertible</td><td rowspan="2">Fully Chiral</td><td>Positive Amphicheiral</td></tr>
<tr><td>Negative Amphicheiral</td></tr>
<tr><td>Invertible</td><td>Reversible</td><td>Fully Amphicheiral</td></tr>
</table>

Fig. 3.23 Five types of knots.

For amphicheirality of 1-knottings with more than one component a good reference is [Kadokami and Kawauchi (2011)].

3.6 Homotopy and Regular Homotopy

Recall the notion of immersion, Definition 1.24. Sometimes, subsets are more naturally described as images of functions that are *not* embeddings, such as an image of an immersion. We weaken our definition of smooth isotopy as follows:

Definition 3.14. Suppose $F\colon X \times I \to Y$ is a smooth function where X and Y are smooth manifolds with $f_t = F|_{X\times\{t\}}$. We say F is a **smooth homotopy between f_0 and f_1.**

Given smooth functions $f\colon X \to Y$ and $g\colon X \to Y$ we say f **and** g **are smoothly homotopic functions** if there is a smooth homotopy F with $f_0 = f$ and $f_1 = g$. The **trace of the smooth homotopy** F is the subset of $Y \times I$, which is image $\mathcal{F}(X \times I)$ where $\mathcal{F}(x,t) = (f_t(x), x)$; the map $\mathcal{F}$ is called the **trace map.** ▮

Example 3.5. Consider the homotopy F_t of a circle C in R^2 indicated in by F_0,$F_{\frac{1}{3}}$, $F_{\frac{2}{3}}$ and F_1 Figures 3.24, 3.25, 3.26 and 3.27. The union for all $a \in I$ of these curves form an infinite stack in $R^2 \times I \subseteq R^3$, Figure 3.28 which is the trace of F, Figure 3.29. ▮

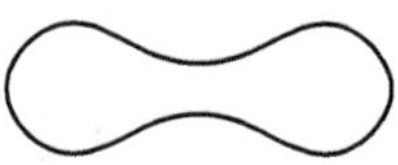

Fig. 3.24 Image $f_0(C)$.

Fig. 3.25 Image $f_{\frac{1}{3}}(C)$.

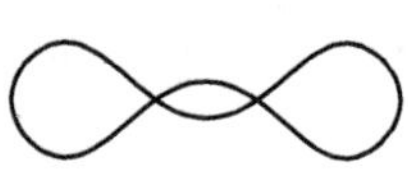

Fig. 3.26 Image $f_{\frac{2}{3}}(C)$.

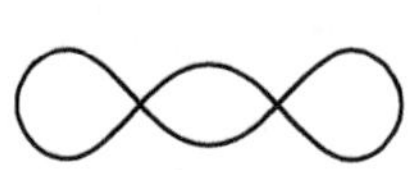

Fig. 3.27 Image $f_1(C)$.

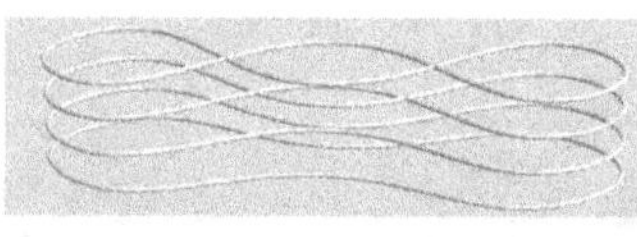

Fig. 3.28 Four curves of form $\mathcal{F}(C \times \{a\})$.

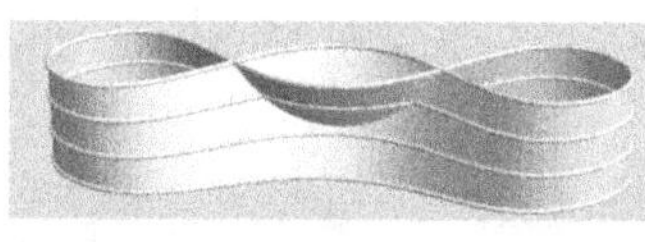

Fig. 3.29 The trace of F of Example 3.5.

Topologically, we classify embeddings via isotopy; topologically, we classify immersions via regular homotopy.

Definition 3.15. Suppose $F\colon X \times I \to Y$ is a smooth homotopy. We say F is a **regular homotopy** if:

(1) f_t is an immersion of $X \times \{t\} \to Y$
(2) the trace map $\mathcal{F}\colon X \times I \to Y \times I$ is an immersion ▮

Identifying X with $X \times \{t\}$ and Y with $Y \times \{t\}$, we can view a regular homotopy as a continuous one-parameter family of immersions $F_t \colon X \to Y$.

Example 3.6. Without condition (2) of Definition 3.15 there are "problems" like the pull-tight problem discussed in Section 3.3. Consider the homotopy g_t of a circle C in R^2 indicated in Figures 3.30, 3.31, 3.32 and 3.33. Informally, we can describe this as "a circle with a loop that is tightened to a point c_0". The peculiarity of this homotopy can be seen in the trace. The trace map $\mathcal{G} \colon C \times I \to R^2 \times I$ fails to be an immersion at $P_0 = (c_0, \frac{3}{4})$, as shown in Figure 3.34 . For smooth manifolds, we avoid a pull-tight for a homotopy if we insist that F is *smooth* and each g_t is an immersion. ∎

Fig. 3.30 Image $g_0(C)$.

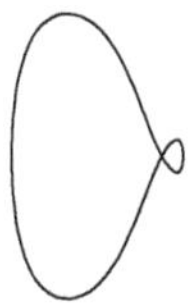

Fig. 3.31 Image $g_{\frac{1}{3}}(C)$.

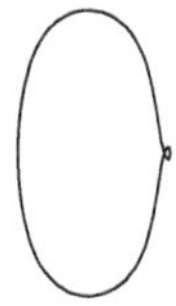

Fig. 3.32 Image $g_{\frac{2}{3}}(C)$.

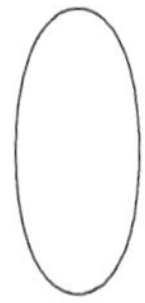

Fig. 3.33 Image $g_1(C)$.

The study of regular homotopies of immersions is a very interesting and important part of geometric topology. An early result is the Whitney–Graustein theorem [Whitney (1937)], which classified *regular* homotopies of smooth planar curves, and later higher dimensional work of Smale [Smale (1958)], [Smale (1959)] and others.

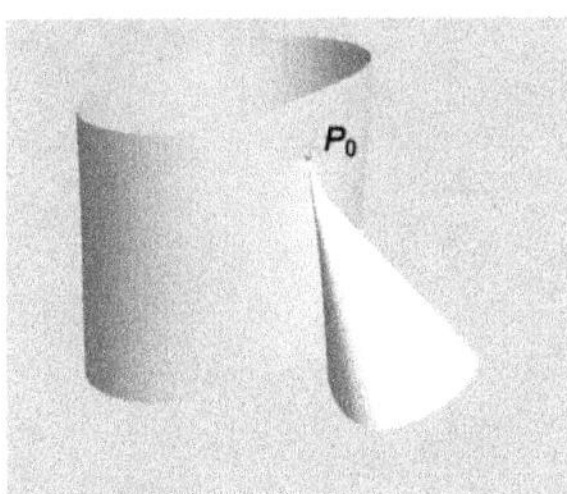

Fig. 3.34 Trace of g_t is not image of immersion at point P_0.

For any map into R^2 we can smoothly shrink its image to a point, so any map is homotopic to a constant map. In particular, any two smooth maps of a circle into R^2 are smoothly homotopic. However, the Whitney–Graustein theorem says that there are infinitely many maps of the circle into R^2, so no two are regularly homotopic. Smale's theorem is often described as saying, "A sphere can be turned inside out in R^3. Readers not familiar with these results are encouraged to view the video Outside In produced in 1994 at the Geometry Center. At the time of writing this book, this video was without an easily available source.

Example 3.7. Regular isotopies of circles in S^2 and R^2 are very different. The embeddings indicated in Figures 3.21 and 3.22 are not regularly homotopic.

However, the equator of S^2 considered clockwise, is isotopic, say by isotopy h_t, in S^2 to the same circle with the direction reversed as can be readily see Figures 3.35 and 3.39. This isotopy is indicated in the sequence of Figures 3.35, 3.36, 3.37, 3.37 and 3.39. ∎

Fig. 3.35 h_0.

Fig. 3.36 $h_{\frac{1}{4}}$.

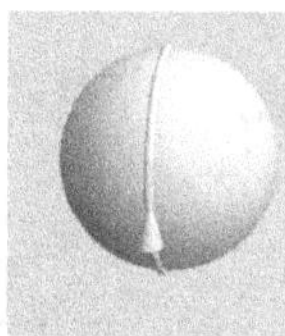

Fig. 3.37 $h_{\frac{1}{2}}$.

Fig. 3.38 $h_{\frac{3}{4}}$.

Fig. 3.39 h_1.

Chapter 4

Topology Topics of 1-Knottings

Unless stated otherwise, all knottings will be smooth. We focus on aspects of 1-knottings of particular significance to higher dimensional knotting. More comprehensive treatments of 1-knottings include [Crowell and Fox (1977)], [Burde *et al.* (2014)], [Lickorish (1997)], [Przytycki (1998)], [Kauffman (2013)], [Livingston (1993)], [Przytycki *et al.* (2024)]. Knot theory began as an attempt to make a table of 1-knottings. The most commonly used early table is in [Reidemeister (1927)]. Modern mathematical techniques combined with machine computation have extended these greatly, [Hoste *et al.* (1998)] and [Hoste (2005)].

4.1 Some Basic Constructions of 1-Knottings

If K is a knotting of a circle in S^3 it might seem complex but, since it is a submanifold, it locally looks simple. If $x \in K$ then there is a small 3-ball B containing x and $B \cap K$ is just a standard disk pair. Now $C = \overline{S^3 - B}$ is also 3-ball, so all the complexity K is somehow expressed in the pair $(C, C \cap K)$. We explore this further.

Remark 4.1. The exterior of the *standard* $S^1 \subseteq S^3$ is homeomorphic to a solid torus $S^1 \times D^2$, Example 2.9. An alternate description of $S^1 \times D^2$ is "a ball with an unknotted hole". We will see that the exterior of the trefoil in S^3 of Figure 4.3 is diffeomorphic to a "ball with a corresponding trefoil-knotted hole", Figure 4.2.

Roughly a "hole in a 3-ball B" corresponds to a long thin tube $H \approx D^2 \times D^1 \subseteq B$ which, when removed, gives a long thin hole. The annulus A corresponding to $\partial D^2 \times D^1$ is "the long tubular side of the hole" and the subset $G \approx D^2 \times \partial D^1$ is a subset of ∂B, consisting of "the disks which cap the hole at the top and at the bottom ".

Now, suppose K is a 1-knotting of a circle with product tubular neighborhood $T \approx S^1 \times D^2$. The knot exterior is $X = S^3 - int(T)$. Write S^1 as union of two closed arcs B_0 and H_0 where $B_0 \subset S^1$ be a small closed arc and $H_0 = S^1 - int(B_0)$. Next, let $B \subseteq T$ correspond to $B_0 \times D^2 \approx D^3$, $H \subseteq T$ correspond to $H_0 \times D^2$ and let $A \subset \partial T$ be the annulus corresponding to $H_0 \times \partial D^2$, Figure 4.1.

Since B is a smooth 3-ball in S^3, $C = S^3 - int(B)$ is also a smooth 3-ball in S^3. We have $X \subseteq C$, in fact $C = X \underset{A}{+} H$. So H "fills up a hole in C". Equivalently, if we remove H from 3-ball C, (then add A back to make a manifold with boundary) we get the knot exterior X. So we may say that the exterior of our knotting is a ball with a knotted hole.

For our second method, we visualize a knot exterior, via inversion in R^3. To keep our focus on R^3 rather than S^3 we use stereographic mapping. If K contains the S_0, the south pole of S^3, then we rotate S^3 just a little so that it no longer contains S_0. We still call the rotated knotting K. (In other words, without loss of generality, we may assume K misses S_0.) Let $K' = P(K)$ be the stereographic image. Next we take a diffeomorphism F which is a translation and scaling such that $K'' = F(K)$ contains $\mathbb{O} \in R^3$ and so that the tubular neighborhood T'' of K'' contains the unit ball $D^3 \subseteq R^3$. If K'', is a (left-handed) trefoil knotting, we would have Figure 4.3. The exterior of $K \subseteq S^3$ is diffeomorphic to the "all points of R^3 on, or outside of, $\partial T''$, plus the point at infinity".

The standard inversion, ι of R^3 will take points inside $\partial T''$ to points outside, and take points outside to points inside. It will take $\mathbb{O} \in R^3$, a point not in our exterior, to "the point at infinity" and brings "the point at infinity", which *is* in our exterior, to $\mathbb{O}$, where "we can see it". So we conclude that the knot exterior of K is diffeomorphic to the points on and inside of $\iota\,(\partial T'')$.

If knotting K'' is (left-handed) trefoil, our tubular neighborhood and standard sphere might look like Figure 4.3. Figure 4.4 is an actual calculation of $\iota\,(\partial T'')$. By cutting a disk out we can visually expose our knotted hole. One technical point: inversion is orientation reversing, so that the exterior of our *left-hand* trefoil knotting of Figure 4.3 is a ball with a *right-hand* knotted hole in Figure 4.4. ∎

Also, we can represent a knotting of S^1 in S^3 using a disk-pair.

Remark 4.2. Given a knotting of an oriented circle $K \subseteq (S^3 - S_0)$ we will obtain a proper knotting of an arc $\alpha_K \subseteq D^3$ with $\partial(\alpha_K) = S_0 \subseteq \partial(D^3)$.

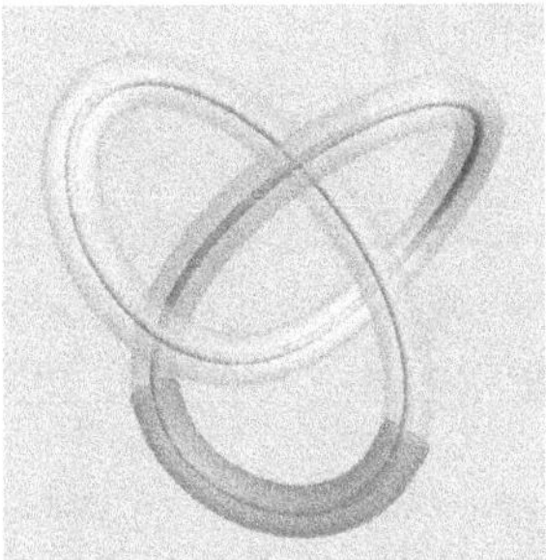

Fig. 4.1 Dark portion of tubular neighborhood is B.

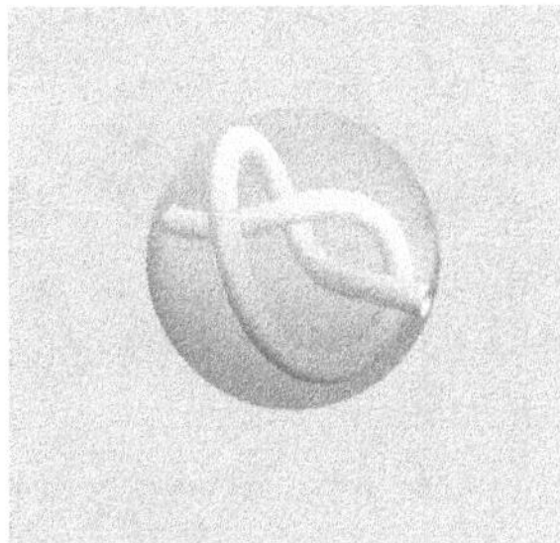

Fig. 4.2 A ball with a knotted hole.

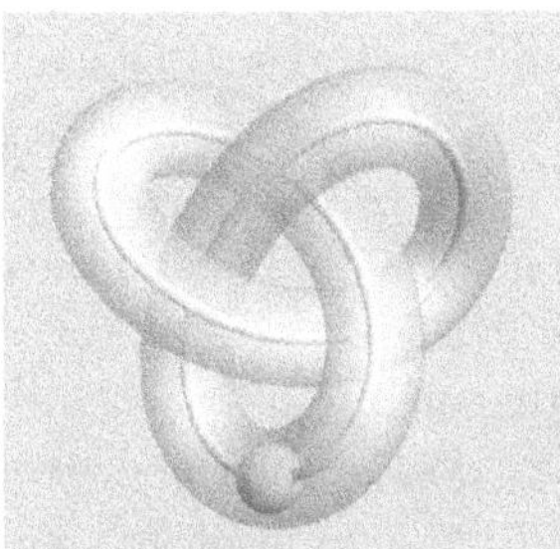

Fig. 4.3 T'' with unit sphere S^2.

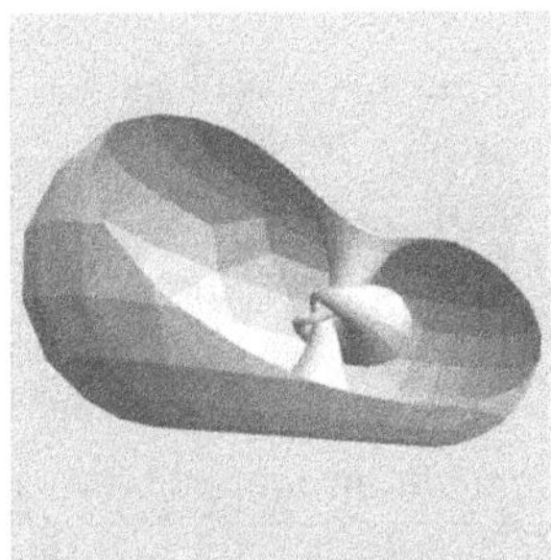

Fig. 4.4 $\iota(\partial T'')$ of Figure 4.3 in S^2 with disk removed to expose knotted hole.

Graphically we go from Figure 4.5 to 4.6. Intuitively, the arc in α_K in a 3-ball, in some sense, "captures all of the self-tangling of K". We will say α_K is a **knotting of an arc associated with the knotting** K. Section 10.7 has a related, slightly different construction.

This knotting α_K is not uniquely determined, but we want the knot-type of α_K determined, where here **knot** is the isotopy class, $[\alpha_K]$, of proper isotopies which are fixed on the boundary.

As in beginning of Remark 4.1 we write S^1 as union of arcs $S^1 = \mathbb{B}_0 \cup \mathbb{H}_0$, but this time we let $\mathbb{B}$ be a small geometric ball in S^3 with center a point $x \in K$ and $\mathbb{B} \cap K = \mathbb{B}_0$. Now $\mathbb{C} = S^3 - int(\mathbb{B})$ is a geometric 3-ball and $\mathbb{H}_0$ is a proper arc in 3-ball $\mathbb{C}$. So, basically, $(\mathbb{C}, \mathbb{H}_0)$ is the ball pair we want. But $\mathbb{C} \neq D^3$ and $\partial(\mathbb{H}_0) \neq S^0$, so we continue.

We find a rotation ρ of S^3 so that $S_0 \in \rho(int(\mathbb{B}_0))$. Let $\mathbb{H}'_0 = \rho(\mathbb{H}_0)$, and, as before, $\mathbb{C}' = \rho(\mathbb{C})$. Using the stereographic mapping we get $\mathbb{C}'' = \mathcal{P}(\mathbb{C}')$ and $\mathbb{H}''_0 = \mathcal{P}(\mathbb{H}'_0)$.

Now $\mathbb{H}''_0$ is a proper 1-disk in a geometric 3-ball $\mathbb{C}''$ in R^3. We can find diffeomorphism f of R^3 (a translation followed by scaling) so that $f(\mathbb{C}'') = D^3$. Now $\beta_K = f(\mathbb{H}''_0))$ is a proper arc in D^3. However, we also need $\partial(\beta) = S^0 \subseteq \partial D^3$.

Since B_0 and $\mathbb{B}$ were chosen small, we have ∂B_0 is a pair of points P and Q that are nearly antipodal in the 2-sphere $\partial\mathbb{B} = \partial\mathbb{C}$. At this point we use the orientation of K which we think of as given by a parameterizations $\phi\colon I \to K$ and we choose our labels for P and Q so that orientation goes from P to Q. That is, $\phi(p) = P$, $\phi(q) = Q$ where $0 \leq p < q \leq 1$.

Following through our construction, let p be the endpoint of β_K corresponding to P and q the point corresponding to Q. Now we can find rotation R of R^3 so that $p = R(P)$ is close to $(1,0,0)$ and $q = R(Q)$ is close to $(1,0,0)$. Let $\beta'_K = R(\beta_K)$, see Figure 4.7.

At this point is should be clear that we can get α_K if we "bend the ends of $\partial(\beta'_K)$ slightly" so that we slide p to $(1,0,0)$, slide q to $(-1,0,0)$ and make sure that our new arc meets $\partial(D^3)$ orthogonally and so the result is a *smooth* proper sub-arc of D^3, Figure 4.8. We avoid these bending details here but only mention there is work to be done in order to get a *smooth* curve that matches all these conditions. (Curves which satisfy all conditions except smoothness are routine to find and are called Hermite splines. However, they are polynomial in nature and can only match with β'_K for a finite number of derivatives.) ∎

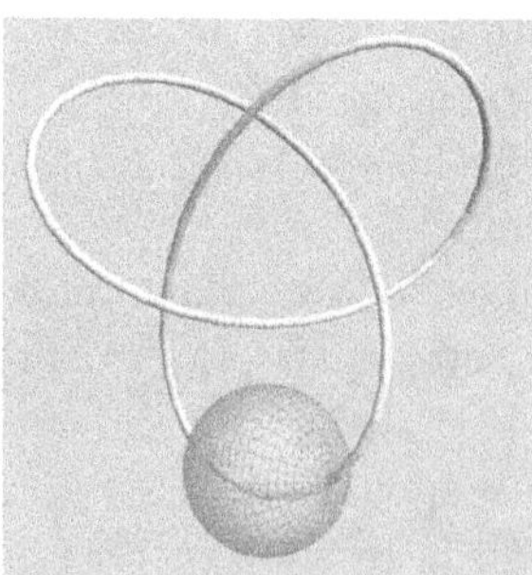

Fig. 4.5 Knotting K and small ball $\mathbb{B}$.

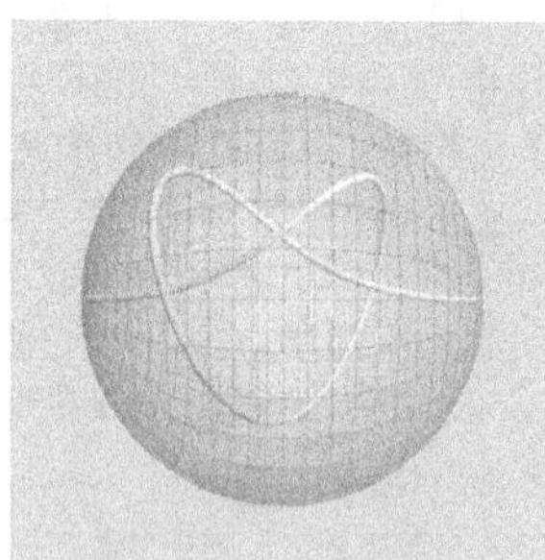

Fig. 4.6 Arc $\alpha_K \subseteq D^3$.

The exterior of a knot does not always determine a knotting. For example, the right-hand trefoil is not equivalent to the left-hand trefoil, yet

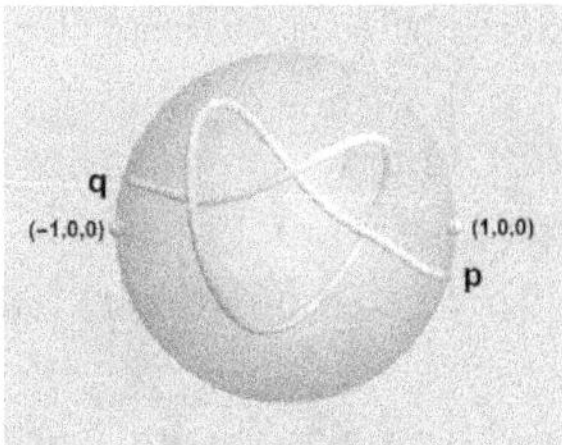

Fig. 4.7 β'_K with endpoints p and q.

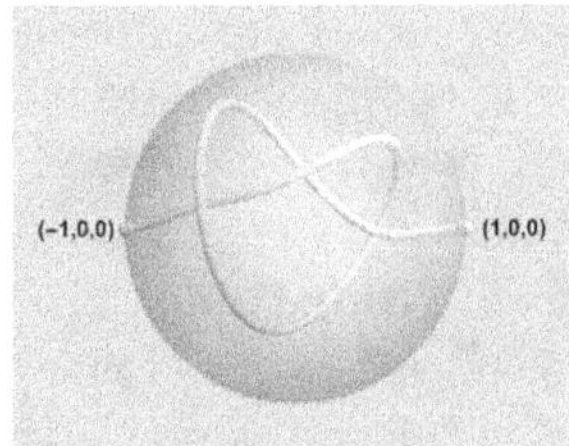

Fig. 4.8 Bend ends of Figure 4.7 and get α_K with endpoints S^0.

since they are mirror images, the complements and exteriors are homeomorphic. But in a strong sense, this is the only problem. In the tame case of knotting of circle [Gordon and Luecke (1989)] showed that except for mirror images, the exterior of a knotting determines the knotting. Note that for links (1-knottings of more than one component), many simple examples exist of non-equivalent knottings with homeomorphic complements. Similarly, for wild knottings of circles, the complement does not determine the knotting—for example, for everywhere-wild knots see [Nanyes (2014)].

Given two 1-knottings of a circle, K and K', we can "glue" these together to get another knotting, the connected sum $K\#K'$, Definition 2.20. For examples see Figure 4.9 and Figure 4.10. Roughly, "we take an unknotted circle, tie K in one small portion, then tie K' in another". We alternatively describe knottings K and K' as *proper ball pairs* (rather than punctured manifolds) and describe $K\#K'$ as obtained by replacing unknotted ball pairs with knotted ones. We go into detail here because it is the key to certain higher dimensional constructions.

Remark 4.3. Begin with two disjoint 3-balls, B and B', centered at two points of a trivial knotting $S^1 \subseteq R^3$, such that $(B, B \cap S^1)$ and $(B', B' \cap S^1)$ are each trivial ball pairs, Figure 4.11. For two 1-knottings, K and K' denote the corresponding knotted arcs in D^3, by α_K and α'_K. Then $K\#K'$ by replacing $B \cap S^1$ with α_K and $B' \cap S^1$ with α'_K. The knottings K and K' are called **summands** of $K\#K'$. For the connected sum of two trefoils, we replace these trivial ball pairs with trefoil arcs, obtaining Figure 4.12, which is isotopic to Figure 4.9. For Figure 4.10 we replace one trivial ball pair with a trefoil arc, the other with its mirror image. These knottings, called the **square knotting** and the **granny knotting** will be of further interest in high-dimensional considerations. ▮

In Remark 4.2 we saw how to represent a 1-knotting of a circle as a proper knotting of an arc in D^3. Similarly we can view a 1-knotting of a circle K is as a proper arc in half-space $\alpha' \subseteq R^3_+$. One way to do this is to use Remark 4.2 to represent K by $\alpha \subseteq D^3$. Next, take a point of $(\partial D^3 - S^0)$ such as $p = (0,0,1)$) and identify $(D^3 - \{p\})$ with R^3_+ using Remark 1.2 by a diffeomorphism f and now let $\alpha' = f(\alpha)$.

Using this, we can say a knotting $K'' \subseteq R^3$ is a connected sum $K \# K'$ if there is a plane $H \subseteq R^3$ that intersects H transversely in two points, such that on one side of H is an arc representing K and, on the other side, is an arc representing K'. We will see this in Figures 11.44-11.47

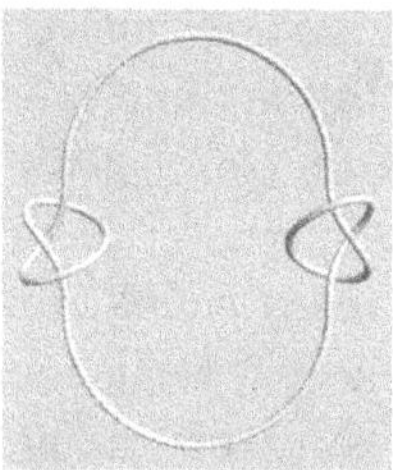

Fig. 4.9 A sum $K \# K$ of trefoil with itself—a "granny knotting".

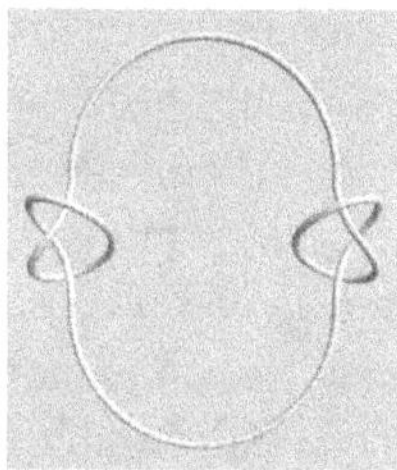

Fig. 4.10 A sum $K \# -K$ of trefoil and mirror image—a "square knotting".

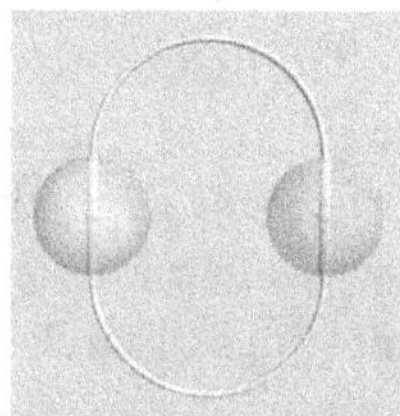

Fig. 4.11 A standard trivial knotting.

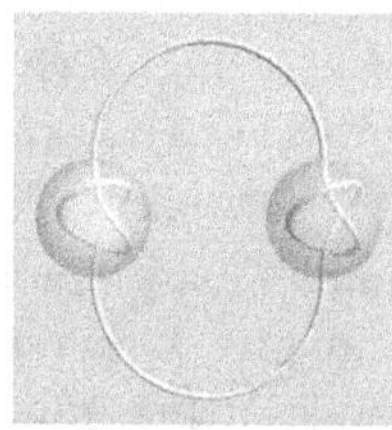

Fig. 4.12 Replace two trivial arcs by non-trivial one.

If we had isotopies, K_t and K'_t, of our knottings, we would have corresponding proper isotopies of our arcs α_t and α'_t which would give an isotopy $K_t \# K'_t$ of the connected sum. So, a connected sum of *knottings* gives rise to a connected sum of *knots*:

Prop 4.1. For any two *connected* 1-knottings, K_1 and K_2, if $[K_1] = [K'_1]$ and $[K_2] = [K'_2]$ then $[K_1 \# K_2] = [K'_1 \# K'_2]$. ∎

Shrink one knotting of a connected sum small enough that we can slide it, following the other completely, and when done restore to original size —in effect interchanging the positions of the two factors as indicated in Figures 10.131 to 10.141. This shows that:

Prop 4.2. For any two knottings K and K' of circles in R^3 (or S^3), $[K \# K'] = [K' \# K]$. ∎

If $\mathcal{T}$ is a trivial knotting, then $\mathcal{K} \# \mathcal{T} = \mathcal{K}$. A **knot** $\mathcal{K}$ **is prime** if the only way to write $\mathcal{K} = \mathcal{K}_0 \# \mathcal{K}_1$ is that one of $\mathcal{K}_0$ or $\mathcal{K}_1$ must be trivial. An important property of knot sum is that we can express any knotting of a circle as sum of prime knots in a unique way up to order. The term "sum" is traditional but somewhat misleading. This "sum" is not like addition in algebra—there are no inverses. There *is* a unit, the trivial knotting T, so that $K\#T \approx T\#K \approx K$ for any knotting K. But there is no K' so that $K\#K' \approx T$. This is intuitively clear from our experience—you cannot untie a knot by tying on another knot. This follows from the additivity of genus $g(K\#L) = g(K) + g(L)$. Also, in algebra, "prime" refers to product and not sum.

Companionship is an important relation between knottings. It provides an interesting way of constructing complex knottings from simpler ones.

Definition 4.1. Suppose K is a 1-knotting of S^1 with tubular neighborhood $T \approx S^1 \times D^2$ and let $P \subset T$ be a knotting of S^1 in T such that P is neither contained in a 3-ball in T nor isotopic to $S^1 \times \{\mathbb{O}\} \subseteq T$. We say that P **is a satellite knotting** which is a **companion of** K and that **the pair** $(S^1 \times D^2, P)$ **is called a pattern.** ∎

If there *is* a 3-ball B with $B \subseteq N(K)$, then we have a split knotting of the knottings, and this will not be considered a satellite.

A connected sum has a satellite/companion structure. Some especially important satellite examples are cables. If $K' \subseteq \partial N(K)$ then K' is called a **cable** about K. A cable about a trivial knotting is called a **torus knotting**, since it lies on a "standard torus".

Torus knottings can be described by parametric equations. If p and q are relatively prime, a (p, q)**-torus curve** is a curve with equations where $0 \leq t \leq 1$:

$$x = (b + a\cos(2\pi qt))\cos(2\pi pt), y = (b + a\cos(2\pi qt))\sin(2\pi pt), a\sin(2\pi pt).$$

The $(0, 1)$-torus curve is a **longitude** of the torus, and the $(1, 0)$-torus curve is a **meridian**, as in Remark 2.6. A torus link is a union of disjoint curves on a torus, all parallel to some (p, q)-torus curve on a torus. The knot corresponding to a torus knotting is called a **torus knot**, Figures 4.13 and 4.14.

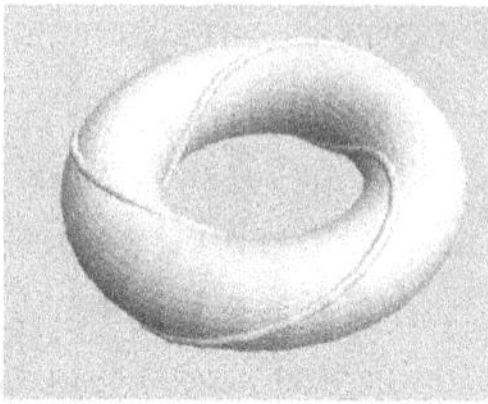

Fig. 4.13 A torus with a (3, 2)-torus curve.

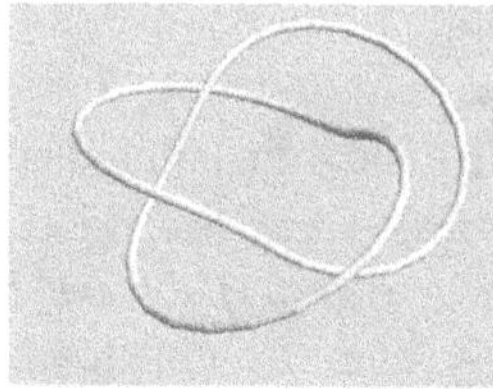

Fig. 4.14 (3, 2)-torus curve is a trefoil knotting.

4.2 Projection, General Position and Diagrams

Traditionally one illustrates a knot by a drawing of a rope displayed on flat surface showing clearly which strand passes above the other as it crosses itself. The corresponding mathematical concept is called a **knot-diagram** where we use the standard projection map, Definition 1.31 which, for R^3, is $\pi(x, y, z) = (x, y)$. For knotting $K \subseteq R^3$, $K_* = \pi(K) \subseteq R^2$ is the **projection of the knotting**, and the **height function** measures the z coordinate .

Definition 4.2. If K_* is projection of knotting $K \subseteq R^3$ and $x \in K_*$, x is an n-**crossing point of** K_* if $\pi^{-1}(x) \cap K$ contains exactly n points where $1 < n$. The set of all crossing points is called the **crossing set**. If $\pi^{-1}(x) \cap K$ is a single point, then x is an **ordinary point of the projection.** ▮

The projection of a 1-knotting K_* is a planar subset—a shadow of K. All height information of points of K has been "lost". Although we cannot reconstruct the *knotting* K we can generally reconstruct the corresponding *knot* $[K]$. Here "generally" refers to knottings in **general position** and "almost" refers to needing a little additional information—the **height relations at the crossings.**

Remark 4.4. The term "general position" occurs in many ways in knot theory as well as other parts of topology. Informally speaking, (and, in a strict mathematical sense, beyond the scope of this book), "most knottings are in general position". You would have to do something special to construct a knotting *not* in general position. Having said that, mathematically generated knottings *are* often special in some ways such as knottings such as K shown in Figures 4.15 or Figure 4.16. General position theorems, such as Proposition 4.3 assure us mathematically that any knotting, even if it has some unusual "non-general" positioning (such as in Figure 4.15), can be mathematically replaced by a "general" equivalent knotting (such as Figure 4.18), by a small change. ∎

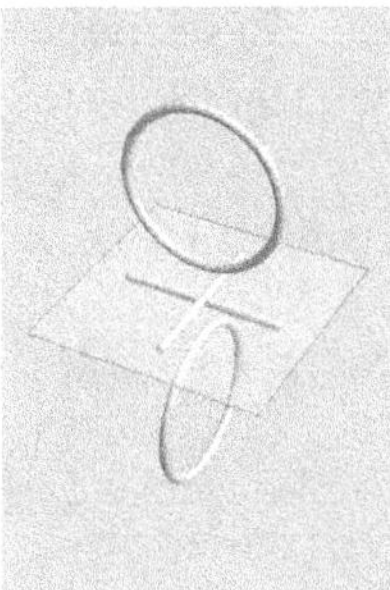

Fig. 4.15 A knotting $K \subseteq R^3$ and its standard projection into R^2.

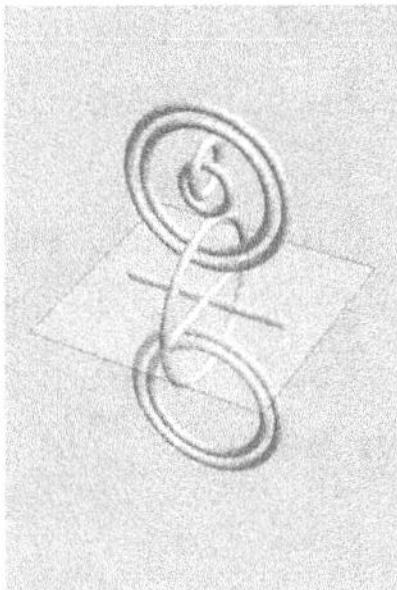

Fig. 4.16 Another knotting whose projection is the same as in Figure 4.15.

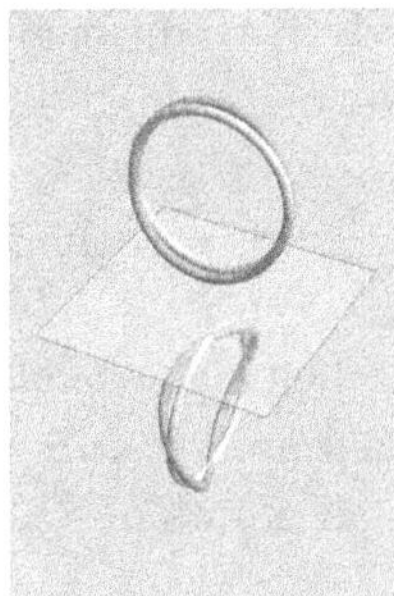

Fig. 4.17 K of Figure 4.15 in dark gray; close equivalent K' in light gray.

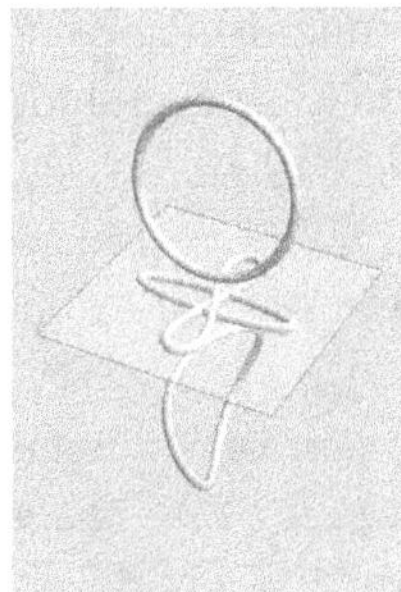

Fig. 4.18 Projection of K' of Figure 4.17 showing general position projection.

Two lines in the plane are in **general position** if they are distinct and not parallel. A term for this is "transverse":

Definition 4.3. Two **intersecting lines in R^2 are transverse** if they intersect in exactly one point. ∎

Definition 4.4. A smooth 1-knotting $K \subseteq R^3$ is in **general position with respect to projection** if:

(1) $\pi|_K$ is at most a 2-to-1 map
(2) no tangent line to $x \in K$ is vertical (vertical means parallel to z-axis)
(3) if $x, y \in K$ with $\pi(x) = \pi(y)$ then the projections of the tangent lines at x and at y are transverse in R^2. ∎

Condition 2 of Definition 4.4 implies that the projection of a 1-knotting in general position will be the image of an immersion (see Definition 1.24) of a 1-knotting.

Prop 4.3 (General position for 1-knottings). Any smooth 1-knotting $K \subseteq R^3$ is equivalent to a knotting K' which is in general position with respect to projection. In fact, this isotopy can be taken to be a small one (in the sense that points are moved by an arbitrarily small amount). ∎

Condition 2 of Definition 4.4 implies that a general position projection of a 1-knotting is the image of an immersion (see Definition 1.24) of a 1-knotting.

To avoid repetitions, for the remainder of this chapter, we will assume that all 1-knottings are smooth and in general position.

The projection terminology, below, of crossing scheme and shards is not standard (nor needed) in traditional theory for 1-knottings. We introduce them as a prelude to similar high-dimensional terminology, Definitions 6.8 and 12.10.

This terminology is for associated, but distinct, structures for K_*, K and M. For example, if $\phi : S^1 \to R^3$ is a knotting of a circle $\phi(S^1) = K$, and $c \in K_*$ is a crossing, then c is a single two-dimensional point, $\widetilde{c} = \pi^{-1}(c)$ is a pair of 3-dimensional points of K and, since $S^1 \subset R^2$, $\widehat{c} = (\pi \circ \phi)^{-1}(c)$ is a pair of 2-dimensional points of S^1.

Definition 4.5. Suppose $K \subseteq R^3$ is a general position 1-knotting of $M \subseteq R^m$ given by ϕ: $M \to K$ with $K_* = \pi(K)$. For $X \subseteq K_* \subseteq R^2$, the set denoted $\widetilde{X} = \pi^{-1}(X_*) \cap K \subseteq R^3$ is called the **lifting of** X. The subset M, denoted $\widehat{X} = (\pi \circ \phi)^{-1}(X) \subseteq R^m$, is called the **pullback of** X. If x is a crossing point, either point of the two points of $\widetilde{x}$ is called a **pre-crossing point of** x **in** K. If $\mathcal{C}$ is the crossing set of K_*, the set $\widetilde{\mathcal{C}} = \pi^{-1}(\mathcal{C}) \cap K$ is the **pre-crossing set of** K_* and $\widehat{\mathcal{C}} = \phi^{-1}(\widetilde{\mathcal{C}})$ is called the **crossing scheme of** K_*. ▮

Definition 4.6. If $\mathcal{C}$ is the crossing set of a projection of a knotting K_*, a component, α, of $K - \mathcal{C}$ will be called a **1-shard of** K_*, $\widetilde{\alpha}$ called a **1-shard of** K and $\widehat{\alpha}$ called a **1-shard of** M.

The closure of α in K_* (that is, adding in the endpoints) is called a **closed 1-shard of** K_*. The closure of $\widetilde{\alpha}$ in K is a **closed 1-shard of** K. The closure of $\widehat{\alpha}$ in M is a **closed 1-shard of** M. ▮

For a 1-knotting a 1-shard is frequently called "an edge" of the projection.

Definition 4.7. If x is a 2-crossing point let $\widetilde{x} = \{a, b\}$ be the pre-crossing set where $h(b) < h(a)$ (i.e. a is above and b below, in the w-coordinate). We say $x_a \in K$ is an **above-point** and $b \in K$ is a **below-point**. The set of all above-points) is called the **above set of** K_* ; the set of all below-points is called the **below set of** K_*.

Similarly if ϕ: $M \to K$ is a homeomorphism with $K = f(M)$ we call $\phi^{-1}(a) \in M$ an **above-point** and $\phi^{-1}(b) \in M$ a **below-point** of M and use the similar terminology for subsets of M corresponding to **above and below sets.** ▮

For 2-crossing points, it is traditional to speak of "over" and "under", Definition 4.7, however, we prefer "above" for "over", and "below" for "under", since the notational use of the first letters a and b is often less awkward than the use of o and u. The above/below as an alternative is used elsewhere—for example, in [Kanenobu and Shima (2002)].

A closed 1-shard $\widetilde{\alpha}$ of a 1-knotting K is a sub-arc. The endpoints of $\widetilde{\alpha}$, say a and b, are pre-crossing points. If it happens that a and b are the above and below-points of the same crossing point, then the projection $\pi(\widetilde{\alpha})$ will be a topological circle—otherwise it will be a topological arc.

A projection of a knotting displayed using the broken line convention is a union of disjoint arcs. Each arc is called an **overpass**. Figure 4.65 shows the three overpasses for a projection of a trefoil knotting. An overpass is made up of above-points and ordinary points of a projection. In our terminology:

Definition 4.8. For knotting K with crossing set $C \subseteq K_*$, let $\mathcal{B} \subset K$ be the below-set. An **overpass of** K is a component P of $K - \mathcal{B}$, and $\pi(P)$ is an **overpass of** K_*. If $\phi\colon M \to K$ is a homeomorphism, then $\widehat{K} = \phi^{-1}(K)$ is an **overpass for the crossing scheme of** K. ▮

In the literature, an overpass is also called an **arc of a diagram**, and a 1-shard is called a **semi-arc**, for example [Elhamdadi and Nelson (2015)].

Another common term for overpass is "**bridge**".A 2-bridge knot has a special knotting. For knots of high crossing number such knots are uncommon. However, all knots with crossing number 7 or less, as well as about half of 8 crossing knots, are of this type.

Definition 4.9. A **2-bridge knot** (also called a **rational knot**) is a one component knot represented by a knotting whose projection has two overpasses. ▮

Example 4.1. In Figure 4.19 $M = S^1$, and knotting K of M represents a trefoil knot. Also shown is the projection on the xy-plane $R^2 \subseteq R^3$. Figure 4.20 shows projection K_* with three crossings P_*, Q_*, and R_*, shown as small balls. In Figure 4.21 we show the six pre-crossing points of $\widetilde{P_*}, \widetilde{Q_*}$, and $\widetilde{R_*}$, labeled $\widetilde{P_a}$ and $\widetilde{P_b}$, with $\widetilde{P_a}$ above $\widetilde{P_b}$, etc. In Figure 4.23 we show pullback points of $\widehat{P_*}, \widehat{Q_*}$, and $\widehat{R_*}$ in $M = S^1$. To graphically emphasize that this figure is two-dimensional, we have changed from using thin tubes to thin lines to represent curves. ▮

Fig. 4.19 The knotting and its projection κ_* with three crossing points in the xy-plane.

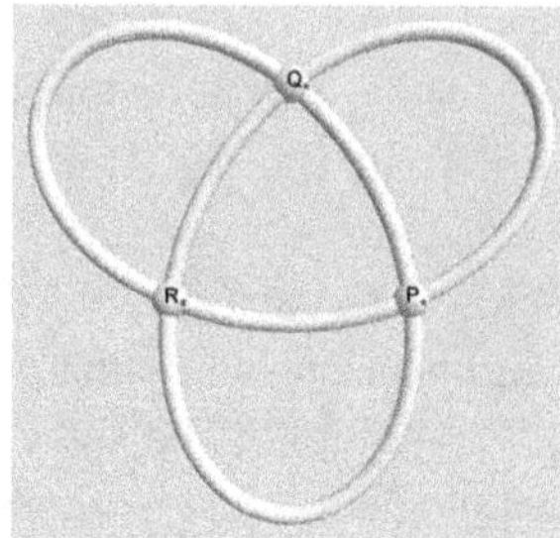

Fig. 4.20 Crossing points labeled as P_*,Q_* and R_*.

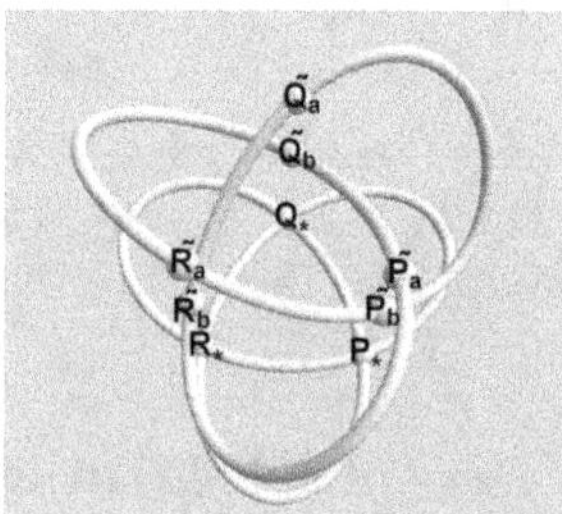

Fig. 4.21 The six pre-crossing points $\widetilde{P_a}, \widetilde{P_b}, \widetilde{Q_a}, \widetilde{Q_b}, \widetilde{R_a}, \widetilde{R_b}$ as subset of K.

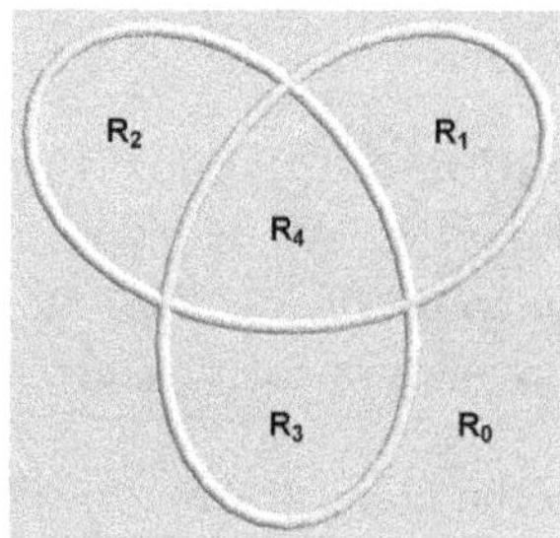

Fig. 4.22 This projection of a knotting has five regions.

Definition 4.10. Suppose K_* is a projection of a 1-knotting. An open component of $R^2 - K_*$ is called an **open region of** K_*. The closure in R^2 of an open region is called a **closed region of** K_*. ▌

The projection of the trefoil knotting of Figure 4.20 has five regions R_0, R_1, R_2, R_3, and R_4 see Figure 4.22. The open regions of a projection are open subsets of R^2 and thus (non-compact) 2-manifolds. The closed regions are not necessarily 2-manifolds. The projection of the knotting with two crossings of Figure 6.33 has four regions in the projection. Three of the closed regions are compact and are topological disks. The fourth closed region (the non-compact outside region) is not a topological disk or half-disk at the crossing points, and so is not a 2-manifold.

The projection $K_* \subseteq R^2$ of a 1-knotting $K \subseteq R^3$ has "lost" information of the z-coordinate— we cannot reconstruct K from K_* alone. But given **crossing information** at crossing points K_* —that is *relative* above/below

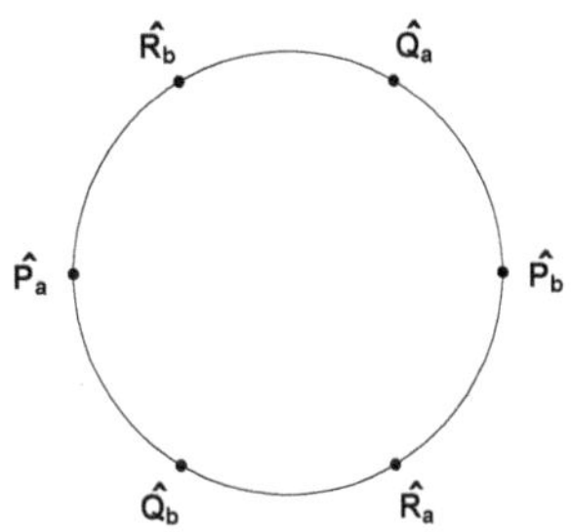

Fig. 4.23 The six pullback points $\widehat{P_a}, \widehat{P_b}, \widehat{Q_a}, \widehat{Q_b}, \widehat{R_a}, \widehat{R_b} \subseteq S^1$.

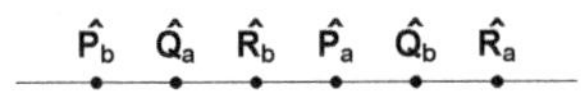

Fig. 4.24 Remove a non-pullback point from S^1 and homeomorphically get a sequence of points on a line.

height information for crossing arcs each crossing point—we can determine the *knot-type* of K:

Prop 4.4. If K and K' are knottings with the same projection and crossing information they are isotopic. ▌

Proposition 4.4 is considered as "obvious" in most traditional discussions of 1 knottings however a proof here will serve as well in higher dimensional settings. Briefly the proof of Proposition 4.4 is: "use a straight-line vertical homotopy". This is a homotopy G_t of K to K' which changes *only* the z-coordinate. by a **vertical isotopy**. We illustrate in the next example.

Example 4.2. The two knottings shown in Figures 4.25 and 4.26 have the same projection onto the plane, see Figures 3.19 and 4.27. Figure 4.28 shows the vertical line segments joining corresponding points of these knottings. The isotopy between these knottings can be visualized as a shrinking of these joining line segments as shown in Figure 4.28. (This subset is, in fact, the track of our isotopy.) Such an isotopy of knottings which have the same projection and compatible height relations at the crossings is called a **vertical isotopy**.

Take any $p = (x, y, z) \in K$. If p_* is *not* a crossing point of K there is a *unique* point $p' = (x, y, z') \in K'$ whose projection is p_*. Our vertical isotopy will be defined on p by $G_t(x, y, z) = (x, y, z + t(z' - z))$ so that $G_0(p) = p$ and $G_1(p) = p'$. Since for each t, G_t projects to an immersion, so must be an immersion and so, G_t is a regular homotopy. We next check that each G_t is an embedding.

If p_* is a crossing point there are *two* points of $\widetilde{p_*}$. One point, $\widetilde{p_a} = (x, y, z_a)$, is an above-point, and one, $\widetilde{p_b} = (x, y, z_b)$, is below. We similarly have two points of K': $\widetilde{p'_a} = (x, y, z'_a)$ an above-point while and one $\widetilde{p'_b} = (x, y, z'_b)$ is below.

We move p_a vertically to p'_a and at the same time move p_b vertically to p'_b as follows: G_t for p_a by $G_t(x, y, z) = (x, y, z_a + t(z'_a - z_a)$ and for p_b by $G_t(x, y, z) = (x, y, z_b + t(z'_b - z_b))$.

It is an algebra exercise to check the homotopy G_t is an isotopy—that since height relations are the same, and thus $z_a > z_b$ and $z'_a > z'_b$, then $z_a + t(z'_a - z_a) \neq z_b + t(z'_b - z_b))$ for $0 \leq t \leq 1$. ∎

Fig. 4.25 A different view of Figure 3.19.

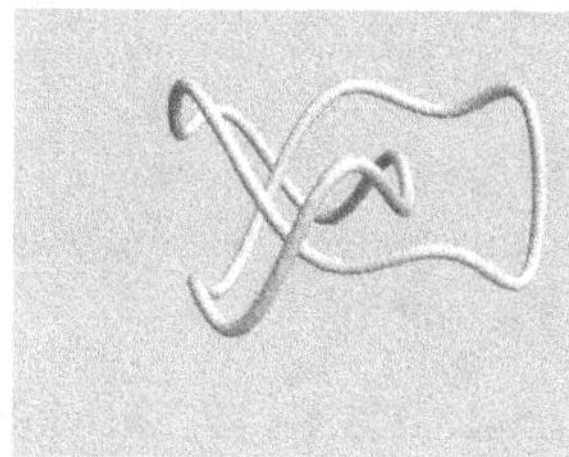

Fig. 4.26 A knotting equivalent to Figure 4.25.

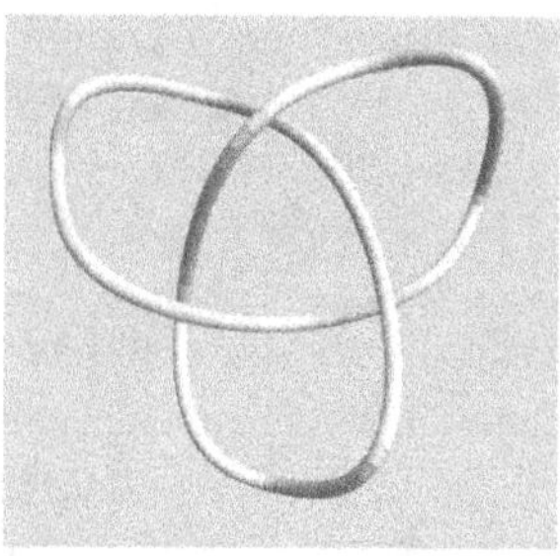

Fig. 4.27 The projection of Figure 4.26.

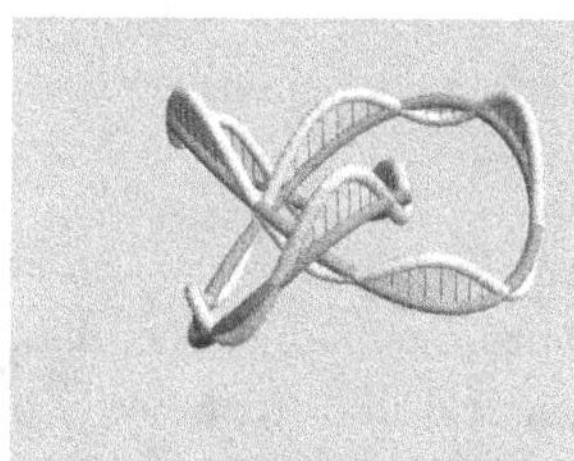

Fig. 4.28 Vertical line segments joining the two knottings.

In general, for an isotopy $h_t(K)$ the projection $h_t(K)_*$ may change topologically as t changes. These changes can be described in terms of **knot-moves**, see Section 4.4. Equivalence class of a projection with no such moves is called a diagram.

Definition 4.11. Suppose we have two knottings $K \subseteq R^3$ and $K' \subseteq R^3$ both in general position with respect to projection. We say **K_* and K'_* are equivalent projections** if there is an isotopy h_t in R^3 that takes K

to K' such that the projection $\pi \circ h_t$ is an isotopy in R^2 of the subset K_*. The equivalence class of projections is called a **knot-diagram for** $\mathcal{K}$. ∎

A single knotting projection, together with relative height information at the crossings, determines a knot-diagram (and so determines a well-defined knot).

Prop 4.5. If K_* is a projection of a knotting of a circle, then the set K_* is also the projection of a trivial knotting K'. ∎

The idea of a proof of Proposition 4.5 is hinted at in Figure 4.29. To define K', take a non-crossing point $(x_0, y_0, 0)$ of a projection K_* and trace a knotting over K_* that continually rises in z but at the very end, with all crossings traversed twice smoothly returns down to $(x_0, y_0, 0)$.

Fig. 4.29 A trivial knotting with same projection as trefoil of Figure 4.19.

Another way of expressing Proposition 4.5 is: any general position projection of a 1-knotting can be made into a projection of a trivial knotting by reversing some of the crossings. For a given projection the minimum number of such crossing changes that result in a trivial knotting is called the **Gordian number of the projection**. The **Gordian number of a knot** $\mathcal{K}$ is the minimum Gordian number of all projections of all knottings K that represent $\mathcal{K}$.

Remove any point of S^1, the resulting space is homeomorphic to R^1. Choose a point that is not a pullback of a crossing point. This gives us a set of points of R^1, as shown in Figure 4.24. Ordering of R^1 gives a sequence of points $\{\widehat{P_b}, \widehat{Q_a}, \widehat{R_b}, \widehat{P_a}, \widehat{Q_b}, \widehat{R_a}\}$. Such a sequence is called a **Gauss code** for the knotting. In practice, integers are used with "below" points written as negative numbers so that our code might be written as $\{-1, 2, -3, 1, -2, 3\}$. Similar codes exist for knottings of several circles by writing the code for each component successively.

Although the knot-diagram cannot always be reconstructed from the code, a slight modification called the **extended Gauss code** will make this possible. An alternate scheme used for computation is the Dowker-Thistlethwaite code [Dowker and Thistlethwaite (1983)].

Remark 4.5. Mirror images and oriented knottings have been discussed in Section 3.5. Given a projection of a 1-knotting one can obtain diagrams for the corresponding mirror images and reverse orientation knottings. Suppose M is a closed 1-manifold with k components. Given a knotting $K \subset R^3$ of M, a diagram for the **mirror image** ꓘ can be obtained by interchanging "above" and "below" at each crossing point of a diagram for K.

An **orientation** choice is most often indicated by a tangential arrow at a point of a 1-shard of that component each component indicating a direction of parametrization. Directions for other shards are oriented compatibly by "sliding across at a crossing point" at the crossings, see Figures 4.30 and 4.32. However, there is an alternate way that will better generalize for diagrams of 2-knottings. Instead of taking a tangent vector at a point of a 1-shard we take a vector normal to the 1-shard and again obtain compatibility by "sliding across at a crossing point" at the crossings, see Figures 4.32 and 4.33. In either case, given an oriented diagram for $\overrightarrow{K}$, a diagram for the inverse $\overleftarrow{K}$ is obtained by reversing the direction of all orientation vectors. ▮

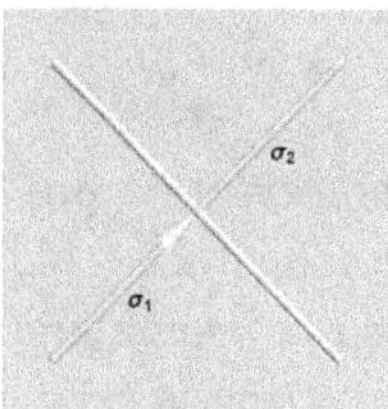

Fig. 4.30 Tangential orientation for shard σ_1.

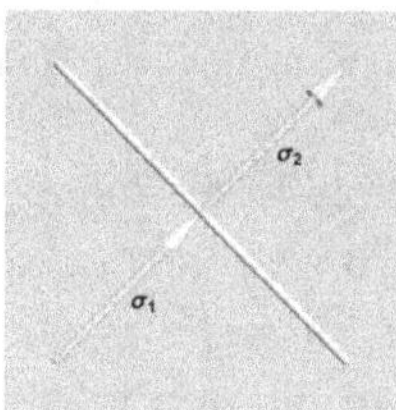

Fig. 4.31 Sliding to get orientation for shard σ_2.

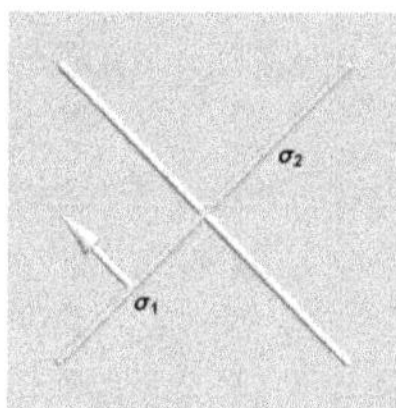

Fig. 4.32 Normal orientation for shard σ_1.

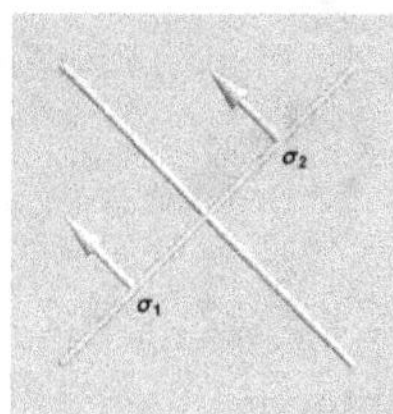

Fig. 4.33 Sliding to get orientation for shard σ_2.

One type of knotting not yet mentioned are certain knottings in $F^2 \times R^1$ where F is a surface.

Definition 4.12. Suppose F^2 is a 2-manifold and K is a knotting of a compact closed 1-manifold in $F^2 \times R^1$. If $\partial F^2 \neq \emptyset$ we require the knotting to be proper. The **projection** $\pi\colon F^2 \times R^1 \to F^2$ is the map $\pi(x,t) = x$; the **height a point** (x,t) is t. ▮

One can in a natural way extend the theorems about knottings in R^3, such as general position to this setting. (One exploration of this is a topic known as virtual knot theory which is of considerable interest in itself, see for example [Kauffman (2012b)], [Fenn *et al.* (2014)] and [Dye (2016)].) We will avoid formal exposition and will limit our attention to a only few particular simple examples.

Some of the papers in this area are [Turaev (2008)] [Cotta-Ramusino and Rinaldi (1991)] [Asaeda *et al.* (2004)] and [Carter *et al.* (2002)].

Example 4.3. In Figure 4.34 we identify $T^2 \times R^1$ with the region in R^3 between two standard concentric tori with a third concentric torus midway between these two representing $T^2 \times \{0\}$. Points outside $T^2 \times \{0\}$ represent positive values in the R^1 factor. Figure 4.35 shows a knotting of two disjoint circles in $T^2 \times R^1$ that lie above $T \times \{0\}$. Figure 4.36 shows the projection of the knotting, T_*, onto $T \times \{0\}$. These two circles meet in a single point p that is a crossing point of T_*. The circle that was higher in $T^2 \times R^1$ is shown as a slightly thicker tube, so that we can visually tell the over/under relation at the crossing p. Here T_* has two 1-shards and one region. ▮

Fig. 4.34 $T^2 \times R^1$ is homeomorphic to the region between two concentric torii. A third torus between these represents $T^2 \times \{0\}$.

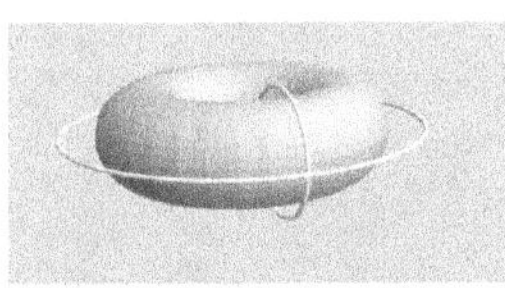

Fig. 4.35 A knotting of two disjoint circles in $T^2 \times R^1$.

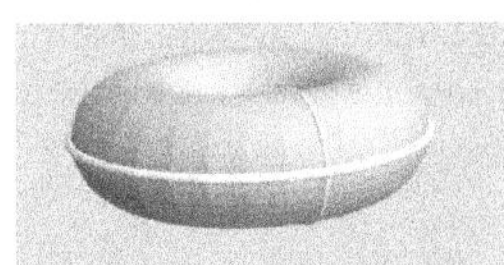

Fig. 4.36 The projection of the knotting to T^2 is a pair of circles with one point in common.

The process of reversing a crossing by "interchanging the above arc and below arc" has analogues in higher dimensions, Section 6.7. Detailed explanation of this process, hardly needed for 1-knottings, will prove helpful in understanding the higher dimensional process.

For the curve in $R^2 \subseteq R^3$ as shown in Figure 4.37 we can "eliminate a crossing" by lifting an arc up in the z-direction as shown in Figure 4.38. If we do this at each crossing, we obtain a knotting of a circle such as Figure 4.39—in this case the knotting represents the (right-hand) trefoil knot, Figure 3.20. Other choices,Figure 4.40, result in a trivial knotting.

This process should be clear enough from the Figures, but to prepare for higher dimensional constructions we add a bit more detail for a single crossing. A crossing c which is the intersection of 1-disks a_* and b_* takes place in a small 2-disk $D_c^2 \subseteq R^2$ with center c as in Figure 4.41. View this crossing as a subset of a 3-ball D_c^3 with center c. By replacing 1-disk with different smooth proper path in D^3, we obtain two *disjoint* proper sub-arcs of this 3-ball—either Figure 4.42 or Figure 4.43.

Fig. 4.37 A planar self-intersecting curve in R^2.

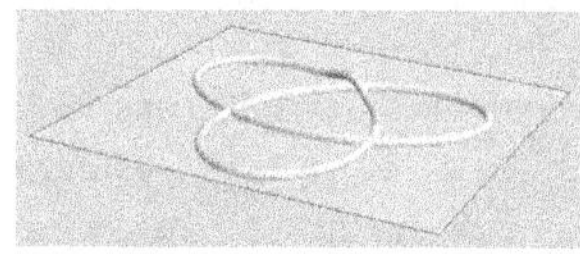

Fig. 4.38 Near one of the self-crossings of Figure 4.37 we lift one of the crossing arcs up in z-direction.

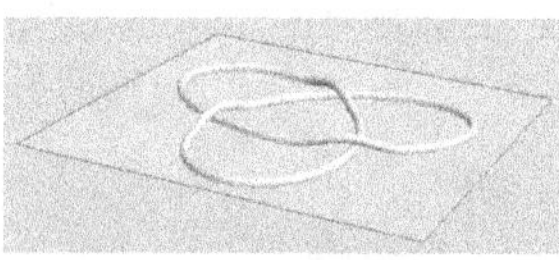

Fig. 4.39 We could do the lifting of Figure 4.38 at each of the crossings and obtain a trefoil knotting.

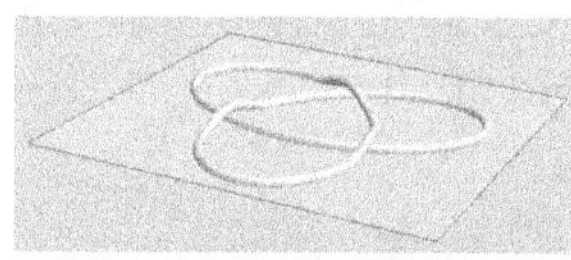

Fig. 4.40 A different set of lifting choices can give a trivial knotting.

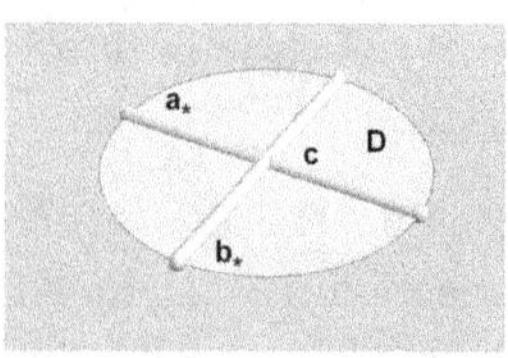

Fig. 4.41 A crossing C in a disk D_c^2 in R^2.

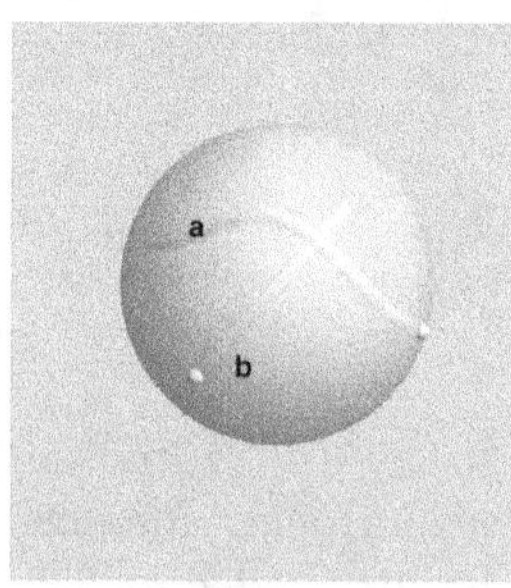

Fig. 4.42 One way to replace a crossing segment.

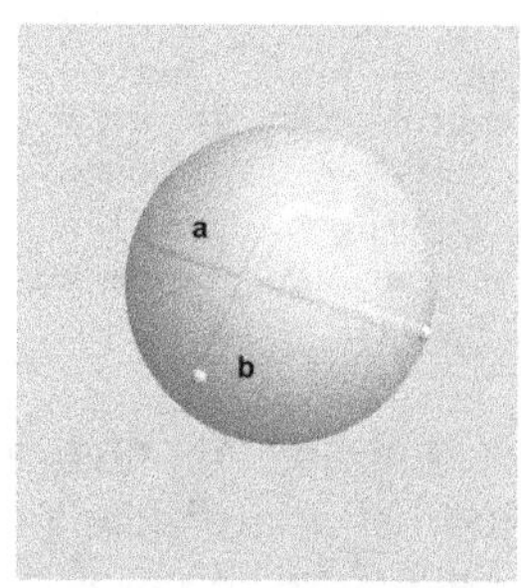

Fig. 4.43 An alternate replacement.

4.3 Twist and Crossing Number at a Crossing

We define a **sign** for each crossing by orienting a 1-knotting and using Figure 4.44 and Figure 4.45.

Although we have introduced an orientation, *for a single component knotting* this sign does *not* depend on choice of orientation. One way to remember the crossing sign is by a right-hand rule:

(1) orient the knotting
(2) at a crossing point grab the overpass with your right-hand with thumb in the direction of the orientation
(3) if your other fingers point in direction of the arc passing below then the sign is positive as in Figure 4.44, otherwise negative, Figure 4.45.

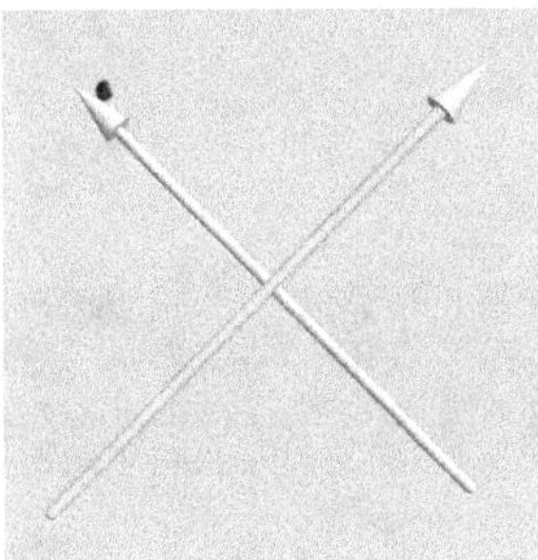

Fig. 4.44 A positive crossing.

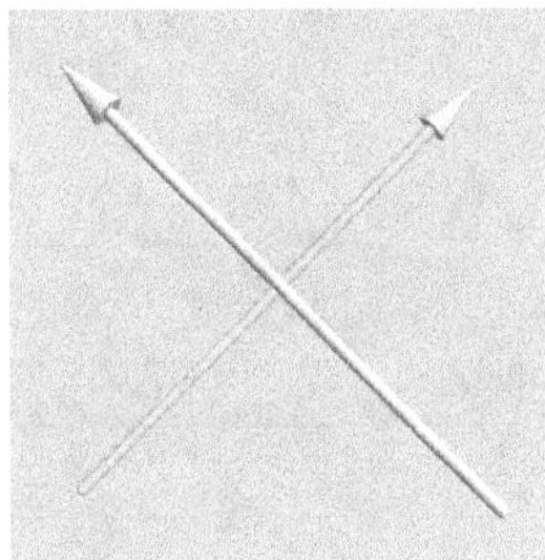

Fig. 4.45 A negative crossing.

Definition 4.13. For a general position projection of a knotting of a single circle, the **writhe of the projection** is the sum of crossing numbers. ▮

Definition 4.14. For a given projection of a knotting of two oriented circles, we define the **linking number** as one-half the algebraic sum of crossing numbers, *counting only crossings between different components.* ▮

If we reverse the orientation of one circle and not the other, the linking number at each crossing will change sign so the linking number will change sign. However, if the linking number happens to be zero, then it will be zero for *any* choice of orientations.

The boundary of an annulus consists of two circles we orient the two in *opposite* directions. By doing this, the linking number of these two circles is well defined.

Definition 4.15. The **twisting number of an annulus** $A \subseteq R^3$ is the linking number of the oppositely oriented components of ∂A. ▮

Example 4.4. For the annulus shown in Figure 4.46, the linking number of the oriented components of the boundary is $+1$, regardless of orientation choices, see Figures 4.47 and 4.48. So, we can say this annulus has a **positive twist**. Similarly, its mirror image has a **negative twist**. The annulus in Figure 4.49 has zero twists, while in Figure 4.50 there are $+2$ twists. These last two examples will provide part of an important observation in higher dimensions, Section 10.4. ∎

Fig. 4.46 An annulus with twist.

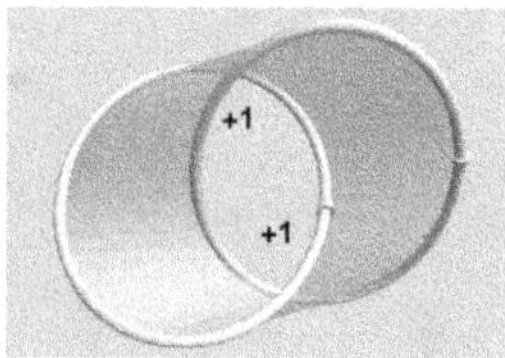

Fig. 4.47 One choice of orientations of boundary gives linking number $+1 = \frac{1}{2}(+1+1)$.

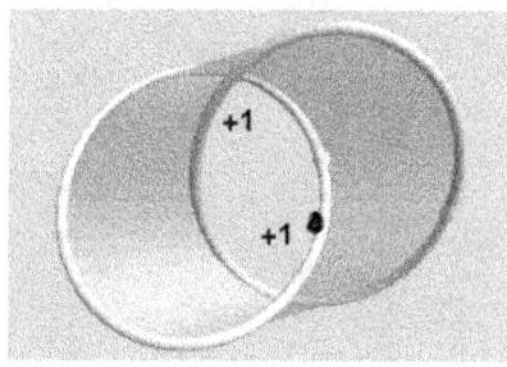

Fig. 4.48 The other choice of orientations of boundary gives linking number $+1$.

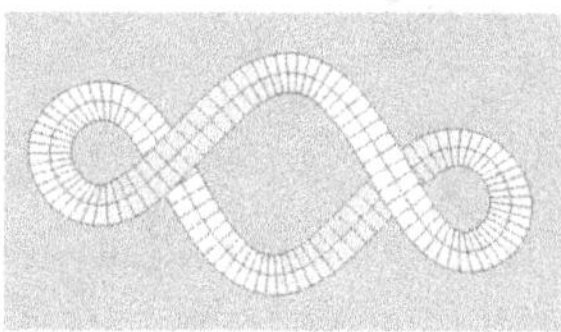

Fig. 4.49 An annulus with zero twists.

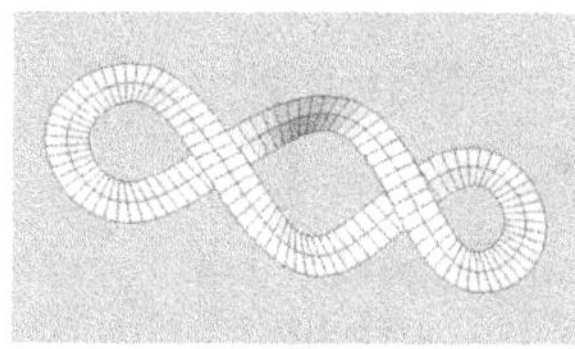

Fig. 4.50 An annulus with $+2$ twists.

Spinners of yarn or rope describe a twist direction related to the crossing sign. Think of a helix wrapped around a semi-transparent cylinder— Figure 4.51 and center attention on the helix portion on the front of the cylinder. The portion of the helix on the left (including the portion on the back side of the cylinder) appears to be shaped like the letter "S", while the helix on the right appears as a letter "Z", Figure 4.52. Spinners call the left a S**-twist**—or a **left-hand twist**. On the right is a Z**-twist** or **right-hand twist**. If we take two helices along each cylinder Figure 4.53, they cross one another—the strand on the front of the cylinder crossing over the strand on the back. Orienting the helices in the same direction along a cylinder, the signs of the "Z" crossings are all negative, and the "S" crossings are all positive, Figure 4.54.

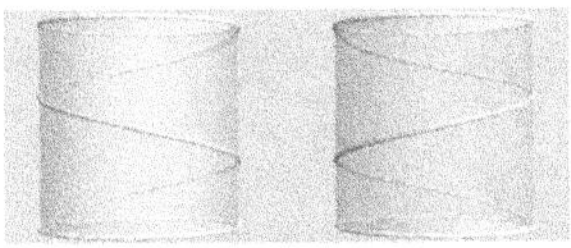

Fig. 4.51 Two ways to wind a helix on a cylinder.

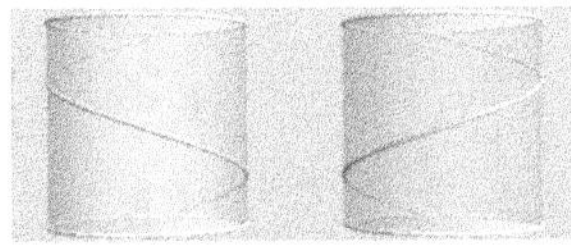

Fig. 4.52 A portion of helix on the left seems to form a letter "S", on the right appears the letter "Z".

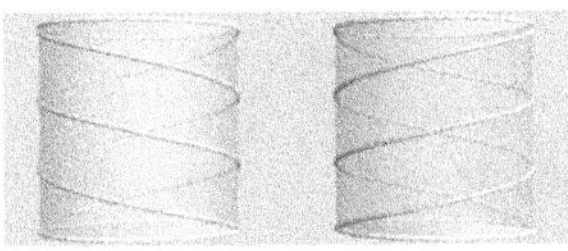

Fig. 4.53 If we use two helices on each cylinder, visually one strand passes over the other.

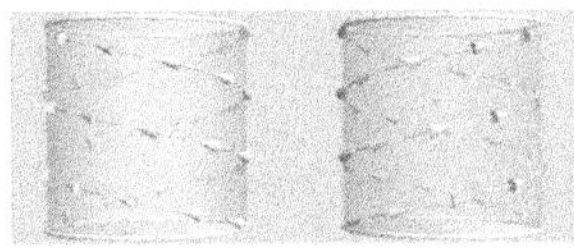

Fig. 4.54 On the left the signs of crossings are -1, on the right +1.

Many numerical aspects of knots can be defined based on diagrams.

Definition 4.16. The **crossing number of a knot** is the minimum number of crossings for a knotting representing the knot. ∎

Definition 4.17. The **unknotting number of a knot** is minimum number of crossings for some diagram representing the knot, which, if reversed, will give a trivial knotting. ∎

4.4 Reidemeister Moves

Knot diagrams represent all 1-knottings. Also, we have diagram changes called "moves" to these diagrams that mathematically generate all isotopies. These changes are called **Reidemeister moves** [Reidemeister (1927)]. These moves show changes, by an isotopy, to a diagram of a knotting in a small region (topologically a 2-disk) $U \subseteq R^2$. The knot-diagram is *unchanged* outside of U. A set U will be called a **support disk** for the knot-move. An example of an isotopy involving three Reidemeister moves is shown in Figures 4.55–4.58.

We arbitrarily label the configuration on the left as **Position** A and the right as **Position** B. Figures 4.59, 4.60 and 4.61 show, respectively, **Reidemeister moves** $\Omega.1$,$\Omega.2$ and $\Omega.3$ as one goes from Position A to Position B. A move that goes from Position B to Position A are also

Fig. 4.55 We begin our isotopy.

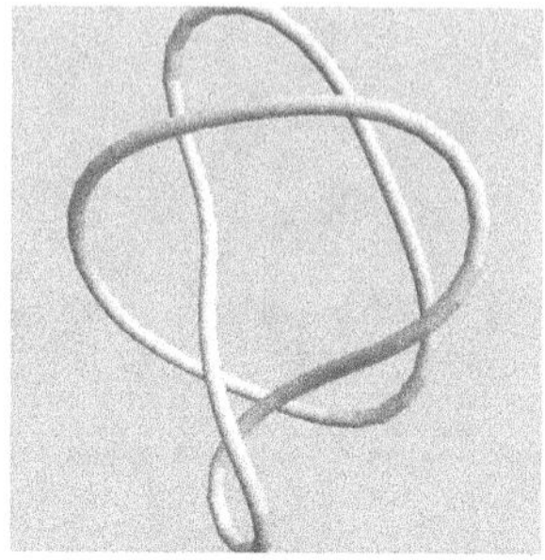

Fig. 4.56 After a type $\Omega.3$ move.

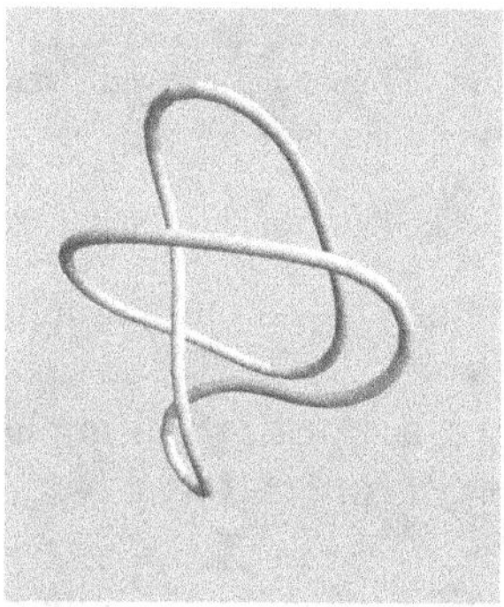

Fig. 4.57 Then a type $\Omega.2$ move.

Fig. 4.58 Finally a type $\Omega.1$ move.

Reidemeister moves denoted $\Omega'.1$,$\Omega'.2$ and $\Omega'.3$. The support disk U in each case not shown, but is taken to be the boundary of the Figure.

In knot theory, it is often stated "there are three Reidemeister moves, and two knots are equivalent if and only if one can change the diagram of one to the diagram of the other by a sequence of these moves". In counting numbers of moves, one needs to make clear when two changes are considered "different". We already have a list of six moves $\Omega.1$, $\Omega.2$ $\Omega.3$ $\Omega'.1$, $\Omega'.2$ and $\Omega'.3$ and this does not take into account the possibly distinguishing differences in the "over/under" information at the crossings. So technically, each of these three moves should be represented by more than one Figure to reflect this. Additionally, it is to be understood that a diagram isotopy that does not involve crossings, such as in Figure 4.62, is also allowed. Many dismiss this as a trivial matter. We consider this as a fourth allowed move $\Omega.0$, and using this collection state:

Prop 4.6 (Reidemeister Moves Theorem). Two knottings K and K' are equivalent knottings if one can change the diagram for K to the diagram for K' by a sequence of Reidemeister moves of type $\Omega.0, \Omega.1$, $\Omega.2$ $\Omega.3$ $\Omega'.1$, $\Omega'.2$ and $\Omega'.3$. ∎

Position A Position B

Fig. 4.59 Move $\Omega.1$ is from Position A to B. Move $\Omega'.1$ is from Position B to A.

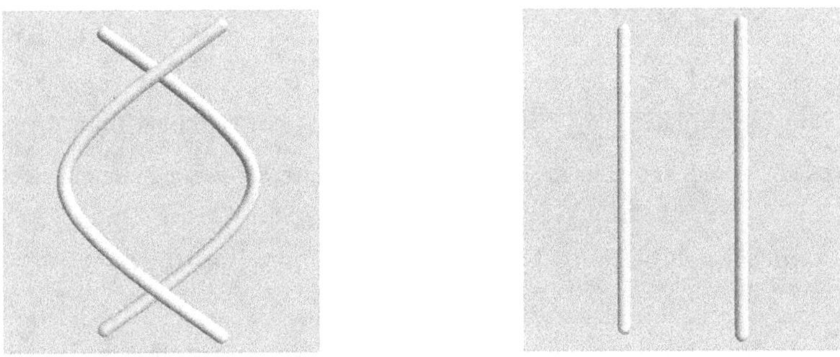

Position A Position B

Fig. 4.60 Move $\Omega.2$ is from Position A to B. Move $\Omega'.2$ is from Position B to A.

Reidemeister's method for proving Proposition 4.6 is sketched at the beginning of Chapter 7. A very combinatorial proof of the Reidemeister moves is in the book [Murasugi (2008)]. Perhaps the earliest proof (without figures) was in Alexander Briggs [Alexander and Briggs (1926/27)].

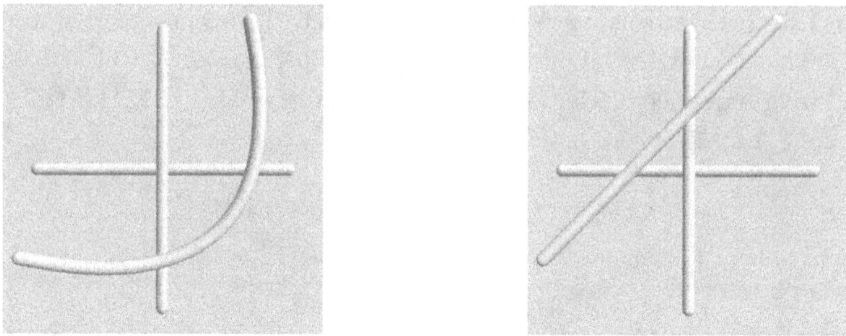

Position A Position B

Fig. 4.61 Move $\Omega.1$ is from Position A to B. Move $\Omega'.1$ is from Position B to A.

Position A Position B

Fig. 4.62 A change $\Omega.0$ of an arc with no crossings, most often not counted as a move.

4.5 Coloring Knotting Diagrams and Coloring of Knots

Example 4.5. We show a simple geometric technique showing the Hopf linking is not equivalent to a split trivial link. For projections we consider: Figure 4.63 and Figure 4.64. We use colorings of Reidemeister moves to show these are not equivalent knottings.

For the knotting of Figure 4.64, we horizontally slide one of the circles rigidly in R^3 so that the new knotting will have a projection with zero crossings. This is a Reidemeister $\Omega.2$ move. However, we show we cannot do this with the Hopf linking.

Distinguish the circles by coloring one red and the other blue. At a crossing c it might be that the above-point and below-point have the same color, then we say c is a **single-color crossing**. Otherwise it is a **two-color crossing** and we have two different types. An Rb**-crossing** has a red above-point and blue below-point; a Br**-crossing** has a blue above-point and red below-point. This technique, which colors each component of a knotting is called component-coloring. It readily extends to colorings with n-colors for knottings with n components.

Consider coloring arcs of the Reidemeister move diagrams either red or blue. An $\Omega.1$ move only involves one single-color crossing. It can be checked that $\Omega.3$ moves do not change the number of crossings of either type (if any). If an $\Omega.2$ has a two-color crossing it has two, and it removes or adds *two* of them of the *same* type.

If we use different colors for the different components of the Hopf linking in Figure 4.63 we will have one Rb-crossing and one Br-crossing. No matter what sequence of Reidemeister moves we use, we will always have an *odd number* of two-color crossings of each type—in particular, never zero. Therefore there cannot be a smooth isotopy that takes the Hopf linking to a trivial linking. ∎

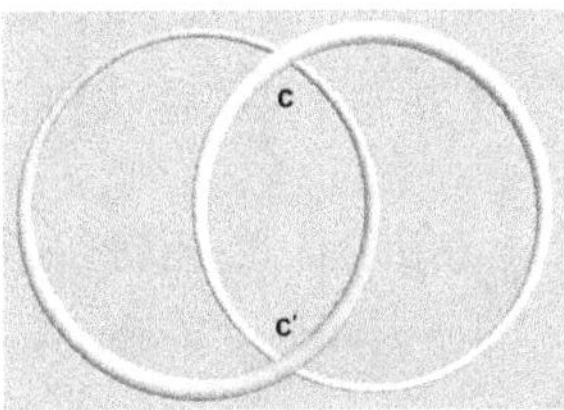

Fig. 4.63 A projection of the Hopf linking.

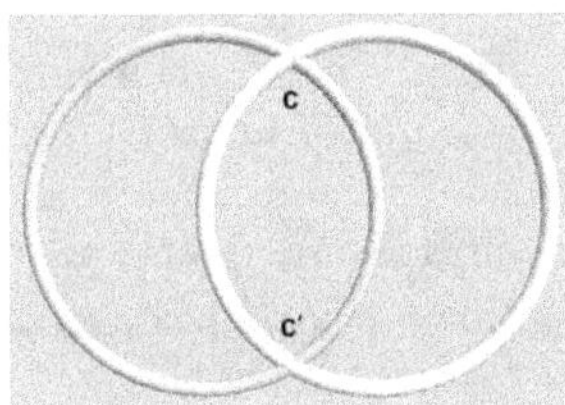

Fig. 4.64 A projection of a trivial linking.

Finer mathematical knot "colorings" of diagrams yield simple but powerful tools that will distinguish many knots. This is different from the component-coloring of Example 4.5—here we will color individual shards (sometimes groups of shards) of a projection. This technique originated in abstract algebra and algebraic topology however our approach avoids the need for such algebraic concepts. For an introduction to algebraic aspects, see [Elhamdadi and Nelson (2015)]. This technique demonstrates the calculational utility of the knot-diagram and its Reidemeister moves.

Mathematically a "color" is just a choice of label. For n colors, we will use labels $L = \{1, 2, \ldots, n\}$.

In most common colorings, 1-shards are required to have the same color if they belong to the same overpass. So, in effect, our coloring is a coloring of overpasses. To simplify our presentation, we will assume a coloring rule is of this type, unless we explicitly say otherwise, as in Remark 4.9.

Trace a knotting projection with colored pencils, lifting the pencil and leaving gaps as we pass under an overpass. We will be allowed to (possibly) change our pencil color *only* when we lift it to make our gap. So, at a crossing, we have at most three different colors possible—one for the overpass and two on either side of the gap. Our rule is: at a crossing, if we know the color of the overpass is b and one of the other colors is a, then the third color must be c.

Notation 4.1. We speak of c as "the color obtained from a if it passes under the overpass colored b". We will use the notation $a \triangleright b = c$. ▌

Definition 4.18. A **diagram coloring** or **diagram labeling** for knotting K is a function λ from the label set L to the set of 1-shards of the knot-diagram for K_* that satisfies a knotting coloring rule so that each shard gets exactly one label. Elements of L are called **labels** or **colors.** ▌

A color can be used more than once in a diagram. Also, there is no requirement to use all given colors. *Any* knot-diagram has at least one diagram coloring—the trivial one:

Definition 4.19. A **trivial coloring** of a knotting K is one that uses the same color for each 1-shard of K_*. ▌

What we really want is a coloring of a *knot*, not just a coloring of a *knotting diagram.*

Definition 4.20. A **knot-coloring rule** R is a diagram coloring rule that extends colorings uniquely to any new knotting under any Reidemeister knot-move. That is, if K' is a 1-knotting obtained from K by an elementary move Ω, then for any R coloring of K there is a unique coloring of K' which agrees with the coloring of K outside the support of move Ω. ▌

Definition 4.21. If R is a knot-coloring rule, K a 1-knotting, the number of distinct colorings of K_* is denoted $N(R,K)$. If $[K] = [K']$ then $N(R,K) = N(R,K')$, and so $N(R,K)$ is called the **number of R-colorings of the knot** $[K]$. If there are no R-colorings, we say $N(R,K) = 0$. ▌

In particular for any two 1-knottings K_0 and K_1, if $N(R,K_0) \neq N(R,K_0)$, then K_0 and K_1 are not equivalent.

The simplest interesting rule is the **three-coloring rule**: there are three colors $L = \{1,2,3\}$ and, at each crossing, either *all three* colors are used or *only one* of the colors. Figure 4.65 shows a knotting of a trefoil with a non-trivial 3-coloring. The fact that this rule extends uniquely if we use any Reidemeister moves can be verified (somewhat tediously) by taking each position (A or B) of each knot-move, coloring it consistent with the three-coloring rule and, then verifying that there is only one possible three-coloring in the other position of that move which agrees with the coloring on the boundary of the support disk. Basic cases are shown in many introductory knot theory presentations, such as [Przytycki (1998)]. We will use an alternate method to verify this using a coloring table.

We can use the three-coloring rule to show the trefoil knotting and the figure-eight knotting are not equivalent. Figure 4.65 shows a knotting of the trefoil is three-colorable using all three colors.

However we will show the diagram for the figure-eight knotting, Figure 4.66, does *not* have a non-trivial 3-coloring. We suppose there *is* a non-trivial 3-coloring, and see what goes wrong. First, suppose we use *different* colors on the overpasses containing crossings x and y. Without loss of generality, we may assume these are colors 1 and 2 on the over passes, as shown. We now have used colors 1 and 2 at crossing x so by the three-color rule we must label the remaining shard at z with color 3. Similarly, at crossing y, we have two colors 1 and 2 and must use color 3 for the remaining shard. So the colors must be as shown in Figure 4.66. However, this violates the 3-crossing rule, since we have exactly two different colors at crossings z and w.

So, the other possibility is that we use the *same* color on the overpasses containing crossings x and y. Without loss of generality, suppose this is color 1. Then, at crossings x and y, we already have the color 1 twice, so the third color must also be 1, and we would have a trivial coloring.

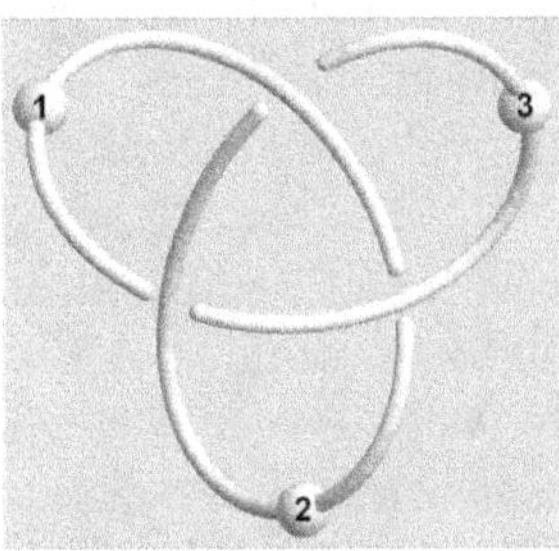

Fig. 4.65 A non-trivial 3-coloring of a trefoil knotting.

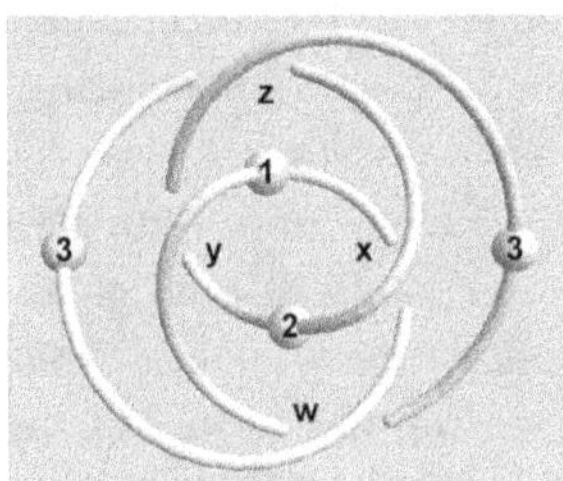

Fig. 4.66 A figure-eight knotting with no 3-coloring.

We can express a coloring rule as a **coloring table** M—a table for $a \triangleright b$. The table for 3-coloring is shown in Figure 4.67 in the style of a multiplication table. Writing M in matrix notation with entries c_{ij}, then c_{ij} is the third color produced from the color i at a crossing with overpass colored by j. The matrix of Figure 4.67 $c_{13} = 2$ tells us that, at an overpass with color 3, if one of the other colors is 1, the third color must be 2.

▷	1	2	3
1	1	3	2
2	3	2	1
3	2	1	3

Fig. 4.67 Table for 3-coloring.

▷	1	2	3	4	5
1	1	3	5	2	4
2	5	2	4	1	3
3	4	1	3	5	2
4	3	5	2	4	1
5	2	4	1	3	5

Fig. 4.68 Coloring table a 5-coloring.

Remark 4.6. We next focus on determining when a diagram coloring rule will be a knot-coloring rule. We will look at the properties of the operation $a \triangleright b$ and then view in terms of the color-change tables, so that given an $n \times n$ table with entries being the integers from 1 to n, we have a calculational method to determine if this is a table for a knot-coloring.

The procedure for each of the three Reidemeister moves is the same. If D is the support disk for the move, call points of our knotting K whose projection lies on ∂D **peripheral points**. Each knot-move corresponds to a pair of knottings: our original knotting K and a "new" knotting K'. An $\Omega.1$ move involves one arc of K, with a pair of peripheral endpoints. Similarly, an $\Omega.2$ move involves two arcs, with two pairs of peripheral endpoints, and an $\Omega.3$ move involves three arcs with three pairs of peripheral points. In each move, peripheral points in K and K' are fixed, but the arcs change. For the rest of this discussion, we focus on the projections of these arcs and peripheral points. A coloring of either K_* or K'_* will give a coloring of the peripheral points.

From Definition 4.20, it follows that the coloring of peripheral points will not change in our move. The reader should keep in mind that: *at a crossing, if the overpass is colored b and one of the other shards colored a, the shard on the other side of the crossing is colored* $a \triangleright b$.

For a Reidemeister $\Omega.1$ move, Figure 4.69 and 4.70, we label one peripheral point in both Figures with color a as shown with an arrow. There are no crossings in Figure 4.70 so the other peripheral point must have color a.

Because of the self-crossing in Figure 4.69 the coloring rule of the other endpoint of the arc must be $a \triangleright a$. For these colorings to agree, comparing K and K', (compare the labeling at tops of Figure 4.69 and 4.70) we require for our rule that:

$$\boxed{\Omega.1 \text{ color requirement : for any color } a, a \triangleright a = a.}$$

For a Reidemeister $\Omega.2$ move, Figure 4.71 and 4.72, we label two peripheral points a and b, as shown with arrows. There are no crossings in Figure 4.72, so the other end points must also be a and b. The coloring at crossing xs, and then y, extend uniquely as in Figure 4.71. In this case, we have a 1-shard that is colored $a \triangleright b$. Again, colorings must agree, comparing K and K', for *all* peripheral points (focus on labelings on at the bottoms of Figures 4.71 and 4.72). So, we require of our rule that:

$$\boxed{\Omega.2 \text{ color requirement : for all colors } a \text{ and } b, (a \triangleright b) \triangleright b = a.}$$

Finally, for a Reidemeister $\Omega.3$ move, Figure 4.73 and 4.74, we label peripheral points with colors a, b and c, as shown. There are three crossings, x, y and z. For Figure 4.73 we can uniquely extend the coloring as shown by considering the crossings y, x, and z in that order. Similarly, we can proceed to label Figure 4.74, as shown, by considering the coloring at the crossings y, x, and z in that order. Again, the colorings must agree, comparing K and K', for *all* peripheral points. So, for this move we require:

$$\boxed{\Omega.3 \text{ color requirement : for all colors } a, b \text{ and } c, (c \triangleright b) \triangleright a = (c \triangleright a) \triangleright (b \triangleright a).}$$ ∎

This motivates a definition found in [Takasaki (1943)]:

Definition 4.22. A set L with operation $a \triangleright b$ is a **kei** if for all $a, b, c \in L$:

(1) $a \triangleright a = a$
(2) $(a \triangleright b) \triangleright b = a$
(3) $(c \triangleright b) \triangleright a = (c \triangleright a) \triangleright (b \triangleright a)$. ∎

A common alternate terminology for "kei" is "involutive quandle"
So, Remark 4.6 has shown:

Prop 4.7. A labeling with labels L with rule diagram coloring rule $a \triangleright b$ is a knot-coloring rule if and only if $(L, \triangleright)$ is a kei. ∎

Remark 4.7. Additionally there are coloring rules called the **Fox n-coloring rules**, or the **Takashita n-coloring rules** after two of the historical co-discoverers, where the entries of the coloring table are $c_{ij} = (2j - i)(mod\ n) + 1$. Figure 4.67 is a table for $n = 3$; Figure 4.68 is a table for $n = 5$. One can verify that such a coloring satisfies Proposition 4.22 by using the formula $i \triangleright j = (2j - i)(mod\ n) + 1$ and verifying conditions (1), (2) and (3). ▌

Remark 4.8. If M is a table of a coloring it is easy and quick to verify that the first two conditions of a kei are satisfied. The third condition is easy but not so quick—it involves a simple calculation, but one might need to be done as many as n^3 times.

The condition $a \triangleright a = a$ means that there are either three different colors or only one at each crossing. In terms of a color rule table M, which uses the colors $1, \ldots, n$ in order for both rows and columns it means that the color $c_{ii} = i$. In standard matrix terminology, this means the diagonal of M is $\{1, \ldots, n\}$.

The condition $(a \triangleright b) \triangleright b = a$ says, "pass color a under b to get color c and, then pass c back under b, the resulting color will be back to a". This implies that each column of M is some permutation of L, since all numbers in L appear one time in each column. Additionally, these are special permutations. For each color i, if i is permuted to j then j is permuted to i. This is called a **permutation cycle of length two**. Labels fixed by the permutation are **cycles of length one**. Geometrically these last correspond to overpasses where the other two 1-shards have the same color. So, the column permutations are composed of cycles of length one or two—something easy to verify in a given table.

There is no simple quick way to tell if the condition $(c \triangleright b) \triangleright a = (c \triangleright a) \triangleright (b \triangleright a)$ is satisfied. However, there is a simple slow way—just try all possible choices of a, b and c. If there are n colors, there will be n^3 choices. For reasonably small values of n, this can quickly be done by making a new table whose entries correspond to $(c \triangleright b) \triangleright a - (c \triangleright a) \triangleright (b \triangleright a)$ and see if you get an $n \times n$ table of zeros. ▌

Another useful knot-coloring rule, closely related to kei coloring, is **quandle-coloring**. Using such colorings, it is possible to distinguish thousands of 1-knottings in established knot tables [Clark *et al.* (2014)]. To define it we need to choose an orientation for the knotting. An orientation of K defines a direction for a tangent vector at any point of K_*, which

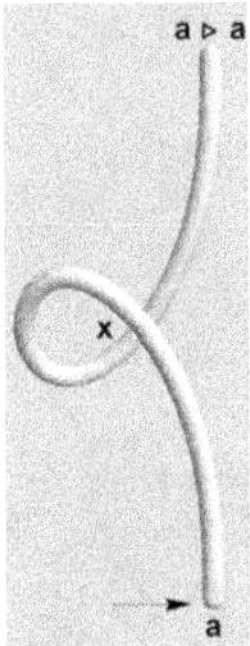

Fig. 4.69 If color is a at endpoint (arrow), the other must have color $a \triangleright a$.

Fig. 4.70 If color is a at endpoint (arrow), the other must be colored a.

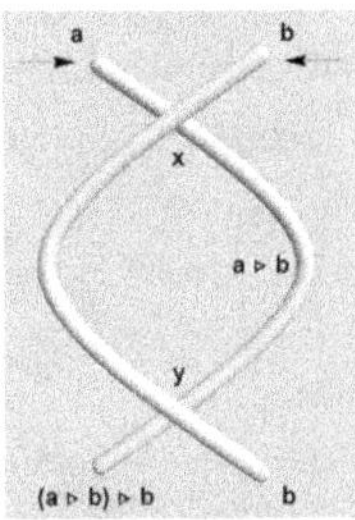

Fig. 4.71 If colors a and b at endpoints (arrows), the move 1-shard and other endpoints must be colored as shown.

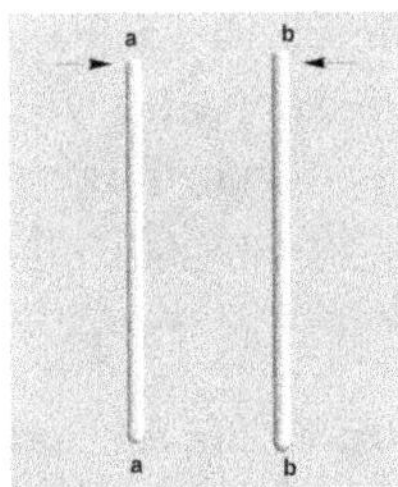

Fig. 4.72 If colors a and b at endpoints (arrows), the other must be colored a and b as shown.

in turn defines a normal vector by rotating the tangent $\pi/2$ in a counter-clockwise direction. We call this an **orientation normal** K_*.

A **quandle-coloring rule** is a knot-coloring rule. We will continue to use the notation $a \triangleright b$. At an overpass colored b, now $a \triangleright b = c$ is the color obtained by passing an arc-colored a under *in the direction of the orientation normal of that overpass.*

If we know the color p of the overpass and one of the other colors at a crossing is q, there are now two possibilities for the third color r or r': we use the color r if the third color is for 1-shard on the *positive* side, but use color r' if we are using this third color on the negative side (we allow the colors r and r' to possibly be the same.) See Figure 4.75 and Figure 4.76 where we have a white overpass and a second color light grey with small black arrows are orientation normals to the overpass. If color is light gray on positive side Figure 4.75 the third color might be medium grey, but if it is on the negative side it might be dark-grey, Figure 4.76.

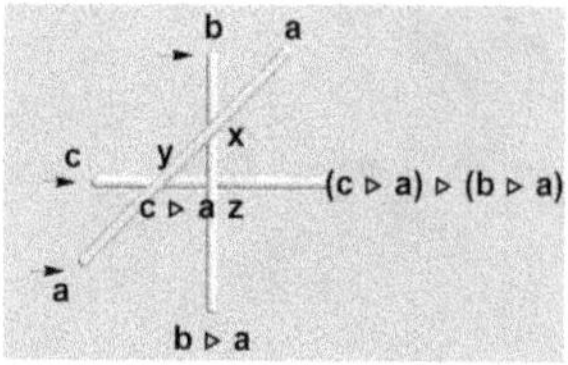

Fig. 4.73 If colors a, b and c at endpoints (arrows), the three move 1-shards and other endpoints must be colored as shown.

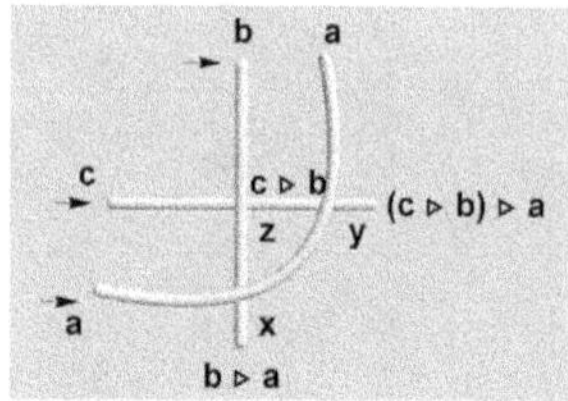

Fig. 4.74 If colors a, b and c at endpoints (arrows), the three move 1-shards and other endpoints must be colored as shown.

A good reference to technical details of computations is [Clark *et al.* (2014)].

If we insist that the quandle-rule results in a knot-coloring rule it, using an analysis as we did with Fox-Takashita colorings, Remark 4.6 (but using *oriented* knot-move figures one can show that quandle-colorings are knot-colorings [Elhamdadi and Nelson (2015)], [Kauffman and Lopes (2018)], where quandle is defined:

Definition 4.23. A **quandle** is a set Q with operation $\triangleright$ that satisfies:

(1) For all $x \in Q$, $x \triangleright x = x$
(2) For all $x, y \in Q$, there is a unique $z \in$Q such that $z = x \triangleright y$
(3) For all $x, y, z \in Q$ $(x \triangleright y) \triangleright z = (x \triangleright z) \triangleright (y \triangleright z)$. ∎

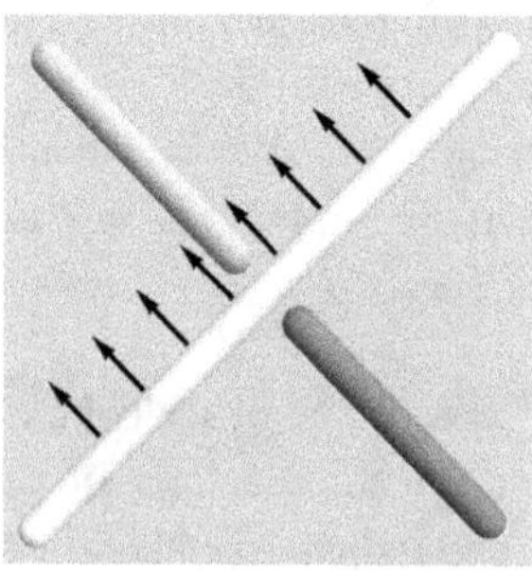

Fig. 4.75 A quandle-coloring for white overpass and light grey on positive side.

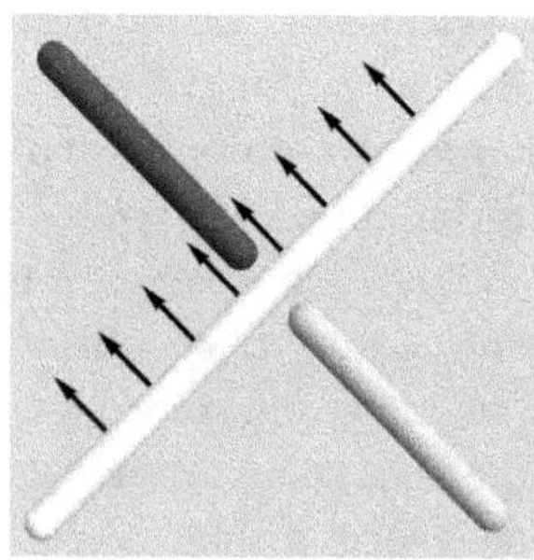

Fig. 4.76 A quandle-coloring for white overpass and light grey on negative side.

The first and third properties of quandle, Definition 4.23, are identical with the first and third properties of kei, Definition 4.22.

$\triangleright$	1	2	3	4	4
1	1	5	4	3	2
2	3	2	1	5	4
3	5	4	3	2	1
4	2	1	5	4	3
5	4	3	2	1	5

Fig. 4.77 Table for a five color quandle, not a Fox coloring.

$\triangleright$	1	2	3	4	5
1	1	4	2	5	3
2	4	2	5	3	1
3	2	5	3	1	4
4	5	3	1	4	2
5	3	1	4	2	5

Fig. 4.78 Table for another five color quandle, not a Fox coloring.

Consider a finite quandle n-coloring defined by a $n \times n$ table M of values for $a \triangleright b$. As in Remark 4.8, M must have value i for diagonal entry cii. Property (2) for a quandle implies that each column of M must be a permutation of $\{1, \ldots, n\}$ but we no longer have a requirement that these permutations are composed of 1-cycles and 2-cycles. Tables in Figures 4.77 and 4.78 are quandle tables but not tables for Fox-Takashita colorings. For example in each the first column has a 1-cycle and a 4-cycle.

Remark 4.9. There are useful coloring rules besides kei and quandles. In this coloring we allow two colors for an overpass—possibly changing at the above-point. So, at a crossing we color four incident 1-shards. There are coloring rules which take colored pair of shards, one from the overpass, and one not and determines the colors of the other two shards. A rule of this type which is a knot-coloring rule is called a **bikei**, and if it is a knot-coloring rule for oriented knottings, it is called a **biquandle**. More details of these ideas can be found in [Fenn (2012)] [Elhamdadi and Nelson (2015)], [Ashihara (2012)]. These methods make more use of shard structures than methods which just use overpass structure. ∎

4.6 Codimension 2 Projections

General position for a 1-knotting $K \subseteq R^3 \approx R^2 \times R^1$, Section 4.2, gives an equivalent knotting which is "nicer" when considering the standard codimension 1 projection to a plane: $R^2 \times \{\mathbb{O}\}$. **Plait position** is a similar idea of a "nicer" position with respect to a codimension 2 projection of R^3 to a line: $\{\mathbb{O}\} \times R^1$. Additionally, there is a different codimension 2 projection where we map the knotting to S^1—this is called **braid position**—in this case, we begin with a knotting in $S^1 \times R^2$ with projection on the first factor.

The height function for a smooth 1-knotting $K \subseteq R^3$ is a smooth real-valued function. The most familiar features of such a function are **critical points** where the tangent line to the knotting is parallel to the xy-plane. There are three types of critical points: local maxima, local minima and horizontal points of inflection. The **plait position** is a knotting that will be "nice" with respect to these features. With a small isotopy we can easily get rid of horizontal points of inflection. (If $x \in K$ is such a point, we need only tip the tangent line slightly away from the horizontal and the point; x will still be an inflection point, but not critical.) However, we must have *some* maxima and minima. If K has n components, then we must have at least n local maxima and n-local minima–the highest and lowest points for each component.

For a local maximum/minimum point of a knotting, the corresponding **critical value** is the height of this point.

Definition 4.24. Any 1-knotting $K \subseteq R^3 \approx R^2 \times R^1$ with height function h is in **plait position** if there is a number C such that all local maximal values of h are above C and all local minimum values are below C. In particular, K is in a **standard plait position** if all local maximal values of h are 1 and all local minimal values are -1. ▮

Prop 4.8. A 1-knotting $K \subseteq R^3$ is isotopic to a knotting in plait position—in fact, one in standard plait position. ▮

Example 4.6. The idea of a proof of Proposition 4.8 is given by Figures 4.79, 4.80 and 4.81.

The knotting of Figure 4.79 is not in plait position since the local minimum at a is above the two local minima at b and c. As shown in Figure 4.80, we can lower critical point's height, without introducing more critical points, by pulling points near the critical point downwards and leaving other points fixed. We cannot necessarily pull straight down, since we must avoid intersections with lower portions of the knotting. We can easily do this by veering slightly as we go down to avoid these lower portions. Once we do this, we are in plait position. To get standard plait position, first pull all the local minima to same level, and similarly, raise all local maxima. Then scale and translate the knotting, so the critical values are -1 for the local minima, +1 for local maxima, as indicated in Figure 4.81. ▮

The subject of braids and braid groups is a large and important part of theory of 1-knottings, with much of the focus on the algebraic topic of

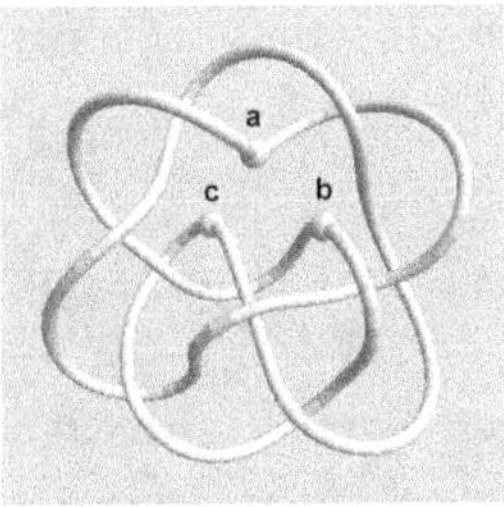

Fig. 4.79 A Knotting not in plait position, see Example 4.6.

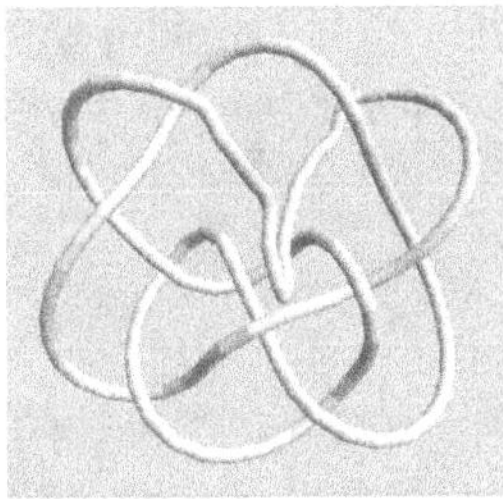

Fig. 4.80 Pulling down one local minimum of Figure 4.79.

Fig. 4.81 Pulling so that it is below all local maxima.

braid groups such as [Birman (1974)], [Kassel and Turaev (2008)],[Birman and Brendle (2005)]. Our focus will be on extensions to higher dimensions as in [Kamada (2002)] and [Kamada (2017)]. The idea is that a 1-knotting is in closed braid position if it "winds around the z-axis and never doubles back".

One way to express this for $K \subseteq (R^3 - axis(z))$ is a 1-knotting is to require that K intersects $H_\theta \subseteq R^3$ transversely for every page $H^2_\theta \subseteq R^3$. Figure 4.82 shows a knotting in closed braid position.

There is an alternate definition that will prove useful in higher dimensions. Consider $\widehat{\pi}\colon (R^3 - axis(z)) \to S^1$ defined $\widehat{\pi}(r,\theta,z) = (1,\theta)$—it takes each open page $\overset{\circ}{H^2_\theta}$ to the point $(1,\theta) \in S^1 \subseteq R^2$. We call $\widehat{\pi}$ the **braid projection.** The transversality condition of closed braid position is equivalent to requiring that the map $\widehat{\pi}|_K$ is an immersion of K to S^1. Thus, this map from K to S^1 is a covering map.

Definition 4.25. We say $K \subseteq (R^3 - axis(z))$ is in **closed braid position with respect to the z-axis** if $\widehat{\pi}|_K$ a covering map from K to S^1. ∎

Prop 4.9 (Alexander's Theorem). Any 1-knotting is equivalent to one in closed braid position. ∎

The original proof of Proposition 4.9 is [Alexander (1923)], an elementary proof is presented in [Adams (1994)] and a more modern proof is found in [Birman and Brendle (2005)] based on proof in [Yamada (1987)] as modified in [Vogel (1990)].

For 1-knottings $K \subseteq S^3$ viewing $S^3 = S^1 * S^1$, the $axis(z)$ corresponds to the second join factor Z. Then closed braid position requires that knottings in $S^3 - Z$ be transverse to all page-disks. We have $S^3 - Z \approx S^1 \times D^2$, and braid projection is the projection on first factor.

Definition 4.26. Let $X \subseteq R^2$ be a finite collection of k distinct points. In $R^2 \times I$ let $P = X \times \{0\}$ and $Q = X \times \{0\}$.

A **braid with k strands**, sometimes also called an **open braid with k strands**, is a disjoint collection A of k proper sub-arcs of $R^2 \times I$,$\{\alpha_i\}$ called **strands** such that $A \cap (R \times \{0\}) = P$, $A \cap (R \times \{0\}) = Q$ and A intersects each $R^2 \times \{t\}$ transversely. ▮

Fig. 4.82 A knotting in closed braid position showing transverse intersection eight pages.

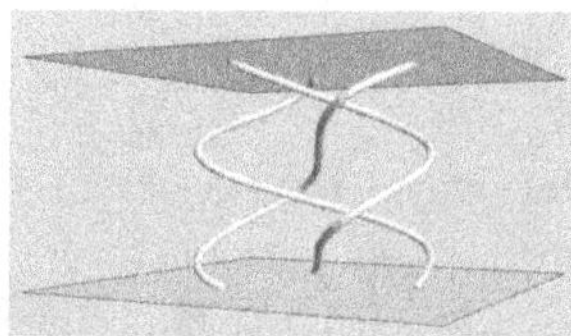

Fig. 4.83 An (open) braid with three strands.

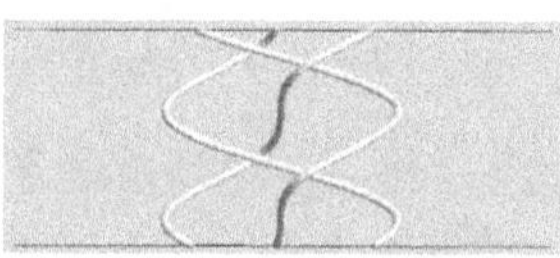

Fig. 4.84 An orthogonal view of braid of Figure 4.83.

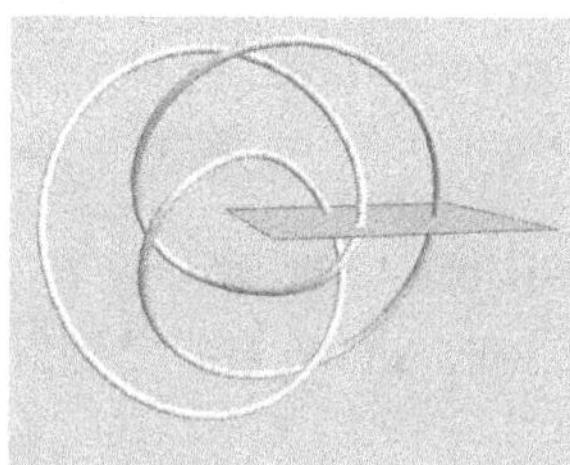

Fig. 4.85 The closed braid corresponding to the open braid of Figure 4.83.

Example 4.7. Figure 4.83 shows an example of an open braid. A closed braid is often obtained from an open braid by "bending the strands of open braid in a circular way and gluing the two ends together", Figure 4.85.

This should be done in a way that avoids introducing whole twists of the braid. The knotting shown in Figure 4.85 represents the figure-eight knot. Figure 4.84 may be helpful in verification of details—one can see more clearly how the four crossings of the projection of open braid correspond to the four crossings in the projection of the figure-eight knotting. ∎

4.7 Siefert Surfaces and Fiberings of 1-Knottings

A basic way of understanding a 3-manifold is to study its 2-dimensional submanifolds. A fundamental tool in understanding the exterior of a 1-knotting K is a **Seifert surface** which can be obtained from an orientable knotting of a surface whose boundary is K.

Definition 4.27. If Σ is a smooth knotting of a compact connected orientable 2-manifold in R^3, or S^3, we say Σ is a **Seifert surface for the 1-knotting** $\partial\Sigma$. ∎

Any knotting of a connected orientable 2-manifold Σ with $\partial\Sigma \neq \emptyset$ is an example of a Seifert surface, such as components shown in Figures 8.44–8.52.

For a Siefert surface in S^3. There is natural version of a Siefert surface Σ' as a subset of the exterior the knotting $\partial\Sigma$. To simplify our next discussions for the remainder of this topic, we consider the case that $\partial\Sigma$ is a single circle. An orientable surface in R^3 has a product tubular neighborhood, and so $\partial\Sigma$ has a product tubular neighborhood $T \approx S^1 \times D^2$. By an isotopy, we can arrange $C = \Sigma \cap T$ to be a smooth annulus corresponding to $S^1 \times I \subseteq S^1 \times D^2$. As a subset of Σ, C will correspond to a collar neighborhood of the boundary. Now $\Sigma' = Ext(K) \cap \Sigma$ is a proper submanifold of $Ext(K)$ diffeomorphic to Σ.

Recall Definition 4.15 of annulus twist. A knotting of a circle $L \subseteq \partial T$ with a smooth *untwisted* annulus between $\partial\Sigma$ and L is called a **longitude of the knotting** $\partial\Sigma$.

Definition 4.28. For any Seifert surface Σ for knotting K, the surface $\Sigma' \subseteq Ext(K)$ is **a corresponding proper Siefert surface in** $Ext(K)$. Going the other way, suppose we have a proper knotting of an orientable surface $S \subseteq Ext(K)$ with $K' = \partial S$ a longitude of T, and an untwisted annulus $C \subseteq T$. Then $S \cup C$ will be **a Siefert surface for** K **corresponding to proper Siefert surface** Σ'. ∎

Every 1-knotting has a Seifert surfaces (in fact many such surfaces):

Prop 4.10. Any smooth 1-knotting is the boundary of a knotting of an orientable 2-manifold. ▮

Proofs, examples and discussions of Proposition 4.10 are found in almost every book on knot theory. The most common presentation is an explicit construction based on a diagram of the knotting. We will take a different approach which is not at all constructive and most directly produces *proper* Siefert surfaces because it is useful in understanding knot exteriors.

Prop 4.11.

(1) Let $K \subseteq S^3$ be a smooth 1-knotting. There exists a smooth map $\phi\colon Ext(K) \to S^1$, which sends each meridian μ_θ of $\partial Ext(K)$ corresponding to $S^1 \times \{\theta\}$ to the point $\theta \in S^1$.
(2) Also, such a ϕ has only non-degenerate critical points, Definition 2.5.
(3) As a consequence, if $\theta \in S^1$ is a regular point of ϕ, then $\sigma_\theta = \phi^{-1}(\theta)$ is a smooth proper orientable 2-dimensional submanifold—and, if it is connected, it will be a proper Seifert surface for $Ext(K)$.
(4) Additionally, if $A \subseteq S^1$ is a closed sub-arc of regular points with $\theta \in A$, then $\phi^{-1}(A) \approx \Sigma_\theta \times D^1$. In other words, $\phi^{-1}(A)$ is a product tubular neighborhood of $\sigma_\theta \subseteq Ext(K)$.
(5) If ϕ has no critical points at all, then $Ext(K)$ is a **fibration over a circle with fiber** σ_θ. In this case K is called a **fibered knotting.** ▮

The proof of Proposition 4.11 (1) is the "hard part". There is no elementary geometric proof, but standard references are [Kervaire (1965a); Levine (1965)] and [Gluck (1962)]. Once we have found these smooth functions, Proposition 4.11 (2) and (4) are versions of Proposition 8.1.

The standard $S^1 \subset S^3$ is a fibered knotting (with fiber D^2) since $Ext(S^3, S^1) \approx S^1 \times D^2$, Remark 4.1.

Good information about the exterior $Ext(K)$ of a knotting can be obtained by cutting along a Seifert surface F and then examining the gluing process that reassembles the exterior. If we have a fibered knotting then $Ext(K)$ is a mapping torus, Definition 2.11, of a gluing diffeomorphism $\phi : F \to F$. It is well understood exactly which 1-knottings are fibered [Harer (1982)]. Examples of fibered knottings are: all torus knots and links (in particular, the trefoil knot and Hopf links), the figure-8 knot, and connected sums of fibered knots. In Section 5.4 we give details of the fibering of a trefoil knotting. In [Rolfsen (1976)] there is a different detailed presentation of the fibering of the trefoil; in [Francis (2007)] is a detailed discussion for the figure-eight fibering.

Remark 4.10. The exterior of a *fibered* 1-knotting K with fiber σ is a union of two handlebodies, each diffeomorphic to $\sigma \times D^1$. To see this let $\phi\colon Ext(K) \to S^1$ be a smooth map with no critical points, Proposition 4.11 (1) and (5). Write $S^1 = H^1_+ \cup H^1_-$ as a union of hemispherical arcs with $H^1_+ \cap H^1_-$ a two-point set. Using Proposition 4.11 (4) we have

$$Ext(K) = \phi^{-1}(H^1_+) \cup \phi^{-1}(H^1_-) \approx (\sigma \times D^1) \underset{\sigma\times\partial D^1}{+} (\sigma \times D^1).$$

By Proposition 5.1, $\sigma \times D^1$ is a handlebody. ▮

4.8 Ideal Knottings

Here are some challenging questions about knots and knottings. Even the smallest progress towards an answer is significant.

Of the infinitely many 1-knottings K that represent a given knot $[K]$, is there a "best choice" or "ideal choice" for K? Which knots have an "ideal knotting". (For the trivial knotting of a circle, the geometric circle is certainly "ideal". Are there others?) If not, what constitutes a "good choice" for K? Is there a concept of "knot complexity" so that given two inequivalent knottings one say one knotting is "more complex" than the other?

A traditional method is to search for a positive numeric $n(K)$ quantity that we can attach to a knotting K so that smaller values correspond to simpler knottings. We then let $N(K)$ be the smallest value of $n(K')$ where $[K'] = [K]$ and use $N(K)$ as our measure of "complexity of K" or "complexity of $[K]$". Such function that assigns a real number to any knotting is called a **functional**. Historically the first example is **crossing number**. Here $n(K)$ is the number of crossings in a general position projection. If K is a trefoil knotting, $n(K)$ could be large, but it is known that $N(K)$=3. For a trivial knotting K, $N(K) = 0$. Beyond this, crossing numbers fail to distinguish knottings. In the traditional knot tables for crossing number nine or less, one finds multiple entries except $N[K] = 0$ or 3 and ordering of these multiple entries being largely random.

There is a concept called **hyperbolic volume**, [Adams (2005)], that is currently the best result. This volume is a real number defined for "most knots" and for any given number V there are only a finite number of knots with hyperbolic volume V. Further discussion is beyond the scope of this book.

Another attempt to define ideal knottings involves a concept called **knot-energy**. Imagine a knotting as a placement in world with no gravity, of a somewhat elastic string of beads of fixed length, where each bead is charged with static electricity. Upon release, the beads would repel one another, (yet constrained by the string length and elasticity) and the string of beads would move. Hopefully, the bead repulsion would prevent the string from crossing itself. If so, the motion of the beads would correspond to an isotopy. One might hope, in time, that this movement would converge to a knotting, which would be a candidate for an "ideal knotting". The strategy to define a real number $E(K)$ which measures the "total energy due to electrical repulsion for a knotting" K in such a way that a knotting with minimal (or perhaps locally minimal) energy is "ideal".

There are a number of candidates for $E(K)$, but very few exact minima found to date. However, well-developed tools are available for finding a minimum of a real-valued function, such as gradient descent and simulated annealing. As with many minimization studies, a problem is that these tools will often find *local* minima, which may not be *global* minima. In practice, for high dimensions, these often work very slowly, and may not lead to exact solutions. However, approximate solutions are of practical value in understanding particular knottings.

Knot-energy for smooth knottings was first defined for piecewise linear knottings in [Fukuhara (1988)] and smooth knottings in [O'Hara (1991)]. Early development is found in [Freedman *et al.* (1994)], [Kim and Kusner (1993)], [O'Hara (1991)] [Kusner and Sullivan (1998)] and [Auckly and Sadun (1997)]. A related notion called **relaxation** is implemented in a widely used software KnotPlot, see [Scharein (1998)] and [Lin and Zhang (2019)]. Two excellent surveys on this topic are [O'Hara (2008)] and [Janse van Rensburg (2005)].

Example 4.8. Figures 4.86 to 4.90 show steps stages of a discrete track of an isotopy produced by KnotPlot. As one goes from left to right, the total energy of the knotting is reduced. Figure 4.86 shows an unfamiliar knotting of a trefoil knot. (This is discussed in Example 5.13 and shown in Figure 5.46.) Figure 4.90 shows a familiar trefoil knotting. The wavy nature of Figure 4.87 is typical of the beginning of a knot-energy isotopy as knot-energy involves a complicated minimization calculation, which is sensitive both locally and globally. ∎

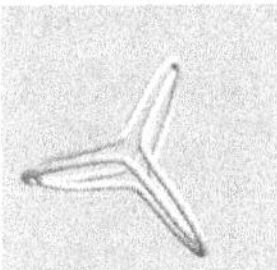

Fig. 4.86

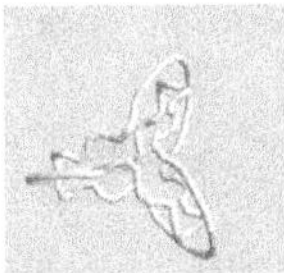

Fig. 4.87

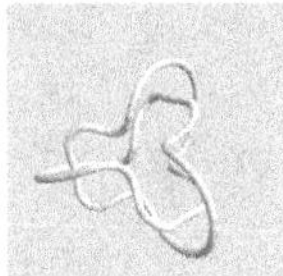

Fig. 4.88

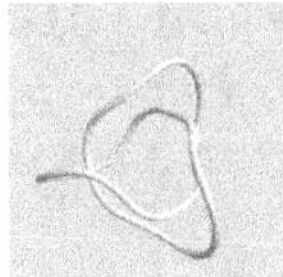

Fig. 4.89

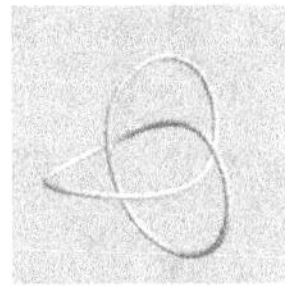

Fig. 4.90

Remark 4.11. Energy considerations are important even for a trivial knotting. If for some energy functional there was a unique global minimum and an effective way of finding an energy minimising flow, it would provide a way that any trivial knotting could untangle itself. For the Möbius energy, it is known that the round planar circle is a global minimum for trivial knottings [Freedman *et al.* (1994)] (as with more general energies [Abrams *et al.* (2003)]) but it is not known if there are any relative minima. Apart from this theoretical question one may hope that in practice knot-energy could be used to find such trivializing isotopies. In a recent preprint [Kauffman (2012a)] it appears this might not to be the case. ▮

4.9 Lattice 1-Knottings

We make an exception to our smooth point of view to consider a very special type of non-smooth submanifold. Our main interest in this topic is in higher-dimensions, Section 11.5, so we state our definitions for R^n:

Definition 4.29. The **integer lattice points Z^n of R^n**, also called **cubic lattice points**, consists of all points whose coordinates are all integers. A **unit lattice m-cube** is an m-cube in some R^n whose edges are all of unit length and whose vertices are integer lattice points. A unit lattice 1-cube is called a **lattice edge**; a unit lattice 2-cube is called a **lattice square.** A **lattice n-knotting** is finite union of unit n-cubes in R^{n+2} which, topologically, is a closed n-manifold. For lattice 1-knotting K, **length of the lattice length of K** is the number of distinct 1-cubes of K.

For n positive integers $p_1, \ldots, p_n$, the $p_1 \times \cdots \times p_n$ **lattice box** is all lattice points in $[0, p_1] \times \cdots \times [0, p_n] \subseteq R^n$. In our discussions we will always have $0 < p_i \leq 9$, so we can abbreviate $p_1 \times \cdots \times p_n$ by the bracketed n-digit number $[p_1 \ldots p_n]$. For example $3\times3\times2$ will be written $[332]$, which we call a **box-number**, as abbreviation of $p\times q\times r$ so for example a standard $3\times3\times2$ lattice box will be denoted $B^{[332]}$. If the box number satisfies $p_1 \geq \cdots \geq p_n$, the box number will be called a **standard box number** and lattice box will be called a **standard lattice box** for that number. Given two standard lattice box numbers $\mathbb{P} = [p_1 \dots p_n]$ and $\mathbb{Q} = [q_1 \dots q_n]$ we say that lattice box $B^{\mathbb{P}}$ is smaller that $B^{\mathbb{P}}$ if, as n-digit integers, $p_1 \dots p_n < q_1 \dots q_n$. ∎

For example, $B^{[322]}$ is a smaller standard lattice box than $B^{[521]}$ since, as integers $332 < 521$, even though each box contains the same number of lattice points, 36.

In the rest of this section we consider only lattice 1-knottings. Lattice 1-knottings have a substantial and varied literature, see [van Rensburg and Whittington (1990)], [Wu (1992)], [van Rensburg (2009)], [Lomonaco and Kauffman (2010)] and [Adams *et al.* (2012)]. Examples of small lattice 1-knottings are found in [Baldridge and McCarty (2010)]. Motivation for this comes from some simple vague questions. What is the simplest non-trivial knotting? What is a "random" knotting? Is the trefoil knotting more likely than the figure-eight knotting? Is the square knotting more likely than a granny knotting? These questions are of interest to certain chemists and molecular biologists for whom lattice models are frequently used. One approach is to use lattice knottings to get some sort of definition for "probability of a knotting":

Definition 4.30. For any box-number $\mathbb{B}$ and 1-knotting k, $k^{\mathbb{B}}$ is the set of lattice 1-knottings in lattice box $B^{\mathbb{B}}$ with some component equivalent (isotopic) to k and the number of elements of this set is written $\|k^{\mathbb{B}}\|$. Also $\|\mathbb{B}\|$ is the **total number of lattice 1-knottings in lattice box $B^{\mathbb{B}}$** which are not split trivial knottings and $P^{\mathbb{B}}(k) = \|k^{\mathbb{B}}\|/\|\mathbb{B}\|$ is called the **probability of k for box-number B.** ∎

By manual calculations, with graphical help from a computer, it is trivial but tedious to search for examples of lattice 1-knottings in small lattice boxes. Some results are as follows. In $B^{[321]}$, the only non-trivial knotting is equivalent to the Hopf linking. In $B^{[321]}$ there is a lattice knotting equivalent to a trefoil knotting as well as link 4.2.1. In $B^{[322]}$ there is a lattice knotting equivalent to a figure-eight knotting as well as 4.3.1 (a split sum of two Hopf linkings. In $B^{[322]}$ one finds, in addition the granny knotting the square knotting 5.1,5.2,5.2.1,6.3.3, and perhaps others. All knots and links in the classical Reidemeister Tables are represented by lattice 1-knottings in $B^{[444]}$.

Example 4.9. Figure 4.91 shows a lattice knotting at square knotting (of length 40) in $B^{[333]}$; Figure 4.92 shows a lattice knotting of a granny knotting (of length 42) in $B^{[333]}$. To help visualize the composite nature of these knottings, each figure indicates a vertical plane which intersects the knotting in two points. This and other examples lead us to conjecture that, for box number [333], that the probability of the square knotting is greater than the probability of the granny knotting. ∎

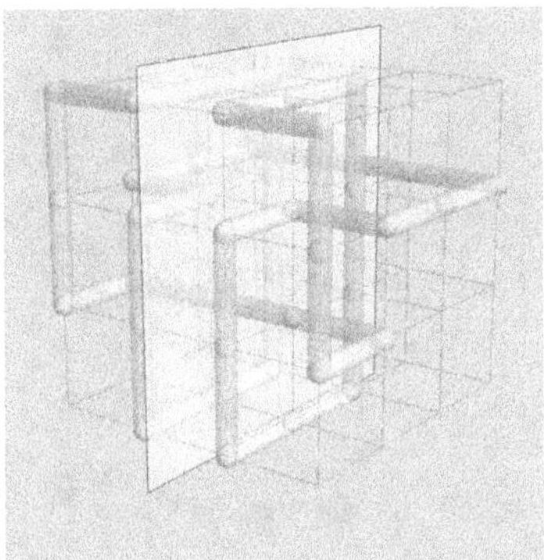

Fig. 4.91 Lattice square knotting.

Fig. 4.92 Lattice granny knotting.

Despite much progress, there is much to learn about this and similar methods. Progress is of great practical interest to chemists and molecular biologists, apart from theoretical interest. A good introduction to these problems is [Stasiak *et al.* (1998)]. We focus attention on some approaches that have higher dimensional extensions.

Part 2
Knotting Surfaces

It is mathematically natural to view the study of surface knotting in R^4 as the extension of knot theory of circles in R^3 into higher dimensions. Smooth or PL knottings circles, or unions of such circles, are unsuitable since they are trivial knottings in four dimensions, see Section 5.1. Surfaces can be tangled and self-tangled in ways similar to knottings of 1-manifolds of R^3. We choose to avoid wild knottings as in Figure 3.2 and restrict our considerations to surfaces that are smooth, with brief consideration to two alternate restrictions: piecewise-linear manifolds or lattice manifolds.

Objects in four dimensions present a visualization challenge even for the experienced. The objects the eye can best see are surfaces. In fact, with the exception of semi-transparent objects, everything we see is a surface. So, four-dimensional visualization employs our visual strengths to tackle one of the most challenging problems.

We study and visualize 2-knottings as images of projection $\pi\colon R^4 \to R^3$, Chapter 6. Another common tool is called slicing, Chapter 8. A **slice** of a knotting is an intersection of the knotting with a 3-dimensional hyperplane. It is often useful to use both tools at once.

Chapter 5

Knotting, Dimension and Codimension

Since knottings of circles in R^3 are so complex, it is clear that knottings of a circle in R^4 are even more complex. However for *knots* (isotopy equivalence classes) it is very simple—there is only one smooth knot of S^1 in R^4, namely the standard S^1. (If we were to consider wild knottings there *are* many knots, but our focus is smooth knottings.) We will show this in Section 5.1. Essentially the reason is that, even though there are more possibilities for constructing circles in four dimensions, compared to three, the overwhelming effect of increase of dimension is to allow more possibilities for isotopies, which allow untanglings not possible in R^3.

As a warm up, we consider a simple low dimensional example of how knotting can be affected when we include a knotting into a higher-dimensional Euclidean space. This is an unusual knotting, since the object we are considering is a union of manifolds of *different* dimensions.

Example 5.1. In R^2, let C be the unit circle and Z be the two points $(0, 0)$ and $(2, 0)$, which we consider to be a 0-sphere, see Remark 1.1. Also consider the 0-sphere Z' consisting of $(0, 0)$ and $(\frac{1}{2}, 0)$. The two subsets $K = C \cup Z$ and $K' = C \cup Z'$ are not equivalent subsets of R^2, see Figures 5.1 and 5.2. Here is a distinction. There is a path in R^2 between the points of Z' that does not meet C— we could use the line segment joining these points. However, *any* path joining the points of Z must intersect C since, intuitively, $(0, 0)$ is inside C and $(2, 0)$ is outside.

We can express this by saying that $C \cup Z'$ is a split trivial knotting in R^2 and $C \cup Z$ is not a split trivial knotting.

Now consider K and K' as subsets of R^3 via the inclusion $R^2 \subseteq R^3$, see Figures 5.3 and 5.4. As subsets of R^3, these *are* equivalent. How did C and Z "become" equivalent? As subsets of R^3, it is obvious that K and K' are isotopic subsets—our isotopy need only move the point $(2, 0, 0)$ to the point $(\frac{1}{2}, 0, 0)$ by a path that misses the circle and keeps all other points of Z fixed. Figure 5.5 shows one possible path and Figure 5.6 shows some intermediate images of the corresponding isotopy. ▌

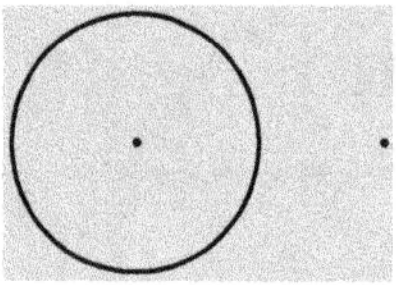

Fig. 5.1 $K \subseteq R^2$.

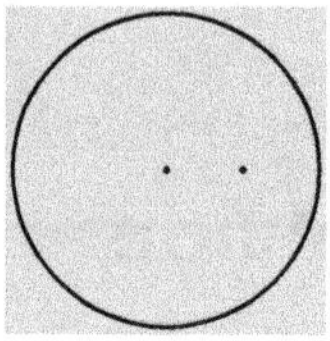

Fig. 5.2 $K' \subseteq R^2$.

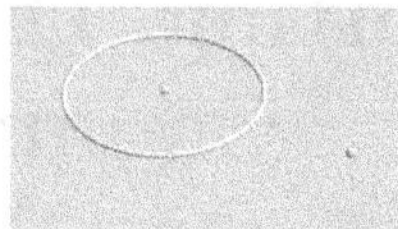

Fig. 5.3 $K \subseteq R^3$.

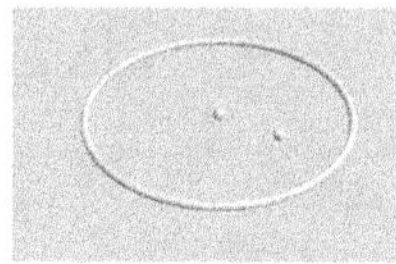

Fig. 5.4 $K' \subseteq R^3$.

Fig. 5.5 An arc in $R^3 - Z$ joining two points of Z'.

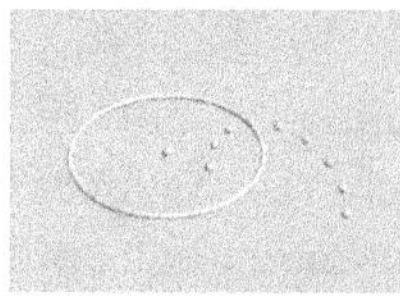

Fig. 5.6 Intermediate images of an isotopy.

For knottings of an n-dimensional manifold in an m-manifold R^m a critical consideration is the *codimension* $(m - n)$. In our definition, an n-knotting is codimension 2 knotting. In many ways the mathematics of n-knottings resembles mathematics of 1-knottings. Knottings of codimension 0 and 1 are fundamental to the study of the structure and classification of manifolds, for example the Schoenflies Theorem and Annulus Theorem below. This makes up a large part of geometric topology but is not generally considered as knot theory. Beyond this section, we will only deal with codimension higher than 2 in a few places such as Example 5.2 and Definition 12.13.

In every dimension, spheres hold a special mathematical place among all the other manifolds of the same dimension. Two basic codimension 1 theorems are:

Prop 5.1. (Smooth Schoenflies Theorem) A smooth simple closed curve in $C \subseteq S^2$ is the boundary of two smooth 2-disks in S^2 whose intersection is C. More generally, a smooth knotting of an n-sphere, $\Sigma \subseteq S^{n+1}$, is the boundary of two smooth $(n+1)$-disks in S^{n+1} whose intersection is Σ. ▌

Because all knottings of an $(n+1)$-disk in S^{n+1} are isotopic, the Schoenflies Theorem implies that all knottings of an n-sphere in S^{n+1} are trivial.

Prop 5.2. (Classical Annulus Theorem) Two smooth disjoint simple closed curves in S^2 form the boundary of an annulus in S^2. More generally, a smooth knotting of two disjoint n-spheres in S^{n+1} is the boundary of a knotting of a smooth $(n+1)$-manifold in S^{n+1} which is diffeomorphic to $S^n \times D^1$. ∎

Non-smooth codimension 1 2-manifolds in R^3 can be quite different—such as the wild knottings of Section 5.5.

Although there are many smooth knottings of closed surfaces in R^3 it turns out that this knotting is essentially captured by 1-dimensional knottings. One simple example is a knotted tube. Graphically, we have been showing knotting of a circle using a thin tube, but we could view this tube as a knotting of the torus $S^1 \times S^1$. Another example is a ball with a knotted hole, Remark 4.1. We can combine these two ideas independently and get very complicated knottings of high-genus surfaces called boundaries of **3-balls with knotted holes and knotted handles**. It turns out that:

Prop 5.3. Every connected knotting of an orientable surface $\Sigma \subseteq R^3$ is the boundary of a 3-ball with (some, possibly none) holes and (some, possibly none) handles. [Bing (1983)]. Every connected knotting of an orientable surface $\Sigma \subseteq S^3$ is the boundary of a 3-ball with (some, possibly none) holes and (some, possibly none) handles *on each side*. The special case, that a smooth knotting of $S^1 \times S^1 \subseteq S^3$ bounds a solid torus on one side, is often called **Alexander's Theorem** [Alexander (1924b)]. ∎

One of the most important theorems on higher dimensional knotting, [Zeeman (1963)] [Zeeman (1960b)] and [Stallings (1963)], is that knottings of S^n in S^m with codimension 3 or more are trivial. The original proofs were in for the piecewise linear and locally flat case. Powerful as this result is, it only deals with knottings of a single sphere. Knottings of a pair of spheres can occur in higher codimension, for example the standard knotting of S^2 and S^2 in $S^5 = S^2 * S^2$ has codimension 3, see Definition 1.38.

An interesting short paper that shows all 2-spheres are trivially knotted in R^5 is [Zeeman (1960c)]; for 3-spheres in R^6, see [Zeeman (1962b)].

5.1 Why Circles Can Be Knotted in 3-Space but Not in 4-Space

The basic idea that "a circle cannot be knotted in four dimensions" was known very early in the mathematical study of knots (Felix Klein). See [Przytycki (2012)] for some of this history. A proof in the locally flat case is found in [Gluck (1963)].

The following example is a higher dimensional version of Example 5.1 and is the kernel of a demonstration that circles unknot in R^4.

Example 5.2. The simplest non-trivial 1-knotting in R^3 is the Hopf linking $H \subseteq R^3$, Figure 4.63, see Example 4.5. It is the union of circles K_0 and K_1 where K_0 is the circle in the xy-plane $\{x^2 + y^2 = 1 \text{ and } z = 0\}$ while K_1 is the circle in the xz-plane with equations $\{(y-1)^2 + z^2 = 1 \text{ and } x = 0\}$. So in the yz-plane given by $x = 0$ we see only all of K_1 and two points, A and B,of K_0.

Intuitively, non-trivial knotted circles exist is because one is not allowed to pass a strand of a knotting through itself. Similarly, the Hopf link is non-trivial is because we do not allow ourselves to pass one circle through the other. If we can show the Hopf link is trivial in R^4 (isotopic in R^4 to a split trivial link), then it is plausible that all 1-knottings are trivial in R^4.

View the Hopf link as a subset of R^4 using $H \subseteq R^3 \subseteq R^4$. Consider the coordinate hyperplane H_{yzw} given by $x = 0$. Clearly $H_{yzw} \cap H = K_1 \cup \{A\} \cup \{B\}$, Figure 5.7. Consider the hemisphere $\mathbb{H}$ in H_{yzw} given by $\{(y-1)^2 + z^2 + w^2 = 1, 0 \leq w \text{ and } x = 0\}$, Figure 5.8. This figure shows all of K_1 since all points have $x = 0$, however only two points of K_0 can be seen—A and B. We have $\partial H = K_1$.

There is an isotopy of K_1 in H_{yzw} from being the equator of $\partial\mathbb{H}$ to being a small circle K_1' tangent to $\partial\mathbb{H}$ which "shrinks K_1 within $\mathbb{H}$". Figure 5.8 shows a discrete track of this isotopy and Figure 5.10 it's projection. That the projection $(K_1')_* \subset R^3$ is a small elliptical curve isotopic to K_1' by a straight line isotopy.

Let $H' = K_0 \cup (K_1')_*$. Clearly H' is a split trivial knotting and we have shown it is isotopic in R^4 to H. ∎3

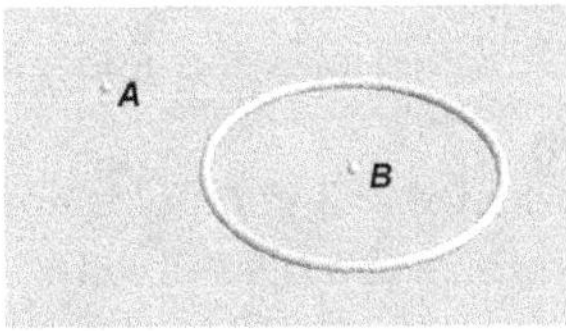

Fig. 5.7 Points of the Hopf linking that lie in the yzw-hyperplane.

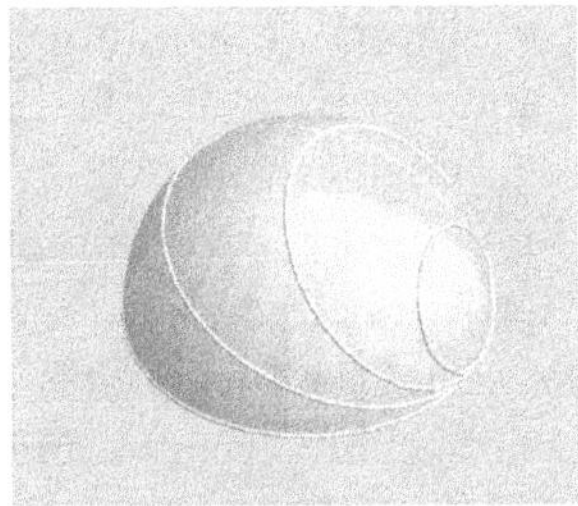

Fig. 5.8 A circle shrinks within the hemisphere $\mathbb{H}$ in yzw-hyperplane which bounds K_1.

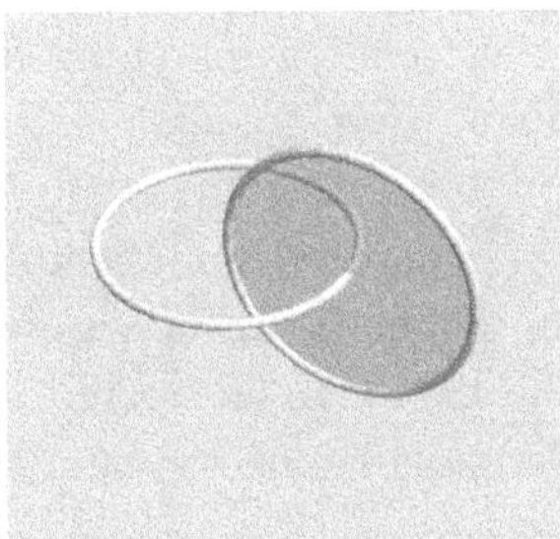

Fig. 5.9 The projection into xyz-space of hemisphere of Figure 5.8.

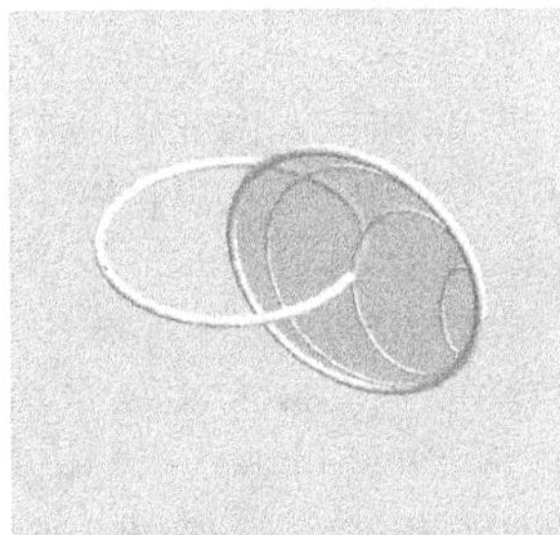

Fig. 5.10 The projection into xyz-space of shrinking circles of Figure 5.8.

5.2 Some Basic 2-Manifolds

Later, in Sections 5.3 and 8.1, we will discuss structures for smooth n-dimensional manifolds. In advance of this we will have need of a few important basic examples: 2-manifolds and n-dimensional handlebodies.

Topological classification of 2-manifolds is addressed in many standard topology texts, for example [Seifert and Threlfall (1980)], [Moise (1977)] and [Hirsch (1976)]. Some other good presentations are [Francis and Weeks (1999)], [Gallier and Xu (2013)] and [Champanerkar *et al.* (2000)].

Closed 2-manifolds fall into two infinite classes: orientable and non-orientable. Orientability is discussed in Section 2.5. The 2-sphere, the torus as well as the many-holed torii Figures 2.44, 5.11, 5.12, and 5.13 are all orientable 2-manifolds. Closed orientable 2-manifolds all embed in R^3. In general, if a closed n-manifold embeds in R^{n+1} it must be orientable. The closed *non-orientable* 2-manifolds do not embed in R^3, so they are less familiar objects. Two simple non-orientable 2-manifolds are the projective plane and the Klein bottle, to be defined and examined in Section 10.9.

The classification of 2-manifolds is frequently done by showing that any 2-manifold can be obtained by taking a $2n$-sided polygon and gluing together pairs of edges as in Section 2.6.

The usual rigorous proofs that closed non-orientable 2-manifolds are not subsets of R^3 involve algebraic topology, for example [Spanier (1966)]. The idea of the proof is to verify that any *closed* surface $M^2 \subseteq R^3$ must separate R^3 into an "inside" and an "outside". This allows us to define two distinct sides of M^2. But if M^2 were non-orientable, it would contain a Möbius band and, this subset of M^2 would inherit this inside/outside distinction. However, famously, the Möbius band is one-sided—so in fact there cannot be a Möbius band contained in M^2. Interested readers searching for a more elementary proof that the projective plane does not embed in R^3 might want to look at [Maehara (1993)].

It is useful to express 2-manifolds in terms of connected sum, Definition 2.20. Any orientable closed manifold except a 2-sphere can be expressed as a connected sum of N torii. Figures 2.43 and 2.44 give one example. A *non-orientable* closed manifold can be expressed as a connected sum of N projective planes. This leads to the definition:

Definition 5.1. A 2-sphere has **genus 0**. Otherwise, for any closed 2-manifold the number N, above, is called the **genus** of the 2-manifold. ∎

So, the torus is the orientable 2-manifold of genus 1; the orientable 2-manifold of genus 2 is as in Figure 2.44. The projective plane is the non-orientable 2-manifold of genus 1; the Klein bottle has genus 2. The genus is clear in Figures 5.11 and 5.12. Some examples such as Figure 5.13 may take some explanation, see Example 5.9 and Figure 5.27.

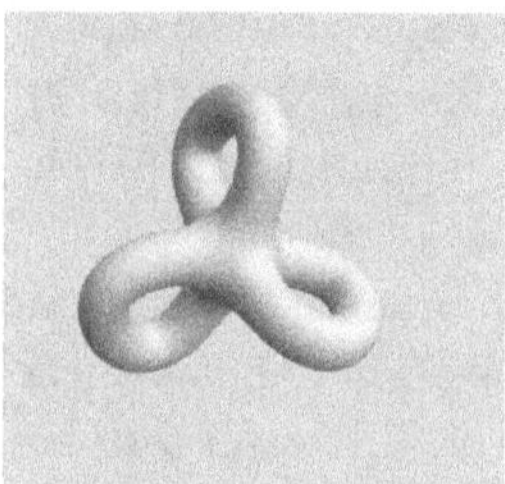

Fig. 5.11 A surface of genus three.

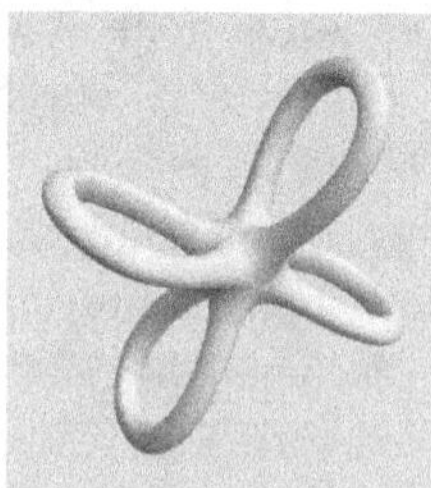

Fig. 5.12 A surface of genus four.

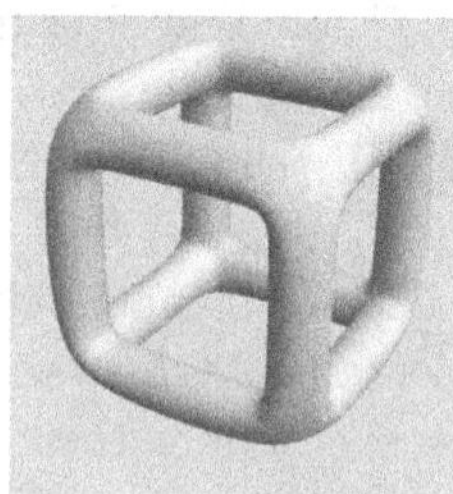

Fig. 5.13 A surface of genus five.

A compact 2-dimensional manifold with k boundary components is diffeomorphic to a closed 2-manifold with k punctures, as defined in Example 2.5. A 2-disk is a once-punctured S^2 and an annulus is a twice-punctured S^2. A once-punctured projective plane is a Möbius band. A thrice-punctured S^2 is often used in geometric topology and is called a **pair of pants** because of embeddings, as shown in Figure 5.14.

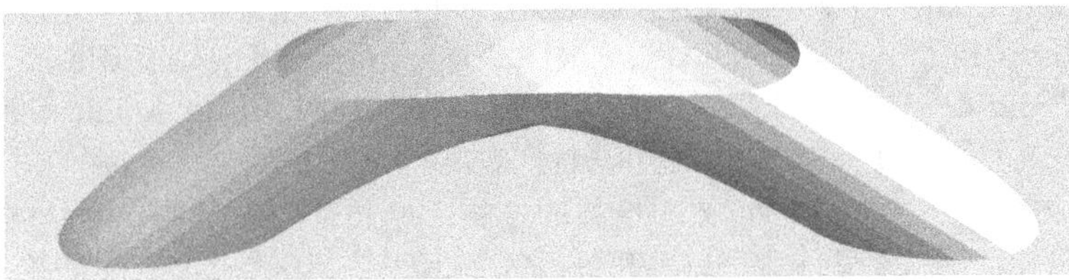

Fig. 5.14 A sphere with three punctures as a "pair of pants".

Some very simple examples of split 2-knottings Definition 3.6 we will soon encounter are: Figure 6.24 a split sum of two spheres, Figure 7.44 a trivial split knotting of a sphere and a torus and Figure 6.160 a projection of trivial split knotting of a torus and a projective plane.

5.3 How to Handle Manifolds

Handles are basic building blocks for manifolds and many aspects of higher-dimensional knottings are can be expressed in terms of handles. Although any n-manifold M^m can be expressed as a union of two basic manifold types each diffeomorphic to R^n or R^n_+, it is often preferable to view M^m as a union of manifolds diffeomorphic to D^n. Such disks will be called **handles**. Existence of such handles can be derived from Morse functions, so it follows that all smooth manifolds can be described using handles, Remark 8.9. In fact, any manifold can be expressed as a handlebody structure in many ways.

For 2-manifolds, there are three types of handles: a 0-handle, a 1-handle and a 2-handle. These correspond to the three ways we can write D^2 as a product of disks: $D^k \times D^{2-k}$ for $k = 0, 1$, or 2. More generally, for an n-manifold Σ there are $(n+1)$-types of handles. In this Part of the book, we will primarily use handles for dimensions 2 and 3. Refer to Section 2.6 for gluing definitions and notations. If M^m is compact, we will only need a finite number of such n-disks.

Definition 5.2. Suppose Σ is a smooth n-manifold and $\Sigma' = \Sigma \cup h$ where h is diffeomorphic to D^n. Write D^n as $D^k \times D^{n-k}$. A **handle gluing map** is a smooth embedding $g\colon \partial D^k \times D^{n-k} \to \partial\Sigma$.

If Σ' is obtained by gluing Σ and h via g (that is $\Sigma' \approx \Sigma \cup_g$) we say Σ' is Σ **plus a** k**-handle** h, denoted $\Sigma \underset{g}{+} h$ or often just as $\Sigma + h$.

The set $A = g(\partial D^k \times D^{n-k}) \subseteq \partial\Sigma$ is called the **attaching set of the handle** h, the set $g(\partial D^k \times \{\mathbb{O}\}) \subseteq \partial\Sigma$ called the **attaching sphere of the handle** h, and g is called the **attaching map of the handle** h. The set $B = g(D^k \times \partial D^{n-k})$ is called the b**-set of the handle** h, and the set $g(\{\mathbb{O}\} \times \partial D^{n-k})$ called the b**-sphere of the handle.**

If we have a j-handle, we refer to j as the **type of the handle.** We optionally use parenthetical superscripts so that $\Sigma + h^{(j)}$ is the notation for Σ plus some j-handle. ∎

The b-set and the b-sphere are subsets of $\partial\Sigma'$, for instance Example 5.8. Focusing on the $(n-1)$-manifolds involved, we obtain $\partial\Sigma'$ by removing the a-set of h and gluing in the b-set of h. In this case one says that one has done **surgery on the** a**-sphere of** h.

For 2-manifolds we have three cases:

- k=0 A is the empty set. In this case, we say h is a **0-handle** and $\Sigma + h$ is just the disjoint union of Σ and a 2-disk h. We allow $\Sigma = \emptyset$. In which case, $\Sigma + h = h$ is just a 2-disk.
- k=1 A is a disjoint union of two closed sub-arcs of $\partial\Sigma$: $A \approx S^0 \times D^1$. The attaching sphere is a pair of points of $\partial\Sigma$, $S^0 \times D^0$ In this case, we say h is a **1-handle.**
- k=2 A is a topological circle: $A \approx S^1 \times D^0$. In this case we say h is a **2-handle.** Here, A is the entire boundary of the disk h, the attaching sphere is the entire attaching set, and g must glue this to an entire component of $\partial\Sigma$.

Example 5.3. The 2-sphere is the union of two smooth 2-disks—hemispheres—glued together along a common equatorial boundary. In terms of handles, we can view this as "a 0-handle plus a 2-handle" and can write $S^2 = h_1^{(0)} + h_1^{(2)}$. ∎

Example 5.4. A different expression of the 2-sphere uses four handles, Figure 5.15: $S^2 = h_1^{(0)} + h_1^{(1)} + h_2^{(1)} + h_1^{(2)}$. We show the dark gray 0-handle ($h_1^{(0)}$, in Figure 5.16), the medium gray 1-handle and the two light gray 2-handles. A 1-handle $h_1^{(1)}$, is added to $h_1^{(0)}$ giving $h_1^{(0)} + h_1^{(1)}$, topologically a cylinder, Figure 5.17. Figure 5.17 suggests "a basket with a handle", where the 0-handle is the "basket", and the 1-handle is the "handle". This accounts, in part, for the usage of the word "handle", however this description does not fit other types of handles.

Another way to think of our 1-handle is: attach an arc (i.e. image of D^1) to the 0-handle, Figure 5.18, then "thicken it up to a two-dimensional disk". The result is shown in Figure 5.20 with attaching set shown Figure 5.19. The boundary of the surface $h_1^{(0)} + h_1^{(1)}$ consists of two disjoint topological circles. We attach two topological disks, one for each circle. These disks are 2-handles denoted $h_1^{(2)}$ and $h_2^{(2)}$. We now have

$$S^2 = h_1^{(0)} + h_1^{(1)} + h_1^{(2)} + h_2^{(2)}.$$

A natural way in which this handlebody structure might be generated is found in Example 8.9.

More generally, one can view the addition of an n-dimensional k-handle to an n-manifold M as a two-step process: first, gluing a k-disk to ∂M, then thickening this disk up to n dimensions. ∎

Fig. 5.15 S^2 expressed using four handles.

Fig. 5.16 First a 0-handle.

Fig. 5.17 A 1-handle has been added to the 0-handle.

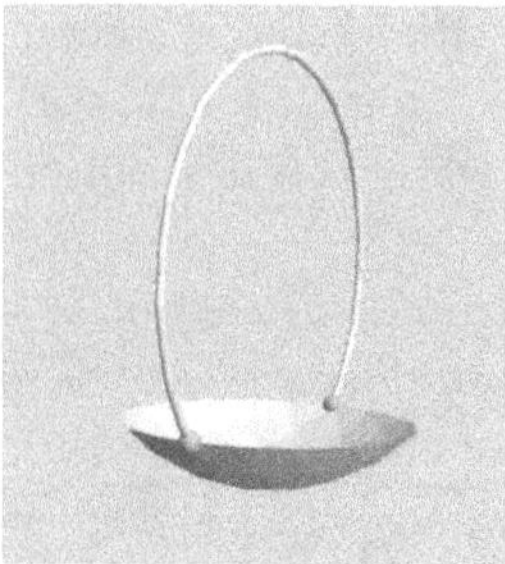

Fig. 5.18 Adding an arc to the boundary of a 0-handle.

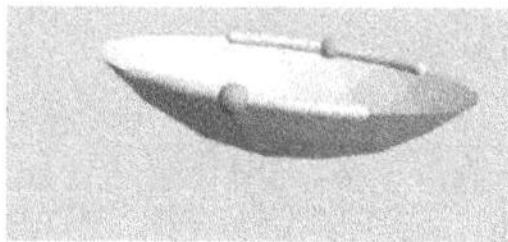

Fig. 5.19 The 1-handle meets the zero handle in two disjoint sub-arcs of the boundary of the 0-handle.

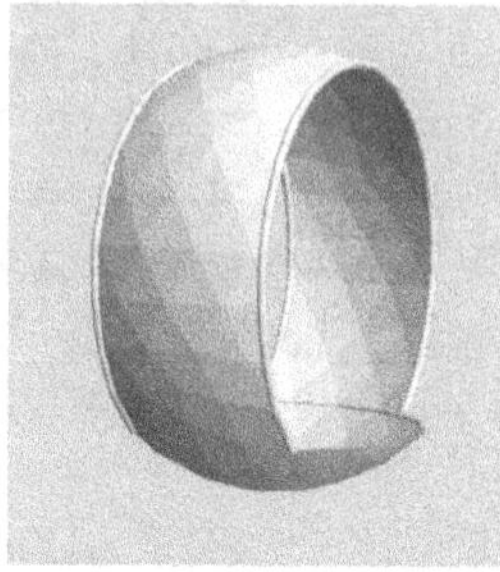

Fig. 5.20 The 1-handle as thickened arc.

Example 5.5. The 2-torus T can be viewed $T = h_0 + h_1 + h_2 + h_3$. A 0-handle h_0 is shown dark gray in Figure 5.21. Two 1-handles h_1 and h_2, medium gray, are added to the 0-handle as shown in Figures 5.22 and 5.23. The remainder of the torus is a 2-disk and will be our 2-handle h_3.

Another expression of the torus is shown in Figures 5.24–5.26. Figure 5.24 shows two 0-handles in dark-grey—one is on the "inside of the hole". We then add four 1-handles—two are attached to only one of the 0-handles, two are attached to both 0-handles. Figure 5.26 shows these handles as subsets of the torus, and so the rest of the torus consists of two topological disks—these correspond to two 2-handles. ∎

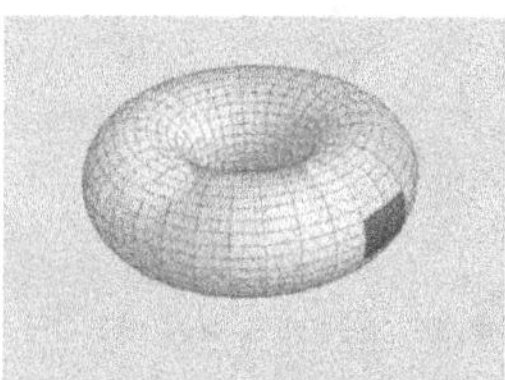

Fig. 5.21 We begin with a topological disk (dark) in a torus, call it a 0-cell.

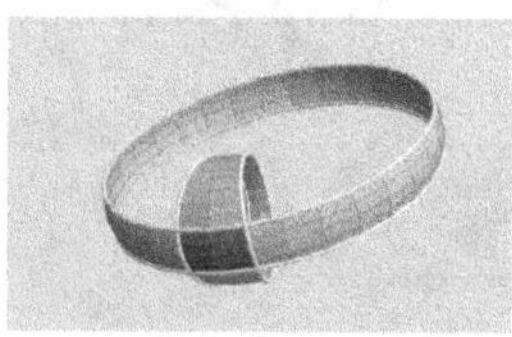

Fig. 5.22 Two other sub-disks will be 1-handles.

Fig. 5.23 The disk that remains (light gray) is a 2-handle.

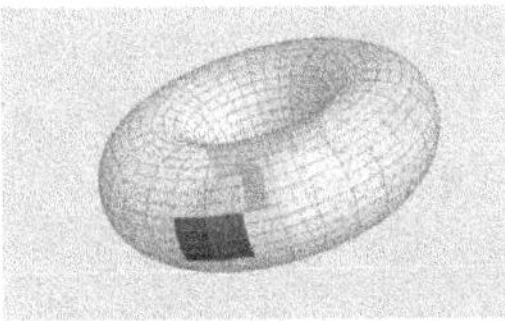

Fig. 5.24 An alternate structure begins with two 0-handles.

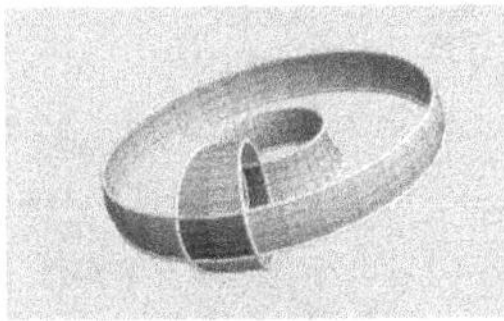

Fig. 5.25 Then we add four 1-handles.

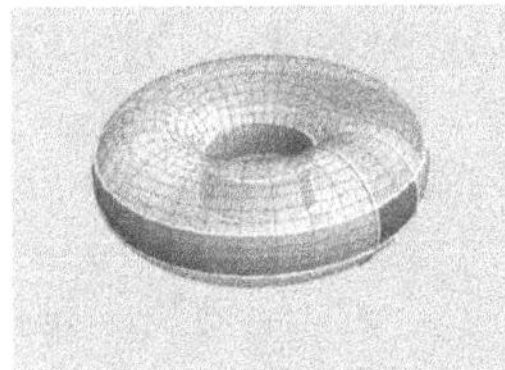

Fig. 5.26 What remains are two 2-handles.

Definition 5.3. A **handle structure**, sometimes called **handle decomposition**) for M is a collection of N handles that fill up the manifold so that $M = (\cdots(((h_1 + h_2) + h_3) + h_4) + \cdots + h_N)$ which we write as $M = h_1 + h_2 + h_3 + h_4 \cdots + h_N$. ▮

Using standard techniques (details omitted) one can show that we may assume without loss of generality that if $M = h_1 + h_2 + h_3 + h_4 + \cdots + h_N$ that the handles are added in order of increasing type (that is if h_i and h_j are handles with $i \leq j$ then the type of h_i is less than or equal to the type of h_j) and that handles of the same type are disjoint sub-disks of M. So, for such a handle structure $M = h_1 + \cdots + h_N$ we change the order of any two handles *of same type* and get another handle structure for M.

If we reverse the order of all handles, we can interpret this as a handle structure called the **dual handle structure** $M = h_N + \cdots + h_1$ where, if h_j was a handle of type k in the original order, this same sub-disk is a handle of type $(n - k)$ in the new order. (If a handle structure is obtained from a Morse function f, the dual handle structure is similarly obtained from $-f$.) A good example is Example 5.12. ▮

Example 5.6. For the dual handle structure of the torus, Example 5.5, we begin by viewing the sub-disk h_3 used as the 2-handle as a 0-handle in the dual structure. Each of the sub-disks h_1 and h_2 meets h_3 in two disjoint sub-arcs of their boundary, and thus can be considered 1-handles added to h_3. Finally, the boundary of the manifold $h_3 + h_2 + h_1$ is a single circle, and thus h_0 can be considered to be a 2-handle added to $h_3 + h_2 + h_1$. ▮

Notation 5.1. In considering a handle structure and its dual for an n-manifold, the same n-ball is considered as a handle in two ways. The order of handle addition is reversed, *and* the type is changed from j to $n - j$. To notationally express the type change, we introduce the following. If (j) represents the type of a handle for an n-manifold, then $[j]$ represents the dual type $(n-j)$. This way, the dual of a handle $H^{(j)}$ is $H^{[j]}$. Suppose $M = H_1 + H_2 + H_3$ where H_1 is a 0-handle, H_2 is a 1-handle and H_3 is a 2-handle. Then, the dual handle structure is $M = H_3 + H_2 + H_1$.

Using the alternate handle-type-indicated-by-superscript notation $M = H_1^{(0)} + H_2^{(1)} + H_3^{(2)}$ our notation allows us to write this $M = H_3^{[0]} + H_2^{[1]} + H_1^{[2]}$ which is also in a handle-type-indicated-by-superscript notation. So *as subsets* $H_1^{(0)} = H_1^{[2]}$. As handles $H_0^{(0)}$ is a 0-handle and $H_1^{[2]}$ is a 2-handle. We will see this notation in use in Examples 5.12 and 5.13. ∎

Definition 5.4. Consider adding a 1-handle $\Sigma' = \Sigma + h^{(1)}$, where Σ is an orientable n-manifold. If Σ' is non-orientable, then we say $h^{(1)}$ is a **non-orientable handle**, otherwise an **orientable handle.** ∎

Example 5.7. There are exactly two 2-manifolds $M = h^{(0)} + h^{(1)}$: an annulus and the Möbius band, depending on whether the 1-handle is orientable or not. For $M = h^{(0)} + h_1^{(1)} + h_2^{(1)}$ we have $C = \partial h^{(0)}$ is a smooth circle. Let $S_1 \subseteq C$ be the attaching 0-sphere of $h_1^{(1)}$ and $S_2 \subseteq C$ be the attaching 0-sphere of $h_2^{(1)}$. We say these 0-spheres are **unlinked** in C, if there is a sub-arc in C between the two points of S_1 that does not contain points of S_2, otherwise they are **linked.** If both handles are orientable with S_1 and S_2 unlinked, M is a pair of pants (i.e. a punctured annulus). If they are linked M is a punctured torus. If one of the handles is non-orientable, then if S_1 and S_2 are unlinked M is a punctured Möbius band, or if linked, a punctured Klein bottle. ∎

Example 5.8. The solid torus $T = S^1 \times D^2$ has a handle structure as a 0-handle plus an orientable 1-handle. Write S^1 in terms of semicircular arcs: $S^1 = H_+^1 + H_-^1$ with $H_+^1 \cap H_-^1 = S^0$. Our 0-handle is $H_+^1 \times D^2$ and the 1-handle is $H_-^1 \times D^2$, whose a-set is the pair of 2-disks $S^0 \times D^2$, the b-set $H_-^1 \times \partial D^2 \subseteq \partial T$ and the b-sphere and be taken to be the meridian $\{S_0\} \times \partial D^2$ where $S_0 = (0, -1) \in S^1 \subseteq R^2$. ∎

Handlebodies are manifolds with simple handle structures:

Definition 5.5. A **handlebody** is an orientable connected n-manifold M with $3 \leq n$ with handle structure that has only p 0-handles and q 1-handles: $M = h_1^{(0)} + \cdots + h_p^{(0)} + h_{p+1}^{(1)} + \cdots + h_{p+q}^{(1)}$. ∎

In some special cases, one can simplify a handle structure by a technique known as **handle cancelation.** which allows us to remove certain pairs of handles (a j-handle paired with a $(j+1)$-handle). In this "cancelation", our pair of handles seems to "vanish", but the process is actually a merging of a handle pair as a ball followed by absorption of that ball into a boundary. A simple example of the process goes like this. Suppose we have a three-dimensional handle structure $M = h_1^{(0)} + h_2^{(0)} + h^{(1)}$. A 1-handle $h^{(1)}$ has as attaching map two disjoint 2-disks, A_1 and A_2. Now further suppose $A_1 \subseteq \partial h_1^{(0)}$ and $A_2 \subseteq \partial h_2^{(0)}$. Then we can "cancel" $h_1^{(0)}$ and $h_1^{(1)}$ and be left with only $h_2^{(0)}$ in the following sense. The subset $B \subseteq M$ corresponding to the union of the two handles, $h_1^{(0)}$ and $h_1^{(1)}$, consists of two 3-balls intersecting only in their boundaries in the 2-disk A_1. So B is a 3-ball. But B intersects $h_2^{(0)}$ only in their common boundary in 2-ball A_2. Using Proposition 2.11 (2) where we view $B \approx A_2 \times I$, we can view $h_2^{(0)} + B \approx h_2^{(0)}$, and so we have $h_1^{(0)} + h_2^{(0)} + h^{(1)} \approx h_2^{(0)}$.

Using such cancellation many times, if needed, we can assume any handlebody has only one 0-handle and then identify a handlebody entirely by the number of 1-handles which is called the **genus of the handlebody**.

Prop 5.4. Any two n-dimensional handlebodies, $n > 2$, with one zero handle and the same number of 1-handles are diffeomorphic. This number is the same as the genus of the boundary of the handlebody. ∎

Proposition 5.4 is not true for other manifolds: a pair of pants and a punctured torus both can be expressed as "a single 0-handle plus two 1-handles".

Example 5.9. View Figure 5.27 as the 3-dimensional handlebody formed by thickening the edges of a cube. We have eight 0-handles (one for each vertex) and twelve 1-handles (one for each edge). Canceling with some of these, we merge them to form a dark 3-ball which is a new 0-handle, and five light colored 1-handles. So, we identify this as a genus 5 handlebody. ∎

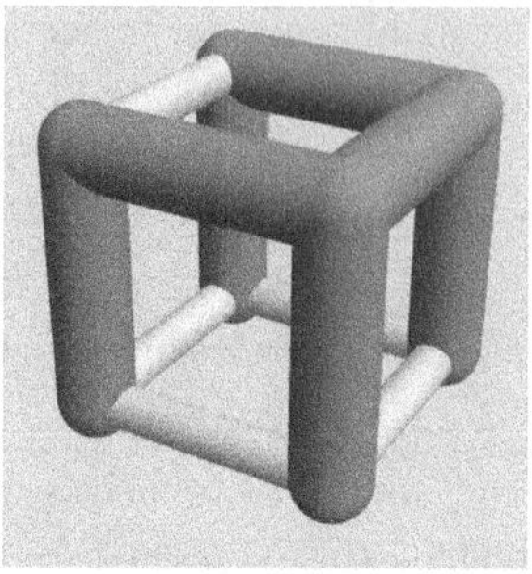

Fig. 5.27 Figure 5.13 shown as ball with five handles.

Handle structures work well with cartesian products with disks:

Remark 5.1. For an n-manifold with a k-handle $N^n = M^n + h^{(k)}$ we can write $N^n \times D^j = M^n \times D^j + H^{(k)}$ as follows. If the attaching map of $h^{(k)}$ is $g\colon D^k \times D^{n-k} \to \partial M$ we can write $(M^n + h^{(k)}) \times D^j$ as $(M^n \times D^j) + (h^{(k)} \times D^j)$ where $H^{(k)}$ the is the k-handle $h^{(k)} \times D^j$ with attaching map $g \times Id_{D^j} : D^k \times D^{(n-k)+j} \to \partial(M^n \times D^j)$. ▮

Example 5.10. If F is a connected 2-manifold with ∂F non-empty and connected, then $F \times D^1$ is a 3-dimensional handlebody since we can write F as a single 0-handle plus some 1-handles. If there is only one 1-handle then F is either an annulus A or a Möbius band M.

View A as a handlebody $h^{(0)} + h^{(1)}$ with $h^{(1)}$ orientable. The solid torus $A \times D^1$ is $(h^{(0)} \times D^1) + (h^{(1)} \times D^1)$, a 3-dimensional handlebody with one 0-handle and one orientable 1-handle.

View M as a handlebody $h^{(0)} + h^{(1)}$ with $h^{(1)}$ *non-orientable* The product $h^{(0)} \times D^1 + h^{(1)} \times D^1$, a 3-dimensional handlebody with one 0-handle and one *non-orientable* 1-handle called a **solid Klein bottle.**

An example of a handle structure, with many handles, for S^3 is (incidently) described in Example 5.12. ▮

Remark 5.2. The exterior of a knotting is generally a complicated 3-manifold. However, it is also delicate in the sense that it becomes a 3-ball if one adds a certain 2-handle to the boundary. This elementary fact will be important in our discussion of Proposition 10.7.

To see this, view the knot exterior as a 3-ball with a knotted hole, Remark 4.1 and view the 2-handle as a thin "cylindrical cork" plugging up one end of the hole. Topologically this is the same as a thick "cylindrical cork". If we make the cork so thick that it fills in the entire hole, we are left with D^3.

That is, we have 1-knotting $K \subseteq S^3$ with tubular neighborhood $T \approx S^1 \times D^2$. Write S^1 as two glued semicircular arcs: $S^1 = H^1_+ + H^1_-$. Then T is expressed as a gluing of two topological 3-balls: $T \approx (H^1_+ \times D^2) + (H^1_- \times D^2)$. Of course $Ext(K) + T \approx S^3$. View $H^1_+ \times D^2$ as a 2-handle H added to $Ext(K)$ and the complement of $Ext(K) + H$ is just the smooth 3-ball corresponding to $H^1_- \times D^2$. By the Schoenflies Theorem, Proposition 5.5, the complement of a smooth 3-ball in S^3 is also a 3-ball, so $Ext(K) + H$ is a 3-ball. ∎

5.4 Fibering a Trefoil Knotting

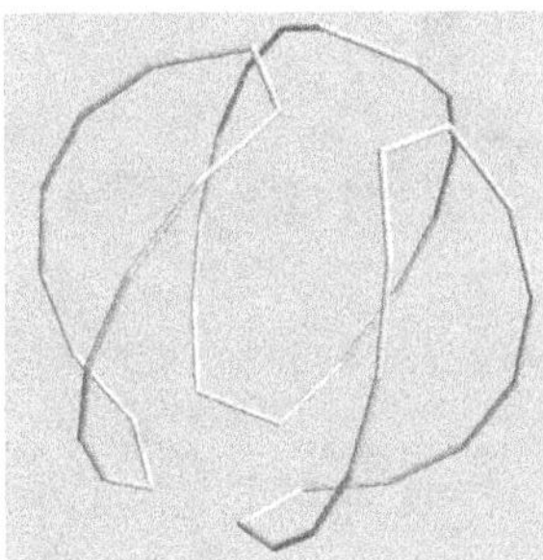

Fig. 5.28 A split trivial link.

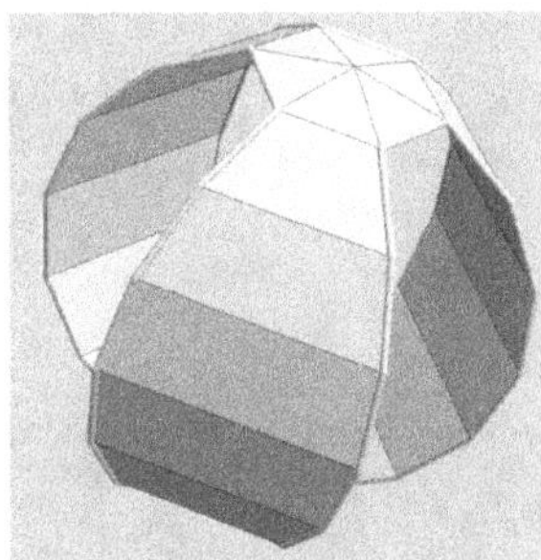

Fig. 5.29 Siefert surface for knotting of Figure 5.28.

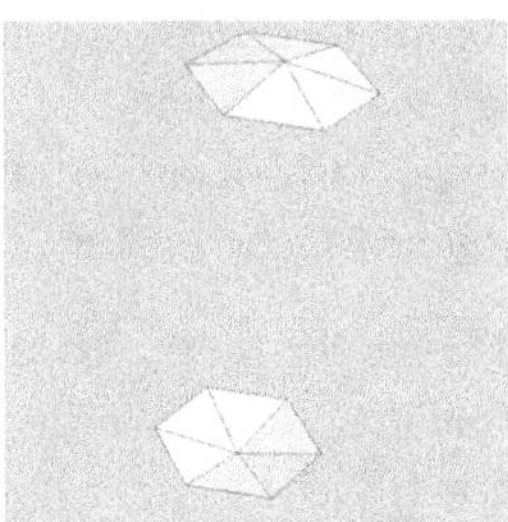

Fig. 5.30 Two 0-handles.

Fig. 5.31 Three disks for 1-handles.

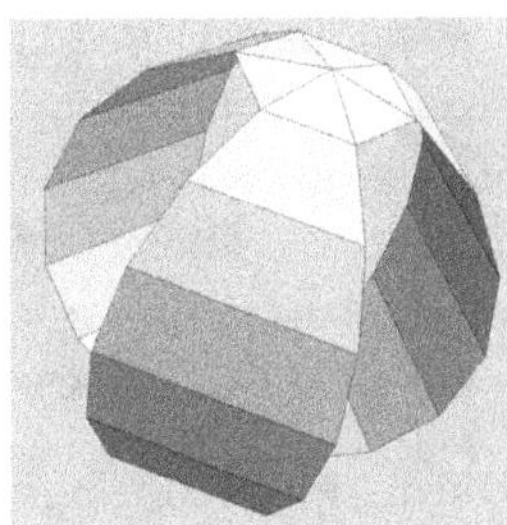

Fig. 5.32 Disks of Figure 5.31 as 1-handles $h_3^{(1)}, h_4^{(1)}, h_5^{(1)}$ added to 0-handles of Figure 5.30.

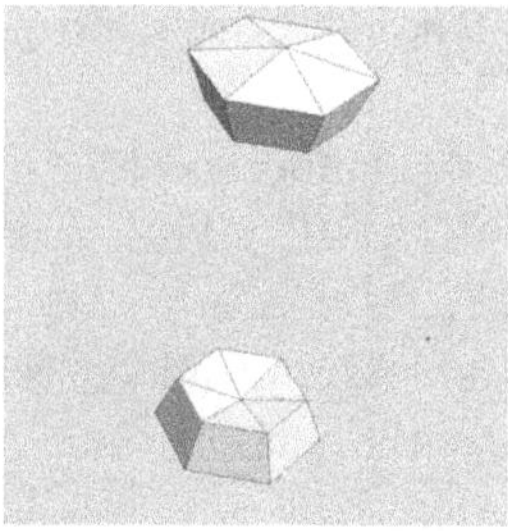

Fig. 5.33 The two 3-disks $H_1^{(0)}$ and $H_2^{(0)}$.

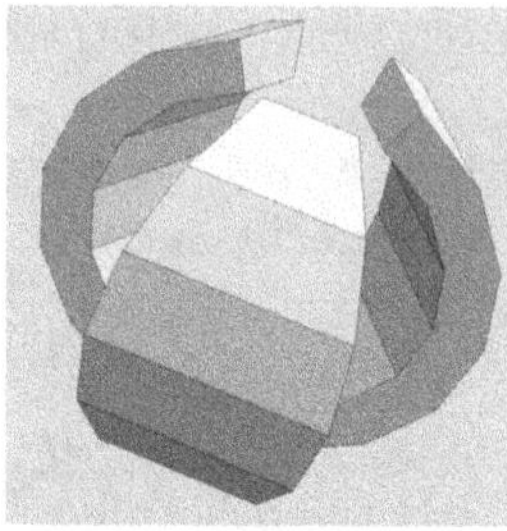

Fig. 5.34 The 3-disks for $H_3^{(1)}, H_4^{(1)}, H_5^{(1)}$.

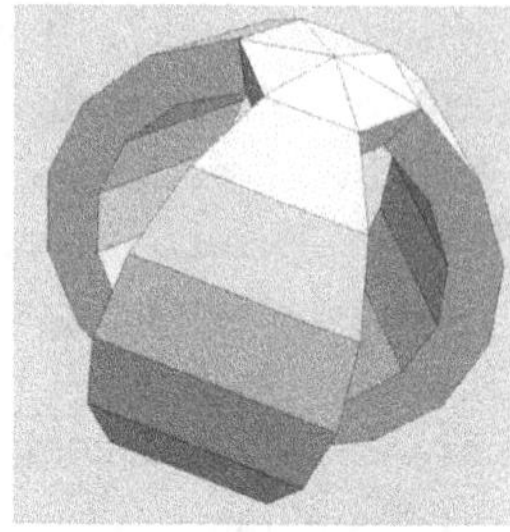

Fig. 5.35 $H_1^{(0)} + H_2^{(0)} + H_3^{(1)} + H_4^{(1)} + H_5^{(1)}$.

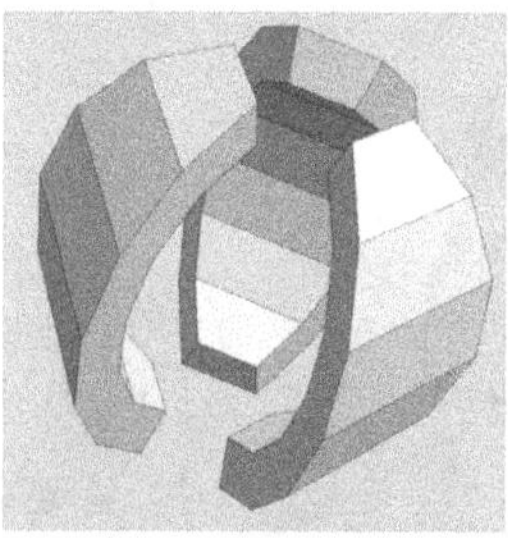

Fig. 5.36 3-disks for $H_6^{(2)}, H_7^{(2)}, H_8^{(2)}$.

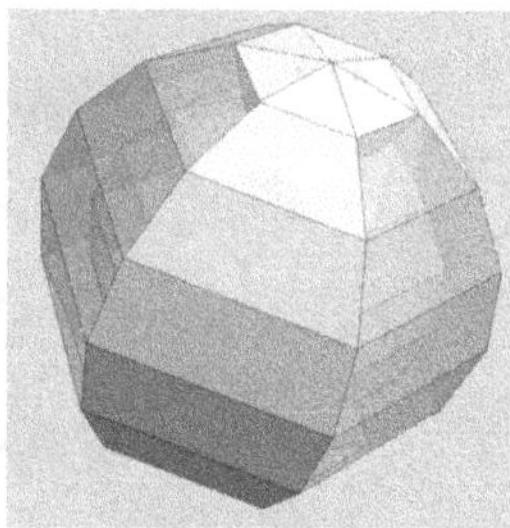

Fig. 5.37 N with handles $H_6^{(2)}, H_7^{(2)}, H_8^{(2)}$ transparent.

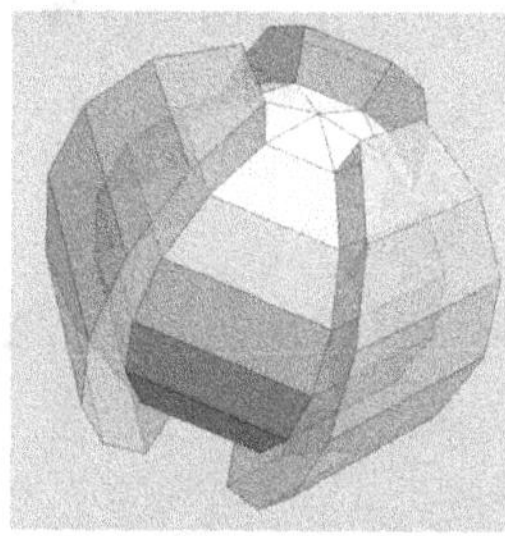

Fig. 5.38 Handles $H_6^{(2)}$, $H_7^{(2)}, H_8^{(2)}$ and $H_9^{(2)}$.

It is not difficult to understand a Siefert Σ surface for a 1-knotting k in R^3, but a fibered knotting, fills S^3 with copies of Σ disjoint except along k. How can we visualize this? Many knottings with very nice Siefert surfaces are not fibered—what goes wrong? Any Siefert surface Σ has a trivial product neighborhood $N \approx \Sigma \times D^1$ which might be considered as a start of a fibration. What could go wrong if we try to "complete this to get a fibration"? The next examples in this section address these questions.

A fibering of a knotting gives us a way of expressing S^3 as a union of handlebodies, Remark 4.10. To discuss this, the following will be useful:

Definition 5.6. Suppose F is a smooth 2-manifold with $\partial F \neq \emptyset$, $M \approx (F \times D^1)$ and E is a smooth 1-dimensional submanifold $E \subseteq \partial M$. We say E **is equatorial in** M, if $(F \times D^1, E) \approx (F \times D^1, F \times \{0\})$. ∎

Example 5.11. A $(2, 2)$-torus link is equatorial as can be seen by taking the tubular neighborhood of an annulus in R^3 with a single full twist.

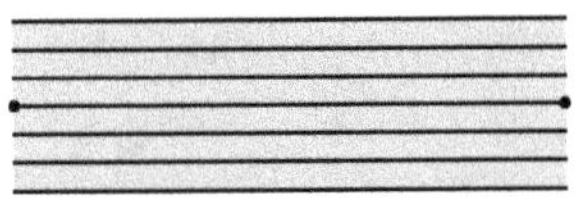

Fig. 5.39 A product $D^1 \times \Sigma^1$, Σ^1 an interval.

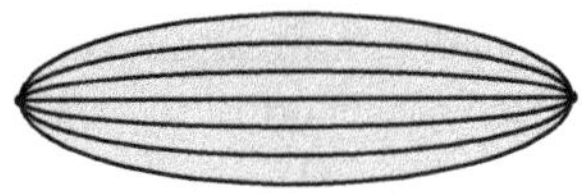

Fig. 5.40 An equatorial smoothing of Figure 5.39.

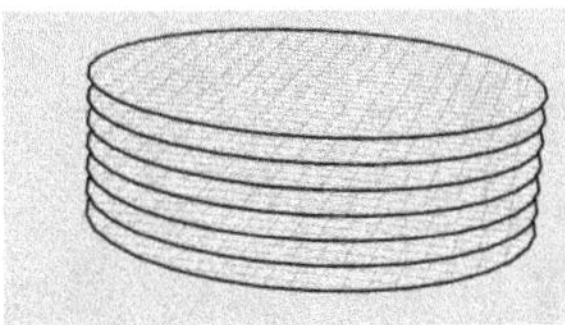

Fig. 5.41 A product $D^1 \times \Sigma^1$, Σ^1 a 2-disk.

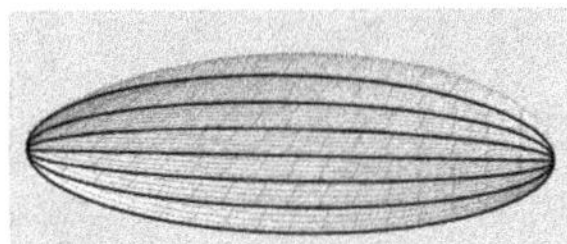

Fig. 5.42 An equatorial smoothing of Figure 5.41 cut in half to reveal details.

Two disjoint meridians are not an equatorial submanifold (it takes a little algebraic topology to verify this). ∎

Remark 5.3. Not all Seifert surfaces Σ for knotting K are candidates to be fibers for a fibering of a knotting, The closed complement X of a tubular neighborhood N of Σ must be a 3-dimensional handlebody, Remark 4.10. *Furthermore, we require that K is an equatorial subset of X and of N.* That is we must have a diffeomorphism of pairs: $(N, K) \approx (X, K)$. ∎

Example 5.12. Consider a split trivial knotting of three circles $K = \partial\Sigma$, Figure 5.28 with Siefert surface Σ, Figure 5.29. We will see that Σ is *not* a fiber of a fibration of K.

We express Σ as a handlebody with two 0-handles Figure 5.30 and three 1-handles Figure 5.31:

$$\Sigma = h_1^{(0)} + h_2^{(0)} + h_3^{(1)} + h_4^{(1)} + h_5^{(1)}.$$

Using Remark 5.1 there is a corresponding thickened handle structure for a tubular neighborhood $N \approx \Sigma \times D^1$ of Σ (Figure 5.32):

$$N = H_1^{(0)} + H_2^{(0)} + H_3^{(1)} + H_4^{(1)} + H_5^{(1)}.$$

Figure 5.33 shows two 3-balls which are 0-handles for N. In Figure 5.34 we show the 3-balls for 1-handles, finally assembled to make N, Figure 5.35.

We now add handles until we fill up S^3. Figure 5.36 shows three 3-balls that serve as 2-handles added to N since each intersects N in an annulus. Add these handles to N and we obtain a submanifold $\mathbb{T} \approx S^2 \times D^1$, Figure 5.37, which we view as a tubular neighborhood of $S^2 \subseteq S^3$:

$$\mathbb{T} = H_1^{(0)} + H_2^{(0)} + H_3^{(1)} + H_4^{(1)} + H_5^{(1)} + H_6^{(2)} + H_7^{(2)} + H_8^{(2)}.$$

The closed complement of $\mathbb{T} \subset S^3$ consists of two 3-balls that serve as 3-handles attached to $\mathbb{T}$—we denote these $H_9^{(3)}$ and $H_{10}^{(3)}$. The three circles that form the attaching maps of the 2-handles happen to be equivalent to K. We will assume that $\mathbb{O}$ is a point of $H_9^{(3)}$ and $H_{10}^{(3)}$ contains the "point at infinity" of R^3, and is everything outside of $\mathbb{T}$. So, we have an expression (Figure 5.38):

$$S^3 = H_1^{(0)} + H_2^{(0)} + H_3^{(1)} + H_4^{(1)} + H_5^{(1)} + H_6^{(2)} + H_7^{(2)} + H_8^{(2)} + H_9^{(3)} + H_{10}^{(3)}.$$

We use this to express S^3 in terms of two handlebodies: the first consists of the first five handles, the second consists of the last five *considered as dual handles*:

$$\begin{aligned} S^3 &= \left(H_1^{(0)} + H_2^{(0)} + H_3^{(1)} + H_4^{(1)} + H_5^{(1)}\right) \cup \left(H_6^{(2)} + H_7^{(2)} + H_8^{(2)} + H_9^{(3)} + H_{10}^{(3)}\right) \\ &= \left(H_1^{(0)} + H_2^{(0)} + H_3^{(1)} + H_4^{(1)} + H_5^{(1)}\right) \cup \left(H_{10}^{[0]} + H_9^{[0]} + H_8^{[1]} + H_7^{[1]} + H_6^{[1]}\right) \end{aligned}$$

An equatorial smoothing of $N \times D^1$ with K is shown in the boundary looks like Figure 5.43. Those same circles, as seen in the complementary handlebody, is shown in Figure 5.44. They happen to be b-spheres of the 2-handles of N. These are clearly non-diffeomorphic pairs (although a proof involves algebraic topology), and so by Remark 5.3 Σ is not a fiber of a fibration of K. ▌

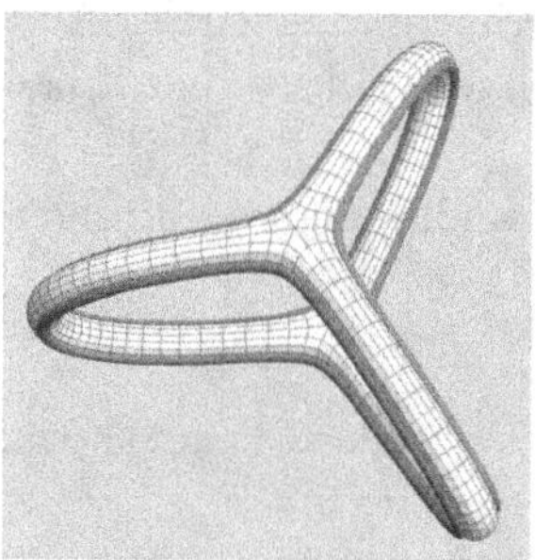

Fig. 5.43 Attaching circles for 2-handles Example 5.12.

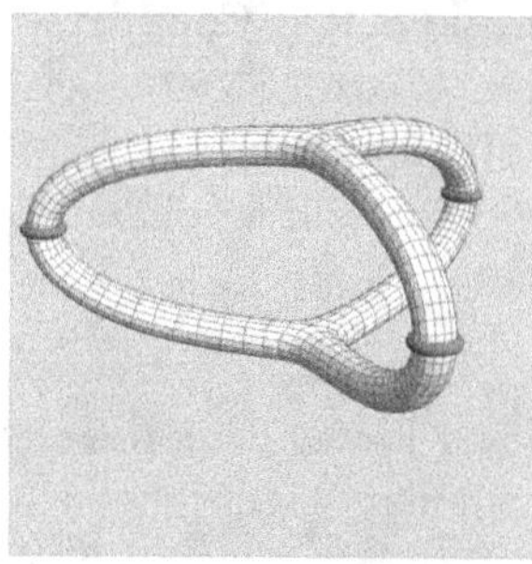

Fig. 5.44 The same attaching circles in the complementary handlebody.

Example 5.13. Figure 5.45 shows a Seifert surface Σ', a once punctured torus, viewed as two 0-handles plus three 1-handles, in Figure 5.45 for trefoil knotting $K' = \partial\Sigma'$. In this case, Σ' happens to also be a Seifert surface of a fibering. (In Example 4.8 we show this is indeed a trefoil knotting.) The

2-manifold Σ' has two 1-handles and three 1-handles shown as connecting strips with a half-twist.

Figure 5.47 shows $\Sigma' \times D^1$, a handle body with two 0-handles and three 1-handles. It is equivalent, as a subset of S^3, to the subset N shown in Figure 5.35. So, even though $\Sigma \not\approx \Sigma'$ we have

$$N = \Sigma' \times D^1 = \Sigma \times D^1 = H_1^{(0)} + H_2^{(0)} + H_3^{(1)} + H_4^{(1)} + H_5^{(1)}.$$

Therefore, the closed complement, X, of $\Sigma' \times D^1$ can be taken to be the handlebody $H_{10}^{[0]} + H_9^{[0]} + H_8^{[1]} + H_7^{[1]} + H_6^{[1]}$. The 3-balls A, B and C, Figure 5.49 are smoothed stretched versions of the 2-handles $H_6^{(2)}, H_7^{(2)}$, and $H_8^{(2)}$ of Figure 5.36. In Figure 5.48 we show a smoother graphic for $K \subset N$ as an equatorial manifold. Note how K goes twice along each 1-handle with a half-twist as it does. To verify we have a fibration we need to show that K corresponds to this same equatorial manifold of X, so that the pair (X, K) will look like Figure 5.48. So we need to show that K goes along each 1-handle of X with a half-twist. The number of times K goes along a handle can be measured as the intersection of K with a b-sphere of the 1-handles of N. The 1-handles of X are the duals of the 2-handles $H_6^{(2)}, H_7^{(2)}$, and $H_8^{(2)}$ and so the b-spheres of these 1-handles are the a-spheres of our 2-handles. In Figure 5.49 the 2-handle A meets N in an attaching annulus A' with a-sphere α but, due to occlusion, all that is visible is a portion of one boundary component β of A'. However, the number of points of intersection of $K \cap \beta$ is the same as the number of points of $K \cap \alpha$. Figure 5.50, shows two points of intersection x and y of $K \cap \alpha$. Continuing this analysis should convince the reader that $(X, K) \approx (N, K)$, thus verifying that the trefoil knotting is indeed fibered. ∎

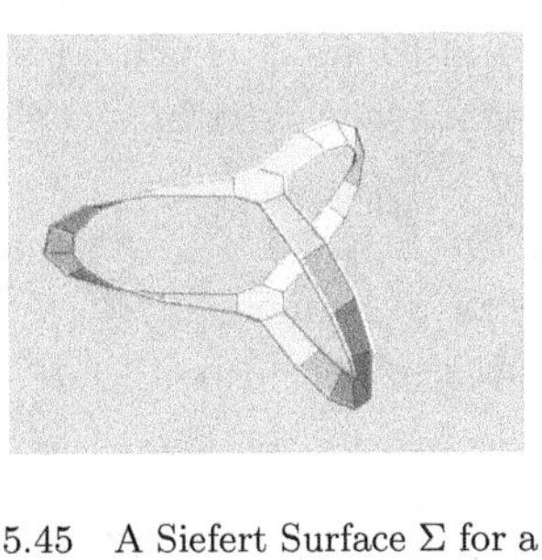

Fig. 5.45 A Siefert Surface Σ for a trefoil knotting.

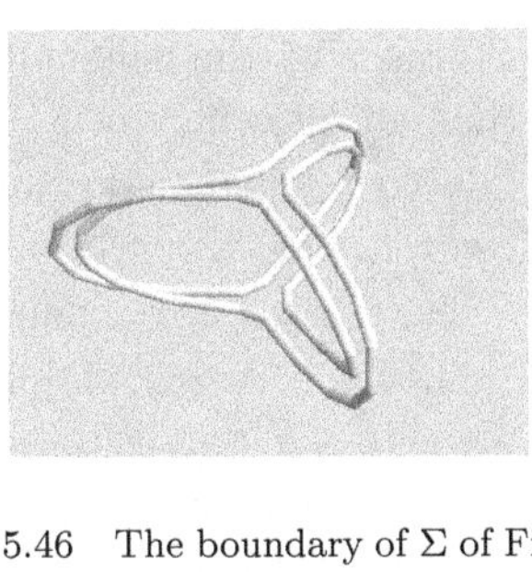

Fig. 5.46 The boundary of Σ of Figure 5.45.

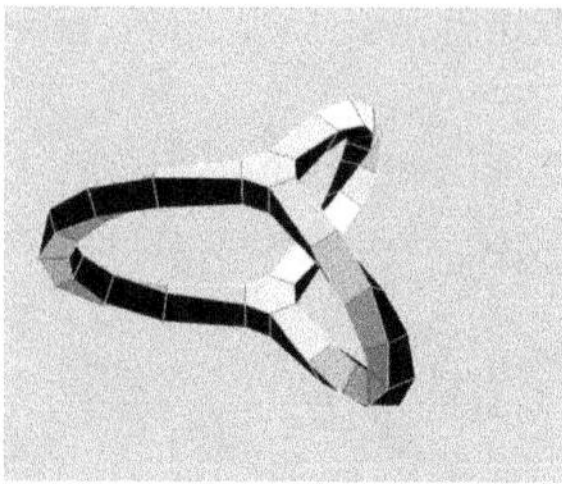

Fig. 5.47 $N = \Sigma' \times D^1$.

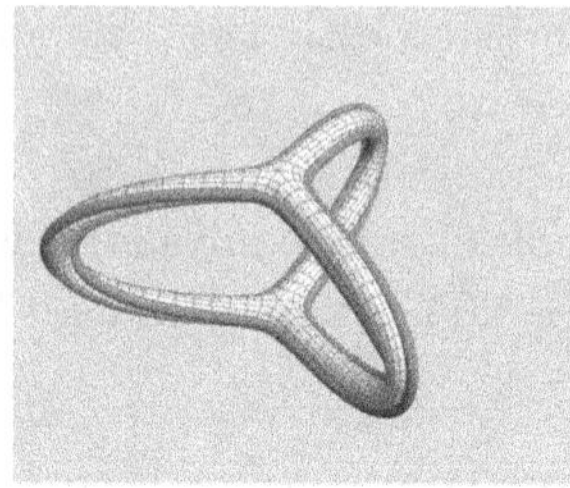

Fig. 5.48 A smooth version of N Figure 5.47 with trefoil knotting as equatorial manifold.

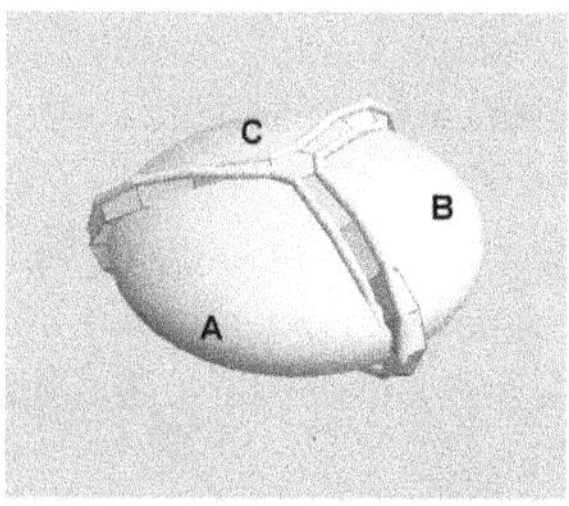

Fig. 5.49 Adding three 2-handles to N.

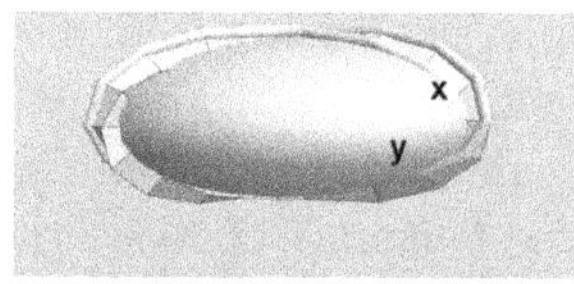

Fig. 5.50 Another view of Figure 5.49.

5.5 Wild Surfaces and Local Knotting

Surfaces in R^3 are codimension 1. As with 1-knottings such as Figure 3.2 we can avoid such wild knottings of 2-manifolds by focusing on *smooth* knottings. There are some well-known wild 2-spheres in R^3. This mathematically contrasts with the fact there are no wild topological circles in the plane. This is by the Schoenflies Theorem, Proposition 5.1 which basically says that all knottings of a circle in R^2 are equivalent, and in fact every such knotting is the restriction of a knotting of a 2-disk.

However, the higher-dimensional *topological* Schoenflies Theorem is false. There are many inequivalent knottings of a 2-sphere in R^3, called "wild spheres". Also, not all topological knottings of a 3-ball in R^3 are equivalent to the standard 3-ball.

The first published example of a wild sphere is known as **Alexander's Horned Sphere**, [Alexander (1924a)]. The two best drawings of the Alexander Horned sphere are in [Bing (1964)] and [Hocking and Young (1961)]. Full discussions on wild examples are found in [Bing (1983)], [Rushing (1973)] and [Rolfsen (1976)].

Example 5.14. One example of a wild 2-sphere in R^3 is based on a knotting of a 1-disk called the **Fox-Artin arc**, [Fox and Artin (1948)], Figure 5.51. This arc knotting is wild since, one can show the complement is not simply-connected, while a tame arc has simply-connected complement. By using a tube that narrows towards a wild point we can obtain a wild sphere in R^3 with two wild points, Figure 5.52. Wilder examples which are everywhere wild embeddings are found, for example, in [Rushing (1973)].

Clearly this wild knotting of 2-sphere bounds a 3-disk $D \subseteq R^3$ and, less clearly, the complement of D is not simply-connected, see Definition 1.20. Thus a purely topological version of the Schoenflies Theorem is false. However smooth manifolds cannot be wild. In fact, the *smooth* version of the Schoenflies theorem *is* true, Proposition 5.5. ∎

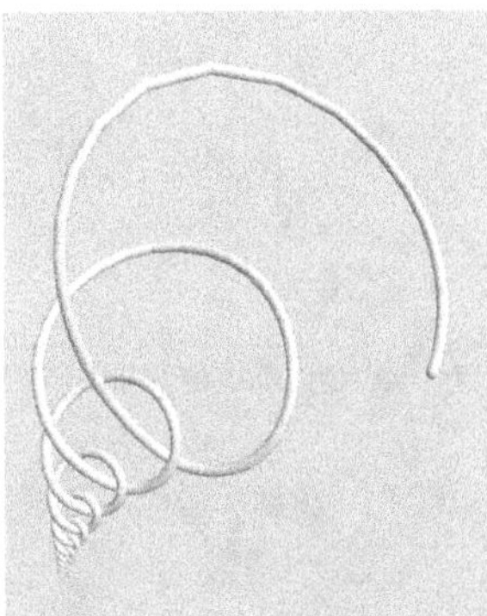

Fig. 5.51 A wild arc in R^3 with one wild point.

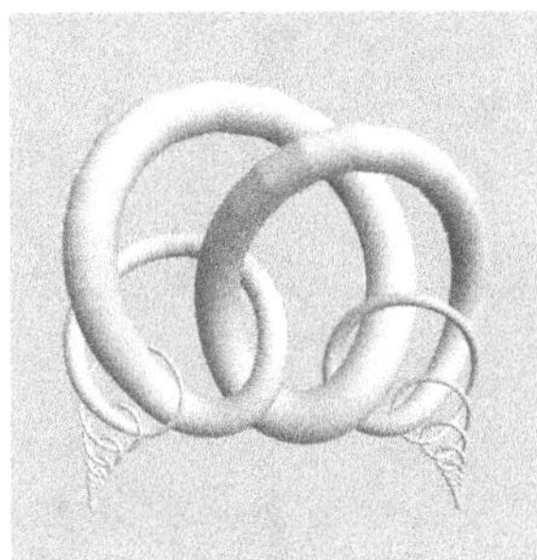

Fig. 5.52 A wild sphere in R^3 with two wild points.

There are wild 2-spheres in R^4 as well. Some can be obtained by spinning and ribbon methods described in later sections of this chapter when applied to wild 1-dimensional knottings. The suspension ΣC of a wild knotting of a circle C see Definition 1.18 is certainly wild and the set of wild points of ΣC is the suspension of the wild points of C. Also, the spinning ΣC see Definition 10.2 of a wild knotting of a circle C is certainly wild and the set of wild points of ΣC is the spinning of the wild points of C.

For an interesting geometric method see [Hinojosa (2007)]. Higher dimensional examples of codimension 2 are found in [Cantrell and Edwards (1965)] and [Tindell (1966)].

When considering higher dimensional knottings, there is a phenomenon called **local knotting** that has no analog for 1-knottings.

Example 5.15. Let K be a trefoil knotting $K \subseteq R^3 \subseteq R^4$ and let $P = (0,0,0,1)$. The cone $C = P * K$, see Definition 1.18, is homeomorphic to the 2-disk D^2. It is a proper sub-disk of R^4_+ since $K = \partial C \subseteq \partial R^4_+$. We view C in Figure 6.1 using projection π_z, defined $\pi'(x,y,z,w) = (x,y,w)$. In this figure, the projection of the line segment in R^4 from P to a point $x \in K$ is the line segment in R^3 from P_* to x_*. If $Q = (0,0,0,-1)$, let $Q * K$ denote cone. Then $\Sigma K = (P * K) \cup (Q * K)$ is homeomorphic to S^2. This knotting of S^2 in R^4 is called the **suspension of the knotting** K.

It can be shown that ΣK fails to be locally flat (and therefore fails to be smooth) at P and at Q (Figures 5.53 and 5.54). One says that P and Q are **local knots** or **local singularities** [Zeeman (1964)], [Suzuki (1969)], [Hudson and Zeeman (1964)] [Fox and Milnor (1966)]. ∎

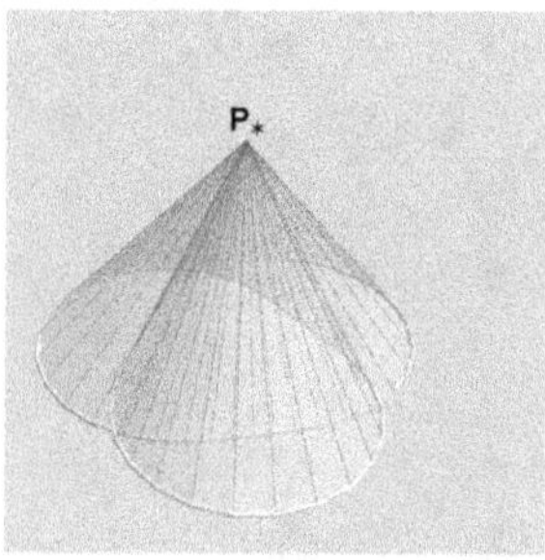

Fig. 5.53 Projection π_z of a cone from $P = (0,0,,0,1)$ to a trefoil knotting K.

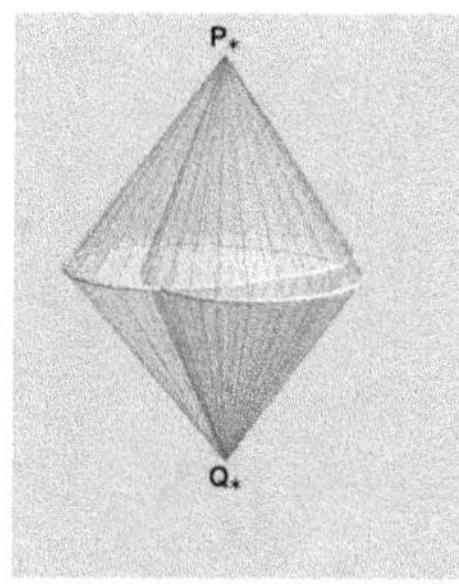

Fig. 5.54 Projection π_z of suspension of a trefoil knotting.

5.6 Codimension 1: Smooth Surfaces in 3-Space and the 3-Sphere

We need only consider *orientable* closed surfaces. It can be shown that closed non-orientable surfaces are never subsets of R^3. Orientable surfaces other than the 2-sphere have many smooth knottings in R^3. These are well understood and are closely related to 1-knottings. A good review of this topic is [Burgess and Cannon (1971)]. One can get interesting knottings by taking a non-trivial 1-knotting K of p components with product tubular neighborhood $N(K)$, Section 2.5. The boundary $\partial N(K)$ will be an interesting knotting of p torii.

Spheres and torii are important special codimension 1 cases.

Prop 5.5 (Schoenflies Theorem for S^3). If $\Sigma \subseteq S^3$ is a smooth knotting of a 2-sphere, then the closure of each component of $S^3 - \Sigma$ is is diffeomorphic to D^3 ∎

By Proposition 2.7 all 3-balls in a manifold are equivalent, so there is only the trivial knotting for 2-spheres in R^3 or S^3. However, there are many non-equivalent knottings of a torus. The boundary of a solid tube about a non-trivial knotting of a circle represents a non-trivial knotting of the torus; tubes about a non-trivial linking of two circles give a non-trivial knotting of a pair of torii. The Torus Theorem has some of the flavor of the Schoenflies Theorem for torii in S^3. As with the Schoenflies Theorem, the proof is non-trivial, and usually involves use of a powerful result known as Dehn's Lemma [Stallings (1971)] [Bing (1983)]

Prop 5.6 (Torus Theorem for S^3). If $\Sigma \subseteq S^3$ is a smooth knotting of a 2-torus then the closure of at least one component of $S^3 - \Sigma$ is homomorphic to $S^1 \times D^2$ ∎

So, for knottings of $S^1 \times S^1 \subseteq S^3$ one is basically considering 1-knottings of a circle. But distinguishing two knottings higher genus can be non-trivial, see for example [Ishii *et al.* (2015)] and [Ishii and Kishimoto (2011)].

Chapter 6

Projections of 2-Knottings

We focus on knottings in R^4, postponing consideration of knottings in S^4 until later. Since the projection $R^3 \to R^2$ is central to visualization and mathematical analysis of 1-knottings, Section 4.2, it is natural to consider projection $R^4 \to R^3$ of 2-knottings.

As with 1-knottings, projection is most often taken to be projection on the last coordinate. Avoiding subscripts, and writing $p \in R^4$ as $p = (x, y, z, w)$, **standard projection** $\pi \colon R^4 \to R^3$ is $\pi(x, y, z, w) = (x, y, z)$. If $K \subseteq R^4$ is a knotting the image $K_* = \pi(K)$ is called the **projection of** K or sometimes **knotting projection.**

For 1-knottings, we want self-intersections of the projection locally modeled on two lines in R^2 intersecting in a single point. Similarly, for projections of 2-knottings, we want the self-intersections to be locally modeled on either two planes intersecting in a line or three planes with a single common point. Additionally, non-immersion points are a new feature to consider, detailed in Section 6.2.

6.1 A Basic Local Example: Crossings and Products

The cartesian product can generate knottings in higher dimensions. For example, if $K \subseteq R^3$ is a trefoil knotting, $D^1 \subseteq R^1$, then $K \times D^1$ will be a knotting of an annulus in $R^3 \times R^1 \subseteq R^4$. Figure 6.1 shows a projection of this product using a non-standard projection $\pi'(x, y, z, w) = (x, y, w)$ as in Example 5.15.

In Figure 6.1 the three crossing points in the trefoil give rise to three crossing lines of self-intersection in the projection of the product 2-knotting. Using cubes, Definition 1.16 we write details of this process of going from a single crossing point to a crossing line.

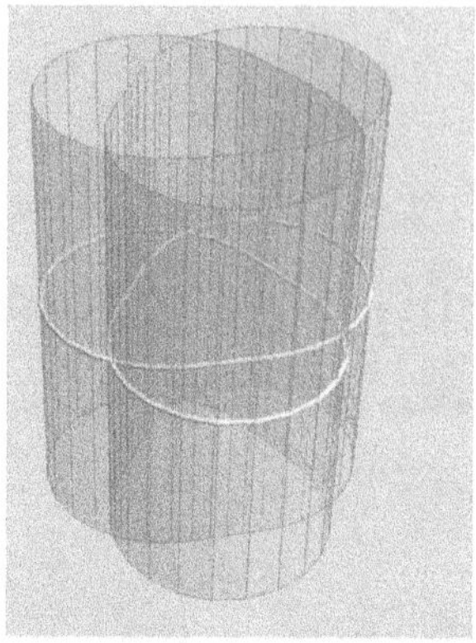

Fig. 6.1 Projection π_z of product of a trefoil knotting and standard $D^1 \subseteq R^1$.

Example 6.1. Just as a cube can be viewed as a stack of squares, $Q^3 = \bigcup_{-1 \leq z \leq 1} Q^2_z$, we can write Q^4 as a stack of 3-dimensional cubes: $Q^4 = \bigcup_{-1 \leq w \leq 1} Q^3_w$. A local model of a single crossing of a 1-knotting is two disjoint proper line segments s_0 and t_0 in Q^3_0 such as Figure 6.2, whose projection in the z-direction is two segments with a single crossing point, Figure 6.3. Viewing a square as a stack of line segments in the w-direction, we have two *disjoint* squares in Q^4: $S = \bigcup_{-1 \leq w \leq 1} s_w$ and $T = \bigcup_{-1 \leq w \leq 1} t_w$. A figure for each Q^3_w in our stack will look just like Figure 6.2 with the 0 subscript replaced by w. The projections of S_* and T_* in the z-direction will be two *intersecting* squares, Figure 6.4. The projections S_* and T_* into R^3 are shown in Figure 6.4. ▌

Let $k = s_0 \cup t_0$ and $K = S \cup T$. Using the product formula in the list, Section 1.1, we have $(Q^4, K) = (Q^3 \times Q^1, k \times Q^1) = (Q^3, k) \times Q^1$. In other words, *the knotting of the product is the product of the knotting.*

From this it follows that $(Q^3, K_*) = (Q^2, k_*) \times Q^1$. In other words *the projection of product knotting is the product of the projection*: the projection of these two squares in Q^3 is the same as the product with Q^1 of the projection of the two line segments in Q^2.

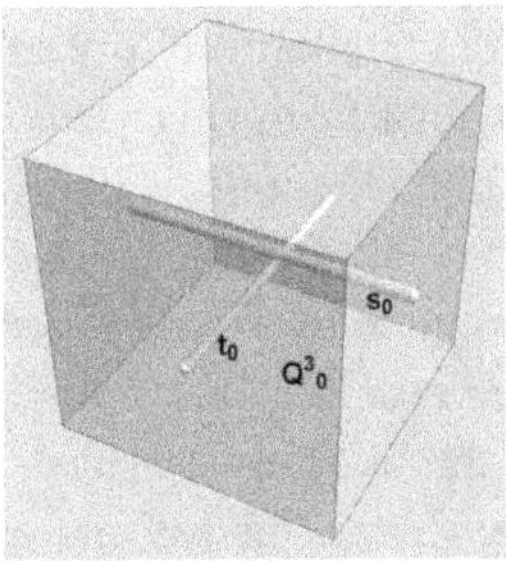

Fig. 6.2 Two disjoint line segments in a cube.

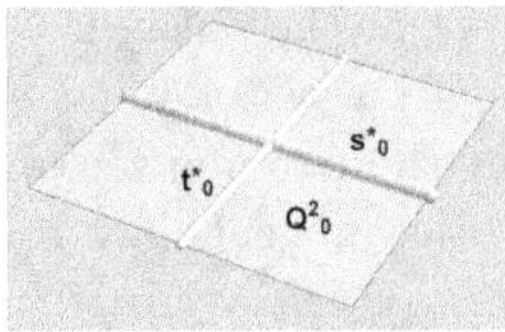

Fig. 6.3 Projection of objects of Figure 6.2.

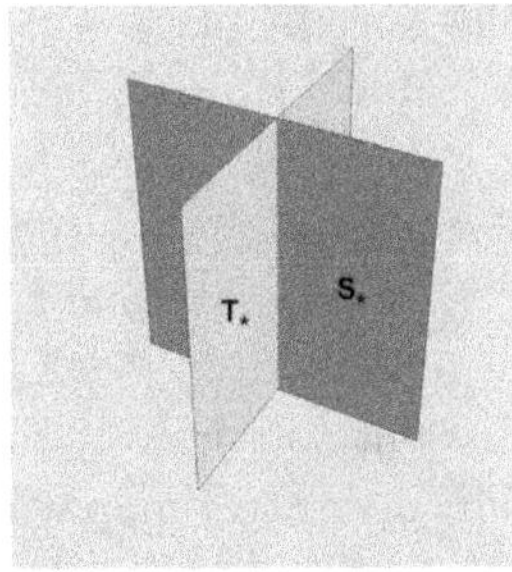

Fig. 6.4 Two disjoint squares S and T of knotting with intersecting projections S_* and T_*.

Another basic local model of a knotting projection of a 2-knotting is a projection of three disjoint planes resulting in a triple point. The three disjoint planes in Q^4 will be S, T and Q_0^2. If we intersect these three planes with Q_0^3, we will have three disjoint subsets s_0, t_0 and Q_0^2, Figure 6.5 and the projection will be the three planes shown in Figure 6.6.

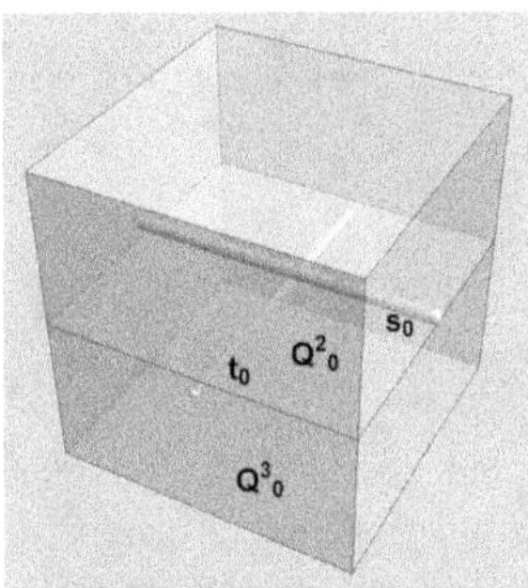

Fig. 6.5 Three disjoint sets: two segments and a plane.

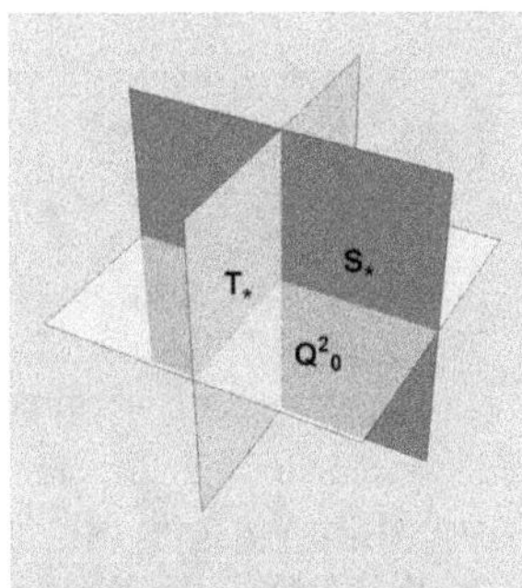

Fig. 6.6 Projection of three squares which are disjoint in Q^4.

6.2 Branch Points

There are two general local models for the image of a projection of a closed smooth 2-manifold $M^2 \subseteq R^4$ with $x \in M^2$. Suppose $x \in U \subseteq M^2$, for small coordinate patch 2-disk U. If the derivative image of the tangent plane at x is a plane then, and U_* will locally look like $R^2 \subseteq R^3$. If not, the image of the tangent plane will be a line and we have a **branch point**, also called a "**pinch point**", [Francis (2007)].

Figure 6.7 shows the projection of a (trivial) smooth knotting K of the 2-sphere in R^4, "a sphere in R^4 with a half twist", which has exactly two branch points, a and b, details will follow in Example 6.3.The tangent plane at any point of $p \in K$ (except at the poles) is a 2-plane spanned a pair of tangent vectors—$\overrightarrow{t_p}$ tangent to a latitude and $\overrightarrow{m_p}$ tangent to a longitude. The derivative map will send these to non-zero vectors *except* at the points A and B. At these points, the tangent lines to the circle C are in the w direction, and so will project to a zero vector. In other words the derivative of the map from K to R^3 is singular only at A and B, see Figure 6.8. This figure shows an image of the tangent plane at a typical point p being a plane but the image of the tangent plane at b is the line as shown.

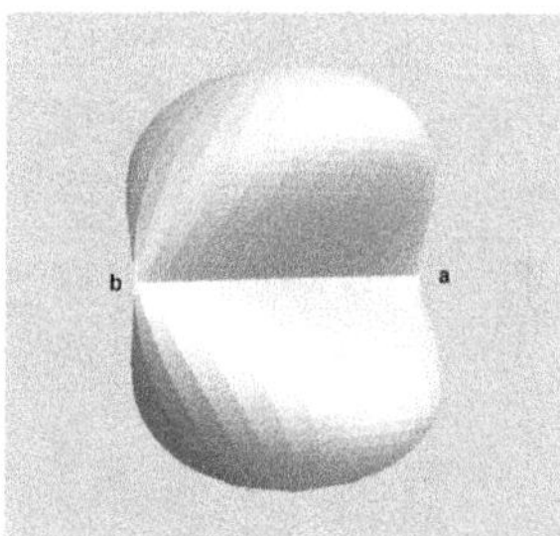

Fig. 6.7 A projection of a sphere with two branch points a and b.

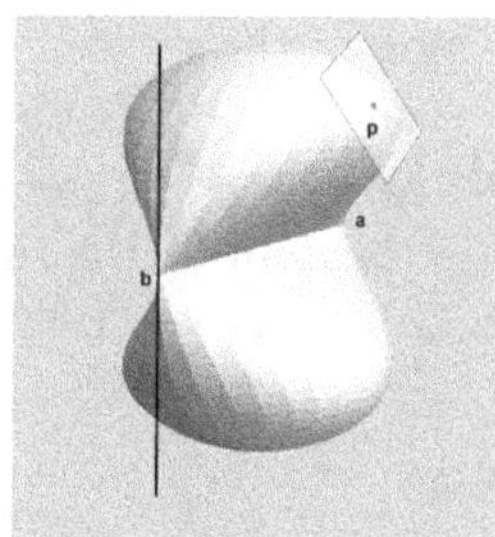

Fig. 6.8 Image of tangent planes at p and b.

Example 6.2. We look first at a lower dimensional version, a "circle τ in R^3 with a half twist", which we visualize as the boundary $\tau = \partial T$ of "a 2-disk with a half twist" $T \subseteq R^3$. View the standard ("untwisted") disk, as a union of D^2 as a union of line segments parallel to the y-axis and two points $D_{-1} = \{-1, 0, 0\}$ and $D_1 = \{1, 0, 0\}$, Figure 6.9. To develop notation, view $S^1 = \partial D^2 \subseteq R^3 = \bigcup_{a \in R^1} H_a^2$ where H_a^2 is the plane perpendicular to the x-axis with equation $x = a$. Now $D^2 = (\bigcup_{-1<a<1} D_a^1) \cup \{D_{-1}, D_1\}$ where $D_a^1 = D^2 \cap H_a^2$. It will be useful to view the two endpoints of a 1-disk D_a^1 $(a \neq \pm 1)$ as a 0-sphere S_a^0 and so $S^1 = (\bigcup_{-1<a<1} S_a^0) \cup \{D_{-1}, D_1\}$, Figure 6.10.

We give D^2 a half twist in the x-direction by rotating each D_a^1 about its midpoint in the plane H_a^2 by an angle of $\pi(1 - a)/2$. Small line segments near D_1 are barely rotated, the diameter D_0^1 is rotated by $\pi/2$ (and so lies on the z-axis) and segments near D_{-1} are rotated by almost π. Let T_a^1 be this rotated version of D_a^1. The result is a new twisted disk $T = \left(\bigcup_{-1<a<1} T_a^1\right) \cup \{D_{-1}, D_1\}$. If we let $\Sigma_a^0 = \partial T_a^1$ then we obtain a 1-knotting of a circle $\tau = \left(\bigcup_{-1<a<1} \Sigma_a^0\right) \cup \{D_{-1}, D_1\}$. We view T in three ways:

z-projection of T, Figure 6.11: The z-projection is the standard projection in R^3. Segment D_0^1 projects to point $\mathbb{O} \in R^2$ and is the single crossing point of the projection of T.

x-projection of T, Figure 6.12: The projection of T has a single crossing point $\{0, 0\}$ in the yz-coordinate plane. The x-projection of each D_a^1 projects of a line segment of the same length containing $\{0, 0\}$. And so, as a varies, the projection of the 0-spheres S_a^0 vary in size as they rotate about $\{0, 0\}$.

y-projection of T, Figure 6.13: This projection of T has no crossing points so it is trivially knotted *topological* planar simple closed curve. However, it is not a *smooth* curve. This projection is an immersion at all points except D_{-1} and D_1, where the tangent lines to T are in the direction of the y-axis. ▮

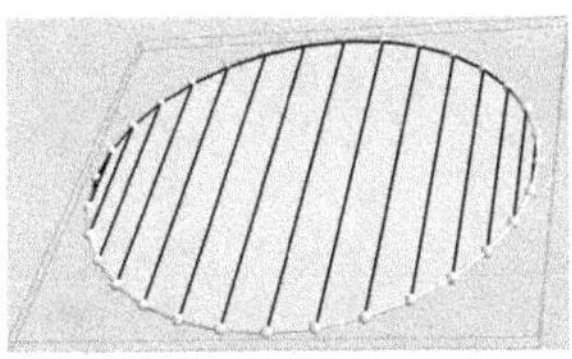

Fig. 6.9 $D^2 = \bigcup_{-1\leq a\leq 1} D_a$, $S^1 = \left(\bigcup_{-1<a<1} S_a^0\right) \cup \{D_{-1}, D_1\}$.

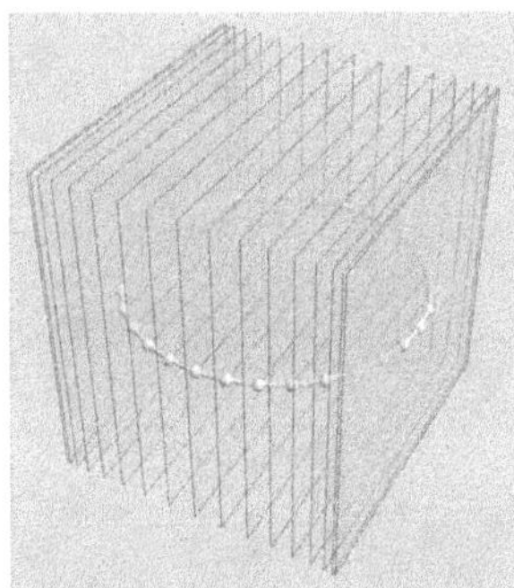

Fig. 6.10 $S^1 = \partial D^2 \subseteq R^3 = \bigcup_{a\in R^1} H_a^2$.

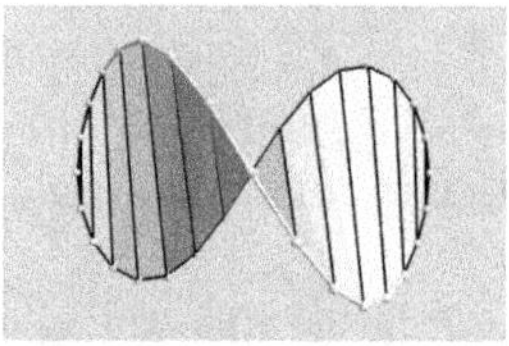

Fig. 6.11 z-projection T.

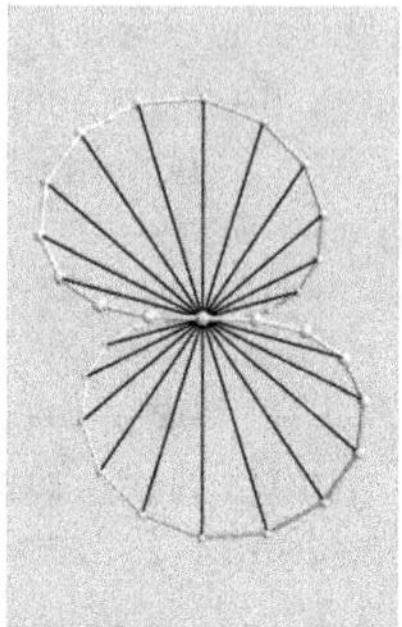

Fig. 6.12 x-projection T.

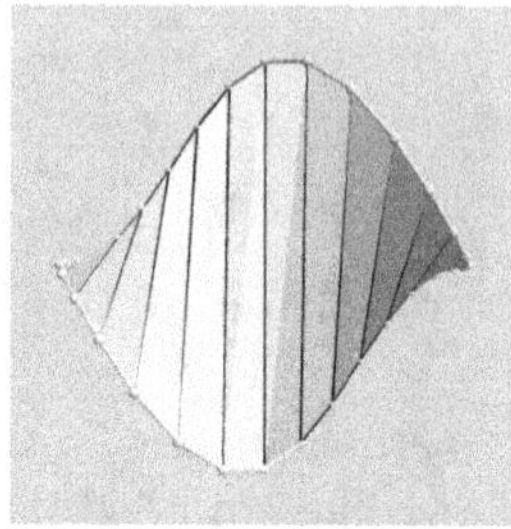

Fig. 6.13 y-projection T.

Example 6.3. We construct a sphere knotting $K \subseteq R^4$ for Figure 6.7 by increasing the dimensions of subsets in the previous construction. (The disks D^1_a used in the previous construction were only needed for visualization—to connect the two points of the 0-spheres S^0_a, so we dispense with analogs of them.) Write $S^2 \subseteq R^3 \subseteq R^4 = \bigcup_{a \in R^1} H^3_a$ where H^3_a is the hyperplane perpendicular to the x-axis with equation $x = a$. Write $S^1_a = S^2 \cap H^3_a$, and express S^2 as a union of 1-spheres S^1_a and two points $S^2 = (\bigcup_{-1<a<1} S^1_a) \cup \{D_{-1}, D_1\}$ where now $D_{-1} = \{-1, 0, 0, 0\}$ and $D_1 = \{1, 0, 0, 0\}$, see Figure 6.14.

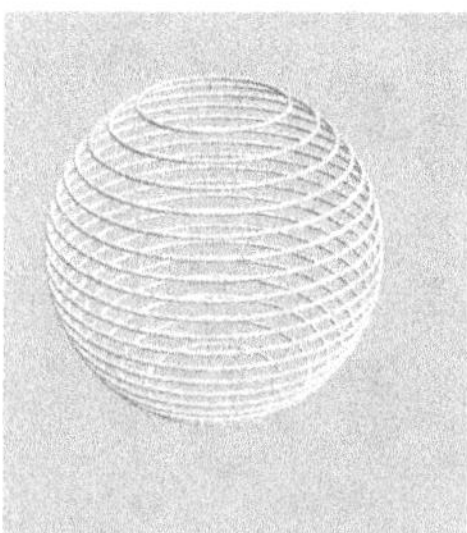

Fig. 6.14 S^2 as union of circles S^1_t.

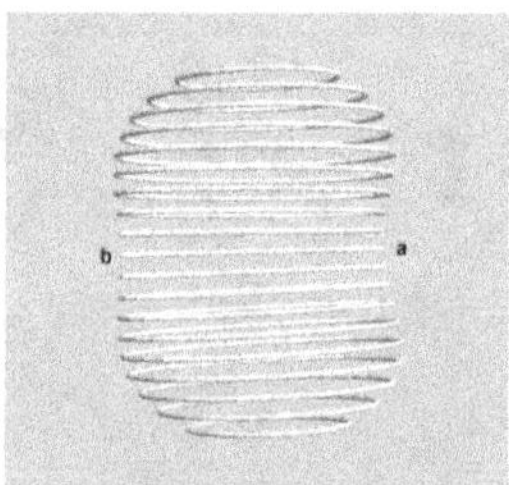

Fig. 6.15 Rotated circles project to ellipse but middle to a line segment.

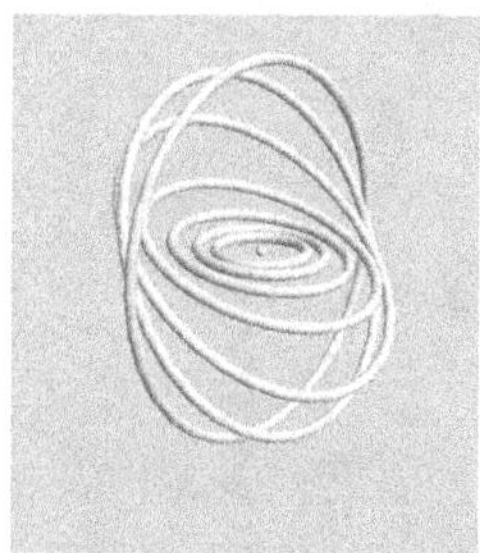

Fig. 6.16 The projection in x-direction of some circles Σ_a.

To get our knotting K we rotate each circle S^1_a *in the three-dimensional hyperplane* H_a along a y-axis (i. e., using z and w coordinates), by an angle of $\pi/2(a+1)$, where $-1 < t < +1$. As with the previous example, this rotation angle varies from 0 to π. The axis of this rotation contains a diameter of the circle. To be clear we call such a rotation a **flip**—as in a "flip of a coin" rather than a wheel-like rotation. Denote this flipped circle by Σ_a^1. Our 2-knotting is now defined as $K = \left(\bigcup_{-1<a<1} \Sigma_a^1\right) \cup \{D_{-1}, D_1\}$. In projection, a flipped circle projects to an ellipse, or if flipped by $\pi/2$, to a line segment, Figure 6.15. Figure 6.16 shows the x-projection of some of these circles.

We view K in four ways:

w-projection of K, Figure 6.17 In the (standard) w-projection, each S_t^1 projects to an ellipse of increasing eccentricity for values of t close to 0—except at $t = 0$ where the rotated circle $C = S_0^1$, rotated by an angle of $\pi/2$ about the x-axis, projects to a line segment between $a = (1, 0, 0)$ and $b = (-1, 0, 0)$.

x-projection of K, Figure 6.18: Topologically this image corresponds to a trivial knotting of a 2-sphere in 3-space. However, as is clear from the irregular shape, it is not the image of a smooth embedding,. This is analogous to Figure 6.12, where the projection of a smooth circle is not a smooth submanifold.

y-projection of K, Figure 6.19: In the y-direction each circle Σ_a is viewed "edgewise" and so projects to a line segment corresponding to a diameter. This diameter appears to rotate in a xz direction. The entire y-projection of K corresponds to our twisted disk T considered in xzw-space, rather than xyz-space.

z-projection of K, Figure 6.20: By symmetry, this subset looks like Figure 6.17 If one compares the wire-frame patterns on these images, one can see the figures are not the same. ∎

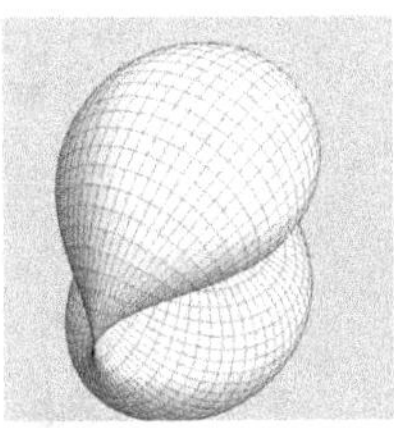

Fig. 6.17 w-projection.

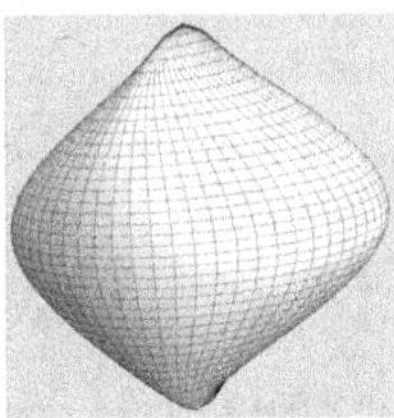

Fig. 6.18 x-projection.

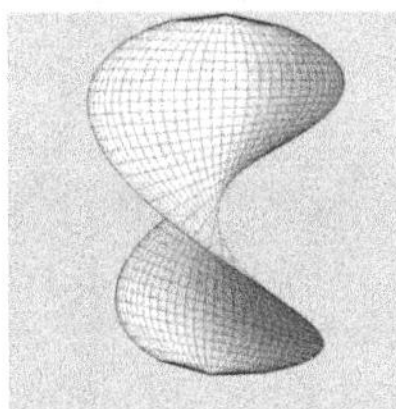

Fig. 6.19 y-projection.

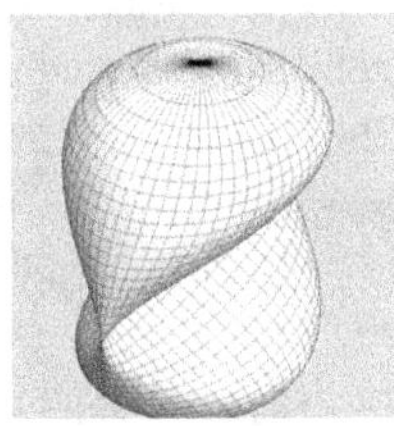

Fig. 6.20 z-projection.

Example 6.3 is the first of many in this book that uses a *sequence* of three-dimensional graphics to depict an object in four-dimensions. In presenting three-dimensional graphics on a page our practice is to use the standard default viewpoint used by the software (in our case, Mathematica) whenever possible. However, often this default view fails to show some critical details, so, without explicit mention, we present our single three-dimensional graphic from an alternate viewpoint. If a sequence of three-dimensional graphics $G_1, \ldots, G_p$ depicts a single object in R^4, then we can use any viewpoint in one G_i but then should use **corresponding viewpoints in the rest.** First of all when we speak of a three-dimensional graphic G on a page we actually refer to a projection where the direction of this projection is called the **viewpoint**. Normally this direction is a three-dimensional vector $\overrightarrow{v}$. If, as is often the case in this book, G is meant to be a three-dimensional hyperplane, $H^3_j(b)$ then the **viewpoint** is a direction in that hyperplane, that is, the 4-dimensional vector $\overrightarrow{V}$ corresponding to $\overrightarrow{v}$ with the j-th component equal to b and we say $\overrightarrow{v}$ and $\overrightarrow{V}$ are **corresponding viewpoints**. For example $\overrightarrow{v} = (1,1,1,0)$ and $\overrightarrow{V} = (1,1,1,-1)$ are corresponding viewpoints; the first in $R^3 \subseteq R^4$, the second in the parallel hyperplane $H^3_4(-1)$. However for the (nonparallel) hyperplane $H^3_2(0)$ the vector corresponding to $\overrightarrow{v}$ is $(1,0,1,1)$.

Consider the Figures 6.17, 6.18, 6.19, and 6.20. These are all compatible projections of the same 2-knotting. It is not easy to find a view-point for the object shown in Figure 6.18 however we chose one the one that seemed best. We were then obligated to use corresponding viewpoints for Figures 6.17, 6.19, and 6.20. Now however, the viewpoints Figures 6.17, and 6.20 were forced to be unavoidably somewhat awkward. (By itself, Figure 6.7 might be a preferred view the object shown in Figure 6.17, but is a poor choice for some of the other projections.)

We alert the reader that there are two distinct, but related, uses of the term **branch point**. One, discussed here, is *branch point of projections*. Definition 6.1, generalized to projections for n-knottings in Section 12.2. The other definition is for a branch point of a *smooth map from one 2-manifold to another* discussed in Section 11.2.

A local model of a branch point *of a projection* is Whitney umbrella:

Example 6.4. Define a smooth map of a square $Q = D^1 \times D^1 \subseteq R^2$ into R^4 by $w(u,v) = (uv, u, v^2, v)$ for $-1 \leq u \leq 1$ and $-1 \leq v \leq 1$. Write $W = w(D^1 \times D^1)$. The standard projection $W_* = \pi(W)$ will be called the **Whitney umbrella**. It is the image of $w(u,v) = (uv, u, v^2)$ and as shown in W_* in Figure 6.21.

If we take a disk for our domain of w the Whitney umbrella projection is shown in Figure 6.22 showing more clearly why it is alternatively called "a cone on a figure-eight". ∎

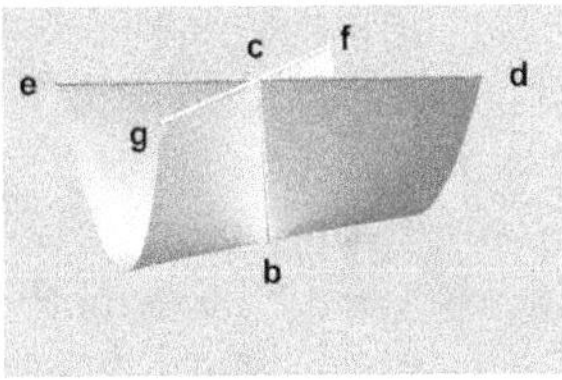

Fig. 6.21 The Whitney umbrella W_* with branch point at b.

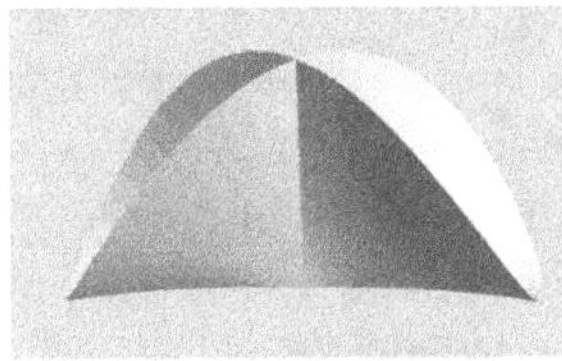

Fig. 6.22 Whitney umbrella as "a cone on a figure-eight curve".

Definition 6.1. For a 2-knotting K, if there is a sub-disk $U \subseteq K$ such that $U_* \subseteq K_*$ is homeomorphic to a Whitney umbrella, the point corresponding to the branch point of the Whitney umbrella is called a **branch point of** K_*; the corresponding point of K, $K \cap \pi^{-1}(b)$, is called a **branch point of** K. ∎

6.3 Projection Terminology for 2-Knottings

We adapt some definitions of Section 4.2. If $\Sigma \subseteq R^4$ is a knotting M^2, we have some embedding $\phi\colon M^2 \to R^4$ with $\phi(M^2) = \Sigma$ and a projection $\Sigma_* = \pi(\Sigma)$. As for 1-knottings, the crossing set $\mathcal{C} \subseteq \Sigma_*$ is central to describing and understanding a 2-knotting. So, we develop terminology for subsets of $\pi^{-1}(\mathcal{C}) \subseteq \Sigma$ and $(\pi \circ \phi)^{-1}(\mathcal{C}) \subseteq M^2$.

Definition 6.2. Suppose $K \subseteq R^4$ is a knotting with standard projection $K_* = \pi(K)$. For $x \in K_*$ write $\widetilde{x} = \pi^{-1}(x) \cap K$ and say $\widetilde{x}$ is the **lifting of** x. More generally for subsets $X \subseteq K_*$, we write $\widetilde{X} = \pi^{-1}(X) \cap K$ and say $\widetilde{X}$ is **the lifting of** X.

If $\widetilde{x}$ contains exactly n points where $1 < n$ we say that x is a n**-crossing point of** K_*. Any of these n points of $\widetilde{x}$ is called a **pre-crossing point of** x **in** K. If K is a 2-knotting and $\phi\colon M \to K$ a homeomorphism where M is a 2-manifold and x is an n-crossing point then any point of $\widehat{x} = \pi^{-1}\phi\circ(x) \subseteq M$ is called a **pre-**n**-crossing point of** x **in** M.

If $\widetilde{x}$ contains exactly one point which is *not* a branch point, x is called an **ordinary projection point of** K_*. A lifting of an ordinary projection point of K_* is called an **ordinary projection point of** K and a pullback of the point called an **ordinary projection point of** M. ∎

A projection of a knotting is a disjoint union of: ordinary points, branch points and n-crossing points. If a projection has a branch point it is not a crossing point but is a is a limit point of crossing points, so, the set of crossing points will not be a closed subset. However, the union of the set crossing points and the set of all branch points *is* closed. That motivates the next definition.

Definition 6.3. Suppose K is a 2-knotting and $\phi\colon M \to K$ a homeomorphism where M is a 2-manifold. The subset $\mathcal{C}$ of all n-crossing points and all branch points of K_* is called the **crossing set of K_*.** The set $\widetilde{\mathcal{C}} = \pi^{-1}(\mathcal{C}) \cap K$ is the **pre-crossing set of K_*.** The set $\widehat{\mathcal{C}} = \phi^{-1}(\widetilde{\mathcal{C}})$ is called the **crossing scheme of K_*.** If x is a branch point of projection K_*, $\widetilde{x}$ is called a **branch point of the pre-crossing set of K_*.** Given $\widehat{x} = (\phi \circ \pi)^{-1}(x)$ is called a **branch point of the crossing scheme.** ▮

Definition 6.4. If x is a 2-crossing point let $\widetilde{x} = \{x_a, x_b\}$ be the pre-crossing set where $h(x_b) < h(x_a)$ (i.e. x_a is above and x_b below, in the w-coordinate). We say $x_a \in K$ is an **above-point** and $x_b \in K$ is a **below-point**. Similarly if $f\colon M \to K$ is a homeomorphism with $K = f(M)$ we call $\widehat{x_a} = f^{-1}(x_a) \in M$ an **above-point** and $\widehat{x}_b = f^{-1}(x_b) \in M$ a **below-point**. More generally for subsets we will say that $A \subseteq R^4$ **is above** $B \subseteq R^4$ if for all $a \in A, b \in B$ with $\pi(a) = \pi(b)$ we have $h(b) < h(a)$, and define A **is below** B similarly. ▮

For 2-crossing points it is traditional to speak of "over" and "under" as for projections of 1-knottings of Definition 6.4, however we prefer "above" for "over" and "below" for "under".

The use of the terminology "crossing scheme" for 2-knottings is new here but the basic idea has been used often in some form or other. As in the case of 1-knottings the scheme alone is just a finite collection of points of M and, by itself, not terribly informative. However, with additional information. such as labeling, it can be useful, see for example Section 6.6.

The above definitions are a little different than those in the literature. In previous papers such as [Roseman (2000)] pre-crossing points were called "double points" and the pre-crossing set called the "double point set". In references such as [Carter (1995)] crossing points are called "double points", and the pre-crossing set is called the "r-decker set". We use the above notation to clarify terminology and, in particular, to avoid the two similar but competing uses of the term "double point". Also, the term "singular set" is used, by some for, crossing set, as in [Kawauchi (2002)]. We avoid this since "singular" is used in many different ways in geometric topology.

Example 6.5. To make our terminology clear, we use the simple example for 2-crossings shown in Figure 6.23. Shown is a projection of a knotting $K = K_0 \cup K_1$ of two disjoint 2-spheres in R^4, with K_1 lying entirely above K_0. Perhaps $K_0 = \{(x, y, z, w) \in R^4\colon (x-1)^2 + y^2 + z^2 = 2 \text{ and } w = 0\}$ and $K_1 = \{(x, y, z, w) \in R^4\colon (x+1)^2 + y^2 + z^2 = 2 \text{ and } w = 1\}$.

Let $M \subseteq R^3$ be a union of two disjoint spheres $M = M_0 \cup M_1$, such as in Figure 6.24, and let $\phi\colon M \to K$ be a homeomorphism (embedding) with $\phi(M_0) = K_0$ and $\phi(M_1) = K_1$. The crossing scheme for this knotting might be as pictured as in Figure 6.25—a circle of above-points $\widehat{A} \subseteq M_1$ and a circle of below-points $\widehat{B} \subseteq M_0$.

A crossing scheme, such as Figure 6.25, depends on the choice of M and ϕ. We can take advantage of this to make our depiction as nice as possible. We can take $\widehat{A}$ and $\widehat{B}$ to be equatorial circles, whereas it is not the case for the corresponding pre-crossing sets $\widetilde{A}$ and $\widetilde{B}$.

A lower dimensional example of this is seen in Figure 4.64 which shows a projection of a split trivial link of two 1-spheres whose crossing set is 0-sphere $\{c, c'\}$. Here the pre-crossing set is a 0-sphere of above-points with a 0-sphere of below-points. ∎

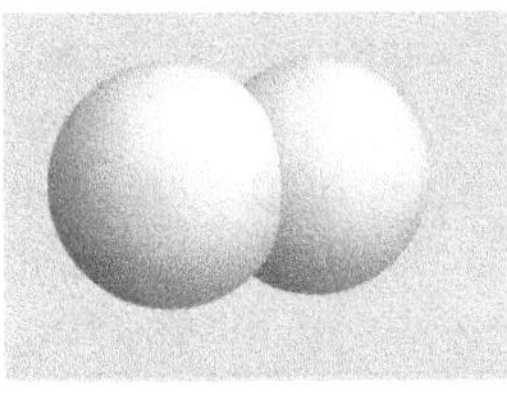

Fig. 6.23 Projection of disjoint union of spheres.

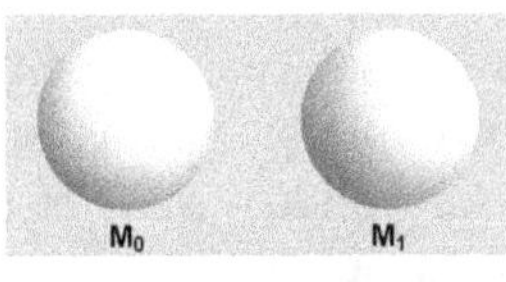

Fig. 6.24 Disjoint spheres M_0 and M_1 in R^3.

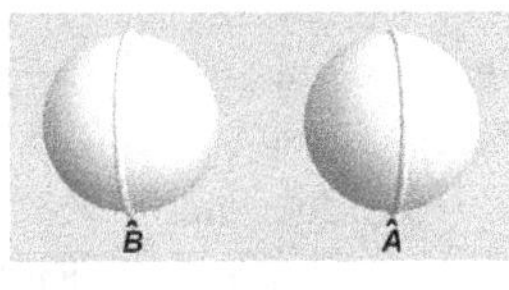

Fig. 6.25 Crossing scheme for Figure 6.23.

Example 6.6. Relabeling branch points of Example 6.3 as b_1 and b_{-1}, we have Figure 6.26. This knotting has a crossing set C which is homeomorphic to a closed interval and the pre-crossing set is a circle as shown in Figure 6.27. The two branch points $\widehat{b_1}$ and $\widehat{b_{-1}}$ separate this circle into an open arc of above-points and an open arc of below-points, Figure 6.27. ∎

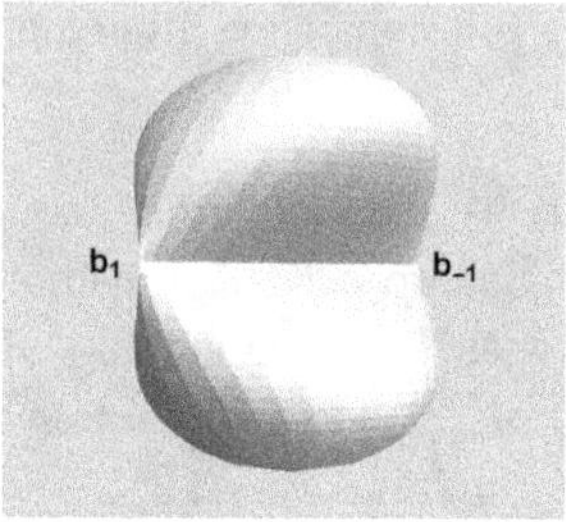

Fig. 6.26 Projection of a knotting with two branch points.

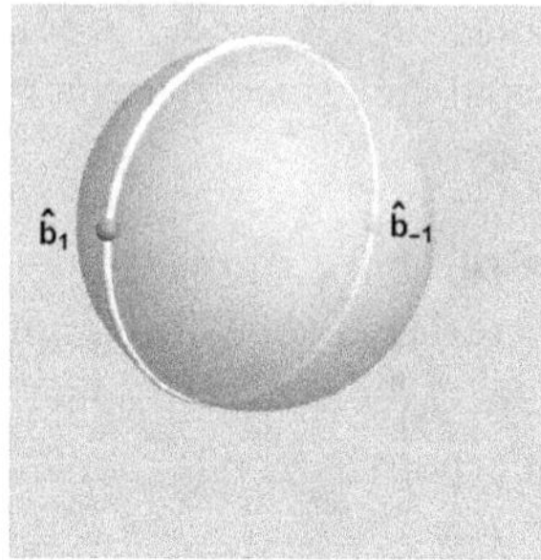

Fig. 6.27 A crossing scheme corresponding to knotting of Figure 6.26 shown with sphere as transparent.

Triple points are an important feature of surface knottings.

Definition 6.5. If x is a 3-crossing point let $\widetilde{x} = \{a, m, b\}$ where $h(b) < h(m) < h(a)$. We say m is the **middle point at** $\widetilde{x}$. ▌

Example 6.7. Shown in Figure 6.28, is a projection of a knotting $K = K_0 \cup K_1 \ \cup K_2$ of three disjoint 2-spheres. First are the two spheres of Example 6.5, the third is $K_2 = \{(x, y, z, w) \in R^4 \colon (x-2)^2 + (y-2)^2 + z^2 = 3^2$ and $w = 2\}$. The crossing set K_* is a union of three circles meeting in two triple points (one hidden from view). The two triple points are $t_1 = (2, 0, \sqrt{5})$ and $t_{-1} = (2, 0, -\sqrt{5})$. Figure 6.29 reveals the crossing set and t_{-1} using transparency for spheres and thin tubes about the crossing set. In each K_i, the crossing set consists of two circles of pre-crossing points which intersect in two points, such as Figure 6.30. It follows that a crossing scheme for this knotting consists of three pairs of intersecting circles as shown in Figure 6.29. ▌

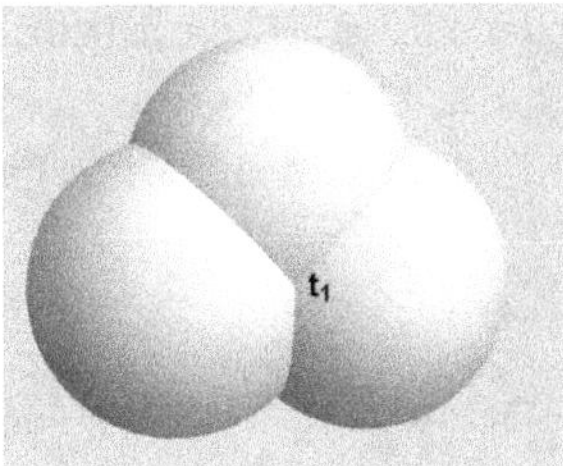

Fig. 6.28 Projection of three spheres disjoint in R^4.

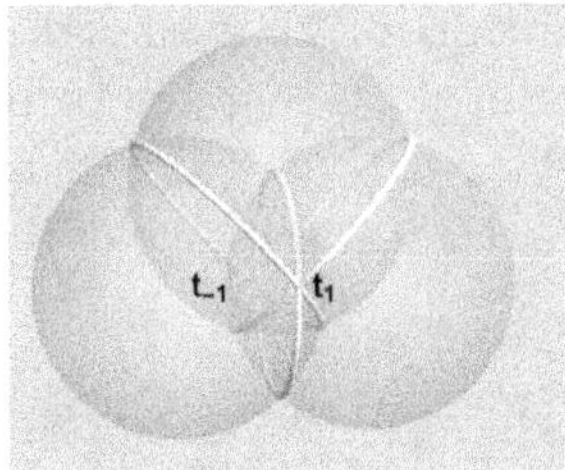

Fig. 6.29 A version of Figure 6.28 with spheres transparent showing intersections of spheres including the second triple point t_{-1}.

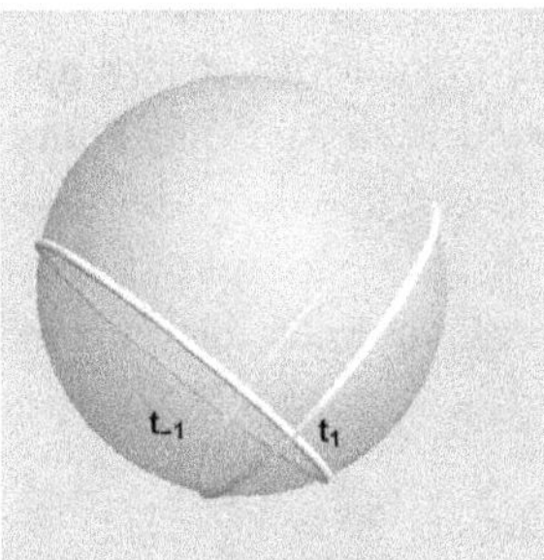

Fig. 6.30 Detail of Figure 6.29 showing just one of the spheres.

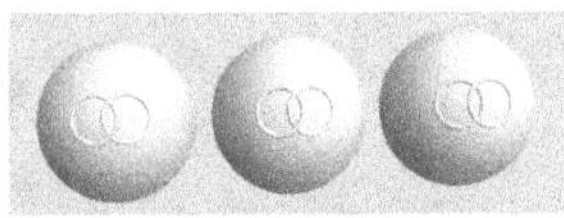

Fig. 6.31 A crossing scheme for the knotting with projection shown in Figure 6.28.

We next extend our notions of general position of Section 4.2 for projections of 2-knottings. (For higher dimensions, see Section 12.2.) First some terminology.

Definition 6.6. A set of **2-planes in R^3 is in general position** if their normal vectors are linearly independent. ▌

In Definition 6.7 we will say a 2-knotting is in general position if the projection is an immersion except at the branch points and locally the projection looks like one of the following four models: a single plane, two or three planes that intersect in general position or (for a branch point) a Whitney umbrella. Mathematically we can avoid knottings not in general position in the sense that any 2-knot *can be* represented by a 2-knotting which *is* in general position. Example 6.8 shows a knotting with a projection we want to avoid.

Definition 6.7. A smooth knotting of a surface $K \subseteq R^4$ is in **general position with respect to projection** if for all $x_* \in K_*$ there is an open subset $x_* \in U \subseteq R^3$ so that $V = K_* \cap U$ and there is a diffeomorphism $f : U \to R^3$ satisfies one of the following:

(1) $f(V) = R^2$
(2) $f(V) = X$ where X is the union of two planes in general position with $f(x_*)$ in the intersection
(3) $f(V) = X$ where X is the union of three planes in general position with $f(x_*)$ in the intersection
(4) $f(V) = X$ where X is the Whitney umbrella $f(x_*) = (0, 0, 0)$ ▮

Example 6.8. Figure 6.32 is a projection of a simple (trivial) knotting of a sphere which is not in general position. The projection clearly has two non-immersion points, but these are not modeled by a Whitney umbrella. A Whitney umbrella can be described as a smooth cone on a *one* crossing knotting of a circle, Figure 6.22. However, in this example the branch points are modeled by smooth cone on a *two* crossing knotting as shown in Figure 6.33. ▮

Fig. 6.32 Example knotting not in general position.

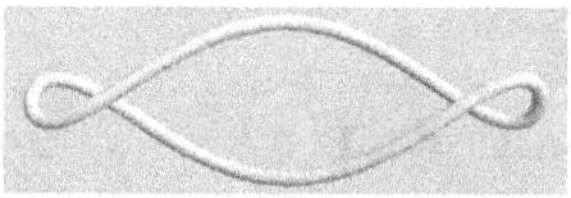

Fig. 6.33 A 1-knotting with two crossings.

Prop 6.1. (**General Position for Surfaces in** R^4) Any smooth 2-knotting is smoothly isotopic to a knotting which is in general position with respect to projection. ▮

There are stronger versions of Proposition 6.1 which say that the isotopy may be chosen to move points by an arbitrarily small amount and that, in a well defined way, most 2-knottings isotopic to K are in general position, see [Roseman (2000)]. Also, a version of the General Position Theorem is true for PL-knotting [Hudson (1969)].

From now on we will work almost exclusively with knottings in general position with respect to projection.

Later, we extend general position for 2-knottings to apply to n-knottings, Definition 12.9, so that branch sets crossing sets and crossing schemes will be defined. To avoid repeating our next definitions of shards we define shards for n-knottings in this section.

Generalizing Definition 4.6, we partition K_*, K and M, into submanifold pieces we call **q-shards**. The prefix of a shard is the **type of the shard**. So a b-shard has type b, a 2-shard has type 2.. The type indicates a local model for points in that shard. So, in a neighborhood of any point a 2-shard of a 2-knotting, K_*, locally looks like two intersecting planes.

Definition 6.8. Suppose K is an 2-knotting of M. with crossing set $\mathcal{C}$ branch set B_*. A **1-shard** of K_* is a component, X of $K_* - \mathcal{C}$. For $1 \leq q \leq 3$, a **q-shard** is a component of the set of q-fold points of $(\mathcal{C} - B_*)$. A **b-shard** is a branch point.

For any of these shards, X, of K_*, a component of $\widetilde{X}$ is a **shard of K of the same type** , and a component of $\widehat{X}$ is a **shard of M of the same type**. A q-shard of M is also called a **q-shard of the crossing scheme**.

For any shard of any type in K_*, K or M a **closed q-shard** is the closure of the shard in K_*, K or M respectively. If we need to emphasize a q-shard is *not* closed we will call it an **open q-shard**. ∎

Every q-shard of K_* is an image of q q-shards of K and of M. In higher dimensions, the projection of the branch set has more structure, see Remark 12.8, but we need only one branch type for 2-knottings. A 1-shard is a component of ordinary points (non-crossing points).

Suppose K is a 2-knotting in general position with respect to projection. Even though K^* is not a 2-dimensional submanifold, it does separate R^{n+1} into pieces called **regions**, and makes each 1-shard two-sided in the sense that [Roseman (1975a)] if $C \subseteq K^*$ is a 1-shard of the projection, C lies in the boundary of exactly two components of $R^3 - K^*$. This, in turn, implies that C is an open *orientable* 2-manifold, even though K might not be.

The term "scheme" introduced here for high-dimensional knottings was used in [Tait (1877)] in a similar fashion to denote a certain coding for projections of 1-knottings.

For an 2-knotting $K \subseteq R^4$ of M in general position with respect to projection, K, K_* or M is a union of disjoint shards or as a union of non-disjoint closed shards. These will be called an **shard structure for K, K_* or M** or **closed shard structure for K, K_* or M.**

For a shard $\sigma \in K_*$, with lifting $\widetilde{\sigma}$ we articulate the idea that one shard of $\widetilde{\sigma}$ is above another. (It follows from connectedness of shards that the following definition does not depend on the choice of test point x_*.)

Definition 6.9. Suppose σ_* is a p-shard of K_*, with two different p-shards $\sigma \subseteq \widetilde{\sigma_*}$ and $\sigma' \subseteq \widetilde{\sigma_*}$. For some $x_* \in K_*$ suppose that $x = \widetilde{x_*} \cap \sigma$ and $x' = \widetilde{x_*} \cap \sigma'$ and that $h(x) < h(x')$, where h is the height function for the knotting. We say σ **lies below** σ', or σ' **lies above** σ, denoted $\sigma < \sigma'$. ▌

Readers familiar with CW-complexes (also called cell complexes) will recognize that the shard structure is modeled after such structures with two major differences. An *open cell* is homeomorphic to an open disk; a *shard* is homeomorphic to an open connected manifold. Secondly, the dimension of an n-cell is n but as mentioned, an n-shard, does not have dimension n.

If we have a smooth embedding of a manifold $\phi\colon M \to R^4$ giving smooth knotting $K = \phi(M)$ in general position with projection, then although the *closed* shards may not be topological manifolds the (open) shards will be *smooth manifolds.* For a 2-knotting, the 1-shards are smooth open 2-manifolds, and the 2-shards are 1-manifolds. The 3-shards and b-shards (isolated points) are 0-manifolds which can be considered to be "smooth". This holds in all three senses: these are smooth manifolds in R^3 as well as smooth sub-manifolds of K and M.

Notation 6.1. We use the notation $K_*^{(q)}$ for the set of q shards of K_* of type q, and similarly, $\widetilde{K^{(q)}}$ for the q shards of K and $\widehat{K^{(q)}}$ for the q shards of M. ▌

Definition 6.10. If K_* is a projection of a general projection n-knotting a component of $R^n - K_*$ is called an **open region of the projection**. A **closed region of a projection** is the closure in R^n of an open region. ▌

A projection K_* partitions R^n into disjoint subsets—the shards and the open regions. An open region is an open subset of R^n and thus an open n-manifold. A closed region may not be a closed manifold. Like a shard, it can be described as a manifold with certain identifications on the boundary.

Example 6.9. We detail the shard structure for schemes of three examples.

In Figure 6.27, a crossing scheme for Figure 6.26, there are two b-shards (branch points), two 1-shards (open hemispherical disks), and two 2-shards (open equatorial arcs). Each *closed* 2-shard is an arc between the two branch points. Each *closed* 1-shard is a closed hemispherical disc. The corresponding *closed* 1-shard in the projection Figure 6.26 is a topological 2-sphere—such as shown in Figure 6.34.

For the crossing scheme of Figure 6.25, there are no b-shards or 3-shards. The two 2-shards $\widehat{A}$ and $\widehat{B}$, each a circle are (open) shards *as well as closed* shards. Also, there are four open hemispherical 1-shards.

In the crossing scheme of Figure 6.31 there are six 3-shards corresponding to the two triple points of the projection, Figure 6.28. In addition, there are twelve 1-shards, twelve 2-shards. ∎

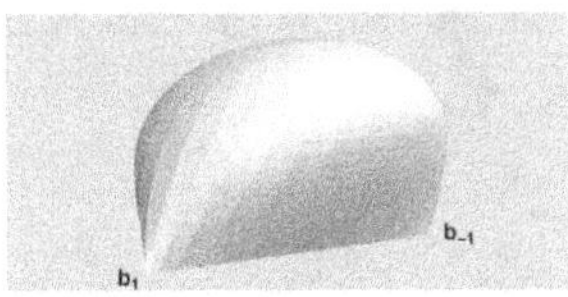

Fig. 6.34 A closed 1-shard of knotting with projection Figure 6.26.

The ideas of general position with respect to projection, shards, crossing scheme, can be extended to include manifolds with non-empty boundary and especially proper knottings. The basic ideas are the same (but increasingly verbose). We will avoid explicit details. The example of disk pair knottings will serve as a guide to this topic.

Consider a proper knotted arc $K \subseteq D^3 \approx D^2 \times D^1$ and projection to first factor $\pi\colon D^2 \times D^1 \to D^2$, Figure 6.35. Here $K \subseteq D^3$ is proper means the two endpoints of K are contained in the bounding sphere $S^2 = \partial D^3$. However, we want more. We want $\pi|_K$ to be a proper map of K which means that $\pi(\partial K) \subseteq \partial D^2 = S^1$, see Definition 2.9. An arc $K \subseteq D^3$ such as in Figure 6.36 *is* proper with $\partial K \subseteq \partial D^3$, but in projection $\pi(K) \nsubseteq \partial D^2$. So for projections of knottings such as this we will assume that our arc is not just proper but also that the endpoints lie on the annulus $\partial D^2 \times D^1$.

Also, we want "nothing unusual to happen". Figure 6.37 shows an example we want to avoid—where both endpoints of our arc project to the same point.

Additional considerations of this type occur in higher dimensions. For example, consider a projection of a knotting of a 2-disk in $K^2 \subseteq D^4 \approx D^3 \times D^1$. We want to avoid the possibility that the projection of branch points, or triple points, to points of K^2_* lie in ∂D^3.

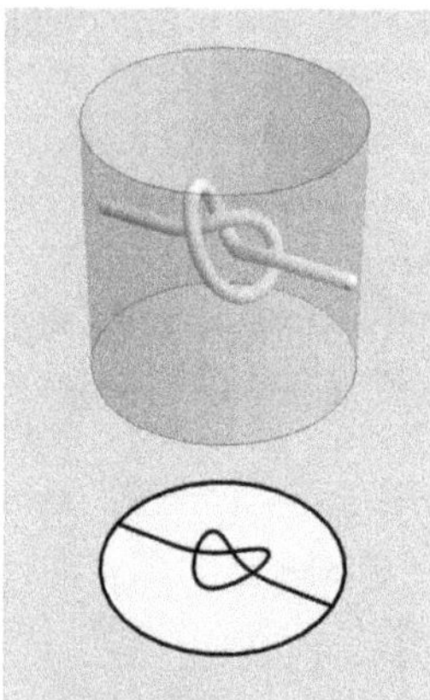

Fig. 6.35 A proper projection of a proper knotting of an arc.

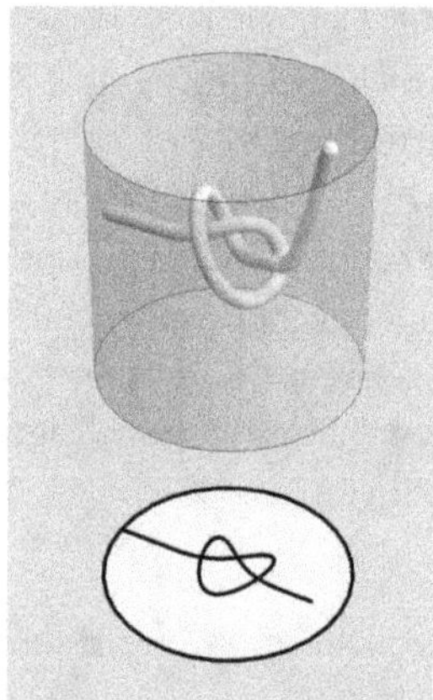

Fig. 6.36 An arc proper in a 3 which does not project to a proper curve in the 2-disk.

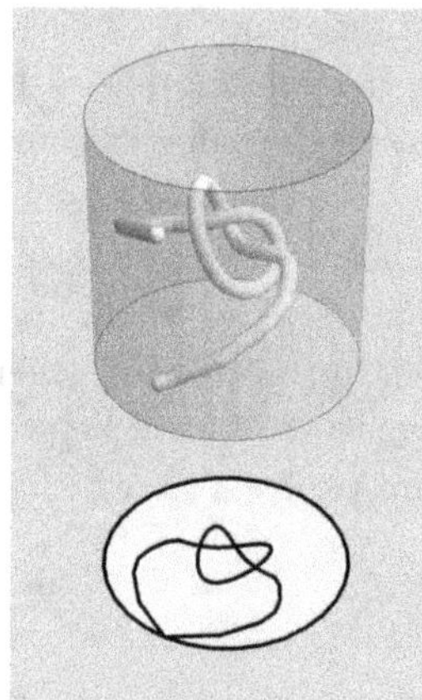

Fig. 6.37 We say this knotting is not in proper general position for projection on the boundary.

Given a finite geometric feature of a knotting K it is natural to investigate the minimum number that occurs for all equivalent knottings. The basic example for 1-knottings is the crossing number. These are generally easy to define but nearly impossible to compute except in a small number of cases. We discuss here such invariants based on projection. There are similar invariants for knot-moves and slicing.

Knottings which have projections with only 2-crossings are of special interest, for example, Proposition 6.1. So we define:

Definition 6.11. A 2-knotting K with a general position projection is **projection-simple** if K_* has no branch points and no triple points. ▮

Remark 6.1. In other words, K_* has crossing set which is a disjoint union of circles. In the literature, such as [Yajima (1964)] a knotted **surface is simple** or **simply knotted** if it can be represented by what we call here a projection-simple. (Note: This is not to be confused with the unrelated term "simple" elsewhere in higher dimensional knot theory such in [Levine (1970)].) Others have called it a **pseudo-ribbon surface diagram** [Kawauchi (2002)] and refer to the corresponding knot as a **pseudo-ribbon**. In Section 12.4 we will consider higher dimensional versions of this type of knot. Looking ahead, a pseudo-ribbon surface would, alternatively, be termed 2-**bulky**, Definition 12.11. ∎

Suppose K^2 is a projection-simple surface. Like Definition 4.16 of crossing number, let p be the smallest number of disjoint crossing circles for any equivalent knotting that has a simple projection. There does not seem to be a common term for this, we call p the **simple crossings number**. It is natural to study simple knottings with small crossing number.

For spheres and torii, any knotting with simple crossing number 2 is a trivial knotting. The spun Hopf link, Example 6.20 and turn-spun Hopf link knottings of a torus pair with crossing number 2. A non-trivial simple knotting of a sphere is the spun trefoil, Example 10.6.

Let B be the below-set of knotting K, the components of $K - B$ are called **sheets** and play the role of overpass for 1-knottings, see for example [Satoh (2009b)]. Technically, a sheet is defined as a component of a broken surface style diagram, see [Kamada (2017)]. [Saito and Satoh (2005)] [Satoh (2010)] [Satoh (2009a)]. The minimum number of sheets for a knotting representing a given knot is called the **sheet number**.

The minimum number of triple points for a knotting representing a given knot is called the **triple point number**. Knots with triple point number zero are projection-simple. Many papers have addressed triple number such as: [Nakamura (2012b)] [Mohamad and Yashiro (2012)] [Mohamad and Yashiro (2009)] [Yashiro (2007)] [Yashiro (2005b)] [Satoh (2005a)] [Satoh and Shima (2005)] [Satoh (2005b)] [Satoh and Shima (2004)] [Satoh (2002)] [Carter *et al.* (2010)] [Hatakenaka (2004)] [Oshiro (2010)]. Related numbers are the triple point cancelling numbers and [Iwakiri (2006)] [Nakamura (2013)] triple linking numbers. [Oshiro (2011)]

6.4 Visualization Methods for a Projection

Most visualization methods for knottings use techniques that reduce dimension, usually by one. The viewer is expected to essentially reconstruct the "lost dimension" information. Generally, a narrative is provided to aid the viewer in this effort. For example, for a non-mathematical two-dimensional perspective drawing of a three-dimensional scene, the lost dimension is the depth of field, and the narrative is "for identical depicted objects if one is smaller on the page, it is because it is further away". In older styles of drawing, a different narrative is used, such as "if one object is drawn to be behind another, it is because it is further away".

For the most part we need not worry about communicating exact value the "lost dimension" as long as we can relatively compare two values in at crossing points and determine the above/below relations. We analyze familiar visualization techniques for 1-knottings and the implied associated narratives—and later extend these to higher dimensions.

Locally, a crossing point of a projection of a 1-knotting looks like two intersecting line segments, Figure 6.38 but a rotation of R^3 reveals they are disjoint—perhaps as in 6.39 where we have added a bounding box as visual aid,revealing that the horizontal segment of 6.38 is above the vertical one.

We next describe several visualization methods for 1-knottings.

Visualization narrative: The projection of knotting is not drawn as thin lines but as thin flat ribbons with edges of contrasting color. This contrast visually reveals which line segment is above and which is below.

Visualization method: Figure 6.40 shows our crossing as a *white ribbon with black edges*. The horizontal line is above since its edges are visible against the vertical ribbon near the crossing, and the vertical edges are not seen, due to occlusion.

Figure 6.41 shows our crossing as a *black ribbon with white edges*. Since the white of the ribbon edge is the same as the white of the background, the effect at a crossing is to "show the segment below as a broken line". These traditional methods have been used by artists for thousands of years, and are effective since they make the above-segment more prominent, and therefore higher, in the "lost dimension".

There is an interesting graphical device for making broken-line drawings. One constructs a flat, thin, white ribbon that lies under the 1-knotting, but very close to it. Visually this is equivalent to adding a white border to the black ribbon. A broken appearance of line is produced by occlusion, assuming a white background.. The crossing in Figure 6.42 seems to have a broken line segment due to a thin white rectangle which lies at a height between the two segments, as revealed in Figure 6.43.

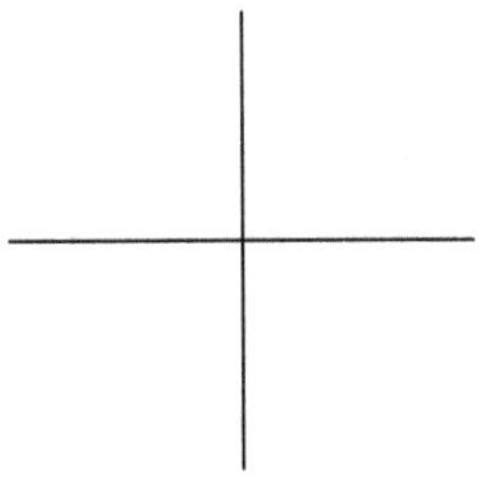

Fig. 6.38 A typical crossing.

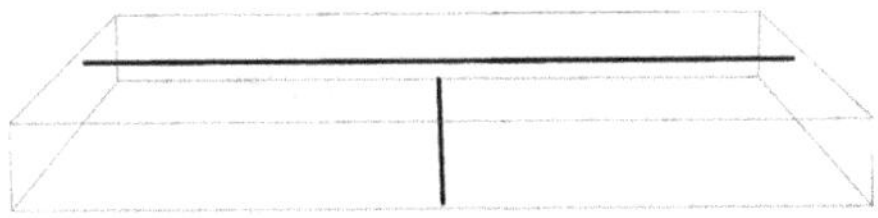

Fig. 6.39 Same figure revealing three dimensional details.

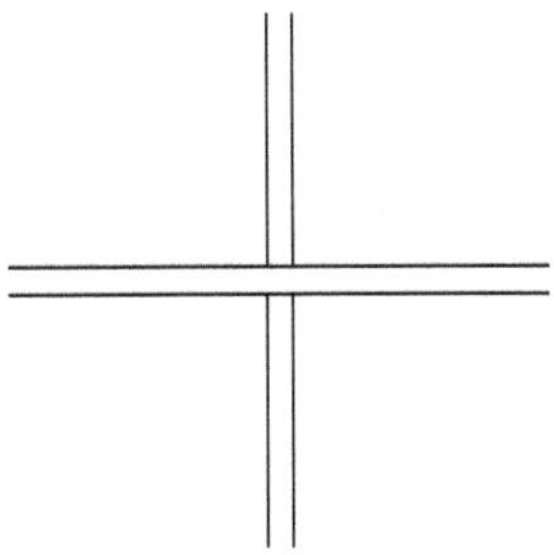

Fig. 6.40 Showing a crossing as white ribbon with black borders.

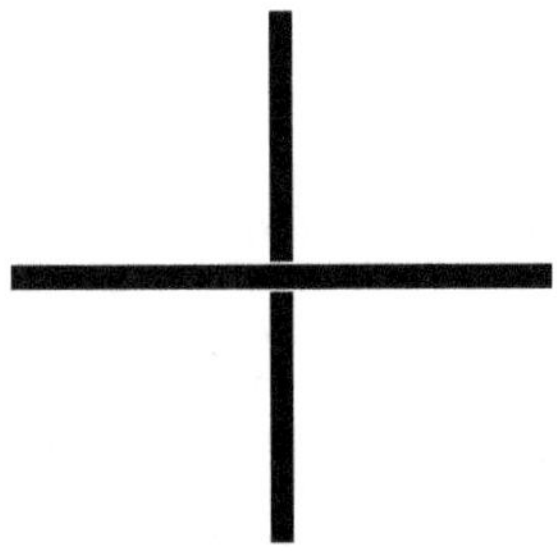

Fig. 6.41 Showing a crossing as black ribbon with white borders.

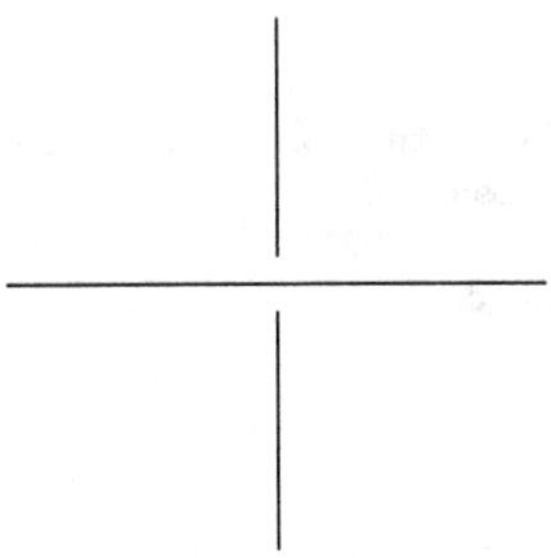

Fig. 6.42 Indicating crossing with "broken line".

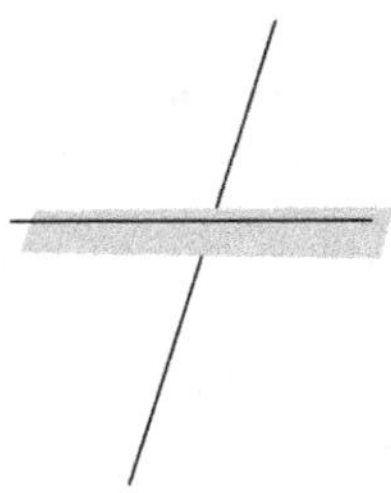

Fig. 6.43 "Broken line" graphics revealed, the white rectangle appears gray here.

This next method is not traditional. Its utility will become evident when we use three-dimensional graphics to consider a certain higher-dimensional visualization called wire-frame coding.

Visualization narrative: Suppose we view a 1-knotting in R^3 as a tube viewed from above in z-direction, as shown in Figure 6.44. Then, higher points the tube, it will appear thicker, since they are nearer.

Visualization method: Artificially maintain this apparent thickness in the projection.

Fig. 6.44 As viewed from the top of the box the portions of the knotting near the viewer will seem larger.

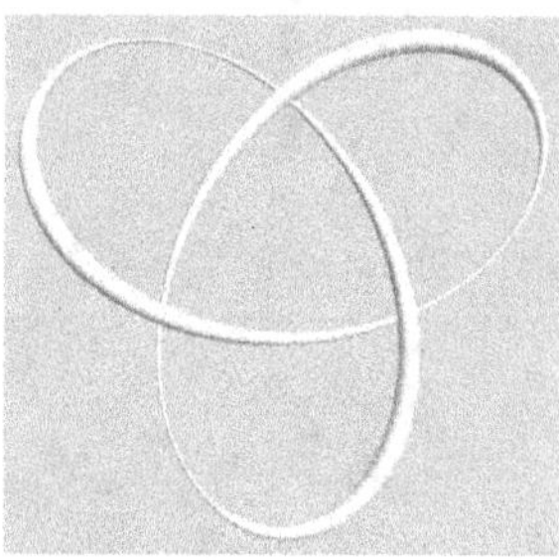

Fig. 6.45 A projection of a trefoil knotting with height of point graphically indicated by the thickness of the tube.

Fig. 6.46 Same object as Figure 6.44 shown from different viewpoint. In this view, at each crossing, we have a thinner tube piercing a thicker tube.

The bounding box of Figure 6.44 indicates the coordinate system. Here the z-direction is vertical on the page. The points near the top of the box are shown as thicker tubing. Our interest is in the projection in the z-direction of this tube, Figure 6.45. We see a projection as a tube about a planar subset of R^3. If we focus on any one crossing we locally see two intersecting tubes where the thin tube actually pierces the thick one. Figure 6.46 is a different view of this projection Figure 6.45 showing more clearly the tube self-intersections.

To be clear—in Figure 6.46 we are not rotating our knotting in R^3—we are rotating our projected planar self-intersecting tube in R^3. This tube-thickness narrative adapts easily to other settings such as the crossing scheme Figure 6.73 and wire-frame coding, Section 6.4.

Remark 6.2. We apply this idea to visualize a discrete track of a planar isotopy C_t, $0 \leq t \leq 1$, of a knotting of a simple closed curve in $R^2 \subseteq R^3$. For example consider an isotopy of the simple closed curve of Example 1.10. Figure 6.47 then displays a discrete trace in five steps of the isotopy. This figure does not convey that we start with the square and end up with a circle. But if we use tube thickness to convey relative values of t, thinner for values of t near 0, thicker for values near 1, then we have Figure 6.48, does visually show that we go from square to circle. A narrative of this idea is to consider isotopy as a change of a simple closed curve beginning at time 0, ending at time 1. We are viewing from the end of the isotopy, from present time 1—the "finish line". The circle is represented by a tube that gets thicker as time goes on, because it corresponds to a time closer to us in the present, so "more recent" tubes are more prominent than those thinner tubes in the "distant past". We will see a less trivial example in Example 7.13. ▮

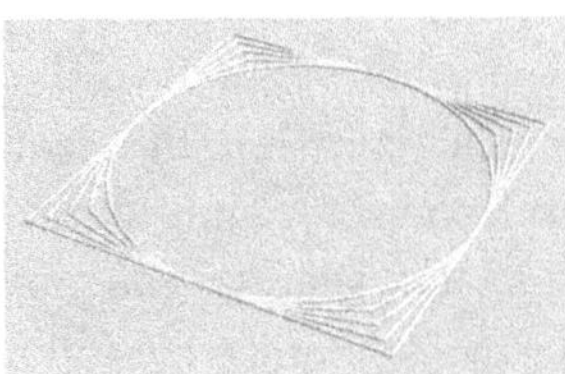

Fig. 6.47 Discrete track visualization, uniform tube thickness.

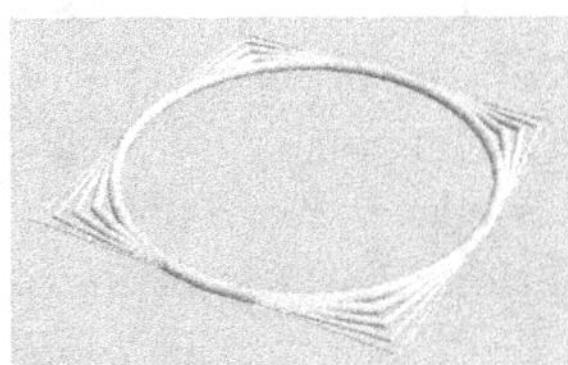

Fig. 6.48 Discrete track, with t value coded as tube thickness.

Visualization narrative: Points close to an observer seem brighter, points far away seem darker and dim.

Visualization method: We extend our tube method and render a tube both thicker *and* whiter if it has high z-coordinate, thinner *and* darker if it has low z-coordinate, Figure 6.49 shows the projection rendered in this way.

Visually such graphic representation of a 1-knotting is hardly needed. However, we will see some utility of this idea when dealing with projections of higher knottings.

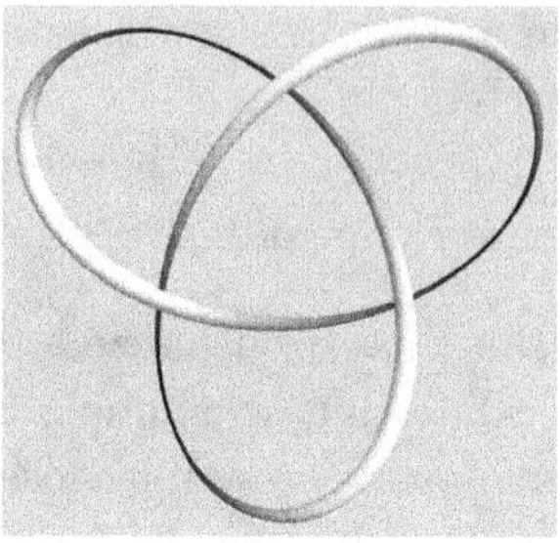

Fig. 6.49 The projection using tube thickness and gray scale.

We turn now to the topic of visualization for a projection of a 2-knotting.

Visualization narrative: Another idea is to generalize the broken-line method used for 1-knotting where we "break an under arc near a crossing point " for a 1-knotting projection. For a 2-knotting we "break an under surface near a crossing arc" for a 2-knotting projection [Carter (1995)].

Visualization method: At an arc of double points, divide the lower surface in two (i.e. "break it") transforming Figure 6.50 into Figure 6.52. At a triple point the lowest surface portion would be locally broken twice broken and appear as four pieces transforming Figure 6.51 into Figure 6.53. This is called the **broken-surface** method.

There are problems with this broken-surface method, related to the fact that the human eye is well adapted to follow lines and line-like features such as edges. Edge detection is a fundamental part of vision and visual navigation. For 1-knotting projections a broken line the eye can instinctively bridge the gap at a crossing *because of* the eye's strong edge-following inclination. It is not as natural to visually fill in the "missing surfaces". Part of the reason for this is that edge detecting visual skills draw attention along the broken edges rather than across them. Following these edges tends to draw attention *away* from the areas of self-intersection rather than *towards* it. Additionally, surface fragmenting produced by breaking makes it difficult to keep in mind the basic form of the projection. Also for computer graphics implementation this method basically requires the difficult computation of all shards.

Fig. 6.50 Projection of two planes. In this example assume that the horizontal plane is highest in the fourth dimension.

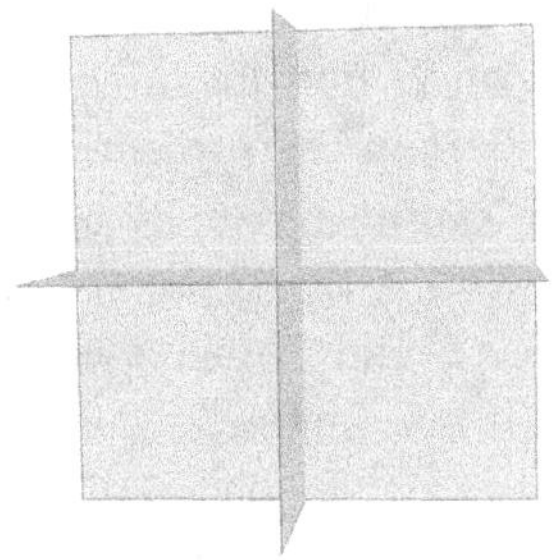

Fig. 6.51 Projection of three planes. In this example assume that the horizontal plane is highest, the plane parallel to the page lowest, the third between.

Fig. 6.52 Broken plane figure corresponding to Figure 6.50.

Fig. 6.53 Broken plane figure corresponding to Figure 6.51.

Visualization narrative: At an arc of crossing points, we have, in terms of the w-coordinate a higher of part the surface passing above a lower part. Suppose the "sun was shining at the far upper end of the w-axis" then the upper part would cast a shadow on the lower part exactly at the crossing points. This shadow coincides with 1 shard of the projection so we spread out the shadow a little on that lower part away from such 1-shards.

Visualization method: We show a small portion of that shadow (only) near the crossing set on the lower part in the projection.

In Figure 6.54 we show a local picture of an arc of crossing points in a projection. The higher part of the surface is A, the lower B. The shadow drawn on lower part B as the dark-grey area on either side of the crossing set. We call this a **crossing shadow**. In Figure 6.55 we artificially remove

A from Figure 6.54 to show, more clearly, this area. Since the shadow is shown on B but not on A, we can visually distinguish the above from the below part in the projection.

We can also visually sort out height relations at a triple point. Figure 6.56 shows a triple point where we suppose that A is the highest part, C the lowest and B in between. Although the triple point of the projection by itself is symmetric, the shadows are not, allowing the viewer to visually construct the height relations. Note that, in Figure 6.56, A has no shadows on it, B has a grey strip of shadows on it, and if viewed separately, would look like Figure 6.55. Finally, C, since it is below both A and B, has two crossing shadow strips which, if viewed separately, appear as in Figure 6.57.

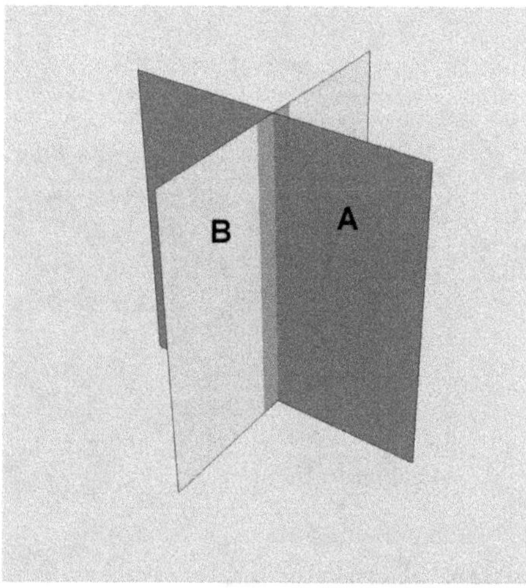

Fig. 6.54 A portion of knotting A passes over portion B. In the projection we indicate by drawing a thick shadow of A onto B.

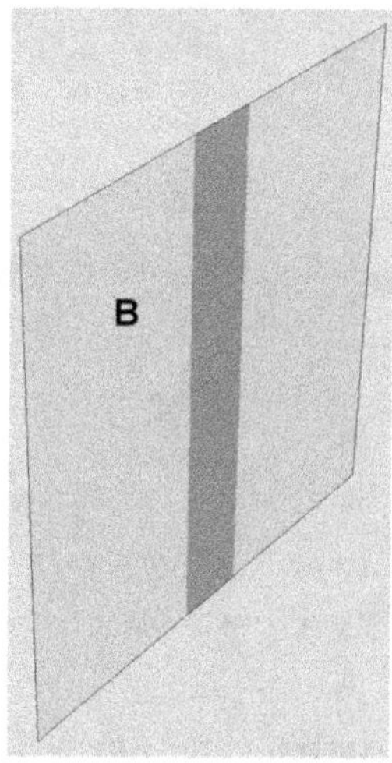

Fig. 6.55 The shadows of A and B as they lie on C.

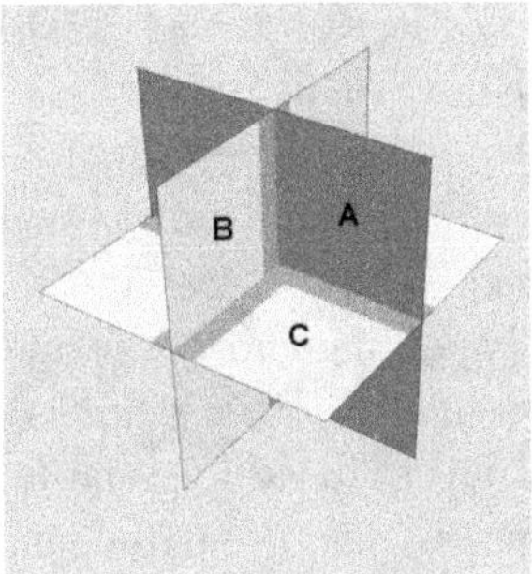

Fig. 6.56 A triple point of a projection with crossing-shadows. We assume A is high C is low with B between.

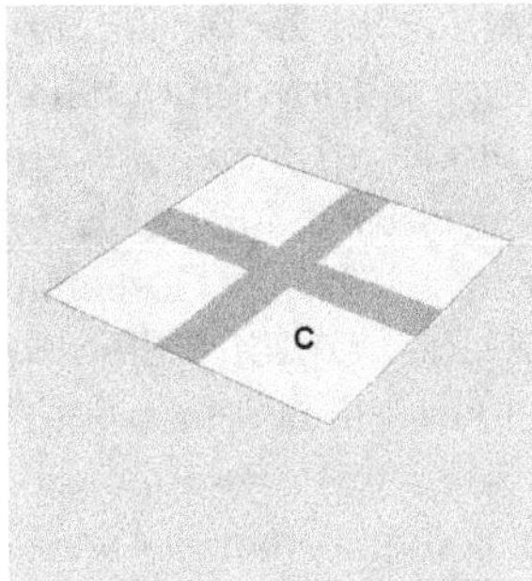

Fig. 6.57 The shadows of A and of B both drawn on C.

Example 6.10. Figure 6.58 shows the use of a shadow on a knotting whose projection is as in Figure 6.26. The crossing set is a line segment between branch points b_1 and b_{-1}. A shadow can be seen on the surface above this segment. This indicates the relative height relation. Along the 1-shard the part of the knotting seen in shadow lies below the other part. For more interesting examples, see Examples 6.19, 6.27 and 6.29. ∎

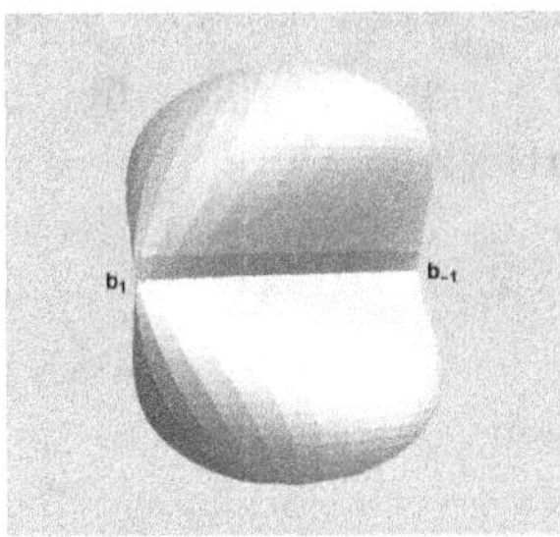

Fig. 6.58 A shadow surface above line between b_1 and b_{-1} indicating that portion of the surface is passing below the other part at the crossings.

A common visualization technique for 2-knottings uses a color gradient (or other color scheme) to indicate the fourth-dimension. For example one can use a blue-to-red color gradient and "paint" the color corresponding to the value of the fourth coordinate onto each point of the projection. Perhaps red points are high, blue points low in w-coordinate and points in the middle some shade of purple. It is relatively easy to do this coloring using computer graphics and in fact this method has been used often, for example the cover of this book where the color gradient is shown as background (graphics for Section 11.3), [Roseman (1997)] and the cover of [Roseman (1993a)]. In software, KnotPlot of Robert Scharein (knotplot.com) uses color gradient, and Geomview (geomview.org) allows implementation in its file formats. Other examples are videos such as [Roseman (1992)], [Roseman (1993b)] [Roseman (1994)], [Dalvit (2017)], and [Dalvit (2016)]. The challenge of visually adding more information for a fourth dimension, both technically and visually, is that color is critically used in three-dimensional computer graphics. In adding the fourth dimension information need to wary of harm to this three-dimensional information.

If we do not have color available (as in the pages of this book) we can use a gray-level gradient. Again, the problem is that shading is routinely used in complex and subtle ways to generate three-dimensional graphics—additional gradient "painting" might conflict with this visually.

A graphics style, **wire-frame coding**, presented here for the first time, allows us to keep most of the information of standard three-dimensional graphics of a projection.

Visualization narrative: Surfaces in computer graphics are typically presented as a union of polygons. The edges of these polygons form a 1-dimensional subset of the surface, usually called a **wire-frame**. We use ordinary rendering of the surface polygons (and thus for the most part do not conflict with color/shading of three-dimensional graphics), but the wire-frame is rendered, artificially, as thin tubes the surface, the thickness of which visually indicates w-coordinates of the surface.

Visualization method: Apply the methods used for tube thickness method Section 6.4 *only* on the wire-frame to graphically show the w-coordinate of the corresponding point of the surface. We have no color in this book, so use a grey-level coloring with light shades meaning high w-values (close to viewer) and darker shades lower w-values.

Example 6.11. In Figures 6.59 and 6.60, A and B are *disjoint* rectangles in R^4, each subdivided into an array of smaller rectangles so that the edges of the smaller rectangles form wire frames for A and B. In projection, $A_* \cap B_* = X \subseteq R^3$ is a line segment with endpoints, p and q. Each of A_* and B_* is shown with wire-frame coding.

Focusing on the points of X in Figure 6.59, the wire-frame for A at q is thicker and lighter than the wire-frame of B, as appears to be the case at all points of X. We conclude that A lies above B in R^4. Figure 6.60 is a different view of the subsets of Figure 6.59. Here we see again that the wire-frame for B_* is thinner and darker than that of A_*.

In Figure 6.61 we have a second example where A and B are *not* disjoint in R^4. At points of X in Figure 6.61 the wire-frame for A at q is thicker and lighter than the wire-frame of B, but the wire-frame for A is thinner and darker at p. Continuity of the difference of the heights of the two sets along X implies that A and B must intersect in R^4. A close inspection indicates an intersection seems to occur above the midpoint of X where the two wire-frames appear to have the same shade and thickness.

One model of two rectangles in R^4 which intersect in a single point, is the projection of a trace of a homotopy of "two skew line segments passing through each other in R^3". Figure 6.62 illustrates this with a modification of our wire-frame coding. We show the projections of our quadrilaterals as transparent. The line segments shown are w-level sets in R^4. The skew lines through p and q that are thick and light are the portions of A and B that are high in R^4. As we consider lower and lower w-values, we get thinner and darker skew lines that get closer together. In the middle of this, we see two intersecting lines, of the same apparent thickness and gray shade, indicating a real intersection of R^4. ∎

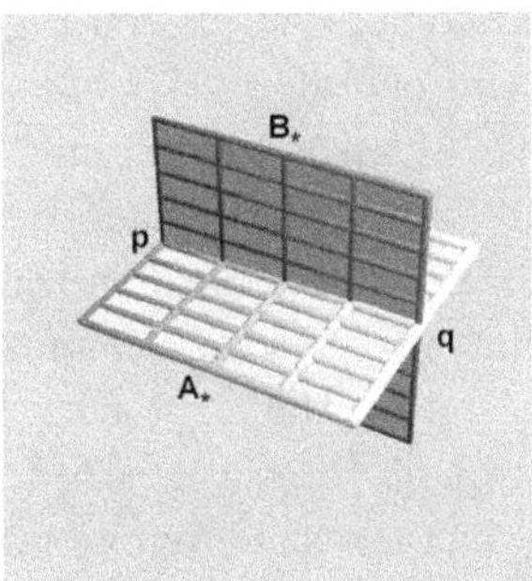

Fig. 6.59 Wire-frame coding indicates A is above B in R^4.

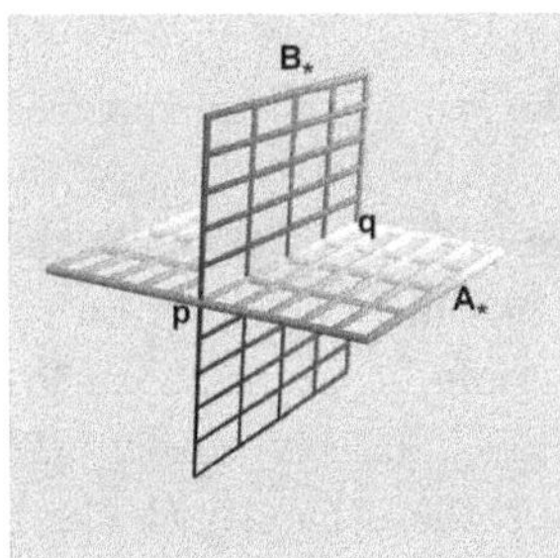

Fig. 6.60 A different view of subsets of Figure 6.59.

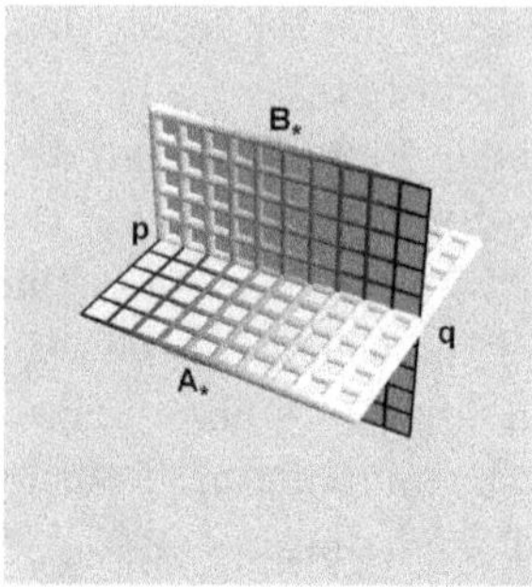

Fig. 6.61 Wire-frame coding indicates A intersects B in R^4.

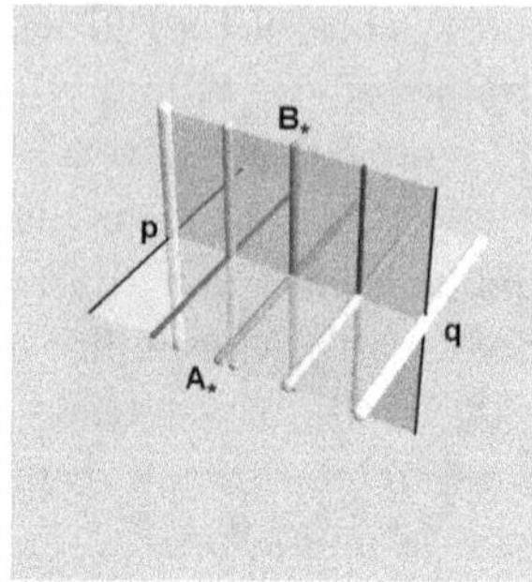

Fig. 6.62 Visualization of intersecting rectangles in R^4 as projection of homotopy.

Example 6.12. In Figure 6.63 we show using wire-frame coding, a projection $\Sigma_* \cup T_* \subseteq R^3$ of 2-knotting $\Sigma \cup T \subseteq R^4$, where Σ is a sphere and a torus T a torus, with crossing set two disjoint circles. We call this a **basic sphere-torus knotting**. The knot-type of this knotting will be called the **basic sphere-torus link**. In Figure 6.64 we make the sphere transparent to reveal an occluded portion of torus projection.

The crossing set of this knotting projection is two disjoint circles—the intersection $\Sigma_* \cap T_*$, more clearly seen if we rotate $\Sigma_* \cup T_*$ in R^3, Figures 6.65 and 6.66. The wire-frame for the sphere seems of uniform thickness and grey-level. So, all w-coordinates for Σ are nearly the w-same value. The points of the torus T must vary considerably in w-coordinate. Referring to Figure 6.63 it is clear that the wire-frame of T_* is thin and dark on the left side and thick and light on the right. Thus points of T_* to the left must be projections of points with low w-values and those on the right with high w-values. At the crossing circle seen in Figure 6.65 the torus wire-frame of the torus is clearly thicker and lighter gray than the same points on the wire-frame of the sphere—therefore at this crossing circle the torus T must be *above* the sphere Σ. On the other hand, in Figure 6.66, the torus wire-frame is clearly thinner and darker than that of the sphere and we conclude that, at this crossing circle, the torus T must be *below* the sphere Σ. ∎

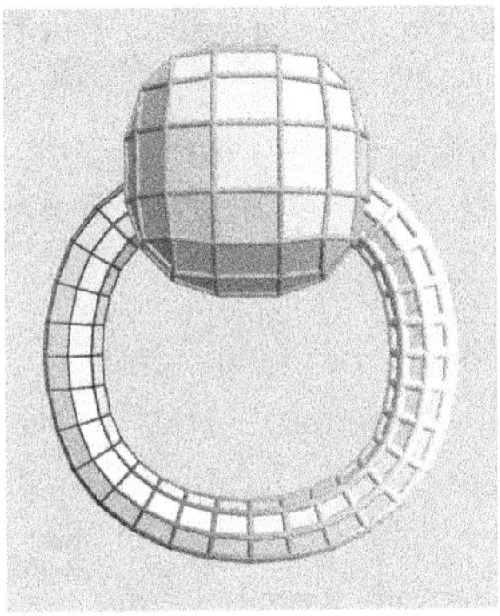

Fig. 6.63 The basic sphere-torus knotting with wire-frame coding.

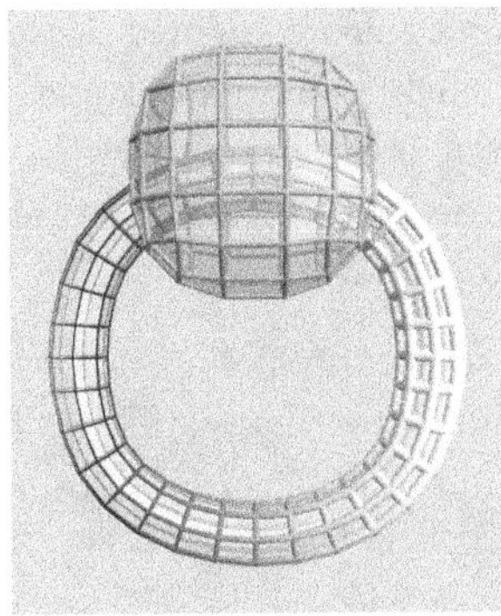

Fig. 6.64 Figure 6.63 with transparency.

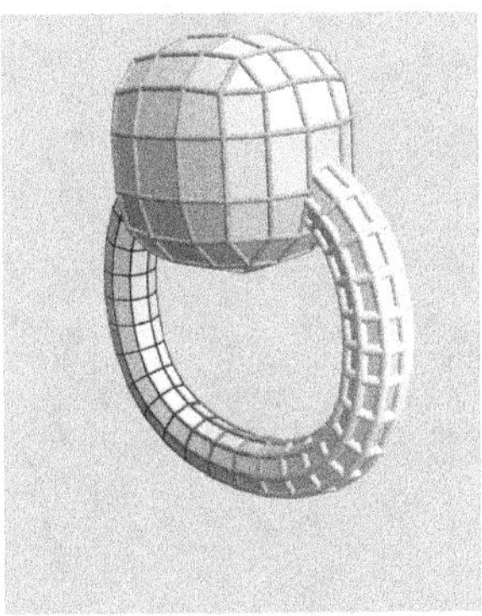

Fig. 6.65 Figure 6.63 showing area near one crossing circle.

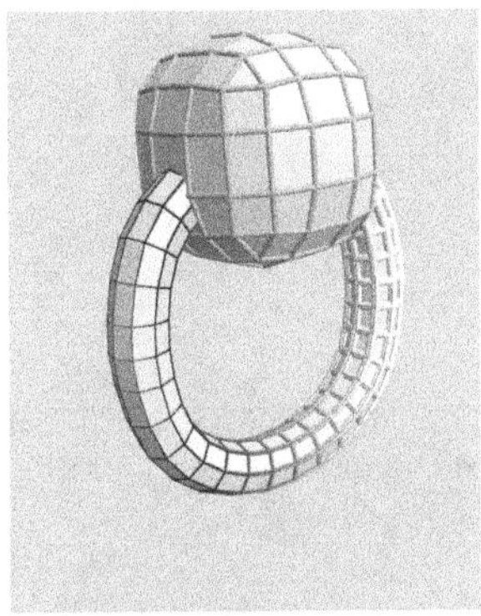

Fig. 6.66 View of Figure 6.63 showing area near other crossing circle.

Cartesian product considerations gives a natural knotting of the torus in R^4, Remark 2.3.

Definition 6.12. The **standard product knotting of the torus in** R^4 is the subset $T^2 = (S^1 \times S^1) \subseteq (R^2 \times R^2) \approx R^4$. ▮

Example 6.13. The standard product torus knotting $T^2 \subseteq R^4$ provides a good introductory visualization exercise. The standard projection $T^2_* \subseteq R^3$ is the cylinder $T^2_* = \{(x, y, z) \colon x^2 + y^2 = 1 \text{ and } -1 \leq z \leq 1\}$. Except for the ends of the cylinder, the points of T^2_* are 2-fold points but clearly not transverse intersections, so T^2 is not in general position with respect to projection.

Rotating T^2 slightly in R^4 by small angle in the xw-direction we obtain $\mathcal{T}^2$, which *is* in general position, see Figure 6.67. The crossing set of $\mathcal{T}^2_*$ has four branch points, two arcs of crossing points. Wire-frame coding near the crossing set shows the relative height relations. We slice $\mathcal{T}^2$ and show the lower half in Figure 6.68 to help visualize Figure 6.67. ∎

Other visualization techniques, not possible in book form, involve animations [Roseman (1992)], [Roseman (1992)] [Roseman (1994)], [Dalvit (2016)], [Dalvit (2017)] or visual environments such as [Francis *et al.* (1995)].

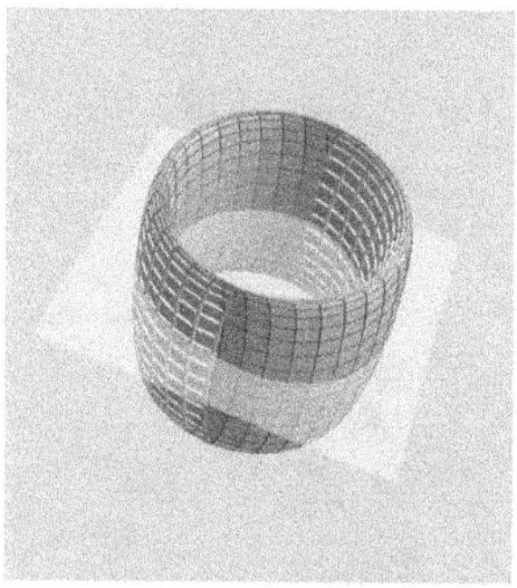

Fig. 6.67 $\mathcal{T}^2_*$, a projection of slightly rotated standard product knotting, and plane P.

Fig. 6.68 Lower half of $\mathcal{T}^2_*$.

6.5 Scheme Knottings

Looking at a crossing scheme such as Figure 6.31 reminds one of the projection a knotting—in this case perhaps some split links or perhaps some Hopf links as in Figure 1.17. In fact, we will show how a crossing scheme $\mathcal{C}$ for a knotting of a 2-manifold M can be considered as a projection of a knotting of a 1-manifold $\overline{\overline{\mathcal{C}}} \subseteq M \times R^1$, called a **scheme knotting**. This manifold and it's possible orientation (see Section 6.6 considered in conjunction with high-dimensional knot-moves are useful, considered from point of view of algebraic topology, a topic beyond the scope of this book.

Suppose we have a 2-knotting $K = \phi(M)$ where $\phi\colon M \to R^4$ a smooth embedding. We first describe a graphical method that visually changes a crossing scheme into a knot-diagram contained in M and later formalize this in Definition 6.13. For each triple point $p_* \subseteq K_*$ point we locally change Figure 6.70 to look like Figure 6.71.

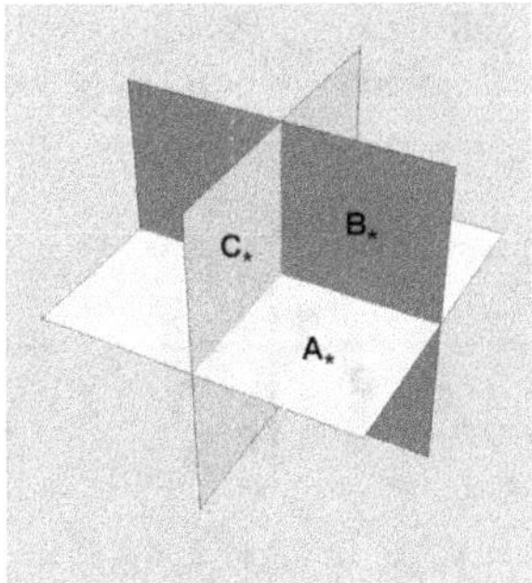

Fig. 6.69 Local figure for triple point of K_*.

At p_* we locally have three disks in K_* as in Figure 6.69. These intersecting disks, A_*, B_* and C_* correspond to three *disjoint* disks $\widehat{A}$,$\widehat{B}$ and $\widehat{C}$ in M, see Figure 6.70. We assume that these are labeled in decreasing height in R^4. That is, $p_a \in \widehat{A}, p_b \in \widehat{B}$ and $p_c \in \widehat{C}$ with $\pi \circ \phi(p_a) = \pi \circ \phi(p_b) = \pi \circ \phi(p_c) = p_*$, and $h(p_c) < h(p_b) < h(p_a)$. In $\widehat{A}$, the line segment b corresponds to the intersection $A_* \cap B_*$ and c corresponds to $A_* \cap C_*$. Since $\widehat{B}$ is higher than $\widehat{C}$ at p_* we graphically indicate this in Figure 6.71 by representing c as a broken line segment. The other two crossing gaps are determined similarly as shown. Continuing in this way for all triple points we will obtain what can be interpreted as a projection of a 1-knotting drawn on M rather than R^2. Here the above/below refers to the R^1 coordinate in $M \times R^1$.

For the crossing scheme of Figure 6.31 we will have trivial split link a pair of circles for each sphere. Figure 6.73 shows this using using an adoption of the tube thickness method of tube thickness of Section 6.4 where tube thickness indicates height of a point.

When graphically presenting a scheme knotting, we sometimes wish to graphically distinguish above 2-shards and below 2-shards. We will use two methods. In Figure 6.71 disk $A \subseteq K$ is highest, and so the two intersecting line segments are portions of *above* shards; C is lowest and those segments correspond to *below* shards. Disk B is in the middle. Points of line segment a are on B and *below* corresponding points of A and so a is a portion of a *below* shard. Similarly, line segment c is a portion of an *above* shard. We use the broken line method to represent this information. Note that graphically on B we draw the below line segment a so as to pass above the above line segment b. A second rendering method is shown in Figure 6.72

where we graphically distinguish above and below shards by additionally rendering below shards with gray dashed style. (If one has use of color it is simpler to have these styles represented by different colors.)

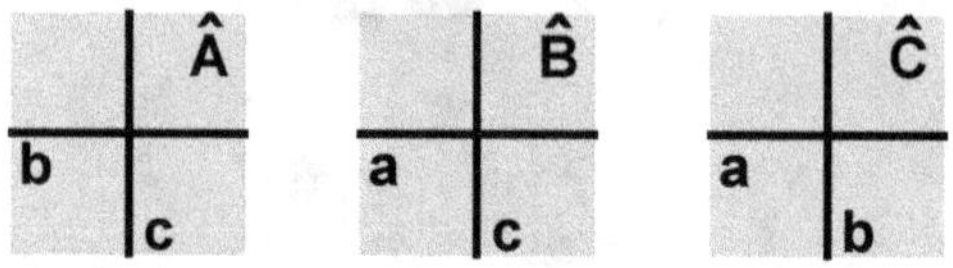

Fig. 6.70 Three disks in M with portions of crossing scheme.

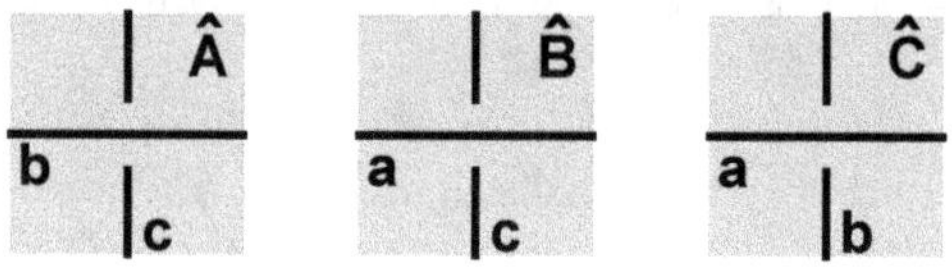

Fig. 6.71 Graphical local resolution of crossing scheme if the above 3-shard is in left square, middle 3-shard center in square, bottom 3-shard in right square.

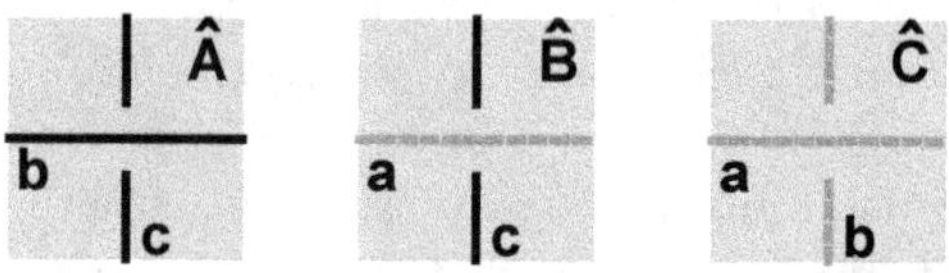

Fig. 6.72 Version of Figure 6.71 with below shards drawn in different style.

Remark 6.3. In general, a crossing scheme is an image of immersed circles. There are a few general things we can say about these immersed circles, if they contain branch points. In general, any branch point b_1 is an intersection of two paired two 2-shards, σ_a and σ_b, in a crossing scheme where σ_a is above, σ_b is below. If we extend these arcs in the crossing scheme, away from b_1, they must eventually join together at a point b_{-1}, which must be another branch point. We call the pair of points, $S = \{b_1, b_{-1}\}$, a **0-dimensional branch-sphere**. For terminology consistency with high-dimensional branch points, we also call S a **branch-manifold**. If γ_a is this extension of σ_a, it must consist of above-points, except for b_1 and b_{-1}. Denote the similar extension of σ_b by γ_b. Let $G = \gamma_a \cup \gamma_b$, and consider the projection G_*. In some knottings, there may be triple points on G_*, for example Figure 7.122, but the path G_* cannot intersect itself, since that would create a point of order four, which is forbidden in a general position projection. So G_* is actually an embedded arc in R^3, and *so* G *is a smoothly embedded circle* in M.

The crossing scheme for a knotting of a surface will be a union of branch points and 1-shards, some above-shards, some below-shards. As we can see in the local models, these all fit together to form a collection of smooth immersed curves in M. Some of these curves will consist entirely in above-points—we call these a-**curves**. Similarly, b-**curves** are those curves containing only below-points. Finally, there are some that contain two branch points which separate above-points and below-points. We call these ab-**curves**. ∎

We describe a smooth 1-manifold $\overline{\overline{\mathcal{C}}} \subseteq M \times R^1$ we call the **scheme knotting** which projects to M with image $\widehat{C}$. (This is basically the same as the unnamed manifolds, denoted Δ_i, less concretely described in [Roseman (2000)] where they were called double points.)

Definition 6.13. Suppose $\phi\colon M \to R^4$ a smooth embedding of a surface with knotting $K = \phi(M)$. Suppose $\widehat{C} \subseteq M$ is the crossing scheme of this knotting. Let $\widehat{x} \in \widehat{C}$ and let $x_* = \pi \circ \phi(\widehat{x}) \in R^3$. We then express $\phi(\widehat{x}) = (x_*, h) \in R^3 \times R^1 \approx R^4$. We will define $\overline{\overline{\mathcal{C}}}_{\widehat{x}} \subseteq \{\widehat{x}\} \times R^1 \subseteq R^k \times R^1$ in three cases:

(1) If $\widehat{x} \in M$ is a branch point of $\widehat{C}$ then $\phi(\widehat{x})$ is a branch point of K which can be written as $(x_*, h_0) \in R^3 \times R^1 \approx R^4$. In this case, $\overline{\overline{\mathcal{C}}}_{\widehat{x}}$ will be the single point $\overline{\overline{\mathcal{C}}}_{\widehat{x}} = (\widehat{x}, h_0) \subseteq M \times R^1$.

(2) If $\widehat{x}$ is a pre-2-crossing point of $\widehat{C}$ then there are two points $\widehat{x_0}, \widehat{x_1} \in \widehat{C}$ with $\pi \circ \phi(\widehat{x_0}) = \pi \circ \phi(\widehat{x_1}) = x_* \in \mathcal{C}$. So we can write $\phi(\widehat{x_1}) = (x_*, h_0)$ and $\phi(\widehat{x_1}) = (x_*, h_1)$. In this case $\overline{\overline{\mathcal{C}}}_{\widehat{x}}$ will be two points of $M \times R^1$: $(\widehat{x_0}, h_1)$ and $(\widehat{x_1}, h_0)$

(3) If $\widehat{x}$ is a pre-3-crossing point of $\widehat{C}$ then there are three corresponding pre-3-crossing points $\widehat{x_0}, \widehat{x_1}, \widehat{x_2} \in \widehat{C}$ with $\pi \circ \phi(\widehat{x_0}) = \pi \circ \phi(\widehat{x_1}) = \phi(\widehat{x_2}) = x_* \in \mathcal{C}$. So we can write $\phi(\widehat{x_1}) = (x_*, h_0)$, $\phi(\widehat{x_1}) = (x_*, h_1)$ and $\phi(\widehat{x_2}) = (x_*, h_2)$. In this case $\overline{\overline{\mathcal{C}}}_{\widehat{x}}$ will be six points of $M \times R^1$:

- $(\widehat{x_0}, h_1)$ and $(\widehat{x_0}, h_2)$
- $(\widehat{x_1}, h_0)$ and $(\widehat{x_1}, h_2)$
- $(\widehat{x_2}, h_0)$ and $(\widehat{x_1}, h_1)$

Define the **scheme knotting of** $\widehat{C}$ to be $\overline{\overline{C}} = \bigcup_{\widehat{x}\in\widehat{C}} \overline{\overline{C}}_{\widehat{x}}$. The **projection of the scheme knotting** $\pi\colon \overline{\overline{C}} \to M$ is the restriction of the projection $R^k \times R^1 \to R^k$. ▌

One can show that $\overline{\overline{C}}$ is a (smooth) manifold. To do this one needs to incorporate the fact that the crossing scheme is the image of immersed 1-manifolds.

Example 6.14. In the Example 6.7 there are only three values of w for points of this knotting: 0 for points of K_0, 1 for points of K_1 and 2 for points of K_2. We adopt the method of tube-thickness Section 6.4 to the scheme projection in place of the standard projection and can present the crossing scheme as a projection of three split links-one on each of M_0, M_1 and M_2, Figure 6.73. ▌

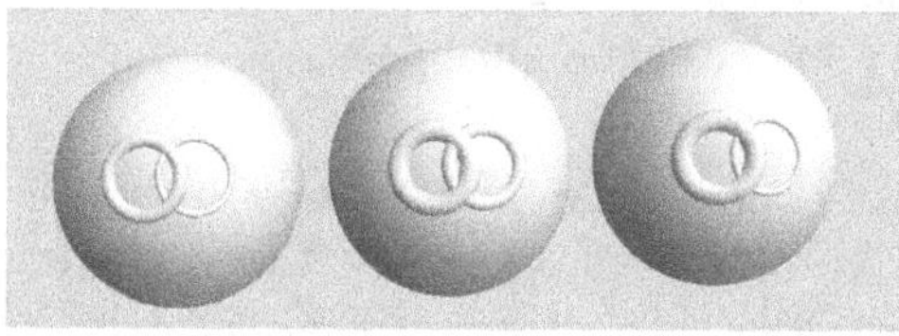

Fig. 6.73 Projection of the scheme knotting using tube thickness method.

To review our definitions and notations for an embedding of manifold $\phi\colon M^2 \to R^4$ with $K = \phi(M^2)$ with crossing scheme $\widehat{C}$ and i denoting subset inclusions we have:

$$\begin{array}{ccccccc}
\overline{\overline{C}} & \xrightarrow{\pi_1} & \widehat{C} & \xrightarrow{\phi} & \widetilde{C} & \xrightarrow{\pi} & C \\
\downarrow i & & \downarrow i & & \downarrow i & & \downarrow i \\
M^2 \times R^1 & \xrightarrow{\pi_1} & M^2 & \xrightarrow{\phi} & K & \xrightarrow{\pi} & K_*.
\end{array}$$

6.6 Orientable Knottings, Orientable Shards

Suppose K is a 2-knotting of an *orientable* manifold. We will show that a choice of orientation of K determines an orientation of each 2-shard $C \subseteq K_*$. This will determine an orientation for the scheme knotting $\overline{\overline{C}}$. This is useful in applications of algebraic topology, which is beyond our chosen scope.

An orientation of K gives a smooth framing F_x of the tangent plane T_x. This determines a based vector at each non-branch point of $x_* \in K_*$ called the **orientation normal**, as follows. Suppose $x \in K$ with $\pi(x) = x_*$. Since x is not a branch point, the image T_x^* under the derivative map is a plane in R^2 with image frame F_x^*. A vector $\overrightarrow{n}_x$ normal to T_x^* such that the ordered triple of the two ordered vectors of F_x^* followed by $\overrightarrow{n}_x$ is a right-handed frame for R^3 will be called an **orientation normal** corresponding to the given orientation of K.

Suppose K is a knotting of M. Figure 6.74 shows a neighborhood in R^3 of a generic point of a 2-shard modeled as two intersecting planes A_* and B_* which have corresponding pullbacks, $\widehat{A}$ and $\widehat{B}$, in M. Here, as in Figure 6.54, we use the shadow method to graphically indicate that $\widetilde{A}$ is above $\widetilde{B}$ in R^4. These planes happen to be orthogonal, but all we need is transversality. Shown are normals to A_* and B_*. The cross-product of these two normals is shown as a gray vector. This direction, upward in the Figure, gives an **orientation of the 2-shard** shown as black intersection line. The smooth nature of the tangent frames for K assures smoothness of this tangent vector for all points of the 2-shards of K_*.

Definition 6.14. Suppose C is a 2-shard and let $\widetilde{C_a}$ and $\widetilde{C_b}$ are the liftings of C to K and $\widehat{C_a}$ and $\widehat{C_b}$ the pullbacks of C to M. An orientation of C can be lifted to K and is called the **lifting of the orientation of C to $\widetilde{C_a}$**. This, in turn, gives rise to an orientation for the pullback $\widehat{C_a}$ called the **pullback of the orientation of C to $\widehat{C_a}$** , see Figure 6.75.

Similarly, we could pullback the orientation of C to $\widetilde{C_b}$ and $\widehat{C_b}$, respectively. **However, we do not use these orientations—we use the reverse orientations.** So, the **lifting of the orientation of C to $\widetilde{C_b}$** is the lifting of the *reverse* orientation of C, and the **pullback of the orientation of C to $\widehat{C_b}$** is the pullback of the *reverse* orientation of C, see Figure 6.76. ▮

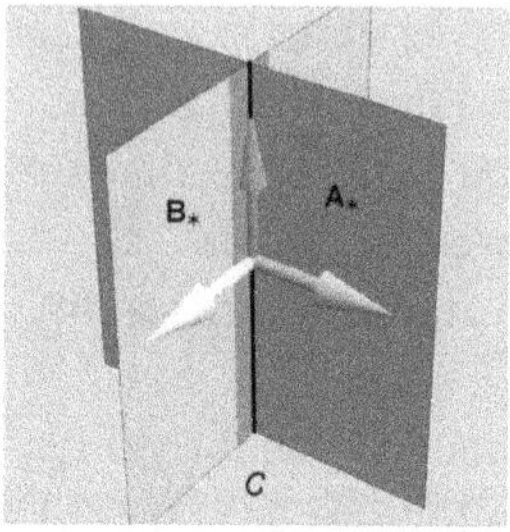

Fig. 6.74 2-shard C as intersection of projections A_* and B_* where $\widehat{A}$ passes above $\widehat{B}$.

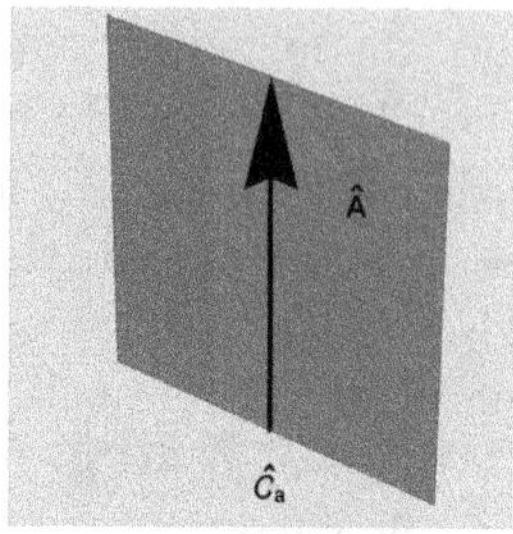

Fig. 6.75 Orientation of pullback of above shard of Figure 6.74.

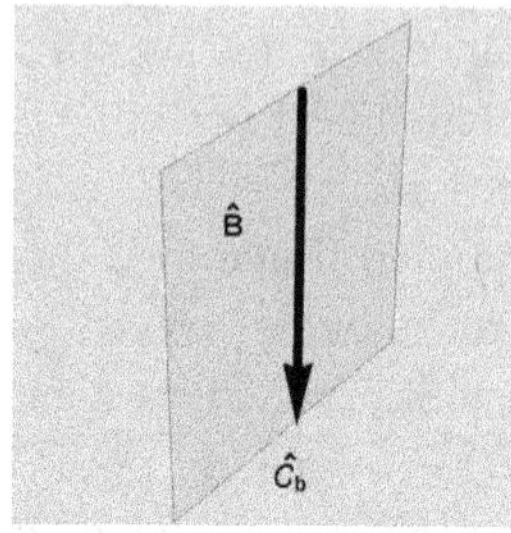

Fig. 6.76 Orientation of pullback of below shard of Figure 6.74.

Example 6.15. Consider the projection of Figure 6.7 of a knotting K of $M = S^2$ with the crossing information given in Figure 6.58.

There is a single 2-shard C in K_* whose endpoints are the two branch points b_1 and b_{-1}. Figure 6.77 shows a frame at a point of this 2-shard C. The thick light colored arrow represents a normal to the portion of the knotting projection passing above. The thin arrow is a normal to the portion of the knotting projection passing below. The darkest arrow represents the cross-product of this ordered pair, which completes this ordered pair into an ordered triple which is a right-handed frame. This vector gives direction to C.

Figure 6.27 shows a pullback of the (unlabeled) crossing set C. The pullback of C consists of an above shard $\widehat{C_a}$ and below shard $\widehat{C_b}$, see Figure 6.78 which meet at the pullback of branch points $\widehat{b_1}$ and $\widehat{b_{-1}}$. The crossing scheme is the union of these and is a topological circle. The orientations are shown as arrows near $\widehat{b_{-1}}$. The orientation of C is from to b_{-1} to b_1 and so the pullback of this an orientation along arc from $\widehat{b_{-1}}$ to $\widehat{b_1}$. However, the pullback of the orientation to $\widehat{C_b}$ is the reverse—along this arc from $\widehat{b_1}$ to $\widehat{b_{-1}}$. Taken together, they give a constant orientation of the crossing scheme. ∎

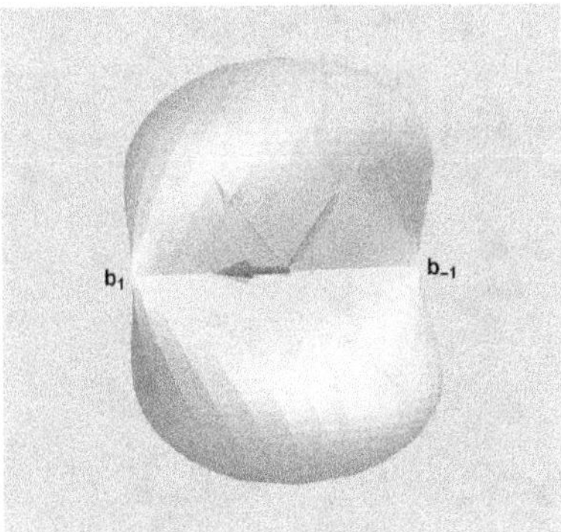

Fig. 6.77 A frame to orient the 2-shard of Example 6.15.

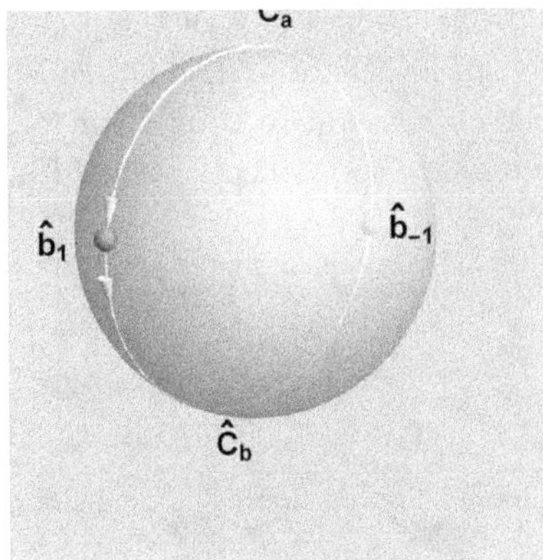

Fig. 6.78 Consistent orientation to the circle pullback.

Example 6.16. Projection of the basic sphere-torus linking, Example 6.12, we have two 2-shards both circles. We will orient both the sphere and torus with outward pointing normals and view the height relations as viewed in Figure 6.79. The thick light gray vector is normal to surface which is higher at these crossings, the thin light gray vector is normal to the surfaces which is lower—consistent with the thickness/thinness of tubes shown on the surface. The cross-product, shown as dark vector, is the cross-product used to orient the 2-shard. Figure 6.80 shows these same frames without using wire-frame coded graphics.

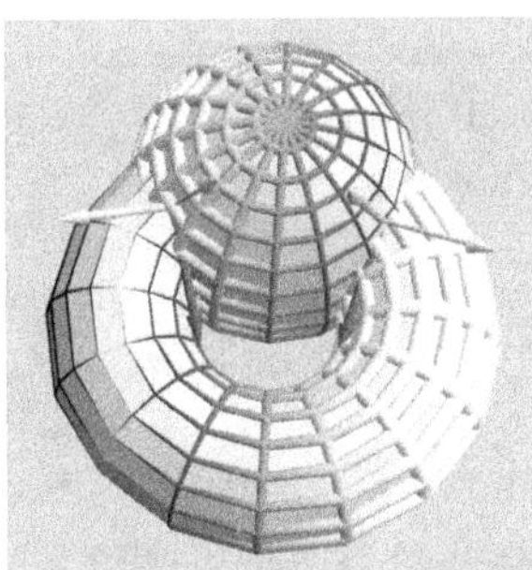

Fig. 6.79 Two frames used to orient the 2-shards, with tube-frame visualization.

The orientations of these circles on the torus component of M are as shown in Figure 6.81. The reader can check that the crossing scheme that lies on the sphere component of M bounds an annulus on M with both boundary components oriented in the same direction. ∎

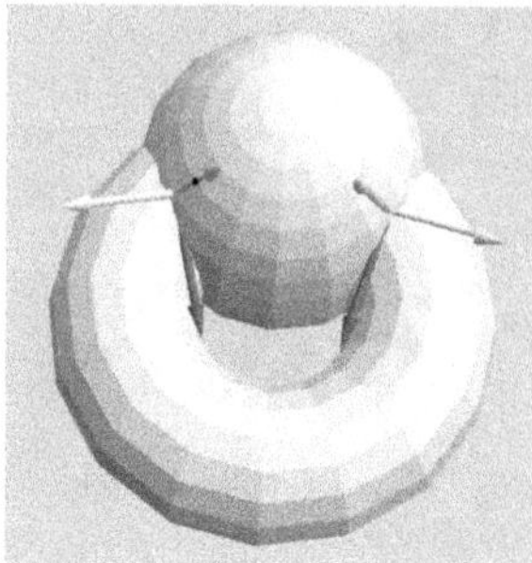

Fig. 6.80 Two frames used to orient the 2-shards.

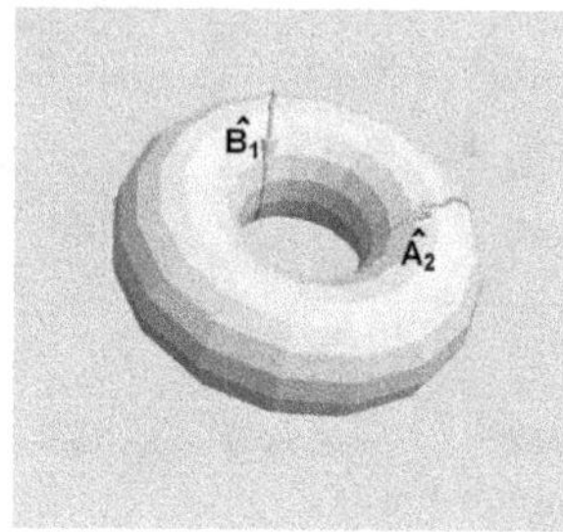

Fig. 6.81 The oriented pullback 2-shards on the torus.

6.7 Constructing a 2-Knotting in 4-Space from an Immersion in 3-Space

A traditional graphic way to construct knottings in R^3, by artists and by mathematicians, is to take one or more simple closed curves in R^2 and change each crossing to a disjoint arc pair—in one of two different ways depending on our above/below choice, Section 4.2.

Can we imitate this and use a self-intersecting surface in R^3 to obtain a 2-knotting in R^4? The answer is "sometimes". There are nice images of surfaces in R^3 which are not the projection of any 2-knotting, [Giller (1982)]. However, if the self-intersection set is not complicated, then we *can* succeed in finding a knotting.

Example 6.17. Consider two annuli A_* and B_* in R^3 which intersect transversely in circle C. We construct *disjoint* annuli A and B in R^4, whose projections are A_* and B_*. We will do this so the above/below information, graphically shown by shadow coding, see Section 6.4, can be either Figure 6.84 or Figure 6.85. In addition we will not change the annuli on the boundary—that is $\partial A = \partial A_*$ and $\partial B = \partial B_*$. The cartesian product based construction, used next, appears often, in various forms, in the rest of this book.

In Section 4.2 we described a construction that changes a pair of intersecting line segments in $R^2 \subseteq R^3$, as in Figure 4.41 to either the disjoint arcs of Figure 4.42 or of Figure 4.43. Our circle $C \in R^3$ has a product tubular neighborhood $T^3 \approx S^1 \times D^2 \subseteq R^3$ *and* product tubular neighborhood $T^4 \approx S^1 \times D^3 \subseteq R^4$. Our basic idea is to use these product structures to make a change such as from Figure 4.41 to Figure 4.42 (or Figure 4.41 to Figure 4.43) at each point of C. For each $\theta \in C$, $0 \leq \theta \leq 2\pi$, write $D^2_\theta = \{\theta\} \times D^2 \subseteq T^3$ and $D^3_\theta = \{\theta\} \times D^3 \subseteq T^4$. For any θ, we will have $\pi(D^3_\theta) = D^2_\theta$. Figure 6.82 shows disks D_θ for several values of $\theta \in C$.

Consider the two intersecting line segments a_* and b_* in D^3 of Figure 4.41. We can assume our annuli in T^4 correspond to : $A_* = S^1 \times a_* \subseteq S^1 \times D^2 \subseteq S^1 \times D^3$ and $B_* = S^1 \times b_* \subseteq S^1 \times D^2 \subseteq S^1 \times D^3$, Figure 6.83.

Replace *intersecting* a_* and b_* in D^3 by *disjoint* a and b in D^3 (but no longer subsets of D^2), as in Figure 4.42, and we will get *disjoint* annuli in T^4: $A = S^1 \times a \subseteq S^1 \times D^3$ and $B = S^1 \times b \subseteq S^1 \times D^3$. In the standard projection, we will have $\pi(A) = A_*$ and $\pi(B) = B_*$, and since a passes above b at c, we will have A passing above B at all points of C as in Figure 6.84. Similarly, had we instead used a and b of Figure 4.43, the annuli produced would have projection as in Figure 6.85. ∎

Fig. 6.82 A tubular neighborhood of C is a solid torus $T^3 \approx S^1 \times D^2$. Shown are some disks D^2_θ where $\theta \in C$.

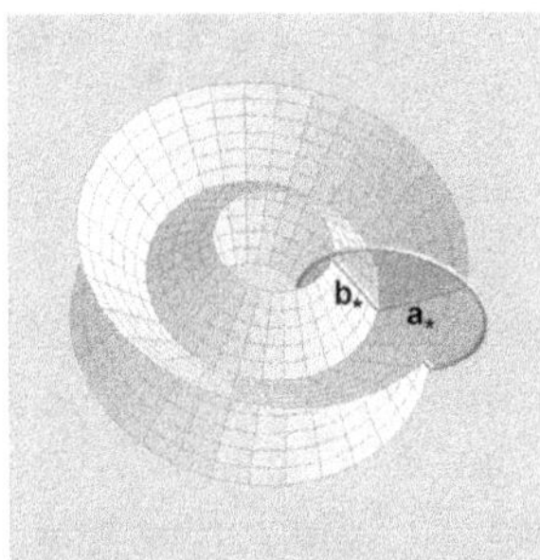

Fig. 6.83 Intersection of a D^2_θ with two annuli gives two crossing line segments in that disk.

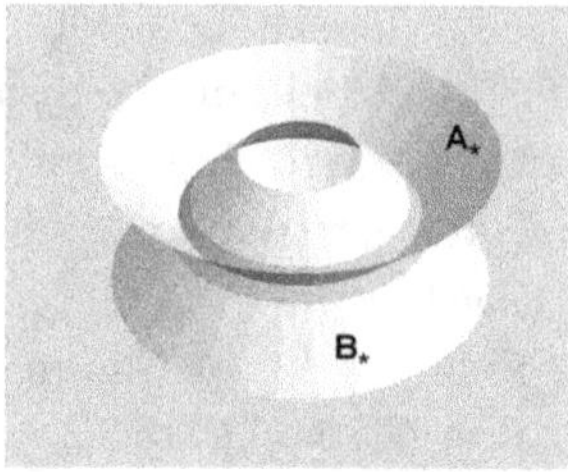

Fig. 6.84 Using crossing-shadows to indicate A passes above B along C.

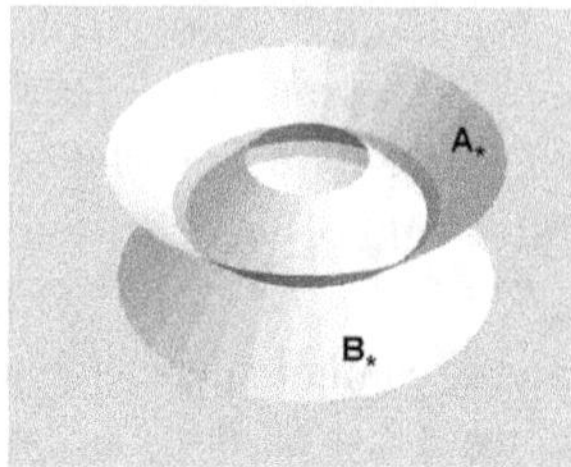

Fig. 6.85 Using crossing-shadows to indicate A passes below B along C.

For the rest of this Section, we consider the image of an immersion of a sphere $X \subseteq R^3$, Figure 6.86. It has 1-spheres of self-crossing points C_1, C_2, C_3, and C_4, see Figure 6.87. Using local construction, Example 6.17, we produce knottings of a 2-sphere in R^4 with projection X.

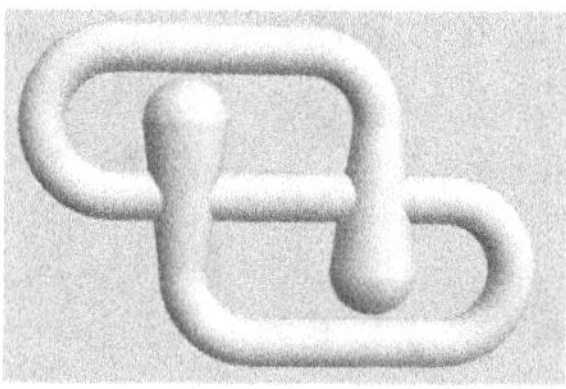

Fig. 6.86 Immersion image of a sphere $X \subseteq R^3$.

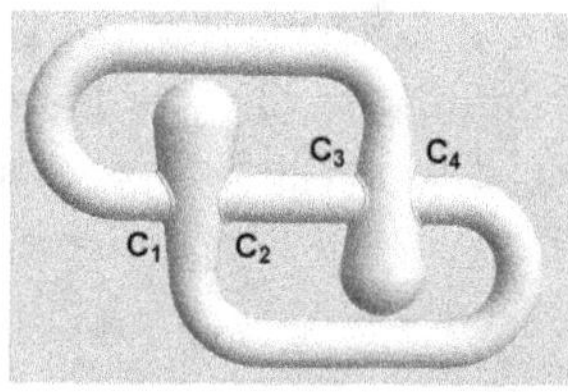

Fig. 6.87 Self-intersections shown in Figure 6.87 are C_1, C_2, C_3, and C_4.

Example 6.18. With four self-intersection circles and two above/below possibilities at each we could consider $2^4 = 16$ knottings. However, some of these knottings represent the same knot. It turns out that there are exactly three knots whose projection is X. In this section we describe these knottings and the associated crossing schemes and shard structures.

There is no standard naming of higher dimensional knots. We will call these three knots: the **spun trefoil**, the **Fox ribbon sphere**, and the **trivial knot**. The spun trefoil is Example 10.6. Example 10.9 shows this to be knotting as in Figure 9.6 whose projection has crossing set of four disjoint circles. The Fox ribbon sphere is often called "Example 10 of Fox" since it appears as "Example 10" of [Fox (1962a)], see Example 9.4. The Fox ribbon sphere was called a miswoven trefoil in [Roseman (1974)], however "Fox ribbon sphere" is now preferred. ∎

Any knotting which projects to X, has no branch points (b-shards) or triple points (3-shards) and each circle C_1, C_2, C_3, and C_4 is a 2-shard. The remainder of X consists of 1-shards $D_1, D_2, D_3, D_4, P_1, P_2, N_1, N_2$, and N_3. Three are visible in Figure 6.88, others are occluded from view and lie inside the dark surface shown in Figure 6.89. Figures 6.91 and 6.92 reveal the shards hidden inside this surface. Each D_i is a disk, Figure 6.90. Each N_i is an annulus, Figure 6.91. Each of the remaining shards P_1, P_2 is a pair of pants. For example, P_1 has, as boundary, circles C_1, C_2, and C_4 as shown in Figure 6.88.

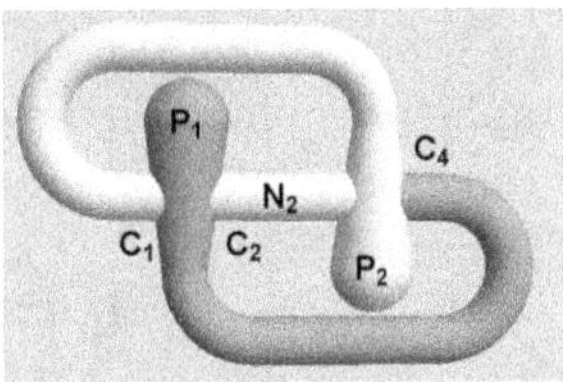

Fig. 6.88 1-shards N_2, P_2 and P_1 (dark-grey).

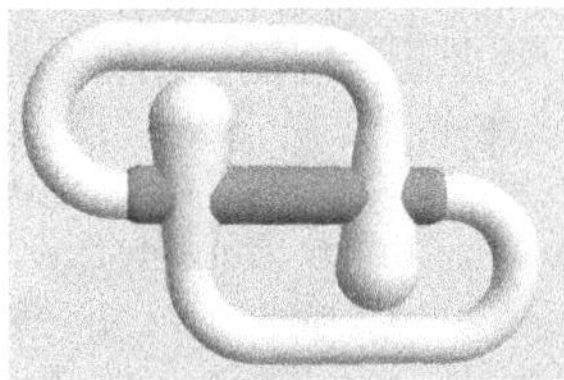

Fig. 6.89 Shard labels inside dark region are shown in Figures 6.90 and 6.91.

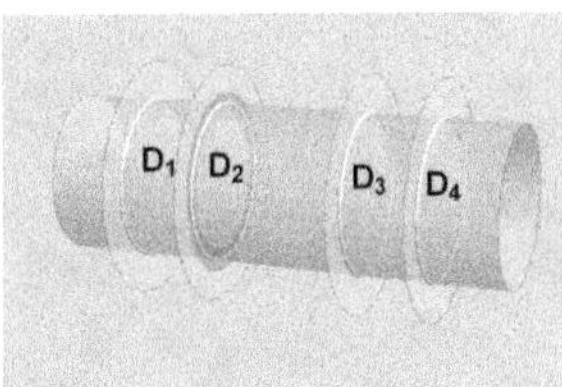

Fig. 6.90 Disks D_1, D_2, D_3, D_4 revealed using transparency.

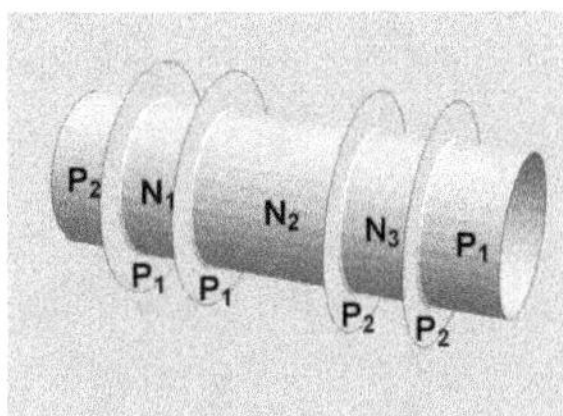

Fig. 6.91 A partial view showing 1-shard labels.

Example 6.19. Two of the 16 possible height relation choices are shown in Figures 6.92 and 6.93. One can distinguish these by focusing on the N_2 shard. Figure 6.93 shows that the closed N_2 passes above the other closed shards on either bounding circle whereas in Figure 6.93 it passes above on one end and below on the other. ∎

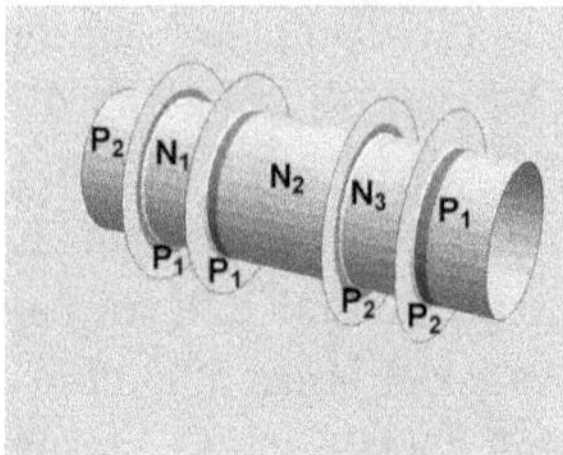

Fig. 6.92 Above/under relations for spun trefoil.

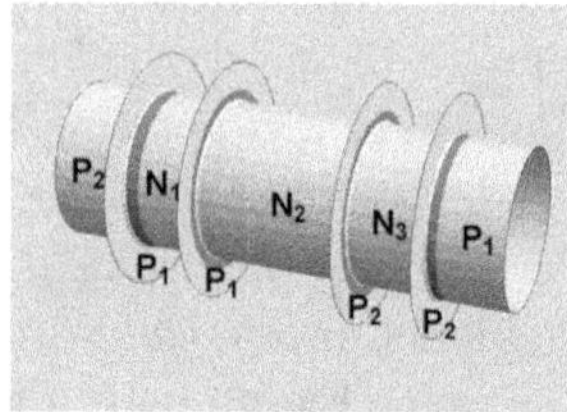

Fig. 6.93 Above/under relations for Fox ribbon sphere.

A crossing scheme for any of these knottings will be a shard structure on a 2-sphere which we view in R^2 via stereographic mapping. Figure 6.94 shows the 1-shards. Here, the south pole of S^2 corresponds to $p \in N^2$, thus p corresponds to the point at infinity and so the annulus $\widehat{N^2}$ of the crossing scheme corresponds to the complement of the two largest circles. In Figure 6.95 we label the 1-shards. In this Figure, we label $\widehat{A_i}$ for above shards and $\widehat{B_i}$ below shards. For the shard N_2, the one 1-shard is above and the other below. One can continue matching and see that this is a labeling for the spun trefoil Figure 6.92.

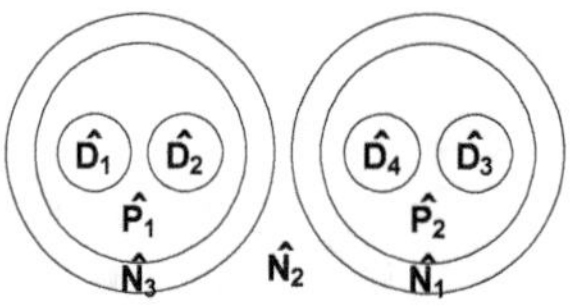

Fig. 6.94 1-shards in S^2.

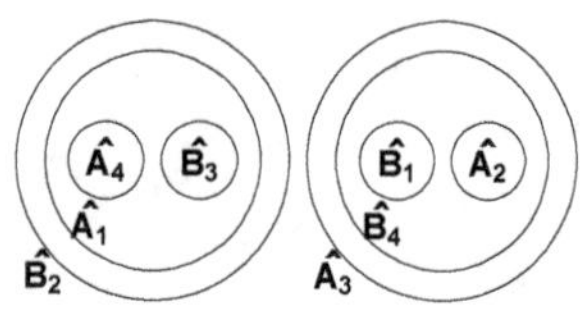

Fig. 6.95 2-shards in S^2.

An instructive 1-knotting whose projection is k_*, Figure 6.96, is analogous to Figure 6.86. Although Figure 6.96 shows *eight* crossing points, we treat these as a union of *four* 0-spheres $C_1, C_2, C_3,$ and C_4 as in Figure 6.97. For each of these 0-spheres we consider possible above/below choices.

Figure 6.98 shows a knotting with particular choices of above/below at the four 0-spheres. Below-crossings are shown as grey (as if it were in a "shadow" of an above-crossing). Points A_i lie above B_i and both project to C_i.

Three particular choices for a knotting which projects to k_* are shown in Figures 6.99, 6.100 and 6.101. (The knotting of Figure 6.99 is same as that of Figure 6.98 from a slightly different perspective and without labels.) The knotting of Figure 6.99 has the knot-type of the square knot. The knotting of Figure 6.100 has the knot-type of the stevedore's knot. The knotting of Figure 6.101 has the trivial knot.

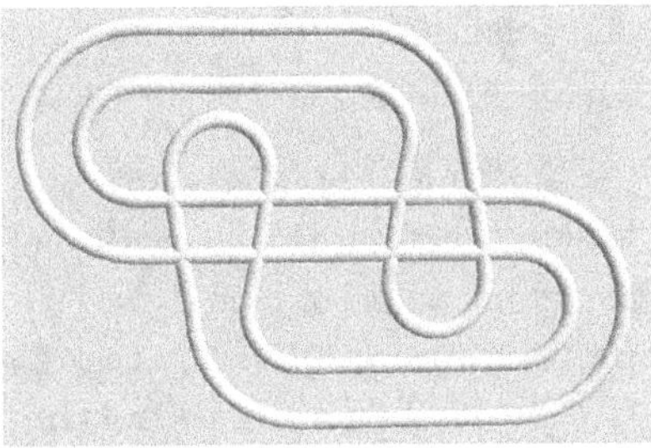

Fig. 6.96 $k_* \subseteq R^2$, an immersed circle.

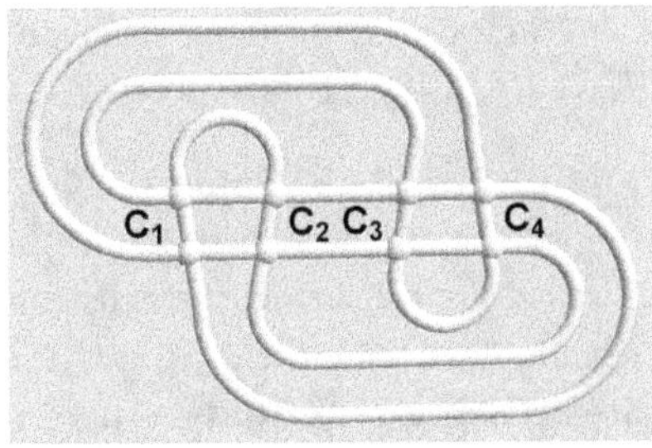

Fig. 6.97 Crossings as four 0-spheres.

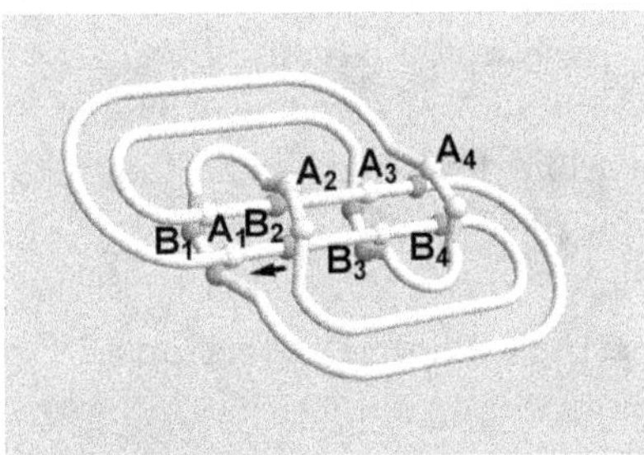

Fig. 6.98 One choice of above/below relations at the four 0-spheres.

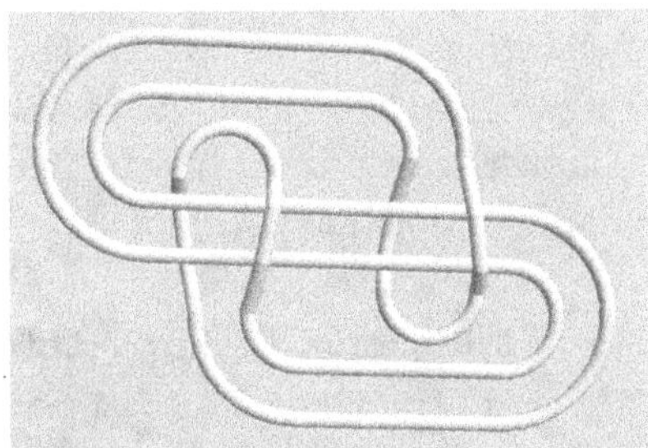

Fig. 6.99 Unlabeled Figure 6.98 from above. The knot-type is a square knot.

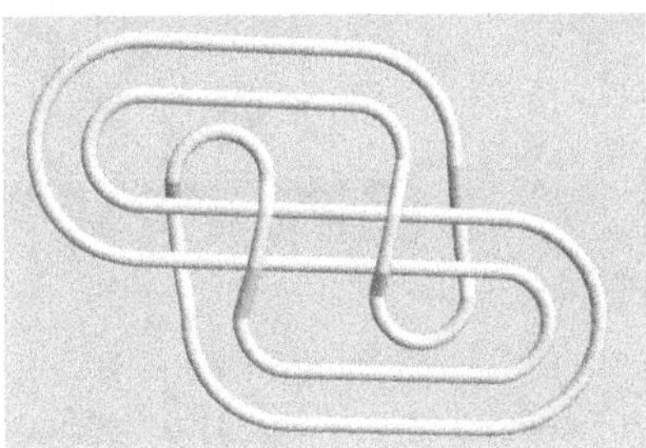

Fig. 6.100 A choice of above/below with knot-type of stevedore's knot.

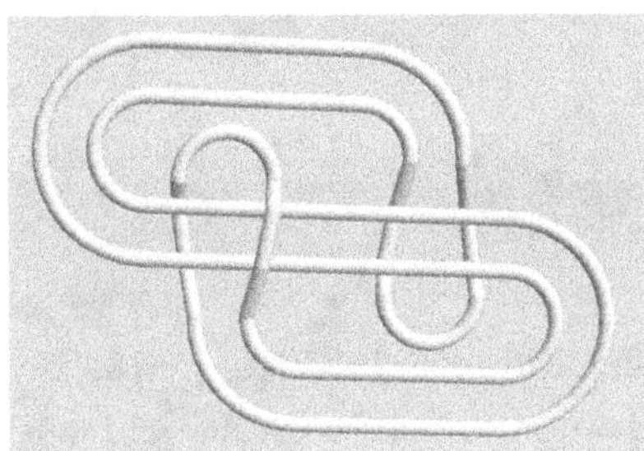

Fig. 6.101 A third choice for above/below with trivial knot-type.

Definition 6.15. A 2-knotting in general position, whose projection K_* is an immersion with crossing set consisting of disjoint union of circles, is called a **pseudo-ribbon knotting**. A 2-knot which can be represented by a psuedo-ribbon knotting is called a **psuedo-ribbon knot.** ∎

Our process of going from an immersion image $X \in R^3$ to a knotting $K \subseteq R^4$ does not have a common definition, so we define:

Definition 6.16. Suppose $X \subseteq R^{n+1}$ is an image of a smooth self-transverse immersion of an n-manifold N with crossing set consisting of 2-crossings (and thus the crossing set is a disjoint union of closed $(n-1)$-manifolds, $C_1, \ldots C_p$). We say X is a **simple immersion of** N and that C_i are **crossing manifolds**. An n-knotting $K \subseteq R^{n+2}$ of N is an *ab*-**knotting for** X if $K_* = X$. ∎

The term *ab*-knotting refers to the fact that the knot-type of K is determined by above/below choices made at each crossing manifold. The 2-knottings we obtain by *ab*-knotting are exactly the pseudo-ribbon knottings. Explanation of the term "pseudo-ribbon" is found in Chapter 9. Another term used mostly in higher dimensions for a such a knotting with limited projection is 2-bulky, see Definition 12.11.

6.8 Additional Examples of 2-Knottings Their Projections

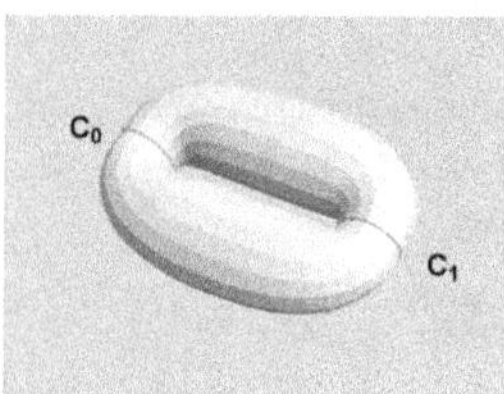

Fig. 6.102 A projection $(K \cup K')_*$.

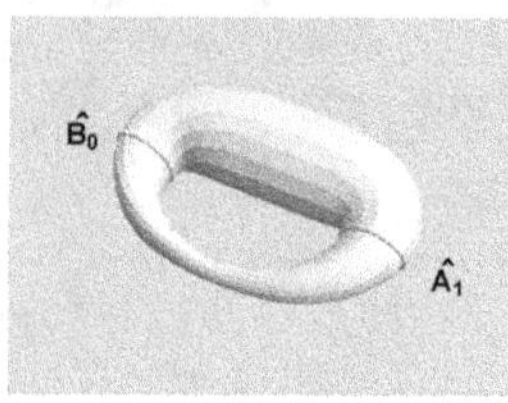

Fig. 6.103 Torus K' from Figure 6.102.

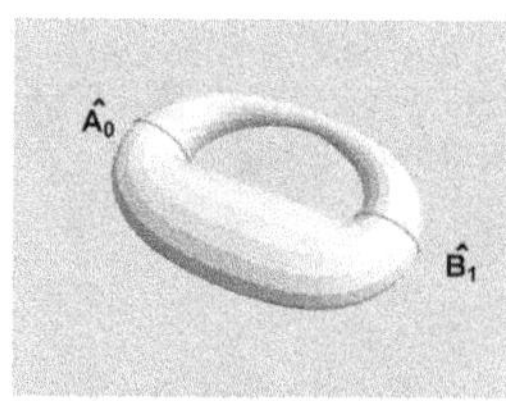

Fig. 6.104 Torus K torus from Figure 6.102.

Example 6.20. Figure 6.102 is a projection of a knotting of two disjoint torii K and K' in R^4. The transverse crossing set consists of two disjoint meridian circles C_0 and C_1. This is called the **spun Hopf link**. In Section 10.2 we explain this name. We indicate w-coordinates with the wire-frame method in Figure 6.105. Also, Figure 6.106 shows Figure 6.105 with transparent faces. So, K passes above K' along C_0 and passes below along C_1. The crossing scheme of this knotting has four 2-shards. We show these as subsets of K and K' in Figure 6.103 and Figure 6.104. The crossing set

C_0 corresponds to circles $\widehat{A_0}$ and $\widehat{B_0}$ where $\widehat{A_0}$ is above-set for $\widehat{B_0}$, Figures 6.103. Using similar labels, we get $\widehat{A_1}$ and $\widehat{B_1}$ for T', Figure 6.104. ∎

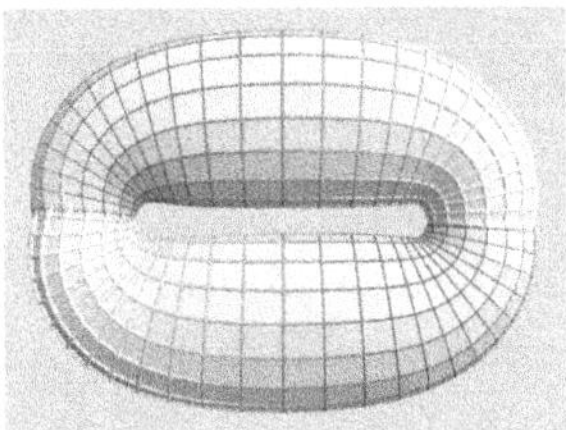

Fig. 6.105 Figure 6.102 with wire-frame coding.

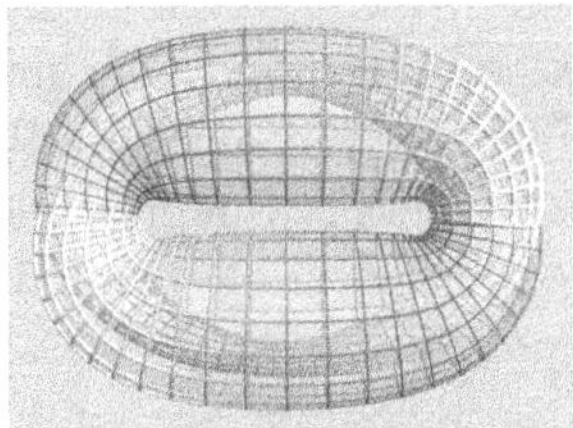

Fig. 6.106 Same as Figure 6.105, but with transparency.

Example 6.21. If p is a point on the inside of both torii and we take an inversion from p, the point p will "be sent to infinity" and the projection in Figure 6.107 will be sent to a subset as shown in Figure 6.105. In the projection, we see two torii which intersect in a pair of circles. Figure 6.108 shows Figure 6.107 cut in half. ∎

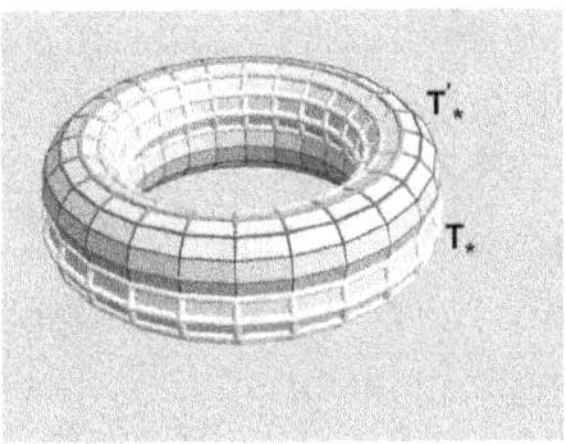

Fig. 6.107 A projection of a knotting of two torii.

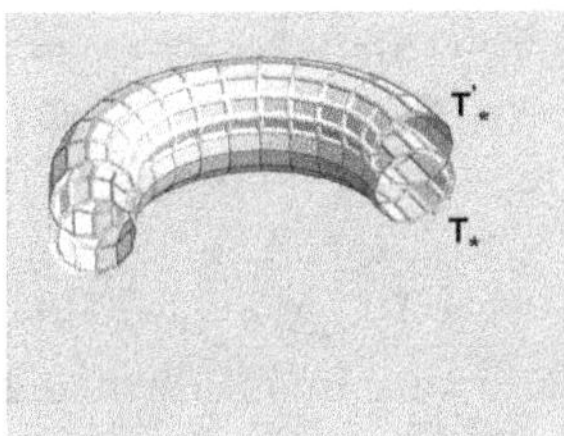

Fig. 6.108 Half of Figure 6.107 with wire-frame coding.

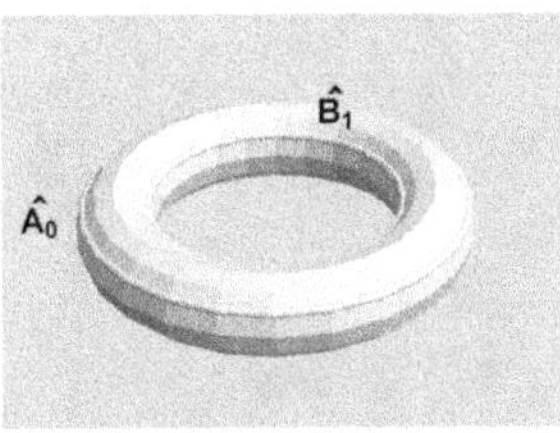

Fig. 6.109 The 2-shards A_0 and B_1 in T for the knotting of Figure 6.107.

Fig. 6.110 The 2-shards A_1 and B_0 in T' for the knotting of Figure 6.107.

Example 6.22. We consider a knotting of M which is a disjoint union of two torii and a 2-sphere. We assume that the spun Hopf linking $\Sigma(H)$ is contained, essentially, in a standard $N = S^1 \times D^3$, and let T and T' denote the two tori whose union is $\Sigma(H)$. Now add the 2-sphere, $S = \{x\} \times \partial D^3 \subseteq N$ for some $x \in S^1$. Our knotting of M is $K = T \cup T' \cup S$. (This is an example of a join spun knotting, Definition 12.3.)

Two views of the projection K_* are shown in Figures 6.111 and 6.112. There are four triple points in this projection. Because of occlusion, it is only possible to clearly see one triple point at one time, but if we render only the wire-frames (traditional wire-frames, not the coded wire-frames) of these surfaces and thicken the crossing set with tubes, we can visualize the entire crossing set clearly, Figure 6.113. Figure 6.114 shows only the crossing set with four labeled triple points. Triple point p is the one shown in Figure 6.111. From the wire-frame coding we can infer that the sphere passes *above* both torii in the fourth dimension at both p and p'. At the triple points q and q' the sphere is shown to pass *below* both torii.

The shards of K_* are: four 3-shards (the triple points p, p', q, and q'); twelve 2-shards (the twelve arcs connecting the triple points as evident in Figure 6.114); fifteen 1-shards—seven corresponding to S and four for each torus T and T'. The 1-shards can be counted by looking at the crossing scheme shown in Figures 6.115 and 6.117. By the symmetry of the knotting, a Figure for the crossing scheme on T' would look the same as for T.

The scheme knotting N is a subset of $M \times R^1 \approx (T \times R^1) \cup (T' \times R^1) \cup (S \times R^1)$. The four circles on S are a projection of the part N that lies in $S \times R^1$. We view this as a knot-diagram on S^2 which corresponds to a knotting in R^3. Tracing through our height relations for K we find this knotting in R^3 is a split sum of two Hopf links, Figure 1.18. Figure 6.116 shows a scheme knotting in $T \times R^1$. In this Figure the torus shown is $T \times \{0\}$ where the original knotting $K \subset R^4$ has positive height and so lies outside the solid torus with projection to T, Figure 6.115. ∎

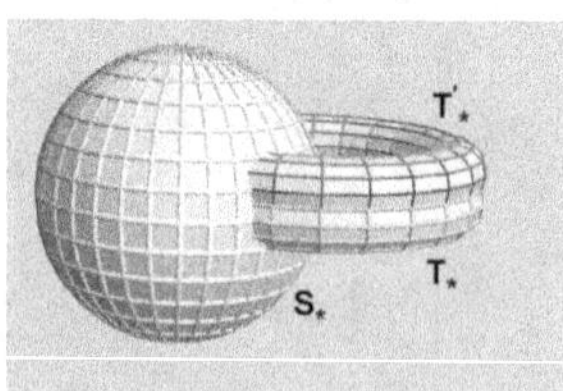

Fig. 6.111 One triple point of K_*.

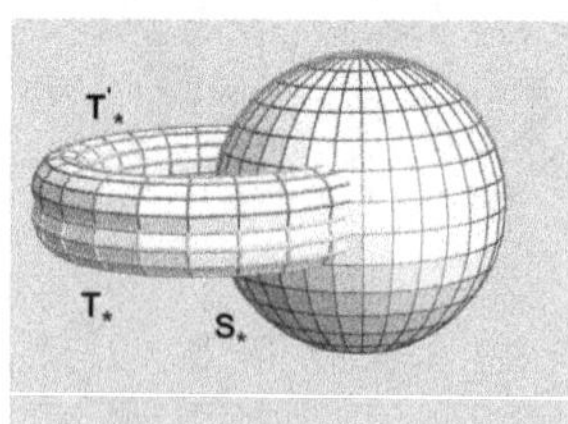

Fig. 6.112 Second triple point of K_*.

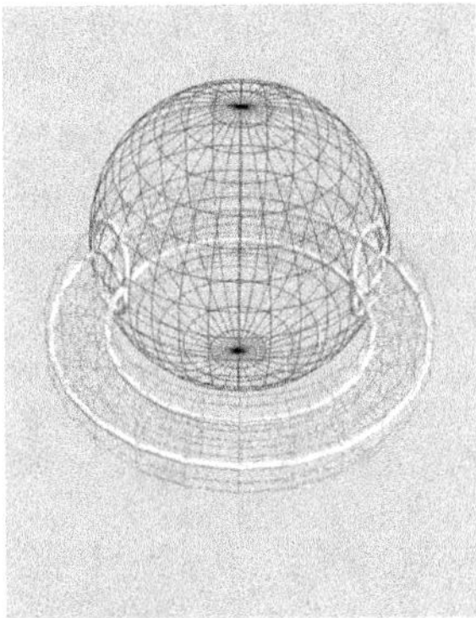

Fig. 6.113 Surfaces as wire-frames, crossing set as thin tubes.

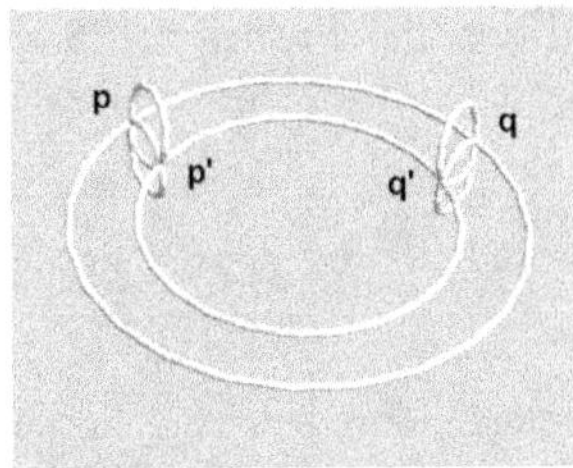

Fig. 6.114 The crossing set only, triple points p, p', q, q'.

Fig. 6.115 The pre-crossing set on T.

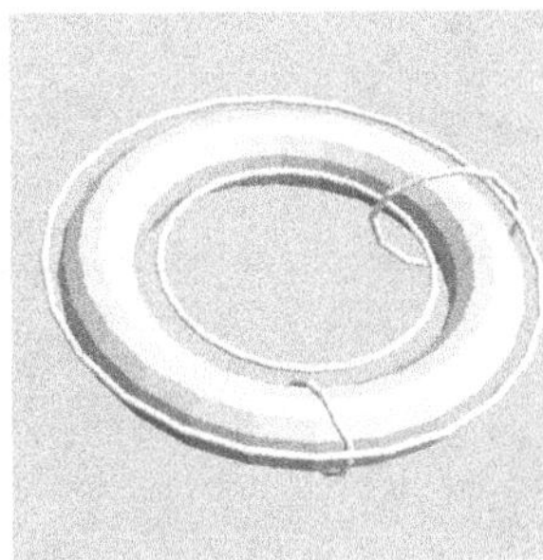

Fig. 6.116 The scheme knotting for $T \times R^1$.

Fig. 6.117 The pre-crossing set which lies on S.

Fig. 6.118 Figure 6.117 rotated slightly.

Example 6.23. Figure 6.119 is a projection of a knotting of a genus two surface with crossing set two disjoint circles C_0 and C_1. The knotting is more completely shown with wire-frame coding in Figure 6.120. In Figure 6.119 we use thick ordinary (non-coded) wire-frame and transparency to reveal details. Figure 6.123 shows a crossing scheme for this knotting. ∎

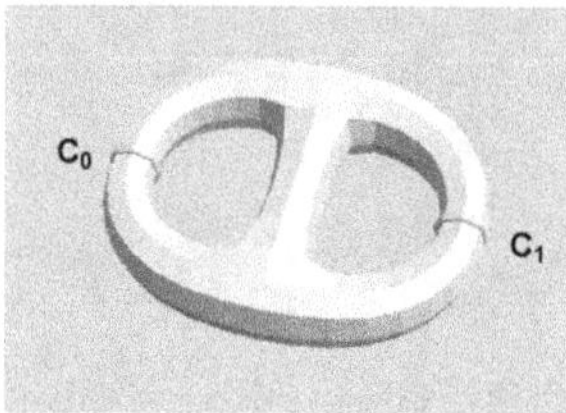

Fig. 6.119 A projection of a knotting of a genus two surface.

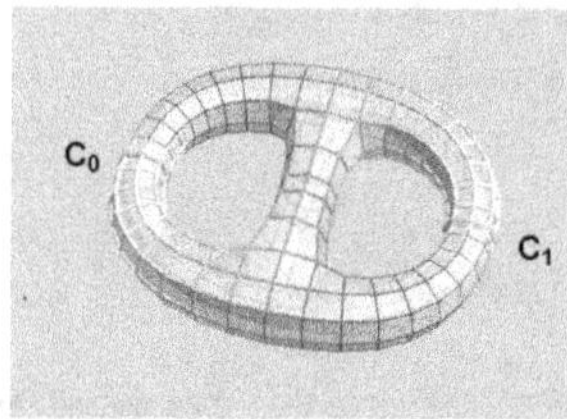

Fig. 6.120 Figure 6.119 with wire-frame coding and transparent faces.

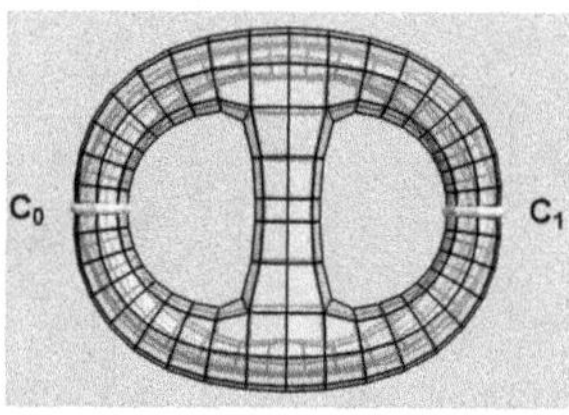

Fig. 6.121 Figure 6.119 with non-coded wire-frame.

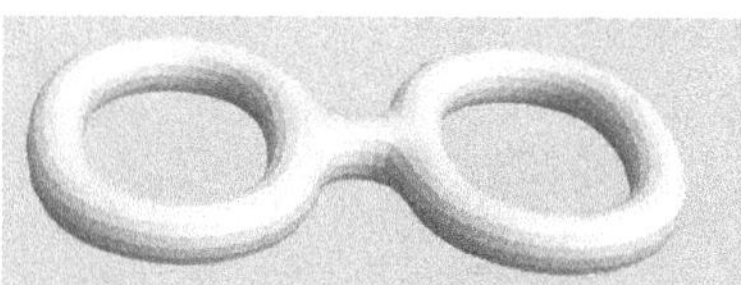

Fig. 6.122 A generic genus two surface.

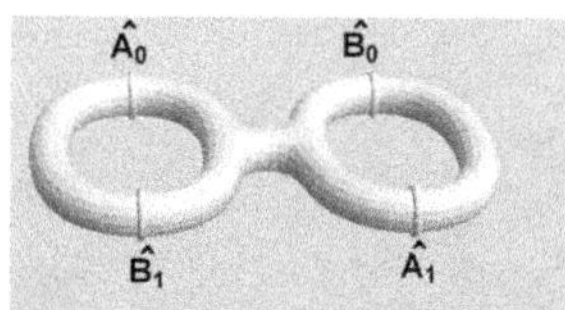

Fig. 6.123 Crossing scheme.

6.9 Some Projections of Non-Orientable 2-Knottings

The two simplest non-orientable closed 2-manifolds are the Klein bottle and the projective plane. We describe knottings and standard projections of these in R^4. Further discussion of these topics is found Section 7.3.

Good background references are [Suzuki (1976)], [Francis (2007)] and [Kamada (1989)]. The technique used in our constructions, generally called **spinning**, will appear in many variations in the rest of this book.

The Klein bottle does not embed in R^3 but it does embed in R^4. The Klein bottle is commonly presented as a self-intersecting surface in $\mathcal{K} \subset R^3$ such as Figure 7.81. More precisely, this is a *projection* of a knotting in R^4 of the Klein bottle.

We will describe a "standard" knotting $\mathbb{K}^2 \subseteq R^4$. The projection $\mathbb{K}^2_*$ has two branch points and an arc crossing points joining them, Figure 6.130. The corresponding knot-type is called a **trivial Klein bottle**, the **standard Klein bottle** or sometimes a **unknotted Klein bottle** in R^4. In Remark 7.10 we show how our choice for standard Klein bottle is equivalent to one with projection $\mathcal{K}$.

A short knotting description for a Klein bottle is: let M^2 be a standard Möbius band in R^3, consider the product $N^3 = M^2 \times I \subseteq R^3 \times I \subseteq R^3 \times R^1 \approx R^4$, now ∂N^3 is a simple 2-knotting of the Klein bottle. By Remark 2.14 we get a description of a Klein bottle as "two Möbius bands glued together on their boundaries". However, this knotting is not in projection general position—in fact $N^3_* = M^2 \subseteq R^3$—so we do not prefer it.

Example 6.24. As a warm-up, view the familiar torus in R^3 as a projection of a knotting $T \subseteq R^4$ using generalized cylindrical coordinates.

Let $M_0 \subseteq R^3_+$ be the circle of radius 1 in the xz-plane, center $(2,0,0)$. Let M_θ be a copy of M_0 put in each page H^3_θ. That is $M_\theta = i_\theta(M_0) \subseteq H^3_\theta$ where i_θ is the page identification for $0 \leq \theta \leq 2\pi$, Figure 6.124. Projection in w sends $H^3_\theta \subseteq R^4$ to the page $H^2_\theta \subseteq R^3$, Figure 6.125. The projection M^*_θ is a geometric circle Figure 6.126. Our torus in R^4 is $\mathbb{T} = \bigcup_{\theta=0}^{\theta=2\pi} M_\theta$. and $\mathbb{T}_*$ is a familiar torus in R^3. Figures 6.127 and 6.128. ∎

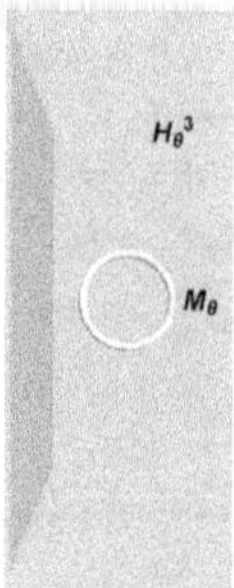

Fig. 6.124 $M_\theta \subseteq H^3_\theta$.

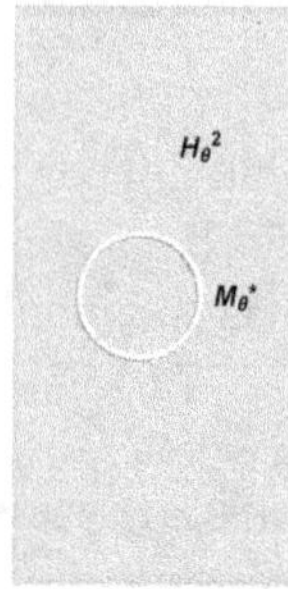

Fig. 6.125 $M^*_\theta \subseteq H^2_\theta$.

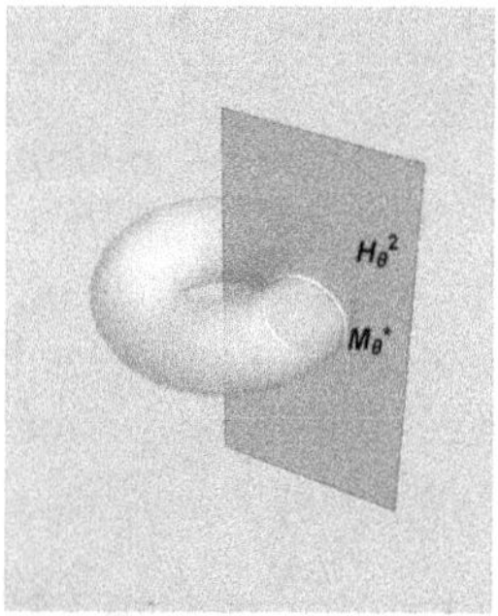

Fig. 6.126 Projection M^*_θ in page H^2_θ.

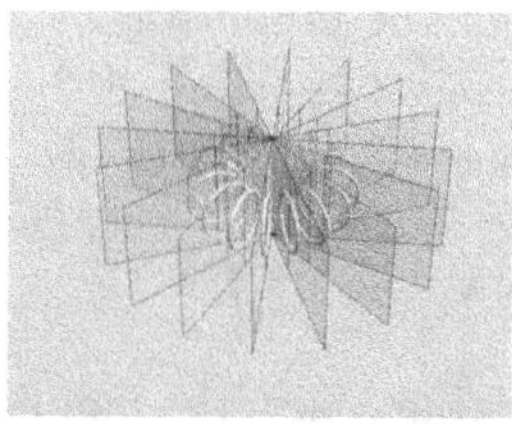

Fig. 6.127 Projection several M^*_θ in pages.

Fig. 6.128 The union of all M^*_θ.

Sometimes a Möbius band is said to be "a rectangle with ends glued with a half twist". Similarly, a Klein bottle is called a "cylinder with ends glued with a half-twist". Instead of "half-twist", we use the term **half-flip**. A **half-flip** is the isotopy h_t, $0 \le t \le 1$ of a geometric circle in R^3 which is rotation *with diameter fixed* of angle π/t. So h_1 is rotation by π and maps the circle to itself. This contrasts to a "twist", which often refers to a rotation of a circle about its center.

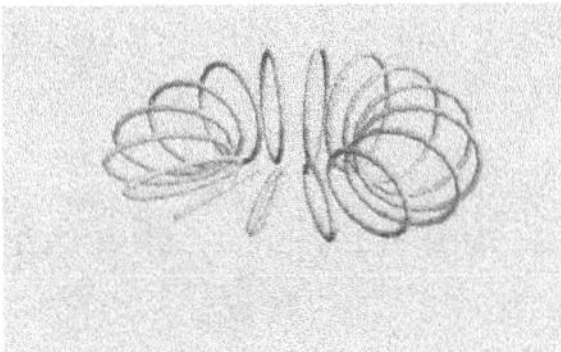

Fig. 6.129 The circles M_θ^*.

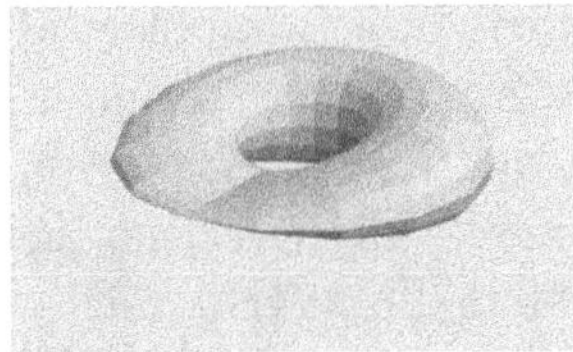

Fig. 6.130 Projection of Klein Bottle $\mathbb{K}_*$.

Example 6.25. We define m_θ to be the half-flip isotopy in R^3_+ of the circle $M_0 \subseteq R^3_+$ parameterized by $0 \leq \theta \leq 2\pi$ where m_θ is the circle M_0 rotated positively about the x-axis by an angle of $\theta/2$. These circles are called **meridians of the Klein bottle**. This rotation will be considered positive if, looking down the x-axis, it appears to be counterclockwise. As a set $m_0 = m_{2\pi}$. We define $M_\theta = i_\theta(m_\theta) \subseteq H^3_\theta$. We take the **standard knotting of the Klein bottle** to be: $\mathbb{K}^2 = \bigcup_{\theta=0}^{\theta=2\pi} M_\theta$.

Figure 6.129 shows the projection of a number of the meridians of $\mathbb{K}^2$. For each θ, $M_\theta \subseteq H^3_\theta \subseteq R^4$ and $M^*_\theta \subseteq H^2_\theta \subseteq R^3$. Circle M^*_0 projects to a circle. The other circles M_θ project to ellipses with M^*_π projecting to a line segment. The resulting projection $\mathbb{K}^2_*$ is shown in Figure 6.130. The crossing set of $\mathbb{K}^2_*$ is a line segment, M^*_π with each endpoint a branch point. Figures 6.131 and 6.132 give more detail for the w-coordinate, using wire-frame coding. ∎

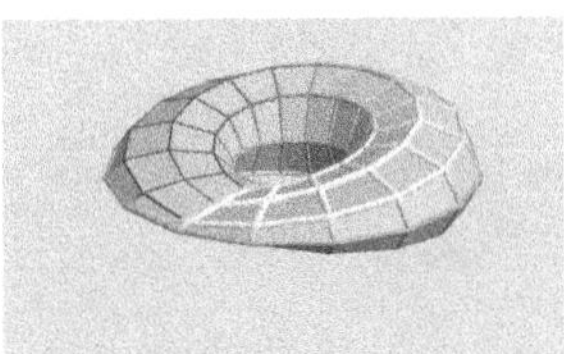

Fig. 6.131 Projection of Klein Bottle with wire-frame coding.

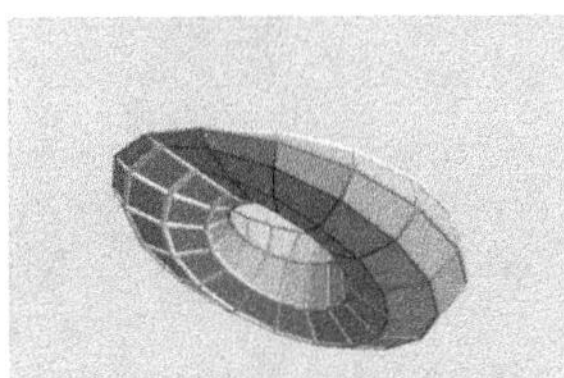

Fig. 6.132 Figure 6.131 from a lower viewpoint.

We used a rotation of angle $\theta/2$ for $0 \leq \theta \leq 2\pi$ but if we rotate by an angle of $-\theta/2$ (that is a negative rather than positive half-flip), we get a different isotopy, m'_θ and a corresponding different knotting $\mathbb{K}'$. We will show in Section 7.3 that these are isotopic.

Remark 6.4. In topological terms one can say: "the Klein bottle is obtained as a connected sum of two proper Möbius bands". We can visualize this using $\mathbb{K}^2_*$ as follows.

Express the circle m_θ is the union of two semi-circles with common endpoints: $A = \{(r, \theta, z, w) \in m_\theta : 2 \leq r \leq 3\}$ and $B = \{(r, \theta, z, w) \in m_\theta : 1 \leq r \leq 3\}$, see Figure 6.133. The subset of $\mathbb{K}^2$ corresponding to the half-flip of A is a Möbius band, similarly for the half-flip of B. Projections are shown in Figures 6.134 and 6.135 respectively. Figure 6.136 is an "exploded view" of both parts to help visualize how they fit together. ∎

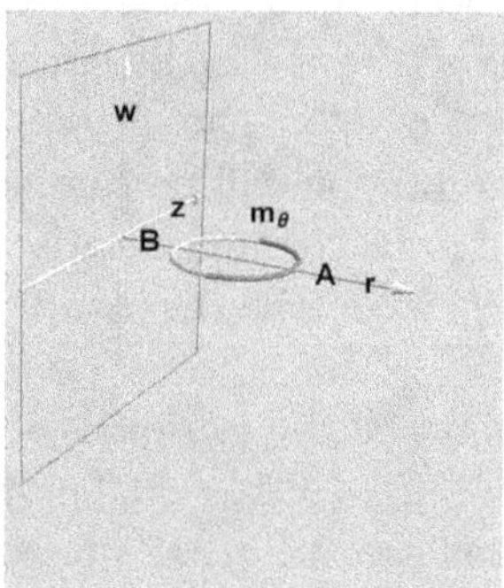

Fig. 6.133 M_θ written as union of two semicircles.

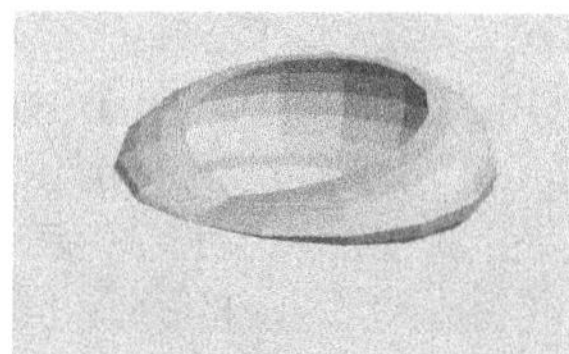

Fig. 6.134 Subset corresponding to A shown separately.

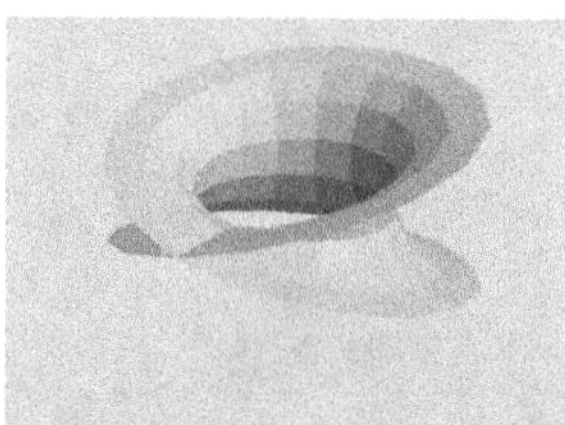

Fig. 6.135 Subset corresponding to B shown separately.

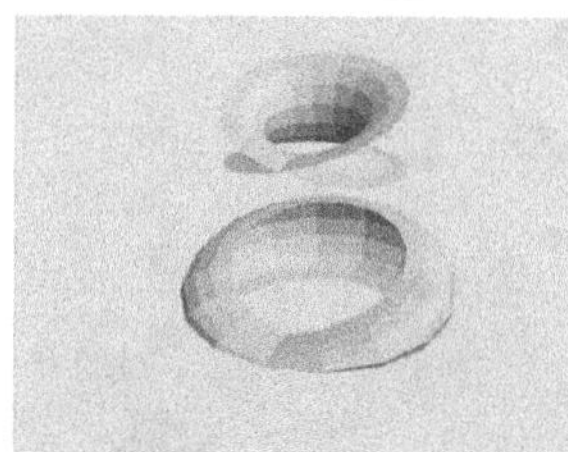

Fig. 6.136 How subsets of Figures 6.134 and 6.135 fit together.

The **projective plane** has a long and interesting mathematical history beginning with non-Euclidean geometry. For us it is a simple basic 2-manifold. As with the Klein bottle, it does not embed in R^3. Unlike the Klein bottle, it has *two* "standard" knottings, not just one.

Example 6.26. Both standard knottings are constructed similarly to the Klein bottle knotting $\mathbb{K}^2$, Example 6.25 only using a slightly different initial circle. Let $M_0 \subseteq R^3_+$ be the circle of radius 1 in the xz-plane with center $(1,0,0)$. Note that M_0 contains the point $(0,0,0)$, Figure 6.137. Denote two isotopies of M_0—a positive half-flip and a negative half-flip. Let m_θ be the circle M_0 rotated an angle of $\theta/2$ about the x-axis. Also, let n_θ be the circle m rotated an angle of $-\theta/2$ and finally let $M_\theta = i_\theta(m_\theta) \subseteq H^3_\theta$ and $N_\theta = i_\theta(n_\theta)$. The **two standard knottings of the projective plane in** R^4 are defined, see Figure 6.138: $\mathbb{P}^2 = \bigcup_{\theta=0}^{\theta=2\pi} M_\theta$ and $\mathbb{P}^2_0 = \bigcup_{\theta=0}^{\theta=2\pi} N_\theta$.

The projective plane is referred to as " a Möbius band and a disk glued together along the common boundary". To see this, we express our circle $m_\theta = A_\theta \cup B_\theta$ and as a union of semicircles, and consider the corresponding subsets for the projective plane, as we did in Remark 6.4 for the Klein bottle. So, $\bigcup_{0\le\theta\le 2\pi} B_\theta$ will be a topological disk and $\bigcup_{0\le\theta\le 2\pi} A_\theta$ is a topological Möbius band, see Figures 6.139, 6.140 and 6.141.

Figure 6.142 shows a different way of expressing this same knotting as a gluing of a disk and a Möbius band. This projection of the Möbius resembles a cap and is known in the literature as a "**cross-cap**". ∎

Fig. 6.137 Projections of some of the circles for projective plane.

Fig. 6.138 Projection of a standard $\mathbb{P}^2$.

Fig. 6.139 Exploded view of projection of projective plane.

Fig. 6.140 Projection of a disk.

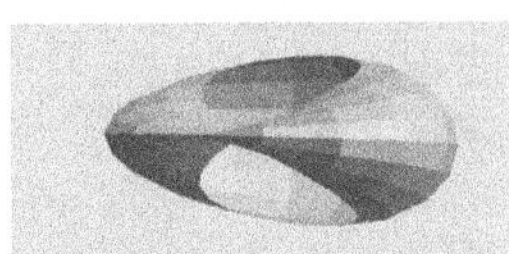

Fig. 6.141 Projection of a Möbius band.

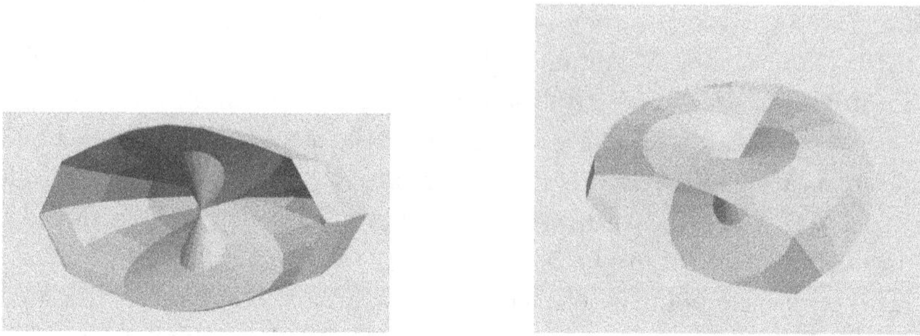

(a) (b)

Fig. 6.142 A disk has been removed from projection leaving a crosscap. (a) One view of a crosscap. (b) Another view of crosscap slightly rotated.

Remark 6.5. It has long been noted, for example [Hosokawa and Kawauchi (1979)] that the two knottings $\mathbb{P}^2$ and $\mathbb{P}^2_0$ are not equivalent, and so there are two candidates for "unknotted projective plane". A right-hand twist seen in a mirror is a left-hand twist, and it follows that $\mathbb{P}^2$ and $\mathbb{P}^2_0$ are mirror images of each other. In general, it can be shown that at most two knottings of P^2 share a given exterior, [Price (1977)]. ▮

The constructions in Section 6.9 are our first examples of a general technique that uses an isotopy of a 1-knotting to obtain a 2-knotting called **spinning**, or more generally **deformation spinning**. These topics will be covered more extensively in Chapter 10.

Definition 6.17. Suppose m_θ is a smooth isotopy parameterized by $0 \leq \theta \leq 2\pi$ of a smooth knotting $M_0 \subseteq R^n_+$ such that $m_{2\pi}(M_0) = M_0$. We define $M_\theta = i_\theta(m_\theta) \subseteq H^n_\theta$. If $\Sigma(m_\theta) = \bigcup_{\theta=0}^{\theta=2\pi} M_\theta$ is a smooth manifold it is called **spinning of the isotopy** m_θ. ▮

Remark 6.6. The subset $\Sigma(m_\theta)$ *will* always be smooth except possibly for points of $M_0 = M_{2\pi}$. Consider two smooth isotopies of a geometric circle in R^2_+ with center $(3, 0)$ that simply expands, then contracts the radius. For the first m_θ we define the radius to be $1 + 0.5\sin(\theta)$. We get a smooth torus Figure 6.143. However, if we define a smooth isotopy m'_θ, so that radius is given by $1 + 0.25(\theta - \pi)^2$, we get the subset shown in Figure 6.144 which fails to be smooth. However we can always smooth this gluing area and obtain a well defined smooth manifold. ▮

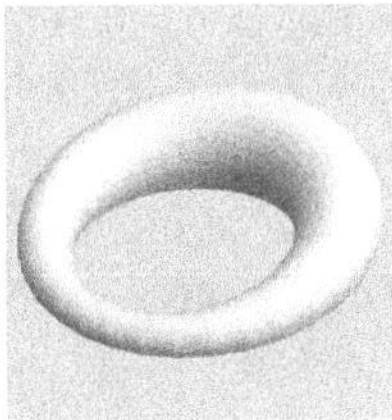

Fig. 6.143 A simple spinning.

Fig. 6.144 A non-smooth spinning. However this is smoothable.

6.10 Some Disconnected 2-Knottings

Example 6.27. Figure 6.145 shows a projection of a simple but useful 2-knotting of a sphere and a torus using the method of crossing-shadows to indicate height relations. One of the crossing circles is not in view in this figure but is revealed in a rotated version, Figure 6.146. If we slice this linking by a hyperplane in R^4 whose projection H shown in Figure 6.147 we get a linking of three circles, Figure 6.148. It is beneficial to view this as a linking of a circle and $S^0 \times S^1$. This sphere-torus linking is equivalent to $10_2^{0,1}$ in Yoshikawa's table [Yoshikawa (1994)], see Figure 8.144 and Example 8.8.

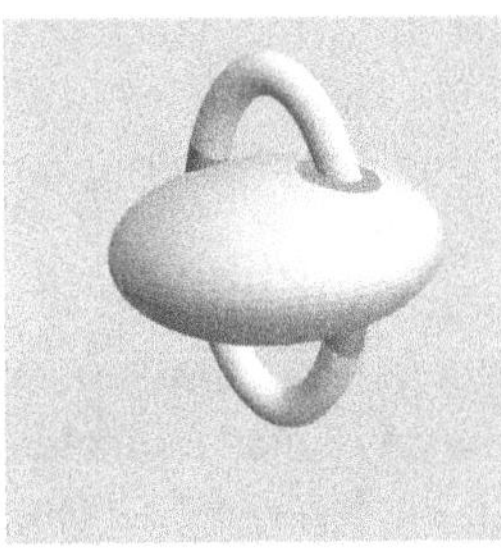

Fig. 6.145 A projection of the linking.

Fig. 6.146 A rotation of Figure 6.145 reveals the fourth crossing circle.

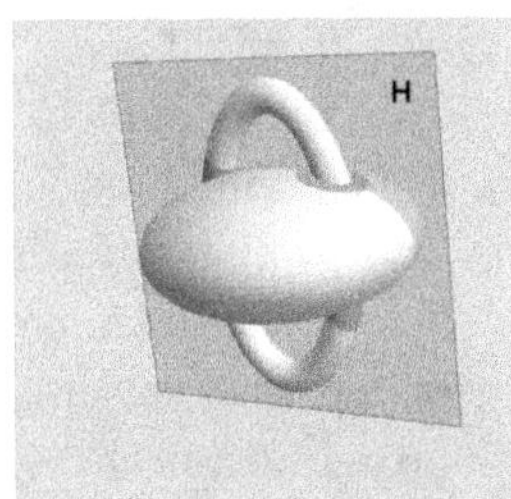

Fig. 6.147 The projection of linking and projection of a slicing plane H.

Informally, this 2-knotting is a version of the basic sphere-torus knotting, Example 6.12, with the torus wrapping about the sphere twice rather than once. Mathematically this knotting is a two-fold branched cyclic cover of the basic sphere-torus knotting. One way to see this is to slice our knotting with a hyperplane, whose projection H, is shown in Figure 6.149. If we cut our link in half, the projection is as shown in Figure 6.150. In H there is a line L as shown. The other side of H is the same, thus we can recognize our knotting as the two-fold cyclic cover of the basic sphere-torus knotting branched along the circle corresponding to L. ∎

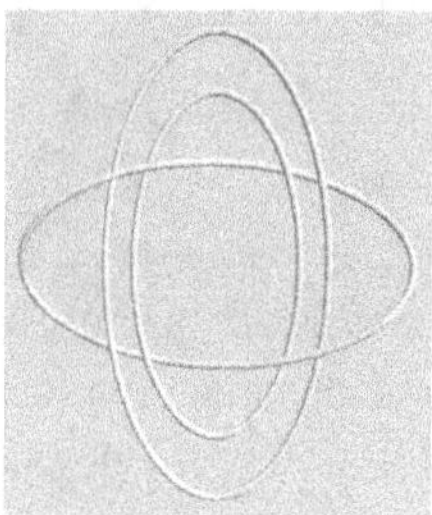

Fig. 6.148 The slice of the linking in the hyperplane corresponding to H.

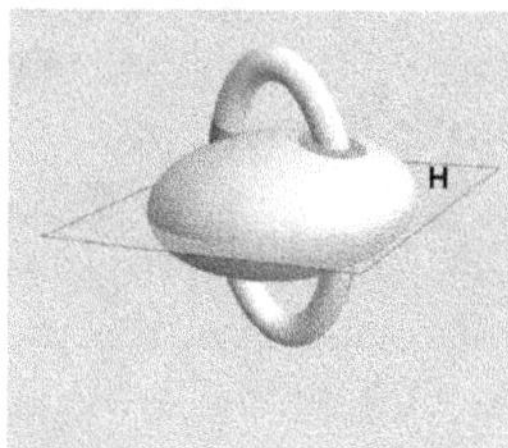

Fig. 6.149 Projection of different slicing hyperplane.

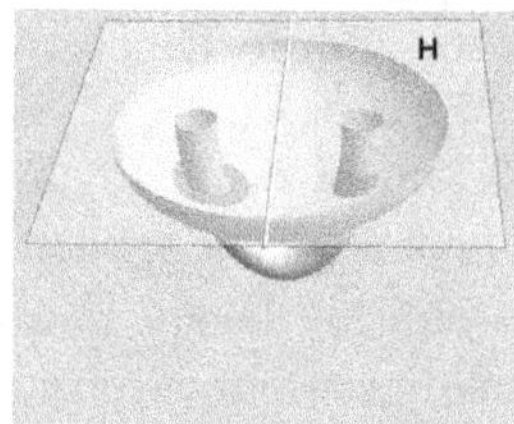

Fig. 6.150 Half of Figure 6.149.

Example 6.28. Figures 6.151 and 6.152 show the projection of a (disjoint) union of a standard projective plane and a standard torus. Call this knotting K. It might seem that K would be a complicated knotting, since the knotting of Example 6.27 is not a split trivial knotting. However, we will see K is a split linking of $\mathbb{P}^2$ and a trivial torus knotting. A projection of the isotopy that reveals this is shown in Figures 6.151–6.160.

The critical change is the transition between Figure 6.154 and 6.155 where one must carefully note the height relations as indicated. This change will be detailed in the next Chapter, Example 7.6, where it is called a **circle-switch**. The transition from Figure 6.157 to 6.160 can be described as two consecutive arch moves as detailed in Example 7.5. ∎

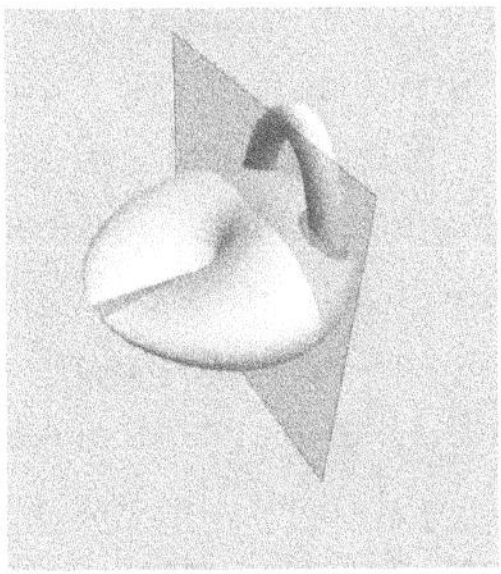

Fig. 6.151 Knotting projection of $\mathbb{P}^2$ and disjoint torus.

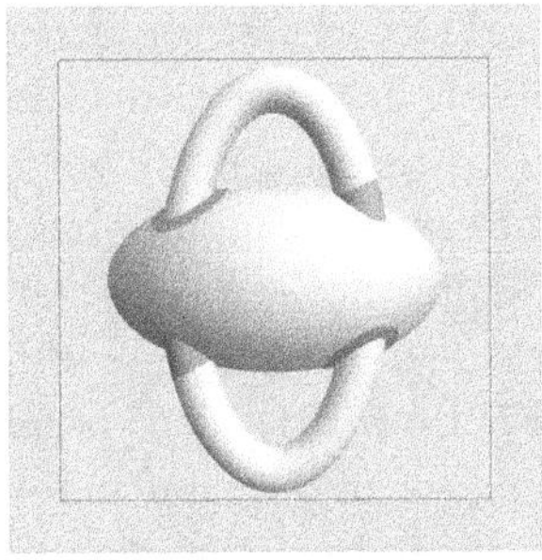

Fig. 6.152 Rear view showing all four crossing circles.

Fig. 6.153 Torus translated forwards.

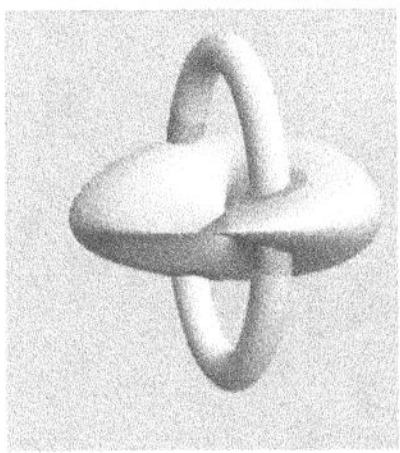

Fig. 6.154 Torus begins half-flip.

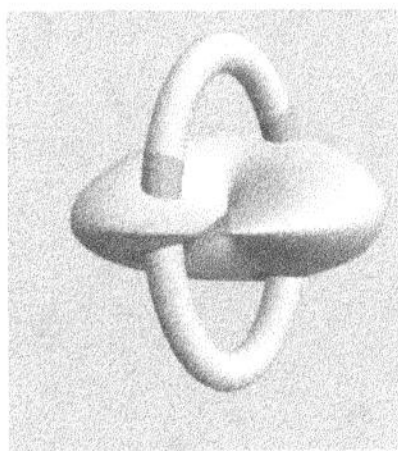

Fig. 6.155 Torus continues half-flip.

Fig. 6.156 Half-flip concluded.

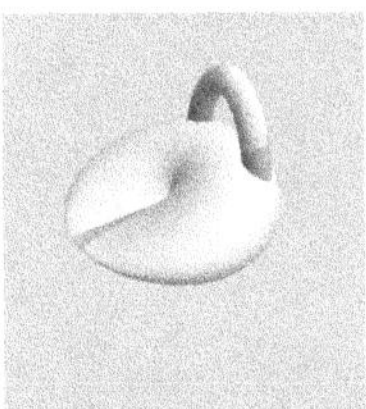

Fig. 6.157 Torus translated backward.

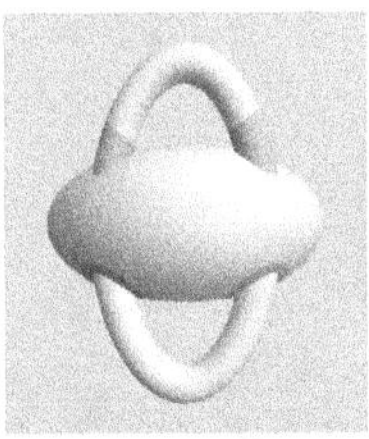

Fig. 6.158 View from back showing all crossings.

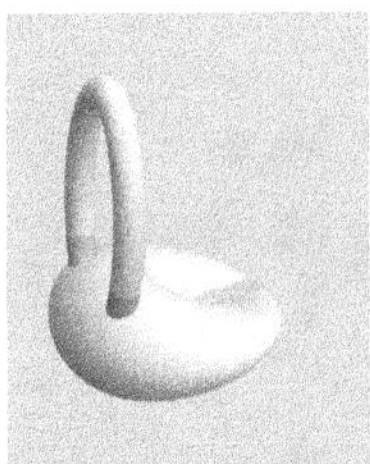

Fig. 6.159 Begin to translate torus upwards.

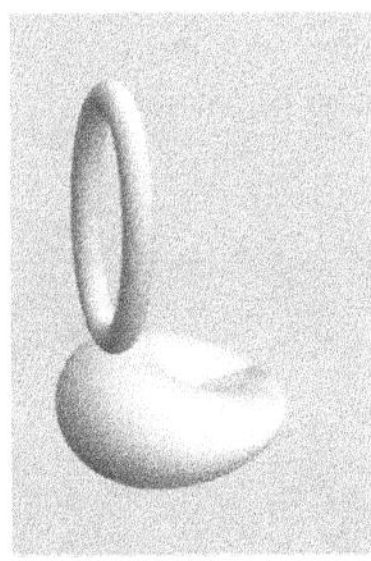

Fig. 6.160 Manifolds now disjoint.

Example 6.29. If K is any 2-knotting and S^2 is the standard $S^2 \subseteq R^4$, we have $K \# S^2 \approx K$. Otherwise, as a rule, a connected sum of two knottings is expected to be more complex than either summand. However, in some cases, this is not true for surfaces that are not spheres. In the next example, we have $\mathbb{P}^2 \# K^2 \approx \mathbb{P}^2$ where $\mathbb{P}^2$ is the standard projective plane and K^2 is a non-trivial knotting of a sphere we call Fox Ribbon Sphere, Section 6.7.

This result was first presented in [Viro (1973)], by different methods. Some details of our isotopy are given in Chapter 7. A projection of the connected sum (with separating plane) is shown in Figure 6.161. Figure 6.162 is a second view showing height relations for four crossing circles. In projection our isotopy moves a tube-like portion of the sphere portion forwards on the other side of the plane as well as slightly towards the viewer, Figure 6.163. For Figure 6.164, we continue to move this tube-like portion across the crossing set of $\mathbb{P}^2$, similar to Example 6.28. Moving our tube-like portion back, we end up with a knotting whose projection is the same set as in Figure 6.165 but with different above/below relations at a pair of crossing circles. At this point, it is not difficult to show that we can eliminate all crossing circles of our sphere—the beginning is shown in Figure 6.166. ∎

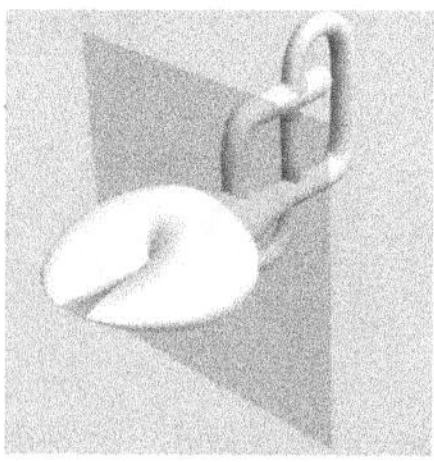

Fig. 6.161 Projection of connected sum $\mathbb{P}^2$ and sphere knotting.

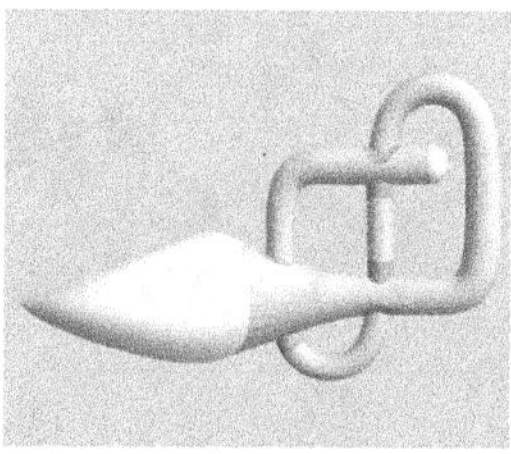

Fig. 6.162 Side view showing crossings on sphere.

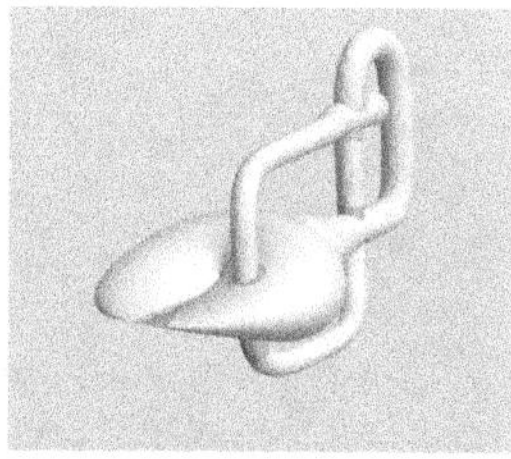

Fig. 6.163 Bringing part of knotting forwards.

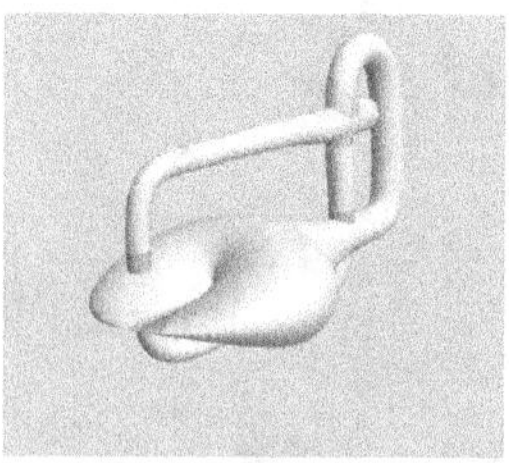

Fig. 6.164 Moving across $\mathbb{P}^2$ crossing set.

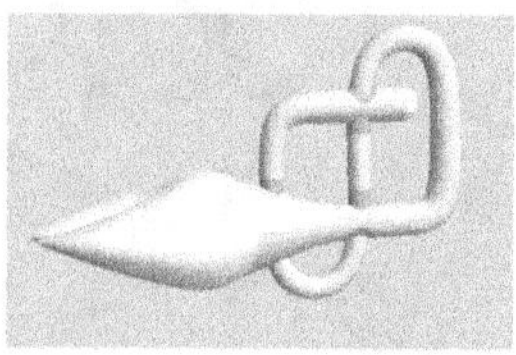

Fig. 6.165 Moving back, note crossings.

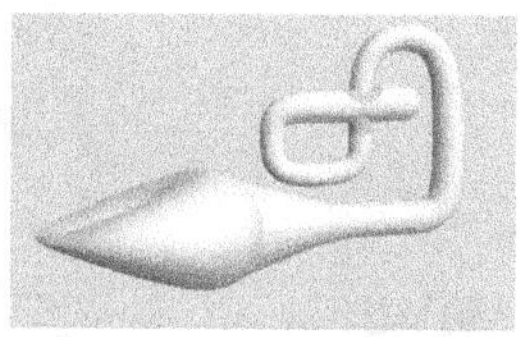

Fig. 6.166 Beginning isotopy to $\mathbb{P}^2$.

Chapter 7

Knot-Moves

Remark 7.1. For 1-knottings,the Reidemeister knot-moves, Section 4.4 have been central to knot theory from [Reidemeister (1927)] to [Kauffman (1990)] and beyond. In [Reidemeister (1927)] a knotting in R^3 was taken to be a polygonal simple closed path K. An **elementary isotopy** was one that introduced or deleted a single vertex v between two consecutive vertices of K, a and b, or deleted a vertex v. A *general* isotopy was a sequence of elementary isotopies. The points a, b and v correspond to a triangle $T \subseteq R^3$ that meets K only in one or two edges. Analysis [Reidemeister (1927)] then shows that in projection elementary moves could be expressed in terms of what is now called Reidemeister moves. The idea is to analyze possible intersections of T_* and K_* and express these in terms of three cases as shown in Figures 7.1, 7.2 and 7.3. In each of these figures, the change in the knotting is the replacement of the dashed edge of triangle T with two solid edges (or the reverse). Figure 7.1 corresponds to smooth moves $\Omega.1$ or $\Omega'.1$, as in Figures 4.59. Figure 7.2 corresponds to moves $\Omega.2$ or $\Omega'.2$, smooth version, as in 4.60 and Figure 7.3 corresponds to moves $\Omega.3$ or $\Omega'.3$, see Figure 4.61. ▮

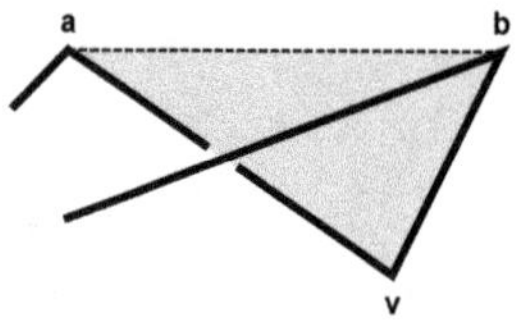

Fig. 7.1 Type 1.

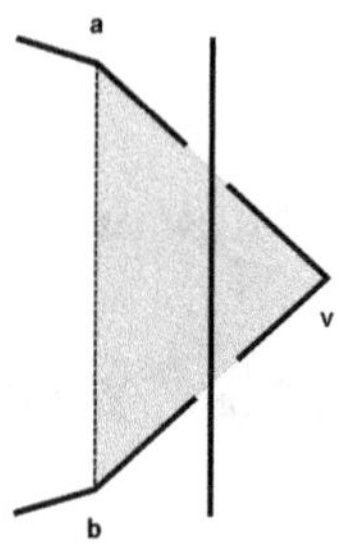

Fig. 7.2 Type 2.

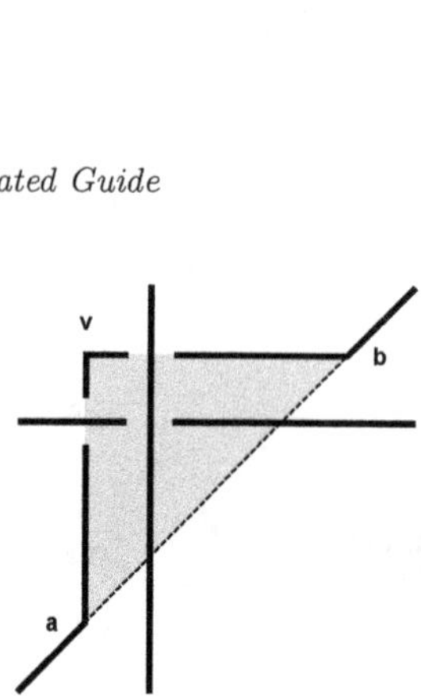

Fig. 7.3 Type 3.

We next outline a proof of the basic theorem for knot-moves Proposition 4.6 for *smooth* 1-knottings which defines isotopy in terms of the trace and uses properties of 2-knottings and their general position projections as presented in Section 6. The advantage of this approach is that it naturally extends to describe knot-moves for higher dimensions.

Remark 7.2. Briefly, Reidemeister moves for 1-knottings correspond to the three possible types of critical points, in the I coordinate, for the shards of the crossing set of the projection of the trace of the isotopy. Definition of these critical points will follow in Definition 7.1, but the essential point is that local models, near these critical points of the trace, are traces of our three knot-moves as seen in Figures 7.8, 7.9 and 7.10. ▮

A simple isotopy $F : S^1 \times I \to R^3 \times I$ is shown in Figures 7.4, 7.5, 7.6 and 7.7 with projections Figures 3.24, 3.25, 3.26 and 3.27. Our isotopy corresponds to an $\Omega.2$ move. The projection of the trace is the trace of the projection, a subset of xyw-space as shown in Figure 3.29. The crossing set $\mathcal{C}$ of this trace projection is a proper arc with endpoints at the end of the isotopy—the two crossing points of $F_1(S^1)$. By critical point here we mean the local minimum of the w-coordinate of the curve $\mathcal{C}$. This corresponds to $t = \frac{1}{3}$, Figure 7.5, where the projection, Figure 3.25, fails to be in general position due to the double point of self-tangency.

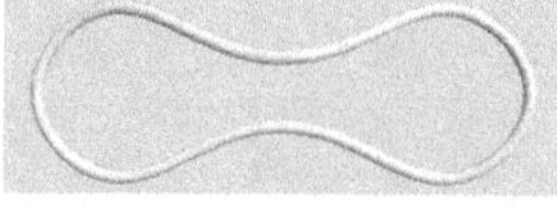

Fig. 7.4 $F_0(S^1)$ of isotopy.

Fig. 7.5 $F_{\frac{1}{3}}(S^1)$ of isotopy.

Fig. 7.6 $F_{\frac{2}{3}}(S^1)$ of isotopy.

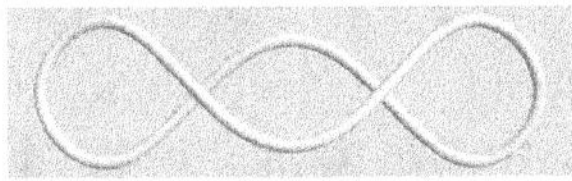

Fig. 7.7 $F_1(S^1)$ of isotopy.

Remark 7.3. We want to consider a trace projection, as defined in Example 3.1, of an isotopy of a 2-knotting that is in "general position".

Our definitions and theorems for **general position** with respect to projection in Section 6, were stated with respect to 2-manifolds with *empty boundary* in R^4. In this case, a general position projection, $\mathbb{K}_*$ has four possible local models: a plane in $R^2 \subseteq R^3$, two different coordinate planes R^3, all three coordinate planes in R^3, or a Whitney umbrella.

We will not go into the details except to list two additional local models needed for a projection of a proper 2-knotting: $R^2_+ \subseteq R^3_+$ or two coordinate half-planes in R^3_+. The first of these models is part of definition of proper submanifold. The second of these models is given by the half-planes $\{y = 0 \text{ and } 0 \leq x\}$ and $\{z = 0 \text{ and } 0 \leq x\}$. In the trace Figure 3.29, the endpoints of the crossing set look locally like this model.With this extension of general position, to include manifolds with non-empty boundary there is no need for changes for any definitions of **shards**. ▮

Definition 7.1. For an isotopy of a 2-knotting with trace projection in general position, $\mathbb{K}_*$ and crossing set $\mathcal{C}$, every *b*-shard (branch point) and 3-shard (triple point) of $\mathcal{C}$ is defined to be a **knot-move critical point of the isotopy**. Additionally, each local maximum and minimum of $\mathcal{C}$ in the I coordinate of $R^2 \times I$ for an 2-shard is a **knot-move critical point of the isotopy**. ▮

The connection between knot-move type and the three types of critical points is as follows. Knot-move critical points that are branch points correspond to Reidemeister move $\Omega.1$ or $\Omega'.1$, Figure 7.8. Knot-move critical points that are local maxima or minima for 2-shards correspond to Reidemeister move $\Omega.2$ or $\Omega'.2$, Figure 7.9. Knot-move critical points that are triple points correspond to Reidemeister move $\Omega.3$ or $\Omega'.3$, Figure 7.10.

The boundary of $R^2 \times I$ corresponds to the beginning of an isotopy or the end. Our general position (in particular our two new local models) will make sure that our knot-move critical points do not occur on this boundary.

Since we are considering smooth *compact* knottings there are only a finite number of local maxima or minima for 2-shards, and so a finite number of knot-move critical points for our isotopy. Each knot-move critical point will correspond to a knot-move based on our local models of knot-move critical point. A refinement of general position, additionally focused on the I direction, that will do this is [Roseman (2000)]:

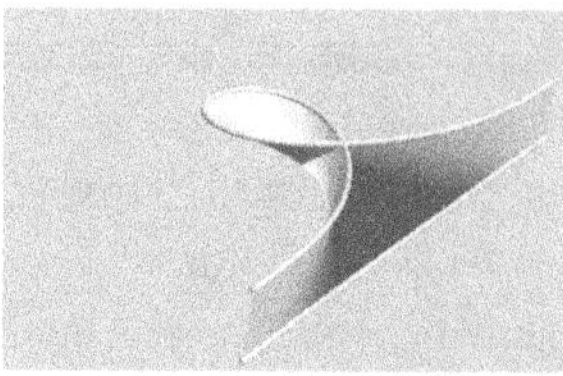

Fig. 7.8 Trace for a Reidemeister $\Omega.1$ move.

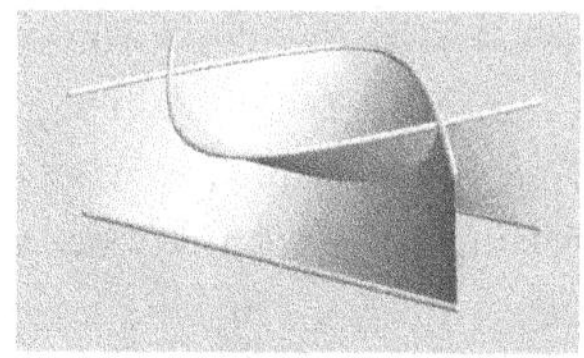

Fig. 7.9 Trace for a Reidemeister $\Omega.2$ move.

Fig. 7.10 Trace for a Reidemeister $\Omega.3$ move.

Definition 7.2. An isotopy with trace $\mathbb{K}$ in general position with respect to trace projection, with crossing set $\mathcal{C} \subseteq R^2 \times I$, is in **general position in the I direction** if:

(1) each trace projection critical point of $\mathbb{K}_*$ should have a separate I coordinate
(2) if $(p, t) \in R^2 \times I$ is a point of an 2-shard γ of $\mathcal{C}$ then it is either a local maximum/minimum or γ meets $R^2 \times \{t\}$ transversely. ∎

In particular, this second condition above says that, on a 2-shard, the height function does not have a horizontal point of inflection.

Prop 7.1. If K and K' are equivalent knottings in general position with respect to projection they are equivalent by a finite sequence of Reidemeister moves that transform K_* to K'_*. ∎

Remark 7.4. Reidemeister moves deal with isotopies of 1-manifolds in $R^3 = R^2 \times R^1$ and the projection on the first factor of $R^2 \times R^1$. Nothing special about R^2 was used other than that it is a smooth 2-manifold. All we have discussed applies equally if we replace R^2 by some other smooth 2-manifold M^2. Reidemeister moves are changes to the projection of a 1-manifold in a sub-disk of R^2.

So suppose we have an isotopy H_t of a 1-manifold M^2 in $M^2 \times R^1$ and π is projection on the first factor. We can define general position with respect to π. We can define Reidemeister moves for H_t to be isotopies whose projection in some sub-disk of M looks like one of the Figures 4.59, 4.60 or 4.61. Finally we can show that: if K and K' are equivalent knottings in general position with respect to projection they are equivalent by a finite sequence of Reidemeister moves that transform K_* to K'_*. We will not make use of this theorem in this book, but it will be useful to speak of these more general Reidemeister moves for some examples, such as Example 10.14 and Remark 10.13. ∎

Knot-moves have been defined for more general knottings. A key to expressing such a theory of knot-moves is to define a projection of some kind. Such an extension to knottings in lens spaces appears in [Cattabriga *et al.* (2013)].

7.1 Elementary Knot-Moves for Surfaces

There are knot-moves for 2-knottings similar to Reidemeister moves [Roseman (2004)]—in fact knot-moves for n-knottings, Section 12.3. In the literature these are frequently called **Roseman moves.** In this introduction we only list and describe these moves. A proof that these moves can describe *any* 2-knotting isotopy [Roseman (2004)] involves examining the trace. Since a trace of an isotopy of a 2-knotting is a 3-dimensional submanifold of a 5-dimensional manifold, we postpone these higher dimensional discussions until Section 12.3.

These moves have been adapted to describe similar geometric changes such as, [Winter (2014)], [Audoux *et al.* (2017)]. Knot-moves represented as sequences of local slices of moves, rather than projections, are often used, [Carter and Saito (1993)], [Kamada (2002)]. Also they are used in conjunction with algebraic topology, [Przytycki and Rosicki (2014)], [Fedoseev and Manturov (2016)], [Matsuda (2022)]. Knot-moves as related to 2-braid charts are found in [Nakamura (2014)]. Other applications are found in [Kharusi and Yashiro (2016)].

These seven moves are shown in the before/after pairs: Figures 7.11 through 7.24. (The wire-frames shown are as in traditional computer graphics an aid to view figures, and not as wire-frame encoding of Section 6.4.) Reidemeister moves of Figures 4.59, 4.60 and 4.61 show *sub-arcs* (1-disks) of a knotting; our higher dimensional moves are local pictures which show projections of 2-disks which are *sub-disks* of a knotting. These disks, disjoint in R^4, have projections with simple intersections and self-intersections in R^3. We arbitrarily label the left figure Position A and the right one Position B.

Figures for a (c)-move and (d)-move are projections of a single disk. Figures for (g)-move involve four sub-disks. The other moves involve two disks. As is the tradition with Reidemeister moves, each move has several possible choices for above/below relations of these sub-disks. This is not graphically indicated and will be discussed later in this section.

If we are considering a knotting of a 2-manifold M, each of these local moves correspond to changes in M of the crossing scheme, shown in Figures 7.25 to 7.31. In these diagrams we graphically show below 2-shards as grey and dashed in the style of Figure 6.72. As with Reidemeister moves, we often make no explicit mention of whether we go from Position A to B or from B to A. In this way, have seven moves rather than fourteen.

Remark 7.5. Here are some comments on each of these graphics:

(a)-move This move, Figures 7.11 and 7.12, produces or eliminates a circle of crossing points. It is related to the $\Omega.2$ Reidemeister move, Figure 4.60 if we think of the two crossing points in the Reidemeister $\Omega.2$ move as a 0-sphere which is produced or eliminated. Figure 7.12, shows there is a "new" region in R^3 whose boundary is two 1-shard disks that share a 2-shard crossing circle. In the scheme Figure 7.25, we see the corresponding 1-shard disks in M, one whose boundary is an above-circle, the other whose boundary is an a below-circle.

(b)-move This move locally looks like a passing of a horizontal a plane past the saddle point of a hyperboloid, Figures 7.13 and 7.14. It seems to move two crossing arcs toward each other, merge then split into two, but in a different manner, then move apart. In the crossing scheme, Figure 7.26, we have two sub-disks of M. Each contains two arcs which are portions of 2-shards. Suppose we label one disk $D_0 \subseteq M$, with arcs of 2-shards $\widehat{s_0}$ and $\widehat{t_0}$; label the other disk $D_1 \subseteq M$ with arcs of

2-shards $\widehat{s_1}$ and $\widehat{t_1}$. Either $\widehat{s_0}$ and $\widehat{t_0}$ are above shards, whereas $\widehat{s_1}$ and $\widehat{t_1}$ are below shards—or $\widehat{s_0}$ and $\widehat{t_0}$ are below shards, whereas $\widehat{s_1}$ and $\widehat{t_1}$ are above shards. In the crossing scheme change for a (b)-move, the pairs of arcs that seem to merge, then split as we go from Position A to Position B, must be either *both* above arcs or *both* below arcs.

(c)-**move** This move, Figures 7.15 and 7.16, either produces or eliminates a pair of branch points. It is related to the Ω.1 Reidemeister move, Figure 4.59. In the crossing scheme a small circle is either produced or eliminated in M, Figure 7.27. This circle is a union of a 2-shard arc of above-points, a (2-shard) arc of below-points and two branch points.

(d)-**move** This move, Figures 7.17 and 7.18, either produces or eliminates a branch point pair. For crossing scheme, see Figure 7.28. This move is useful but redundant, see Remark 7.7. Comparing this figure with Figure 7.17, we see that the two 2-shards whose arcs are shown in the right side of Figure 7.28, are paired with one above and one below. If a 1-shard meets a below 1-shard it must be at an endpoint of each shard, which is a branch point. These must be separated in a compatible way, so that after an elimination of a pair, one will be left with two arcs in the crossing scheme, one above the other below.

(e)-**move** This move, Figures 7.19 and 7.20, either produces or eliminates a pair of triple points. Figure 7.20 shows three "new" regions in R^3 each of which is bounded by three 1-shards of the projection. In the crossing scheme, six crossings are introduced, eliminated, Figure 7.29.

(f)-**move** This move, Figures 7.21 and 7.22, either produces or eliminates a single triple point. Figure 7.21 shows one "new" region in R^3 bounded by two 1-shards, with associated crossing scheme, Figure 7.30.

(g)-**move** The move, Figures 7.23 and 7.24, is related to Ω.3 Reidemeister move, Figure 4.61. The model for the Position A figure consists of the three coordinate planes of R^3 with the fourth plane P corresponding to $x + y + z = 1$. For Position B, we replace P by plane P' with equation $x + y + z = -1$. Distinctive to this move is a tetrahedral region whose boundary consists of four 1-shards. Figure 7.33 used transparency to reveal this region in Position B. A different tetrahedral region for Position B is revealed in Figure 7.32. The associated crossing scheme is shown in Figure 7.30. ▮

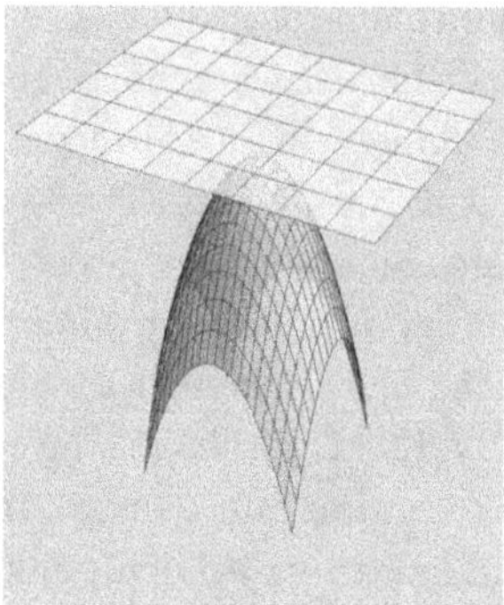

Fig. 7.11 (a)-move, Position A.

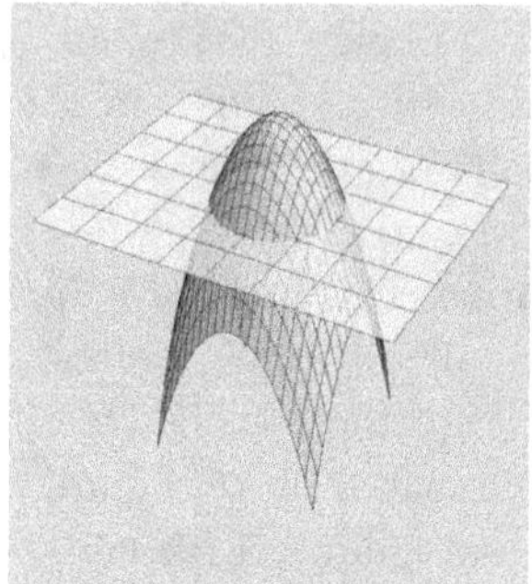

Fig. 7.12 (a)-move, Position B.

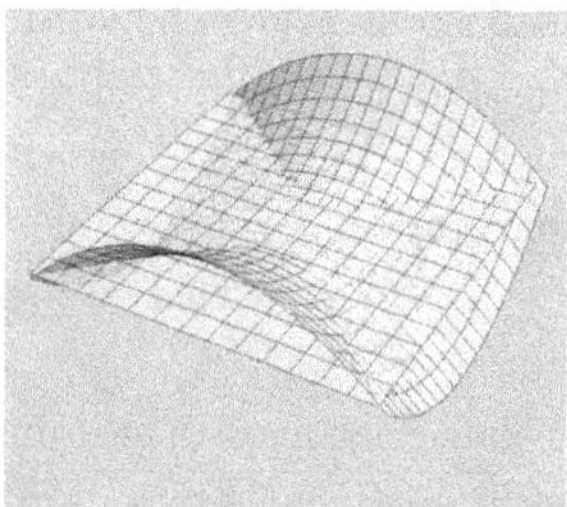

Fig. 7.13 (b)-move, Position A.

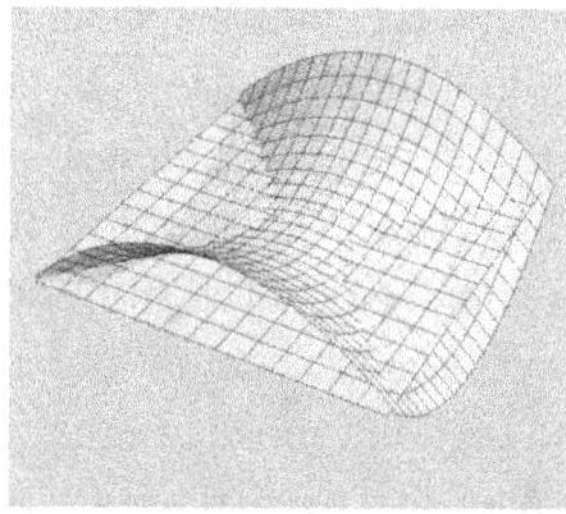

Fig. 7.14 (b)-move, Position B.

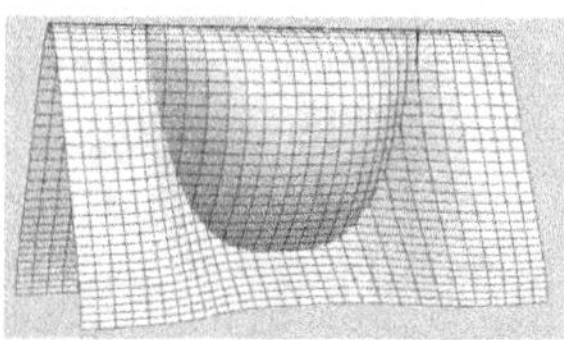

Fig. 7.15 (c)-move, Position A.

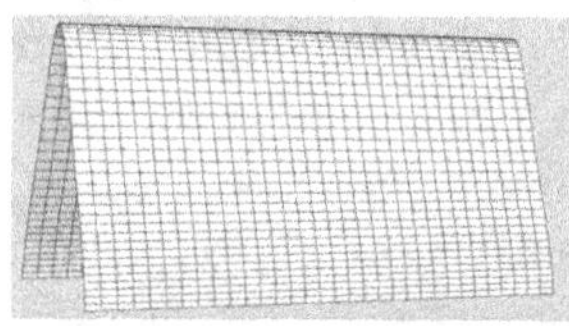

Fig. 7.16 (c)-move, Position B.

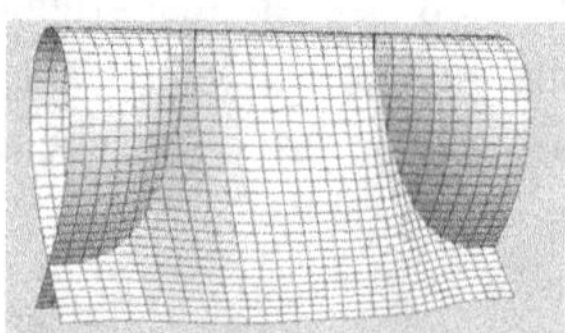

Fig. 7.17 (d)-move, Position A.

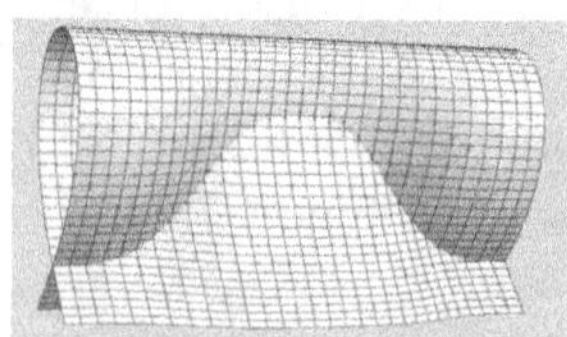

Fig. 7.18 (d)-move, Position B.

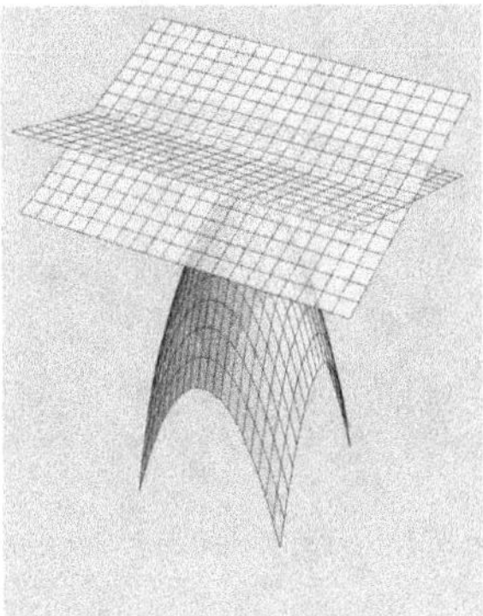

Fig. 7.19 (*e*)-move, Position *A*.

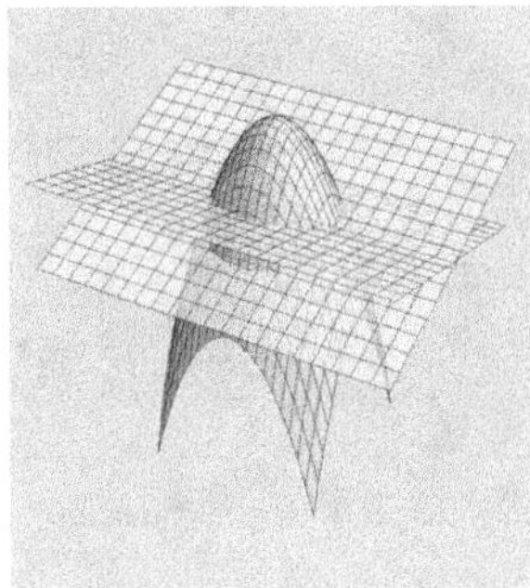

Fig. 7.20 (*e*)-move, Position *B*.

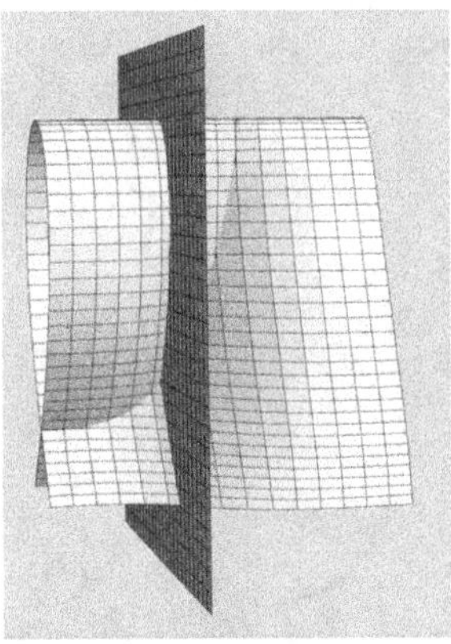

Fig. 7.21 (*f*)-move, Position *A*.

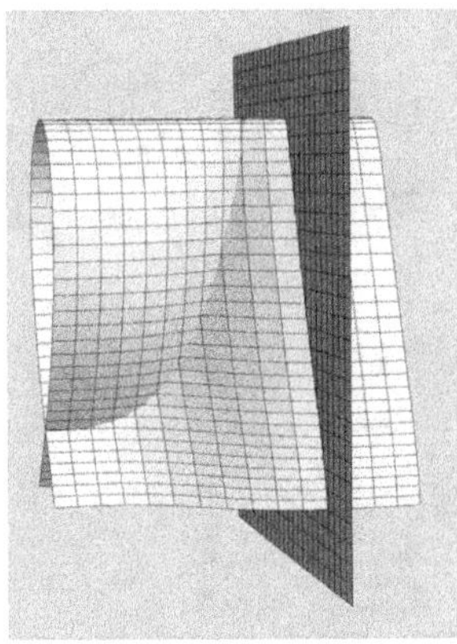

Fig. 7.22 (*f*)-move, Position *B*.

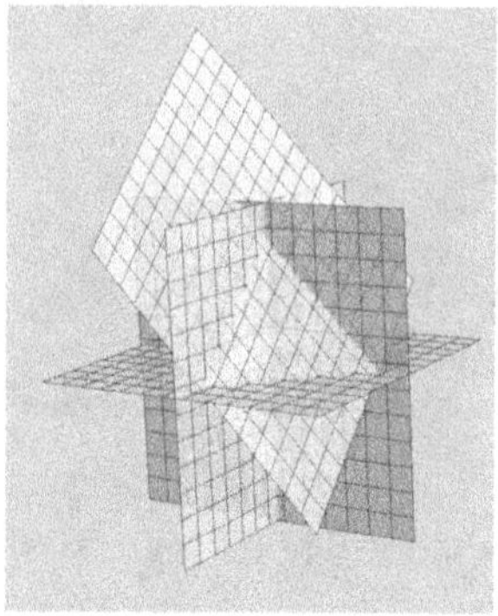

Fig. 7.23 (g)-move, Position A.

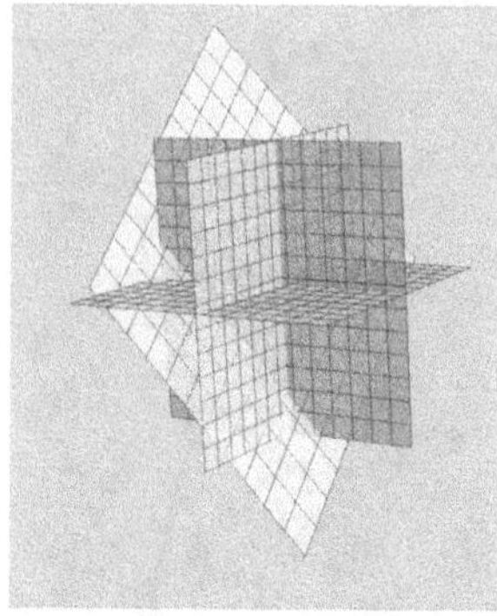

Fig. 7.24 (g)-move, Position B.

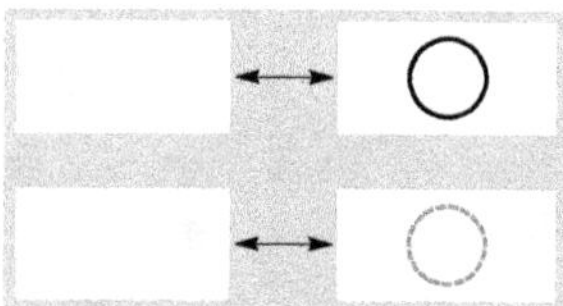

Fig. 7.25 Local scheme change (a)-move.

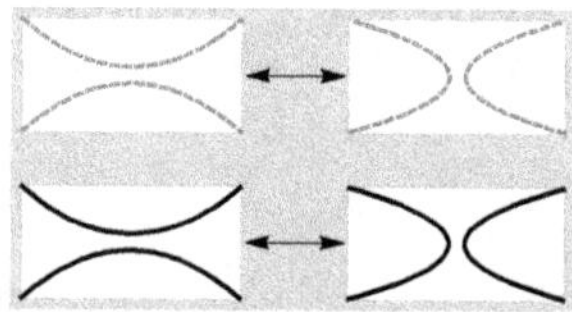

Fig. 7.26 Local scheme change (b)-move.

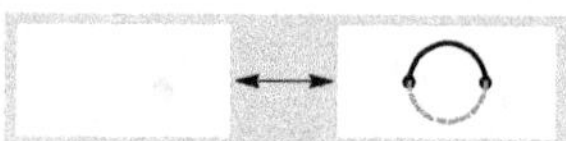

Fig. 7.27 Local scheme change (c)-move.

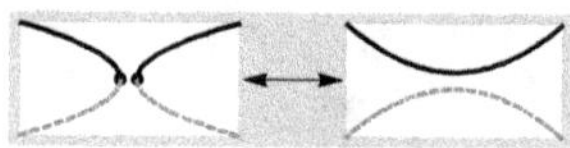

Fig. 7.28 Local scheme change (d)-move.

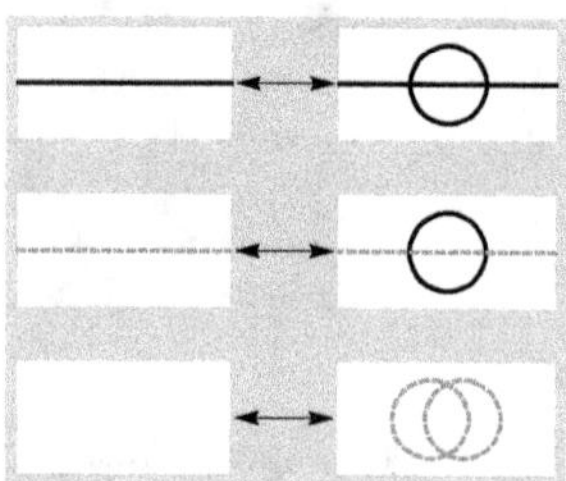

Fig. 7.29 Local scheme change (e)-move.

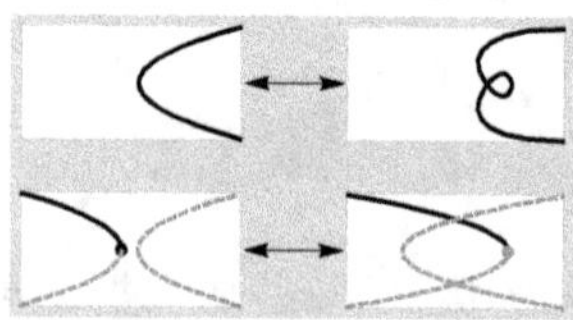

Fig. 7.30 Local scheme change (f)-move.

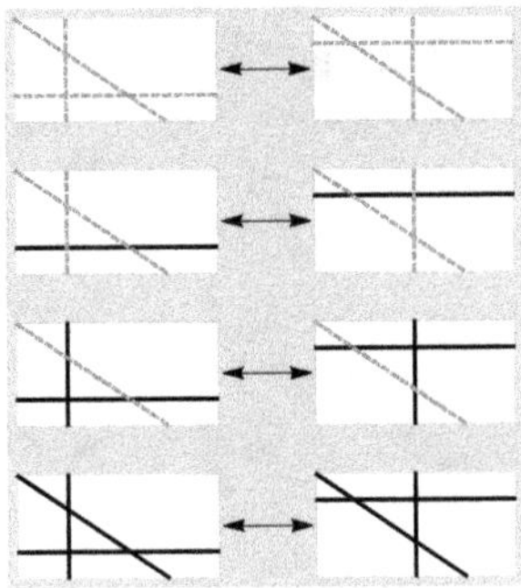

Fig. 7.31 Local scheme change (g)-move.

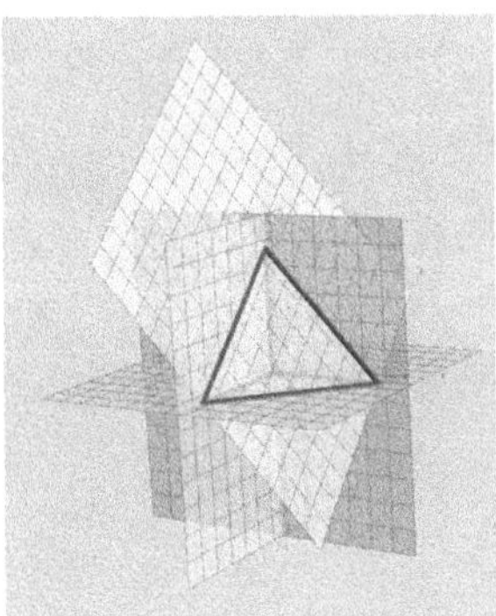

Fig. 7.32 Elementary (g)-move, Position A showing tetrahedral region.

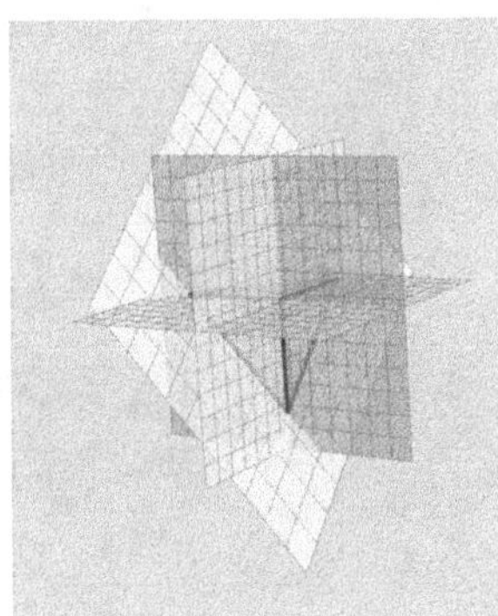

Fig. 7.33 Elementary (g)-move, Position B showing tetrahedral region.

Example 7.1. An isotopy that illustrates two knot-moves is shown, in projection, in Figures 7.34 to 7.36. In R^4 the sphere lies above the torus, and so the crossing scheme might be as shown in Figure 7.37. The transition from Figure 7.34 to Figure 7.35 is a (b)-move. The transition from Figure 7.35 to Figure 7.36 corresponds to an (a)-move. So, our knotting represents a split trivial link. This projection (with different above/below relations) is the same as the non-trivial basic sphere-torus link, see Example 6.12. ∎

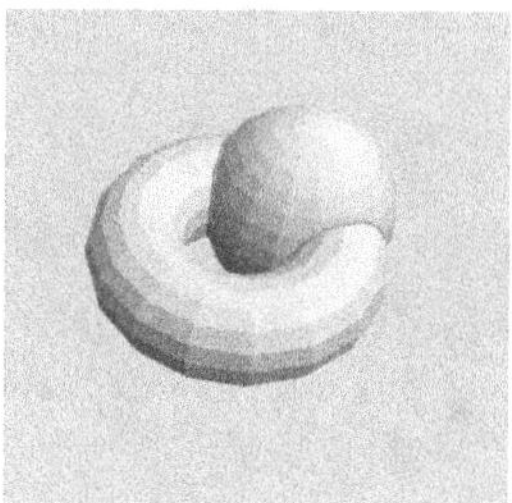

Fig. 7.34 Projection of initial knotting, two crossing circles.

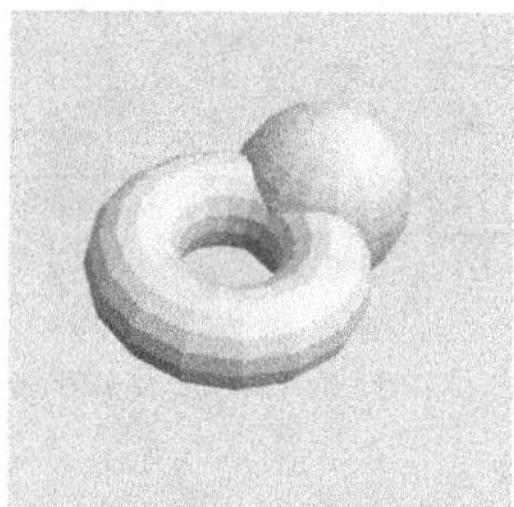

Fig. 7.35 After (b)-move there is one crossing circle.

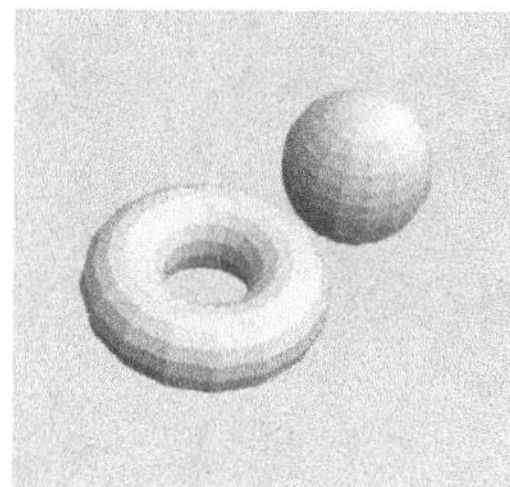

Fig. 7.36 After an (a)-move there are no crossings.

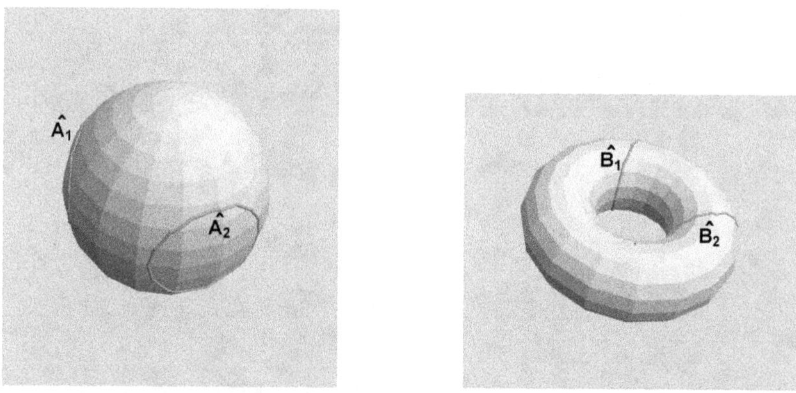

(a) (b)

Fig. 7.37 (a) Portion of crossing scheme on the sphere—two above circles: $\widehat{A}_0$ and $\widehat{A}_1$. (b) Portion of crossing scheme on the torus—two below circles: $\widehat{B}_0$ and $\widehat{B}_1$.

Example 7.2. Consider a knotting that has projection shown in Figure 6.86 height relations as detailed in Figure 7.38. We will show this knotting represents a trivial knot.

A (b)-move changes the projection Figure 7.41 to Figure 7.42. The move has the effect of merging the two circles of self-intersection to one circle. Figures 7.38, 7.39 and 7.40 show crossing schemes for this move respectively where $\widehat{A_0}$ and $\widehat{B_0}$ correspond to C_0. Figure 7.39 shows isotopy step when circles merge. As the crossing circles C_3 and C_4 merge to form C_0 we see that the corresponding above circles $\widehat{A_3}$ and $\widehat{A_4}$ merge as do the below circles $\widehat{B_3}$ and $\widehat{B_4}$. The remainder of the isotopy is a sequence of moves, each of type (a)—Figures 7.43 and 7.44. In the end we have a knotting whose projection has no crossings and so must be trivial. ∎

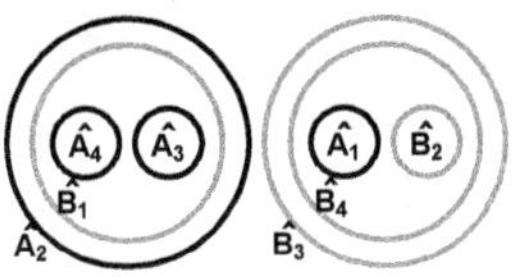

Fig. 7.38 Initial crossing scheme.

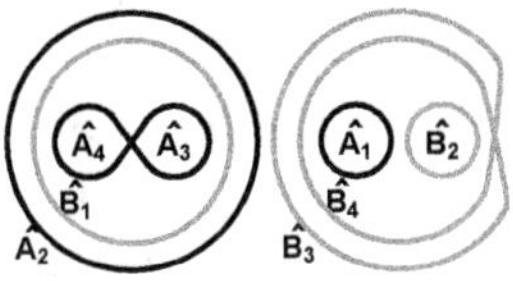

Fig. 7.39 Circles merge at midpoint of (b)-move.

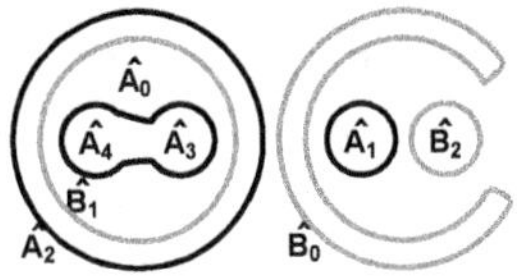

Fig. 7.40 Scheme after a (b)-move.

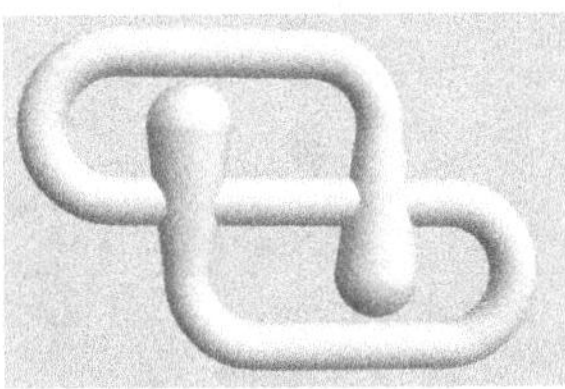

Fig. 7.41 Projection of isotopy beginning.

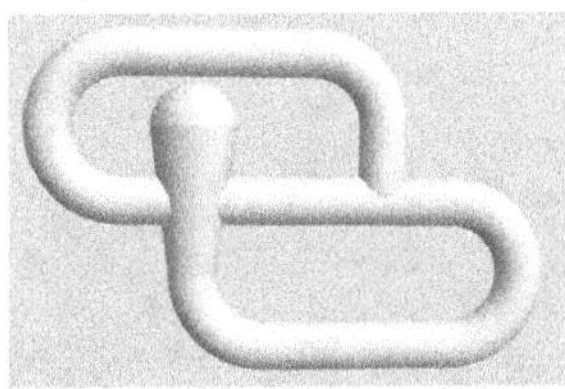

Fig. 7.42 After a (b)-move.

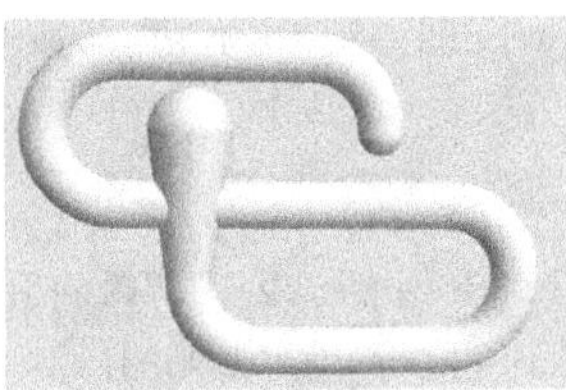

Fig. 7.43 An (a)-move has removed newly merged circle of self-intersection.

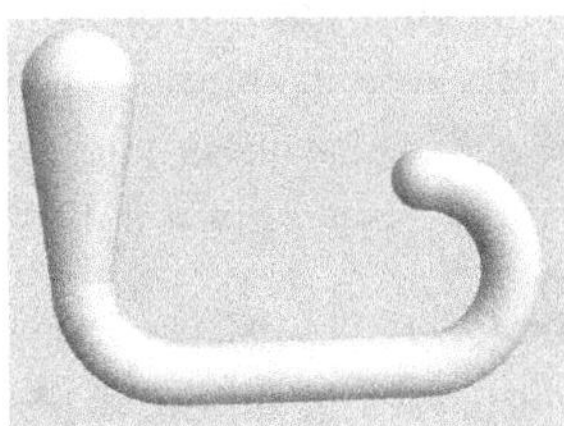

Fig. 7.44 After two more consecutive (a)-moves.

Example 7.3. The knotting shown in Figure 6.26 is a trivial knotting since one (c)-move will give a projection with empty crossing set. ∎

Remark 7.6. The figures for these seven first appeared, in hand drawn form, in [Roseman (1998)]. These Figures all relate to *smooth* knottings however there are minor technical graphics inaccuracies regarding smoothness near branch points, corrected in our current figures. In [Roseman (1998)] the three-dimensional graphics shows locally a branch point as apex of a geometric cone resting on a plane, touching along an element of that cone, Figure 7.45. This set is not the image of smooth map at the branch point. The image of the tangent plane at a branch point is a line. The elements of the cone should be in the direction of this line but clearly these elements have many directions. So, technically this graphic is not correct. Correct graphics for branch points are shown in our current Figures for moves of type (c), (d) and (f). Also, branch points for the crossing scheme were shown as in Figure 7.46 where the branch point is a large dot. At that dot the curve is not smooth but since it should be modeled on a Whitney umbrella, the crossing scheme must be a smooth curve at a branch point. This is shown correctly in Figures 7.27, 7.28 and 7.30. ∎

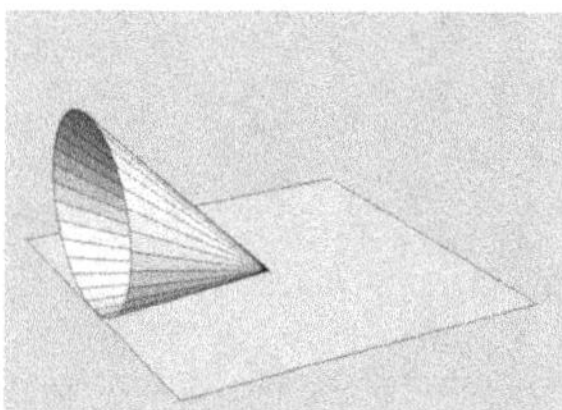

Fig. 7.45 Old style graphic for branch point in projection.

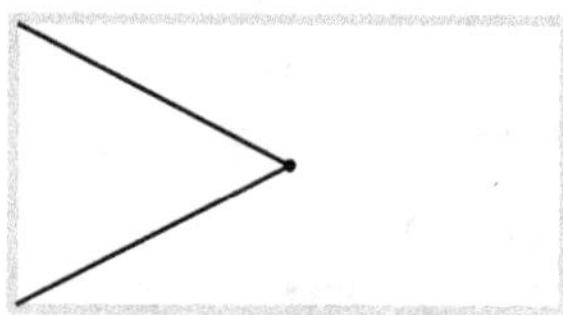

Fig. 7.46 Old style graphic for branch point in crossing scheme.

Given that any isotopy of general position 2-knottings can be expressed with the seven elementary moves, it is natural to ask: do we really need all seven? The answer is "no" see Remark 7.7. Also: what isotopies can be done if one uses only a subset of moves? This is addressed in Remark 7.8

Remark 7.7. Our list of moves follows from the Remark 12.12. This list is complete but not minimal. As was pointed out in [Yashiro (2005a)] one can express a d-move as a c-move followed by a b-move. This is shown in Figures 7.47 to 7.50. Figure 7.50 is equivalent to Figure 7.18. In [Kawamura (2015)] it is further noted that a c-move can be expressed as an a-move followed by a d-move—as shown in Figures 7.51 to 7.54—and further noted that there are no other such combinations within the moves. In both these Figure sequences, an isotopy projection is shown between the second and third step. In short, both c and d moves are moves that remove (or create) a pair of branch points, we need one or the other, but need not use both. ∎

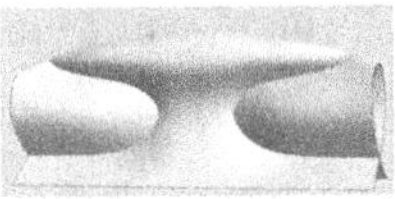

Fig. 7.47

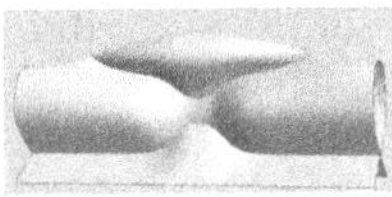

Fig. 7.48

Fig. 7.49

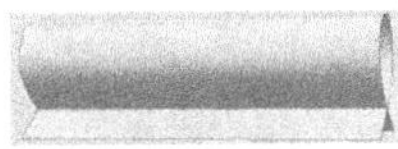

Fig. 7.50

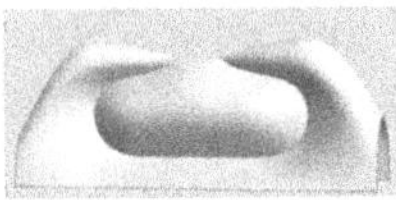

Fig. 7.51

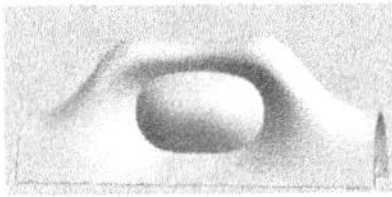

Fig. 7.52

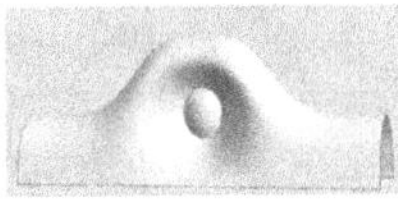

Fig. 7.53

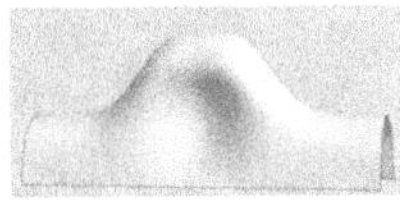

Fig. 7.54

Remark 7.8. By Proposition 7.2 any orientable 2-knot can be represented by a knotting that has no branch points. However we still need to consider branch points when studying isotopies orientable 2-knots. A projection of a 2-knotting is an immersion and the natural equivalence of immersions is regular homotopy, Definition 3.15. In other words there are equivalent 2-knottings K and K' such that the projections of each is an immersion, but no isotopy of these knottings has a projection which is a regular homotopy. In terms of knot-moves this means that any isotopy between K and K' must use some of the knot-moves which involve branch points: moves of type (c), (d) or (f). The first example was for knottings of a 2-torus [Satoh (2001)] with a different proof in [Oshiro and Tanaka (2015)] and with the case of 2-spheres settled in [Takase and Tanaka (2016)]. We discuss a related Klein bottle example in Remark 7.13. ∎

Consider knot-moves which involve triple-points: type (e), (f), and (g). One can ask: if K_* and K'_* are projections of equivalent knottings with neither projections having triple points is there always an isotopy between them that does *not* use any of these? The answer to this is "no", see [Jablonowski (2012)]. Also, if equivalent knottings have the same number of triple points, can we go from one to the other without needing to use some of these moves?. The answer is "no", [Kawamura *et al.* (2016)], where examples are given that specifically require moves of type g.

Remark 7.9. A projection K_* of a 2-knotting K can be a complex subset of R^3. What elementary knot-moves are possible for K_* and how can we describe them? The neat local models for knot-moves shown in Figures 7.11 to 7.24 may not be immediately evident or easy to describe. For two cases, we use what we call **move-guides**. A move-guide is a 2-disk $G \subseteq R^3$ with $G \cap K_* = \partial G$ that, in some sense, serves as a guide for the move. One way to search for possible moves for a projection is to search for possible move-guides.

A move-guide for a b-move is shown in Figure 7.55. This is a 2-disk G such that $G \cap K_* = \partial G$ and such that ∂G contains exactly two points, a and b, which are 2-fold crossing points. The pullback $\widehat{\partial G} \subseteq M$ of ∂G consists of two disjoint arcs α and β. A requirement for a b-move is that either both endpoints of α are above-points or both are below-points. For the knotting projection in Figure 7.56 a disk is shown in dark gray which intersects the projection in its boundary. If the height relations are as shown in Example 7.1 the disk will satisfy the above/below condition and will be a move-guide for the b-move. But if this is a projection of the basic sphere-torus knotting Example 6.12 then the above/below conditions are not met.

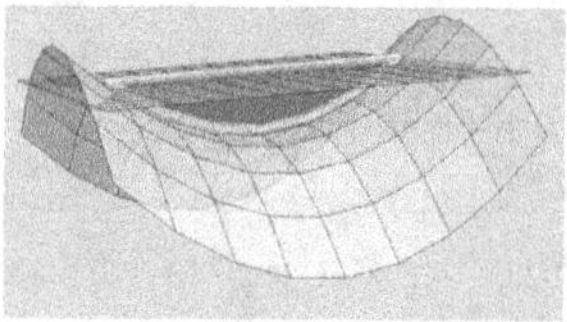

Fig. 7.55 Move-guide for (b)-move.

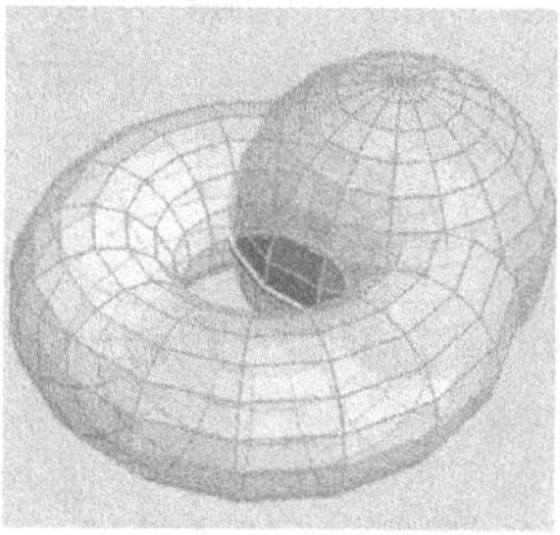

Fig. 7.56 A disk whose boundary lies on projection.

A d-move which creates two branch points corresponds to a move-guide G with $G \cap K_* = \partial G$ whose boundary meets a single 1-shard of K_*. Figure 7.57 is a version of Figure 7.18 using transparency to reveal an example of such a disk G—in dark gray.

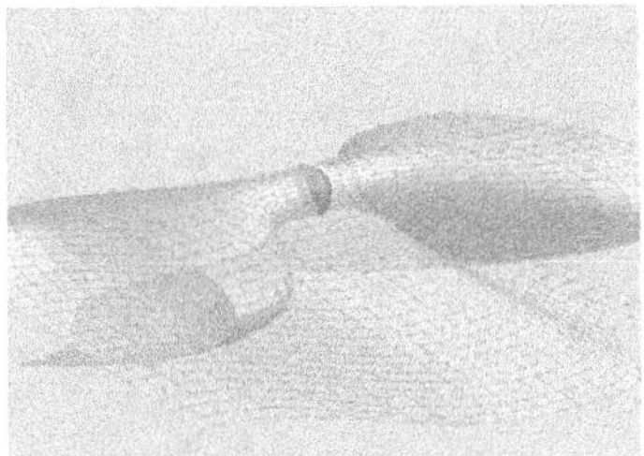

Fig. 7.57 Move-guide for (d)-move.

A d-move which *cancels* two branch points needs to be considered with special care. An alternate local model of two branch points contained in a disk image of Figure 7.17 is Figure 7.58. We say these are **mirrored branch points**. Here the mirror M is a plane perpendicular to a line segment joining the branch points and half-way between them. This is in contrast to the branch points in Figure 7.59 which we call **mismatched branch points**. Locally the crossing set as seen in Figure 7.17 has two components. A second consideration are above/below relations at these components. For a (d)-move, these above/below relations must be compatible. Figure 7.60 uses crossing-shadows to show an example of compatible above/below relations. These shadows are also mirror images of each other. In this case we say the branch points have **mirrored height** relation. Figure 7.61 shows a crossing shadow rendered local version of Figure 7.17. Graphically the shadows "merge" appropriately when the branch points cancel. *So, for two branch points to cancel by a (d)-move they must be mirrored with mirrored height relation.* ∎

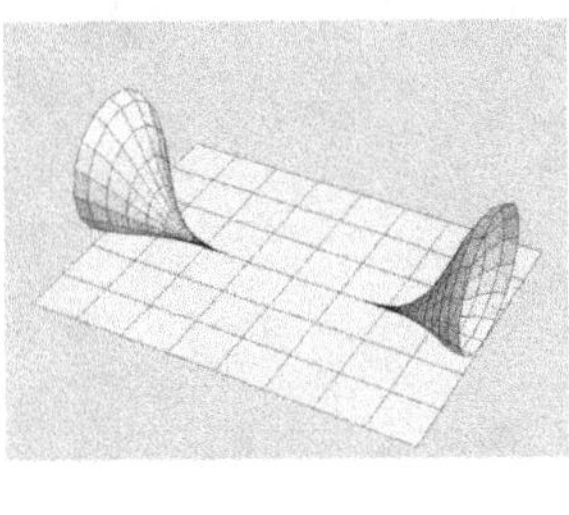

Fig. 7.58 Two mirrored branch points.

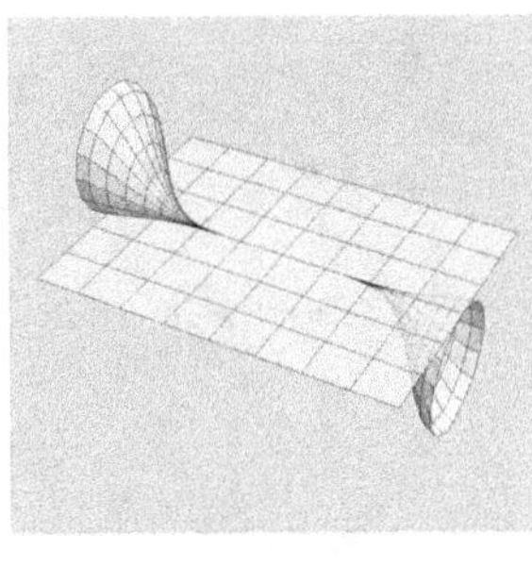

Fig. 7.59 Two mismatched branch points.

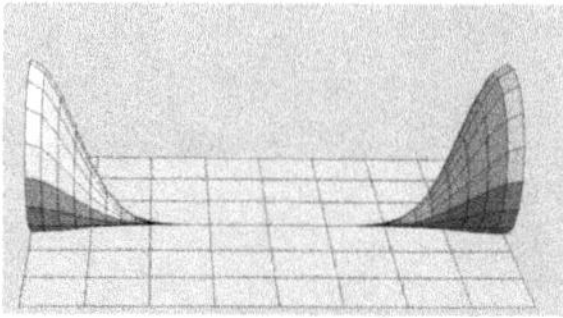

Fig. 7.60 Crossing-shadows near canceling branch points.

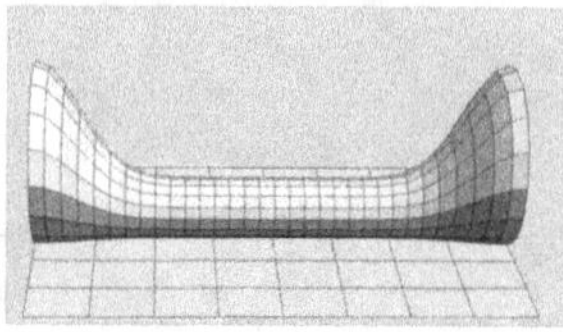

Fig. 7.61 Crossing shadow after cancelation.

We take a closer look at a (d)-move shown with wire-frame coding in Figures 7.62 and 7.63 are wire-frame codings for Figures 7.17 and 7.18. Figure 7.62 is a projection of a disk with two branch points. It is important to distinguish this from a similar projection of a disk with two branch points shown in Figure 7.65 which can *not* be canceled by a knot-move. In the terms of Remark 7.9, in each case the branch points are mirrored but in Figure 7.62 the height relations are also mirrored while in Figure 7.65 they are not.

We next consider a pair of examples that compare and contrast cancelling and non-cancelling branch points. Both Figure 7.62 and Figure 7.65 can be viewed as a projection with two branch points of an isotopy of an arc $\alpha \subseteq R^3$. At each isotopy step, the projection of α has either one crossing or none. The isotopy shown in Figure 7.64 corresponds to 7.62, that of Figure 7.66 corresponds to 7.65. A description of the Figure 7.64 isotopy is: first rotate α vertically a quarter of a turn, then reverse direction and rotate to original position. For the isotopy of Figure 7.66 we can say: rotate α vertically one complete turn.

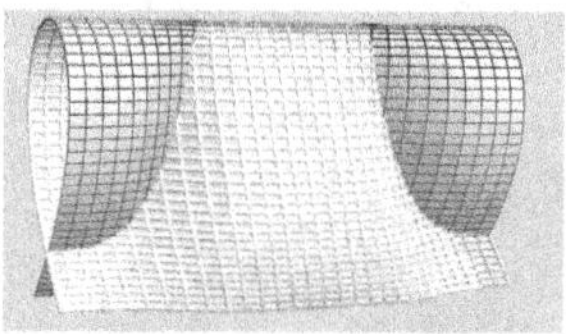

Fig. 7.62 A wire-frame coding for Figure 7.17.

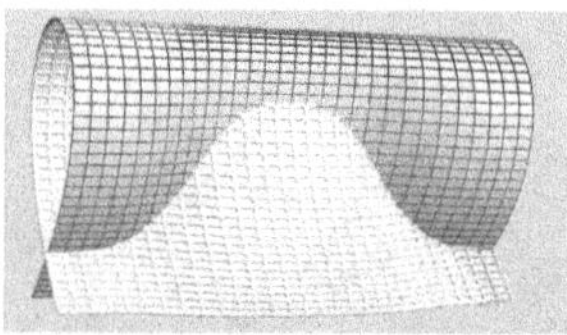

Fig. 7.63 A wire-frame coding for Figure 7.18.

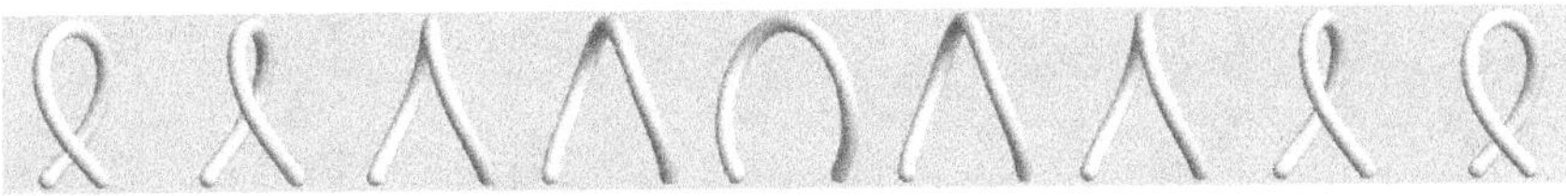

Fig. 7.64 Isotopy track of arc α with trace Figure 7.62—a quarter vertical twist then reverse quarter twist.

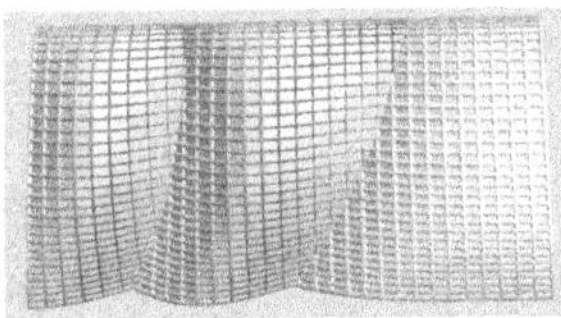

Fig. 7.65 Branch points that do not cancel.

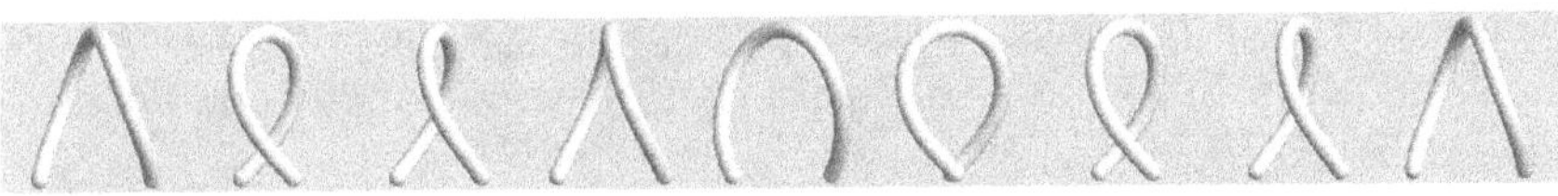

Fig. 7.66 Isotopy track of arc α with trace Figure 7.65—a full vertical twist.

The only knot-moves that reduce the number of branch points are the (c)-move and (d)-move. However, we can avoid using the (c)-move for this purpose, if we wish. Any (c)-move that cancels two branch points and eliminates an arc of crossing points as in Figures 7.67 can be replaced by a (d)-move which cancels these branch points and turns an arc of crossing points into a circle of crossing points, see Figures 7.68, 7.69 and 7.69.

Fig. 7.67 A (c)-move can "cancel" the two branch points shown.

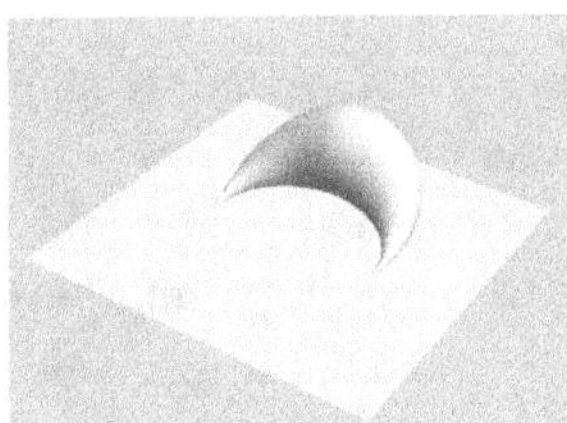

Fig. 7.68 Alternatively we could isotop beginning as shown.

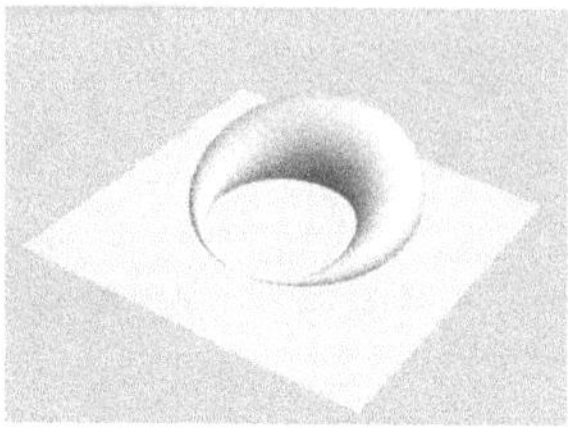

Fig. 7.69 Then continuing to bring branch points closer.

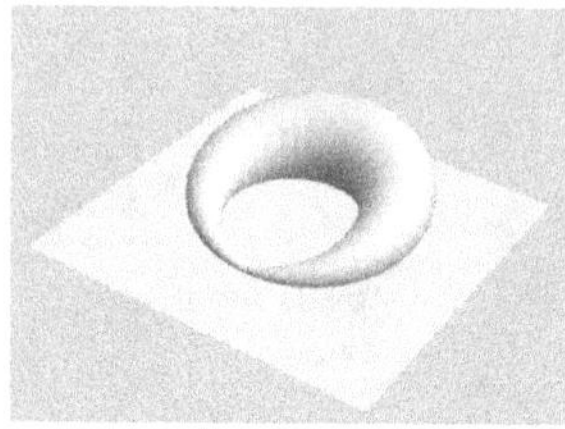

Fig. 7.70 And now "cancel" branch points with (d)-move.

It is of interest in to explore what can or cannot be done by using only *some* of these moves. In [Satoh (2001)] it was shown that the torus whose projection is spin of a 1-crossing knotting and standard embedding of a torus although isotopic cannot be transformed without using moves that involve branch points. Examples for spheres are found in [Takase and Tanaka (2016)]. In [Oshiro and Tanaka (2015)] other quandle tools are used to explore this topic.

A **2-foam** is an object in R^4, of some current interest, similar to a 2-dimensional submanifold but at some places, locally may look like a book with three pages. Extensions of knot-moves have been defined and studied in [Carter (2015)], [Carmi and Moskovich (2017)]

7.2 Some Non-Elementary Moves

For finding examples of isotopies it is useful to have simple isotopies in addition to the elementary knot-moves. All elementary moves involve changes for *disks*. These non-elementary moves involve an *annulus*.

Example 7.4. A handy simple isotopy of an annulus is the one which in projection changes between Figure 7.71, a projection of an annulus with a circle of crossing points and Figure 7.73, a projection of an annulus with an arc of crossing points between two branch points. Our move is simply a (d)-move as applied to the annulus. In other words like Figures 7.67 and 7.70, but taken in reverse order for an annulus rather than a disk. Figure 7.72 shows an intermediate position. This process can be roughly described as "grab each end of the annulus in Figure 7.71 and pull it tight in R^4".

Details of this (d)-move are shown in Figures 7.74, 7.75 and 7.76, using transparency to reveal occluded details. To aid viewing, we have drawn arc γ that is the projection of an arc in R^4 which joins the two boundary circles of the annulus. In Figure 7.74 the projection of γ forms a small loop which is the boundary of a move-guide for a (d)-move as in Remark 7.9. ∎

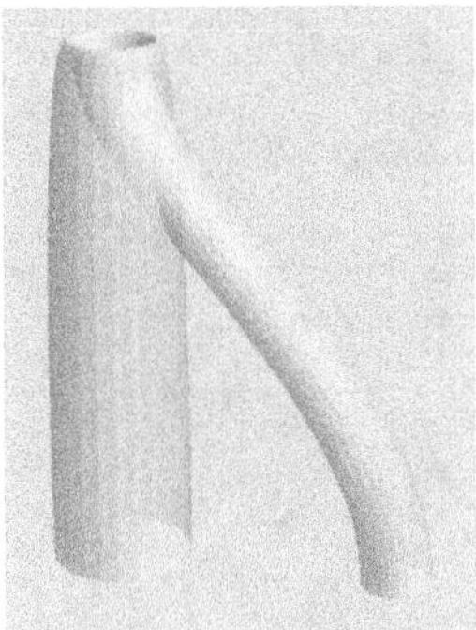

Fig. 7.71 Annulus with circle of crossing points.

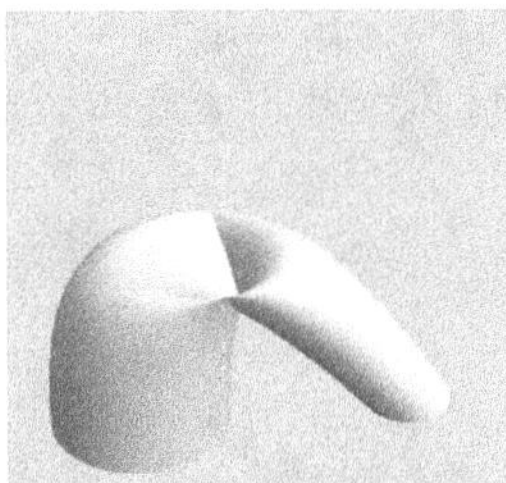

Fig. 7.72 Intermediate position.

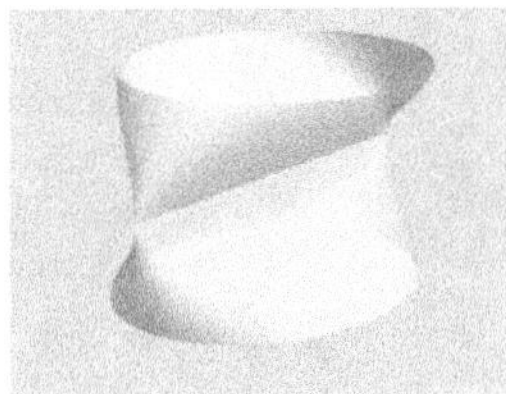

Fig. 7.73 Annulus with arc of crossing points.

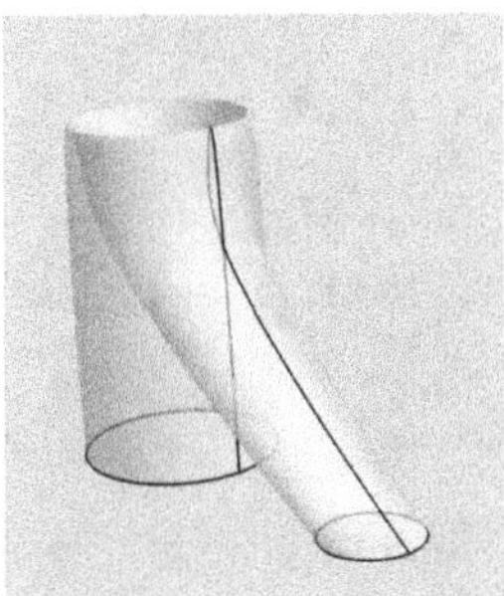

Fig. 7.74 Showing γ forms a loop.

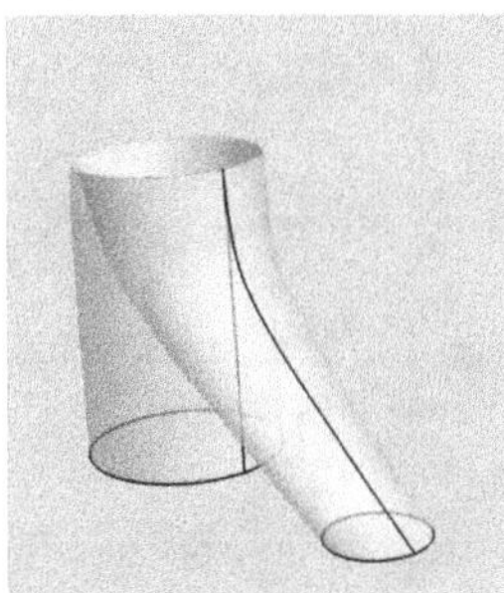

Fig. 7.75 Note the loop in γ no longer appears.

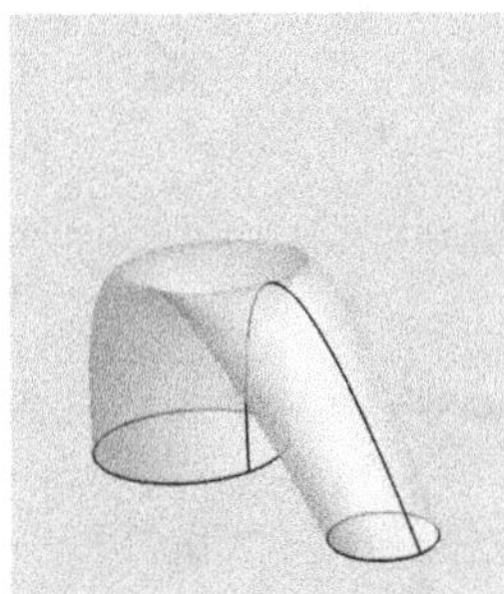

Fig. 7.76 Another view of Figure 7.72.

Remark 7.10. Using Example 7.4 we can transform the Klein bottle projection of Figure 6.130 to the more familiar graphic, Figure 7.81.

There are two basic configurations for a crossing arc with two branch points as end-points. These correspond to two different ways of putting together two Whitney umbrellas, Figure 7.77 or Figure 7.78. Figures 7.79 and Figure 7.80 show this assembly graphically. A 2-knotting which exhibits Figure 7.77, is the projection of the Klein bottle Figure 6.130. An example of Figure 7.78 is the knotting of the projective plane Figure 6.138. ∎

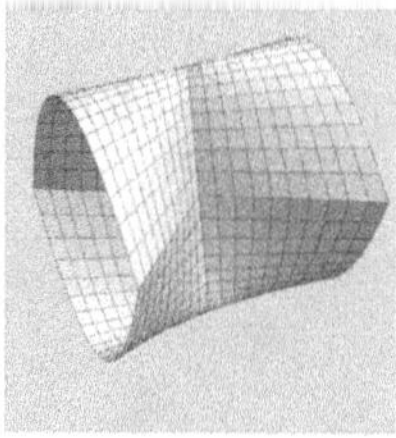

Fig. 7.77 This union of umbrellas is annulus projection.

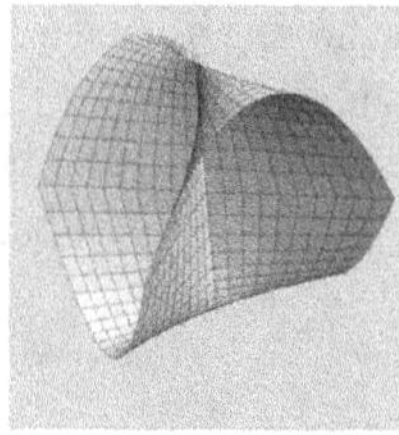

Fig. 7.78 This union of umbrellas is Möbius band projection.

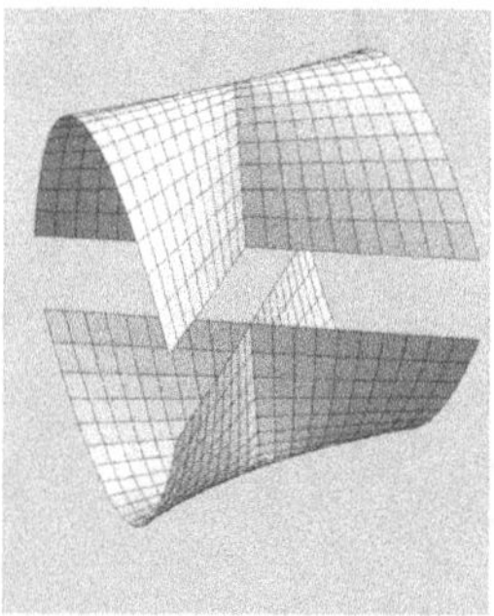

Fig. 7.79 Mirrored branch points.

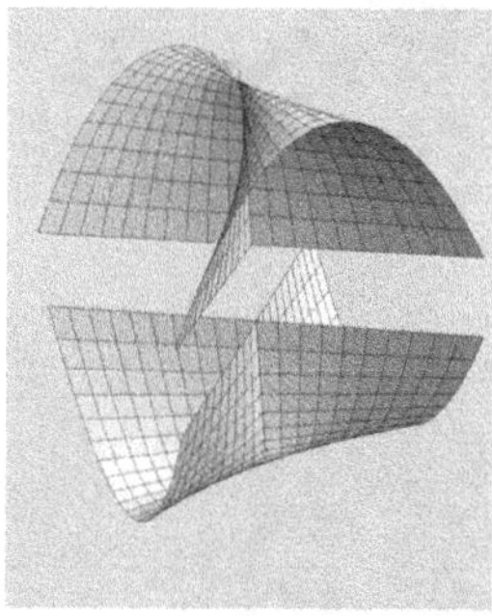

Fig. 7.80 Mismatched branch points.

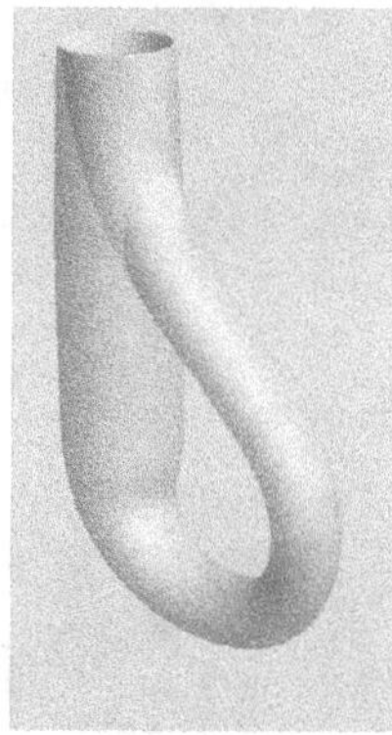

Fig. 7.81 A traditional graphic for the Klein bottle.

Example 7.5. The **arch-move** is another non-elementary move, see Figures 7.82, 7.83 and 7.84, involves a disk D and an annulus A projecting with two crossing circles. In the crossing scheme, we have two non-trivial circles on A—either both above or both below. In Figure 7.82 the crossing-shadows indicate both circles are above, so the annulus passes above the disk in R^4. The move is a combination of a single (b)-move followed by an (a)-move and has been used in Example 7.1 and Example 7.2. ∎

Fig. 7.82 Beginning of arch-move.

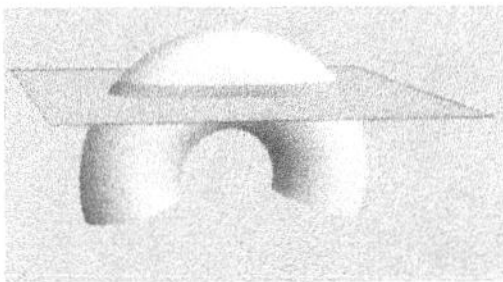

Fig. 7.83 Figure 7.82 after elementary (b)-move.

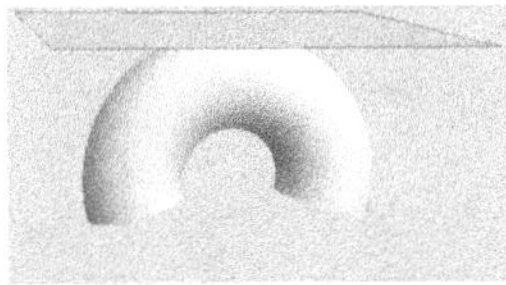

Fig. 7.84 End of arch-move.

Example 7.6. A third move we call a **circle-switch** involves one sub-annulus N and two sub-disks D and E, see Figures 7.85, 7.86 and 7.87. The crossing set consists of two circles— $C = N_* \cap D_*$ and $K = N_* \cap E_*$ and one line segment $D_* \cap E_*$. The shadow coding at $D_* \cap E_*$ indicates that D passes above E in R^4. Also, the annulus N_* above D along C and below E along K. Our move, starting with Figure 7.85 consists of an (e)-move which creates two triple points, see Figure 7.86, followed by a second (e)-move which deletes these two triple points ending with Figure 7.87. Briefly this move moves N between D and E in R^4. Note that the height relations along C and K have been switched in the process.

This switching in the crossing scheme corresponds to N—Figures 7.88 and 7.89 where we switch positions in N of an a-circle $\widehat{A_C}$ and a b-circle $\widehat{B_K}$. This is similar to a classic Reidemeister $\Omega.3$-move. In Figure 4.61, focus on the line segment that has one above-point and one below-point— it lies between the other two segments in R^3. Note how the above and below-points "switch positions" in the Gauss code under this move. ∎

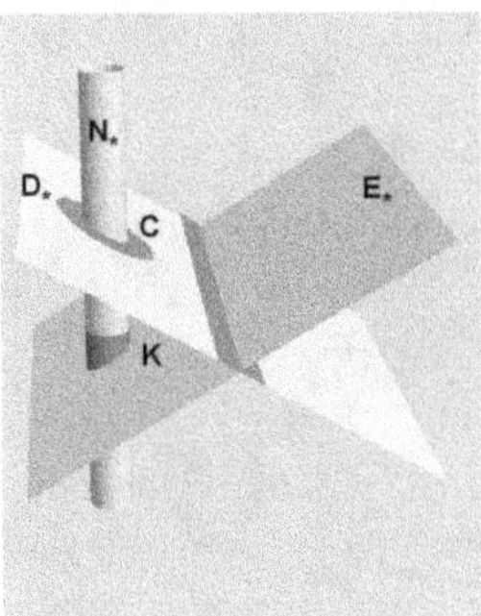

Fig. 7.85 An initial position for circle-switch.

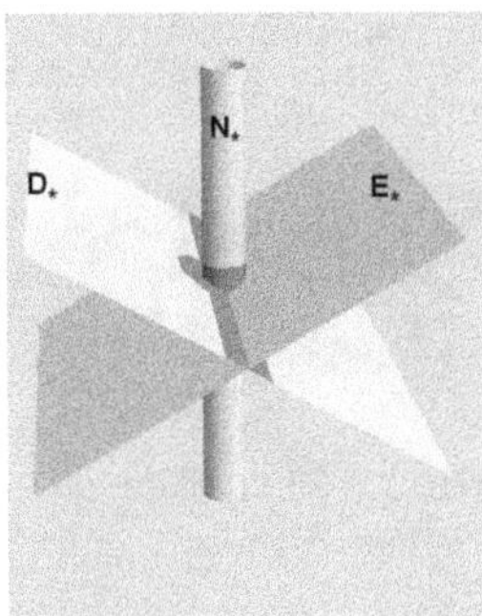

Fig. 7.86 Mid position of move.

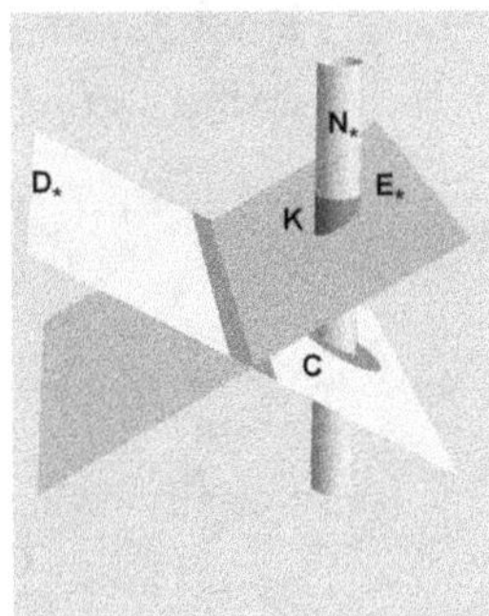

Fig. 7.87 The completed move.

Fig. 7.88 The portion of crossing scheme in N based on Figure 7.85.

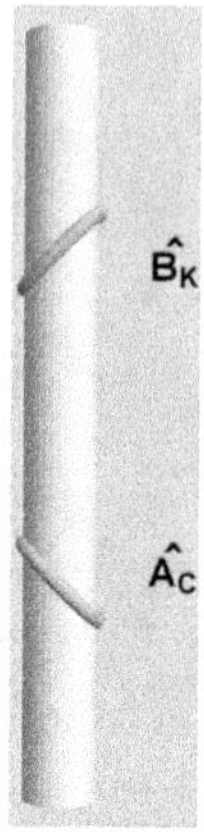

Fig. 7.89 The portion of crossing scheme in N based on Figure 7.87.

Example 7.7. A **tube-move** is just a useful special case of a (b)-move, which introduces a circle of crossing points that might not bound a disk, see Figures 7.90 and 7.92. Figure 7.91 corresponds to a critical point Figure. ∎

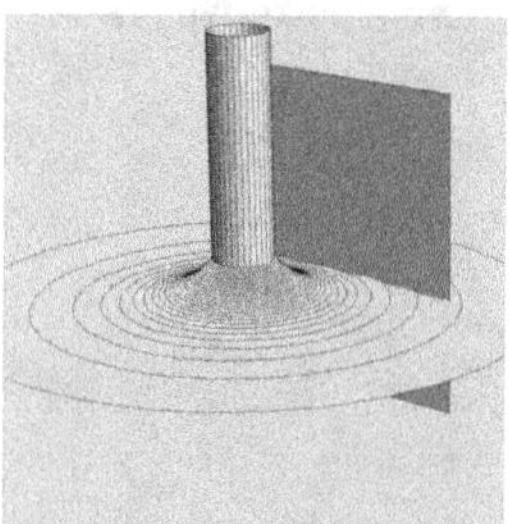

Fig. 7.90 Slide-move position A.

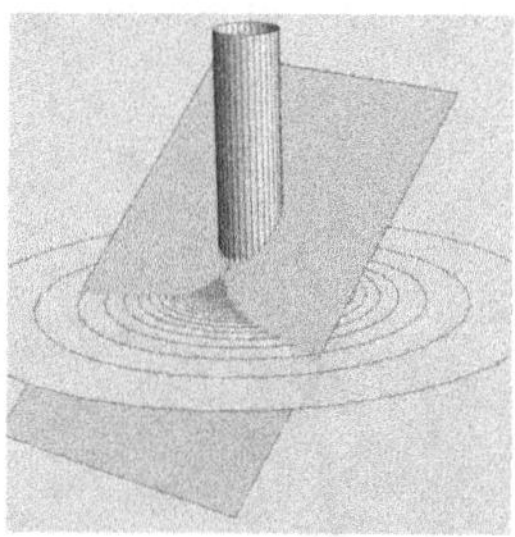

Fig. 7.91 Slide-move position B.

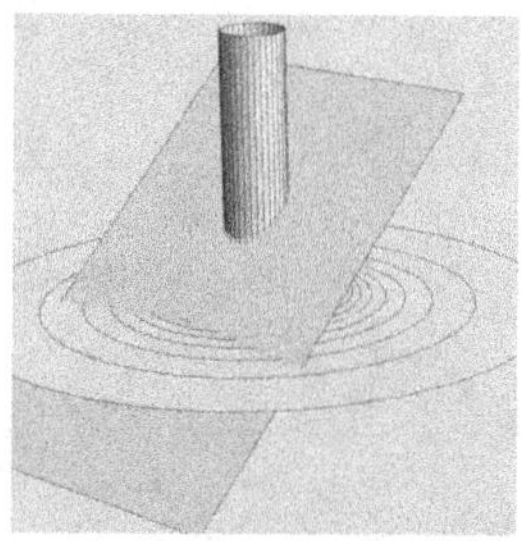

Fig. 7.92 Slide-move position B.

7.3 Some Isotopies of the Standard Klein Bottle

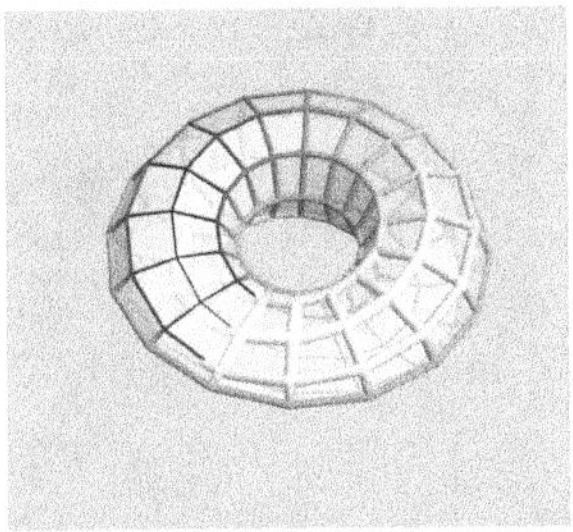

Fig. 7.93 Wire-frame coding for Klein bottle with positive half-flip.

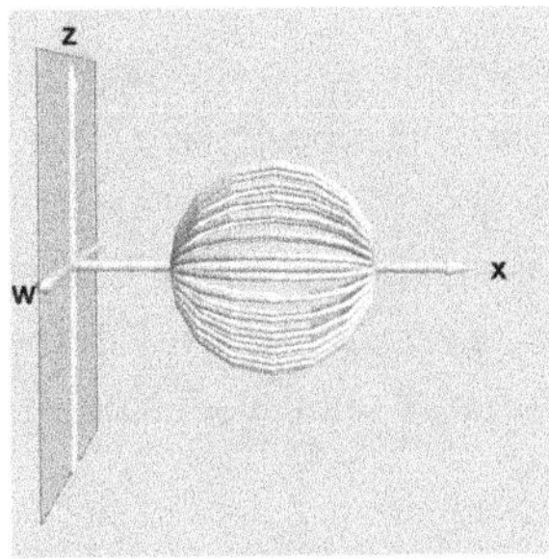

Fig. 7.94 Meridians M_θ^* for Figure 7.93, presented on same page for comparison to show flip.

Remark 7.11. In Definition 6.25 of $\mathbb{K}^2$ a *positive* half-flip of a circle was used. If we change the direction of the half-flip we get a knotting $\mathbb{K}_0^2$. We will show that $\mathbb{K}^2$ and $\mathbb{K}_0^2$ are isotopic knottings. This contrasts with a seemingly similar situation for the projective plane, in which two non-isotopic knottings are produced, see Remark 6.5.

This isotopy is unusually simple in that it needs no elementary moves at all. That is to say: the projection of the isotopy of this knotting is the isotopy of the projection of this knotting. This isotopy of the projection is shown using wire-frame coding as a sequence of steps in Figures 7.93, 7.95, 7.97, 7.99 and 7.101. Figure 7.93 is projection of $\mathbb{K}^2$ and Figure 7.101 is projection of $\mathbb{K}_0^2$.

Notationally, we are describing an isotopy K_ϕ in R^4, where our parameter is $0 \leq \phi \leq \pi$ with $K_0 = \mathbb{K}^2$ and $K_\pi = \mathbb{K}_0^2$. A peculiarity of this construction is that each individual knotting K_ϕ of our Klein bottle of is itself defined by an isotopy (of a circle), so K_ϕ is defined by a continually changing isotopy. This can be explicitly constructed using a continuous rotation in the xz-plane by angle ψ of our flipping axis.

The track of the isotopy used to define $\mathbb{K}^2$ is a sphere. A discrete track is shown in Figure 7.94. It is a union of circles which, at various steps of the half-flip isotopy where the flipping axis has direction of the x-axis. Although the circles intersect in the isotopy track, each circle corresponds to a separate meridian of $\mathbb{K}$. The projection of each of these meridians in Figure 7.93 corresponds to its projection in R^3_+ to the xz-half-plane, with one circle projecting to a line segment, the rest to ellipses as seen in Figure 6.129. From the mutual intersection of these circles, it is visually clear that the x-axis is the axis of the half-flip. Figure 7.96 shows a discrete track of an isotopy used to produce Figure 7.95—the axis of this flip makes angle of $Pi/4$ with the positive x-axis. This will be $K_{\pi/4}$ in our isotopy. Continuing to rotate our flipping axis in the xz plane, we similarly get Figure 7.97 that corresponds to a half-flip of m_θ about an axis parallel to the ray z-axis, Figure 7.98 and corresponds to $K_{\pi/2}$ of our Klein bottle isotopy. Our isotopy continues similarly to obtain $K_{3\pi/2}$ and concludes with K_π. ∎

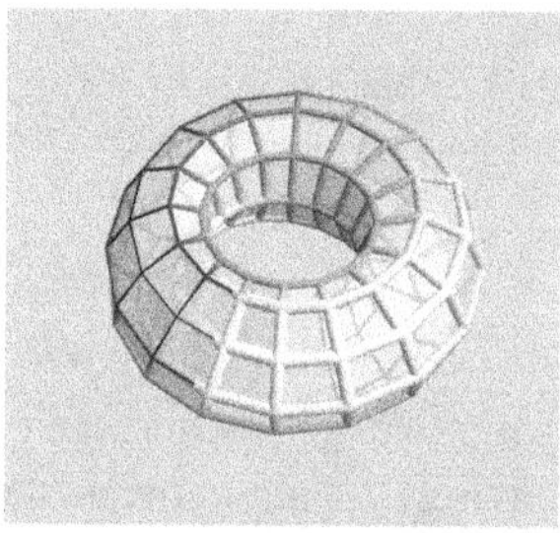

Fig. 7.95 A second step of isotopy.

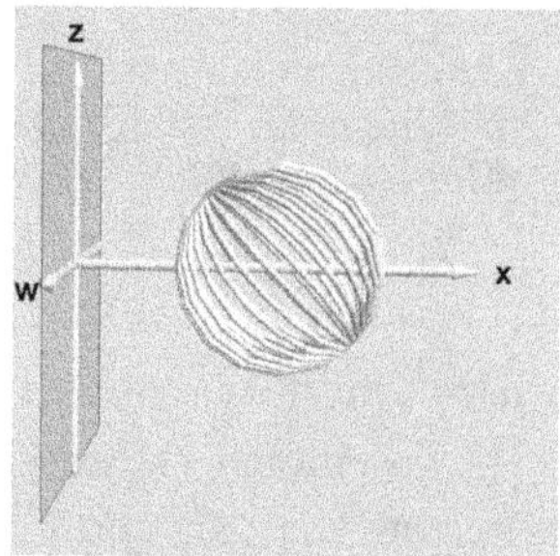

Fig. 7.96 Meridians M_θ^* for Figure 7.95. Note different axis of flip.

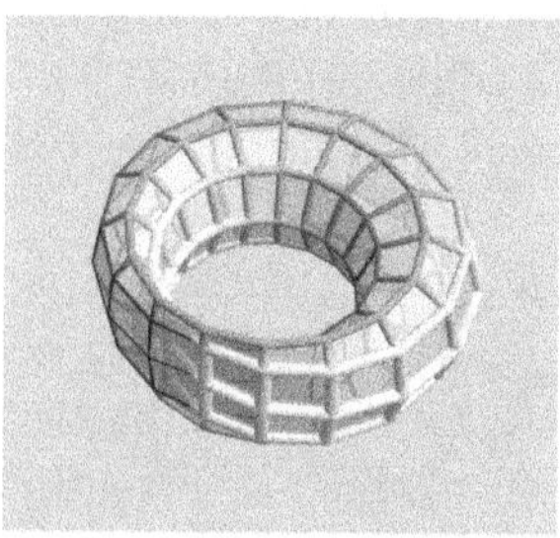

Fig. 7.97 A third step of isotopy.

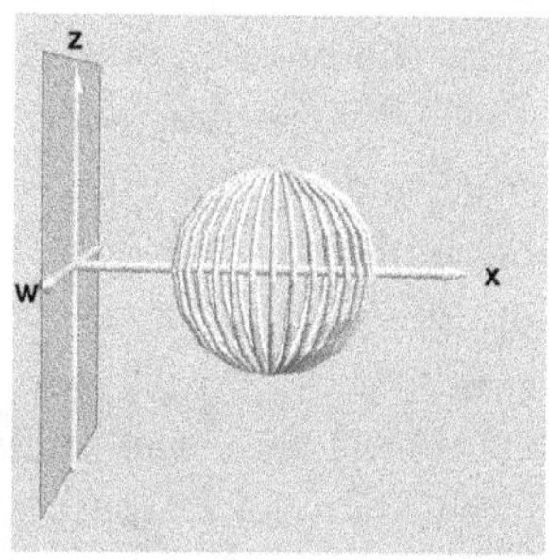

Fig. 7.98 Meridians M_θ^* for Figure 7.97. Note different axis of flip.

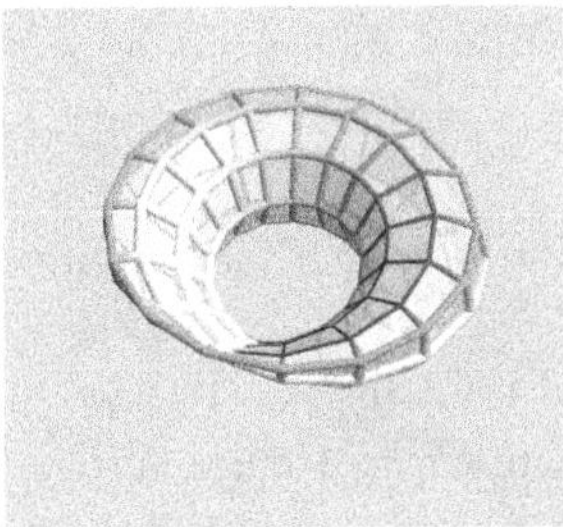

Fig. 7.99 A fourth step of isotopy.

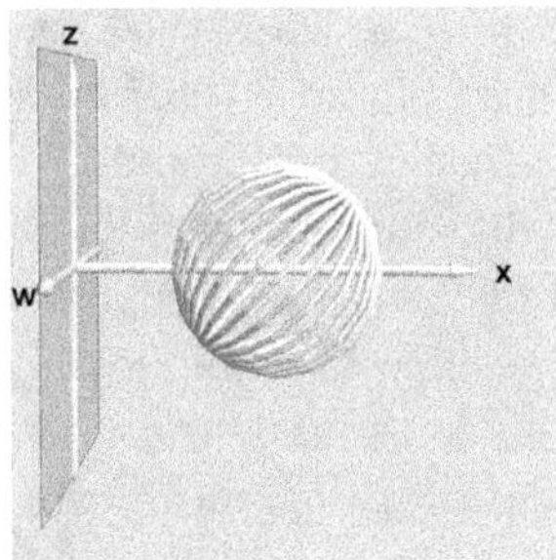

Fig. 7.100 Meridians M_θ^* for Figure 7.99. Note different axis of flip.

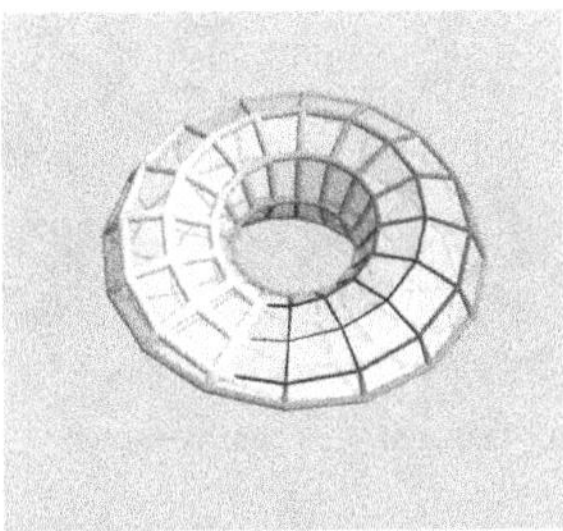

Fig. 7.101 Final state of isotopy. Wire-frame coding for Klein bottle with negative half flip. Compare height relations with Figure 7.99.

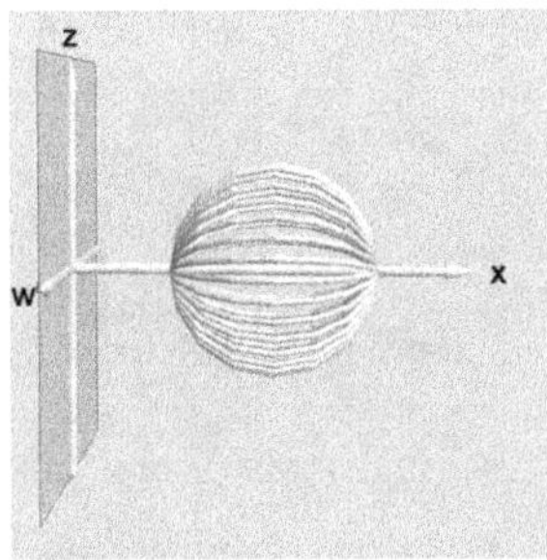

Fig. 7.102 Meridians M_θ^* for Figure 7.101.

Example 7.8. Another knotting of the Klein Bottle commonly called a **figure-eight knotting of the Klein bottle** is derived from a circle M_0 in $R^3_+ \approx H^3_0$ whose projection has one crossing and an isotopy $m\theta$ parameterized by $0 \le \theta \le 2\pi$ with steps indicated in Figure 7.103. If we project this isotopy in the w-direction it will appear to be a one-crossing curve that is slowly rotated so that the crossing point is fixed. The rotation of the projection in the xz plane is *counter-clockwise* ends up with a rotation of $m_{2\pi}$ mapping the circle to itself with orientation reversed. This isotopy is different from a half-flip. In a half flip there are two points that do not move; a half turn has no fixed points.

As in previous construction we define meridians $M_\theta = i_\theta(m_\theta) \subseteq H^3_\theta$ and obtain a 2-knotting of the Klein bottle $\mathbb{F} = \bigcup_{\theta=0}^{\theta=2\pi} M_\theta$, Figure 7.104. (A meridian A has been given a direction—, note that as we come around to B we have a reversed orientation.) The projection of this knotting is shown in Figure 7.106. We get a second knotting which we denote $\mathbb{F}'$ by instead using an isotopy m'_θ which uses a *clockwise* rotation, Figure 7.105. (We will soon see these knottings are isotopic.)

A nice feature of this Klein bottle projection is that it is an immersion with a single circle of crossing points.

This knotting is equivalent to the standard knotting $\mathbb{K}$. To see this do an elementary (d)-move using move-guide as shown in Figure 7.107. The result of the move will be to alter the crossing circle and change it to a crossing arc between two endpoints. If one takes this arc and makes it small and fairly straight one can see that we essentially have our standard projection as in Figure 6.130. ∎

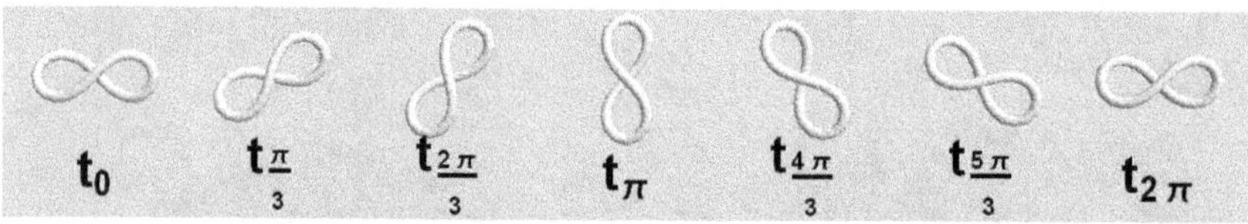

Fig. 7.103 Steps of a half-turn isotopy of a knotting with one crossing.

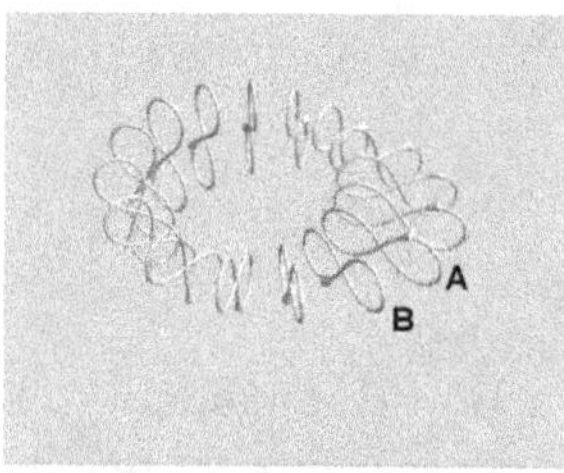

Fig. 7.104 Projection meridians for Klein bottle $\mathbb{F}$.

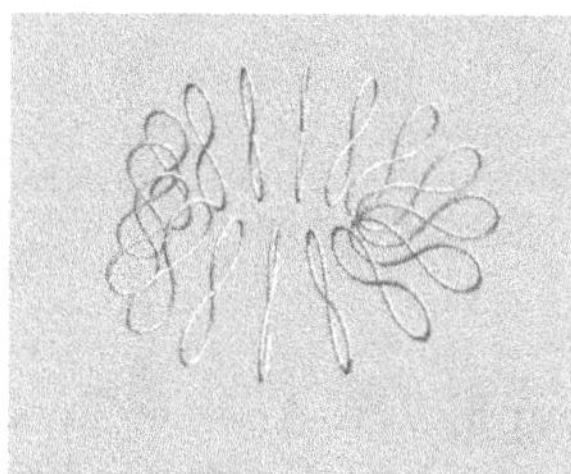

Fig. 7.105 Projection meridians for Klein bottle $\mathbb{F}'$.

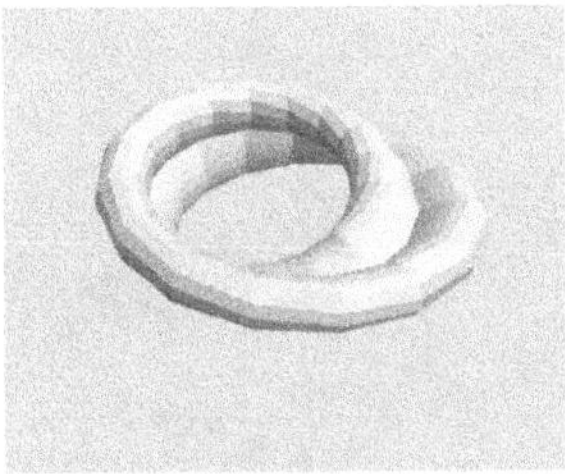

Fig. 7.106 A projection of the Klein bottle $\mathbb{F}$.

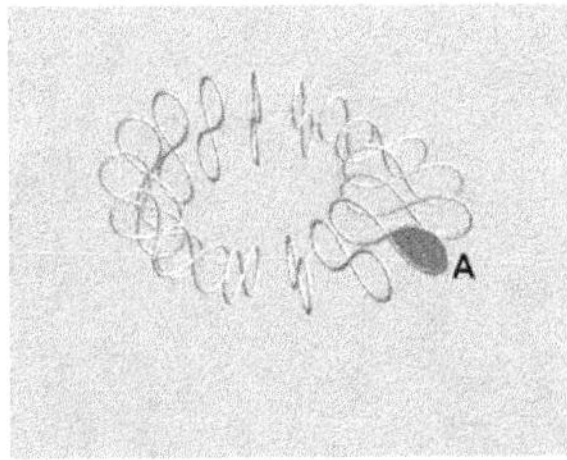

Fig. 7.107 A move-guide about projection of meridian A used for (d)-move.

Remark 7.12. There is an interesting mathematically natural way to implement the isotopy M_θ of Example 7.8. For our knotting of a circle with one crossing we use a circle that is nearly planar and is subset of geometric sphere, Figure 7.108. We take an isotopy m_θ which rotates this circle about the x-axis—Figures 7.108 to 7.111 show the beginning of this isotopy. A discrete track of the entire isotopy is shown in Figure 7.112.

The projection of the circle in Figure 7.108 has one crossing. The projection of Figure 7.108 has one small loop and one large one. The projection of Figure 7.108 fails to be an immersion, and finally 7.111 has no crossing. The projection of the resulting Klein bottle is shown in Figure 7.115. The crossing set is an arc with each endpoint a branch point. This is clearly equivalent to the projection of $\mathbb{K}$, Figure 6.130. ▮

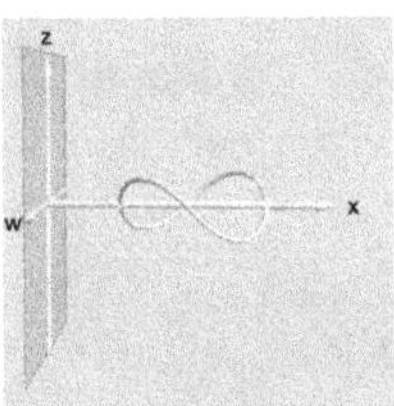

Fig. 7.108

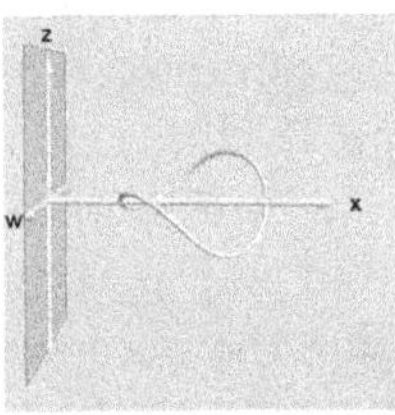

Fig. 7.109

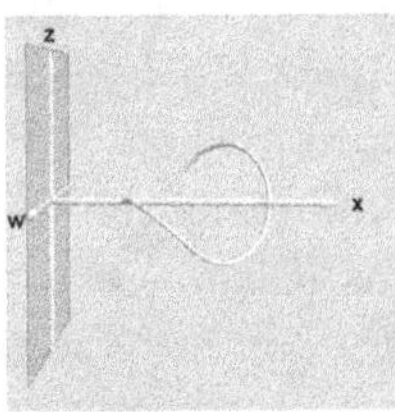

Fig. 7.110

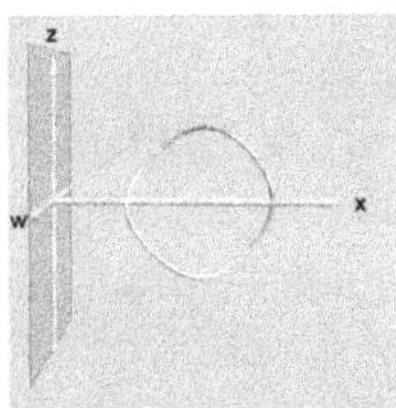

Fig. 7.111

Remark 7.13. The projection of a figure-eight knotting of a Klein bottle has been the focus of attention in the study of regular homotopies of surfaces in R^3. The first graphic seems to be [Banchoff (1976)]. Of interest, is the fact that the version with right-hand twist is not regularly homotopic to the version with a left-hand twist [Pinkall (1985)] (see [Sequin (2013)] for nice graphical exposition).

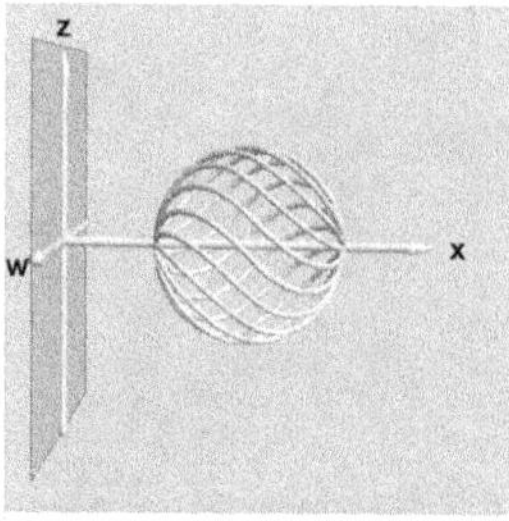

Fig. 7.112

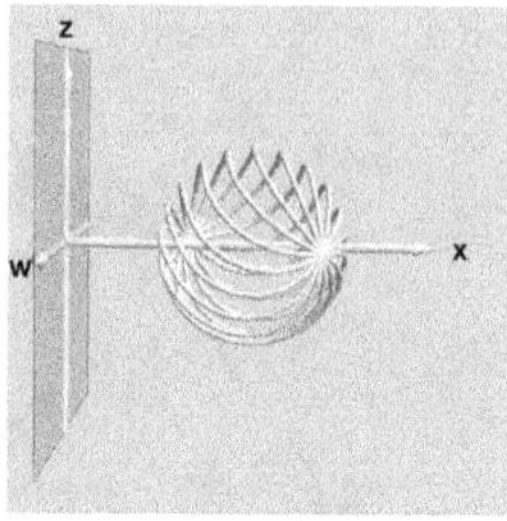

Fig. 7.113

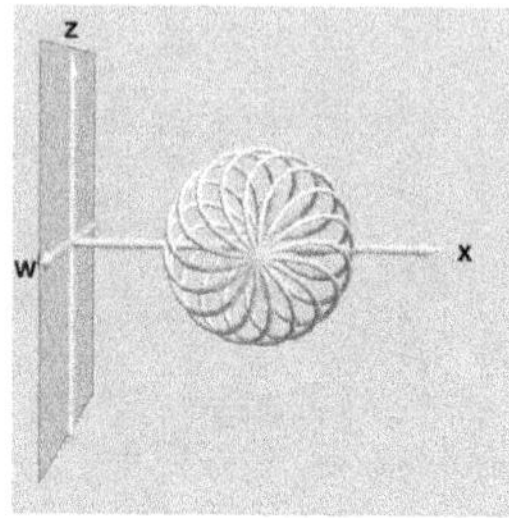

Fig. 7.114

Fig. 7.115 Produced by isotopy of Figure 7.112.

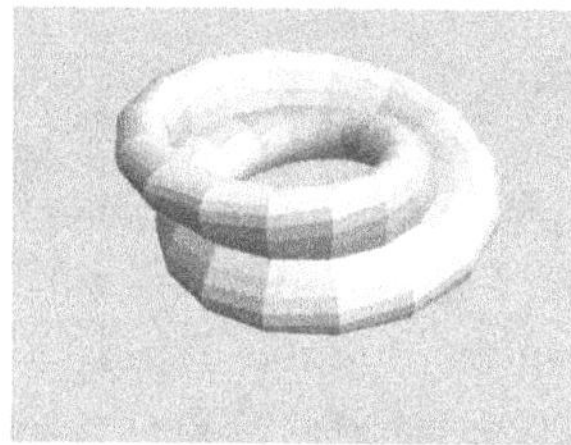

Fig. 7.116 Produced by isotopy of Figure 7.114.

In the style of our previous isotopies, we can use a rotation of our isotopy track to show $\mathbb{F}$ and $\mathbb{F}'$ are isotopic.

As before we chose our meridian curves m_θ so that the track of the isotopy lies on a sphere such as Figure 7.117. This time we augmented our visualization by of Remark 6.2 to display our discrete track. For example, Figure 7.117 is a coded version of Figure 7.114. This way, we can visualize the sequencing (and thus the rotation direction) of the m_θ—the tube thickness increases as θ increases. So our isotopy from $\mathbb{F}$ and $\mathbb{F}'$ will be parameterized by $[0 \leq \psi \leq \pi]$ as we rotate the isotopy track by angle of ψ this time in the wz-plane, Figures 7.117 to 7.121. If we compare the tube thickness pattern in Figure 7.117 with that in Figure 7.121, we can see that these represent meridian rotations in opposite directions.

So these two immersions are projections of equivalent (trivial) knottings of the Klein bottle, providing yet another example of two equivalent knottings whose projections are immersions that are not non-regularly homotopic. ∎

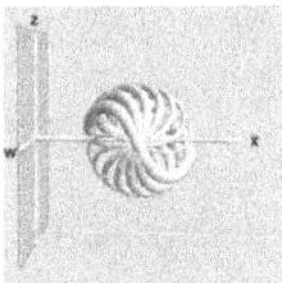

Fig. 7.117

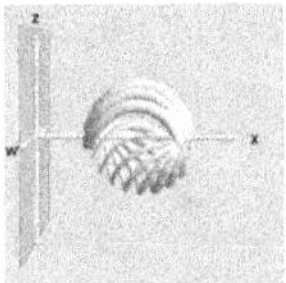

Fig. 7.118

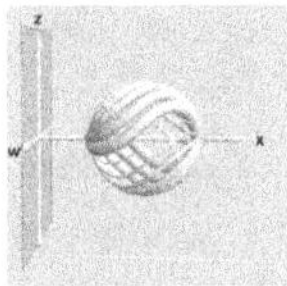

Fig. 7.119

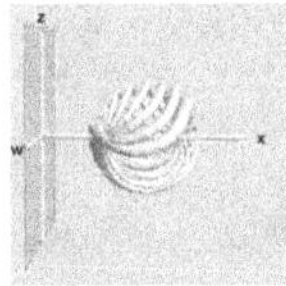

Fig. 7.120

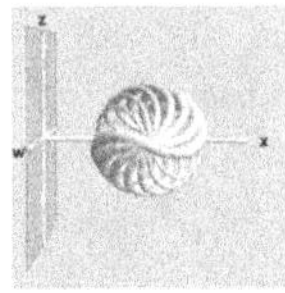

Fig. 7.121

7.4 Getting Rid of Branch Points—When Possible

The existence of branch points for a projection of a knotting is generic, but if we are concerned with knots rather than knottings of orientable manifolds, we can "avoid" branch points: any knotting of an *orientable* surface in R^4 is smoothly isotopic to one whose projection is an immersion (i.e. has no branch points). Similar elimination of branch points can be done in higher dimensions, Proposition 12.3.

Prop 7.2. Any 2-knotting K of an orientable surface is smoothly isotopic to a knotting K' in general position with respect to projection such that K'_* and contains no branch points. ▮

A proof of Proposition 7.2 can be found in [Giller (1982)] based on manipulations of sequences of slices (described by Giller as "difficult"). The approach we outline in Remark 7.14 is based on the projection and knot-moves, [Roseman (2000)]. The restriction to orientable surfaces is essential. Every knotting of a projective plane in R^4 must have branch points in any projection, [Giller (1982)]. There are well-known general position immersions $\mathbb{P}^2 \looparrowright R^3$, see [Francis (2007)] for some excellent illustrations, and so these cannot be projections of a knotting of $\mathbb{P}^2$ in R^4.

Remark 7.14. We outline a proof of Proposition 7.2. and illustrate with a simple example—a knotting $A \cup B$ of two spheres $\phi : M \to (A \cup B) \subseteq R^4$ where M is a union of two disjoint 2-spheres, with projections shown in Figure 7.122. This projection has two branch points b_{-1} and b_1 and two triple points s and t. We will assume that B lies below A in R^4. We describe our isotopy in terms of moves. It is helpful to consider crossing schemes to describe these moves. In our case, the crossing scheme for a knotting of a ***pair*** of spheres can be viewed as *pair* of planar diagrams (using stereographic projection for each sphere, Figures 7.127 to 7.132.

By Remark 6.3 we know that in the crossing scheme of there is an *embedded* circle $\widehat{\gamma} = \widehat{\alpha} \cup \widehat{\beta} \subseteq M$ where $\widehat{\alpha}$ is an above-arc, $\widehat{\beta}$ is a below arc and $\widehat{\alpha} \cap \widehat{\beta} = \{\widehat{b_1}\} \cup \{\widehat{b_{-1}}\}$, see Figure 7.128.

In any 2-knotting projection branch points can be paired as endpoints of embedded arcs of crossing points. In our example, $\pi \circ \phi(\widehat{\gamma}) = G_* \subseteq R^3$ is an embedded arc in the crossing set with endpoints of crossing points b_1 and b_{-1}, Figure 7.124. (In the general case, we would have a number of such arcs.) We will use G_* to pair branch points which we will then cancel.

To aid our construction, as a guide, we consider an arc $H \subseteq K_*$ near G_* and parallel to it. At the end of H near b_1 we will make a (c)-move as in Figure 7.124 where, to keep the graphic simple, we temporarily omit A_*. This will introduce two new branch points, B_1 and B_{-1}. We will do our (c)-move in such a way that B_1 is close to b_1 and b_1 and B_1 are *mirrored branch points with mirrored height relations.* This move creates a small "bubble" in the projection. Now slide our second new branch point B_{-1} along H towards b_{-1} as indicated in Figure 7.125. This stretches the "bubble" along H. This isotopy consists of two consecutive (f)-moves since H pierces sphere A_* twice. (In the general case, the number of moves will depend on the number of triple points along G_*.) The total picture then, complete with image A_* made transparent is Figure 7.126, (another view without transparency is Figure 7.123) with crossing scheme in Figures 7.129 and 7.130. We know all three pairs of points $\{b_1, b_{-1}\}$, $\{B_1, B_{-1}\}$ and $\{b_1, B_1\}$ are mirrored with mirrored height relations—it follows that the pair $\{b_{-1}, B_{-1}\}$ is mirrored with mirrored height relations. Finally, we cancel each of the pairs $\{b_1, B_1\}$ and $\{b_{-1}, B_{-1}\}$ using (d)-moves leaving us with no branch points and crossing scheme Figure 7.131 and 7.132. ∎

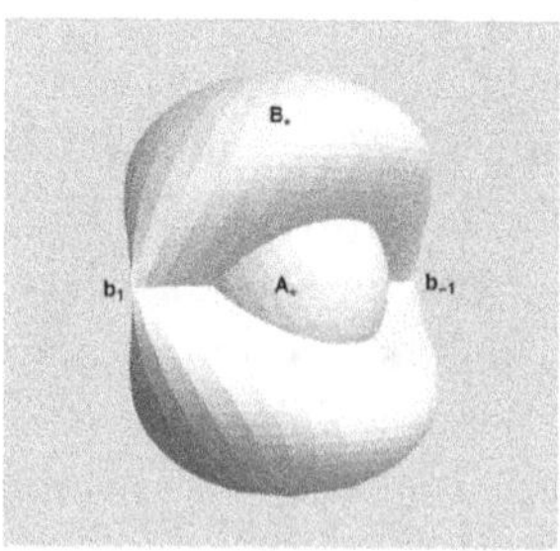

Fig. 7.122 Projection $(A \cup B)_*$.

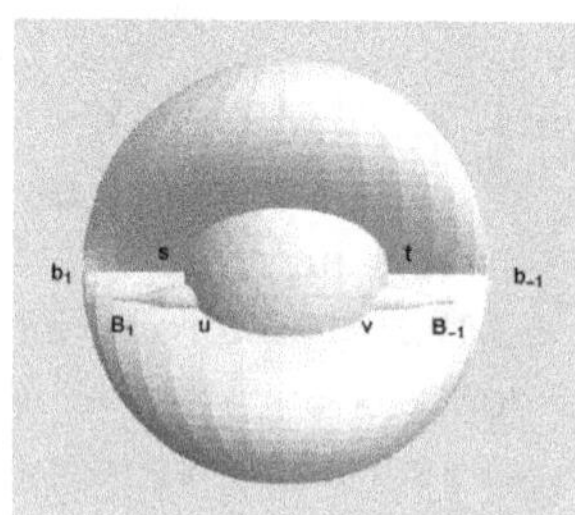

Fig. 7.123 Prepared for cancelation.

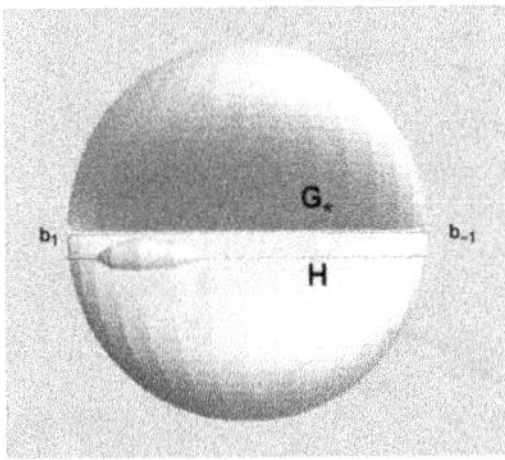

Fig. 7.124 Arc H after a (c)-move.

Fig. 7.125 Stretching along H.

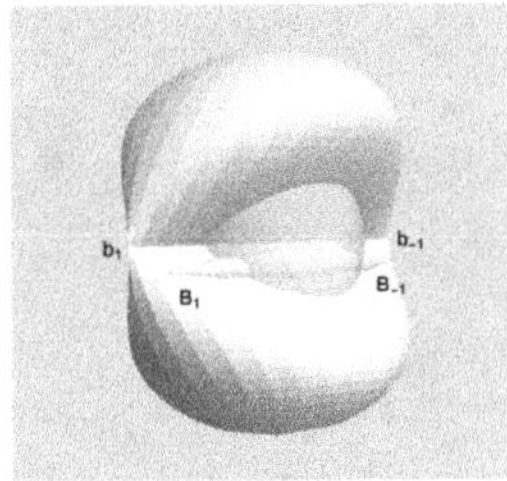

Fig. 7.126 Resulting change to $A \cup B$.

For more discussion of branch point cancelling as well as consideration of the case for non-orientable surfaces see [Carter and Saito (1992)].

One consequence of Remark 7.14 is that every orientable 2-knotting $K \subset R^4$ of manifold M has a product tubular neighborhood, see Remark 12.9.

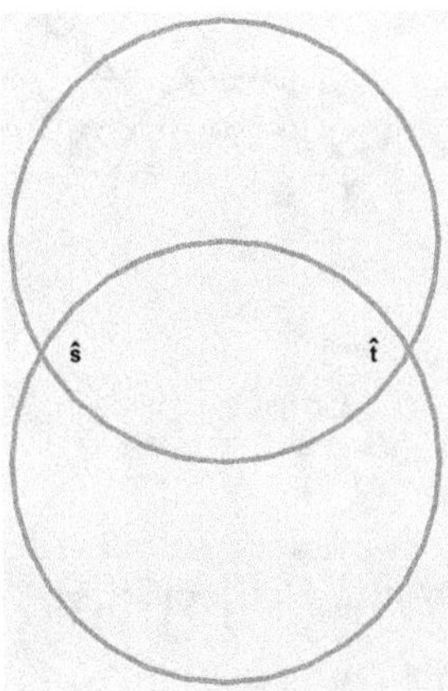

Fig. 7.127 Crossing scheme on $\widehat{A}$ for Figure 7.122.

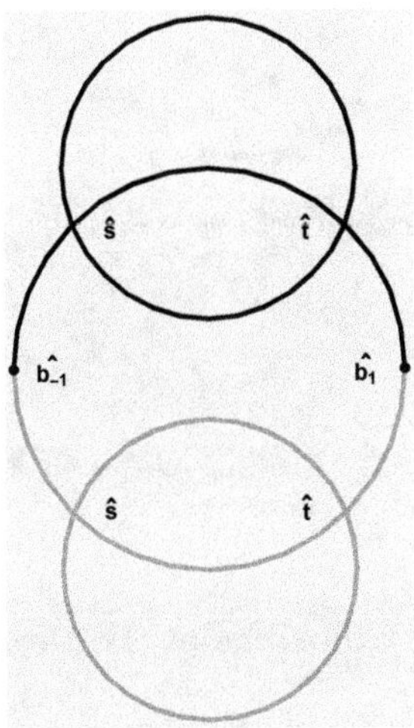

Fig. 7.128 Crossing scheme on $\widehat{B}$ for Figure 7.122.

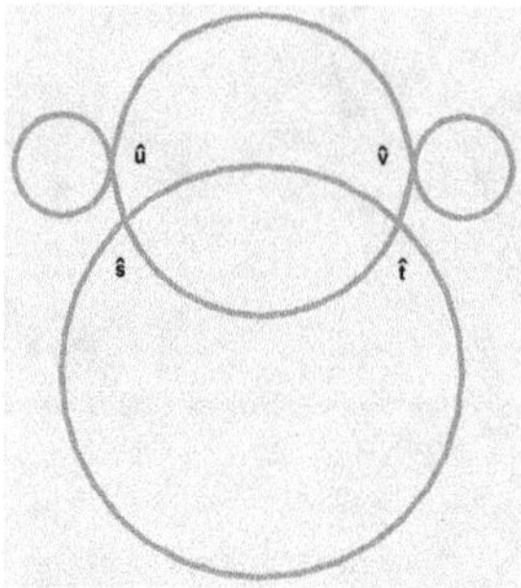

Fig. 7.129 Crossing scheme on $\widehat{A}$ for Figure 7.123.

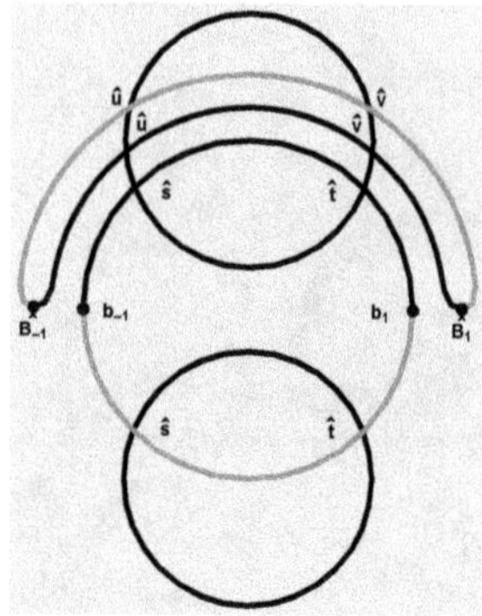

Fig. 7.130 Crossing scheme on $\widehat{B}$ for Figure 7.123.

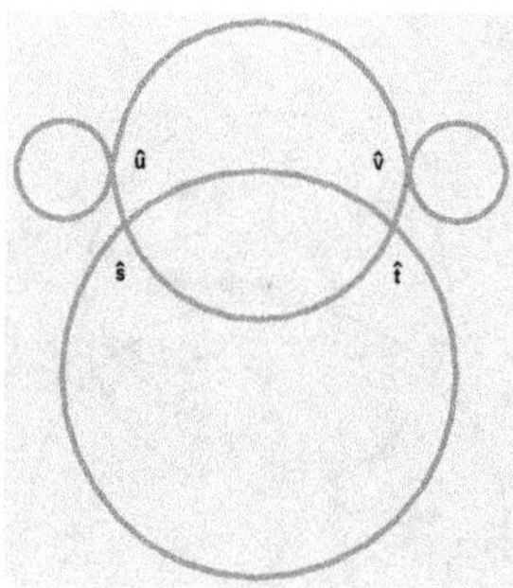

Fig. 7.131 Crossing scheme on $\widehat{A}$ after the two (d)-moves.

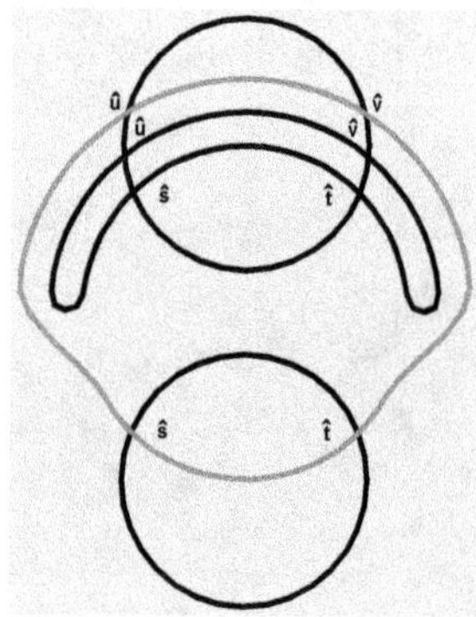

Fig. 7.132 Crossing scheme on $\widehat{B}$ after the two (d)-moves.

7.5 Colorings at Triple Points of 2-Knotting Projections

In this section we show how to do component-coloring, as done for 1-knottings in Example 4.5, for 2-knottings and how to apply it to study triple points of 2-knottings with more than one component.

Suppose we have a 2-knotting $K \subseteq R^4$ of a 2-manifold M in general position with respect to projection. Suppose $M_1, \ldots M_q$ are the components of M and that $q \geq 2$. As in other terminologies, we speak of a label as a "color" but could just as well use letters or integers.

In this Section, we use **component coloring**, color each component of K with a different color. The local picture of a triple point of $p_* \in K_*$

is three intersecting planes, Figure 6.6. We have three points p_+, p_0, p_- in K which project to p_* and if h is the height function (given by the last coordinate) we label so that $h(p_-) < h(p_0) < h(p_+)$ Each plane corresponds to a 2-disk of K. We label these $D_{p_-}, D_{p_0}, D_{p_+}$ called the **bottom disk** the **middle disk** and **top disk**, where $p_- \in D_{p_-}$, $p_0 \in D_{p_0}$ and $p_+ \in D_{p_+}$.

Definition 7.3. Suppose we have a component coloring of 2-knotting K. Suppose p_* is a triple point of knotting projection K_* where the top disk has color p, the middle disk color q and bottom disk color r. We say p_* **has component-color type** pqr. If all three colors are different, we say the point is colorful triple point. A triple point of component-color type pqp with $p \neq q$ is called p_* **a sandwich-point of** K_*. ∎

Prop 7.3. If K and K' are isotopic 2-knottings of three or more components, the number of colorful triple points of a given pqr-type triple points are either both even or both odd. Also the number of sandwich points of type pqp are either both even or both odd. ∎

We show Proposition 7.3 it is enough to prove it in the special case that the isotopy is a single move. We examine component-coloring the figures for the knot-moves where all triple points have component-color type pqr, where *all three colors are different.* Only the e-move, f-move and g-move make changes in the triple points of a projection. The f-move involves only two sub-disks of K, so the triple point of this type must have only one or two different colors. So there will be no change in the number of colorful triple points of any type. In Figures 7.11 to 7.24, one finds that *if a knot-move makes a change that either a pair points of the same color type pqr is added or a pair of points of the color type pqr is deleted.* A move of type (e) is the one that can add/delete pairs. The triple point involved with move (f) cannot have three different colors. A move of type (g) does not change the number of triple points of any type.

The argument for sandwich points is similar. Note that a move of type f will produce single triple points of only of type ppp, ppq or qqp but *never* pqp, a sandwich triple point.

For 2-knottings of *orientable* manifolds, we can further define an oriented component-color type of a triple point.

Definition 7.4. Suppose K is a knotting of an oriented 2-manifold with component colors in general position with respect to projection, and $p_* \in K_*$ is a triple point. In the local model of the triple point let $\overrightarrow{n_+^*}$ be a vector based at p_* normal to D_+^* in the direction consistent with the induced orientation of D_+^*. Similarly, define normal vectors $\overrightarrow{n_0}$ and $\overrightarrow{n_-}$. Also suppose that the colorings of these disks are, in order, p, q, and r. If the ordered triple $\{\overrightarrow{n_+}, \overrightarrow{n_0}, \overrightarrow{n_-}\}$ is a right-handed frame in R^3, then we say the triple point p_* has oriented color type $+pqr$, if negative, then p_* has oriented color type $-pqr$.

7.6 Coloring 1-Shards of 2-Knottings

Knot colorings, Section 4.5, can be extended to 2-knottings in a natural way, and extended to n-knottings as well, Section 12.2. We focus on Fox colorings since it needs minimal back ground. Quandle colorings can be done with oriented 2-knottings, giving a finer tool, but requiring more algebra, [Carter *et al.* (2004)]. As for 1-knottings one can verify these are colorings of a 2-knot rather than just colorings of a 2-knotting by verifying the coloring rules are consistent with the 2-dimensional knot-moves.

Definition 7.5. For knotting K, let $\mathcal{B} \subset K$ be the below-set. An **overpass of** K is a component P of $K - \mathcal{B}$, and $\pi(P)$ is an **overpass of** K_*. If $\phi\colon M \to K$ is a homeomorphism, then $\widehat{K} = \phi^{-1}(K)$ is an **overpass of the crossing scheme of** K. ▌

Every 2-shard of the projection has an above-shard and below-shard in the crossing scheme. The above 2-shard will be a subset of an overpass.

For Fox coloring an n-color is an integer, $\{1, \ldots, n\}$.

Definition 7.6. An n**-color** is an integer, $\{1, \ldots, n\}$. A **Fox** n**-coloring** for a projection of a 2-knotting in general position is an assignment of an n-color to each overpass. one that satisfies:

(1) **Overpass Color Rule** The two 1-shards that join to form part of an overpass must have the same color.
(2) **Color Sum at Crossing Rule** For every 2-shard, twice the overpass color equals the sum of the other two colors plus one (mod n). ▌

As with Fox coloring for 1-knottings, a Fox n-coloring satisfies all three properties of a kei, Definition 4.22.

Definition 7.7. Suppose R is a knot coloring rule and K is a 2-knotting in general position with respect to projection. The number of distinct colorings of K_* is denoted $N(R, K)$ and is called the **number of** R**-colorings of the knotting** K. If there are no R-colorings we say $N(R, K) = 0$. ▌

Any coloring of a projection of a knotting gives a coloring of the crossing scheme as follows. Every 1-shard in a crossing scheme belongs to a unique overpass, so a coloring of overpasses gives a coloring of 1-shards. Consider a 2-shard γ in the projection. It corresponds to two 2-shards, $\widehat{\gamma_a}$ and $\widehat{\gamma_b}$, (an above-shard and a below shard) in the crossing scheme $\widehat{\gamma}$. The above 2-shard is a subset of some overpass—we use the color of this overpass for *both* 2-shards.

Locally a 2-shard of the crossing scheme meets two 1-shards, Figure 7.133. Suppose the above 2-shard is colored b and the 1-shard we see part of on the left is colored a. The color for the other 1-shard is labeled $b \triangleright a$.

Prop 7.4. Suppose R is a knot coloring rule and K_0 and K_1 are two equivalent 2-knottings in general position with respect to projection. Denote the knot $\mathcal{K} = [K_0] = [K_1]$. Then $N(R, K_0) = N(R, K_1)$ and we call $N(R, \mathcal{K})$ the **number of R-colorings of the knot $\mathcal{K}$.**

In particular for any two 2-knottings K_0 and K_1 if $N(R, K_0) \neq N(R, K_0)$ then K_0 and K_1 are not equivalent. ∎

We outline a proof of Proposition 7.4 in the same style as for 1-knottings Section 4.5. That is, we verify that if K and K' are knottings that differ by a single 2-knot move that a coloring of K_* corresponds uniquely to a coloring of K'_*. We could do this by coloring the knot-move Figures 7.11 to 7.24 much as we did for 1-knottings in Figures 4.69 to 4.74, but we find it easier to follow if we view our coloring on the crossing scheme.

The crossing scheme contains the shard structure and relative heights of matching shards so, as we will see, colorings of K_* can be examined on the crossing scheme of K.

Remark 7.15. Figure 7.133 shows a small portion of a crossing scheme with $\widehat{C_b}$, a below 2-shard, indicated by dashed line. There is a corresponding above shard $\widehat{C_a}$, which belongs to some overpass. Suppose we color this overpass with color a. We indicate this by labeling the dashed line as $\overline{a}$. (Generally, in our graphical notations, $\overline{x}$ indicates "a line colored x".) On either side of $\widehat{C_b}$ is a 1-shard. Suppose we label the 1-shard on the left with color b. There is a 1-shard on the right of line $\overline{a}$ and the color of this, as determined by the color-rule is denoted $b \triangleright a$ which we think of as "the color b becomes if it passes under an overpass with color a". ∎

Remark 7.16. A triple-point (3-shard), τ_* of K_* corresponds to three 3-shards in the crossing scheme: an above-shard $\widehat{\tau_a}$, a middle-shard $\widehat{\tau_m}$, and a below shard $\widehat{\tau_b}$. In Figure 6.72, $\widehat{\tau_a} \in \widehat{A}$, $\widehat{\tau_m} \in \widehat{B}$, and $\widehat{\tau_b} \in \widehat{C}$, where above 1-shards shown as solid black, below 1-shards dashed gray. It follows that all four 1-shards, seen partially in $\widehat{A}$, must belong to a single overpass.

In $\widehat{B}$ the two 1-shards seen above the dashed line must belong to one overpass, and the two below to another overpass. (Because of the shard structure away from this triple point it is possible these two overpasses turn out to be the same; as we will see in Figures 7.138, 7.139 and 7.141.)

We now consider local coloring of Figure 7.134. This corresponds to $\widehat{C}$ of Figure 6.72. In the crossing scheme we see four (dashed) 2-shards $\widehat{W_b}, \widehat{X_b}, \widehat{Y_b}, \widehat{Z_b}$ that meet at $\widehat{\tau_b}$. Without loss of generality we may assume that $\tau_a = \widetilde{W_b} \cap \widetilde{X_b}$. This means that these three all lie in disk $\widehat{A}$ and in fact all shards seen in disk $\widehat{A}$ must belong to a some overpass. We suppose this overpass has been assigned color a. So, we color $\widetilde{W_b}$ and $\widetilde{X_b}$ color a and so label these with $\overline{a}$ in Figure 7.135.

Therefore $\widehat{\tau_m} = \widetilde{Y_b} \cap \widetilde{Z_b}$. Assume the above shard for $\widetilde{Y_b}$, $\widetilde{Y_a}$, corresponds to a solid black line segment in $\widehat{B}$. All in $\widehat{B}$ above the dashed line lies in some overpass. Suppose $\widetilde{Y_a}$ lies in this overpass and suppose we give this overpass a color b. We record this in Figure 7.135 by labeling $\widetilde{Y_a}$ with $\overline{b}$. Now, as in Remark 7.15, the overpass in $\widehat{B}$ which lies below the dashed line will be colored $b \triangleright a$. So, we label $\widehat{Z_b}$ as $\overline{b \triangleright a}$.

In Figure 7.134, color a 1-shard (of $\widehat{C}$) with color c as shown. We are now ready to color the other three 1-shards. First, color two of the shards $c \triangleright a$ and $c \triangleright b$ as shown. Having done this we have two colorings for the fourth 1-shard: $(c \triangleright b) \triangleright a$ and $c \triangleright a) \triangleright (b \triangleright a)$. Since a coloring can have only one color per overpass, we must have: $(c \triangleright b) \triangleright a = (c \triangleright a) \triangleright (b \triangleright a)$, which we know we have since a Fox-coloring satisfies the kei properties, and this is precisely condition (3) of Definition 4.22. ∎

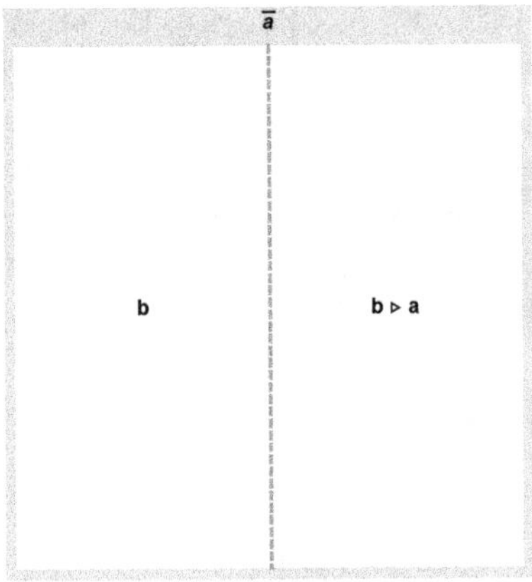

Fig. 7.133 a is above b so 1-shard on right is $b \triangleright a$.

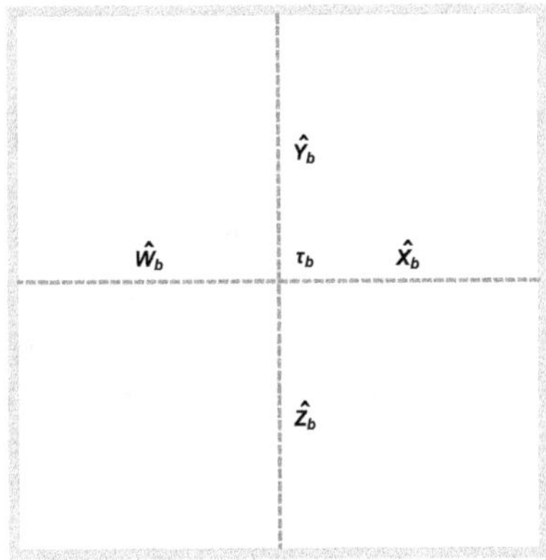

Fig. 7.134 In a neighborhood of τ_b.

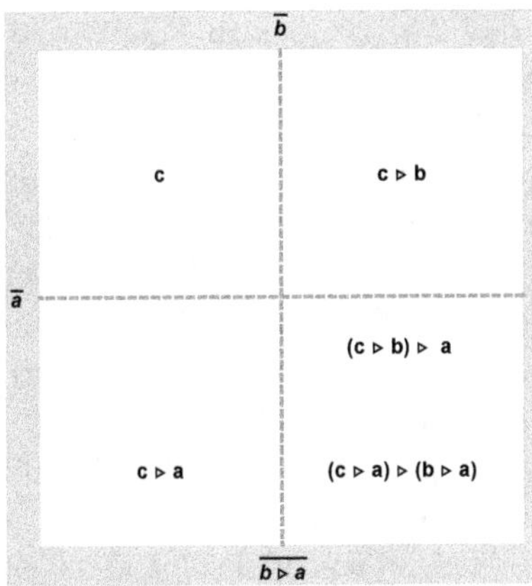

Fig. 7.135 $(c \triangleright b) \triangleright a = (c \triangleright a) \triangleright (b \triangleright a)$.

Remark 7.17. Now we are ready to consider colorings knot-moves, Figures 7.25 to 7.31. We assume our coloring satisfies the three kei conditions, Definition 4.22. For each move we look at Position A, assign a color to some 1-shard of each disk and observe there is only way to complete the coloring of other 1-shards seen in that disk. We then transfer these colors to the corresponding 1-shards in Position B and observe there is only one way to complete the coloring of that disk.

a-move: In Figure 7.25, one disk in Position A is colored a, the other b. Then in Position B, there is only one coloration possible, Figure 7.136.

b-move: In Figure 7.26, one 1-shard in the first disk of position A is colored a, a 1-shard in the second disk colored b as shown. There are no below 2-shards in the second disk so all 1-shards must have color b. In the first disk we must have the colors as indicated. In Position B, the second disk all 1-shards must be labeled b. In the first disk, we see locally two 1-shards from Position A appear to have merged into one shard, so the two colors inherited from Position A must agree and we see that we need $(a \triangleright b) \triangleright b = a$, which we know is true by part (2) of Definition 4.22, Figure 7.137.

c-move: In Figure 7.27, the disk in Position A is colored a. We transfer this color to Position B, where we have a "new" 1-shard (inside the circle). Here, using the below 2-shard, this 1-shard must have color $a \triangleright a$. Using the above shard is must be colored a. So, we need $a \triangleright a = a$ which we do have, Definition 4.22 condition (2), Figure 7.138.

d-move: In Figure 7.28, the disk in Position A is colored a and we deduce that all 1-shards we see in both positions must be colored a. As with (c)-moves, we conclude that we need to use $a \triangleright a = a$, Figure 7.139.

e-move: In Figure 7.29, we chose colors a, b and c for the 1-shards in Position A, as shown. We will need additional color $b \triangleright a$ for the second disk, Position A. In Position B for the bottom disk note the labeling of one of the 2-shards as $\overline{b \triangleright a}$. The "central" 1-shard in this disk is bounded by two below 2-shards, giving it two possible colors $(c \triangleright a) \triangleright (b \triangleright a)$ and $(c \triangleright b) \triangleright a$ for those to be the same color we need $(c \triangleright a) \triangleright (b \triangleright a) = (c \triangleright b) \triangleright a$, which is assured by Definition 4.22 property (3), Figure 7.140.

f-move: In Figure 7.30, we chose colors a,and b for the 1-shards in Position A, as shown. Then we need to color a 1-shard in position A with $b \triangleright a$, as shown. In Position B note the need to label a 2-shard as $\overline{b \triangleright a}$. Then the "central" 1-shard in the lower disk has two colors $b \triangleright a$ and $(b \triangleright a) \triangleright (b \triangleright a)$ which should be the same—and they are since $(b \triangleright a) \triangleright (b \triangleright a) = b \triangleright a$ is just Definition 4.22 property (1) applied to the color $b \triangleright a$, Figure 7.141.

g-move: Finally, a proof for knot-moves of type (g) can be done using Figure 7.31 and four colors and the method used in other six knot-moves. There are four 3-shards involved and for each considerations such as in Remark 7.16 are required. We leave tedious details for the (patient) reader. ∎

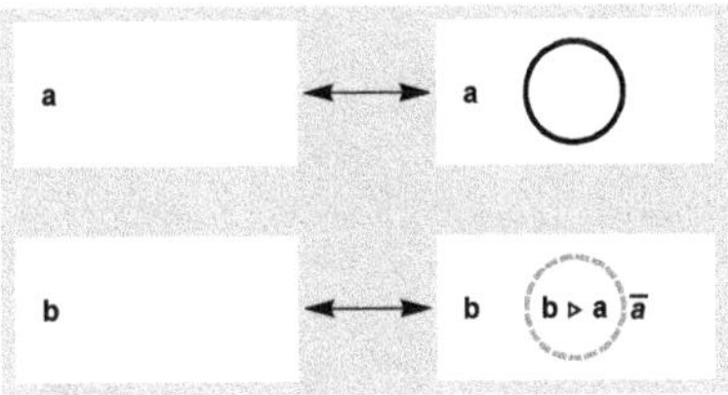

Fig. 7.136 Using (*a*)-move.

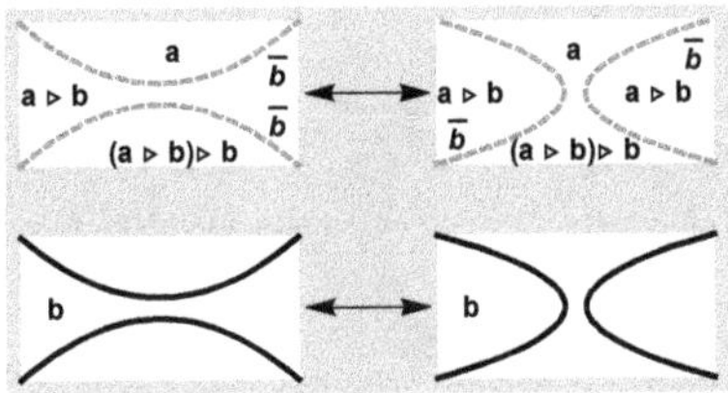

Fig. 7.137 Using (*b*)-move, need $(a \triangleright b) \triangleright b = a$.

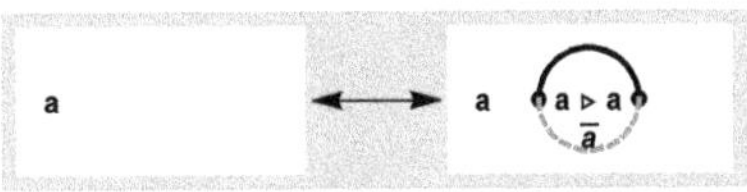

Fig. 7.138 Using (*c*)-move, need $a \triangleright a = a$.

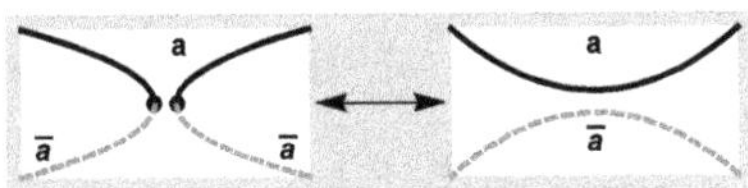

Fig. 7.139 Using (*d*)-move, need $a \triangleright a = a$.

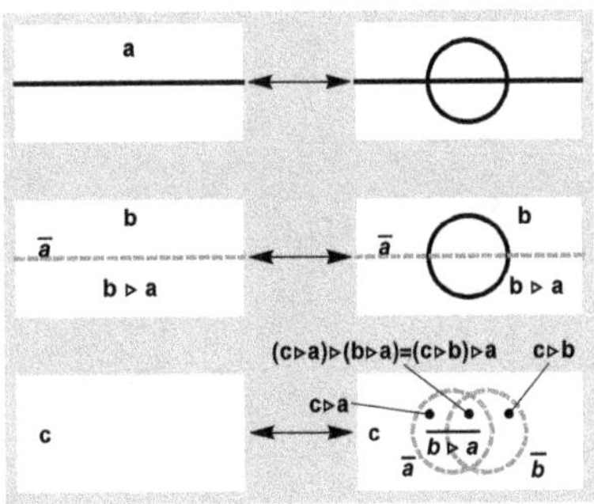

Fig. 7.140 Using (*e*)-move, need $(c \triangleright a) \triangleright (b \triangleright a) = (c \triangleright b) \triangleright a$.

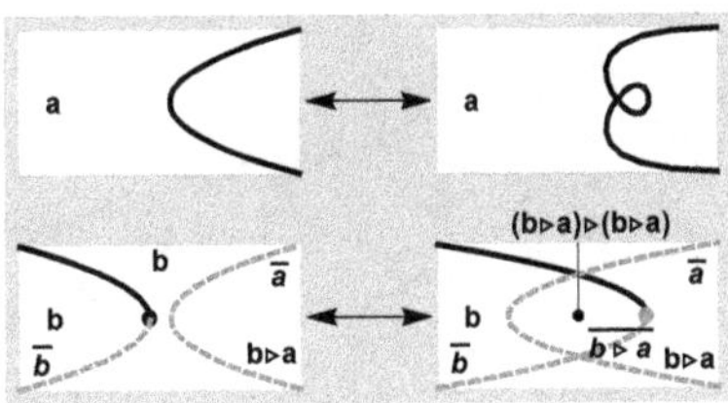

Fig. 7.141 Using (*f*)-move, need $(b \triangleright a) \triangleright (b \triangleright a) = b \triangleright a$.

Chapter 8

Slicing Surfaces in 4-Space

We understand a surface $\Sigma \subseteq R^3$ by intersecting it with a family of parallel 2-planes $P_1, \ldots, P_M$. A contour plotting of $z = f(x, y)$ which uses planes with equation $z = a$. Figure 8.3 is a contour plot for the function whose graph is shown in Figure 8.1. Contours are curves in the *domain* of f. The analogous technique for surfaces in R^4 is called **slicing**.

We explore a surface $\Sigma \subseteq R^4$ by intersecting it with a family of parallel 3-planes $P_1, \ldots, P_M$, obtaining a sequence $C_i = \Sigma \cap P_i$. Most often P_i is with equation $w = a$. Using standard projection $\pi : R^4 \to R^3$ we get a sequence of subsets of R^3. If C_i is a collection of topological circles, this 1-knotting is called a **slice knotting**, and the corresponding knot **slice knot**.

Remark 8.1. As with a contour curve of a surface in R^3, there are some rare but significant circumstances where a slice of a surface is *not* a 1-knotting. Figure 8.1 shows a single saddle (hyperbolic) singularity at a point p. Figure 8.2 is the same surface as seen from above with a tangent plane at p also shown. Our focus is on the *topological changes* of slices as we change from below a critical point to above it. Level curves below level $q = f(q)$ are pairs of circles, above this level the q contour is single circle, Figure 8.3. The contour curve at q is a self-intersecting "figure-eight curve". ∎

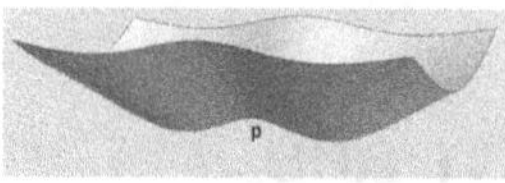

Fig. 8.1 A surface with two local minima and one saddle point at p.

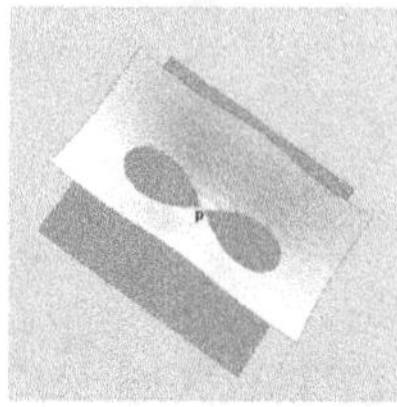

Fig. 8.2 A view from above of the tangent plane at the saddle point p.

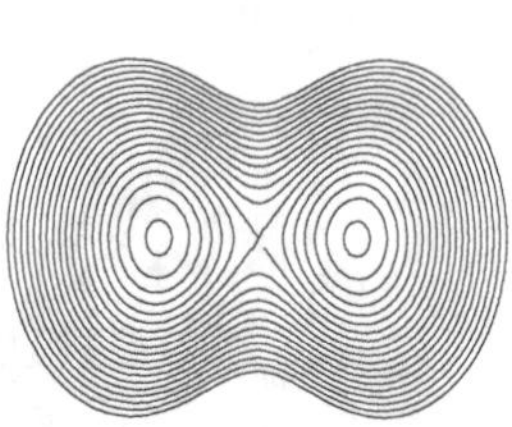

Fig. 8.3 A contour graph for surface of Figure 8.1.

Most often slicing a knotting in R^4 refers to slicing by w-hyperplanes but there are occasions to use other hyperplane families.

Definition 8.1. If K is a 2-knotting and $H_w(a)$ the hyperplane $w = a$ write the intersection as $\widetilde{K}|a = K \cap H_w(a)$. The image of this set via standard projection into R^3 is written $K|a = \pi\left(\widetilde{K}|a\right)$ and called the **slice of K at level a** Then $\widetilde{K}|a$ is called the **lifting of the slice** $K|a$. For the x, y and z directions we use notations $K|^x a$, $K|^y a$ or $K|^z a$ for the **x-slice, y-slice** and **z-slice**, (and so $K|a = K|^w a$).

Furthermore, if $[a, b]$ is a closed interval then the standard projection of $K \cap (R^3 \times [a, b])$ is called the **section of K between a and b** or $[a, b]$-**section**. It is denoted $K||_a^b$ with $\widetilde{K}||_a^b = K \cap (R^3 \times [a, b])$ called the **lifting of the section**. As with slicing we can take sections using different coordinates and use notations $K^x||_a^b$, $K^y||_a^b$, $K^z||_a^b$, and $K^w||_a^b$.

Finally, we define the **section of K below a** as $K||_{-\infty}^a = \pi\left(K \cap \left(R^3 \times (-\infty, a]\right)\right)$ and the **section of K above a** as $K||_a^\infty = \pi\left(K \cap \left(R^3 \times [a, \infty)\right]\right)$. ∎

The relations between slicing and projections in these notations is: the knotting K is the union of all slice liftings: $K = \bigcup_{a \in R} \widetilde{K}|_a$; the projection of a knotting K is the union of all slices: $K_* = \bigcup_{a \in R} K|_a$.

Similarly for sections, if $a_0 < a_1 < \cdots < a_r$, we have $K = \widetilde{K}||_{-\infty}^{a_0} \cup \widetilde{K}||_{a_0}^{a_1} \cup \cdots \cup \widetilde{K}||_{a_r}^{\infty}$ and so $K_* = K||_{-\infty}^{a_0} \cup K||_{a_0}^{a_1} \cup \cdots \cup K||_{a_r}^{\infty}$.

The definitions and notations of slices and sections extend to n-knottings, where **standard slicing** is done with respect to the last coordinate, x_{n+2}.

Presenting a 2-knotting as a sequence of slices is appealing since each slice looks like a familiar 1-knotting. However, there are hidden pitfalls. It might seem tempting to think that, if a given slice of a surface knotting is complex, that the surface knotting might be complex, or if all slices are simple 1-knottings, that the surface knotting might be somehow also "simple". This is not the case. An analogous situation is slicing a 1-knotting by planes obtaining a sequence of configurations of points in a plane such as in Example 4.6. Any given set of points in a plane is easy to understand by itself but fitting these together to assemble a 1-knotting is challenging.

In the literature, often a slice is considered to be a 3-dimensional "picture". So, the sequence of slices (or corresponding liftings) is called a "3-dimensional motion picture" [Carter (1995)]. This visualization via slicing is called the "motion picture method", [Carter (1995)] [Kamada (2002)].

For visualization we look at "thin" sections we call **slabs**. We first look at some examples and delay the definition of slab until, Definition 8.7.

8.1 Critical Points and Slicing 2-Knottings

In cartography the basic geographic features are summits, basins and passes. In multivariable calculus these correspond to local maxima, local minima and saddle points—familiar features studied using only first and second derivatives. Morse theory is a powerful extension of these ideas to all dimensions. In this chapter, we restrict attention to 2-knottings. Morse Theory assures us that we can analyze any 2-knotting by using only a finite number of slices with consecutive slices related in a simple way. This view of 2-knottings traces back to [Fox and Milnor (1966)] and [Hosokawa (1967)]. Morse theory can be applied to geometric topology in several ways, as explained in many fine books such as [Milnor (1963)], [Milnor (1965)], [Matsumoto (2002)], [Nicolaescu (2011)] and [Knudson (2015)].

Using Morse Theory we can replace *any* smooth function by a "better" approximating function called a **Morse function** that has simple critical points. These singularities are locally modeled on familiar multi-variable quadratic functions. In particular, any smooth knotting is smoothly isotopic one whose w-coordinate function (height function) function is a Morse function. **Slices** of 2-knottings that are 1-knottings are those that *avoid* critical points.

It might seem that the next definition is highly specialized but it can be shown that "most" real-valued functions *are* Morse functions. Recall the definition of degenerate critical point, Definition 2.5.

Definition 8.2. A smooth real-valued function is called a **Morse function** if it has no degenerate critical points. ▮

Example 8.1. Examples of non-degenerate critical points are local maxima, local minima and saddle points for ordinary functions of two variables. A basic result of Morse theory is that *degenerate* critical points are delicate and contrived in that almost any slight change of the function will remove degenerate critical points and give a function with only non-degenerate critical points, if any. For the **parabolic sheet** $f(x,y) = x^2$, the critical points are all points of the y-axis, Figure 8.4 with contour plot, Figure 8.5. All these points are *degenerate* critical points since the Hessian is $\begin{vmatrix} x & 0 \\ 0 & 0 \end{vmatrix}$, which is a matrix of zeros for points on the y-axis. Almost any small change to this function will remove all critical points. For example, if we add a linear function $ax + by$ for almost any choices of a and b (in effect skewing the graph slightly) then $x^2 + ax + by$ will have no critical points and a contour graph of $x^2 + ax + by$ will look like Figure 8.6.

Fig. 8.4 Plot of $f(x,y) = x^2$ with tangent plane at a point of y-axis.

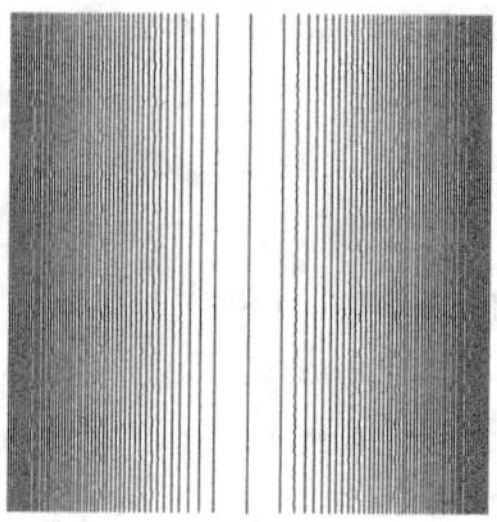

Fig. 8.5 Contour plot $f(x,y) = x^2$.

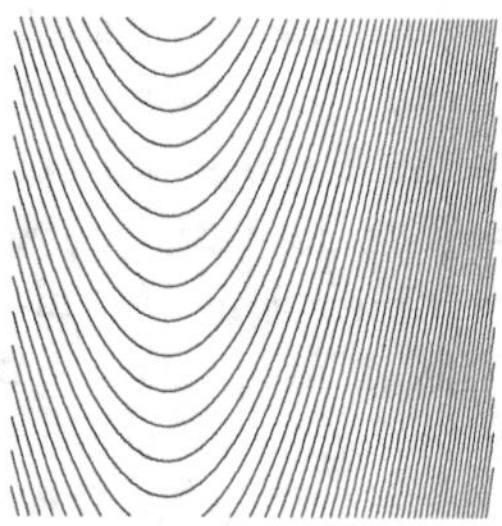

Fig. 8.6 Contour plot $f(x,y) = x^2 + (ax + by)$.

Another example called a **monkey saddle**, is given $f(x,y) = x^3 - 3xy^2$ This has a single critical point at the origin and it is degenerate there— the Hessian is $\begin{vmatrix} 6x & -6y \\ -6y & -6x \end{vmatrix}$ which is a matrix of zeros at $(0,0)$. See Figure 8.7 for a plot and Figure 8.8 for a contour plot. Adding almost any small linear function we get a new function $x^3 - 3xy^2 + ax + by$ which has two *non-degenerate* critical points with contour curves such as Figure 8.9. ▮

Fig. 8.7 Monkey saddle: plot of $f(x, y) = x^3 - 3xy^2$ with tangent plane at $\mathbb{O}$.

Fig. 8.8 Contour plot of monkey saddle.

Fig. 8.9 Contour plot of $f(x, y) = x^3 - 3xy^2 + (ax + by)$.

The following is a special case (for 2-manifolds) of what is known as the **Morse Lemma**. It says that for a surface, a non-degenerate critical point can be locally modeled on a local minimum, local saddle point, or a local maximum. A standard name for each of these cases is "index", where index is the number of minus signs in the corresponding formula:

Prop 8.1. If (p, q) is an isolated non-degenerate critical point of h, then with an appropriate choice of local coordinates (x, y), we can express h as one three **index** types:

- **Index 0:** A local minimum at a and $h(x, y) = h(p, q) + (x^2 + y^2)$
- **Index 1:** A saddle point at a and $h(x, y) = h(p, q) + (x^2 - y^2)$
- **Index 2:** A local maximum at a and $h(x, y) = h(p, q) + (-x^2 - y^2)$ ∎

In higher dimensions, for a smooth function $h\colon M^n \to R^1$ if a is an isolated non-degenerate critical point with critical value b, then with appropriate choice of local coordinates $(x_1, \ldots, x_n)$, we can express h as $h(x, y) = h(a) - x_1^2 - \cdots - x_k^2 + x_{k+1}^2 + \cdots x_n^2$ and k is called the **index of the critical point**.

The important connection between *regular* values and slices is:

Proposition 8.1. *If a is a regular value for a knotting of a 2-manifold K, the slice $K|a$ is a 1-manifold. If $[a, b]$ is a closed interval of regular values, the lifted section $\widetilde{K}||_a^b$ is a 2-manifold diffeomorphic to the product of $[a, b]$ with $\widetilde{K}|a$. In fact $K|_t$ is an isotopy of $K|_a$ in R^3 parameterized by $t \in [a, b]$ and $K||_a^b$ is the track of this isotopy.* ∎

In higher dimensions, a version of Proposition 8.1 for an n-knotting K is as follows. If a is a regular value the slice $K|a$ is an $(n-1)$-manifold. If $[a,b]$ is a closed interval of regular values then the lifted section $\widetilde{K}||_a^b$ is an $(n+1)$-manifold which is a diffeomorphic to the product of $[a,b]$ and $K|a$, $K|_t$ is an isotopy and $K||_a^b$ is its track.

We illustrate slices, slabs, projections, slice sequences and associated slab sequences for three (equivalent, as will be seen) knottings A, B and C: Example 8.2, Example 8.3 and Example 8.4.

It is important to note that in each graphic we have scaled the graphic so that it nearly fills the space allotted in order to make details clear. As a consequence however, consecutive graphics generally have slightly different scales. Especially for slabs when one wants to mentally assemble objects in consecutive graphics, one generally needs to mentally scale one to fit more exactly.

Example 8.2. Figures 8.10 to 8.18 show nine slices of a 2-knotting we will call **Knotting A**. The slices are presented with increasing height values labeled $\{a, b, c, d, dx, e, ex, f, g, h\}$. The slices at dx and ex each contain one critical point of index 1. All other slices shown are for regular values. As is traditional, critical points of index 0 are implied but not shown. A minimum occurs just below level a and a local maximum above level g. There is a second local minimum m_2 between level a and level b and a local maxima one between f and g.

A viewer is expected to interpolate and fill in all other slice information. Since this is our first example, we supply details. Somewhere below level a is the first non-empty slice—a single point which is a minimum point of the knotting. As we slice above that point, we obtain a small (unknotted) circle which expands until, at level a, we have Figure 8.10. Somewhere between level a and b a local minimum occurs. As we slice above this local minimum, a small second circle appears which expands to that shown until, at level b. we have Figure 8.11. As we slice upwards from level b to c and then d, the circles move slightly and expand, Figures 8.12 and 8.13. As we progress from level d to dx, two parts of the larger circle begin to come together and at level dx actually meet at a single point, forming a figure-eight curve Figure 8.14. This singular point is a saddle point of the height function. Above level dx the figure-eight splits into two circles. These separate and shrink as we slice towards level e, Figure 8.15. As we continue slicing to level ex, portions of these two previously split circles come together and finally merge into a figure-eight curve in level ex, Figure 8.16. The changes in the figure-eight curves of level dx and ex behave differently in R^3. At level dx, the formation of the crossing appears to happen *through* the second circle, while in level ex it is *away from* this circle. Between level ex and level f, the figure-eight crossing is resolved into a larger single circle that is pinched slightly, Figure 8.17. Between level f and g, the smaller circle becomes smaller and, at some level before level g, has shrunk to a point corresponding to a local maximum M_1. This leaves a single circle in level g, Figure 8.18. It is understood that above level g, this circle will appear to shrink, and become a point at a second local maximum.

A second way of viewing Knotting A is via corresponding thin sections, called slabs (defined later), for these slices, Figures 8.19 to 8.27.

To relate our slice figures to the slab figures for Knotting A, recall that, by definition, each slab has a top slice and a bottom slice. (For local maxima/minima one of these slices will be empty). For example Figure 8.21 shows $K||_b^c$—two annuli whose boundary is four circles. Two of these circles are $K|b$, shown in Figure 8.11; the other two are $K|c$ shown in Figure 8.12. Similarly, the five boundary circles of slab $K||_d^e$ of Figure 8.23 is the union of the circles shown in Figure 8.13 and Figure 8.15.

Technical notes: For Knotting A, there is actually a non-trivial slice K_h not shown. This is not shown in order to allow Figures 8.10 to 8.18 to include $K|dx$ and $K|ex$, and to form a 3×3 array of graphics—K_h looks like a small version of the curve seen in Figure 8.18. In order to fit Figures 8.19 to 8.27 into a 3×3 array, we include the slab $K||_g^h$ and $K||_h^\infty$ which has slice at h as one boundary curve.

This slab method allows viewing of *all slices*, not just a finite number. Also, Figure 8.19 shows a slab below a. The existence of our index 0 critical point m_1 is visual and does not need to be inferred. Figure 8.20 shows slab $K||_a^b$ —a local minimum m_2 occurring as well as a slight expansion of the circle of slice K_a. In Figures 8.21 and 8.22 the annuli are traces of an isotopy of this circle pair. Each annulus is a union of non-intersecting circular slices. Figures 8.23 and 8.24 show our saddle points clearly.

Additionally, we can use slabs to show more details of the two saddle points of our Morse function. Figures 8.28 and 8.29 show these slabs $K||_d^e$ and $K||_e^f$ of Figures 8.23 and 8.24 respectively, with the slices at the critical points (at levels dx and ex) added. ∎

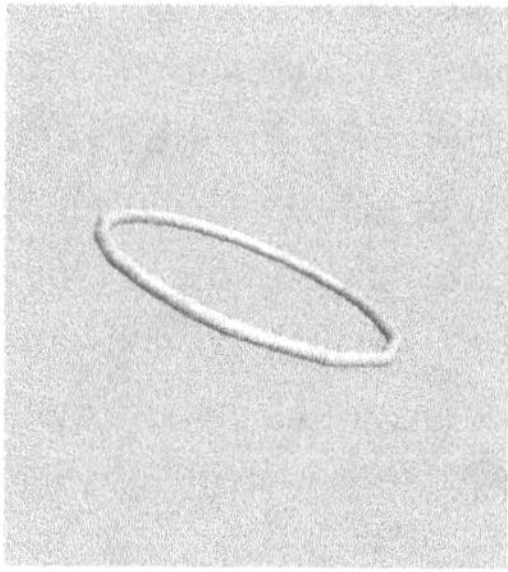

Fig. 8.10 Slice $K|a$, Knotting A.

Fig. 8.11 Slice $K|b$, Knotting A.

Fig. 8.12 Slice $K|c$, Knotting A.

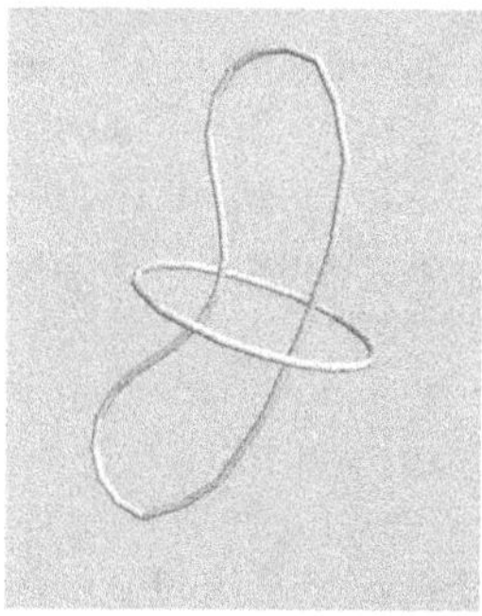

Fig. 8.13 Slice $K|d$, Knotting A.

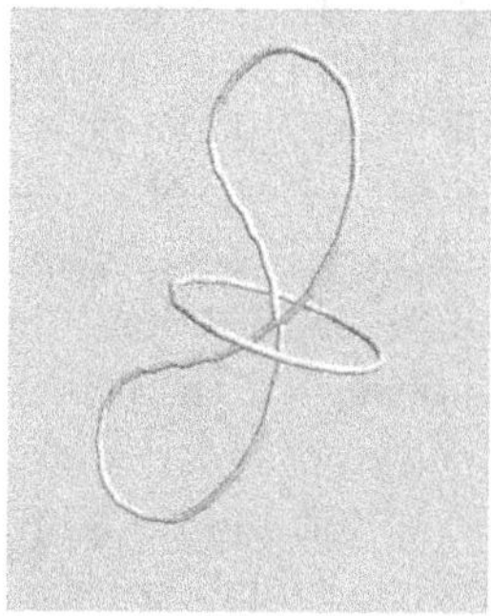

Fig. 8.14 Slice $K|dx$, Knotting A.

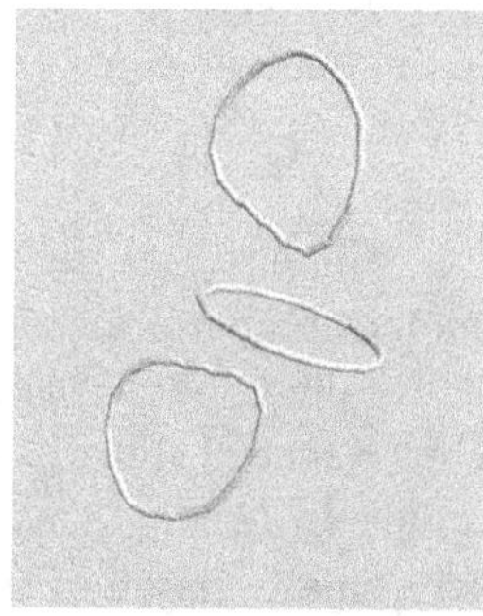

Fig. 8.15 Slice, $K|e$, Knotting A.

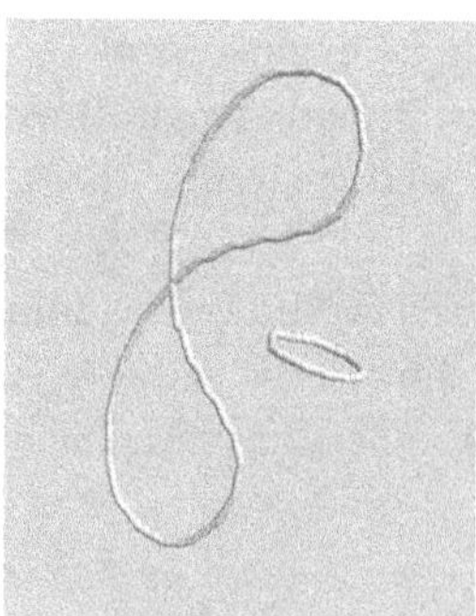

Fig. 8.16 Slice $K|ex$, Knotting A.

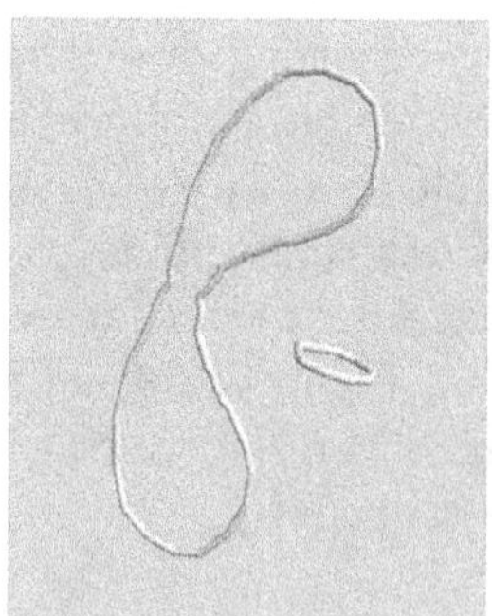

Fig. 8.17 Slice $K|f$, Knotting A.

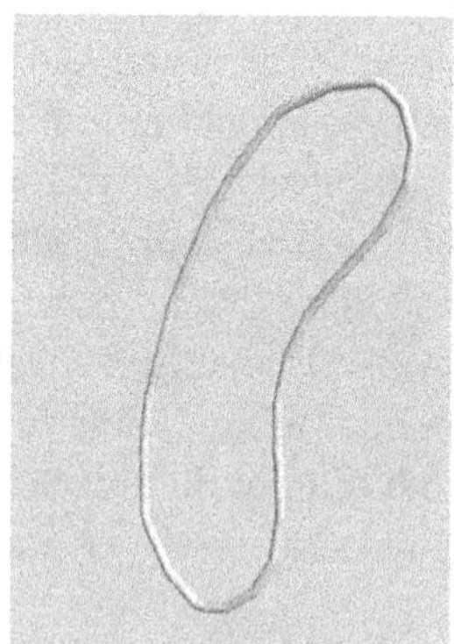

Fig. 8.18 Slice $K|g$, Knotting A.

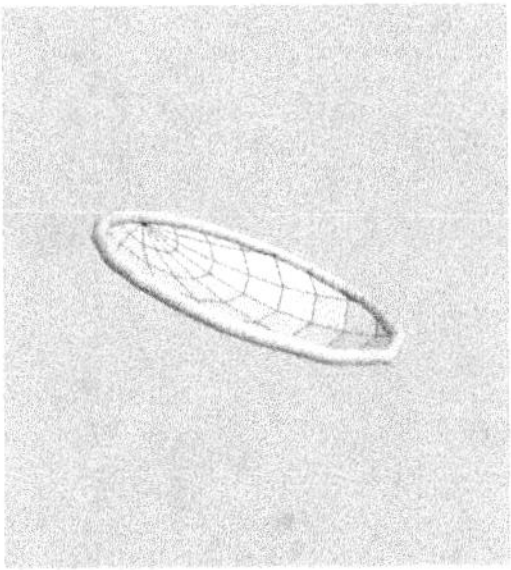

Fig. 8.19 Slab $K||_{-\infty}^{a}$, Knotting A.

Fig. 8.20 Slab $K||_{a}^{b}$, Knotting A.

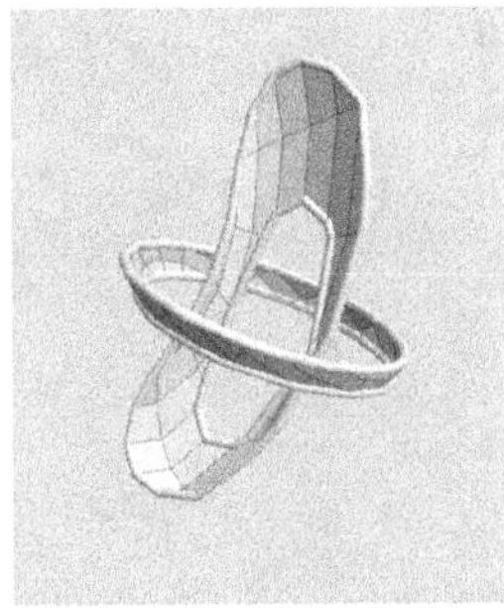

Fig. 8.21 Slab $K||_{b}^{c}$, Knotting A.

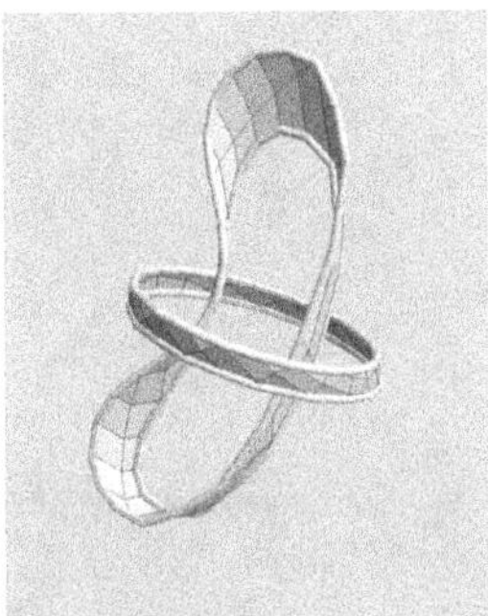

Fig. 8.22 Slab $K||_{c}^{d}$, Knotting A.

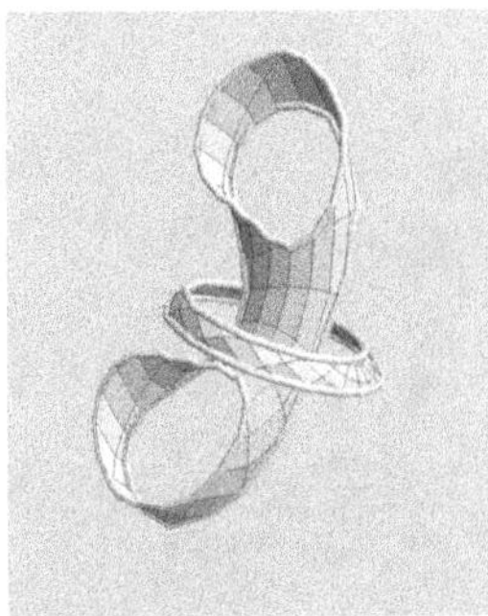

Fig. 8.23 Slab $K||_{d}^{e}$, Knotting A.

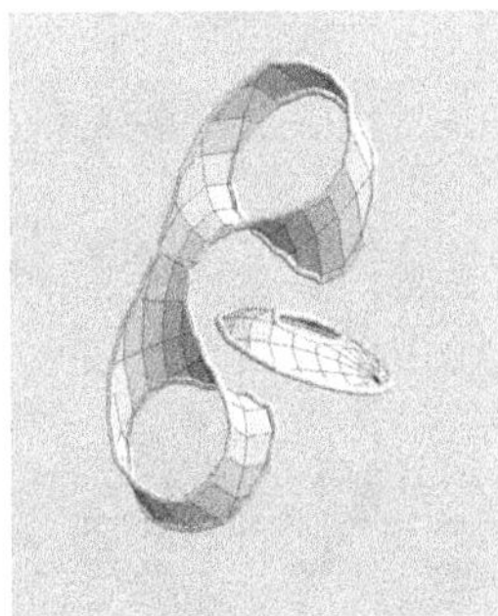

Fig. 8.24 Slab $K||_{e}^{f}$, Knotting A.

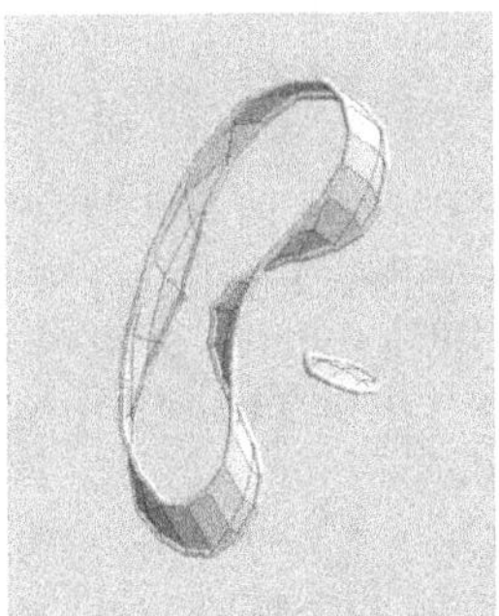

Fig. 8.25 Slab $K||_{f}^{g}$, Knotting A.

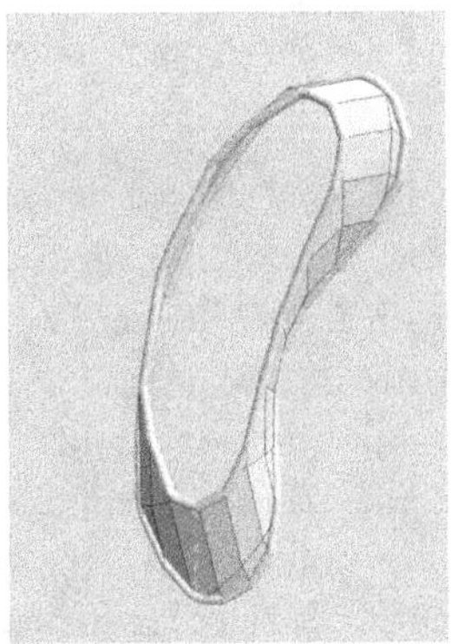

Fig. 8.26 Slab $K||_{g}^{h}$, Knotting A.

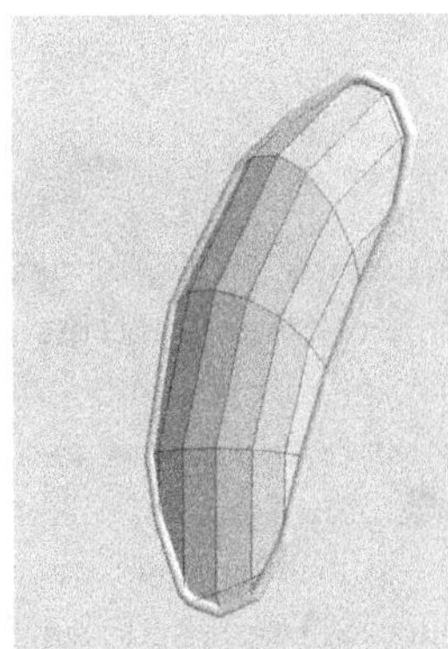

Fig. 8.27 Slab $K||_{h}^{\infty}$, Knotting A.

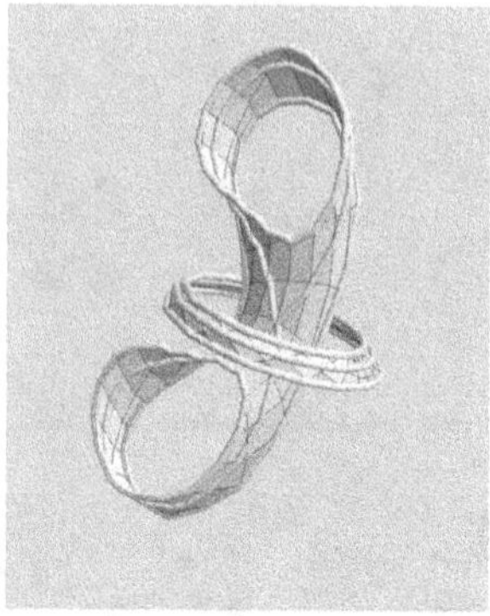

Fig. 8.28 Figure 8.23, showing slice $K|dx$.

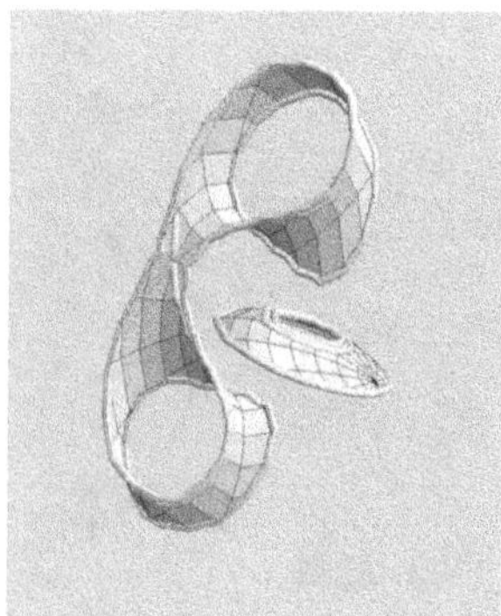

Fig. 8.29 Figure 8.24, Knotting A, showing slice $K|ex$.

Remark 8.2. We next show that Knotting A is equivalent to the basic sphere-torus knotting, Section 6.12. In fact, we show that the projection consists of an embedded torus intersecting an embedded sphere, with the intersection being two crossing circles that lift to meridians of the torus, with one an above 2-shard, the other a below 2-shard.

In Figure 8.30, we show the union of all slabs of Figures 8.19 to 8.27 with slices represented as thin tubes. Each of these are individually embedded in

R^3 by projection, but different slabs can (and, as we see, do) intersect, and their union is the projection of our knotting. It is clear that the projection, shown in Figure 8.30, is a torus and a sphere intersecting in two disjoint circles. To verify that Knotting A is equivalent to the basic sphere-torus knotting it would be enough to verify that, at one crossing circle, the torus passes over the sphere, while at the other circle it passes under. With care, the reader could match up the slab sequence, which we know are ordered with height, to the total image of Figure 8.33 and deduce that is indeed the case. However, we will next use an easier and more interesting graphical way to make this verification.

These curves are like the familiar contour curves for a function of two variables, since the height function is constant on each of these curves. However, the domain of our function (the height function) is our surface viewed as a subset of the knotting projection rather than a planar subset.

We use coding such as in Section 6.4, and use thickness and color to indicate the w-height of a slice. Each point of a slice has the same w value, so each slice will have uniform thickness and grey level. Slices that are high would be thicker lighter gray tubes, and low slices darker and thinner. We call this visualization method **contour coding**. A similar color coding of contour lines is sometimes used for functions of two variables.

We apply this coding to our figure and rotate to get a better look at the two crossing circles, Figures 8.31 and 8.32. The lightening of shade and the thickening of consecutive contours gives an idea of how the surface rises in the fourth coordinate. The darker thinner contours on the sphere near this circle verify that the sphere is below the torus along this circle for Figure 8.31. Similarly, the sphere is above the torus on the crossing circle of Figure 8.32. As in Example 6.11, due to graphical lighting issues, determination of above/below may be unclear at *some* points of a crossing circle—however, we need only find *one* point along each circle where we are confident from the coding of the above/below relation.

A wire-frame is an artifact of computer graphics. As a subset of the knotting, it may not be related to the knotting at all. Contour coding has one advantage over wire-frame coding since the tubes we use *are* related to the knotting. A disadvantage to using contour coding is that if we want to change the knotting (say by a rotation in R^4) we would have to recalculate all the contours—perhaps even alter the slicing levels. ▮

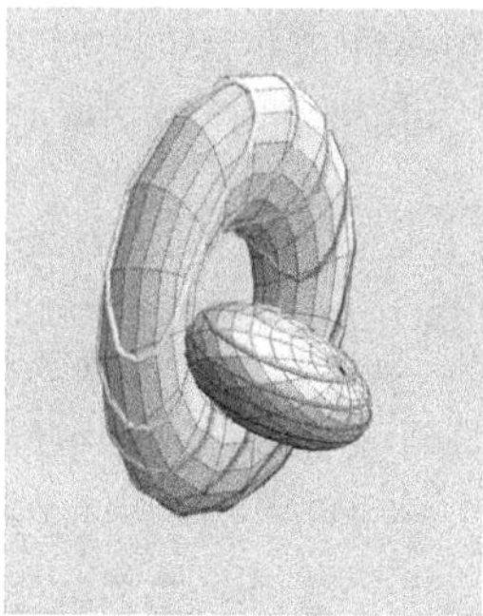

Fig. 8.30 All slabs of Knotting A with contour coding.

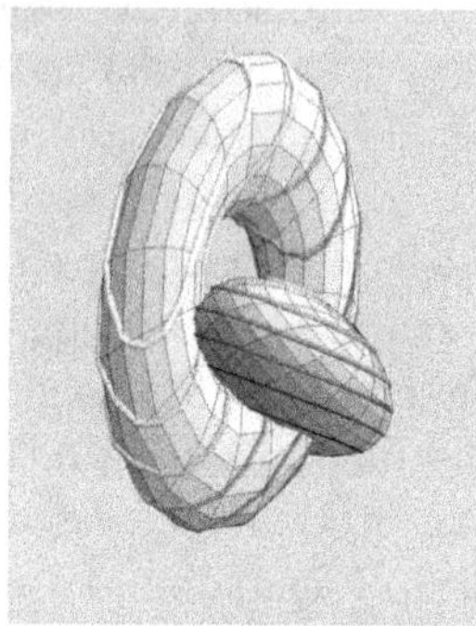

Fig. 8.31 View of one crossing circle showing sphere passes below torus.

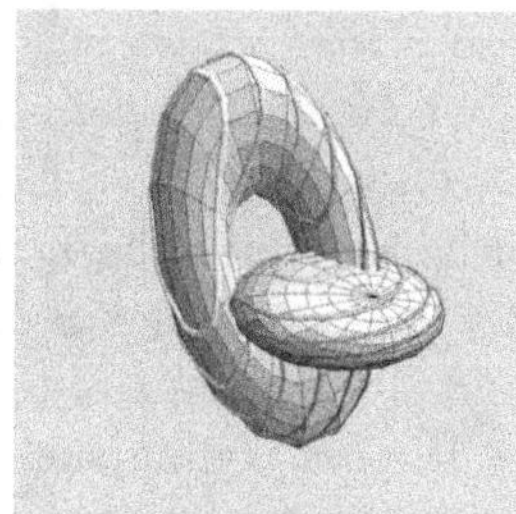

Fig. 8.32 View of the other crossing circle showing sphere passes above torus.

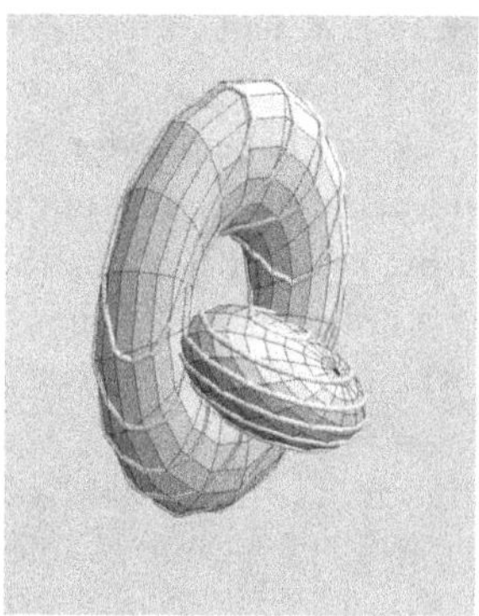

Fig. 8.33 All slabs of Knotting A with contour coding.

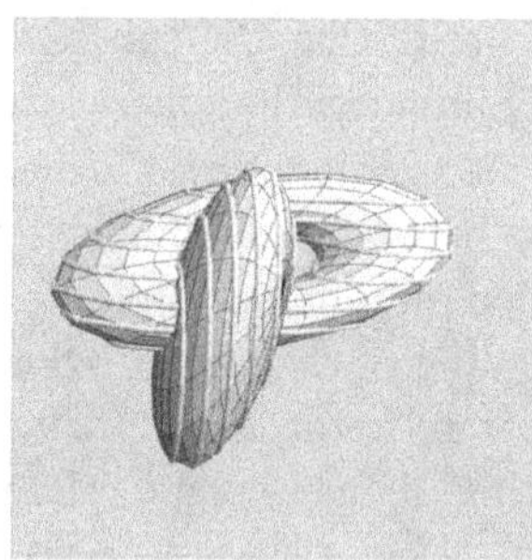

Fig. 8.34 A rotation of Figure 8.34.

Remark 8.3. All of the slices of Knotting A shown in Figures 8.10 to 8.18 that are 1-knottings (that is, ignore the figure-eights) are split trivial links. Yet, as we will show, the basic sphere-torus is not a trivial split 2-link. So, the non-triviality of this example is not revealed in any of the (trivial) slices but must therefore be due to how the slices relate to each other. Further examples show that is not unusual. ▌

Example 8.3. A second example, **Knotting B**, is shown in Figures 8.35 to 8.43. We label values $\{a, b, c, d, dx, e, ex, f, g, h\}$ as before. However, numerically, the values are different. Figures 8.44 to 8.52 show the corresponding sequence of slabs with details of saddle points in Figures 8.53 and 8.54. Figure 8.55 shows a projection of Knotting B. Knotting B is obtained by a slight rotation of Knotting A in R^4, a fact not readily apparent from the slices or slabs. This illustrates that substantial changes to slices and slabs can occur with simple changes of a knotting.

Comparison with the projection of Knotting A, Figure 8.30 reveals that Knotting A is equivalent to Knotting B. Each projection is a union of an almost standard torus in R^3 with an almost standard sphere: with crossing set a union of two disjoint circles. Closer inspection, see Figures 8.56 and 8.57, reveals that the height relations at the crossing circles are basically the same in these two knottings, showing that these knottings are equivalent.

The most striking difference between slices of these two knottings is seen in the slice of Knotting B at level e Figure 8.40. This slice is a not trivial 1-knotting, while Knotting A has only trivial slices, see Remark 8.3. ▌

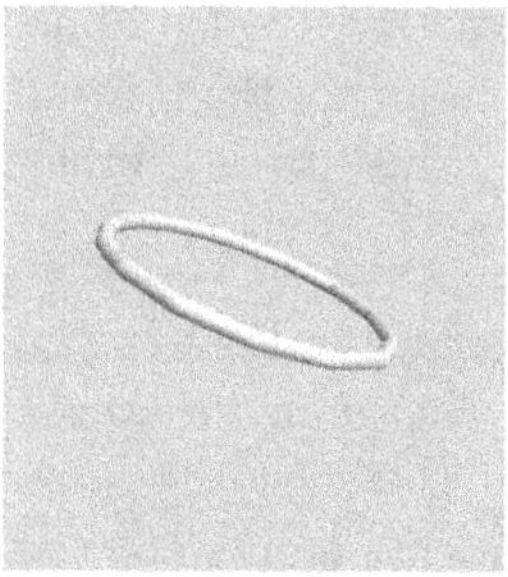

Fig. 8.35 Slice $K|a$, Knotting B.

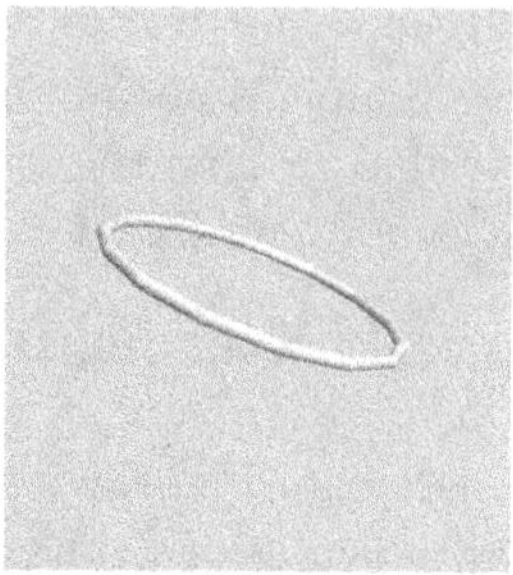

Fig. 8.36 Slice $K|b$, Knotting B.

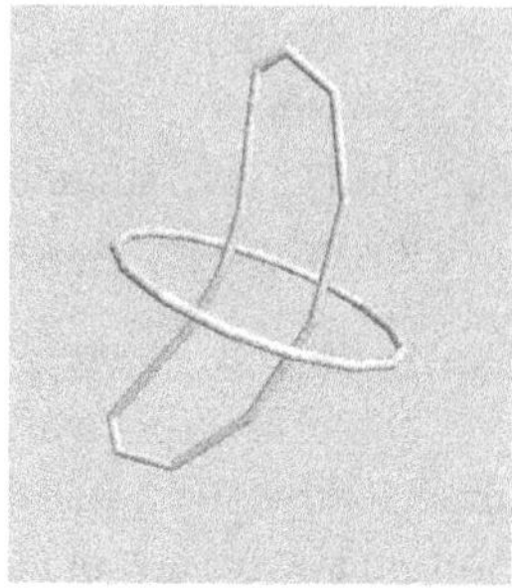

Fig. 8.37 Slice $K|c$, Knotting B.

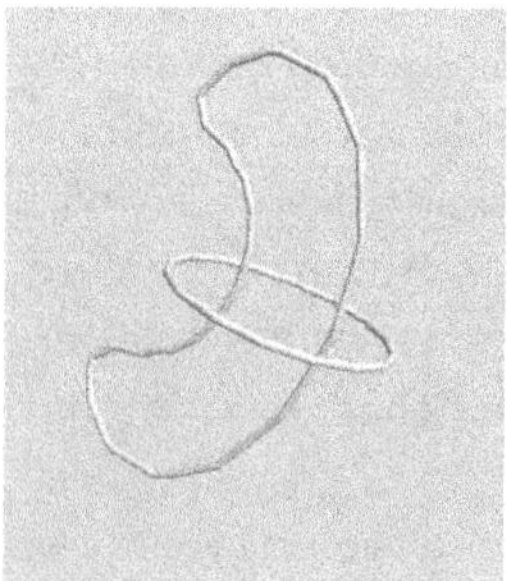

Fig. 8.38 Slice $K|d$, Knotting B.

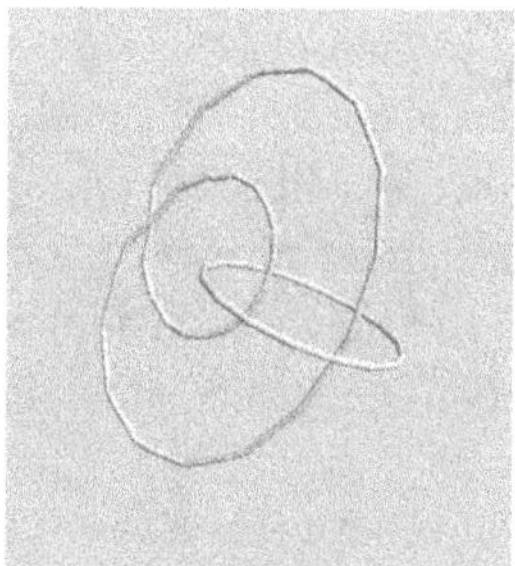

Fig. 8.39 Slice $K|dx$, Knotting B.

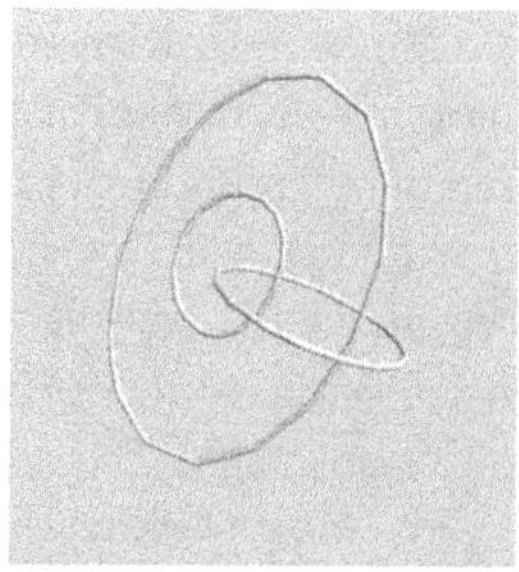

Fig. 8.40 Slice $K|e$, Knotting B.

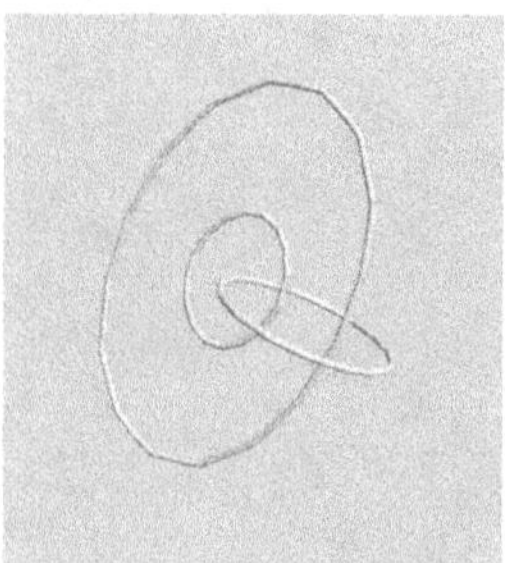

Fig. 8.41 Slice $K|ex$, Knotting B.

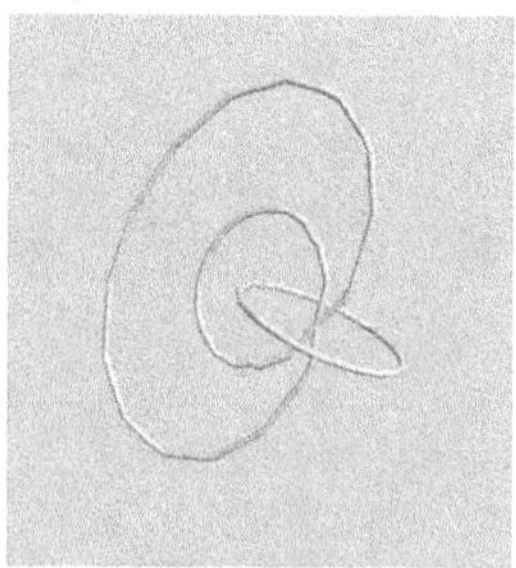

Fig. 8.42 Slice $K|f$, Knotting B.

Fig. 8.43 Slice $K|g$, Knotting B.

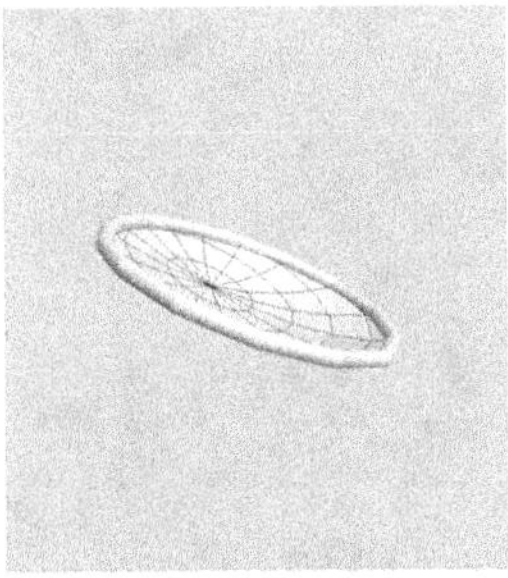

Fig. 8.44 Slab $K||_{-\infty}^{a}$, Knotting B.

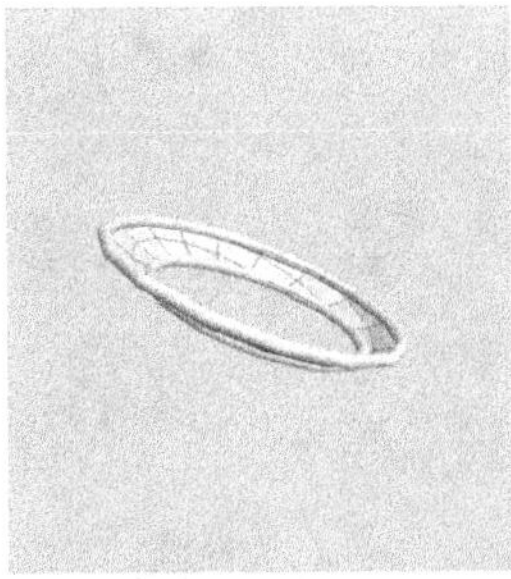

Fig. 8.45 Slab $K||_{a}^{b}$, Knotting B.

Fig. 8.46 Slab $K||_{b}^{c}$, Knotting B.

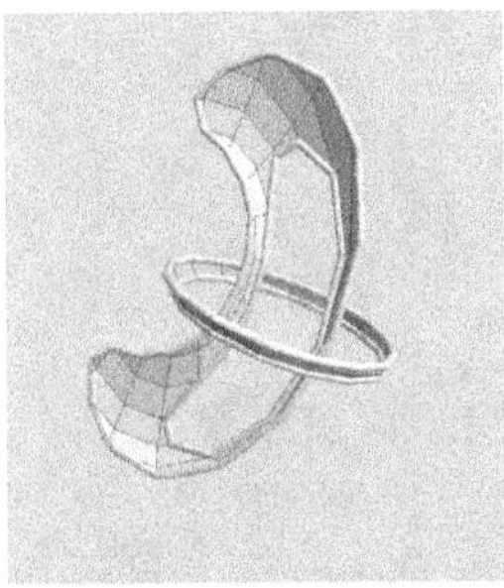

Fig. 8.47 Slab $K||_{c}^{d}$, Knotting B.

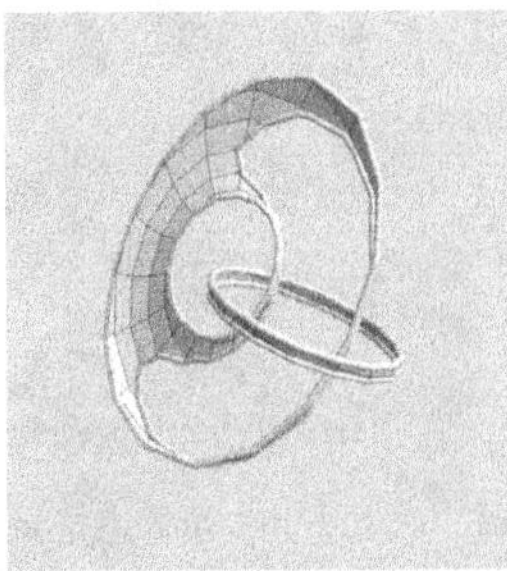

Fig. 8.48 Slab $K||_{d}^{e}$, Knotting B.

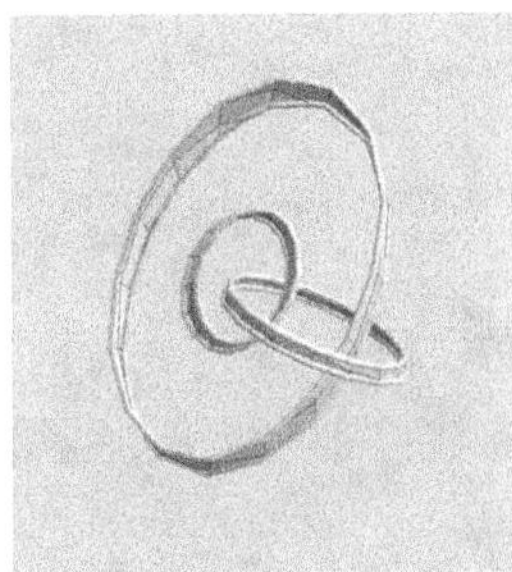

Fig. 8.49 Slab $K||_{e}^{f}$, Knotting B.

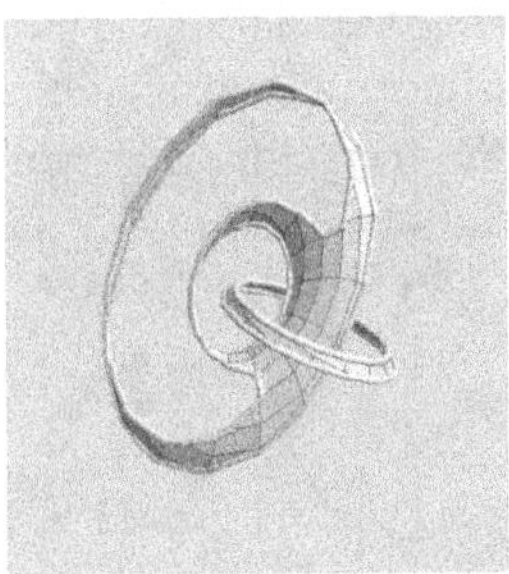

Fig. 8.50 Slab $K||_{f}^{g}$, Knotting B.

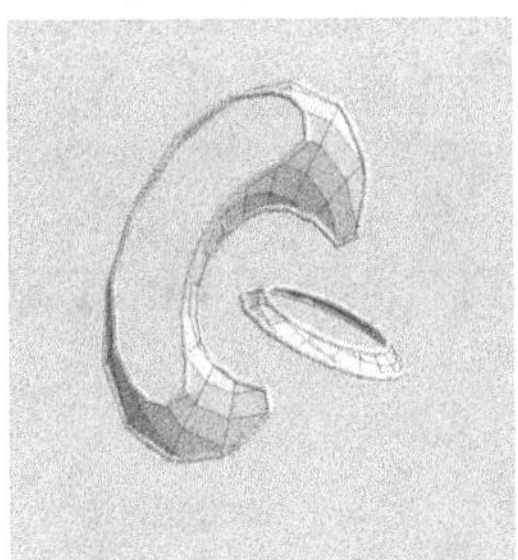

Fig. 8.51 Slab $K||_{g}^{h}$, Knotting B.

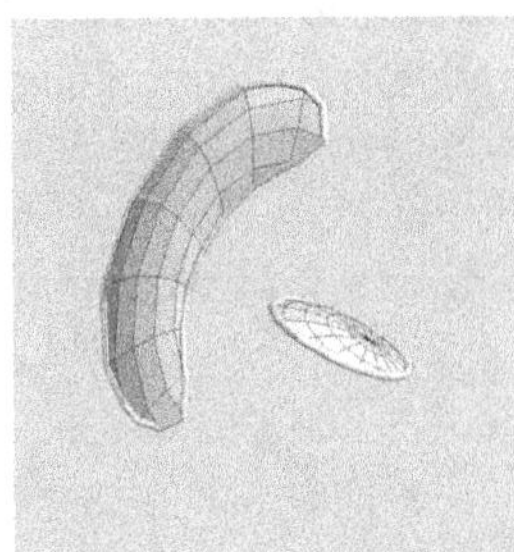

Fig. 8.52 Slab $K||_{h}^{\infty}$, Knotting B.

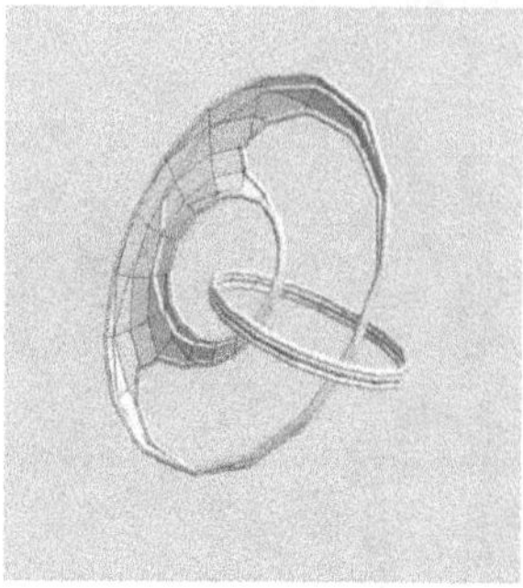

Fig. 8.53 Basic sphere-torus slab 4, Knotting B, showing slice $K|dx$.

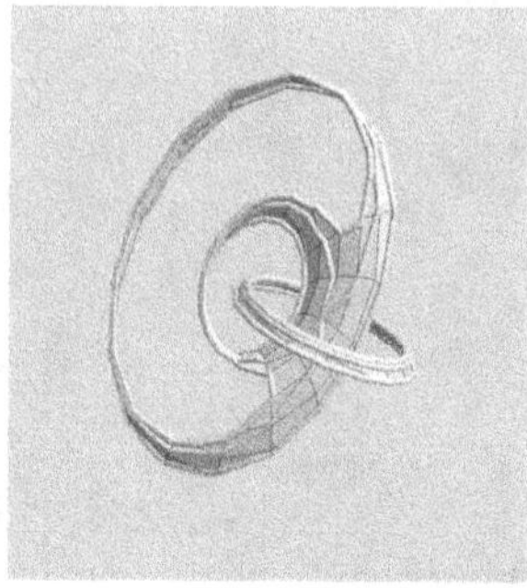

Fig. 8.54 Basic sphere-torus slab 6xb, Knotting A, showing slice $K|ex$.

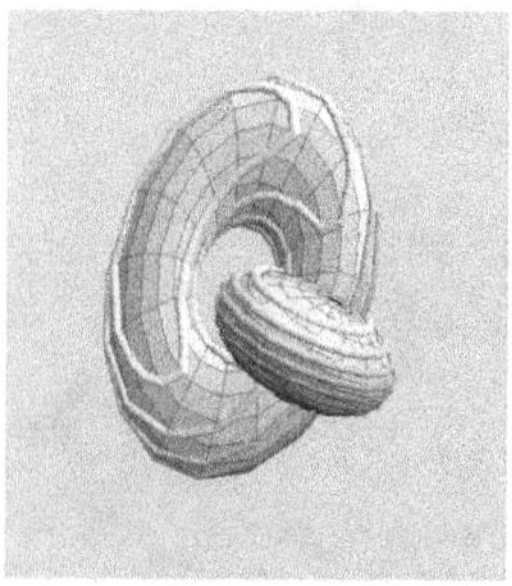

Fig. 8.55 Knotting B, all slabs assembled.

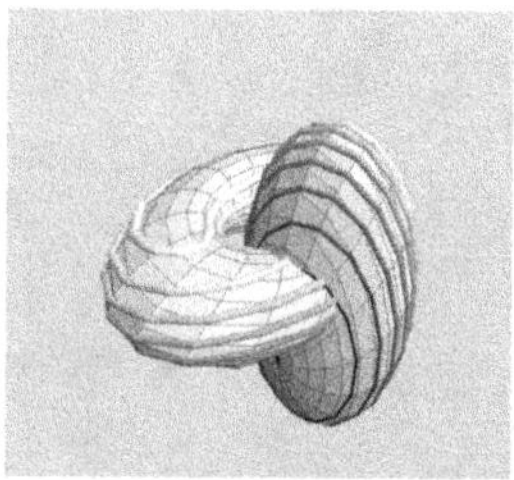

Fig. 8.56 Knotting B contour coding viewing one crossing circle.

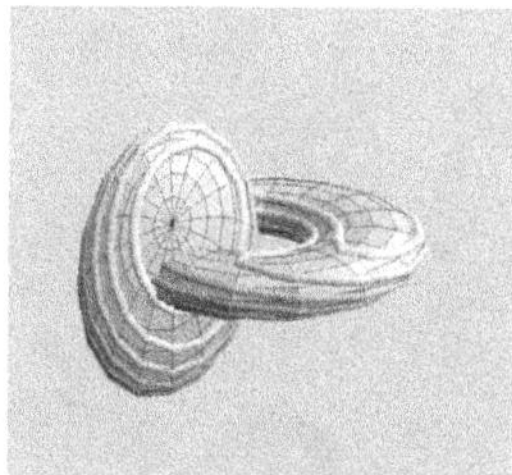

Fig. 8.57 Knotting B contour coding viewing one crossing circle.

Example 8.4. Knotting C with slices shown in (Figures 8.58 to 8.66) and corresponding slabs (Figures 8.67 to 8.75) is a 2-knotting with saddle points at the *same* critical level. A projection with contour coding is shown in Figure 8.76. It is another knotting of the basic sphere-torus knotting as can be seen by comparing Figures 8.77 with Figures 8.30 or 8.55. ▮

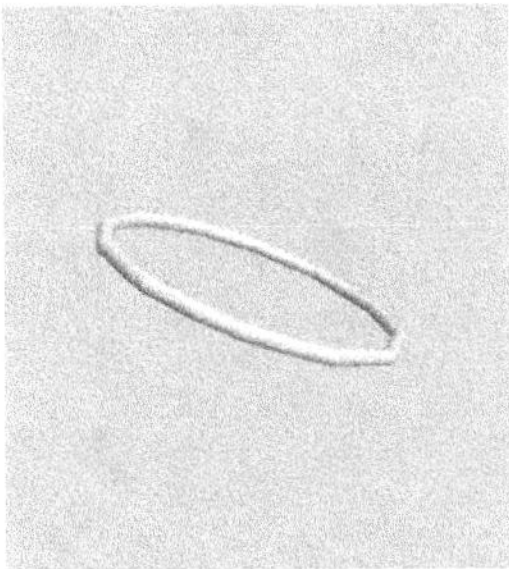

Fig. 8.58 Slice $K|a$, Knotting C.

Fig. 8.59 Slice $K|b$, Knotting C.

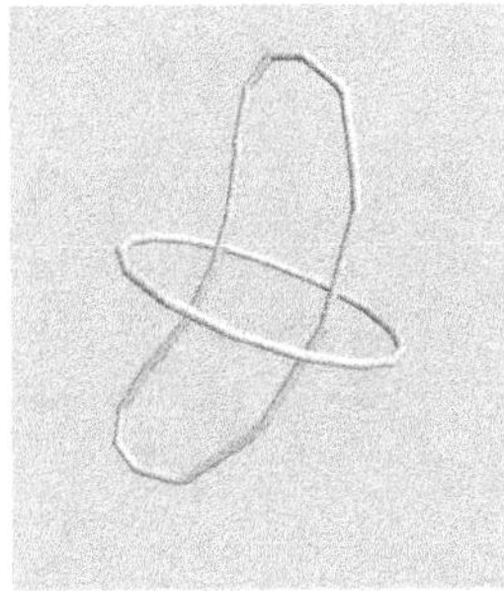

Fig. 8.60 Slice $K|c$, Knotting C.

Fig. 8.61 Slice $K|d$, Knotting C.

Fig. 8.62 Slice $K|e$, Knotting C.

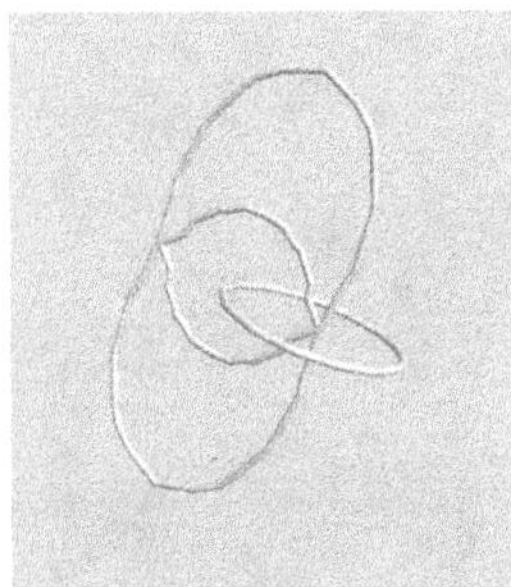

Fig. 8.63 Slice $K|ex$, Knotting C.

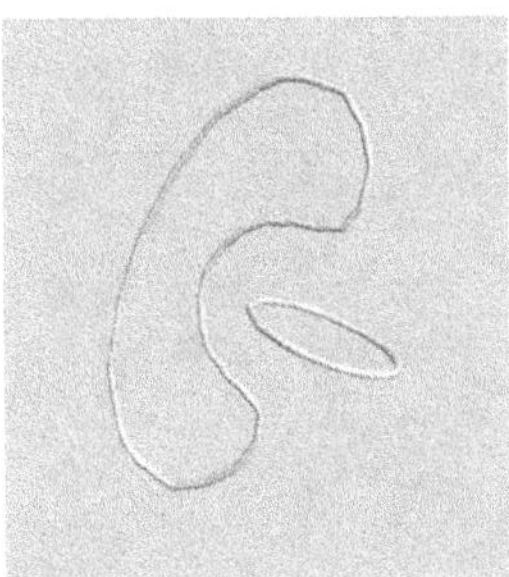

Fig. 8.64 Slice $K|f$, Knotting C.

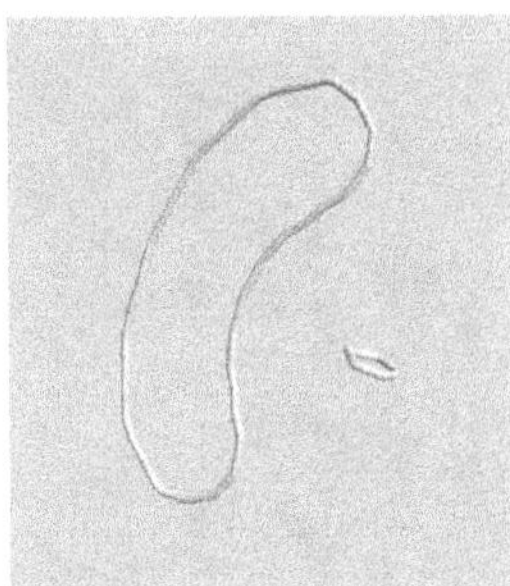

Fig. 8.65 Slice $K|g$, Knotting C.

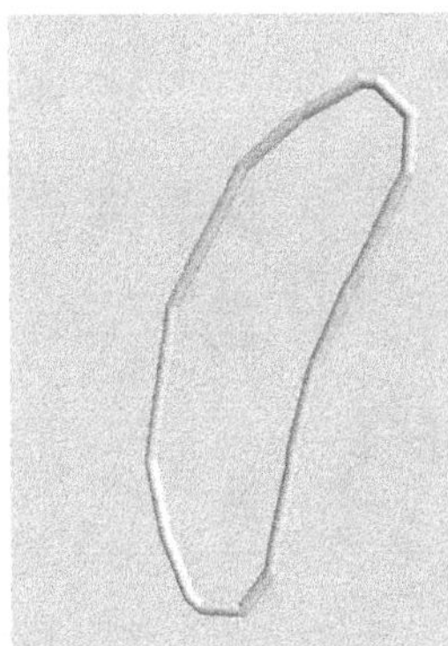

Fig. 8.66 Slice $K|h$, Knotting C.

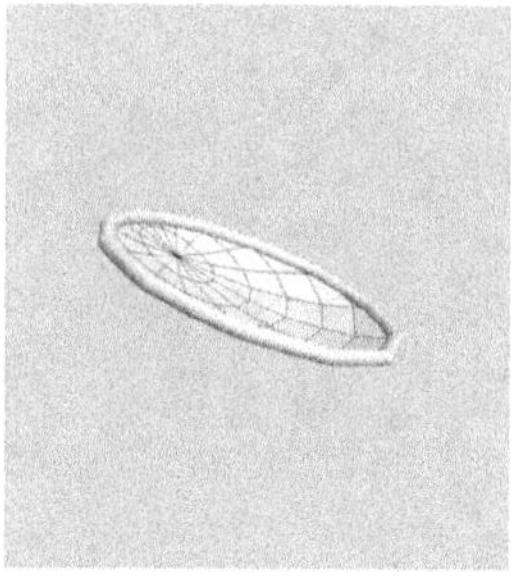

Fig. 8.67 Slab $K||_{-\infty}^{a}$, Knotting C.

Fig. 8.68 Slab $K||_{a}^{b}$, Knotting C.

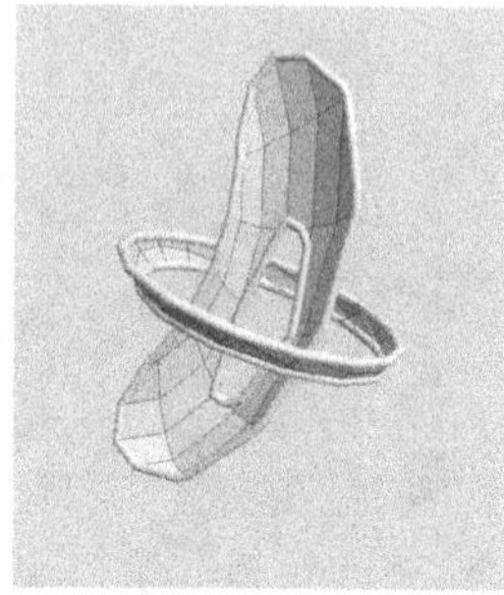

Fig. 8.69 Slab $K||_{b}^{c}$, Knotting C.

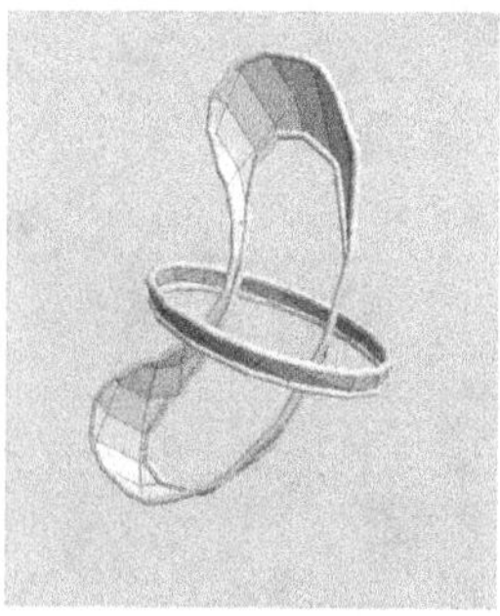

Fig. 8.70 Slab $K||_{c}^{d}$, Knotting C.

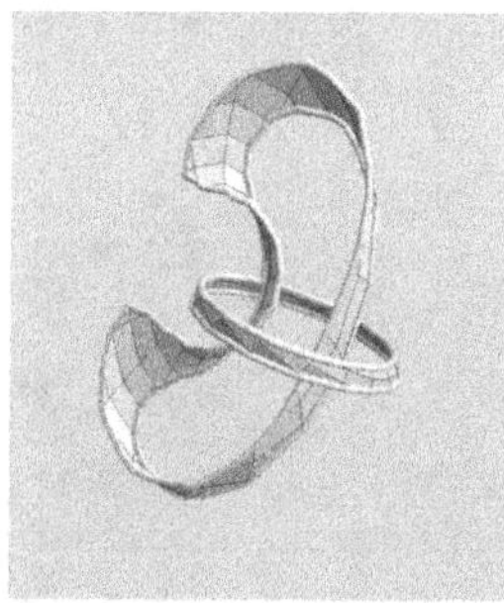

Fig. 8.71 Slab $K||_{d}^{e}$, Knotting C.

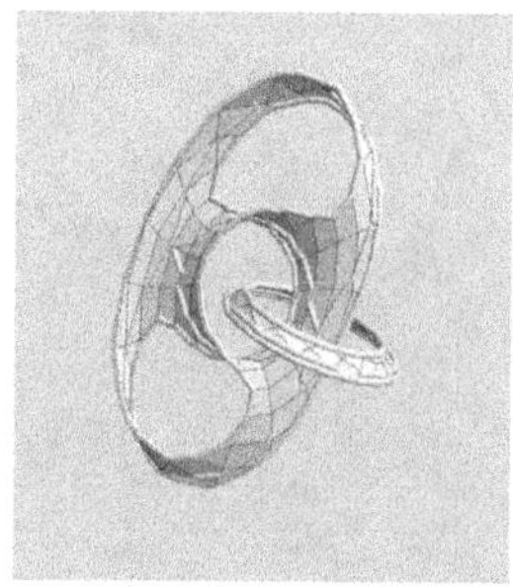

Fig. 8.72 Slab $K||_{e}^{f}$, Knotting C.

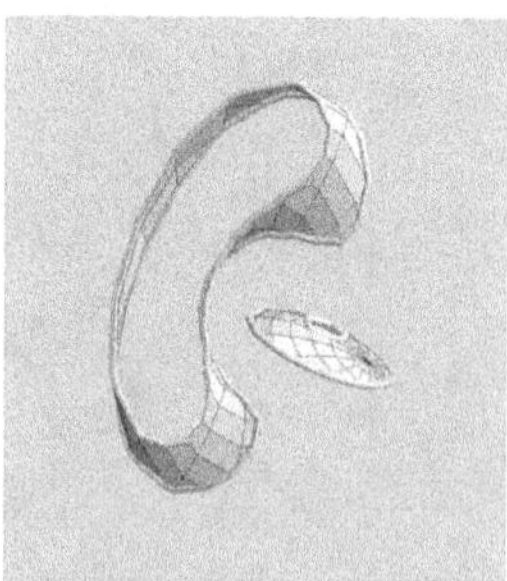

Fig. 8.73 Slab $K||_{f}^{g}$, Knotting C.

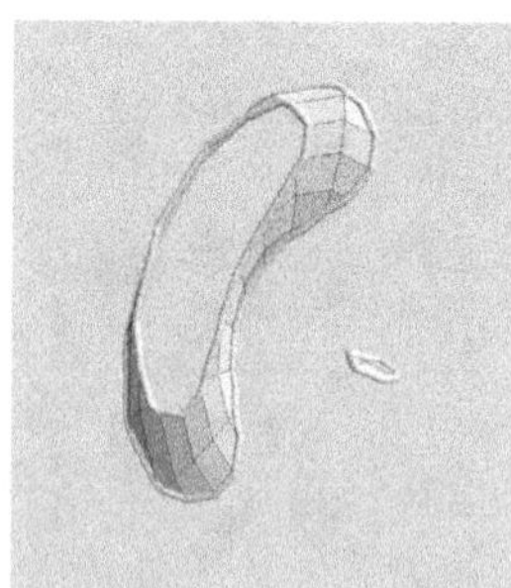

Fig. 8.74 Slab $K||_{g}^{h}$, Knotting C.

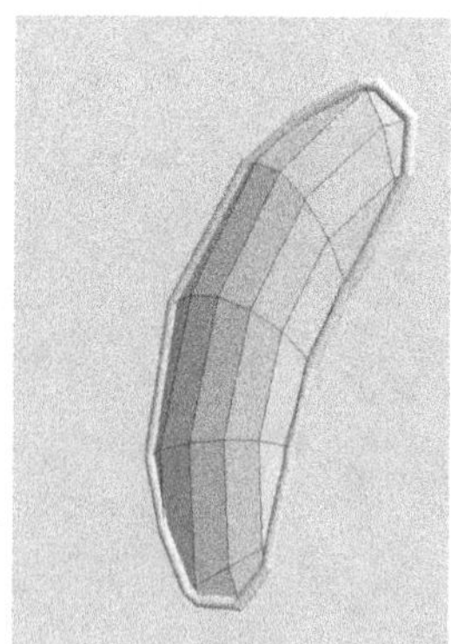

Fig. 8.75 Slab $K||_{h}^{\infty}$, Knotting C.

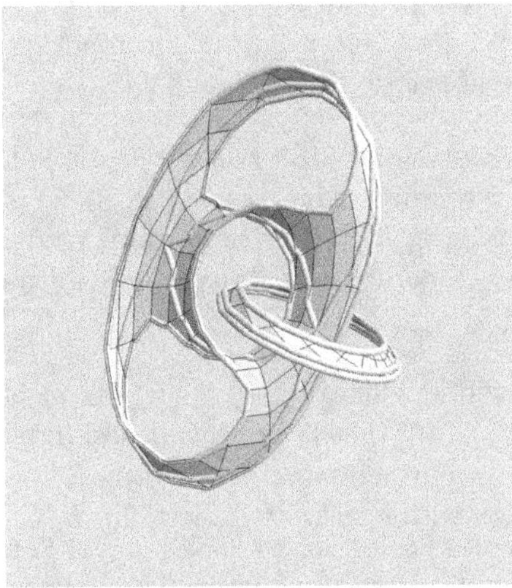

Fig. 8.76 A view of the two saddle points on slab 5, Knotting C.

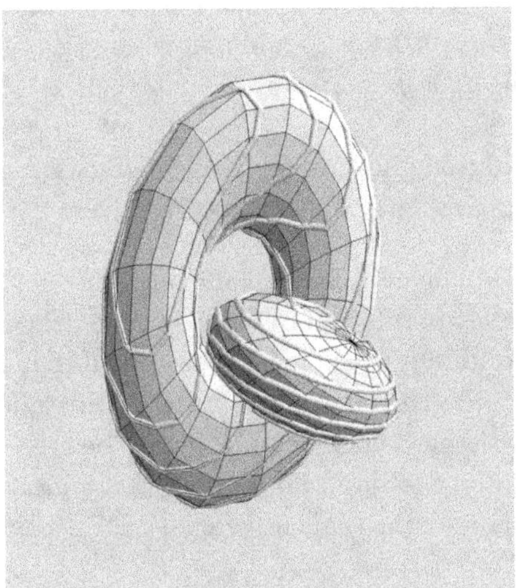

Fig. 8.77 All the slabs from Knotting C.

An area of mathematics known as Morse theory proves, among other things, that slicing examples we have been looking at are "typical". So, when we slice, we can get at most one familiar topological change between slices or slabs, close in height, Proposition 8.2.

Definition 8.3. A knotting $K \subseteq R^n$ is in **general position with respect to slicing** if the last coordinate function $x_n : K \to R^1$ is a Morse function. More generally, a knotting $K \subseteq R^n$ is in **general position with respect to slicing in the x_i-th coordinate** if the coordinate function $x_i : K \to R^1$ is a Morse function. ∎

Prop 8.2. Any smooth 2-knotting $K \subseteq R^4$ is smoothly isotopic to a smooth knotting $K' \subseteq R^4$ that is in general position for slicing with K' arbitrarily close to K. Furthermore, this knotting K' can also be chosen to be in general position with respect to projection. ∎

General position with respect to projection and general position with respect to slicing are independent notions. For example, $S^2 \subseteq R^3 \subseteq R^4$ is clearly in general position with respect to projection—in fact, with no crossing points. However, the w height function is the constant 0 and so *all* points are degenerate critical points. The sphere $S = \{(x, y, z, w) \in R^4 : y^2 + z^2 + w^2 = 1 \text{ and } x = 0\}$ has only two non-degenerate critical points—a minimum at $(0, 0, 0, -1)$ and a maximum at $(0, 0, 0, 1)$. However, the projection is the disk $y^2 + z^2 \leq 1$ and $x = 0$. It is clearly not a general position image.

Although critical points of Morse theory is focused on (smooth) manifolds, a slice *at* a critical level of a 2-knotting is *never* a 1-manifold.

Remark 8.4. We can usefully explain our "fix" for this topologically using some PL models to "flatten out the surface near the critical point". In the process, we obtain a topological 2-disk in K that is contained in the slicing hyperplane and geometrically represents our saddle point. We will also see that this disk is a 1-handle which could be seen as attached to the below-slab or to the above-slab. For smooth knottings, a critical point is associated with a horizontal tangent plane— a PL counterpart will be a horizontal polygon.

To apply our PL models to smooth knottings we can consider our PL models as piecewise smooth manifolds which, when corners are rounded, correspond to smooth manifolds.

For PL local maxima or minima (index 0 or 2) we can flatten by truncation. A local minimum is modeled by an inverted cone with the minimum at the vertex—we flatten by making a slight horizontal truncation near the cone tip. A model of a PL saddle point might look like Figure 8.78—with a flattened version like Figure 8.79. Figure 8.80 shows a PL version of three slices near a saddle point. We replace this middle level (two intersecting line segments) with a horizontal square with line segments attached to each corner, as shown in Figure 8.81. ∎

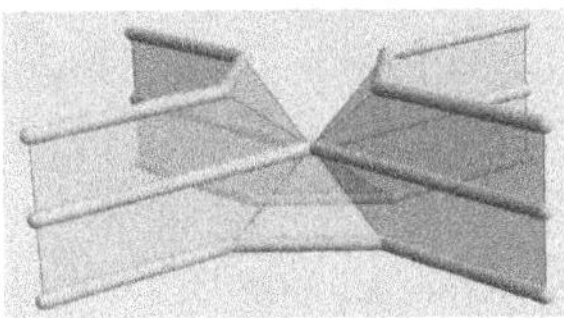

Fig. 8.78 A PL saddle point.

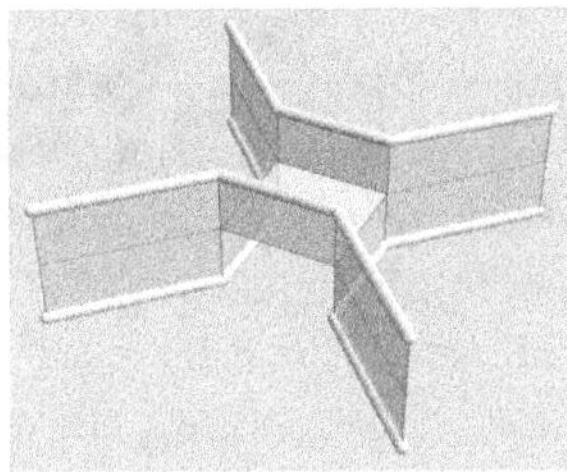

Fig. 8.79 Flattened version of PL saddle.

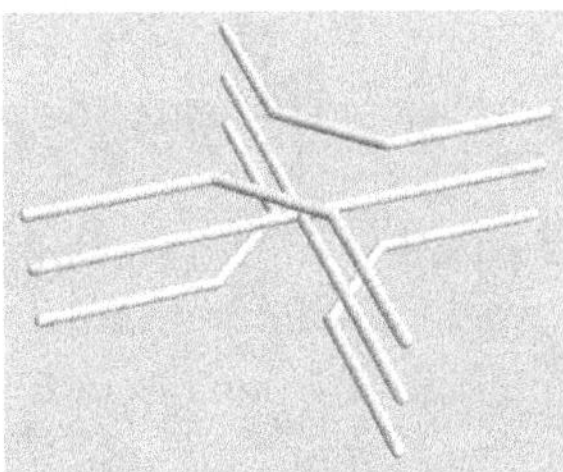

Fig. 8.80 Local picture of slices near saddle point.

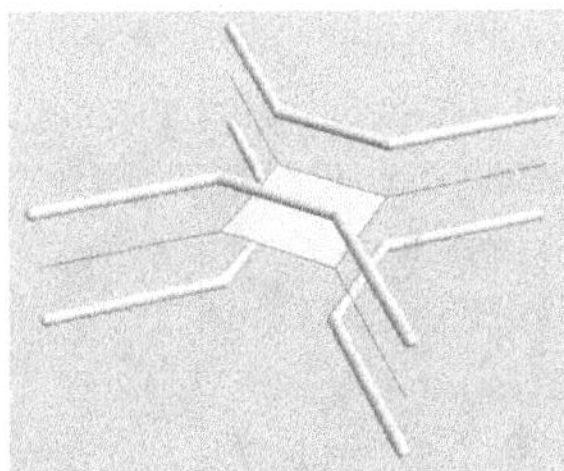

Fig. 8.81 Replace crossing of Figure 8.80 by horizontal square.

Example 8.5. Figure 8.88 shows a (trivial) PL knotting of a torus in R^3. Here our height function is the z-coordinate. There are four critical rectangles (in dark-grey) at heights 0, 2, 4 and 6. Most examples of slices and sections we illustrate are for 2-knottings $K^2 \subseteq R^4$, but this particular example is a subset of R^3 which needs no projection. So, we will look at liftings such as $\widetilde{K}|a$ and $\widetilde{K}||_a^b$, rather than projections $K|a$ and $K||_a^b$, etc.

The entire torus in Figure 8.88 can be considered as either $\widetilde{K}||_{-\infty}^6$ or $\widetilde{K}||_0^6$. The rectangle at level $z = 0$ is a (local) minimum and serves as a model of a critical point of index 0; the rectangle at level $z = 6$ serves as a model of a critical point of index 2. The rectangles at level $z = 2$

and $z = 4$ will correspond to critical points of index 1. (If we add to a horizontal rectangle those polygons of $\widetilde{K}$ that share an edge we get a larger PL sub-disk of $\widetilde{K}$ that more closely resembles a saddle as in Figure 8.89.)

For this PL knotting *every* section $\widetilde{K}||_a^b$ is a 2-manifold including slices at critical values for a or b. Note that the topology of a section only changes at critical values. For example, Figure 8.88 topologically shows a 2-torus, Figure 8.82 shows a 2-disk, Figure 8.83 shows a larger 2-disk, Figure 8.84 shows an annulus, Figure 8.85 shows a larger annulus, Figure 8.86 shows a punctured 2-torus and Figure 8.87 shows a larger punctured 2-torus.

We have so far considered *sections* of our PL knotting. Next, we consider *slices* and focus for the moment on the slice at $z = 2$, $K|_2$. This is a union of a horizontal square $C \cup P$ where P consists of two disjoint arcs, Figure 8.91. Since $\widetilde{K}|_2 = \widetilde{K}||_{-\infty}^2 \cap \widetilde{K}||_2^{\infty}$, we can view this slice as a subset of either of these sections. First we view $\widetilde{K}|2 \subseteq \widetilde{K}||_{-\infty}^2$. Figure 8.84 shows $K||_{-\infty}^2$ is a disk whose boundary is one topological circle c. The intersection $A_c = c \cap C$ consists of two disjoint arcs (opposite sides of C), Figure 8.93. and C intersects this in two arcs labeled A_c. Next view $K|_2 \subset K||_2^{\infty}$, Figure 8.92. Here $\widetilde{K}||_2^{\infty}$ is an annulus whose boundary c' is a disjoint union of topological circles. The intersection $B_c = c' \cap C$ consists of two disjoint arcs (opposite sides of C), Figure 8.94—but note these are the *other* opposite pair of edges of C.

To put this another way, we write the four edges of C as two opposite pairs of sides A_c and B_c. We can construct two 1-manifolds from Figure 8.91. Using sides A_c we get $P \cup A_c$ which is topologically one circle Figure 8.93. Using sides B_c we get $P \cup B_c$ which is two disjoint circles, Figure 8.94. As we go up in height, at value b, the slice changes from one circle to two. if we pass b going down, slices change from two circles to one. ∎

Fig. 8.82 $\widetilde{K}||_{-\infty}^0$.

Fig. 8.83 $\widetilde{K}||_{-\infty}^1 = \widetilde{K}||_0^1$.

Fig. 8.84 $\widetilde{K}||_{-\infty}^2 = \widetilde{K}||_0^2$.

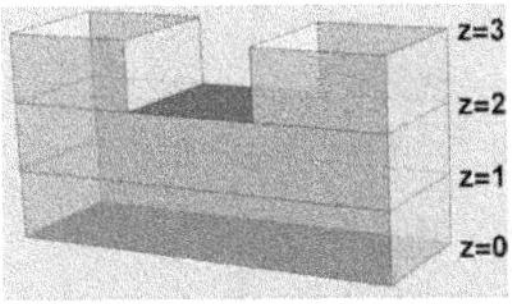

Fig. 8.85 $\widetilde{K}||^{3}_{-\infty} = \widetilde{K}||^{3}_{0}$.

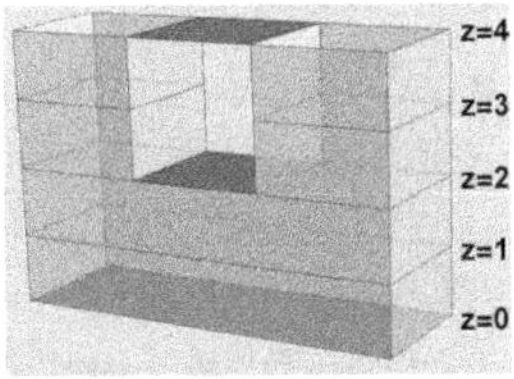

Fig. 8.86 $\widetilde{K}||^{4}_{-\infty} = \widetilde{K}||^{4}_{0}$.

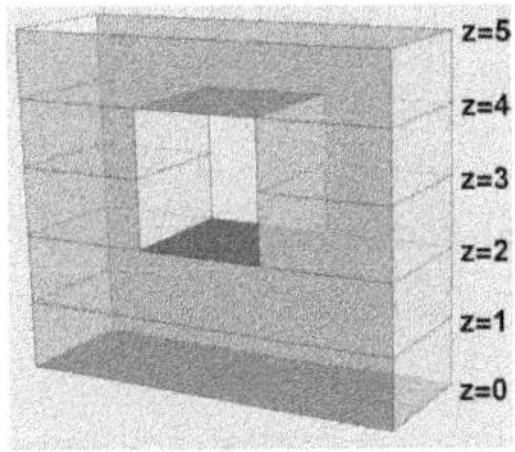

Fig. 8.87 $\widetilde{K}||^{5}_{-\infty} = \widetilde{K}||^{5}_{0}$.

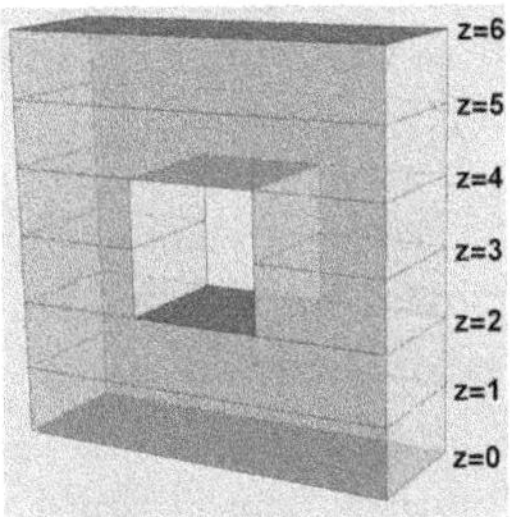

Fig. 8.88 $\widetilde{K}||^{6}_{-\infty} = \widetilde{K}||^{6}_{0}$.

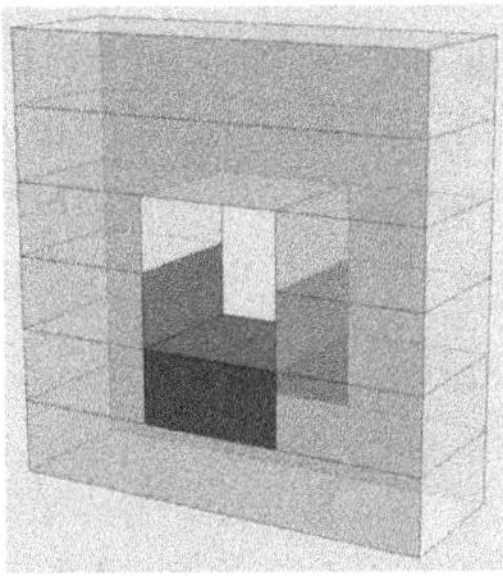

Fig. 8.89 Another topological disk representing a 1-handle.

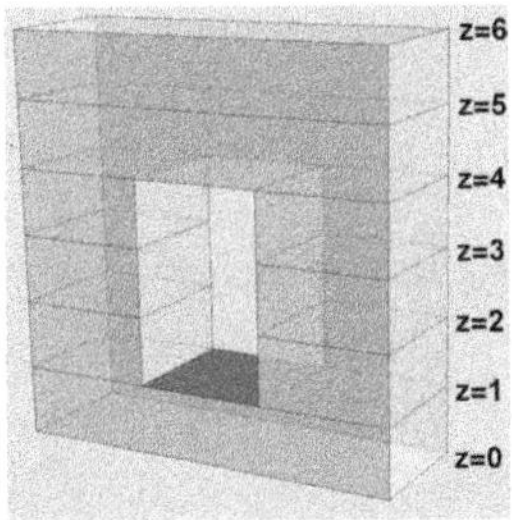

Fig. 8.90 Lowering a critical point.

Fig. 8.91 Slice at $z = 2$: two topological arcs and a disk.

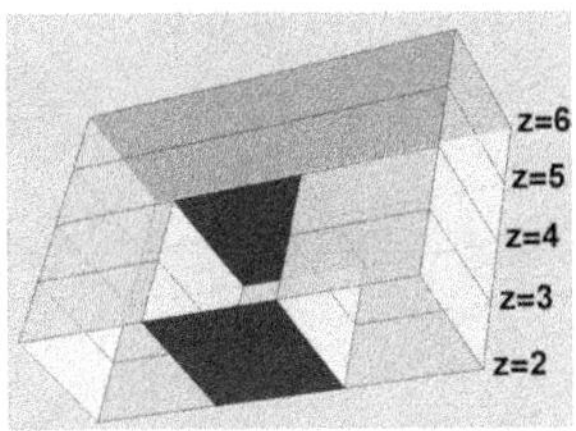

Fig. 8.92 $K||^{6}_{2}$ from below.

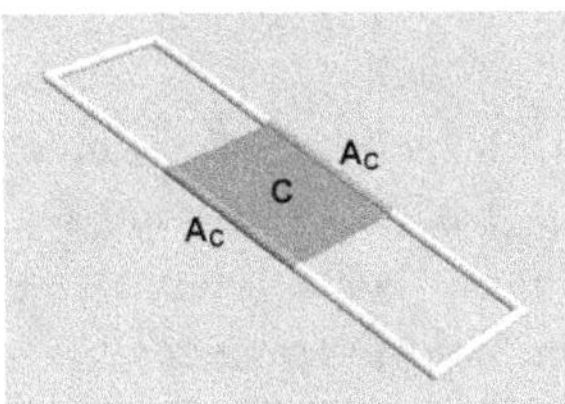

Fig. 8.93 A view of Figure 8.91 as one circle and a band.

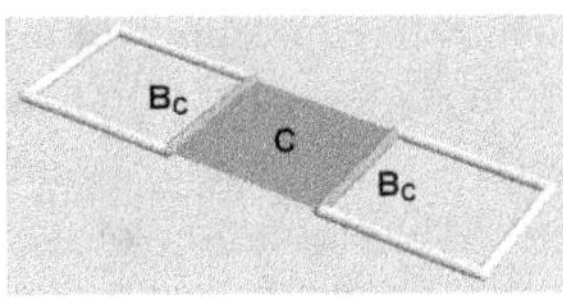

Fig. 8.94 A view of Figure 8.91 as two circles and band.

Definition 8.4. A slice of a smooth knotting that contains a horizontal polygon C representing a critical point of index 1 with critical value c, is called a **band representation of the critical point**. The subset C is called a **band**. Let κ be the slice $\kappa = \widetilde{K}|_c$. The intersection $A_C = \kappa \cap C$ is called the **attaching set of the band** or the a**-set of the band**. Now $\partial R - \kappa$ consists of two open topological arcs. The closure $B_C = \overline{\partial C - \kappa}$ is a union of two topological arcs called the b**-set of the band**. ▌

Definition 8.5. The band, considered as a subset of $\widetilde{K}||^c_{-\infty}$, is a 1-handle of $\widetilde{K}||^c_{-\infty}$ where the attaching set of the band is the same as the attaching band of the handle. A 1-disk corresponding to the core of this 1-handle will be called a **core of the band**, or the **attaching disk of the band**. We graphically use a short line segment at a saddle point as a reminder of this band a-set Figure 8.95. We call this line segment a **slash**. ▌

Remark 8.5. We next remove the band from our Figure but keep the slash as a reminder. The result is as a graphic shorthand for the saddle, which first appeared in [Lomonaco (1981)]. We can topologically reconstruct a saddle from the slash at a crossing by reversing this process as illustrated locally by the sequence: Figure 8.97, then Figure 8.96 and finally Figure 8.95. ▌

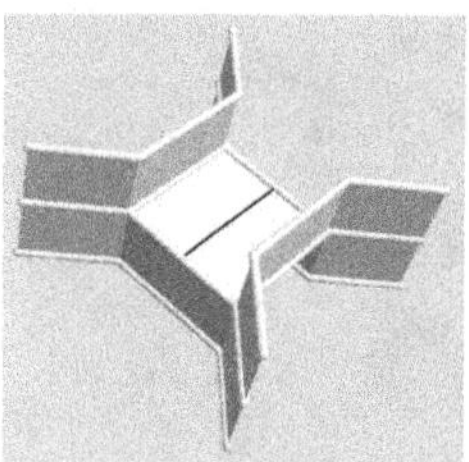

Fig. 8.95 Black line segment is the a-disk of the band.

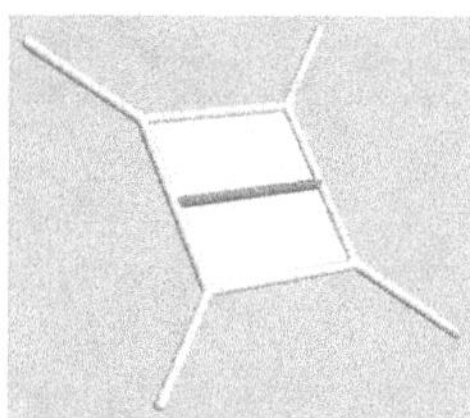

Fig. 8.96 The portion of saddle contained in slicing hyperplane.

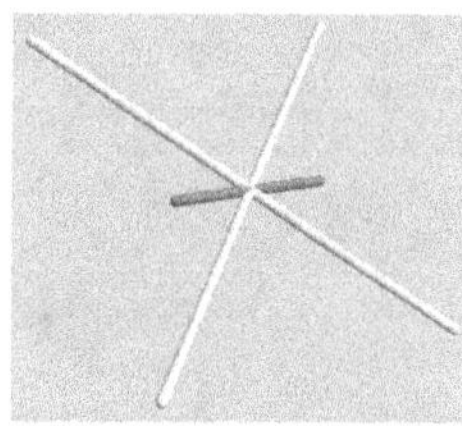

Fig. 8.97 Simplify further to the slash representation in slicing hyperplane.

Both a band and a 1-handle are 2-disks with attaching set A diffeomorphic to a disjoint pair of 1-disks. However, for a band, A is a subset of a 1-knotting in R^3 (or a 3-dimensional hyperplane). For a 2-handle A is a subset of the boundary of a knotting of a 2-manifold in some R^n.

A band is a topological 2-disk. Graphically a small rectangle. However, on occasion, it is convenient to view a band as a long thin strip—a "thickening of the band core"— such as Figures 8.98, 8.99 and 8.100.

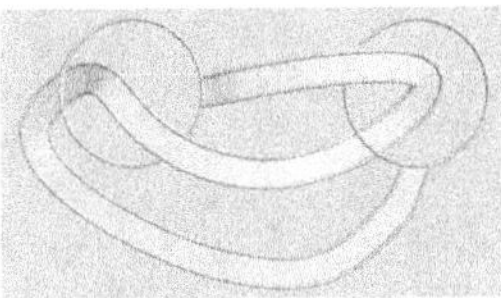

Fig. 8.98 Two circles and one long band.

Fig. 8.99 Two circles and a different long band.

Fig. 8.100 Four circles and three bands.

Remark 8.6. For example, we can use Figure 8.101 to describe Knotting C. We are using PL knottings (rather than smooth knottings) in these graphics in appreciation of the beauty of the style as used in [Yoshikawa (1994)].

Suppose in Knotting C the two saddle points have critical value 1. The slice $w = 1$ is the diagram of Figure 8.101, with slashes removed. This consists of two circles that intersect in a pair of points, and a disjoint third circle. It is isotopic to the slice Figure 8.63. A w-slice just above this is obtained from Figure 8.101 by slashing the crossing points in the direction of the slash and rejoining the pieces by small line segments parallel to the slash, resulting in Figure 8.102. (In this Figure, we show slashes as a reminder of our graphic, but these are not part of the slice.) Figure 8.102 shows a split trivial knotting of two circles, and corresponds to Figure 8.64.

A w-slice below $w = 1$ is the result of slashing the two crossings of Figure 8.101 in the direction *perpendicular* to the previous cuttings and rejoining with small line segments perpendicular to the slash, obtaining Figure 8.103. This Figure (again, with slashes removed) is also a split trivial knotting. This corresponds to the slice Figure 8.62.

Similarly Knotting A can be described as Figures 8.104 to 8.108 and Knotting B can be described as Figures 8.109 to 8.113. ∎

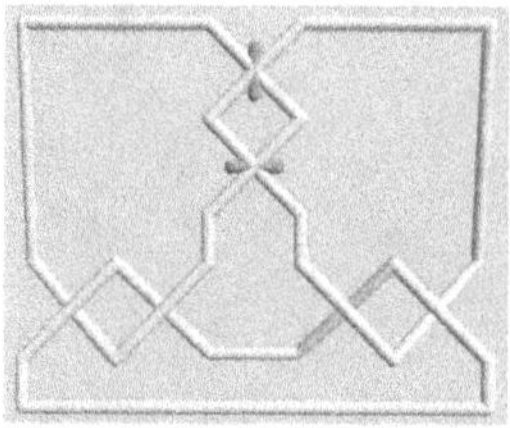

Fig. 8.101 Notation for a 2-knotting with two saddle points at $w = 1$.

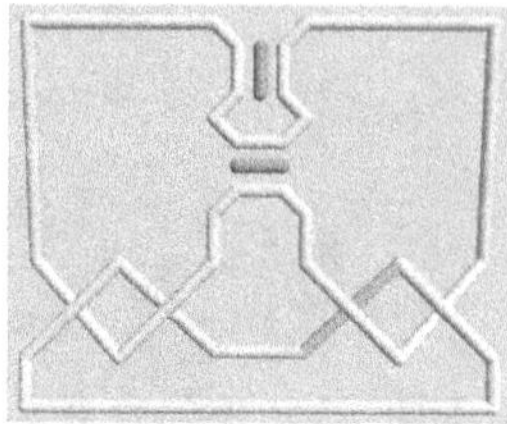

Fig. 8.102 Slice slightly above $w = 1$ slice.

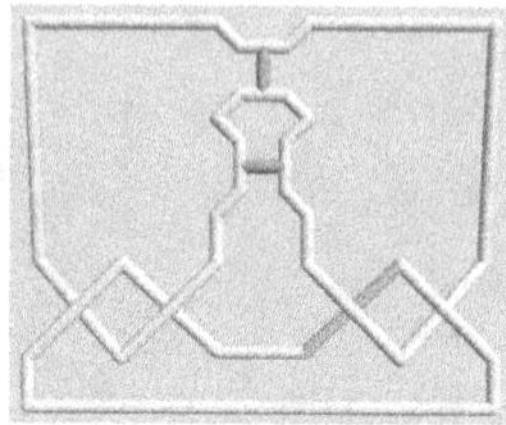

Fig. 8.103 Slice slightly below $w = 1$ slice.

We next define **slab** of knotting K, Definition 8.7. It is a projection $K||_a^b$ of a "thin" section—thick enough to be more interesting than a slice, but not so thick that it is difficult to view. A projection K_* is a union of slabs so slabs help in the visualization of a projection.

The simplest possible slab is a knotting of disjoint annuli. If $[a, b]$ contains no critical value then the section lifting is a product $\widetilde{K}||_a^b$ is a union of annuli and if the interval is small enough, and $K||_a^b$ contains no branch points, the projection will be an embedding.

Suppose K_* has a branch point p and $p \in K||_a^b$ then, of course, p will be a branch point of $K||_a^b$, the crossing set of $K||_a^b$ will be non-empty and so $K||_a^b$ will fail to be a submanifold of R^3. In this case we can require that the crossing set be "as simple as possible" namely an arc. Our term for this is "umbrella shaft" (which in an actual umbrella, is the part that connects the umbrella tip to the umbrella handle):

Definition 8.6. If $K \subseteq R^4$ is a knotting in general position with respect to projection with a non-empty boundary such that crossing set C is an arc with one endpoint on $(\partial K)_*$ and the other point a branch point, we will say C is an **umbrella shaft of K_*.** ∎

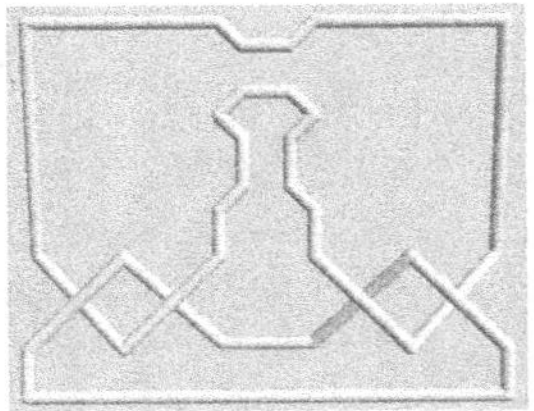

Fig. 8.104 Below both saddle points.

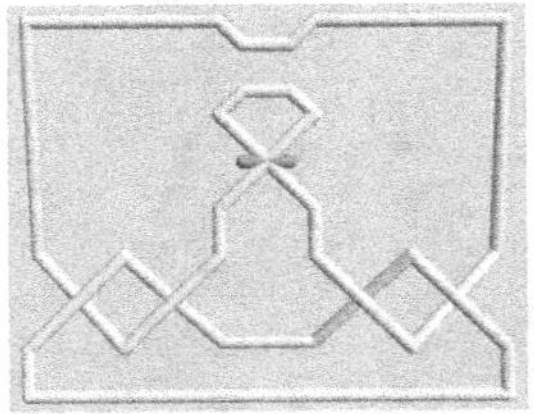

Fig. 8.105 First saddle point.

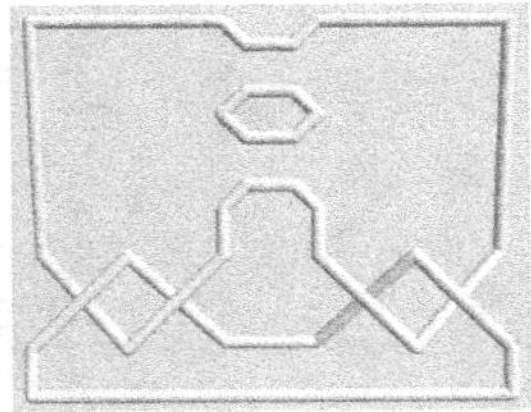

Fig. 8.106 Between two saddle points.

Fig. 8.107 Second saddle point.

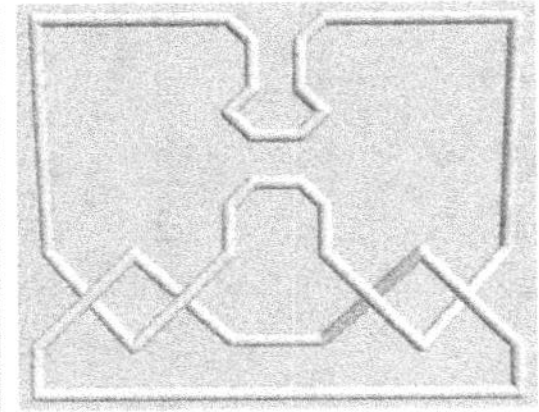

Fig. 8.108 Above both saddle points.

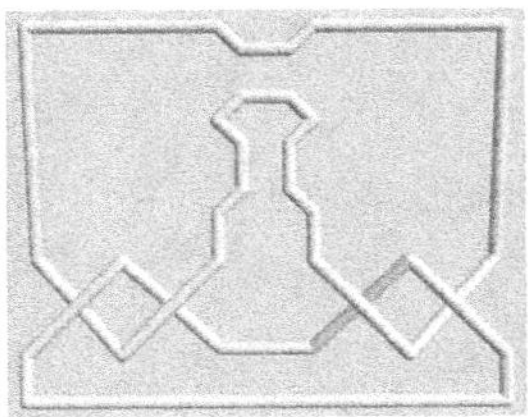

Fig. 8.109 Below both saddle points.

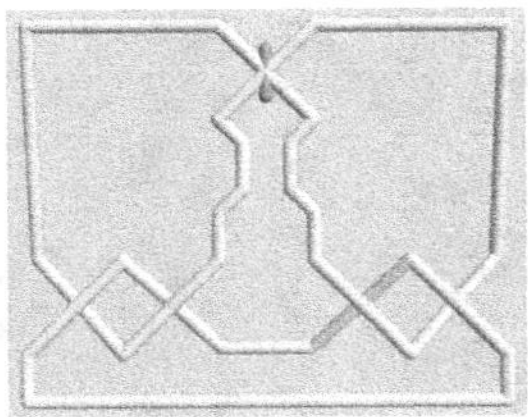

Fig. 8.110 First saddle point.

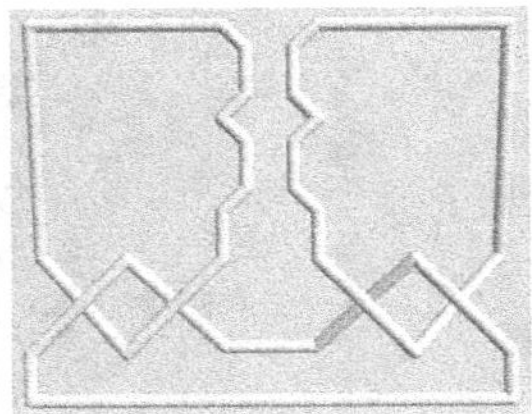

Fig. 8.111 Between two saddles.

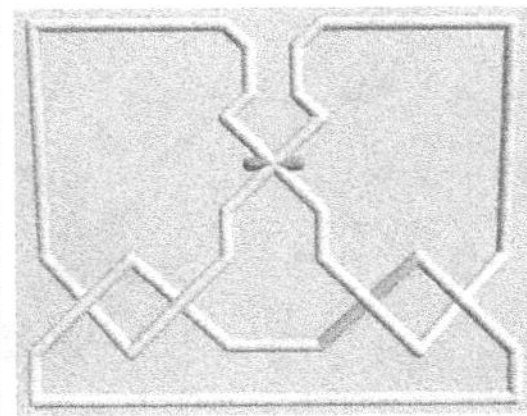

Fig. 8.112 Second saddle point.

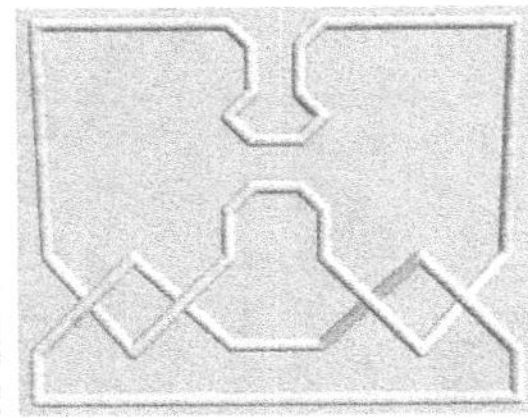

Fig. 8.113 Above both saddle points.

Remark 8.7. Figure 8.114 shows a local model of an umbrella point. Figure 8.115 is an example of an projection of an annulus with an umbrella point b, with x a crossing point on projection of the boundary—here the shaft is a line segment of crossing points between b and x. This Figure is a little inaccurate since there is a cone, with cone point b in the projection of the annulus. So, this is not a projection of a *smooth* annulus as explained in Remark 7.6. As we will see, the projection near an umbrella point is often small and subtle as in Example 8.6, Figures 8.126 and 8.127. ▮

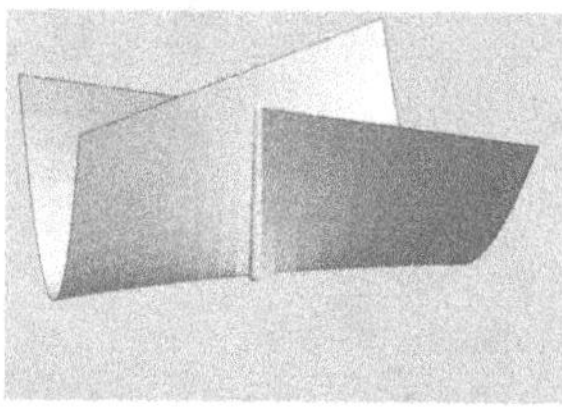

Fig. 8.114 Local model of an umbrella shaft.

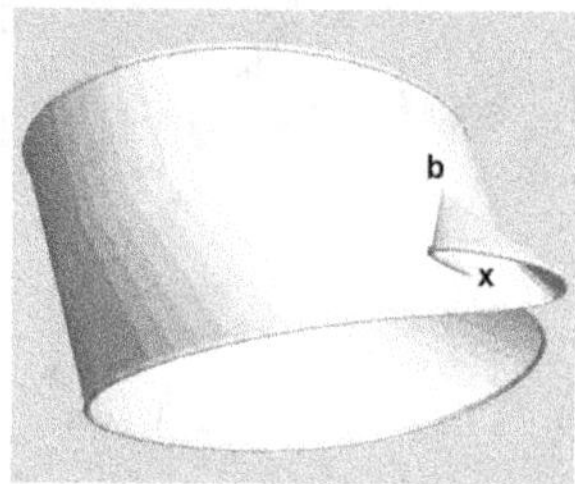

Fig. 8.115 A cylinder with an umbrella shaft.

Definition 8.7. Suppose $K \subseteq R^4$ is a 2-knotting, $[a, b]$ a closed interval with section lifting $\widetilde{K}||_a^b$ with at most one critical value of the height function c, $a < c < b$. The projection $K_{|}|_a^b$ is called a **slab of** K if the crossing set is a (possibly empty) disjoint union of umbrella shafts. ▮

We can extend our general position arguments to show that for any knotting (perhaps adjusted by a small isotopy) there are some $a_1 < \cdots a_m$ so that projection K_* can be written as a union of slabs $K_* = \bigcup_i K||_{a_i}^{a_i+1}$.

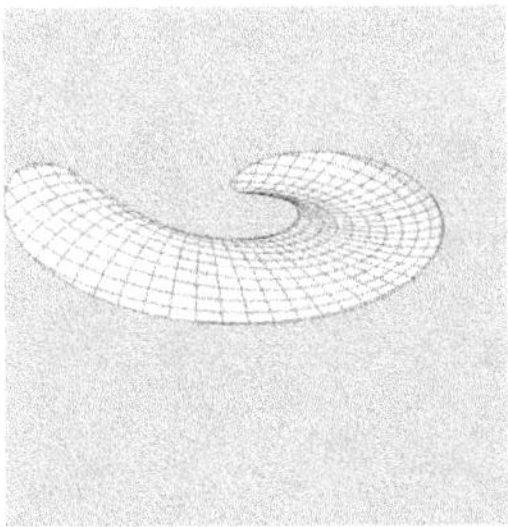

Fig. 8.116 Standard Klein bottle lower w-slab.

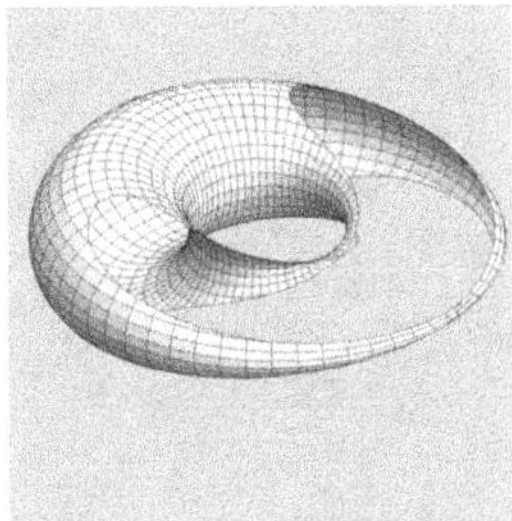

Fig. 8.117 Standard Klein bottle middle w-slab.

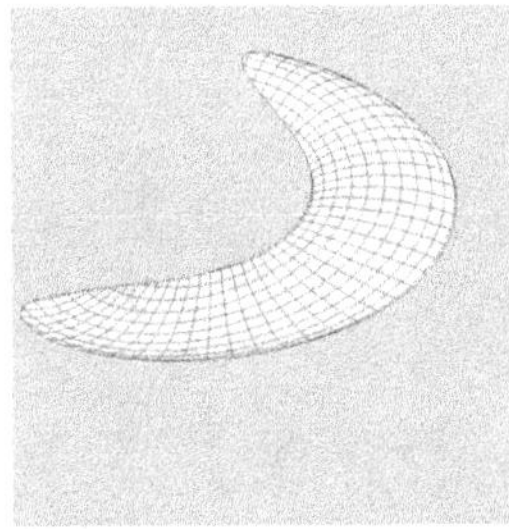

Fig. 8.118 Standard Klein bottle upper w-slab.

Remark 8.8. Consider slabs for the standard knotting of the Klein bottle $\mathbb{K}^2 \subseteq R^4$, Example 6.25. Using standard slabs (slicing in w-direction) we need only three slabs, Figures 8.116, 8.117 and 8.118. The standard projection $\mathbb{K}^2_*$ has two branch points, see Figure 6.131. We find this in the middle slab, Figure 8.117 as tips of two very short umbrella shafts. ∎

Example 8.6. One can get a idea about how to slice a knotting in one direction by analyzing a projection in a different direction. By looking closely at the standard w-projection $\mathbb{K}^2_* \subseteq R^3$ we can see what slices we should make in (say) the x-direction to avoid critical points and only capture at most one critical value between slices. There are four critical points for $\mathbb{K}^2$ in the x-direction indicated by four tangent planes orthogonal to the x-axis at x-values, Figure 8.119. If we avoid slicing at these critical levels, we will divide $\mathbb{K}$ into four slabs—two disks and two pairs of pants. The projections of these are shown in Figures 8.122, 8.123, 8.124 and 8.125. The projection of $\mathbb{K}$ in the x-direction has many branch points. There are eight branch points of $\pi_x(\mathbb{K})$ some of which are visible in Figures 8.120 and 8.121. A slight rotation of this projection shows one branch point more clearly, symmetry of projection hints at others. In the slabs, some branch points are shown in Figures 8.126 and 8.127 —each has (very short) umbrella staff. ∎

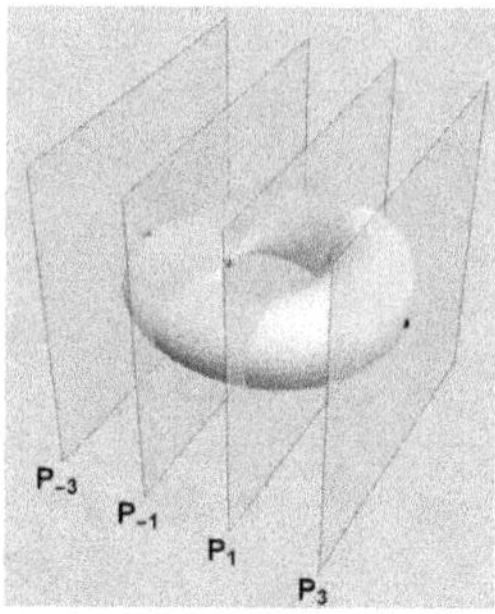

Fig. 8.119 Klein bottle, four critical points.

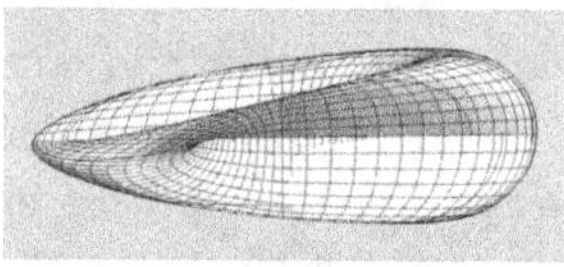

Fig. 8.120 x-projection $\mathbb{K}^2$.

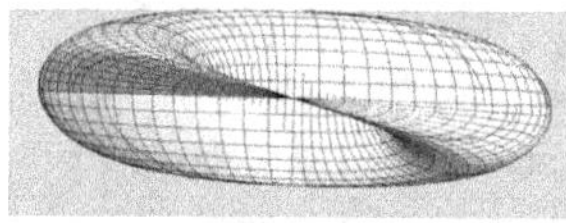

Fig. 8.121 x-projection $\mathbb{K}^2$ rotated $\pi/4$.

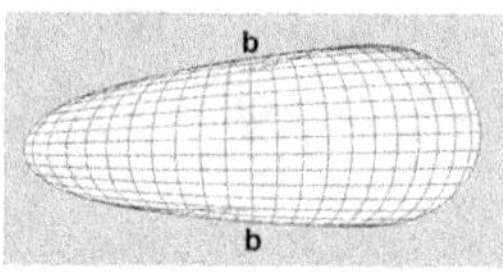

Fig. 8.122 Standard Klein bottle, bottom x-slab.

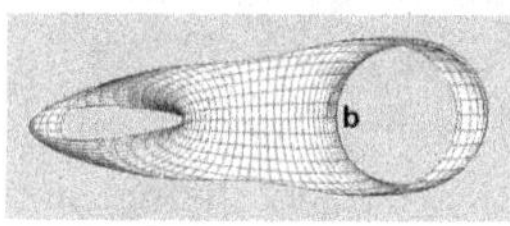

Fig. 8.123 Standard Klein bottle, second x-slab.

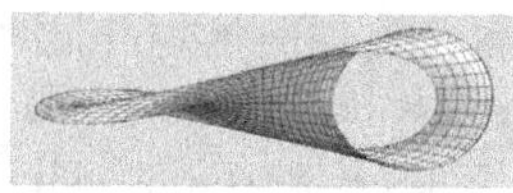

Fig. 8.124 Standard Klein bottle, third x-slab.

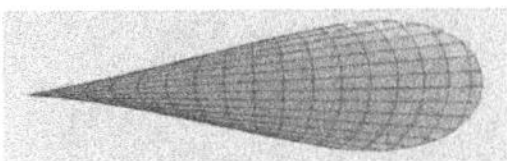

Fig. 8.125 Standard Klein bottle, top x-slab.

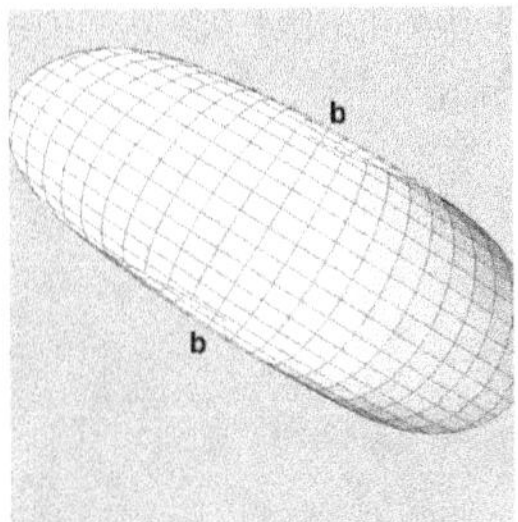

Fig. 8.126 Two branch points of the bottom slab.

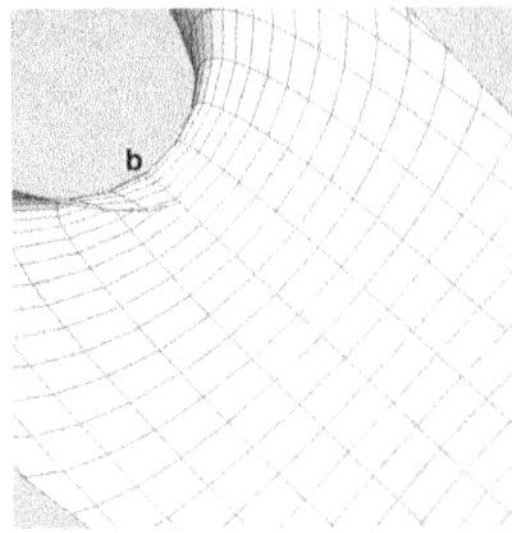

Fig. 8.127 Close-up of a second slab branch point.

Example 8.7. We use the slab method to view a 2-knotting K of the standard projective plane $\mathbb{P}^2$ Figure 6.138). Our slabs will be in the x-direction. The entire projection (and thus the union of all our slabs) in the x-direction is shown in Figure 8.128—it has six branch points, two symmetrically on each axis, and one central triple point. Topologically, this is a well-known image of the projective plane called the **Roman surface**; see, for example, [Francis (2007)] or [Gray *et al.* (2006)].

We divide $\mathbb{P}^2$ into three slabs, similar to our sectioning of the Klein bottle. Each slab will contain a single critical point and two branch points. In these figures branch points are all labeled b. The bottom slab Figure 8.131 is a projection of a 2-disk and contains a local minimum and two branch points. Topologically, the lifting of the middle slab $S \subseteq K$ is a twice-punctured projective plane (that is, a once-punctured Möbius band). One can describe S as a 2-disk with two added bands (1-handles), where each band has a half-twist. The projection of each band is a Whitney umbrella. The top slab Figure 8.132 is a projection of a 2-disk and contains a local maximum and two branch points. All figures, Figure 8.128 to 8.132 are from the same viewpoint of yzw-space. The two projected disks of Figure 8.131 and 8.132 intersect in a line segment between two branch points. This is not obvious from the figures. To help understand this we show the two disks in Figure 8.133 viewed along the direction of this intersection line segment.

Figure 8.129 shows a projection of some of the generating circles for $\mathbb{P}^2$, which may help follow the other figures. One can see the half twist in the two bands of the center slab is due to the flipping motion of the generating circle. Since we have not indicated the height information, our projection K_* serves to illustrate both $\mathbb{P}^2$ and $\mathbb{P}^2_0$ knottings. The direction of the half-twist of the bands will determine which standard projective plane we represent, $\mathbb{P}^2$ or $\mathbb{P}^2_0$, see Remark 6.5. ∎

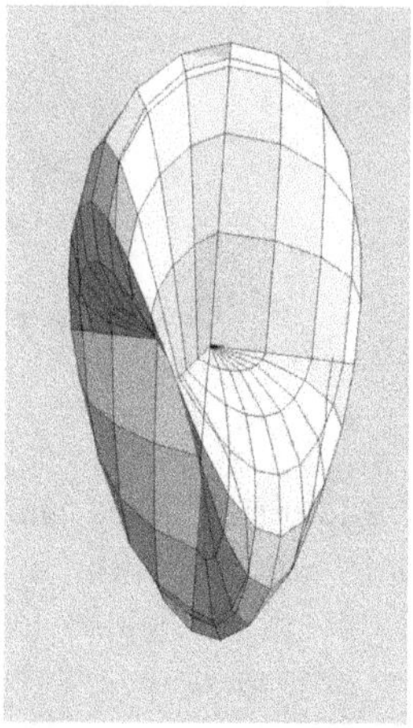

Fig. 8.128 The projection of $\mathbb{P}^2$ is a version of the Roman surface.

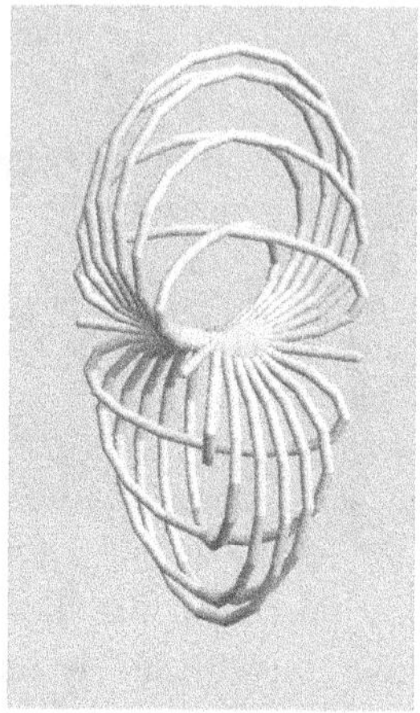

Fig. 8.129 x-projections of some generating circles for $\mathbb{P}^2$.

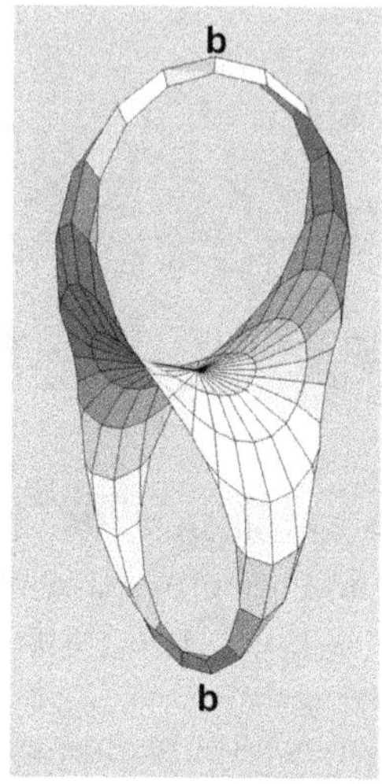

Fig. 8.130 Middle x-slab $\mathbb{P}^2$.

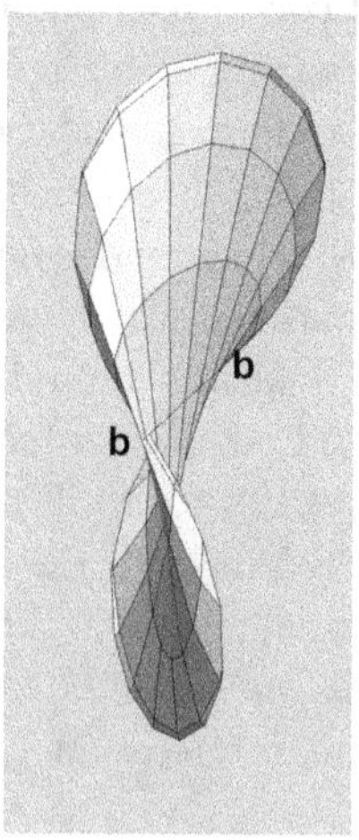

Fig. 8.131 Bottom x-slab $\mathbb{P}^2$.

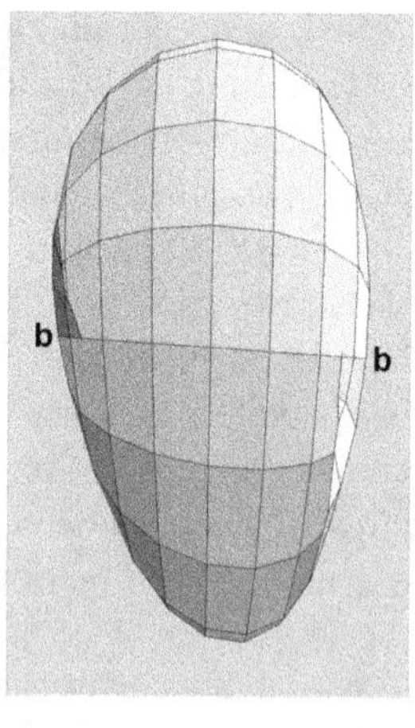

Fig. 8.132 Top x-slab $\mathbb{P}^2$.

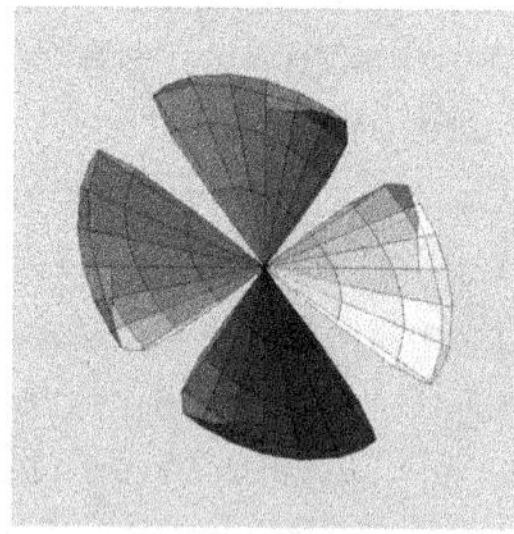

Fig. 8.133 Top and bottom slabs seen from direction of intersection interval.

In a few cases, a small number of critical points limits the knotting possibilities. For example, for a 2-knotting K whose only critical points are one local maximum and one local minimum, K must be a trivial knotting of S^2. In fact, more can be said [Scharlemann (1985)] [Howie and Short (1985)] [Thompson (1987)]: a 2-knotting with only four critical points is equivalent to either an unknotted sphere or an unknotted torus.

By [Bleiler and Scharlemann (1988)] a non-orientable 2-knotting with four critical points is equivalent to one of the standard projective planes.

An important connection, in all dimensions, between Morse functions and handle structures for sections is: A handle of index j, in all dimensions, corresponds to a non-degenerate critical point of index j. In particular:

Proposition 8.2. *Suppose M^n is a smooth manifold, $f\colon R^1$ smooth function, with b a critical value with $f^{-1}(b)$ a single point. Also, suppose we have numbers with $a < b < c$. Then $\widetilde{M}||^c_{-\infty}$ is diffeomorphic to $\widetilde{M}||^a_{-\infty}$ plus a j-handle, whose attaching set is a subset of $\widetilde{M}|_a$.* ∎

Remark 8.9. For any manifold $M^n \subseteq R^k$ we can adjust, if needed, so that the height function is a Morse function. It follows that every n-manifold has a handle structure.

Consider the height function (z-coordinate) of the sphere of Figure 8.134. This function has four critical points: one minimum below the lower plane (that will correspond to a 0-handle) Figure 8.135, two local maxima above the higher plane (those will correspond to two 2-handles) Figure 8.137, and a single saddle point between the two planes (which corresponds to a 1-handle) Figure 8.136. This is equivalent to the handle structure of Example 5.4. ∎

Fig. 8.134 A 2-sphere in R^3 with two slicing planes.

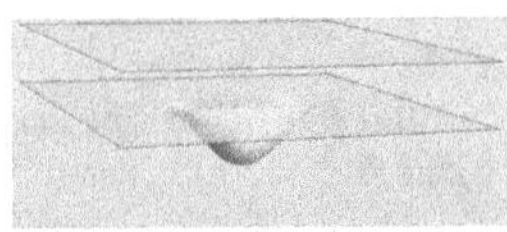

Fig. 8.135 Portion of Figure 8.134 below lower plane gives a 0-handle.

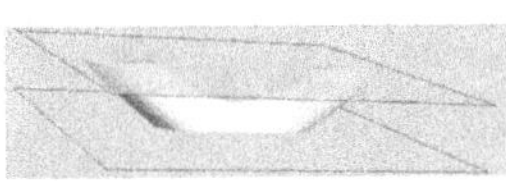

Fig. 8.136 Portion of Figure 8.134 between two planes gives a 1-handle.

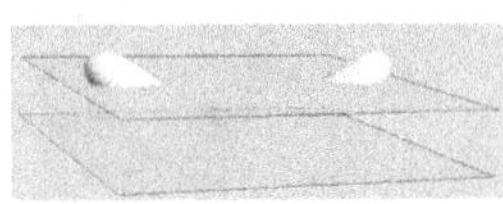

Fig. 8.137 Portion of Figure 8.134 above upper plane gives two 2-handles.

8.2 Morse Theory and Moving Critical Points

Generally, the number and nature of critical points change for a knotting during an isotopy. We explore the possibility of finding isotopies that allow us to control these changes.

A basic technique of Morse theory is called **reordering of critical values** or **reordering of handles**. For a 2-knotting K, the idea is find an equivalent knotting K' so that the number and indices of the critical points of the two are the same, but the set of critical *values* of these points are altered in a controlled way. This is frequently used in conjunction with a second technique called **handle cancellation**, which reduces the number of handles in a decomposition.

That isotopies can change the relative heights of critical points is not remarkable. Consider the isotopic surfaces in R^3 shown in Figure 8.138. These are clearly isotopic in R^3 and all have three local minima and two saddle points. This isotopy changes the relative heights of both the local minima and the saddle points. It is interesting to view the level contours of these three positions in Figure 8.139. Even though the surface has changed little, the contour level patterns are quite different. The isotopies we will need are more complex than this example.

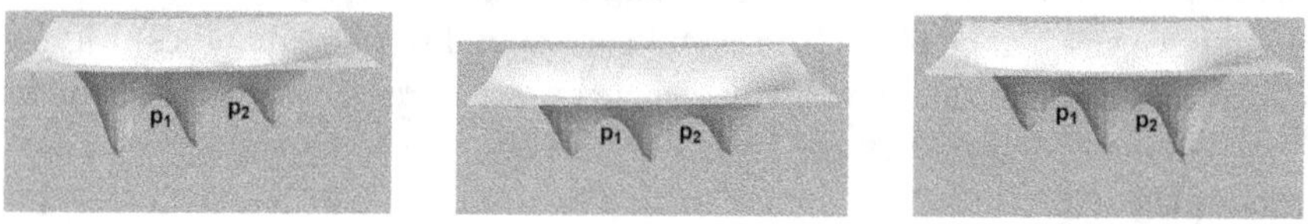

(a) (b) (c)

Fig. 8.138 A surface in R^3 with three local minima and two critical points of index 1 at p_1 and p_2. (a) critical values increase left to right. (b) critical values same for points of same index. (c) critical values decrease left to right.

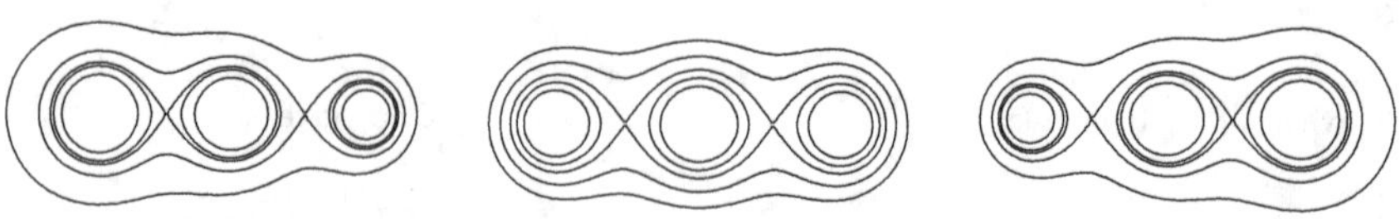

Fig. 8.139 Contour curves corresponding to Figure 8.138.

Remark 8.10. Suppose we have a section of a knotting $K|_a^b$ that contains one critical c point of index 1 and critical value A. Using our PL local model we can see how we can find a new knotting that changes the value of the critical point to any value A' with $a < A' < b$ without introducing any other critical points and not moving points on the slices $K|a$ and $K|b$. Here, we change the height of the critical rectangle while shrinking or expanding vertical rectangles appropriately, see Figures 8.140 and 8.141. Figures 8.88 and 8.90 illustrates a lowering of a saddle point from $z = 2$ to $z = 1$. ▮

Fig. 8.140 Lowering the critical rectangle from Figure 8.141.

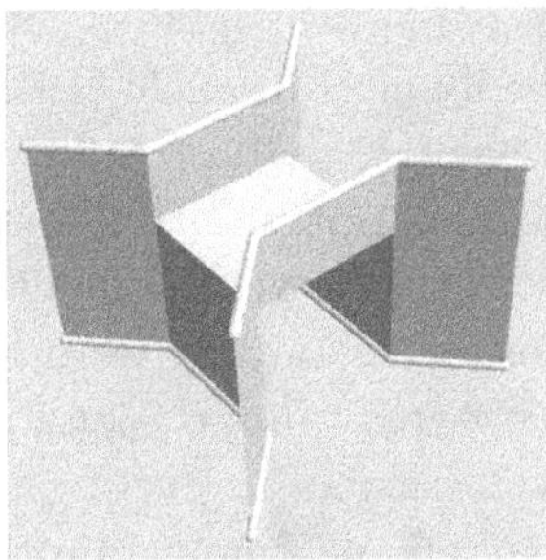

Fig. 8.141 Raising the critical rectangle from Figure 8.140.

The basic lemma that allows us to change the heights of some critical points *without introducing new ones* by an isotopy is Proposition 8.3. If we had a critical point of index 0 in K that was higher than a saddle or local maximum, we could pull it down below the level of these singularities and not introduce any new singularities. Also for critical points of index 0, we could find an isotopy that would switch the two critical values.

Prop 8.3. Suppose K is a knotting in general position with respect to slicing and $a \in K$ is a critical point of index i_a of the height function with critical value A, $b \in K$ is a critical point of index i_b of the height function with critical value B. Furthermore suppose $i_a \leq i_b$ and $B \leq A$. Then there exist open subsets of K, U_a and U_b and a smooth isotopy F of $K \subseteq R^4$ such that:

(1) F is the identity on $K - (U_a \cup U_b)$.
(2) $F(U_a)$ contains only one critical point a of index i_a and value A'
(3) $F(U_b)$ contains only one critical point b of index i_b and value B'
(4) $A' \leq B'$ ▮

The idea of the proof of Proposition 8.3 is similar to the process used for plait presentations of 1-knottings, Section 4.6. Briefly we altered a 1-knotting by an isotopy so that pulled down the local minima raised up the local maxima. We might not be able to pull directly down or raise directly up since we needed to avoid intersecting with other parts of our knotting. However once aware of this problem one could do our pulling/raising with slight perturbations so as to "dodge" such intersections. For our 2-knotting that we can similarly take care of our critical points of index 0 (local minima) and index 2 (local maxima). For critical points of index 1 (saddle points) we can do a similar pulling/raising. Details can be found [Kawauchi (1996)], [Kawauchi *et al.* (1982, 1983)].

In general position for slicing, each critical value corresponds to a distinct critical point. However several authors have defined positions that require many critical points to have a common value.

For any 2-knotting we can scale and translate it so it lies in $R^3 \times [0, 2]$. Next we apply Proposition 8.3. We lower all local minima to level 0, raise all local maxima to level 2 much as we do for a plait form for a 1-knotting, Section 4.6 and finally adjust critical points of index 1 to have value 1.

Definition 8.8. A 2-knotting is in **self-indexed position** (also called **hyperbolic splitting** [Yoshikawa (1994)]) if the height function is Morse and critical points of index i have critical value i. ∎

A version of Proposition 8.4 is found in many places such as [Lomonaco (1981)], [Kawauchi *et al.* (1982)] and [Yoshikawa (1994)]. Nice short proofs for an extended version known as **normal form** are in [Kawauchi (1996)] and [Kamada (2002)].

Prop 8.4. Any 2-knotting K is isotopic to a self-indexed K'. ∎

A 2-knotting K in self-indexed position, K yields a single diagram (with slashes) of the $w = 1$ slice, such as Figure 8.101, which captures the corresponding knotting. Because all saddle points have critical value 1, the only interesting slice is at $w = 1$, denoted $\widetilde{K}|_1$. All other non-empty slices are split trivial links. In [Yoshikawa (1994)] such a diagram is just called a diagram for the knotting. We prefer to call it a **Yoshikawa diagram**.

Remark 8.11. A Yoshikawa diagram looks like a 1-knotting diagram where some crossings have been replaced by slashed crossings, see Definition 8.5. The *ch*-index is the number of crossings plus the number of slashes. For any integer n there are only a finite number of 2-knots expressible as Yoshikawa diagrams with cf-index less than n. In principle, one can use this as a basis for constructing a table of all 2-knottings. It is tedious but possible to enumerate all Yoshikawa diagrams of index n or less that are distinct as labeled planar subsets, and then determine which diagrams represent isotopic knottings. This is the hard part of the task, but for small *ch*-index, it is possible. A list of all distinct diagrams of *ch*-index 10 or less is in [Yoshikawa (1994)]. This table of knots has been reproduced in several books [Kawauchi (1996)], [Kamada (2017)] and some corrections of algebraic calculations noted in [Ishii *et al.* (2018)].

Figures 8.142 and 8.143 represent the two standard embeddings of the projective plane. ∎

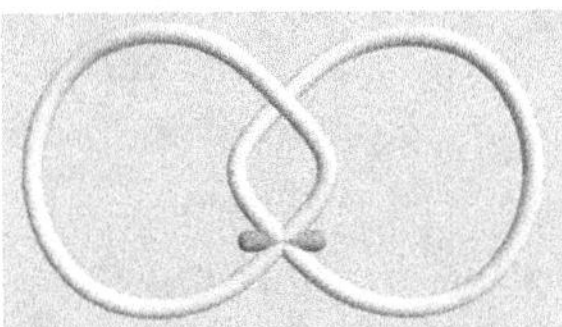

Fig. 8.142 Yoshikawa diagram for negative standard $\mathbb{P}^2_0$.

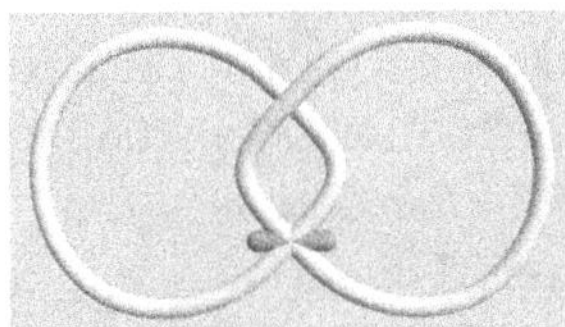

Fig. 8.143 Yoshikawa diagram for positive standard $\mathbb{P}^2$.

Example 8.8. If we take a diagram for sphere-torus link of Example 6.27 and we push one singularity up and one down as indicated for Figure 8.144 one obtains Figure 8.145, the slice linking shown in Figure 6.148. ∎

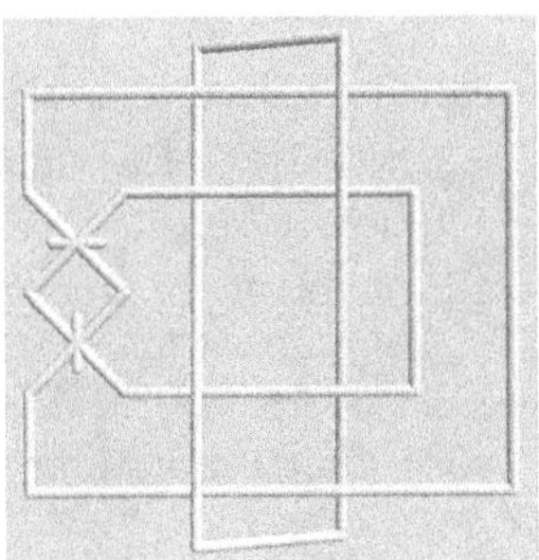

Fig. 8.144 Figure for link $10_2^{0,1}$ of Yoshikawa's table.

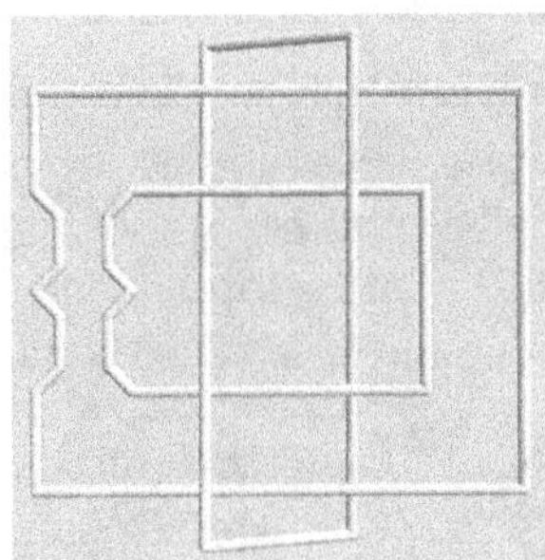

Fig. 8.145 Diagram for Figure 8.144, heights of singularities changed.

8.3 Slice Knots and Knot Cobordism

Slices of 2-knottings help understand 1-knottings. After all, our fundamental relation between knottings is isotopy. By definition, an isotopy K_t of a circle is a set of slices of the trace, which is a knotting of $S^1 \times I \subseteq R^3 \times I \approx R^4$.

In **knot cobordism** theory, or sometimes **knot concordance** theory, one studies n-knottings that are related by being consecutive slices of $(n+1)$-knottings. The study began in [Fox and Milnor (1966)]. A good survey is [Livingston (2005)]. However, this large, important and interesting topic involves algebra beyond our intended scope.

As usual, all knottings and isotopies are to be considered smooth unless stated otherwise and the next definitions adhere to this. In later remarks, we briefly discuss the important matter of what happens if we consider removing "smoothness".

Every 1-knotting k is a slice of some 2-knotting K. Suppose $\kappa \subseteq R^3$ is a knotting with Seifert surface $\Sigma \subseteq R^3$ for κ, then κ is a slice of $D(\Sigma) = \partial(\Sigma \times D^1)$. Our slice is by $R^3 \subseteq R^4$ where we have $\Sigma \times D^1 \subseteq R^3 \times D^1 \approx R^4$. However the question of whether k is a slice of a *particular* 2-knotting of some given 2-manifold M leads in interesting mathematical directions. For example, a 1-knotting which is a slice of a trivial 2-knotting, called **doubly-null-cobordant**, is a class often considered as in [Terasaka and Hosokawa (1961)] and [Sumners (1971)]. Concerning slices of non-spheres one sometimes considers other manifolds. The **slice genus** is the minimum genus of surface M that k can be a slice of a 2-knotting of M, [Rudolph (1983)] [Lewark and McCoy (2019)] [Gilmer (1982)] [Florens and Gilmer (2003)] [Dye *et al.* (2017)].

Definition 8.9. A 1-knotting $\kappa \subseteq R^3 \subseteq R^4$ is called a **slice knotting** if it is a slice of a 2-knotting $K \subseteq R^4$ where K is a knotting of a union of disjoint 2-spheres. A knot represented by a slice knotting of a single circle is called a **slice knot**; a knot represented by a slice knotting of more than one component is called a **slice link.** ▮

So, a 1-knotting k of S^1 in $R^3 \subseteq R^4$ is a slice knotting if there is a proper knotting K of D^2 in H^4_+ such that $K \cap R^3 = k$.

We will provide many examples of slice knots in Chapter 9. Slice knots have some distinctive properties in terms of algebraic topology. Using algebraic topology one can show that a trefoil knot is not a slice knot, for example [Fox (1962a)]. Being a slice knot is an uncommon property. The only 7 knots in the Reidemeister 1-knotting table that are slice knots are: 6.1,8.8, 8.9, 8.20, 9.27, 9.41 and 9.46.

If we are considering a knotting k of S^1 in S^3, rather than R^4, we identify $S^3 = \partial D^4$ and say k is a **slice knotting** if there is a proper knotting K of D^2 in D^4 such that $K \cap S^3 = k$.

More generally (also considering higher dimensions) we consider:

Definition 8.10. Suppose K and K' are two n-knottings of S^n in S^{n+2}. A **cobordism between** K **and** K' is a proper knotting M of $S^n \times I$ in $S^{n+2} \times I$ such that $M \cap (S^{n+2} \times \{0\})$ corresponds to K and $M \cap (S^{n+2} \times \{1\})$ corresponds to K'. In this case, we say K **and** K' **are cobordant knottings.** ∎

With this definition, we can say a knotting is slice if and only if it is cobordant to a trivial knotting. The trace of a isotopy of S^n in S^{n+2} is a *level preserving* proper submanifold of $S^n \times I$ in $S^{n+2} \times I$. So, a cobordism is what we get if we drop the level preserving condition for isotopy.

The study of slice knots and cobordisms becomes very different if we remove "smoothness" from our definitions.

If we take any locally flat 1-knotting κ of a circle the suspension $K = \Sigma(\kappa)$ will be a locally flat knotting of a 2-sphere whose intersection with a hyperplane is κ. Similarly, any PL knotting of a circle is a slice of the suspension of that knotting.

We can make similar definitions for slicing and cobordism for topological, PL or locally flat submanifolds. For example

Definition 8.11. Suppose K and K' are two n-knottings of S^n in S^{n+2}. A **topological cobordism between** K **and** K' is a locally flat knotting M of $S^n \times I$ in $S^{n+2} \times I$ such that $M \cap (S^{n+2} \times \{0\})$ corresponds to K and $M \cap (S^{n+2} \times \{1\})$ corresponds to K'. In this case, we say K **and** K' **are topologically cobordant knottings**. A locally flat knotting that is locally cobordant to a trivial knotting is called a **topologically slice knot.** ∎

In the early 1980's it was discovered that topological and smooth cobordism are not the same.

The first examples of knottings that are topologically slice but not smoothly slice were first given in [Freedman (1982)] with a corrected and simplified proof found in [Garoufalidis and Teichner (2004)]. In fact, given an example, one can construct exotic R^4 (a manifold homeomorphic to R^4 but not diffeomorphic to R^4), [Gompf and Stipsicz (1999)]. An example of a topological slice knot which is not smoothly slice is a companion of a right-hand trefoil called the positive Whitehead double. Interested readers can find much information in: [Freedman and Quinn (1990)], [Livingston (2005)], [Feller and McCoy (2016)], [Piccirillo (2020)].

Chapter 9

Ribbon Constructions for Surface Knottings

Ribbon 2-knotting are easiest 2-knottings to understand since they are very similar to ribbon 1-knottings. A 2-ribbon is a type of immersed 2-manifold in R^3 that can be used to produce certain 2-knottings. We will show two constructions—*ab*-ribbon knottings and symmetric-ribbon knottings. The boundary of a 2-ribbon is a 1-knotting called a **ribbon 1-knotting**. Every ribbon 1-knotting is a slice knotting. A long-standing unresolved question is whether every slice knotting is a ribbon knotting. In higher dimensions an n-ribbon is often restricted to immersions of n-disks rather than more general n-manifolds. We will use the terms 2-ribbon and ribbon 1-knotting rather than (as in [Fox (1962b)]) "ribbon" and "ribbon knotting".

Not all 2-knots are 2-ribbon knots. Any 2-ribbon knot can be represented by a knotting whose projection has no triple points. However, there are many 2-knots that do not have this property, such as Example [Satoh and Shima (2005)]. Such knots are sometimes called **pseudo-ribbon knots**, Remark 6.1.

Every *orientable* 2-manifold M^2 can be written as a double, DN^2, of a 2-manifold N^2 where $\partial N^2 \neq \emptyset$, see Remark 2.14. The 2-knottings we obtain from an immersion of N^2 are knottings of $D(N^2)$. So, S^2 is a double of D^2, and a ribbon 2-disk gives a knotting of S^2. The 2-torus is a double of $S^1 \times D^1$, so we get 2-torus knottings from an annular ribbon surface.

Definition 9.1. Suppose $F\colon N^2 \to R^3$ is a smooth immersion of a 2-manifold such that each component has a non-empty boundary and

(1) $F|_{\partial N^2}$ is a smooth embedding;
(2) F is at most 2-to 1;
(3) the self-intersection set $\Sigma(F)$ is a finite disjoint union of arcs α_i. Each α_i called **ribbon self-intersection**, **ribbon singularity** or **ribbon crossing**;

(4) F is self-transverse along each α_i;
(5) For each i, $F^{-1}(\alpha_i)$ is a disjoint union of two arcs P_i and Q_i;
(6) P_i is a proper sub-arc of N^2 and is called a **slice** or **ribbon slice**;
(7) Q_i is contained in the interior of N^2 and is called a **slot** or **ribbon slot**.

Then $F(N^2)$ is called a **2-ribbon** or **ribbon surface**, the 1-knotting $F(\partial N^2)$ is called a **ribbon 1-knotting**, and the knot-type is called a **ribbon 1-knot**. If $N^2 = D^2$, then $F(N^2)$ is called a **ribbon 2-disk**. The manifold N^2, together with the labeled slices and slots, is called a **slot-slice diagram of the 2-ribbon.** ▮

Definition 9.1 has a natural extension to higher dimensions Definition 9.2 in several different ways, Section 12.4. The next definition is the most common extension. Later in this section, we use 3-ribbons of this kind.

Definition 9.2. Suppose $F\colon D^n \to R^{n+1}$ is a smooth immersion of a disk such that:

(1) $F|_{\partial D^n}$ is a smooth embedding;
(2) F is at most 2-to 1;
(3) the self-intersection set $\Sigma(F)$ is a finite disjoint union of $(n-1)$-disks α_i. Each α_i called **ribbon self-intersection**, **ribbon singularity** or **ribbon crossing**;
(4) F is self-transverse along each α_i;
(5) For each i, $F^{-1}(\alpha_i)$ is a disjoint union of two $(n-1)$-disks P_i and Q_i;
(6) P_i is a proper 2-disk of D^n and is called a **ribbon slice**;
(7) Q_i is contained in the interior of D^n and is called a **ribbon slot**.

Then $F(D^n)$ is called an n-**ribbon**, and D^n, together with the labeled slices and slots, is called a **slot-slice diagram of the ribbon.** ▮

A ribbon is not a submanifold of R^{n+1} but it is a subset of R^{n+1}. The equivalence of ribbons is just equivalence as subsets:

Definition 9.3. If R_1 and R_2 are n-ribbons in R^{n+1} they are **equivalent ribbons** if there is an orientation preserving diffeomorphism h of R^{n+1} such that $h(R_1) = R_2$. ▮

Definition 9.1 Conditions (4), (5), (6) and (7) imply that the self-intersections of a 2-ribbon all locally look like Figure 9.1, called the **local model of a 2-ribbon crossing**. In detail, define two rectangles in

R^3: $R = \{(x,y,z)\colon |x| \leq 1, |y| \leq 1 \text{ and } z = 0\}$ and $r = \{(x,y,z)\colon |y| \leq 1/2, |z| \leq 1 \text{ and } x = 0\}$. Additionally, suppose A is the two edges of R such that $y = \pm 1$ and a the edges of r with $y = \pm 1/2$. We want these four edges to correspond to the boundary of the 2-ribbon.

Example 9.1. The ribbon surface shown in Figures 9.2 has one ribbon crossing and a corresponding ribbon 1-knotting of three circles as shown in 9.3. ▮

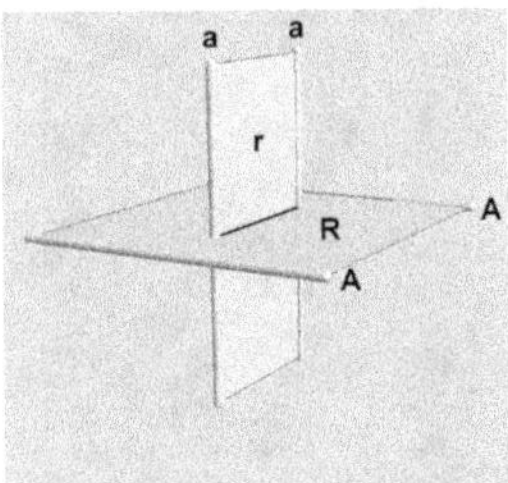

Fig. 9.1 Local model of ribbon crossing.

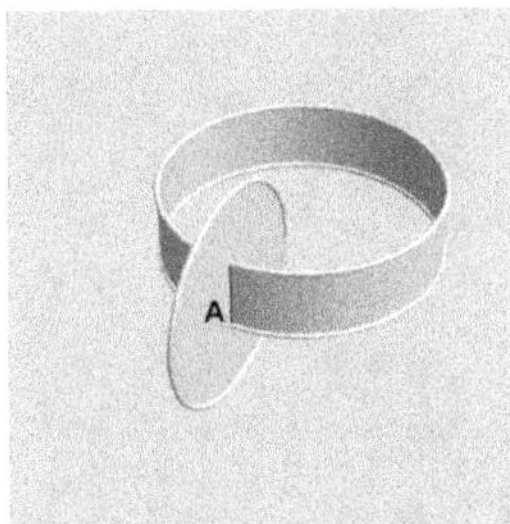

Fig. 9.2 The ribbon surface with one ribbon self-intersection at A, see Definition 9.1.

Fig. 9.3 The edge of the ribbon surface of Figure 9.2 is a non-trivial knotting of three circles.

9.1 Constructing 2-Knottings from 2-Ribbons

We present two constructions that convert 2-ribbons to 2-knottings: *ab*-ribbon knottings and symmetric ribbon knotting Definition 9.4. A ribbon surface $F\colon N^2 \to R^3$ can be "thickened a bit" in R^3 to get an immersion $\phi\colon N^2 \times D^1 \to R^3$ so that $\phi|_{N^2\times\{0\}} = F$. This immersed 3-manifold is called a is a **tubular neighborhood of the immersion** F. (This thickening can be obtained as a union of short line segments normal to the immersed surface where we shrink length near the ribbon boundary.) Now $\partial(N^2 \times D^1) \approx D(N^2)$. Then $\phi|_{\partial(N^2\times D^1)}$ a smooth immersion of $D(N^2)$. We can control this process by making the thickening "not too big", so that the self-intersections of $D(F)$ will be disjoint circles.

We could then proceed as in Section 6.7 with above/below decisions at each self-intersection circle to produce *ab*-knottings $K \subseteq R^4$, see Definition 6.16. Since the projection $K_* = F|_{\partial(N^2\times D^1)}$. We will call these knottings ***ab*-ribbon knottings** for R.

Example 9.2. Figure 9.4 is a more interesting ribbon 2-disk with two ribbon singularities A_1 and A_2, slot-slice diagram shown in Figure 9.5 and ribbon 1-knotting shown in Figures 9.7 and 9.8. A thickening of the ribbon of Figure 9.4 might look like Figure 9.6. This immersed sphere in R^3, whose self-intersections are disjoint circles, is basically the same as in Figure 6.87. In Section 6.7 we have discussed in detail knottings obtainable from this immersion considered as a projection. ▌

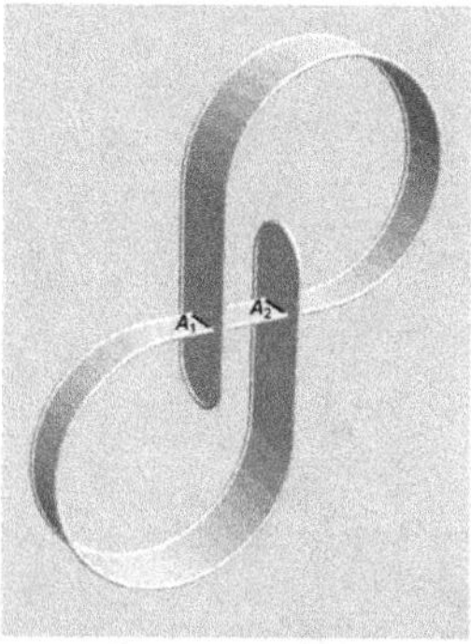

Fig. 9.4 A ribbon with self-intersections at A_1 and A_2.

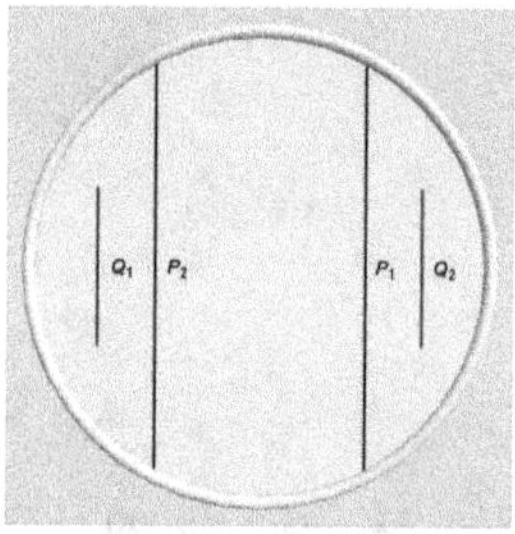

Fig. 9.5 The two slots and two slices for the ribbon of Figure 9.4.

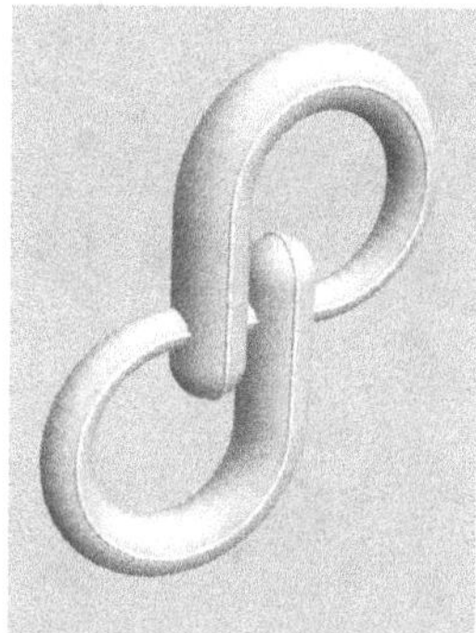

Fig. 9.6 A thickening of ribbon of Figure 9.4.

Fig. 9.7 The ribbon knotting corresponding to the ribbon shown in Figure 9.4.

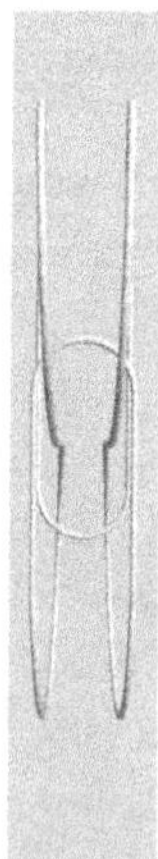

Fig. 9.8 The knotting of Figure 9.7 viewed from different point showing it is "square knotting".

Example 9.3. Using the ribbon surface (a disk and an annulus) shown in Figure 9.3 the immersion we would get is a union of an embedded sphere and an embedded torus whose intersection is a pair of circles just like the immersion in Figure 6.63 which is the projection of either a split trivial knotting or the basic sphere-torus knotting. ∎

We next describe another knotting obtained from our ribbon surface that has interesting slices (but will not be in general position with respect to projection).

We construct a 2-knotting of $D(N^2)$ whose slice is ∂R. Concrete construction details will follow in Remarks 9.1 and 9.2. Briefly, one views F as an immersion into R^4 which is then altered. First with ∂R fixed push the other points of $F(N^2)$ straight upwards into R^4 pushing a little further near the slice compared to the slot. This results in a proper knotting, call it N_+, of N^2 in R^4_+. Symmetrically, we could push $F(N^2)$ downwards and get a proper knotting, call it N_-, of N^2 in R^4_-. Our 2-knotting of $D(N^2)$ is now $N_+ \cup N_-$ and will be called a **symmetric ribbon 2-knotting**. Such a 2-knotting is far from uniquely determined, since there are lots of ways to "push". A formal definition follows, Definition 9.4,

For those who wish, an outline of a concrete construction for a symmetric ribbon knotting is given in the next two Remarks. It might be a good idea for the reader to scan Examples 9.4 and 9.5 so as to have in mind some examples of what the construction might look like for several different 2-ribbons. In Remark 9.1, we show how two intersecting rectangles in R^3 such as in Figure 9.9 give rise to disjoint annuli in R^4 whose projections are like Figure 9.13. We complete the construction in Remark 9.2.

Remark 9.1. View each rectangle as a union of line segments parallel to y-axis. A plane $H^2_x(a)$ given by $x = a$ meets R in line segment L_a and shorter line segment l_a in r, Figure 9.9. At the intersection, l_a is a subinterval of L_a: l_a corresponds to a slot while L_a corresponds to a slice. Points on L_a have constant z-coordinate Z_a; points on l_a have constant z-coordinate z_a.

Each plane $H^2_x(a)$ is a projection of a 3-dimensional hyperplane $H^3_x(a) \subseteq R^4$. In each $H^3_x(a)$ we construct two circles. Circle $C(a)$ has diameter equal to the length L_a and lies in a plane parallel to the yz-coordinate plane at w-height $w = Z_a$. Circle $c(a)$ has diameter equal to the length l_a and lies in a plane parallel to the wy-coordinate plane at w-height $w = z_a$. In Figure 9.12 we show these circles in three hyperplanes corresponding to Figure 9.9. Figure 9.10 shows the standard projection of nine pairs of circles. The projection of all of them is a pair intersecting cylinders intersecting in a pair of circles, as in Figure 9.13.

Note that Figure 9.14 shows the projection of the portion of the cylinders that lie in R^4_+—is a local view of a ribbon singularity—topologically two 2-disks intersecting in a 1-disk. Similarly for Figure 9.15 and R^4_-. ∎

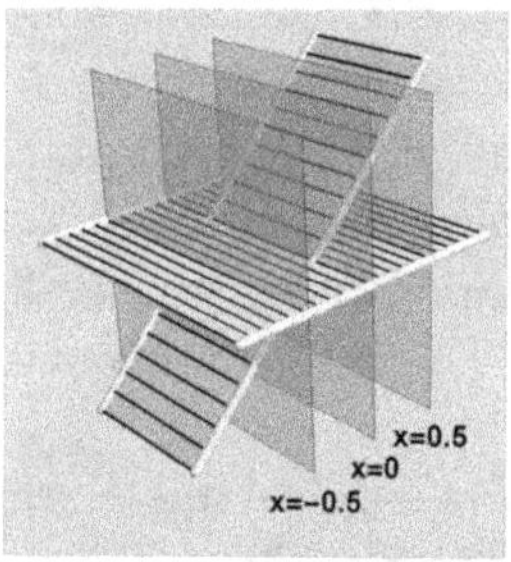

Fig. 9.9 Local model of a ribbon self-intersection.

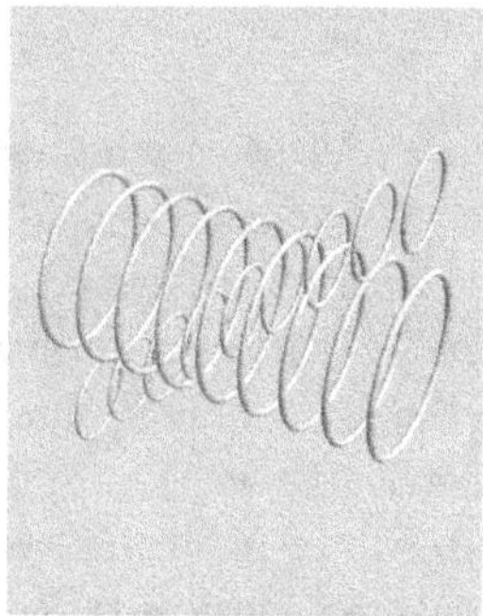

Fig. 9.10 Projections of circle pairs from nine hyperplanes.

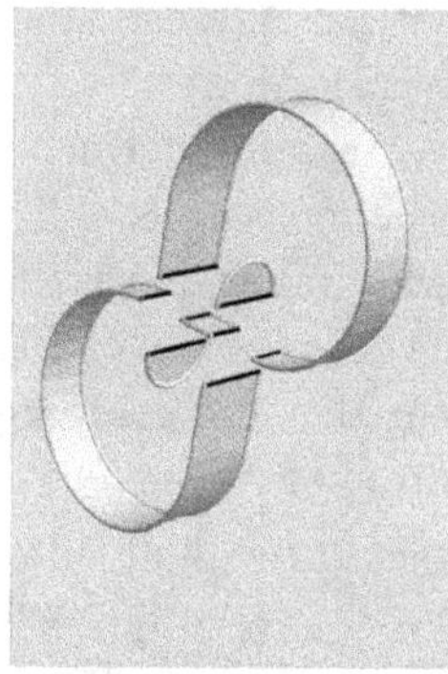

Fig. 9.11 Ribbon with crossing models removed.

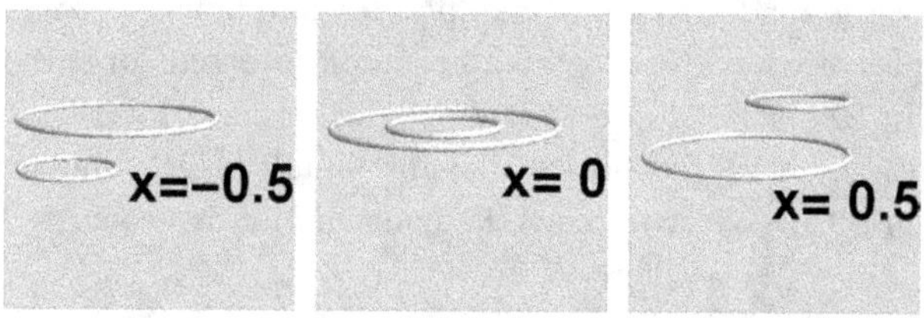

Fig. 9.12 Above each plane of Figure 9.9 is a three dimensional hyperplane with each containing two circles each of which project to a line segment of the ribbon.

Fig. 9.13 Projection of two annuli corresponding to a ribbon crossing.

Fig. 9.14 Portion in R^4_+ will be part of N^2_+.

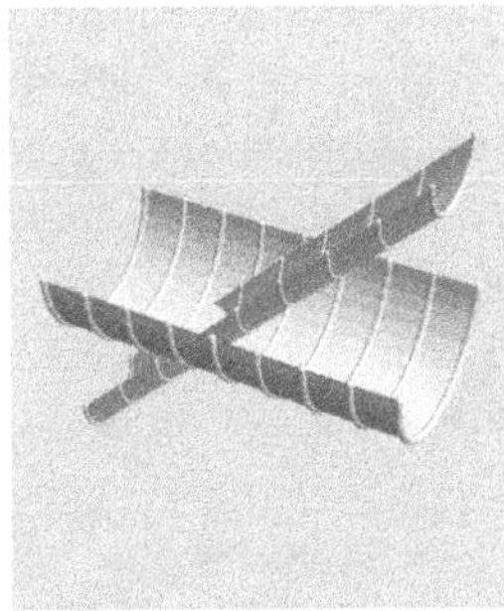

Fig. 9.15 Portion in R^4_- will be part of N^2_-.

Remark 9.2. Using Remark 9.1 we get portions of N_+ and N_- for each ribbon singularity. Let $Q \subseteq N^2$ be the subset corresponding to all rectangles we are using for local model. What remains is to deal with the rest of N^2 which we express as a compact manifold $M = \overline{N^2 - Q}$. Figure 9.11 shows M for the ribbon of Figure 9.4. Write boundary $\partial M = X \cup Y$ where X is a union of disjoint sub-arcs of ∂R shown as thin tubes and Y is a union of disjoint sub-arcs s_i shown in black which correspond to as semicircles S_i in R^4_+ in the construction of Remark 9.1.

We want an embedding M_+ of M into R^4_+ whose projection is M. In other words M_+ will be the graph $g : M \to R^1$ where g gives the w-coordinate. We know exactly what we want for w-coordinates on ∂M_+. Points of Y will be points of ∂M_+ so that when we slice our new 2-knotting we will get ∂R. So the w-coordinate of such a point will be 0. However, for points of X we must match up with our progress from Remark 9.1 so we want to use the semicircle S_i for the w-coordinate for ∂M_+ along s_i. So if we let $G = g(\partial M)$ we will need $G = \partial M_+$.

There are many ways proceed to get this M_+, so what follows is only one way to do this. By the Collar Neighborhood Theorem, Proposition 2.9 we can write $M = M_0 \underset{\partial M_0 = \partial M \times \{0\}}{+} (\partial M \times I)$ or more briefly $M = M_0 \cup C$ where $C \approx \partial M \times I$. Let $M_1 = M_0 \times \{1\} \subseteq R^3 \times R^1_+ \approx R^4_+$, we view M_1 as the graph of the constant function $g(x) = 1$ defined on M_0.

Consider a special case where M lies in the xy coordinate plane. View the z projection of our construction so that our subsets are in xyw-space.

Figure 9.16 shows example of $M = M_0 \cup C$ in xyw-space as be subset in the xy plane. Figure 9.17 shows G and M_1.

We finish by defining g on C, which agrees on one end with G and ∂M_1 on the other end, Figure 9.18. Define: $g\colon \partial M \times I$ by $g(x,t) = g(x)(1-t)+t$. Then $g|_{\partial M \times \{0\}} = g(x)$ and so $g(\partial M \times \{0\}) = G$. Also $g|_{\partial M \times \{1\}} = 1$ so $g(\partial M \times \{1\}) = \partial M_1$.

Finally, this construction will clearly be piecewise smooth. To obtain a smooth knotting one needs to round the corners. ∎

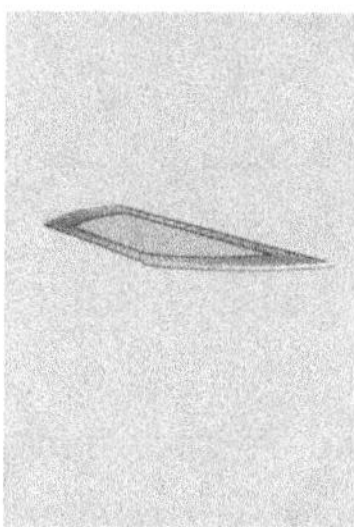

Fig. 9.16 If M was in xy-plane.

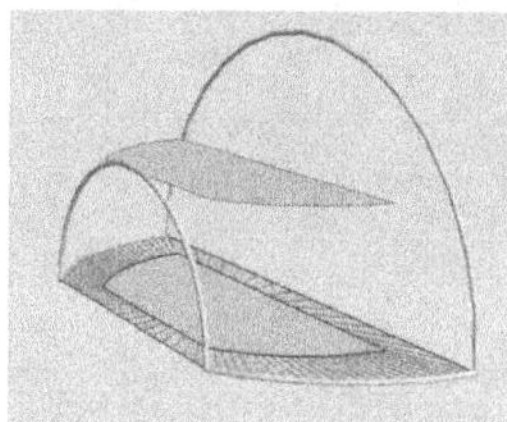

Fig. 9.17 G shown with thin tubes. Also M_1 shown.

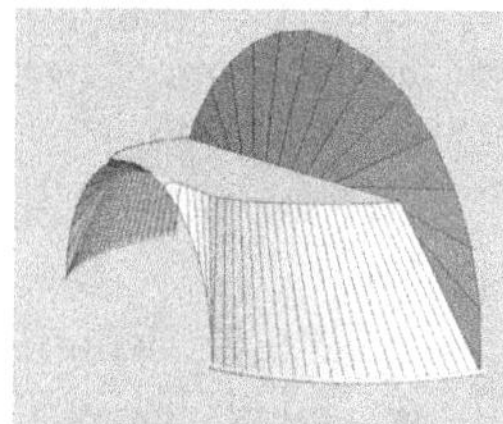

Fig. 9.18 New knotting M_+.

Definition 9.4. Given a ribbon knotting $R \subseteq R^3$ of N^2 a knotting denoted $S(R) = N_+ \cup N_-$ constructed using Remarks 9.1 and 9.2 is called a **symmetric ribbon 2-knotting associated with** R. ∎

For any ribbon R, ∂R is a slice of the associated symmetric ribbon 2-knotting, so we have:

Prop 9.1. If K is a ribbon 1-knotting then K is a slice 1-knotting. ∎

Unresolved as of this writing is the converse, first mentioned in print in [Fox (1962b)] and often called the **slice-ribbon problem**—Is every slice knot of smooth knotting of a sphere in R^4 a ribbon knot? This is known for some classes of knots: [Lisca (2007)] for two bridge knots and [Greene and Jabuka (2011)] for 3-stranded pretzel knots. That is, if a slice knot is either two-bridge or a 3-stranded pretzel then it is also ribbon.

Definition 9.5. A **ribbon 2-knot** is the knot class of a symmetric ribbon 2-knotting. ∎

We next look at some more examples of ribbons.

Example 9.4. A second ribbon with two ribbon self-intersections is shown in Figure 9.19; the boundary of this ribbon is shown in Figure 9.20. The corresponding 1-knot is called the **stevedore's knot**, 4.1 in standard tables. So, we call our 2-ribbon the **stevedore's ribbon** and the symmetric ribbon knot the **stevedore's 2-knot**. Since the stevedore's knot is not equivalent to the square knot, it follows that the square ribbon and the stevedore's ribbon are not equivalent ribbons in R^3, see Definition 9.3.

The difference between the 2-ribbons of Figure 9.4 and Figure 9.19 is (visually) a half-twist. For both of these, the immersed sphere we get as a boundary of a tubular neighborhood is the same—basically like Figure 6.87. The spun trefoil and the stevedore's 2-knot, obtained from these ribbons are not equivalent 2-knots. This is shown in [Fox (1962a)], where the spun trefoil is Example 9 and the stevedore's 2-knot is Example 10. ∎

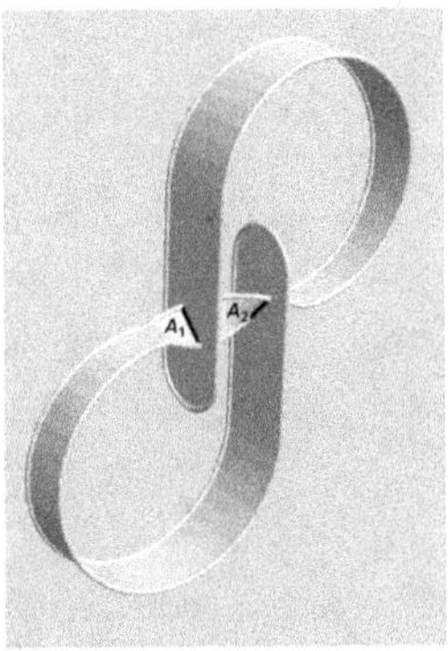

Fig. 9.19 A ribbon for Example 9.4.

Fig. 9.20 The boundary of ribbon of Figure 9.19 is called the stevedore's knot.

Example 9.5. A more complex ribbon Y_0 is shown in Figure 9.21 with boundary ribbon knot Figure 9.22. If we label ribbon crossings as in Figure 9.23 we get a slot-slice diagram Figure 9.24. Figure 9.25 shows an immersion corresponding to this ribbon and Figure 9.26 shows one possibility for an *ab*-construction which happens to differ from the symmetric ribbon 2-knotting. Note this example is discussed in [Kanenobu and Komatsu (2017)] and refers to earlier work in [Yasuda (2009)]. One can generalize Example 9.5 constructing ribbons such as Figure 9.27 with immersed spheres as in Figure 9.28. ∎

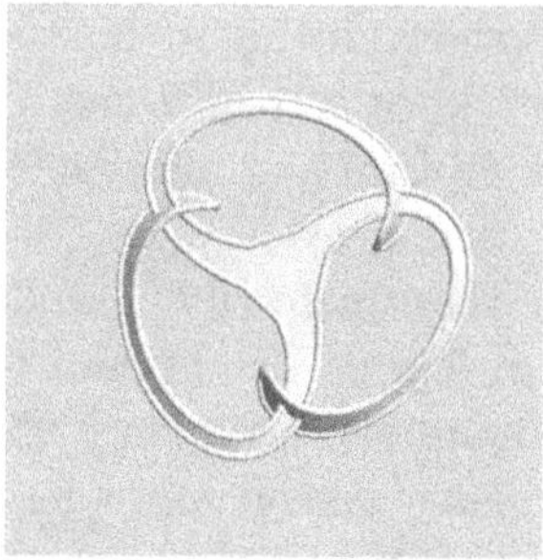

Fig. 9.21 The ribbon R_{Y_0}, see Example 9.5.

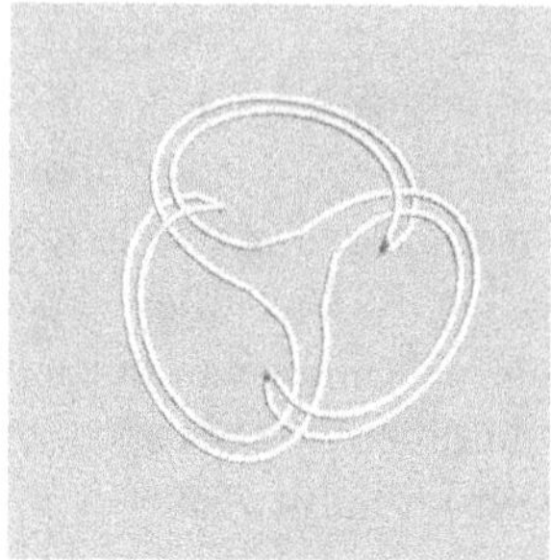

Fig. 9.22 The ribbon knot ∂R_{Y_0}.

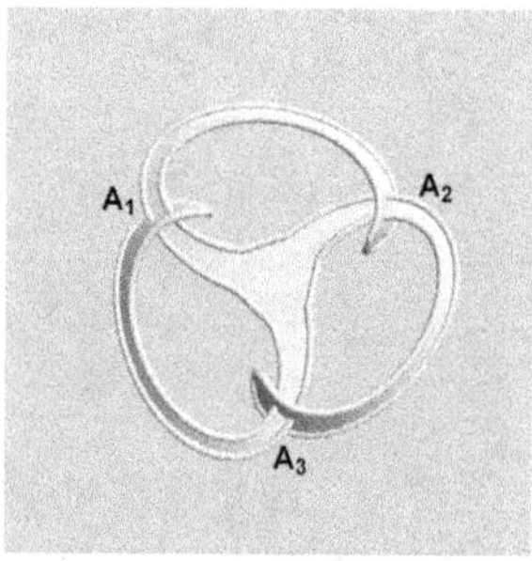

Fig. 9.23 The self-intersections of R_{Y_0} labeled $A1, A_2, A_3$.

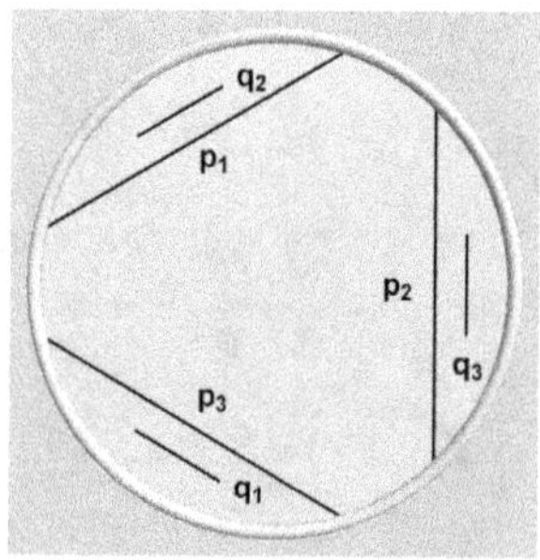

Fig. 9.24 A corresponding slot-slice diagram corresponding to Figure 9.23.

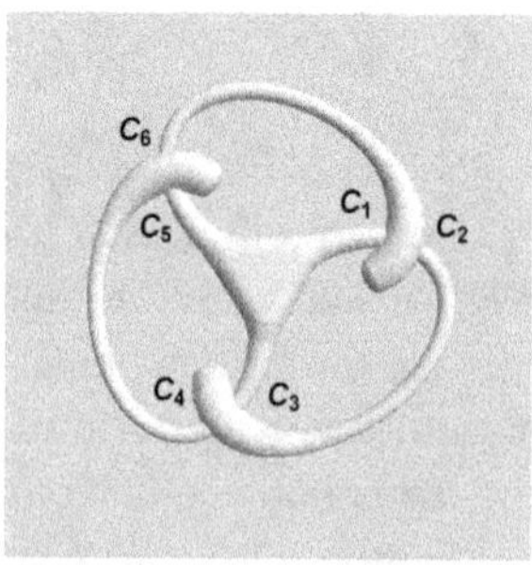

Fig. 9.25 An immersed sphere with six circles of self-intersection $C_1, \ldots, C_6$.

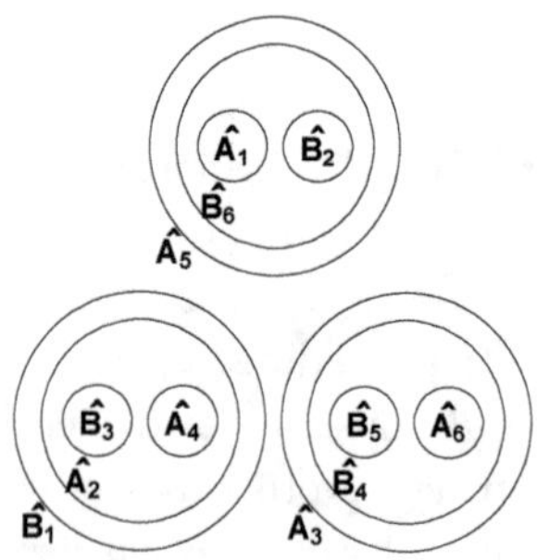

Fig. 9.26 A crossing scheme corresponding to Figure 9.25.

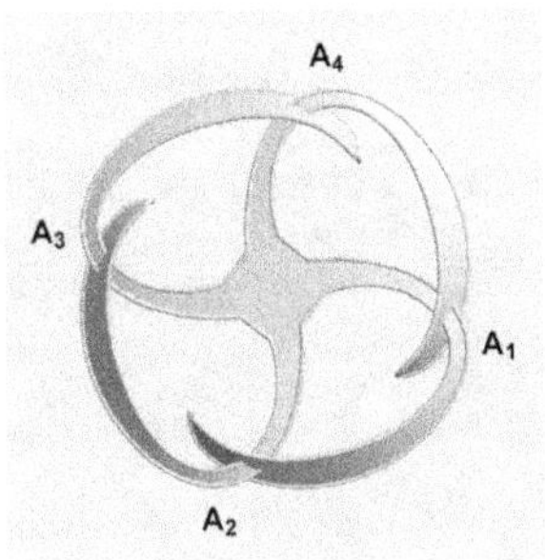

Fig. 9.27 Ribbon R_{X_0} with four arcs of self-intersection A_1, A_2, A_3, A_4.

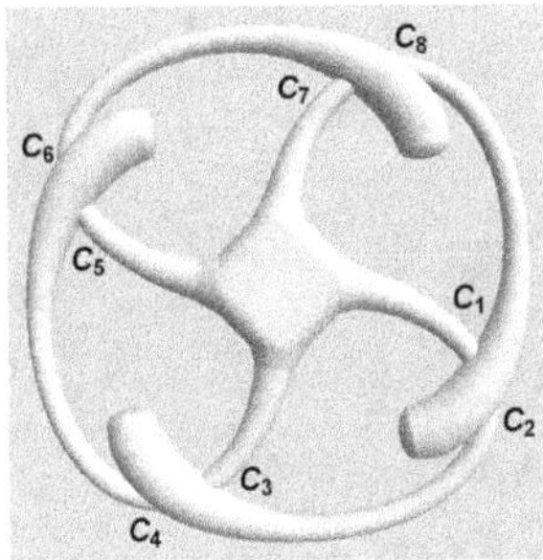

Fig. 9.28 An immersed sphere with eight circles of self-crossing $C_1, \ldots, C_8$.

A general construction of such ribbons and immersed spheres involves branched cyclic covers. We illustrate with the immersion of a sphere, Figure 6.86.

Example 9.6. Figure 9.29 shows an immersed sphere $S \subseteq R^3$ and a rectangle meant to represent a half-plane P. We view R^3 with cylindrical coordinates (r, θ, z) with $-\pi < \theta \leq \pi$ and visualize the z-axis as rising out of the page towards the viewer. The half-plane P has equation $\theta = \pi$ and intersects S in an arc A and a circle C.

The construction of the two-fold cover of S branched along is often describe as follows. In cylindrical coordinates we have half-space $H = \{(r, \theta, z) \colon r \neq 0, pi/2 < \theta < \pi/2\}$ and $R^3 - P$ is homeomorphic to open half-space by halving the angle, $h(r, \theta, z) = (r, \theta/2, z)$. The image $h(S - P)$ will be (non-compact) surface in H. So, we consider $H \subseteq R^3$ and the closure S' of $h(S - P)$ in R^3. The points added in this closure will form the boundary of S'. This boundary consists of three circles:two circles, each corresponding to C, and one circle corresponding to A, as in Figure 9.30. Let S'' be the image of S' under a rotation, in cylindrical coordinates, of π. Then the union $S' \cup S''$ will be our immersed surface which is a two-fold branched cyclic cover of S (Figure 9.31). This is basically the immersion of Figure 9.6 while the three-fold and four-fold covers correspond to Figures 9.26 and 9.28, respectively. ∎

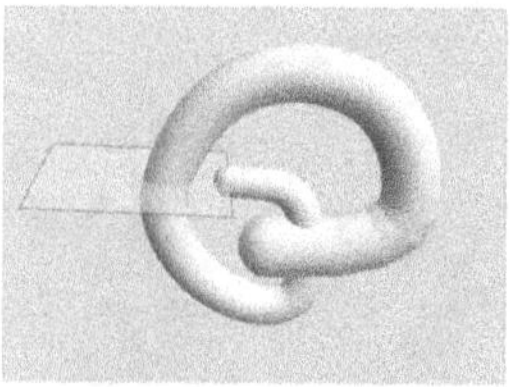

Fig. 9.29 An immersed sphere in $S \subseteq R^3$ and a half-plane $\theta = \pi$.

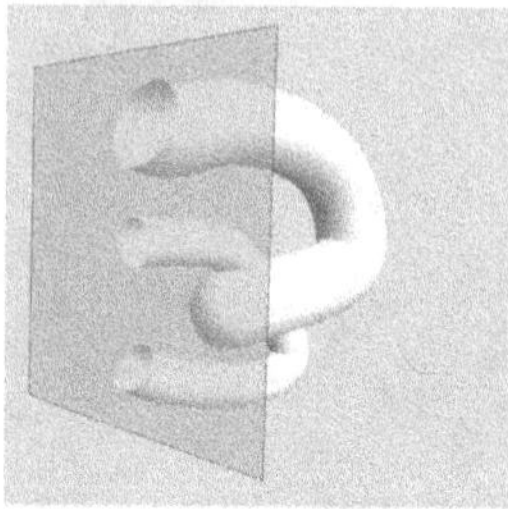

Fig. 9.30 The surface S'.

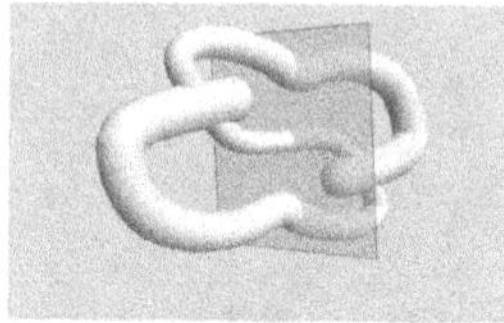

Fig. 9.31 The branched cyclic cover $S' \cup S''$.

9.2 Constructing 2-Ribbons

A style of ribbon construction as in Figures 9.35–9.39 is based on handlebody structure. A connected 2-manifold F with $\partial F \neq \emptyset$ is a 2-dimensional handlebody. For a ribbon $R = F(N^2)$, one can incorporate the slot/slice structure a the handlebody structure to get a useful way of constructing and analyzing ribbons. For example, we can arrange so that all slots lie in the zero handles and the slices all lie in the 1-handles. It is simplest to describe if $N^2 = D^2$ using the fact that a proper arc in D^2 divides D^2 into two sub-disks, two disjoint proper arcs divides D^2 into three sub-disks, etc.

Example 9.7. Consider the 2-ribbon disk $R \subseteq R^3$, Example 9.2. We have two slices P_1 and P_2, Figure 9.5. We express D^2 as "three 0-handles plus two 1-handles. Disjoint tubular neighborhoods H_1 and H_2 of the slices of R will correspond 1-handles, Figures 9.32. The closure of the complements $\overline{D^2 - (H_1 \cup H_2)}$ is a union of three sub-disks of D^2—these are the 0-handles D_1, D_2, D_3. Figure 9.33 shows image of the 1-handles in the ribbon.

The expression of ribbon in terms of the handlebody structure is not unique. Figure 9.34 shows an expression of the same ribbon with only one 1-handle shown in dark gray. ∎

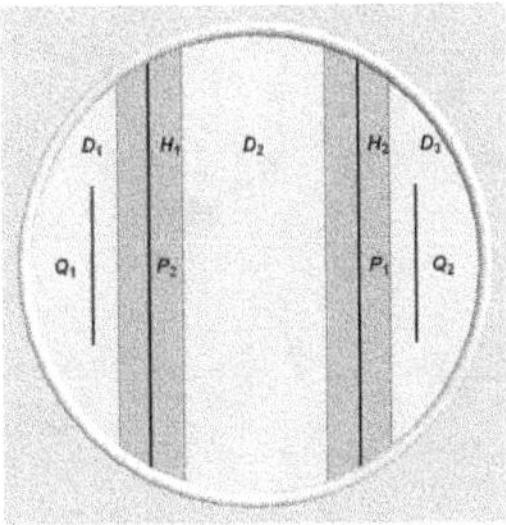

Fig. 9.32 Two disjoint tubular neighborhoods H_1 and H_2 of slices.

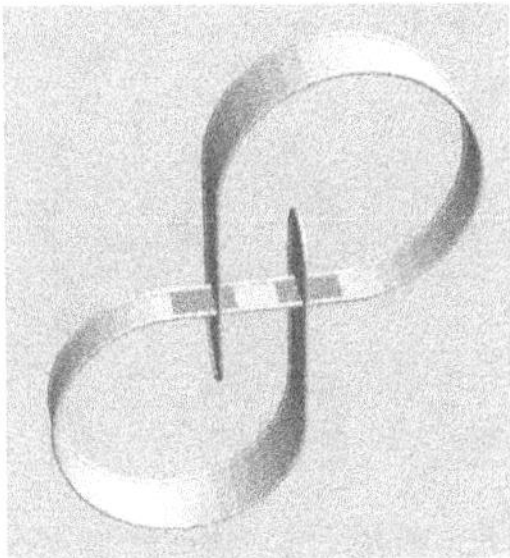

Fig. 9.33 Images of H_1 and H_2 in ribbon.

Fig. 9.34 Another possibility with only one 1-handle.

Example 9.8. So, we can describe any 2-ribbon in R^3 as follows: Take a collection of disjoint 2-disks D_i in R^3 (these will be our 0-handles). Now add in disjoint bands B_j (these will be our 1-handles). For each band, each of the two ends is glued onto a boundary of some 2-disk. Otherwise B_j and only intersect D_i in a ribbon singularity.

Figures 9.35, 9.36, 9.38 and 9.39 are examples of ribbons expressed in this way. Because bands can be twisted and tangled in R^3 there are many non-equivalent 2-ribbons that correspond to a 2-knot. For example, the ribbons of Figures 9.35, 9.38 and 9.39 give rise to a single symmetric ribbon 2-knot. The argument that linking and knotting of bands do not affect the 2-knots we construct is much like the argument that circles unknot and unlink in R^4 discussed in Section 5.1. A separate argument shows that a *full twist* in a band has no effect on the 2-knots produced, although *half-twists* can matter, as we have seen in the stevedore's ribbon, Example 9.4. ∎

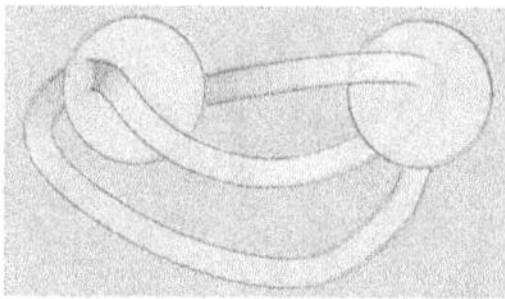

Fig. 9.35 A ribbon with two 0-handles and one 1-handle.

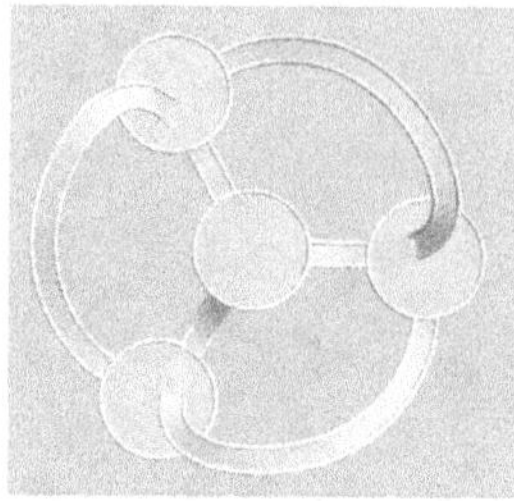

Fig. 9.36 A ribbon with four 0-handles and three 1-handles.

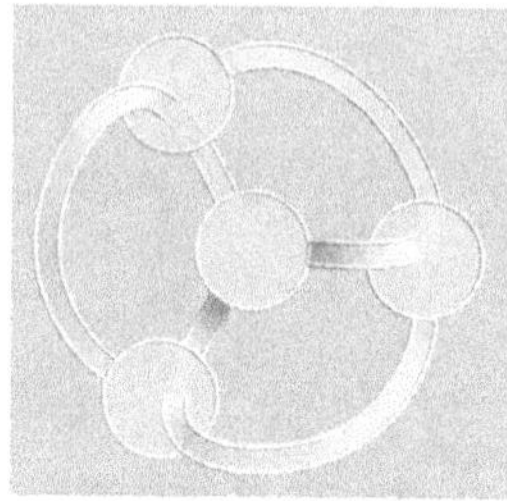

Fig. 9.37 A different ribbon with four 0-handles and three 1-handles.

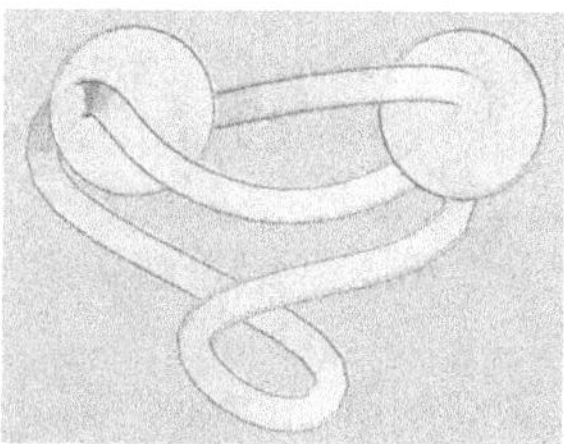

Fig. 9.38 There may be twists in the 1-handles.

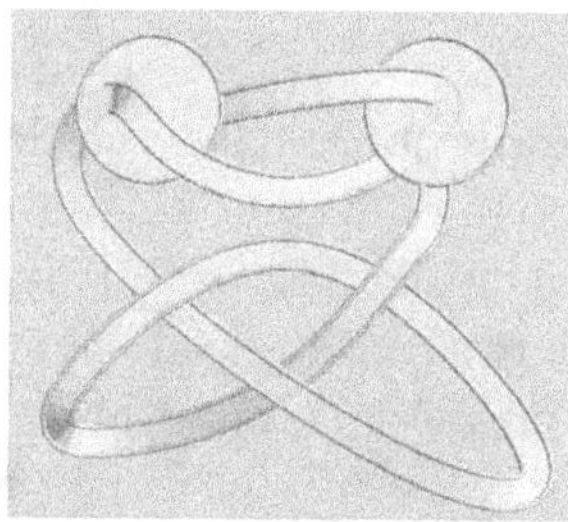

Fig. 9.39 1-handles can be tangled and self-tangled.

Using a handle structure simplifies the process of making a list of simple ribbon knots. We can describe the ribbon with a simple code that captures how the 1-handles intersect the 0-handles that allows one to construct a table of ribbon knots according handle structure and ribbon crossing numbers [Yasuda (2001)], [Yasuda (2018)] and [Kanenobu and Komatsu (2017)].

We can view n-ribbons similarly and express D^n as a handlebody where slots lie in the zero handles and slices lie in the 1-handles. In this description an n-ribbon is a union of disjoint n-disks sometimes called "bases" (corresponding to 0-handles) and disjoint union of connecting "n-pipes" (corresponding to 1-handles) with the disks and pipes intersecting is limited ways. The corresponding $(n-1)$-ribbon knotting is then described boundary of such a ribbon.

For a n-handlebody structure if components of the attaching map of a 1-handle lie on different 0-handles the effect "fuse" the 0-handles. As a consequence, in the literature, a 1-handle is called a **fusion**. In these terms a ribbon is sometimes expressed as bases and fusions. This is called a **ribbon presentation.**

The **fusion number of a ribbon knotting** is the smallest number of fusions of a ribbon R such that $\partial R = \Sigma$. Alternatively we could consider the **base number for a ribbon knotting** which is the smallest number of bases.

A simple method of constructing some ribbons uses a proper 1-knotting $K \subseteq R^3_+$. **Standard projection** $R^3_+ \to R^2$ is the restriction of standard projection; a smooth proper knotting $K \subseteq R^3_+$ **is in general position with respect to projection** if $K - \partial K$ is in general position with respect to the standard projection π. (We exclude endpoints of components of K since properness requires a vertical tangent to $R^2 \in R^3$.)

If K' is a mirror image of K in R^3, then $K \cup K'$ is a ribbon 1-knotting with ribbon R where R is the union of all line segments joining a point of K and its mirror image. A more sophisticated version is called a symmetric union of knots, see [Kinoshita and Terasaka (1957)].

Example 9.9. Figure 9.40 shows a proper arc in R^3_+ which has two crossing points in projection. The mirror ribbon R corresponding to this is shown (with R^2 rendered transparently) in Figure 9.41. Note ∂R represents the square knot. This ribbon was discussed earlier as Example 9.2. Figure 9.42 shows a trefoil knotting in R^3_+ whose projection has three crossings. The mirrored ribbon we produce, see Figures 9.43 and 9.44, is a ribbon annulus. The corresponding symmetric ribbon 2-knotting is a knotting of a 2-torus. ∎

Fig. 9.40 A proper arc in R^3_+.

Fig. 9.41 Mirror ribbon of Figure 9.40 arc.

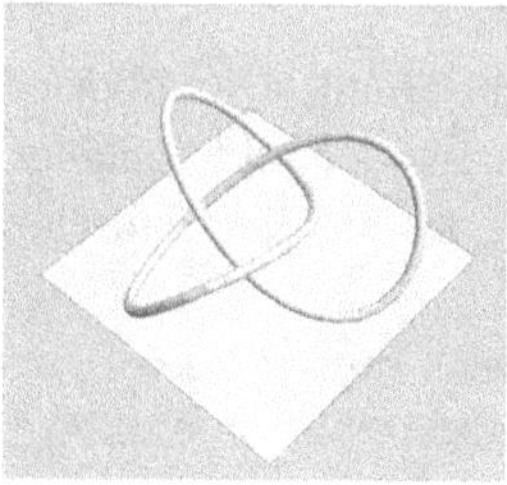

Fig. 9.42 A trefoil knotting in R^3_+.

Fig. 9.43 Mirror annulus ribbon for Figure 9.42.

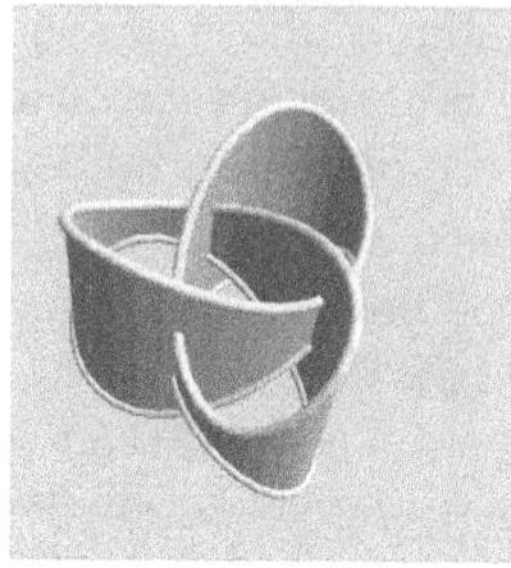

Fig. 9.44 Ribbon of Figure 9.43, without R^2.

More generally we can define:

Definition 9.6. For any $x \in R^n$ write $\mathbb{I}_x$ for the line segment between x and its mirror image in R^n; for any $X \subset R^n$ write $\mathbb{I}_X = \bigcup_{x\in X} \mathbb{I}_x$, we call this the **symmetric extrusion of** X. Suppose M^n is a proper knotting in R^{n+2}_+. **Standard projection** $R^{n+2}_+ \to R^{n+1}$ is the restriction of standard projection A smooth proper knotting $M^n \subseteq R^{n+2}_+$ **is in general position with respect to projection** if $M^n - \partial M^n$ is in general position with respect to the standard projection π. ▮

We get a 3-ribbon R' in a natural way from a 2-ribbon R as the symmetric extrusion of symmetric ribbon 2-knot: $R' = \mathbb{I}_K$ where $K = S(R)$.

If K is a symmetric 2-knotting for ribbon $R = F(N^2)$ then $\mathbb{I}_K$ is a 2-ribbon $R' = F'(N^2 \times D^1)$ where F' sends the interval $x \times D^1$ linearly onto $\mathbb{I}_x$ and $F'(N^2 \times \{1\}) = N_+$. This is an immersion and the $F(\partial(N^2 \times D^1)) = K$ so it is a 2-knotting on the boundary on the image of $\partial(N^2 \times D^1)$.

For a ribbon crossing point where F(p)=F(q) and p is in the slice and q is in the slot, the w coordinate of p will be larger than for q so that $\mathbb{I}_q \subsetneq \mathbb{I}_p$. Since this is the case for all points in a given crossing of K, we conclude that the self-intersections of $F'(N^2 \times D^1$ are 2-disks with one interior and one proper. (Note that $R \times D^1 \subset R^4 \times R^1$ is not a 3-ribbon since $R \times \{1\}$ is a copy of R and has self-intersections.).

If D is a slot/slice diagram for R, $D \times D^1$ will be a slot/slice diagram for R'. An expression of ribbon R as 0-handles $h^{(0)}_i$ and 1-handles $h^{(1)}_j$, naturally gives an expression of R' as 0-handles $H^{(0)}_i$ and 1-handles $H^{(1)}_j$.

In a paper [Yajima (1964)] a 2-knotting of a sphere to be **simple** if it had a projection in general position with only double points. In terms used later in this book a projection-simple sphere is a **2-bulky knotted 2-sphere**. A third name for this is **pseudo-ribbon surface-link**, [Kawauchi (2002)].

In Remark 6.1, we defined a simple 2-knotting as one whose projection only has double points (no triple points, no double points).

Prop 9.2. Every projection-simple 2-sphere is a 2-ribbon knot. ▮

Proof. We give a rough idea of the proof, for details see [Yajima (1964)], with an alternate proof in [Kanenobu and Shima (2002)]. Of interest is the fact that it is constructive, based largely on the crossing scheme. The strategy is as follows. We are given a projection-simple sphere K. We find a 3-ribbon $R \subseteq R^4$ that $K_* = (\partial R)_*$ where ∂R has the same crossing scheme as K. Then $\partial R \approx K$ by a vertical isotopy, see Example 4.2. Our ribbon R will be modeled as 0-handles and 1-handles such that each handle *individually* will project to a 3-ball in R^3. We will build $R_* \subset R^3$. These projected handles are 3-balls bounded by 2-spheres constructed by assembling 1-shards of K_*.

A crossing scheme $\widehat{C}$ of a simple knotting of a $K = F(S^2)$ has no branch points or 3-shards so consists of only 2-shards $\widehat{C^{(2)}}$. So it is a collection of disjoint pairs of circles in S^2 such as Figures 6.94 or 9.26.

We illustrate key parts of the proof by showing how can we find a ribbon R if we were only given Figure 9.26 and told it was a crossing scheme $\widehat{C}$ for some 2-knotting K. We use this crossing scheme to construct a projection R_* as in Figure 9.25. View this Figure as a projection of a ribbon formed from four 0-handles and three 1-handles. In Figure 9.25 there are four "obvious" candidates for our projected 0-handles. All happen to have boundaries made up of images of four 1-shards—one pair of pants and three disks. There is a "central" 0-handle, a 3-ball in R^3 which has circles C_1, C_3 and C_5 in its boundary. Another 0-handle is a ball which has C_1, C_2 and C_6 in its boundary, etc. The 3-balls corresponding to the three 1-handles are hidden from view in the Figure. The boundary of each 1-handle consists of an annulus and two disks. One of them has on its boundary the two circles C_1 and C_2, etc. We next indicate how to get these handles from Figure 9.26.

In general, for any collection of disjoint circles in a 2-manifold M^2, a circle which bounds a 2-disk in M^2 which contains no other circle is called an **innermost circle**.

In searching for clues on how to find our ribbon, we focus on innermost 2-shards of $\widehat{C}$. Note that, in Figure 9.26, for every above/below pair of 2-shards, one is an innermost and one is not.

The 3-ribbon we aspire to build has projection as in Figure 9.25 It will have four 0-handles and three 1-handles. Our 3-ribbon is a higher dimensional version of the 2-ribbon of Figure 9.37.

We begin by constructing a 0-handle for R_*. Consider Figure 9.45. The union of three closed 1-shards $\widehat{C_1} \cup \widehat{C_2} \cup \widehat{C_3}$ is shown in gray. It is a smooth sub-disk with boundary a 2-shard denoted $\widehat{A_2}$ in Figure 9.26. Also shown in gray in Figure 9.45 is closed 1-shard $\widehat{C_8}$ whose boundary is the 2-shard $\widehat{B_2}$. Now $\widehat{A_2}$ and $\widehat{B_2}$ are above/below shards for a circle of crossings $C_2 \subseteq K_*$. (For lack of space we have omitted the labels for the 2-shards found in Figure 9.26. The reader will have to refer to these 2-shard labels to follow the details of our discussion.) A 1-shard $\widehat{C_i}$ of the crossing scheme corresponds to a 1-shard C_i of K_*. The union of the 1-shards $X = C_1 \cup C_2 \cup C_3 \cup C_8$ is a 2-sphere in R^3—two disks glued together at common boundary. So X bounds a 3-ball call it $H_1^{(0)} \subseteq R^3$ since it will correspond to a 0-handle for a 3-ribbon R we intend to construct. As a guide to following this, if the K_* happened to look like Figure 9.25 and we viewed this projection as having three "arms", this ball would correspond the "arm" on the lower right.

Our second 0-handle. corresponds to the central portion in Figure 9.25. Consider the four 1-shards in gray in Figure 9.46. Recall these shards are really subsets of S^2 and not the plane, so $\widehat{C_5}$ is really a disk with three holes (i.e. a pair of pants). Inspection of the above/below pairings on the boundary of these shards reveals that the union of the corresponding 1-shards in K_*, that is $C_5 \cup C_2 \cup C_4 \cup C_6$, is a 2-sphere in R^3. Our second 0-handle for R_*, $H_2^{(0)} \subseteq R^3$, is the ball bounded by this sphere.

Next, we find a 1-handle using Figure 9.47. Examining the above/below pairings of the bounding 2-shards for 1-shards $\widehat{C_7}, \widehat{C_4}$, and $\widehat{C_8}$, we conclude that the shards C_7, C_4, and C_8 fit together to form a 2-sphere in R^3. We let $H_1^{(1)}$ denote the 3-ball bounded by this sphere.

So far we have 0-handles $H_1^{(0)}$ and $H_2^{(0)}$ with a 1-handle $H_1^{(1)}$ which is attached to these 0-handles with attaching set $C_4 \cup C_8$. If we continue in this way we can use these methods to eventually obtain four 0-handles and three 1-handles. The union of these is not a ribbon. We have $H_1^{(1)}$

We now push up/down as in Figures 6.96 to 6.101.

Our method worked because our crossing scheme of our example was not complicated. For the general case, more work is required to simplify a given crossing scheme.

In our example, every 2-shard C_j corresponds to a pair of circles $C_j^{(2)}$, one of which is an innermost 2-shard and the other is not. Reflection on our argument reveals that this helped us find 2-*spheres* in K_* rather than some other 2-manifold. We can arrange for this to be so. In [Yajima (1964)] it is shown that *any* simple 2-knotting of a sphere K is equivalent to a knotting whose scheme consists of pairs of 2-shards, one of which is an innermost 2-shard and the other is not.

Additionally, there is a general guide for constructing our 3-ribbon R using an embedded graph G—for deciding which 3-balls should become 0-handles and which should become 1-handles as well as how to assemble R. The graph $G \subseteq S^2$ has one point in each 1-shard and a smooth arc for each 2-shard circle $\widetilde{C_j^{(2)}}$, transverse to the 2-shard and joining the points on either side as in Figure 9.48. For a 2-sphere, the graph G is a tree (that is, a graph that contains no embedded circles). ∎

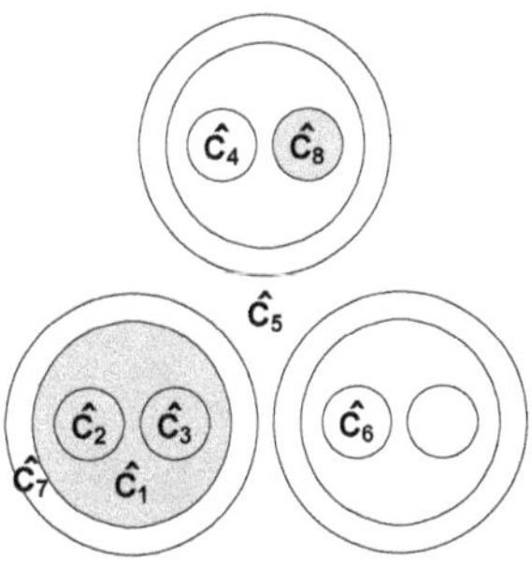

Fig. 9.45 1-shards used for $\widetilde{H_1^{(0)}}$.

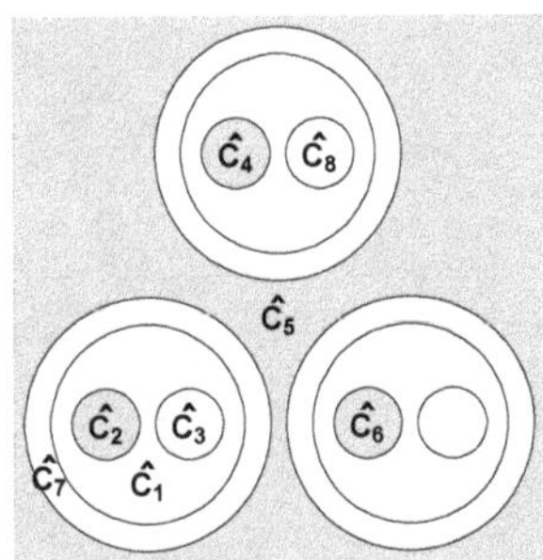

Fig. 9.46 1-shards used for $\widetilde{H_2^{(0)}}$.

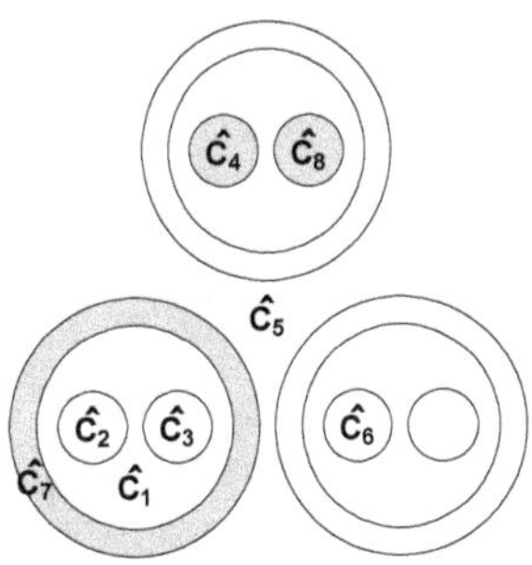

Fig. 9.47 1-shards used for $\widetilde{H_1^{(1)}}$.

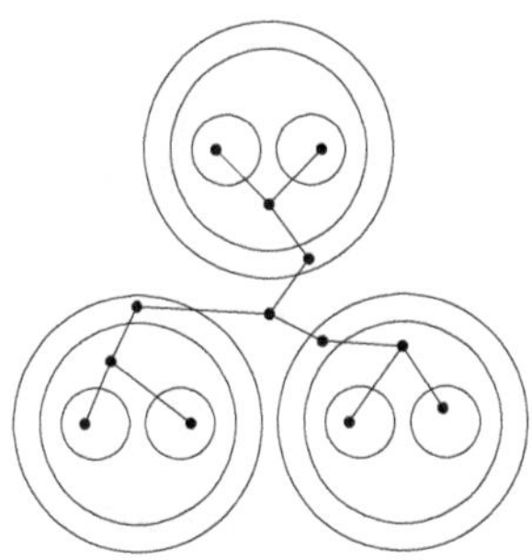

Fig. 9.48 A 1-shard graph G.

Chapter 10

Spinning Constructions for Surface Knottings

The oldest construction of higher dimensional knotting is **spinning**, [Artin (1925)], Section 10.2. Spinning has many variations all of which use a 1-dimensional knotting $\kappa \subseteq R^3$, represented as proper knotting (D^3, k_α), to produce a 2-dimensional knotting $K(\alpha) \subseteq R^4$ by a process which locally is a product $(D^3, k_\alpha) \times D^1 \approx (D^4, k_\alpha \times D^1)$, see Section 10.1.

The simplest construction Section 10.2 is described using generalized cylindrical coordinates and uses a 1-knotting to produce a 2-knotting whose components are spheres and torii. Other spinning constructions create 2-knottings of any genus surface orientable or not.

A knot is an equivalence class of a knotting. It is important to note that all spinning constructions have the property that if 1-knottings α and β are properly isotopic in D^3 then the 2-knottings $K(\alpha)$ and $K(\beta)$ are isotopic 2-knottings in R^4. So, **a spinning creates a 2-knot from a 1-knot**. In higher dimensions, a more general spinning creates an $(n+1)$-knot from an n-knot, Example 12.1.

10.1 Basic Construction: Cartesian Products and Disk Pairs

Fundamental to higher dimensional knotting is cartesian products of pairs. Using set pair notation, Notation 1.4, we begin with an elementary example in R^3 which serves as a guide to higher dimensional constructions:

Example 10.1. Consider the proper path P in the square Q^2 as shown in Figure 10.1—in pair notation (Q^2, P). This is a $(2,1)$ ball pair. Taking a cartesian product with D^1 we obtain a $(3,2)$ ball pair: $(Q^2, P) \times D^1 \approx (Q^3, R)$. Here R is a proper 2-disk $R = P \times D^1 \subseteq Q^2 \times D^1 \approx Q^3$, Figure 10.2. Now focus on the *boundaries* of this disk pair. The subset $C = \partial R$ is a topological circle which lies on the boundary of the cube ∂Q^3 (a topological 2-sphere), see Figures 10.2 and 10.3. Radial projection maps a cube boundary to a sphere, and we can view Figure 10.2 as Figure 10.4. Using stereographic on C, we get an image in the plane, Figure 10.5. The curve ∂R is a union of two copies of P. **Note that, more exactly, it is made by connecting a copy of P and the *mirror image* of P.** ▮

Fig. 10.1 A path P in the square Q^2.

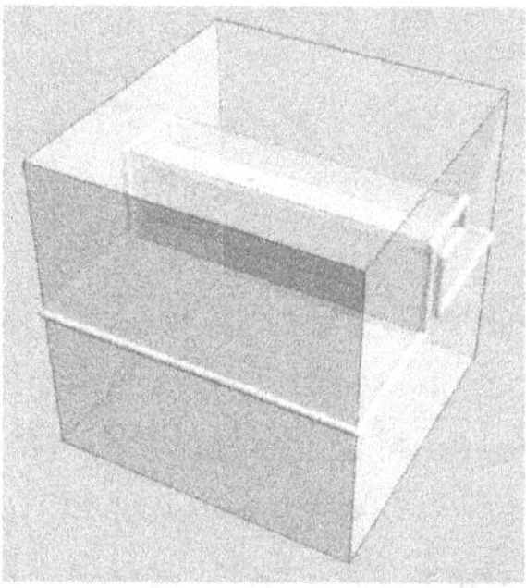

Fig. 10.2 The product of this disk pair with D^1 is a disk pair.

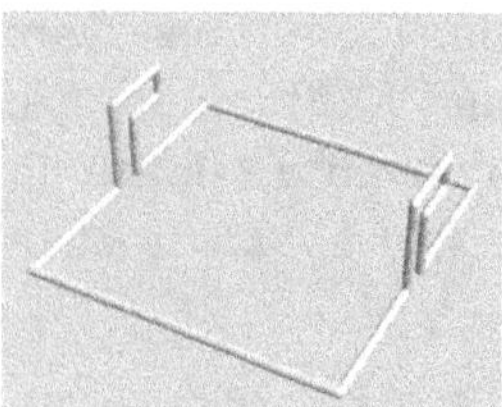

Fig. 10.3 The boundary $C = \partial R$ of the 2-disk $R = D^1 \times P$ by itself.

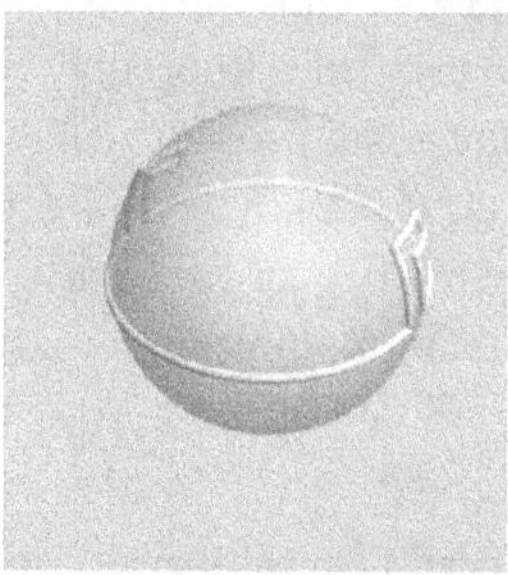

Fig. 10.4 Project C radially to S^2 to a curve on S^2.

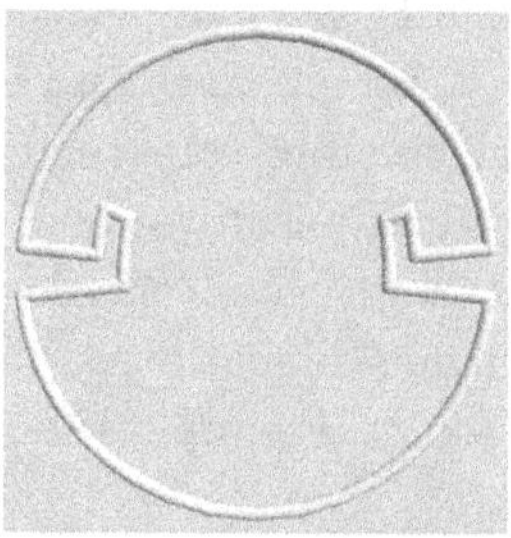

Fig. 10.5 Stereographic mapping gives image of C in R^2.

The product of D^1 and a $(3, 1)$-ball pair gives a similar example in R^4:

Example 10.2. Suppose $P \subseteq Q^3$ is a proper arc as in Figure 10.6 and let $R = P \times D^1 \subseteq Q^3 \times D^1 \approx Q^4$. This proper arc represents a trefoil knotting of a circle. Clearly R is a proper 2-disk in Q^4. So, $(Q^3, P) \times D^1$ is a $(4, 2)$-ball pair (Q^4, R).

If we look at the boundaries and apply stereographic mapping, we will see a 1-knotting as shown in Figure 10.11. The projection P_* is shown in Figure 10.7. In Figure 10.8 we show the projection R_* in the projection $Q^4_* = Q^3$. For clarity, we show a second view of R_* without the cube, Figure 10.10. The boundary $C = \partial R$ is a topological circle and ∂Q^4 is a topological 3-sphere so topologically C is a 1-knotting in S^3.

From the projection of this knotting, as seen in Figure 10.9 it is apparent that C is a connected sum of two trefoil knottings—but C is not a connected sum of a trefoil knotting with itself. Figure 10.11 was produced as follows—two parallel copies of P were put into R^4, in fact into opposite sides of Q^4, the endpoints connected to form C, which was then projected radially to S^4 and then stereographically projected to R^3. As in the lower dimensional Example 10.1, we can see that C is the connected sum of T with its *mirror image* $-T$. So C has the type of the square knot $T\# -T$ and not the granny knot $T\#T$. ∎

Fig. 10.6 Proper knotting R of an arc in 3-cube Q^3.

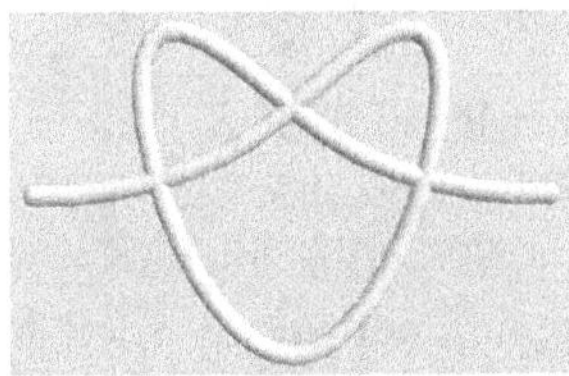

Fig. 10.7 The standard projection R_* of knotting of Figure 10.6.

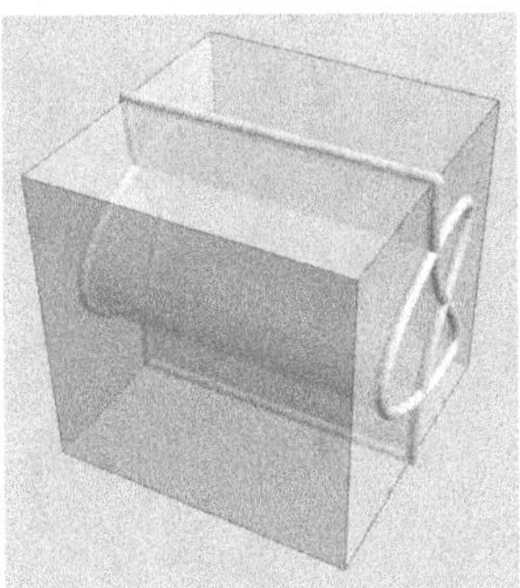

Fig. 10.8 The projection R_* inside a cube $Q^3 = Q^4_*$.

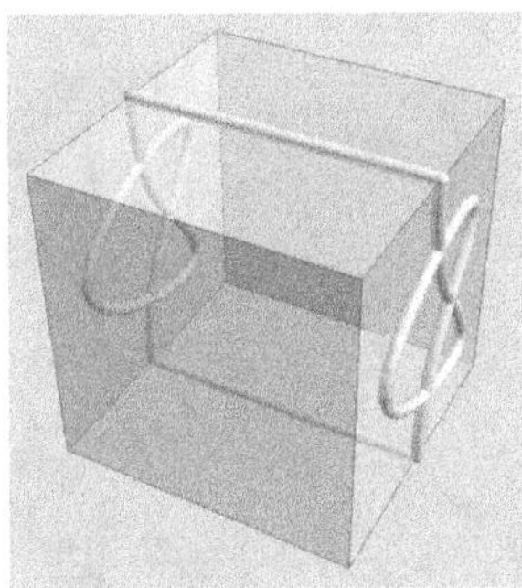

Fig. 10.9 The projection of the boundary C_*.

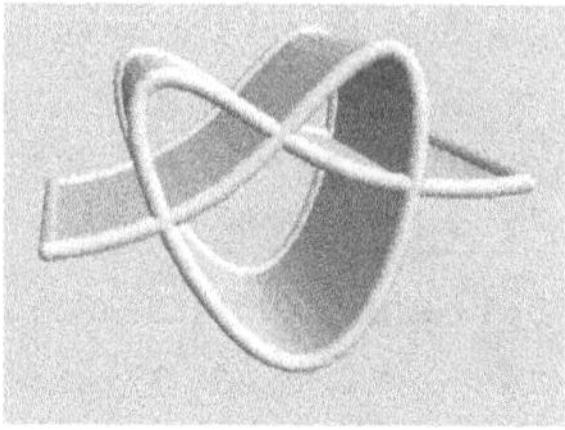

Fig. 10.10 A different view of R_*, and C_* without the cube.

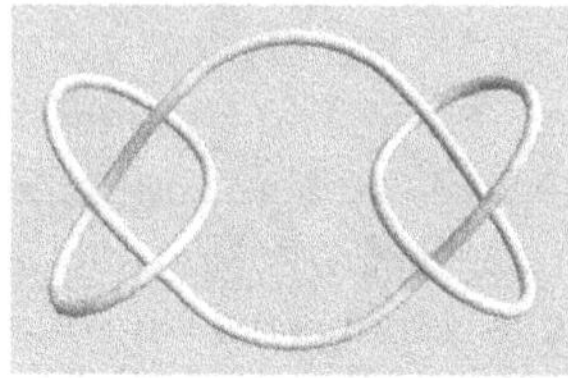

Fig. 10.11 Stereographic mapping of boundary of sphere pair $(\partial Q^4, \partial R)$.

We can think of the product knotting of Example 10.2 as changing a standard proper 2-disk in D^4 into another:

Definition 10.1. Let $D^4 = D^1 \times D^1 \times D^2$ and for $x \in D^1$ write $D^3_x = D^1 \times \{x\} \times D^2$. Let $\alpha_0 \subset D^3_0$ be a proper 1-knotting with $\partial\alpha_0 = \partial D^1 \times \{\mathbb{O}\}$ and $\alpha_x \subseteq D^3_x$ be a copy of α_0. Finally, let $D = D^1 \times D^1 \times \{\mathbb{O}\}$ and let $\Sigma(\alpha) = \bigcup_{x \in D^1} \alpha_x$. We say $\Sigma(\alpha)$ is the **local spinning of** α **for** D. ▮

Example 10.3. If $K \subset Q^3$ is a knotting of a left-hand trefoil, Figure 10.12, the projection of $K \times D^1 \subseteq Q^3 \times D^1 \approx Q^4$ is shown in Figure 10.13. Identify $\partial Q^4 \approx S^3$ and take stereographic mapping to R^4. We can see $\partial(K \times D^1)$ is a split knotting of a left-hand trefoil knotting and a right-hand trefoil knotting, Figure 10.14. ▮

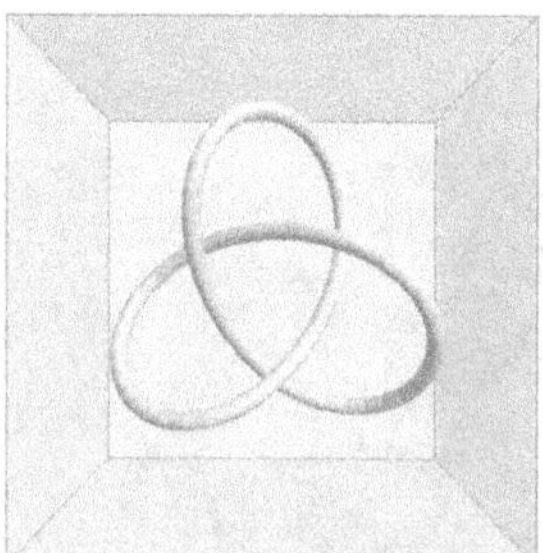

Fig. 10.12 Left-hand trefoil knotting K in 3-cube Q^3.

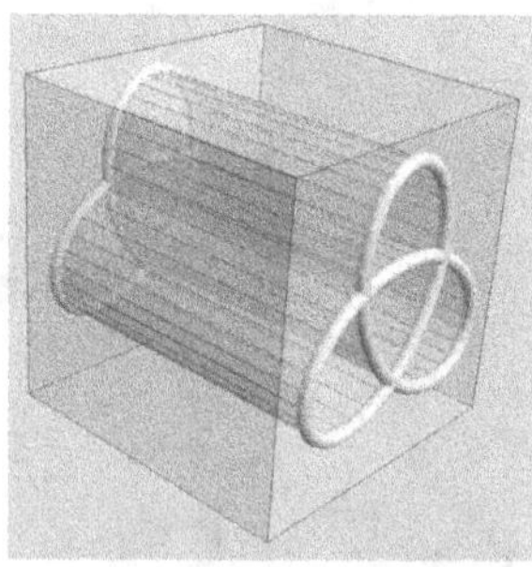

Fig. 10.13 The projection $K \times D^1$ inside a cube $Q^3 = Q^4_*$.

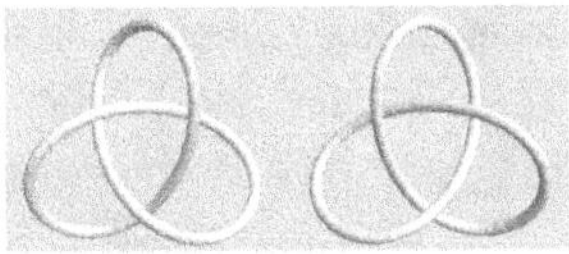

Fig. 10.14 $\partial(K \times D^1)$ is left-hand trefoil and right-hand trefoil.

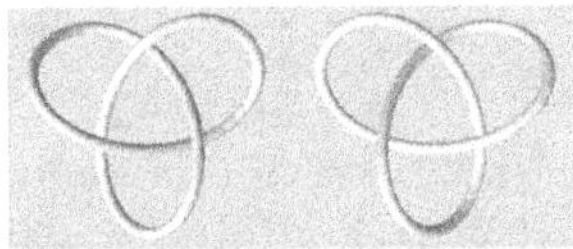

Fig. 10.15 Figure for Remark 10.1.

Roughly, Figure 10.8 is "Figure 10.7 times D^1". A brief expression of this is: **"the projection of the product is the product of the projection"**—symbolically put:

$$(P \times D^1)_* = P_* \times D^1.$$

In this expression, we understand in context that "$*$" is notation for two different standard projections—from $Q^3 \to Q^2$ and from $Q^4 \to Q^3$.

10.2 Spinning a Knotting Using Generalized Cylindrical Coordinates for 4-Space

Using generalized cylindrical coordinates for R^4 spinning can be defined for any subset of $X \subseteq R^3_+$ to produce a subset $\Sigma(X) \subseteq R^4$:

Definition 10.2. If $p \in R^3_+$ with coordinates $p = (x_p, y_p, z_p)$ define $\Sigma(p) = \{(r, \theta, y_p, z_p) \in R^4 \text{ with } r = x_p\}$. The **spinning of** X is $\Sigma(X) = \bigcup_{p \in X} \Sigma(p)$.
Alternatively, for any θ with $0 \leq \theta \leq 2\pi$ and page $H^3_\theta \subseteq R^4$, we define $X_\theta \subseteq H^3_\theta$ by $X_\theta = \{(r, \theta, y, z) \text{ with } r = x\}$. Then we can write $\Sigma(X) = \bigcup_{0 \leq \theta \leq 2\pi} X_\theta$. ∎

If $p = (x, y, z)$ with $x \neq 0$ then $\Sigma(p)$ will be a circle of radius x in R^4; if $x = 0$ then $\Sigma(p)$ will be a single point. The circle $\Sigma(p)$ projects to a geometric circle in the z or w direction; to a line segment in the x or y directions. In the alternate definition each X_θ is a homeomorphic copy of X in the page H^3_θ.

A torus is a cartesian product, so we begin with torus knottings in R^4.

Definition 10.3. For a 1-knotting $\alpha_0 \subseteq \mathring{R}^3_+ = \mathring{H}^3_0$ if α is a single circle we call $\Sigma(\alpha)$ a **spun torus knotting for** α. If it has more than one component it is called a **spun torus linking for** α. ∎

If K_t for $0 \leq t \leq 1$ is a smooth isotopy between 1-knottings K_0 and K_1 then $\Sigma(K_t)$ is a smooth isotopy between knottings $\Sigma(K_0)$ and $\Sigma(K_1)$. So, for a 1-knotting K of a circle, the corresponding knot $[K]$ produces a well defined knot of a torus in R^4, namely $[\Sigma(K)]$.

Example 10.4. If α is a trefoil knotting as in Figure 10.16, its projection α_* to the half-plane will be the familiar projection with three crossing points (2-shards) and six arcs (1-shards). The projection of the spinning is the spinning of the projection, Remark 1.6, as in Figure 10.17. The crossing set of $\Sigma(\alpha)_*$ consists of three disjoint circles which are the spinning of the three crossing points (2-shards) of α_*. The six 1-shards of $\Sigma(\alpha)_*$ are annuli obtained as the spinning of the six 1-shards of α_*. Figure 10.18 shows half of the projection to reveal certain details. ▮

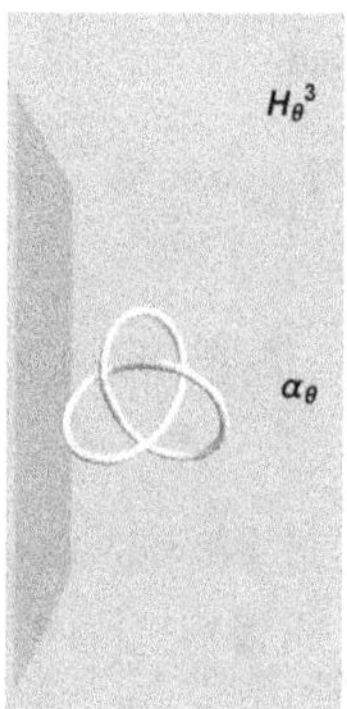

Fig. 10.16 A trefoil knotting in half-space.

Fig. 10.17 Projection of spinning of trefoil.

Fig. 10.18 Projection spun torus cut in half.

It is clear that in general:

Prop 10.1. For any 1-knotting α, the knot-diagram of the spinning is the spinning of the diagram. ▮

Remark 10.1. The right-hand trefoil is not equivalent to the left-hand trefoil, Remark 3.5. However, the spinnings of these knottings are equivalent. An isotopy of R^4, which is rotation of π with x and w fixed, sends a spinning of a right-hand trefoil to the spinning of a left-hand trefoil, seen in projection in Figures 10.19, 10.20 and 10.21.

We can view Figure 10.14 as the slice of our knotting by the yzw-coordinate plane P which we can view as a union of the pages H_0^3 and H_π^3

with the left-hand trefoil contained in H_0^3, the right-hand trefoil in H_π^3. The restriction of our rotation will give a rotation in P that interchanges these pages and takes Figures 10.14 to 10.15. ▮

Fig. 10.19 A spun right trefoil projection half transparent.

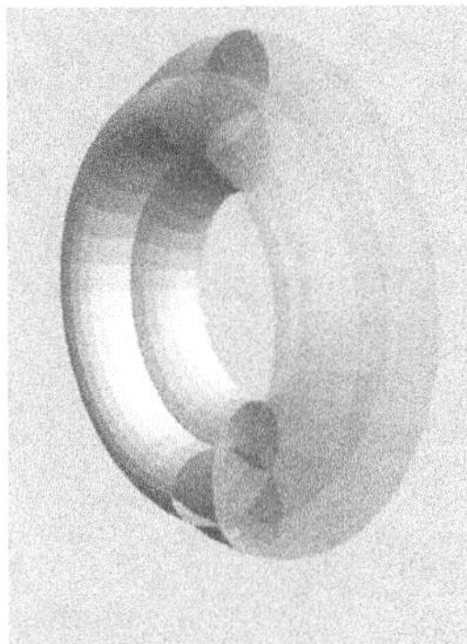

Fig. 10.20 Rotated $\pi/2$ with x and w fixed.

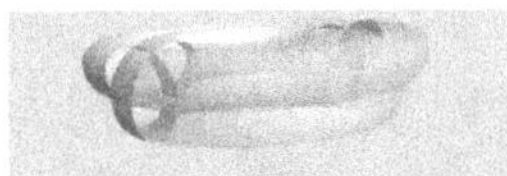

Fig. 10.21 Rotated π with x and w fixed.

Remark 10.2. From Proposition 10.1 it follows that for any knot-coloring rule R, α can be colored by R if and only if $\Sigma(\alpha)$ can be colored by R. So the spinning of a trefoil knotting produces a non-trivially knotted torus, the spinning of the Hopf link gives a non-trivial linking of torii. Many of the 2977 1-knots of 12 crossings or less can be distinguished by quandle colorings [Dionísio and Lopes (2003)] [Clark *et al.* (2014)] and these colorings transfer to produce many distinct knotted spun torii. ▮

We produce knotted *spheres* by spinning using proper knotted *arcs* in R_+^3. We start by producing an trivially knotted sphere from a trivially knotted arc.

Example 10.5. For $S^2 \subseteq R^3$, let $M_\theta = S^2 \cap H_\theta^2$. Then M_θ is a (trivially knottted) proper sub-arc of H_θ^2, Figure 10.22. Since $H_\theta^2 \subseteq H_\theta^3$ we can consider $M_\theta \subseteq H_\theta^3$ where it is also a proper arc, see Figure 10.23. The standard sphere $S^2 \subseteq R^3$ is the union of longitude arcs M_θ, $0 \leq \theta \leq 2\pi$ which all meet at the poles $N_0 = (0,0,1)$ and $S_0 = (0,0,-1)$, each in a page H_θ^2, Figures 10.24, 10.25 and 10.26. Similarly we consider R^4 as a union of three dimensional pages H_θ^3. Each H_θ^3 projects in the w-direction to H_θ^2. ▮

The union of longitude arcs M_θ, $0 \le \theta \le 2\pi$ gives an embedding of the sphere in R^4.

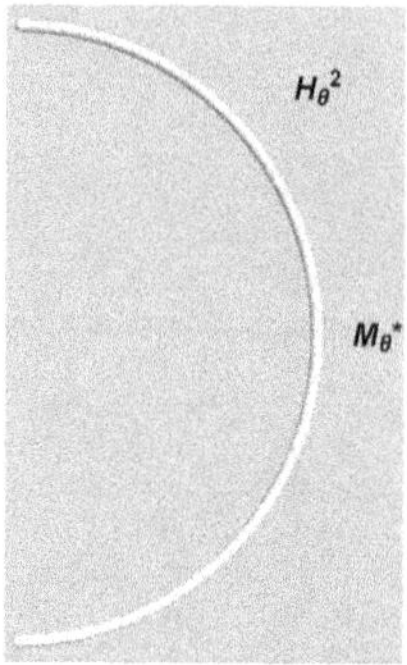

Fig. 10.22 Longitude and page of Figure 10.24.

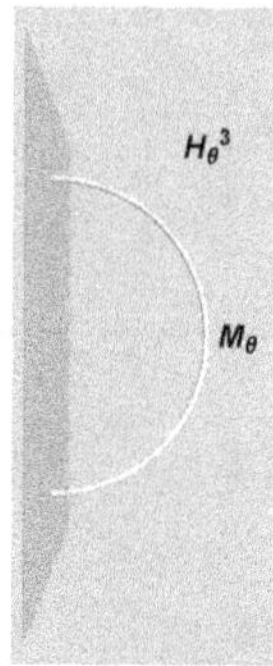

Fig. 10.23 Corresponding figure of meridian and half-space.

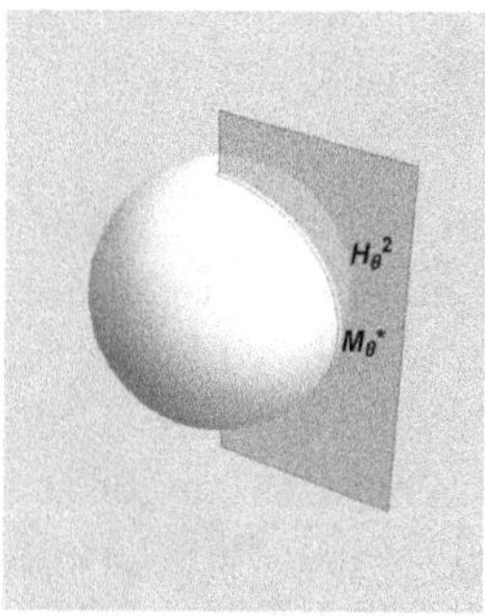

Fig. 10.24 Projection M_θ^* in page H_θ^2.

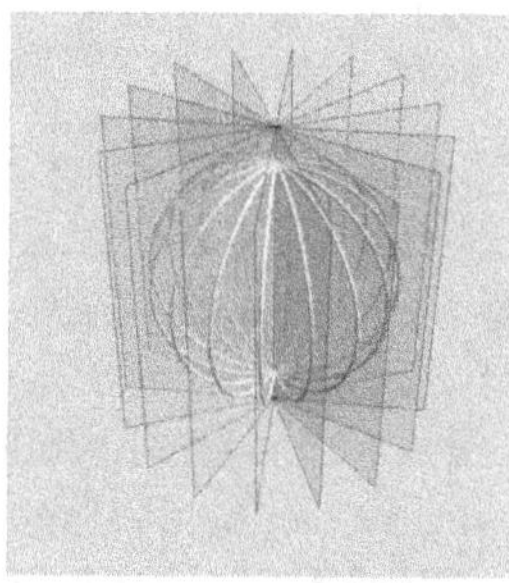

Fig. 10.25 Projection several M_θ^* in pages.

Fig. 10.26 Union of all arcs is sphere.

Definition 10.4. Let $\kappa \subseteq R^3$ be a knotting of a circle with $\alpha \subseteq R^3_+$ an associated proper 1-disk. The spinning of α is called the **spinning of** k, the **spun sphere of** κ, or the **Artin spinning of** κ. Artin spinning is called **simple spinning** in [Friedman (2005)]. ∎

(In the literature, and in this book, the present tense "spin" and the past tense "spun" are used interchangeably.)

Example 10.6. The knotting of an arc shown in Figure 10.27 is obtained from a trefoil knotting. Using this we obtain a **spun trefoil knotting** of a sphere in R^4. The corresponding knot is called a **spun trefoil.**

Figure 10.28 shows a projection. Figure 10.29 shows half and reveals details. If we label the crossing points of 10.27 as shown in Figure 10.30, the crossing scheme of the spun trefoil will consist of six "parallel" circles three above 2-shards and and three below 2-shards, Figure 10.31. ∎

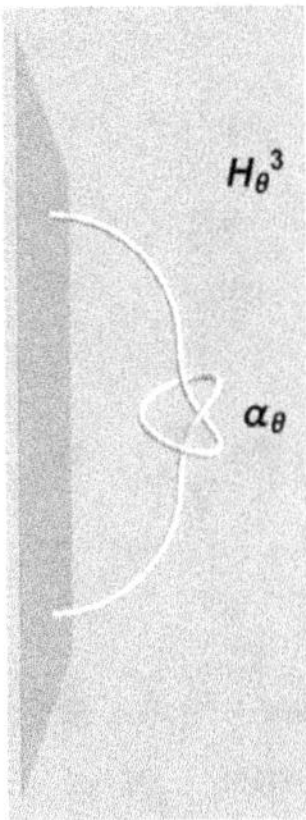

Fig. 10.27 A trefoil knotting of an arc in half-space.

Fig. 10.28 Projection of spinning of trefoil arc.

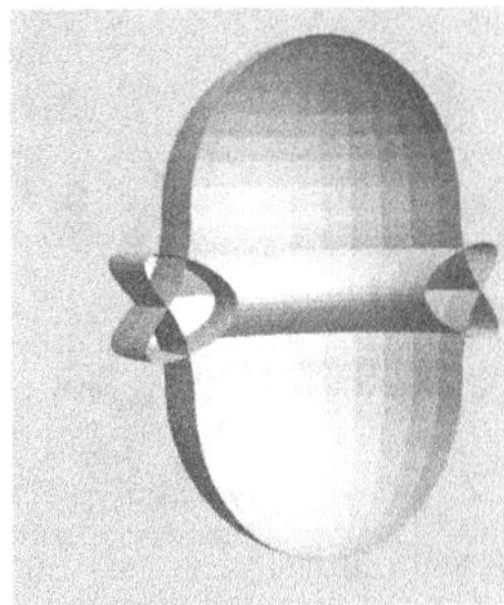

Fig. 10.29 Projection spun sphere cut in half.

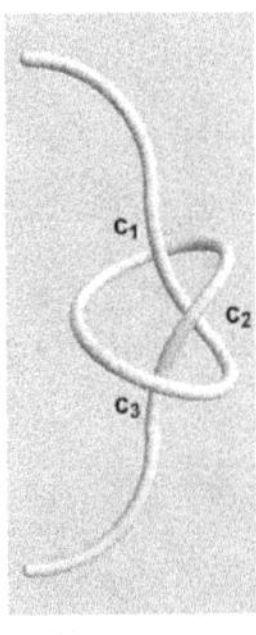

Fig. 10.30 Three crossings of trefoil arc.

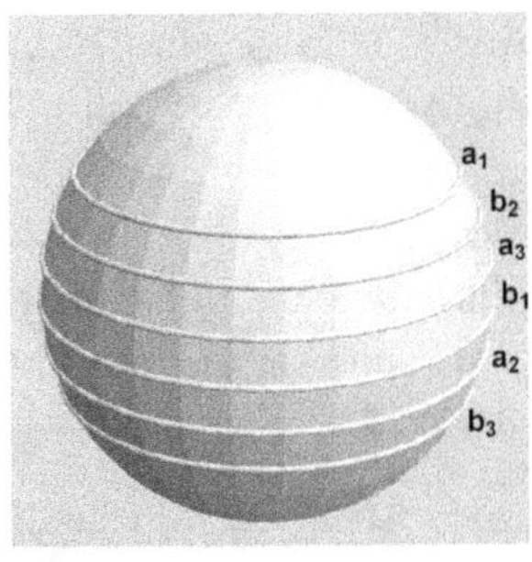

Fig. 10.31 Corresponding six circles of the crossing scheme.

Example 10.7. Spinning can be done with any proper 1-knotting in R^3_+. Spun links were first fully considered in [Zeeman (1960a)] followed by [Epstein (1960)]. The spin of the 1-knotting of Figure 10.32 is a 2-knotting of two components—a sphere *and* a torus equivalent to the basic sphere-torus knotting Example 6.12. The spin of the 1-knottings shown in Figure 10.33

is a linking of 2-spheres. Translate a Hopf linking to $\overset{\circ}{R}{}^3_+$, spin it and we get a knotting of two torii, the knot-type, the **spun Hopf link.** ▌

Fig. 10.32 Spin this 1-knotting and get 2-knotting of sphere and torus.

Fig. 10.33 Spin this 1-knotting and get a linking of two spheres.

We can be assured of a plentiful supply of non-equivalent spun spheres. As with torii of Remark 10.2, the projection of the spinning of α is the spinning of the projection of α_*, Remark 1.6 and the shard structure of $\Sigma(\alpha)_*$ is related in a simple way to that of α_*. However, there are some examples of two 1-knottings, k and k', which are not equivalent, and yet $\Sigma(k)$ and $\Sigma(k')$ are equivalent, as in Examples 10.3 and 10.10.

Example 10.8. Figure 10.34 shows a wire-frame coding of half of $\Sigma(K)$. On the boundary is a slice knotting and a close look at the wire-frame coding *on the edge* reveals details of the thickness coding see Section 6.4 of the slice knotting. Figures 10.37 and 10.38 show that this is a connected sum with a left-hand trefoil on the left and a right-hand trefoil on the right. (Focus on the relative tube thickness at each self-crossing along the edge.) ▌

Remark 10.3. As a visualization exercise, we show that the spinning of a right-hand trefoil $\Sigma(K)$ and the spinning of the left-hand trefoil $\Sigma(-K)$ are equivalent 2-knottings (although K and $-K$ are *not* equivalent 1-knottings).

Suppose the slicing hyperplane for Figure 10.34 is $x = 0$. We will parameterize H_θ by $0 \leq \theta \leq \pi$. Let $R(\theta)$ be a rotation of R^4 in the yz-direction (i.e. with x and w fixed). In projection $\pi \circ R$ looks like rotation $r(\theta)$ by θ in R^3 about the x-axis.

An isotopy H_θ between $\Sigma(K)$ and $\Sigma(-K)$ can be obtained from a rotation by π of $\Sigma(K) \subseteq R^4$, as shown in projection in Figures 10.34, 10.35

and 10.36. The projection in Figure 10.34 is the same subset of R^3 as Figure 10.36. Comparing Figure 10.39 and 10.40 we can see the height relations each crossing have reversed height relations. So, we can identify Figure 10.36 as spinning $\Sigma(-K)$. Thus $\Sigma(K)$ and $\Sigma(-K)$ are equivalent knottings. ∎

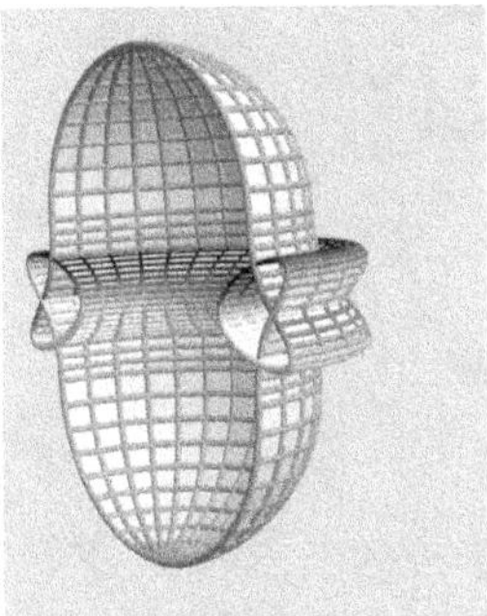

Fig. 10.34 Figure 10.29, wire-frame coded.

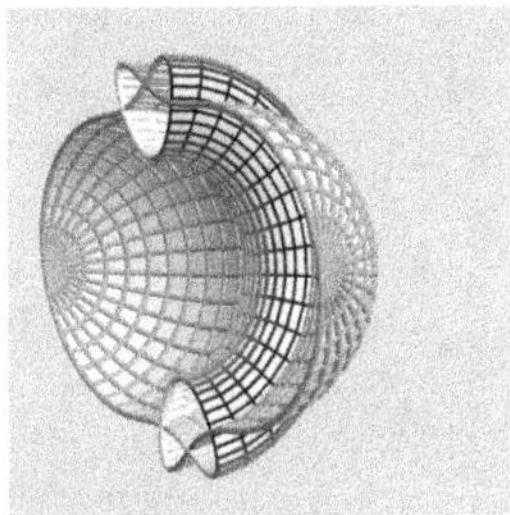

Fig. 10.35 Figure 10.34 after rotation of $\pi/2$.

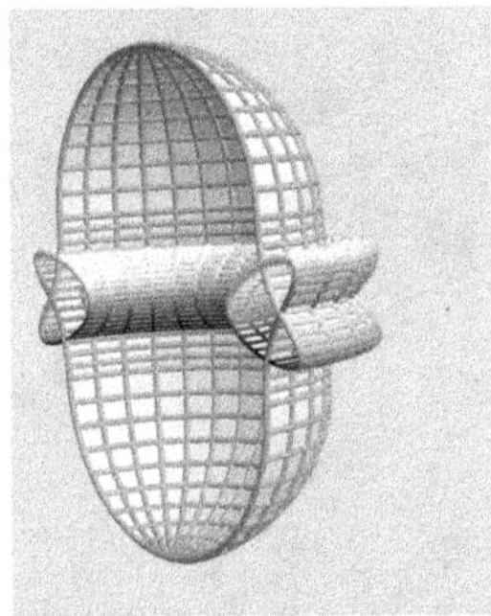

Fig. 10.36 Figure 10.34 after rotation of π.

Fig. 10.37 Detail of left side of Figure 10.34—a left-hand trefoil.

Fig. 10.38 Detail of right side of Figure 10.34—a right-hand trefoil.

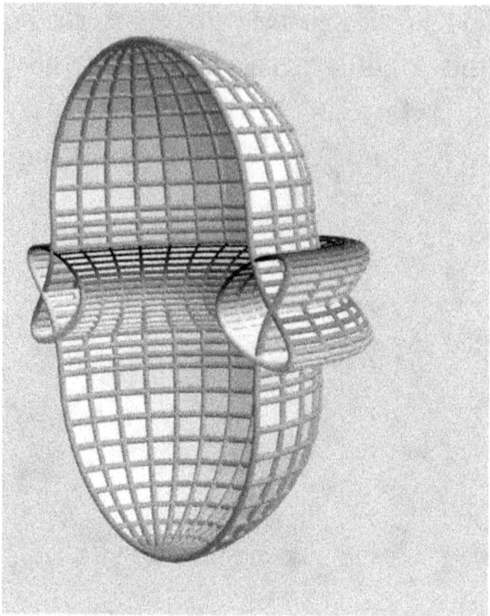

Fig. 10.39 Enlarged Figure 10.34, before rotation.

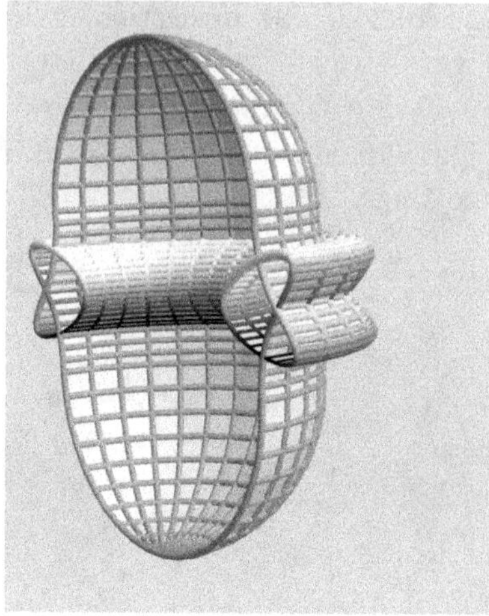

Fig. 10.40 Enlarged Figure 10.36, equivalent to $\Sigma(-K)$.

Prop 10.2. Any spun 1-knotting, $\Sigma(\alpha)$, is a ribbon 2-knot.

Proof. By Definition 10.2 $\Sigma(\alpha)$ is a union of *geometric* circles (and two points, circles of radius 0) in R^4: $\Sigma(X) = \bigcup_{p\in\alpha} \Sigma(p)$. Let D_p be the *geometric* 2-disk in R^4 with $\partial D_p = \Sigma(p)$, then $R = \bigcup_{p\in\alpha} D_p$ will be a ribbon with boundary $\Sigma(\alpha)$.

Here are some details if α is a single proper arc, other 1-knottings are similar. Consider $p = (a, b, c) \in \alpha \subseteq R^3_+$, and let $P\colon R^3_+ \to \partial R^3_+$ be the projection $P(x, y, z) = (0, y, z)$. Define $H(p) = a$, $H(p)$ is the "height" of p (with respect to ∂R^3_+).

We consider a non-standard projection—in the x-direction. We may assume that α is in general position with respect to π_x in that $\pi_x(\alpha)$ is an immersed arc with transverse self-crossings. The projection of α of Figure 10.27, $\pi_x(\alpha)$, has four crossings, see Figure 10.41.

The spinning of each one-point subset $\Sigma(\{p\})$ is a geometric circle contained, in $\Sigma(\alpha)$, lying in the 2-plane of R^4 given by $z = b$ and $w = c$ with center $(0, 0, b, c)$ and radius $a = H(p)$. The union of all these circles, together with the two endpoints of α is $\Sigma(\alpha)$.

The projection π_x will send each circle $\Sigma(\{p\})$ to a line segment and will project each D_p "edgewise" to the same line segment. In fact $\pi_x(\Sigma(\{p\}))$ will be the line segment with endpoints (a, b, c) and $(-a, b, c)$. We will view this as a geometric 1-disk, D'_p, in R^3 with radius $H(p)$ and center $(0, b, c)$. If $p = (0, b, c)$ is an endpoint of α, we will just say $D_p = \{p\}$.

We note that $R' = \bigcup_{p \in \alpha} D'_p$ is a 2-ribbon in R^3, Figure 10.43. It is the projection of our 3-ribbon R. ∎

Example 10.9. Each of these four crossing points in Figure 10.27 corresponds a slot/slice of our ribbon as shown in Figure 10.43. Figure 10.42 shows α and its mirror image. Suppose p is a crossing point of α then The ribbon knotting for Figure 10.43 is equivalent to the simpler ribbon of Figure 9.41. One way to see this is to note there is a *proper* isotopy between the arc of Figure 10.27 in R^3_+ and the arc of Figure 9.40. As a consequence we see that the spun trefoil has a projection as a ribbon knotting with four circles of self-intersection as in Figure 9.6. ∎

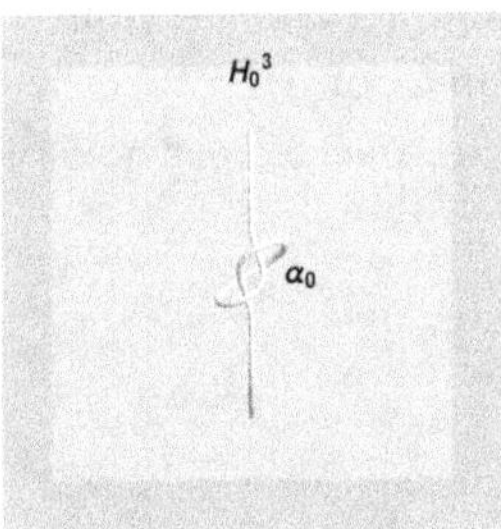

Fig. 10.41 Rotated view of Figure 10.27.

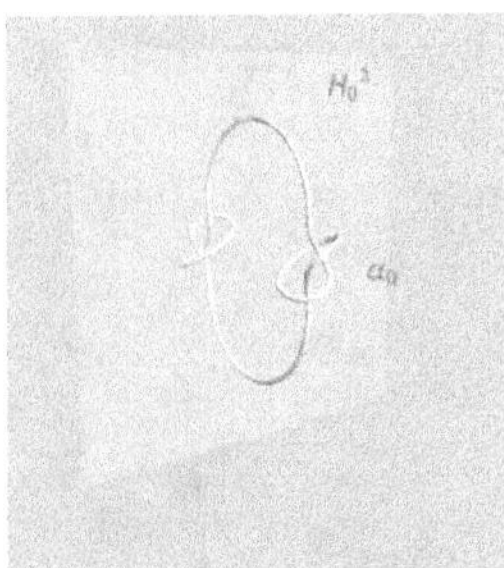

Fig. 10.42 Arc, mirror image, cut by plane.

Fig. 10.43 $R' = \pi_x(R)$.

10.3 Spinning Along a Submanifold

Using generalized cylindrical coordinates makes the spinning construction very concrete—coordinates for a 1-knotting determine coordinates for a 2-knotting. However, we are limited to producing 2-knottings for the two 2-manifolds, spheres and torii. We next extend the spinning construction so that it applies for 2-knottings of other 2-manifolds. Extensions of this construction to higher dimensions, is found in Remark 12.18.

Example 10.10. The spun trefoil torus shown in projection in Figure 10.48 is the union of two surfaces: a spun trefoil arc, Figure 10.47 and a second piece, Figure 10.46. If we had spun an *unknotted* arc, we would have gotten an annulus as in Figure 10.45. Assembling the two pieces of Figures 10.45 and 10.46 we obtain a trivial knotting of a torus, Figure 10.44.

Reassemble and view the spinning process as going from Figure 10.44 to Figure 10.48. We start simple smooth 2-knotting K (in our case a trivial knotting of a torus). We capture the notion of spinning by a simple closed path $C \subseteq K$ such as in Figure 10.44. We have two product tubular neighborhoods of C. In K, C has a product tubular neighborhood— $N \approx C \times D^1$. As mentioned in Example 2.8 C has a product tubular neighborhood in R^4, $T \approx C \times D^3$. Writing in pairs, $(T, N) \approx (C \times D^3, C \times D^1) = C \times (D^3, D^1)$ where (D^3, D^1) is the standard (trivial) ball pair. We obtain our spinning of the trefoil arc α by replacing for each $x \in C$, $\{x\} \times (D^3, D^1)$ with $\{x\} \times (D^3, \alpha)$. Putting all this together we have $N' = C \times \alpha \subseteq C \times D^3 \approx T \subseteq R^4$. We have N and N' diffeomorphic. Our new torus knotting is $(K^2 - N) \cup N'$.

The projection of several of these trivial ball pairs is shown in Figure 10.49. Each $\{x\} \times D^3$ projects to a disk with center x_* normal to circle C_*. All the trivial arcs fit together to form an annulus Figure 10.49 previously viewed by itself as Figure 10.45. Similarly, all the copies of α fit together to form an annulus knotting—Figures 10.50, 10.51, and 10.52 previously viewed, by itself, as Figure 10.47. ∎

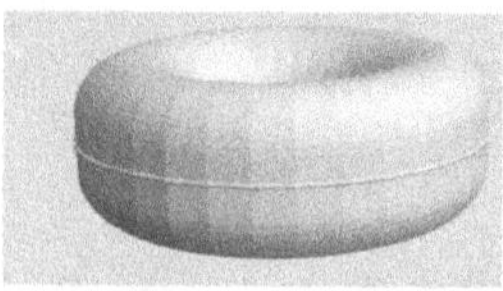

Fig. 10.44 A circle on a torus.

Fig. 10.45 A tubular neighborhood of circle is annulus.

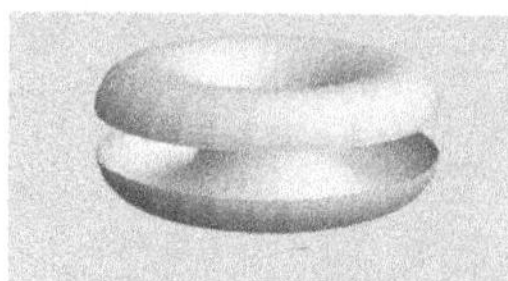

Fig. 10.46 Torus with annulus tubular neighborhood removed.

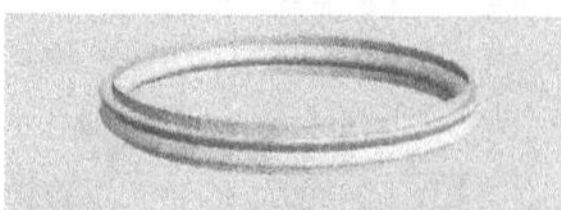

Fig. 10.47 Spinning of trefoil arc.

Fig. 10.48 Spun trefoil torus assembled.

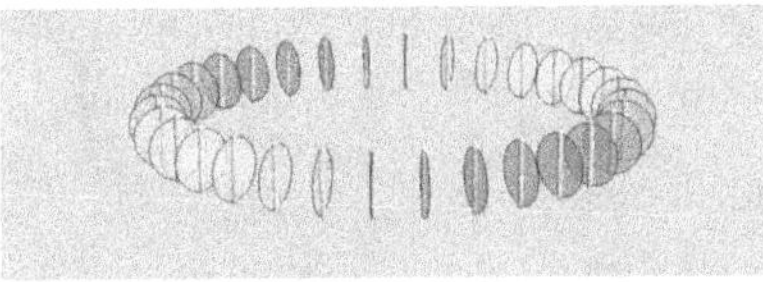

Fig. 10.49 Projection of a number of trivial disk pairs.

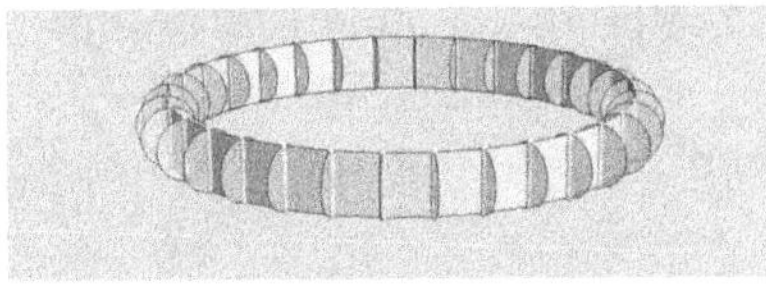

Fig. 10.50 All projected trivial arcs form annulus.

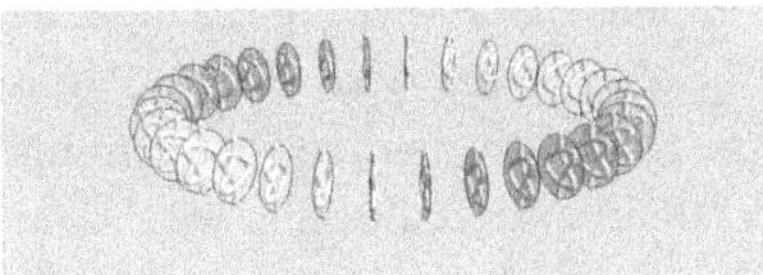

Fig. 10.51 Replace a number of trivial $\{x\} \times D^1$ by trefoil arc in each $\{x\} \times \alpha$.

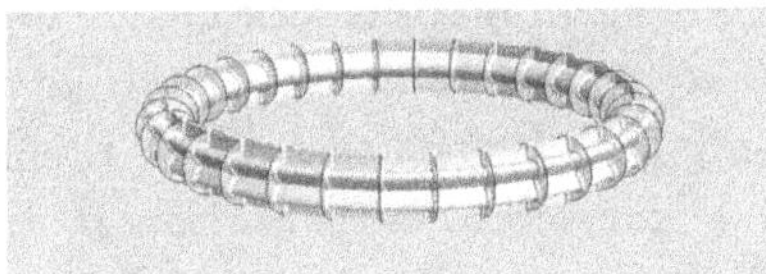

Fig. 10.52 All projected trefoil arcs form knotting of an annulus.

A re-reading of our reassembly reveals (1) no mention was made that K was a torus (it could be any closed orientable 2-manifold), (2) that the knotting K was standard (it could have been any knotting), or (3) that C was a particular curve on the manifold we wish to knot. So, our general construction is:

Definition 10.5. Suppose we have knottings $M \subseteq K \subseteq R^4$ where M is a closed 1-manifold with a product tubular neighborhood N in 2-manifold K and $\kappa \in R^3$ 1-knotting of a circle with associated ball pair (D^3, α). Let T a product tubular neighborhood of $M \subseteq R^4$. So we have $(T, N) \approx M \times (D^3, D^1)$. Let $N' = M \times \alpha \subseteq M \times D^3 \approx T$. The new knotting is called the **knotting of K obtained by spinning κ about M** and is defined $K' = (K - N) \cup N'$ The new knotting K' is denoted $(K, M) \bigotimes \kappa$ or $(K, M) \bigotimes \alpha$ ∎

Of course, K and K' are homeomorphic subsets. We do not need C to be a single circle—that is, it could be some other closed (possibly disconnected) 1-submanifold of K and, in this case, we could spin different knotted arcs on different components. Equivalently, we could do spinning on each component in an iterated manner, see discussion in Section 10.10.

Notation 10.1. The use of the notation $\bigotimes$ is unrelated to the tensor product. It was used for spinning constructions in [Cappell (1970)] (where his $S^1 \bigotimes \alpha$ corresponds to our $(S^3, S^1) \bigotimes \alpha$ and later [Roseman (1989)].

Locally all spinning is a product knotting. As a mnemonic, one can view $\otimes$ as a simplified version of $\circlearrowleft\!\!\!\!\times$ which combines $\circlearrowleft$, a reference to spinning and $\times$, a reference to product. ▮

Example 10.11. There are many interesting knottings of 1-manifolds in 2-manifolds. Any orientable 2-knotting, has product tubular neighborhood, Remark 12.9. Consider the simple closed curve on the torus in Figure 10.53. (This corresponds to the diagonal of $S^1 \times S^1$, Definition 1.4.) We find a generalization of this discussed for higher-dimensions in Section 12.4. If we use this to spin our trefoil arc α we obtain a 2-knotting with projection as in Figure 10.54. It turns out this knotting is not equivalent to the spinning of Example 10.4 see Section 10.4. Figures 10.55 and 10.56 illustrate some more interesting possibilities for genus 2 manifolds. ▮

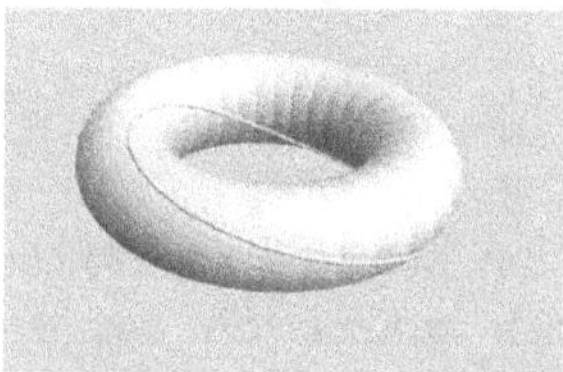

Fig. 10.53 An interesting simple closed curve on a torus.

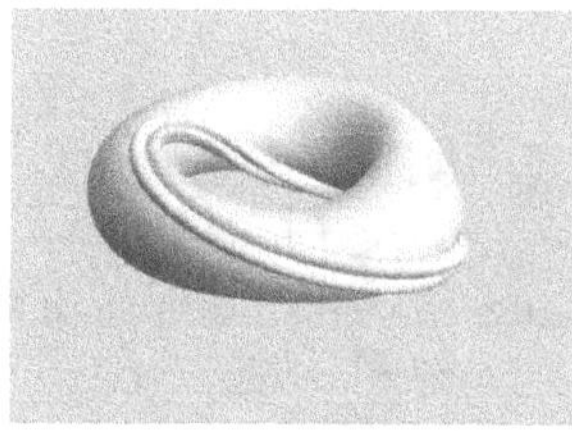

Fig. 10.54 Spun trefoil along curve of Figure 10.53.

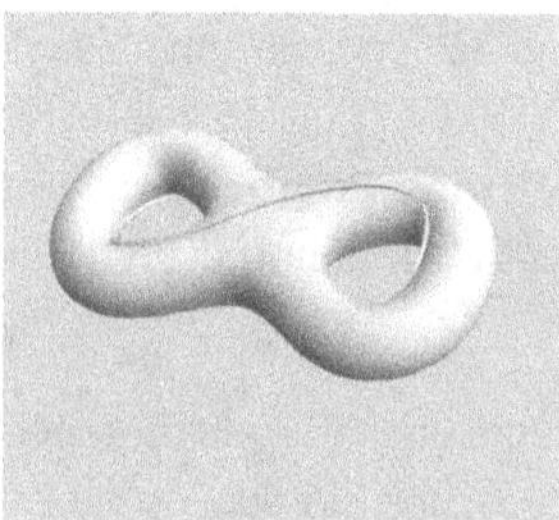

Fig. 10.55 A non-trivial curve on genus two surface.

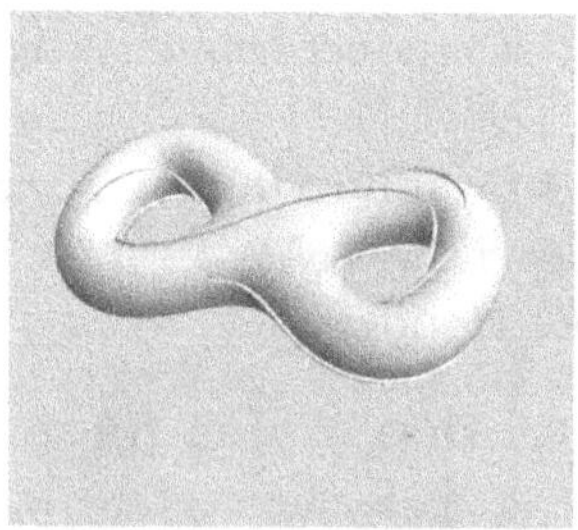

Fig. 10.56 Two disjoint non-trivial simple closed curves on genus two surface.

The spinning process produces a 2-knot from a 1-knot—not just a 2-knotting from a 1-knotting. If we spin equivalent 1-knottings, we obtain equivalent 2-knottings since an isotopy κ_t of a 1-knotting κ gives a 2-knotting isotopy $(K, M) \bigotimes \kappa_t$. Also, if M_t is an isotopy of $M \subseteq K$ then $(K, M_t) \bigotimes \kappa$ is an isotopy.

Remark 10.4. In review, a knotting of a torus as replacing an unknotted annulus by a knotted one has the following steps:

(1) Begin with the standard knotting of the torus $K \subseteq R^4$.
(2) Take a $(1,3)$-disk pair—a proper knotting κ of D^1 in D^3.
(3) Consider a simple closed curve, call it M (we took a longitude for M).
(4) Consider the tubular neighborhood $N \approx M \times D^1 \subseteq K$ of $M \subseteq K$.
(5) Consider the tubular neighborhood $N' \approx K \times D^2$ of $K \subseteq R^4$.
(6) Consider $N \subseteq N'$. We obtain our spinning by replacing the knotting $N \subseteq N' \subseteq R^4$. At each $x \in M$ we have an unknotted $(1,3)$-disk pair: $\{x\} \times D^1 \subseteq (\{x\} \times D^1) \times D^2 \subseteq K \times D^2 \subseteq R^4$.
(7) Replace each such unknotted ball pair by knotted ball pair κ. Product structures of N and N' assure continuity as we vary points $x \in M$. ∎

If K' is smoothly isotopic to K, M' is smoothly isotopic to M in K and κ' is smoothly isotopic to κ in D^3, then $\Sigma(M, K, \kappa)$ will be smoothly isotopic to $\Sigma(M', K', \kappa')$.

This construction gives the Artin-spinning if k is a 1-knotting k if M is a single circle in $K = S^2 \subseteq R^4$. By the Brower-Jordan Theorem, any circle M is isotopic in S^2 to any other. In particular, $K \subseteq M$ is isotopic to a very small circle $M' \subseteq K$, that is, contained in a 4-ball in R^4, our knotting just the connected sum of K and the Artin spinning of κ. That is $\Sigma(M, K, \kappa) \approx K \# \Sigma(\kappa)$.

Remark 10.5. Spinning along a submanifold gives many interesting examples in higher dimensions as in Example 12.5. There is one lower-dimensional example that will prove instructive. Begin with $S^0 \subset S^1 \subset R^3$ and some proper knotting κ of a 1-disk which represents a 1-knotting K of a circle. Our spinning construction will naturally produce a 1-knotting which is, in fact, $K \# -K$. Because S^0 is a 0-dimensional manifold, we say $K \# -K$ is the **0-spinning of** κ or the 0-spinning of K.

A tubular neighborhood N of $S^0 \subset S^1$ is a pair of arcs and a tubular neighborhood N' of $S^0 \subset S^3$ is a pair of 3-balls. So (N', N) is diffeomorphic to a disjoint union of standard disk pairs as in Figure 4.11. Replacing each standard pair with a knotted pair, we obtain a new knotting that is a connected sum. However, there is a question of which connected sum. Because N and N' are not connected one could do the construction in two different ways—one which gets $K\#K$ or $K\# - K$, as in Figures 4.9 and 4.10. We will take $K\# - K$ for our 0-spinning since it is consistent with Artin-spinning, see Figures 10.11 and Figures 10.34, 10.37 and 10.38. ▮

10.4 Turn-Spinning

Turn-spinning is a knotting construction for torii that has extensions to higher genus 2-manifolds. It can be considered as a special type of spinning along a curve in the standard torus.

A rotation in $f\colon R^3 \to R^3$ is determined by a directed line representing an axis and an angle of rotation about that axis.

Definition 10.6. A **turn of** R^3 is an isotopy $t_\theta\colon R^3 \to R^3$ of rotations parameterized by $[0, 2\pi]$ so that t_θ is rotation about the axis of angle θ.

Let $K \subseteq R^3_+$ be a 1-knotting. A **turn of** K **in** R^3_+ is a proper isotopy $f_\theta \subseteq R^3_+$ where $0 \leq \theta \leq 2\pi$ of K which is a restriction of a turn of R^3. ▮

Example 10.12. We can define a turn for any knotting by finding an axis so that a turn for R^3 about that axis will keep the image of K in R^3_+. Figure 10.57 shows an axis appropriate for turning a trefoil knotting as in Figure 10.57 If we identify the page H^3_θ with R^3_+ and let $K_\theta = f_\theta(K)$, then $T(K) = \bigcup_{0\leq\theta\leq 2\pi} K_\theta$ will be a smooth knotting of a torus called the **1-turn spinning of** K, or the **turn-spinning of** K, see Figures 10.58 and Figures 10.59. If K is a knotting of q circles, then $T(K)$ will be a knotting of q torii. ▮

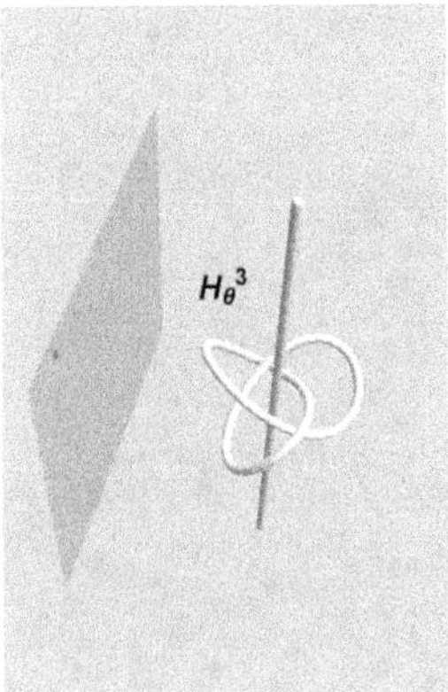

Fig. 10.57 An axis for a turn of a trefoil knotting in R^3_+.

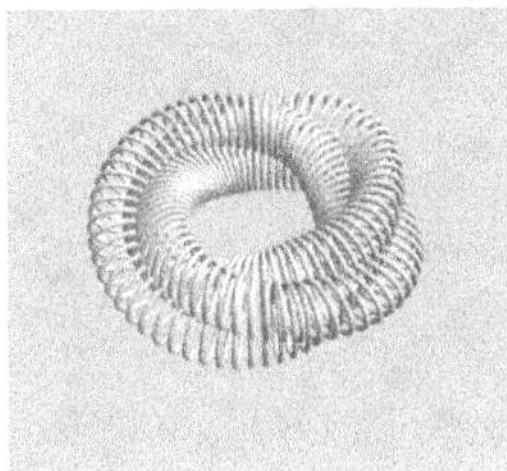

Fig. 10.58 Projection of several K_θ.

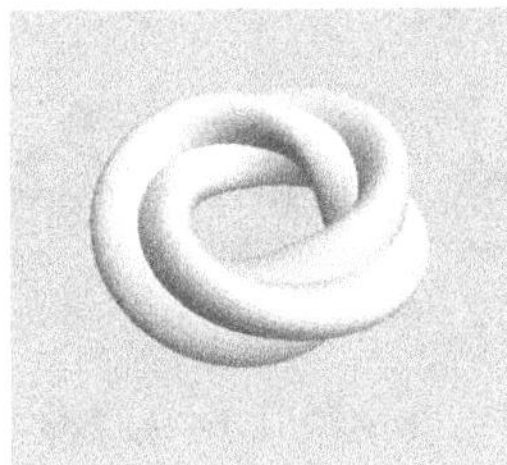

Fig. 10.59 Projection of turned trefoil knotting.

Example 10.13. A turn-spinning of $k \subseteq R^3$ is a knotting obtained by spinning k (with zero twist) along a curve on the standard torus $T \subseteq R^3$ such as is shown in Figure 10.53. If we have an isotopy K_t of a circle, $0 \leq t \leq 1$ then $T(K_t)$ is an isotopy between $T(K_0)$ and $T(K_0)$. For any 1-knotting K_0 of a circle consider a shrinking isotopy K_t so that K_1 can be described as a tiny knotting along a standard circle. For example, a standard trefoil knotting is isotopic to the knotting shown in Figure 3.8. If we turn this knotting, the tiny trefoil arc seems to rotate along an unknotted circle. A projection of this turned knotting is seen in Figure 10.54. ∎

Example 10.14. By repeating the turning isotopy, we can define a two-turn knotting. Figure 10.60 shows a projection of a two-turn trefoil knotting. If we isotop this trefoil knotting by shrinking as we considered in Remark 10.13, we obtain a projection as in Figure 10.61 which is a trefoil knotted arc spun along the curve shown in Figure 10.62. ∎

Just as for spun knots, there are many distinct non-trivial turn-spun knottings. The projection of a turn-spun knotting $T(K)$ is a turn-spinning of the projection of K. The j-shards of $T(K)_*$ correspond in an obvious way to the turn-spinning of the j-shards of K_*. For example, the projection of the trefoil Figure 4.20 has six 1-shards which are arcs. The 1-shards of the projection the turn-spun trefoil Figure 10.59 are six annuli clearly one for each arc. Using this correspondence one can see that a knot-coloring of

Fig. 10.60 A twice turned trefoil knotting.

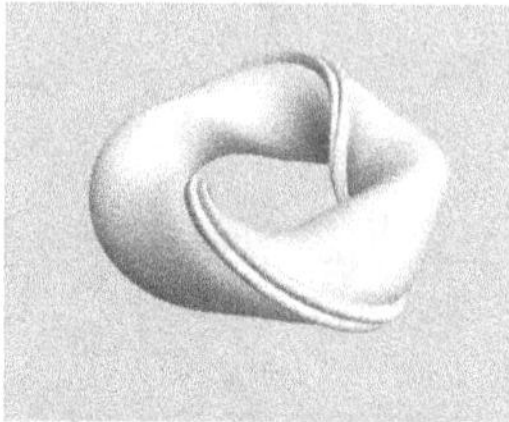

Fig. 10.61 Projection of knotting equivalent to Figure 10.60.

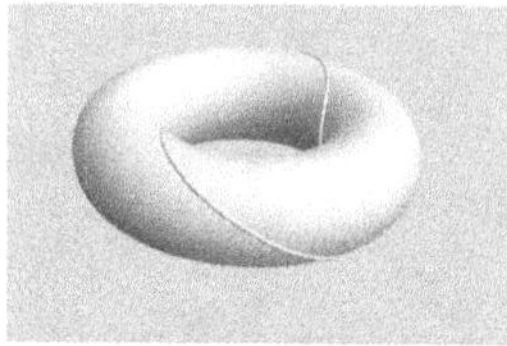

Fig. 10.62 Figure 10.61 is a spinning of trefoil arc along this curve.

K_* transfers to a knot-coloring of $T(K)_*$ and vice versa. For example, since the trefoil is 3-colorable, so is the turn-spinning of the trefoil. Furthermore, in [Boyle (1993)] it was shown that turning and spinning yield distinct knots and in particular that turn-spun trefoil is not equivalent to the spun trefoil.

For any knotting the projection of a turn-spinning is the same as the turn spinning of the projection. In particular the crossing schemes can be taken to be identical, and the spinning of a knotting and the turn-spinning of the same knotting are non-equivalent knottings but not distinguishable using colorings of any kind.

In [Shima (1997)] there is a comprehensive discussion of knotted tori with crossing set three disjoint circles, and in [Shima (2003)] it was noted that the turn-spun trefoil was also not ribbon. So, Proposition 9.2 "spun knots are ribbon knots" does not extend to knottings of surfaces which are not S^2, see also [Kawauchi (2002)].

It might seem that multiple turns would produce even more complicated examples than a single turn. However, we next see that **any two-turn 1-knotting of a circle K is equivalent to the torus obtained by Artin spinning K** (that is, zero turns). The proof of this follows immediately from Proposition 10.3. Proposition 10.3 is basically a version of a well-known geometric "trick" that goes by several names: "the Dirac's string trick", "the belt trick" or "the plate trick". Each of these can be roughly described as "two twists is equivalent to zero twists".

Prop 10.3. Let $T^2 \subseteq R^3 \subseteq R^4$ be the standard torus and λ a longitude of the T^2 as pictured in Figure 10.44 and let γ be a $(1,2)$-torus curve, as shown in Figure 10.62. There is a smooth isotopy F_t of T^2 in R^4, $0 \leq t \leq 1$, that $F_1(T^2) \subseteq R^3 \subseteq R^4$ and $F(\lambda) = \gamma$.

Proof. For visualization purposes, we view $T^2 \subseteq R^3 \subseteq R^4$ as the product of a circle and the perimeter of a square so that T is a union of four annuli, Figure 10.63.

We describe our isotopy in steps which are isotopies for t in the subintervals: $[0, \frac{1}{5}]$, $[\frac{1}{5}, \frac{2}{5}]$,$[\frac{2}{5}, \frac{3}{4}]$,$[\frac{3}{5}, \frac{4}{5}]$, and $[\frac{4}{5}, 1]$. Our Figures will show a projection of an isotopy in R^4 so technically subsets of R^3 are standard projections of themselves. For example, we view $T^2 \subseteq R^3 \subseteq R^4$ as projection $F_0^*(T^2)$. We depict λ as the center line of the top annulus.

Step 1: For $t \in [0, \frac{1}{5}]$, F_t is an isotopy in $R^3 \subseteq R^4$ of this torus from that shown in Figure 10.63 to position $F_{\frac{1}{5}}^*(T^2)$ shown in Figure 10.64.

Step 2: For $t \in [\frac{1}{5}, \frac{2}{5}]$, keeping most of the torus in R^3 we use a vertical isotopy of the torus in R^4 to raise a small portion directly up in the w-coordinate until at $t = \frac{2}{5}$ a wire-fame coding of the projection is as shown in Figure 10.65. During this Step, the *projection image* of the torus in R^3 does not change.

Step 3: For $t \in [\frac{2}{5}, \frac{3}{5}]$ the isotopy of the torus does not change the w-coordinate of any point, but, in projection goes from Figure 10.65 to Figure 10.67. Here $F_{\frac{2}{5}}^*(T^2)$ and $F_{\frac{3}{5}}^*(T^2)$ are embedded torii in R^3, however midway through this isotopy we must have some double points of the projection. For example, the projection $F_{\frac{1}{2}}^*(T^2)$ might look like Figure 10.66. Based on the wire-frame coding we can infer that this is a projection of an embedded torus in R^4.

Step 4: For $t \in [\frac{3}{5}, \frac{4}{5}]$ our isotopy of the torus in R^4 is vertical and a reverse the lifting of Step 2. In particular, $F_{\frac{4}{5}}(T^2) \subset R^3 \subseteq R^4$. We do not show $F_{\frac{4}{5}}^*(T^2)$ but it looks like Figure 10.67 but with uniform wire-coding.

Step 5: The torus $F_{\frac{4}{5}}^*(T^2)$ clearly bounds a solid torus in $X \subseteq R^3$. For $t \in [\frac{4}{5}, 1]$ we take an isotopy of the torus in $R^3 \subseteq R^4$ which takes the center of the solid torus X to the standard unit circle. At the end, our torus will look like Figure 10.68. Following our curve in this sequence of Figures we can see that our curve representing λ in $F_0(T^2)$ in Figure 10.63 has been taken to a curve that is equivalent to γ in $F_1(T^2)$. ∎

10.5 Deform Spinning

In the definition of spinning, we begin with a smooth 1-manifold $M \subseteq K^2 \subseteq R^4$ and a knotting $\kappa \subseteq R^3$, and we get a new knotting $K' \subseteq R^4$. We have tubular neighborhood A of $M \subseteq K$ and tubular neighborhood N of $M \subseteq R^4$ so that at each $x \in M$ we have a normal trivial ball pair

Fig. 10.63 $F_0^*(T^2)$ is isotopic to standard torus in $R^3 \subset R^4$.

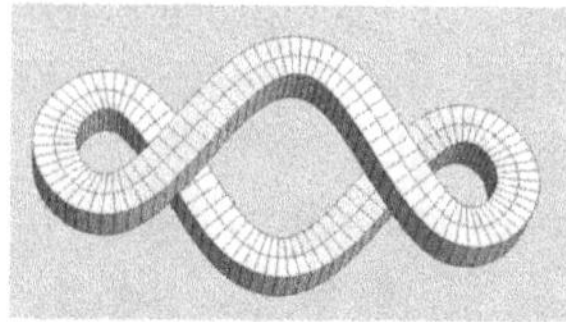

Fig. 10.64 $F_{\frac{1}{5}}^*(T^2)$ is isotopic to Figure 10.63.

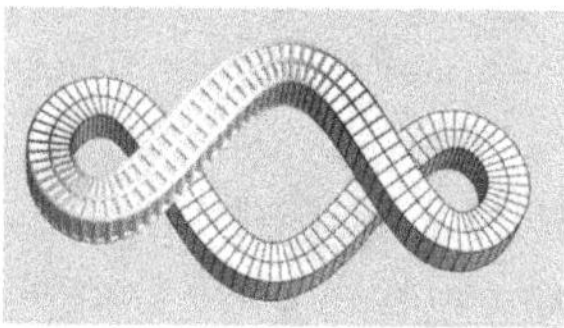

Fig. 10.65 $F_{\frac{2}{5}}^*(T^2)$ Wire-frame coding of knotting with portion lifted up from $R^3 \subseteq R^4$.

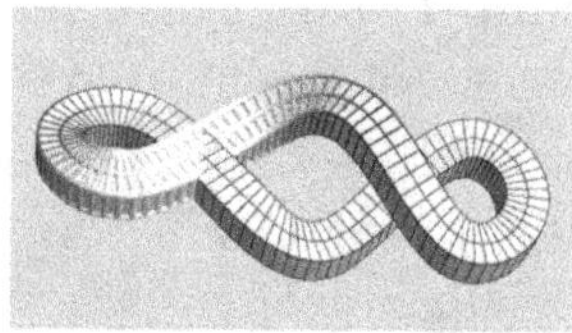

Fig. 10.66 $F_{\frac{1}{2}}^*(T^2)$ A projection of knotting which has double points.

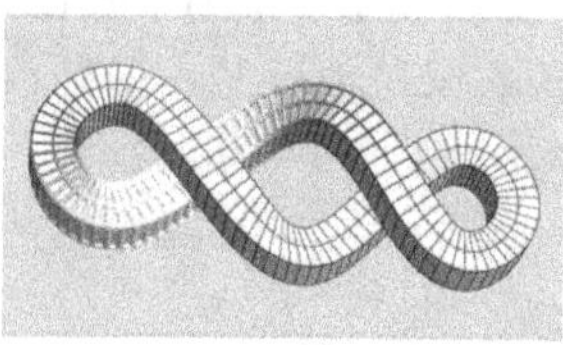

Fig. 10.67 $F_{\frac{3}{5}}^*(T^2)$ Projection at end of Step 3.

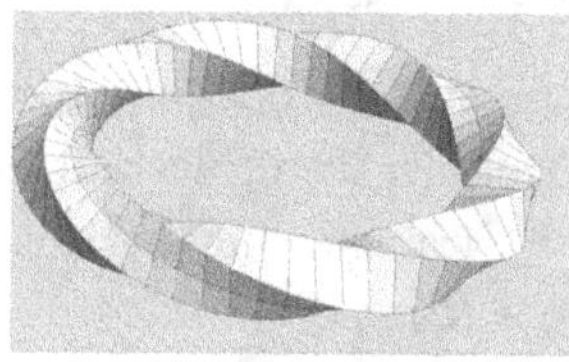

Fig. 10.68 $F_1^*(T^2)$ End of isotopy in R^4.

$\{x\} \times (D^3, D^1)$. We get K' by replacing $D^1 \subseteq D^3$ by a knotting of a 1-disk which represents the knotting κ of S^1. *The subtlety here is the choice of knotting of a 1-disk to represent* κ. In our spinning definition, we took *one choice* (D^3, α) to represent κ for all $x \in M$—we replaced each $\{x\} \times (D^3, D^1)$ with $\{x\} \times (D^3, \alpha)$.

But suppose for each $x \in M$ we have an individual knotting (D^3, α_x) each of which represents κ and let $\mathcal{A}$ denote the collection of these—$\mathcal{A} = \{\alpha_x\}_{x \in M}$. Now consider $A'' = \bigcup_{x \in M} \{x\} \times \alpha_x \subseteq M \times D^3 \approx N \subseteq R^4$. Without conditions on how α_x depends on x, the set A'' is potentially very strange—it may not be compact nor a manifold even topologically. So we impose the condition that A'' is a smooth proper submanifold of N diffeomorphic to A. This will give us a smooth 2-knotting $K' = (K - A) \cup A''$ with $K \approx K'$. Examples of such α_x are found in Sections 10.6 and 10.7.

Definition 10.7. Suppose we have $M \subseteq K^2 \subseteq R^4$, K a closed orientable 2-manifold, M a smooth 1-manifold and a knotting $\kappa \subseteq R^3$ and a collection of proper 1-disk knottings $\mathcal{A} = \{\alpha_x\}_{x\in M}$ such that $A'' = \bigcup_{x\in M}\{x\} \times \alpha_x \subseteq M \times D^3 \approx N \subseteq R^4$ is a smooth proper submanifold of N diffeomorphic to $M \times D^1$. Suppose A is a tubular neighborhood of $M \subseteq K$

Then we will say that **$\mathcal{A}$ is a deformation of κ** and that $K' = (K - A) \cup A''$ is a **deform-spun knotting**, or that **K' is the deform-spinning of $\mathcal{A}$**, with notation $(K, M) \bigotimes \mathcal{A}$. ∎

A standard definition [Litherland (1979)] is more useful for mathematical investigation, however Definition 10.7 is easier to visualize. Not every n-knotting of a sphere is a deform-spun spinning, [Budney and Mozgova (2009)].

10.6 Twist-Spun Spheres

Twist-spinning originated by R. H. Fox, see [Fox (1966)] is a type of deformation that can be done for any knotting of a circle. The first substantial paper, a response to questions about the construction raised by Fox, was [Zeeman (1965)] which accounts for the fact that the name Zeeman is most often associated with the concept. Important aspects of this construction will be considered in Section 10.11.

Example 10.15. We illustrate with a trefoil knotting. We represent κ by a proper arc knotting α_0 as in Figure 10.69. A single twist deformation is obtained by a gradual rotation by angle θ about the x-axis (which keeps the endpoints of α_0 fixed), where $0 \leq \theta \leq 2\pi$, as shown in Figures 10.69 to 10.76. If we construct the corresponding proper arcs β_θ in R^3_+ we obtain Figures 10.77 to Figures 10.84. Put a copy of β_θ in each page $H^3_\theta \subseteq R^4$ we would obtain the 1-twist-spun trefoil—a smooth 2-knotting for the 2-sphere.

We can do any number of twists. For p twists, we rotate by an angle of $p\theta$ (that is rotate p times as fast). The result is called the **p-twist spinning of κ**. A general knotting of this type is called a **p-twist spun sphere**. The number of twists can be either positive or negative. ∎

Example 10.16. Similarly a twisting of an entire circular knotting, Figures 10.85 to 10.92, can be spun to generate a knotting of a torus. In [Boyle (1988)] it is shown that for $n = 1, 2$, that the twist-spun torii (for trefoil or

Fig. 10.69 α_0.

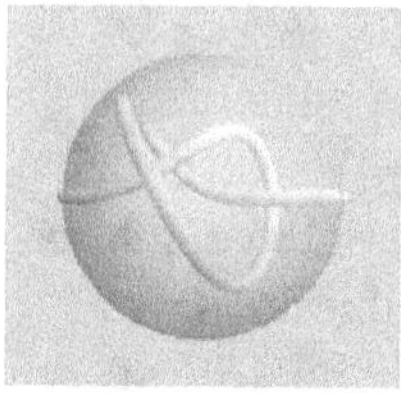

Fig. 10.70 $\alpha_{\pi/4}$.

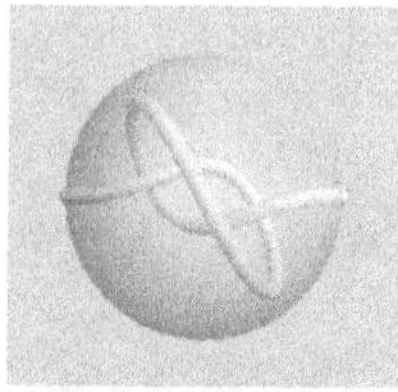

Fig. 10.71 $\alpha_{2\pi/4}$.

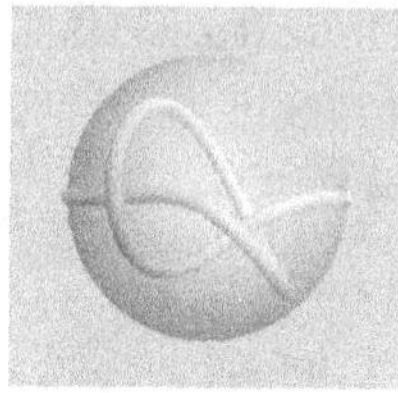

Fig. 10.72 $\alpha_{3\pi/4}$.

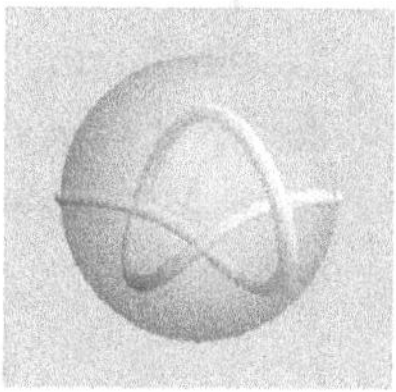

Fig. 10.73 $\alpha_{4\pi/4}$.

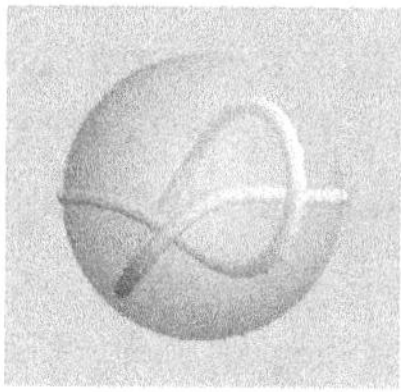

Fig. 10.74 $\alpha_{5\pi/4}$.

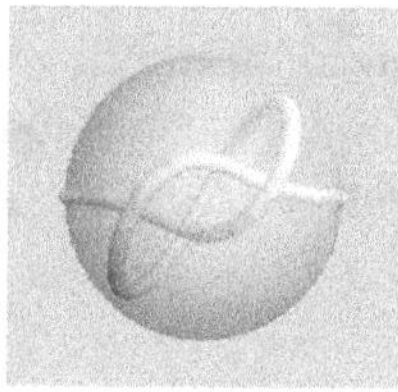

Fig. 10.75 $\alpha_{6\pi/4}$.

Fig. 10.76 $\alpha_{7\pi/4}$.

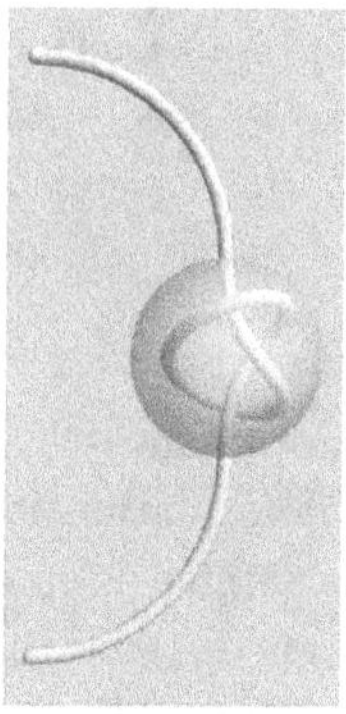

Fig. 10.77 β_0.

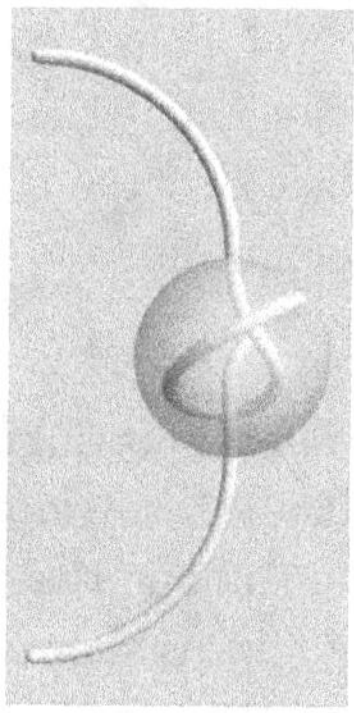

Fig. 10.78 $\beta_{\pi/4}$.

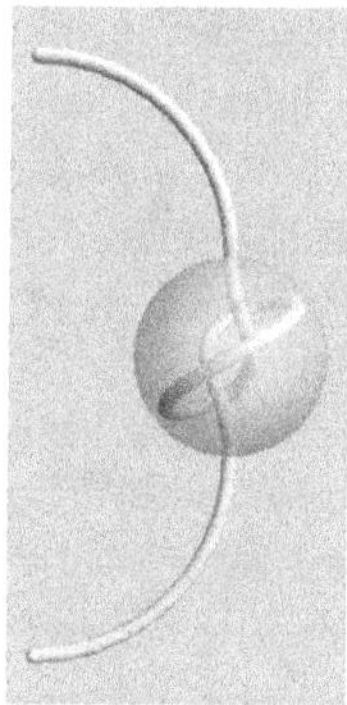

Fig. 10.79 $\beta_{2\pi/4}$.

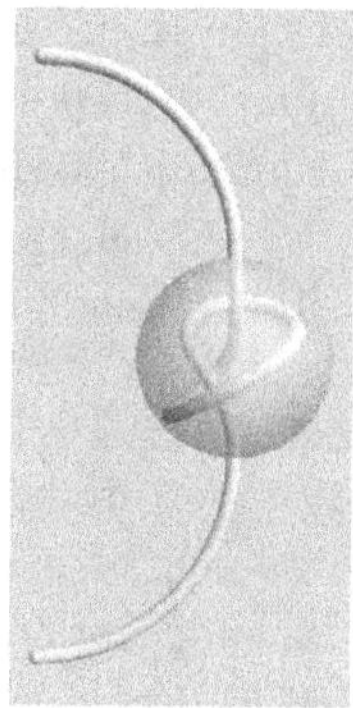

Fig. 10.80 $\beta_{3\pi/4}$.

figure-eight knottings) are connected sums of an n-twist-spun sphere and an unknotted torus, but for other n this is not the case. ▮

Notation 10.2. We write the knotting of a sphere obtained from knotting k by spinning with p twists as $\Sigma^{\circlearrowleft}(k,p)$. There is no standard notation for this. We will reuse this notation for higher-dimensional versions, Example 12.6. ▮

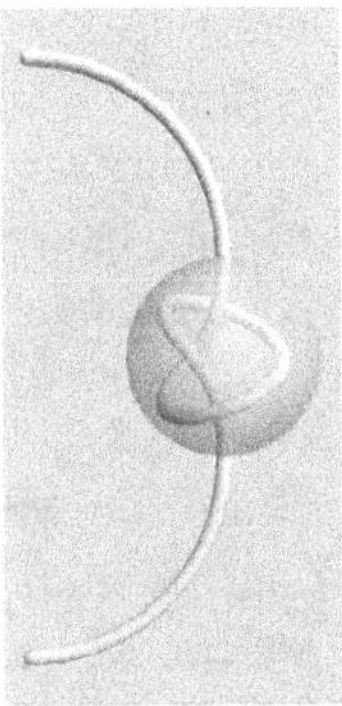

Fig. 10.81 $\beta_{4\pi/4}$.

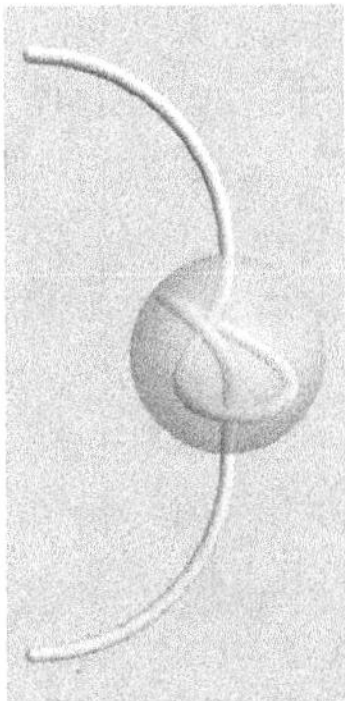

Fig. 10.82 $\beta_{5\pi/4}$.

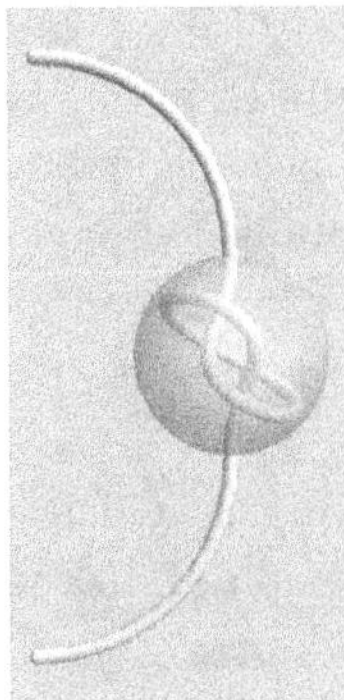

Fig. 10.83 $\beta_{6\pi/4}$.

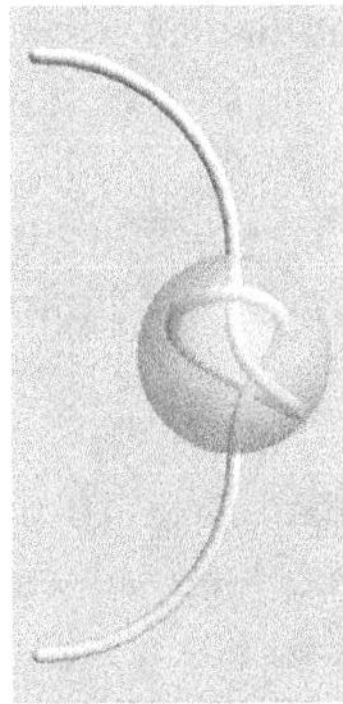

Fig. 10.84 $\beta_{7\pi/4}$.

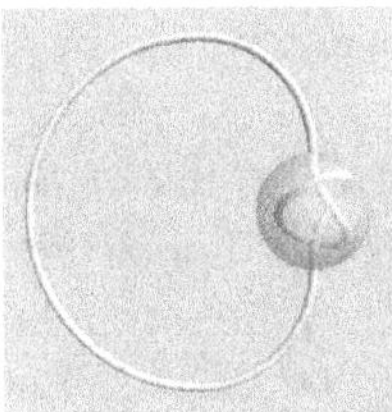

Fig. 10.85 β_0.

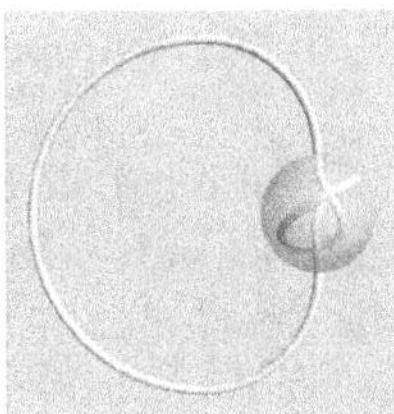

Fig. 10.86 $\beta_{\pi/4}$.

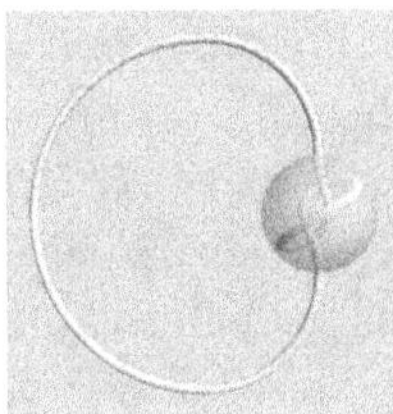

Fig. 10.87 $\beta_{2\pi/4}$.

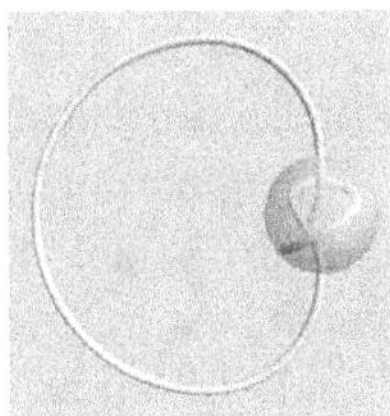

Fig. 10.88 $\beta_{3\pi/4}$.

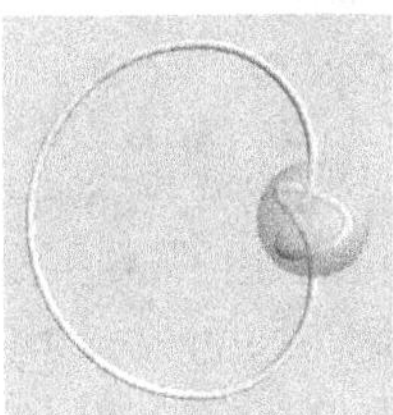

Fig. 10.89 $\beta_{4\pi/4}$.

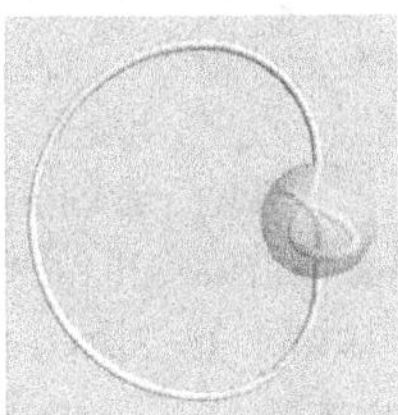

Fig. 10.90 $\beta_{5\pi/4}$.

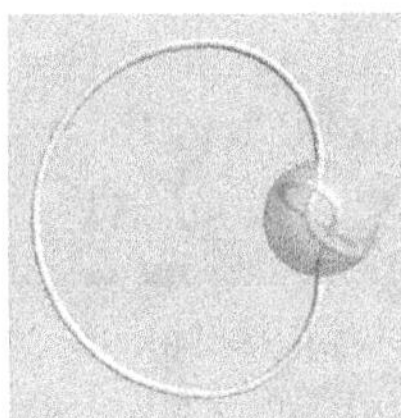

Fig. 10.91 $\beta_{6\pi/4}$.

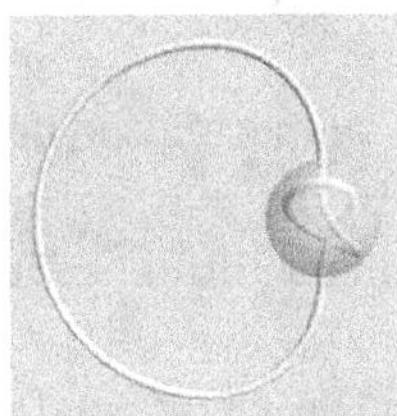

Fig. 10.92 $\beta_{7\pi/4}$.

10.7 Rolling a 1-Knotting

Knot rolling was introduced in the short cryptic paper [Fox (1966)]. This paper concerned a particular variation of spinning. Only one example was given, described by a sequence of knotting projections of a figure-eight knotting.

Fox's own description reads: " This operation (whose name derives from its resemblance to the operation of "rolling a stocking") is somewhat difficult to describe in totally precise terms, and I will content myself herewith referring to Figure II, in [3], where it is shown how to roll a figure-eight knot 4_1". The reference might have been to some earlier manuscript version since Zeeman's published paper, reference "[3]", contains no such figure nor is the figure-eight knot even mentioned. These details aside, Fox does clearly describe a proper isotopy of an arc that represents the knot-type 4_1.

Remark 10.6. Rolling is a deformation. Our description starts off by recalling the construction of disk pairs which represent knottings, Remark 4.2. For a smooth 1-knotting $K \subseteq S^3$ of a circle and $x \in K$ we can find a (small) 3-ball B_x with center x so that $(B_x, B_x \cap K)$ is equivalent to the standard pair (D^3, D^1) as in Figure 10.93. Now $D_x = S^3 - int(B_x)$ is also diffeomorphic to D^3. Suppose $\phi_x \colon D_x \to D^3$ is such a diffeomorphism. Let $\alpha_x = \phi_x(D_x \cap K)$. Then the pair (D^3, α_x) is a proper knotted arc in a 3-ball representing K.

Imagine ourselves at $x_0 \in K \subseteq S^3$. Label our ball B_0 and think of B_0 as a tiny vehicle with a transparent spherical windshield ∂B_0 and ourselves seated, at x_0, in this vehicle so that the nearby points of K are straight ahead. What we see outside our vehicle corresponds to (D^3, α_{x0}). Our vehicle moves smoothly along K, keeping our seat confined to K and, as we are seated, we are facing in the direction of K.

We travel one complete time about K. We begin at $x = 0$ and end at $x = 2\pi$. For each point x, we get B_x and corresponding (D^3, α_x). Comparing the beginning and the end, it is clear that, *as subsets* $B_0 = B_{2\pi}$. But we may not have $(D^3, \alpha_0) = (D^3, \alpha_{2\pi})$. The problem is that our vehicle might rotate, as it travels along K, and so, our seat, although still centered, and still facing in the direction of K, might be tilted when compared to when we started, and that will affect our view. At this point we need to have a more detailed description of the process of going from a knotting of a circle to a proper knotting of an arc.

Chose an orientation for K this will give us a unit tangent vector $\overrightarrow{v}_x$ at each point $x \in K$ which varies smoothly as we change x. The tubular neighborhood circle in R^3 is a product. This gives us a smoothly changing set of two orthogonal unit vectors, both orthogonal to v_x. Label them n_x and b_x. So, we have a frame at x $F_x = (\overrightarrow{v}_x, \overrightarrow{n}_x, \overrightarrow{b}_x)$ where we order these vectors so that each frame $(\overrightarrow{v_x}, \overrightarrow{n_x}, \overrightarrow{b_x})$ agrees with the standard orientation of R^3. If the torsion and curvature of K were not zero at x we could use a Frenet frame at x. In Figure 10.93 the orientation of K indicated by arrow along path determining the direction of tangent vector $\overrightarrow{v_x}$.

Let $\psi_x\colon R^3 \to R^3$ be a rigid motion that translates x to $\mathbb{O} = (0,0,0)$ and rotates and sends F_x to the standard frame at $\mathbb{O}$, Figure 10.94 and let $K_x = \psi_x(K)$. Next, let $\iota\colon (R^3 - \mathbb{O}) \to R^3$ be inversion in a unit sphere, Definition 1.27. Let L_x be the image $\iota(K_x - \mathbb{O})$; L_x will be a knotting of R^1 tangent to the x-axis, Figures 10.95 and 10.96.

Essentially the arc knotting we want α_x is $L'_x = D^3 \cap L_x$. As in Remark 4.2, adjustments near endpoints of α_x can be made so that $\partial\alpha_x = S^0 = \partial D^1 \subseteq D^3$. Also, since inversion is orientation reversing we will use a reflection R in the xy-plane and so choose $\alpha_x = \big(D^3, R(L'_x)\big)$ as a **standard representative of** K. ∎

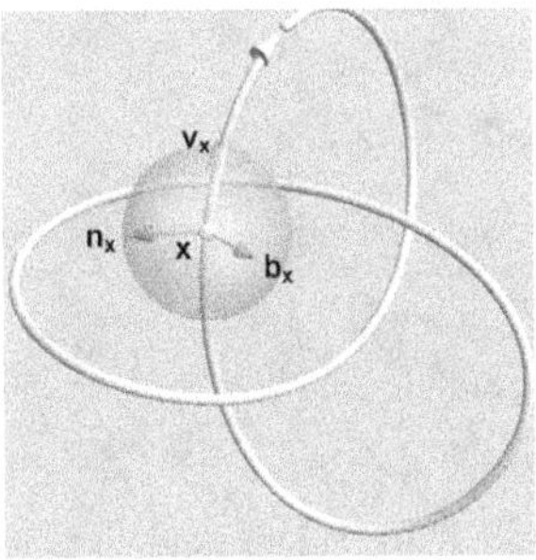

Fig. 10.93 A small ball and frame at $x \in K$.

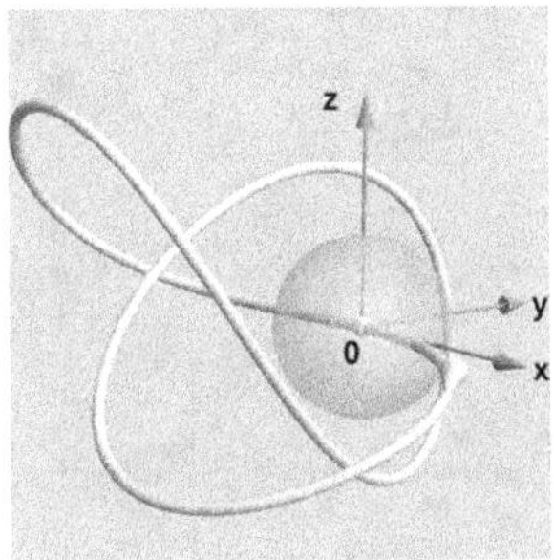

Fig. 10.94 Result of rigid motion which takes frame to standard frame.

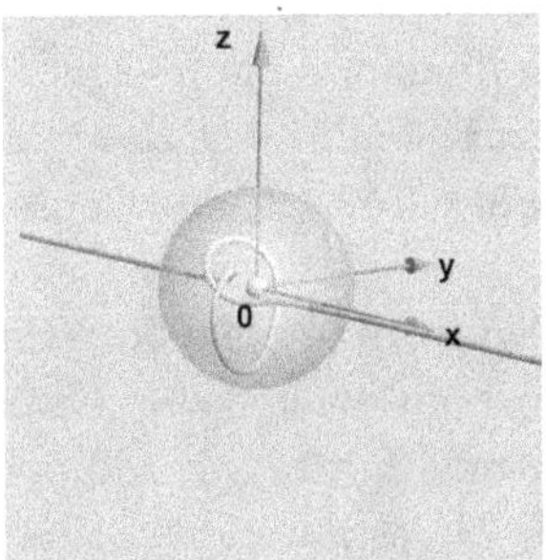

Fig. 10.95 Standard inversion sends knotting to knotting of line.

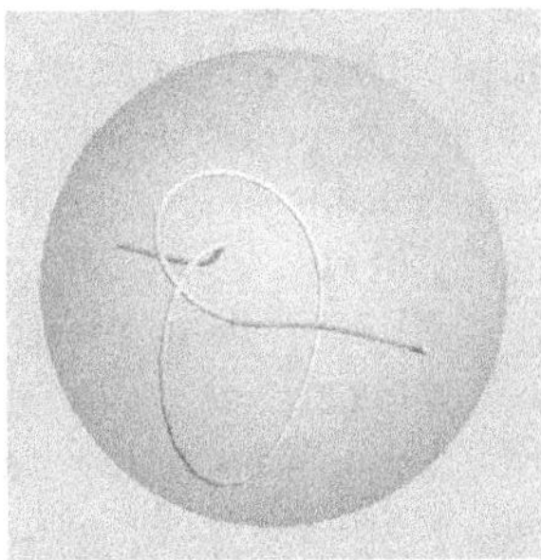

Fig. 10.96 Portion of line knotting in unit ball.

Definition 10.8. Given an oriented knotting of a circle $K \subseteq R^3$ we will define **a rolling of** K as a deformation-spin $\mathcal{A} = \{\alpha_x\}$ as defined above using a framing that corresponds to an untwisted trivialization of a tubular neighborhood of K. For twist see Remark 2.9. If the frame corresponds to n-twists, the deformation is called the **n-twist roll.** Knottings of spheres and torii constructed with such a deformation are called **roll-spun sphere** and **roll-spun torus.**

The knotting corresponding to going along the circle ρ times is called a **ρ-roll-spun sphere (or torus).** Doing this with τ twists one can have a **ρ-roll-τ-spun sphere (or torus).** ▮

If we have an isotopy of our original knotting K with a knotting K' one can extend this to an isotopy of a tubular neighborhood which will give a smooth change of framings and using this framing on K' it can be shown all possible choices will yield equivalent 2-knottings.

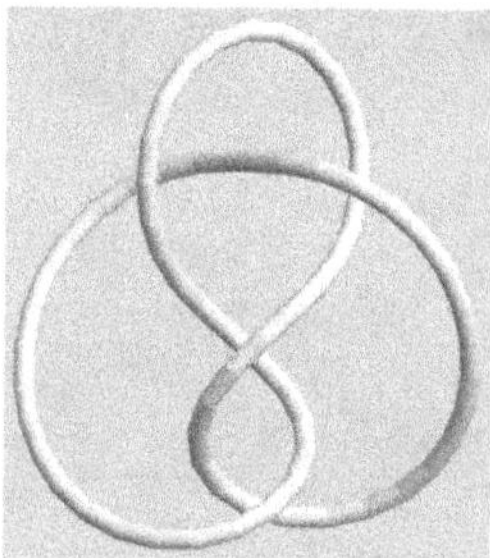

Fig. 10.97 Figure-eight knotting.

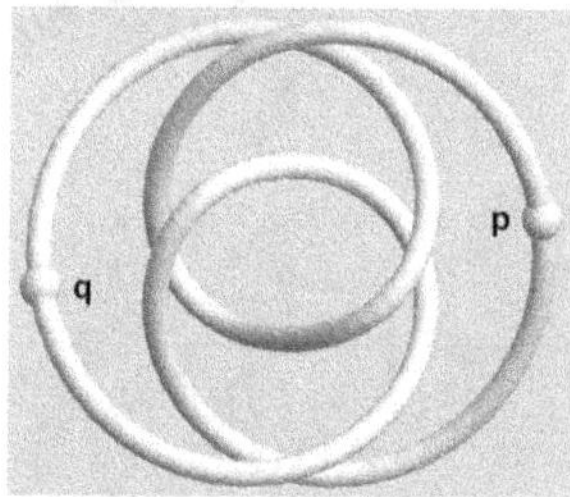

Fig. 10.98 Figure-eight knotting as symmetric closed braid.

Example 10.17. We examine the process of rolling for a figure-eight knotting. We will use the knotting K shown in Figure 10.98 since this was the original example of [Fox (1966)], however we will note that, due to use of a symmetry of this knotting, the example is not typical. An equivalent commonly used equivalent knotting is shown in Figure 10.97. The knotting of Figure 10.98 can be thought of as a two-fold cover as explained in Example 4.7. Since there is a rigid rotation of R^3 that sends K to itself and point p to point q, the knotted arc pairs (D^3, α_p) and (D^3, α_q) are identical.

Figures 10.99 to 10.118 show the arc knottings as we go from p to q along K. Figure 10.99 (rotated $\pi/2R$ clockwise) corresponds to Figure I of [Fox (1966)]. Similarly, Figure 10.103 corresponds to II, 10.107 corresponds to III,10.111 corresponds to IV and 10.118 corresponds to V. Since $(D^3, \alpha_p) = (D^3, \alpha_q)$ this deformation defines a smooth annulus and thus a 2-knotting of sphere or torus. However, this is not a complete roll as we defined it—one should continue past q until we return back to p. ▮

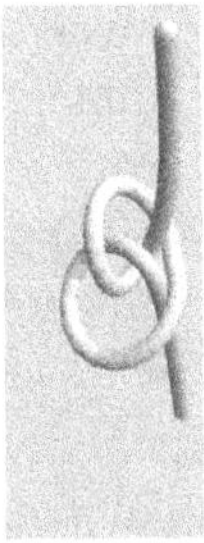

Fig. 10.99

Fig. 10.100

Fig. 10.101

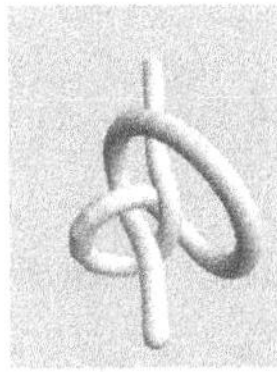

Fig. 10.102

Fig. 10.103

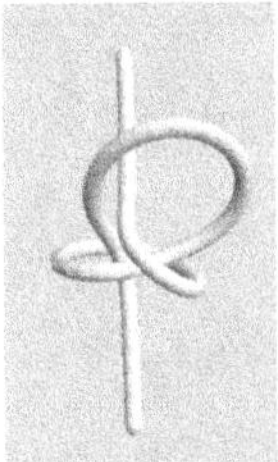

Fig. 10.104

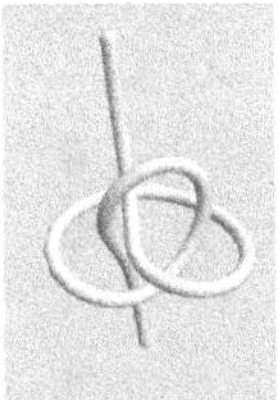

Fig. 10.105

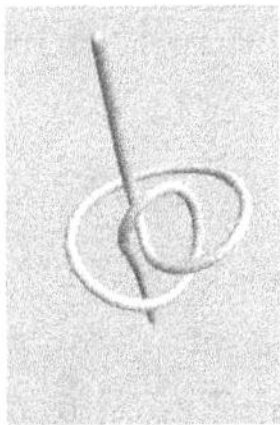

Fig. 10.106

Fig. 10.107

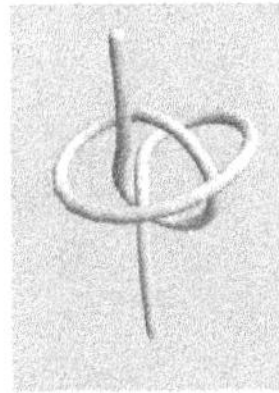

Fig. 10.108

Fig. 10.109

Fig. 10.110

Fig. 10.111

Fig. 10.112

Fig. 10.113

Fig. 10.114

Fig. 10.115

Fig. 10.116

Fig. 10.117

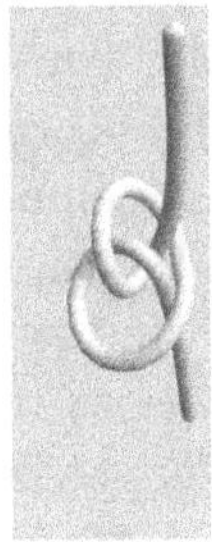

Fig. 10.118

Example 10.18. Generally twist-spinning and rolling are not equivalent deformations [Fox (1966)], however there are rare exceptions. Rolling a torus knotting is equivalent to a twist-spinning of that knotting (with the number of twists depending on choices of framing). Consider a trefoil knotting K. We say in Example 2.15 that K can be expressed as a three-fold branched cover and so we should expect a single rolling with blackboard framing to be a deformation that repeats itself three times during one roll. (One can calculate directly that the blackboard framing of trefoil knotting of Figure 2.54 has three twists. This also follows from fact that it is a three-fold cover of the knotting of Figure 2.51 whose blackboard framing has one twist.) Figure 10.119 shows a complete roll of a trefoil. Figure 10.120 shows the first third of the roll and so is equivalent to a twist. ▮

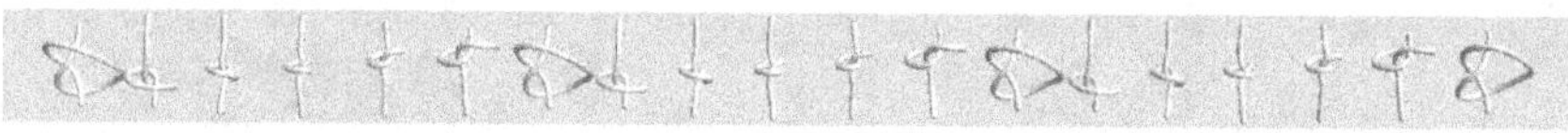

Fig. 10.119 A discrete track for a rolling of a trefoil knotting.

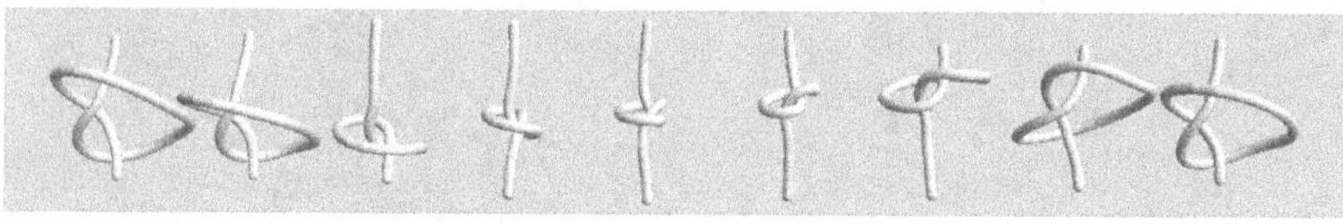

Fig. 10.120 The first third of Figure 10.119.

Example 10.19. The rolling of a composite knotting is interesting and useful, as we will see in Section 10.10. We consider in detail a rolling of the connected sum $\kappa = \tau\#\phi$, Figure 10.121, where ϕ is a knotting for the figure-8, τ a trefoil knotting. We show a discrete track of this isotopy using consecutive points $\{x_0, \ldots, x_{32}\}$ of along the knotting κ. Using the blackboard framing we obtain a sequence of arc knottings $\{\alpha_0, \ldots, \alpha_{32}\}$. In a composite knotting, there is no mathematically well-defined location of the summands. However, in these figures, since they are all produced by inversions of the same knotting, we can keep track of the two original portions of the circle and refer to them as the trefoil portion, and the figure-eight portion of the corresponding arcs.

If we take a point x on the trefoil arc in Figure 10.121, the trefoil portion is near x, and the figure-eight portion is relatively far away. A geometric inversion centered about x will look like Figure 10.122. Points near x are sent far away, and far away points are brought close. This way the figure-eight portion becomes small, and the trefoil becomes (relatively) large. One might describe this arc as a small figure-eight knotting tied along a larger trefoil knotting.

Let us start with α_4 which we describe as a smaller figure-eight above a larger trefoil, then follow through the Figures 10.123 to 10.130. All of these Figures have the same viewpoint in R^3. It is not possible to find a single point of view so that all figures are easy to see, but we wanted to portray an accurate set of illustrations.

A clearer deformation for a connected sum of figure-eight and trefoil knottings constructed "by hand" is shown in Figures 10.131 to 10.140.

In any case, we informally describe our isotopy as a sequence of "**shrinks, swells and slides**", where "shrinks", "swells", and "slides" are simple isotopies, and "sequence" means concatenation of isotopies. We describe this deformation as follows: From Figures 10.131 to 10.132 we shrink the figure-eight; from Figures 10.132 to 10.140, we slide the figure-eight knotting along the trefoil knotting; from Figures 10.140 to 10.141 we swell the figure-eight knotting; from Figures 10.141 to 10.142 we shrink the trefoil knotting; from Figures 10.142 to 10.151 we slide the trefoil knotting along a figure-eight knotting; from Figures 10.151 to 10.152 we swell the trefoil knotting.

In the other figures knowing what to look for, one can now see shrinking, swelling and sliding in Figures 10.123 to 10.130 with the slight complication that the blackboard framing is not untwisted, so there is twisting going on as well. ∎

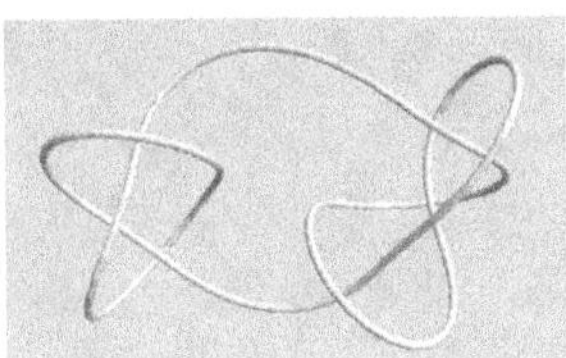

Fig. 10.121 A connected sum $\kappa = \tau \# \phi$ of a trefoil τ shown at left and a figure 8 ϕ shown on right.

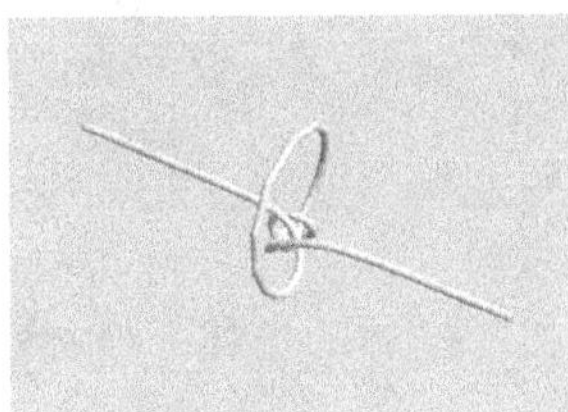

Fig. 10.122 A knotting of an arc made by inverting the knotting of Figure 10.121 in one of its points.

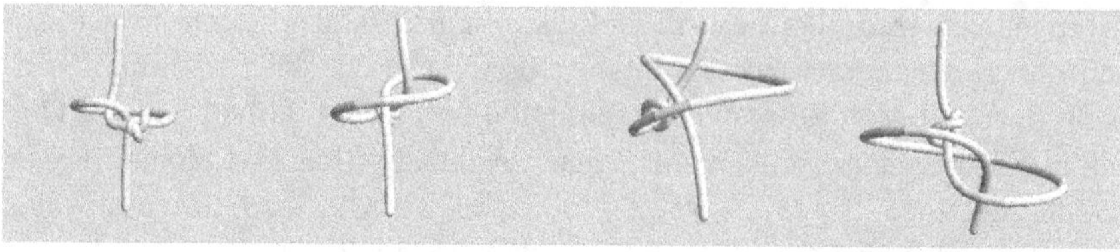

Fig. 10.123 Steps $\alpha_0, \alpha_1, \alpha_2, \alpha_3$, of Example 10.19.

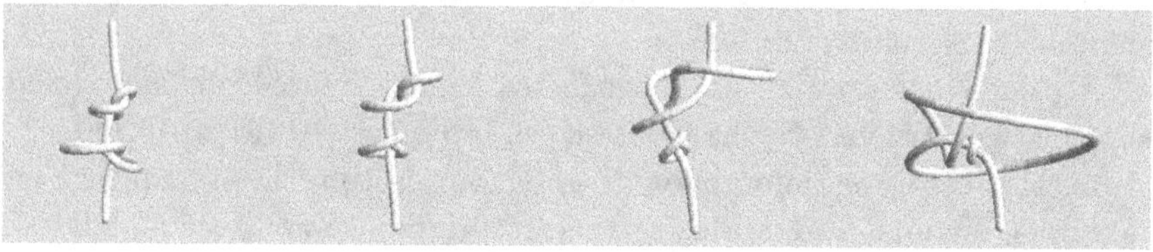

Fig. 10.124 Steps $\alpha_4, \alpha_5, \alpha_6, \alpha_7$, of Example 10.19.

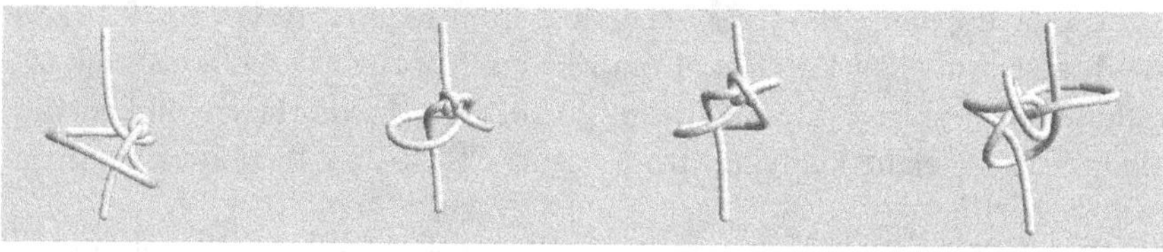

Fig. 10.125 Steps $\alpha_8, \alpha_9, \alpha_{10}, \alpha_{11}$, of Example 10.19.

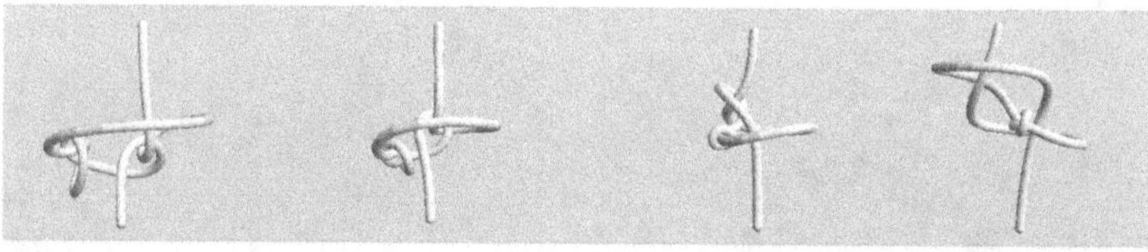

Fig. 10.126 Steps $\alpha_{12}, \alpha_{13}, \alpha_{14}, \alpha_{15}$, of Example 10.19.

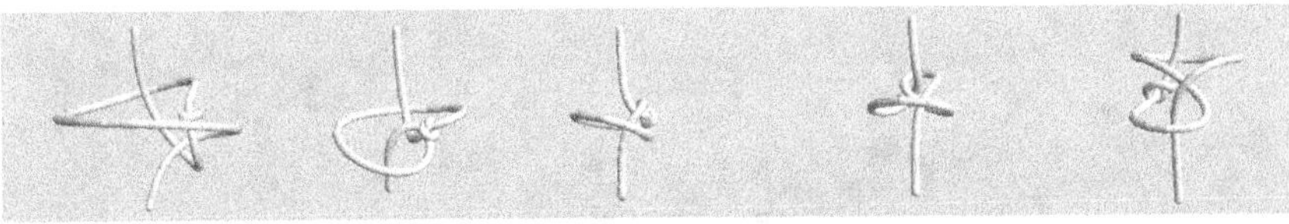

Fig. 10.127 Steps $\alpha_{16}, \alpha_{17}, \alpha_{18}, \alpha_{19}, \alpha 20$, of Example 10.19.

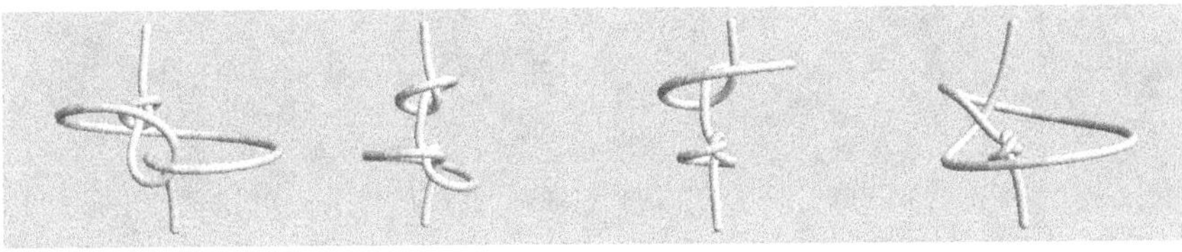

Fig. 10.128 Steps $\alpha_{21}, \alpha_{22}, \alpha_{23}, \alpha_{24}$, of Example 10.19.

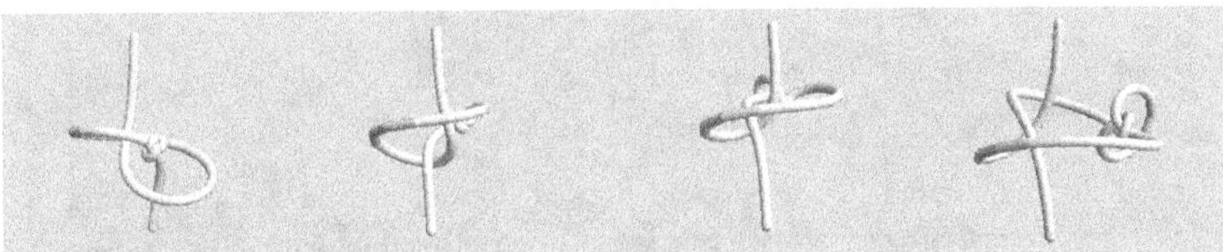

Fig. 10.129 Steps $\alpha_{25}, \alpha_{26}, \alpha_{27}, \alpha_{28}$, of Example 10.19.

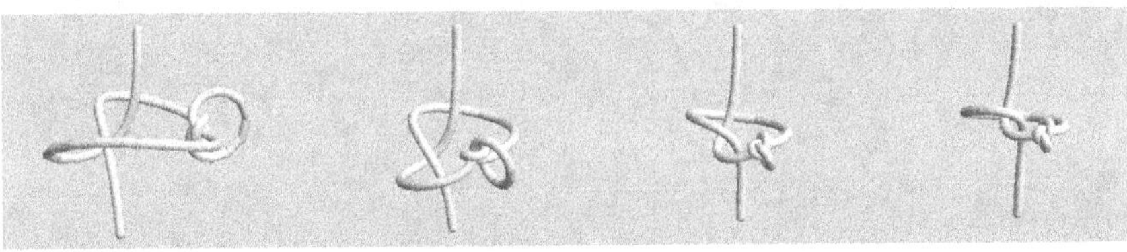

Fig. 10.130 Steps $\alpha_{29}, \alpha_{30}, \alpha_{31}, \alpha_{32}$, of Example 10.19.

Fig. 10.131

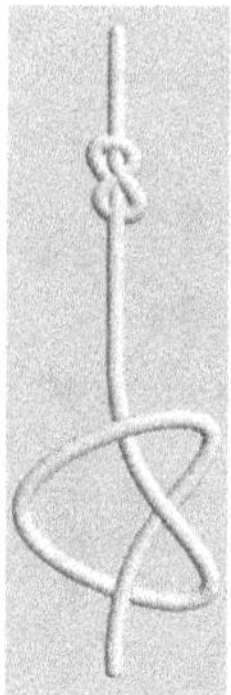

Fig. 10.132

Fig. 10.133

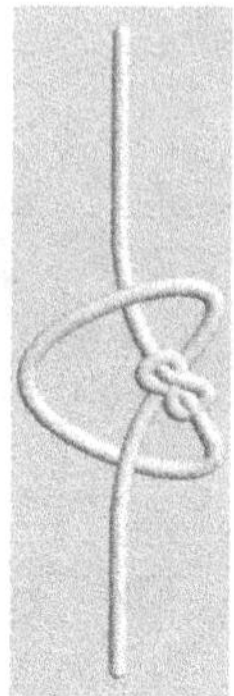

Fig. 10.134

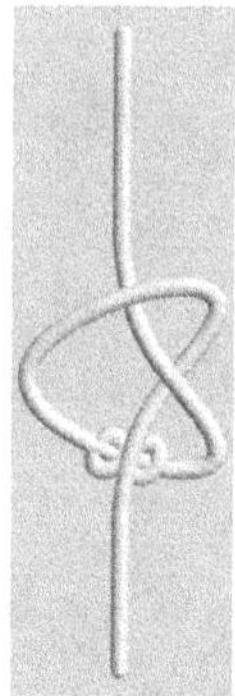

Fig. 10.135

Fig. 10.136

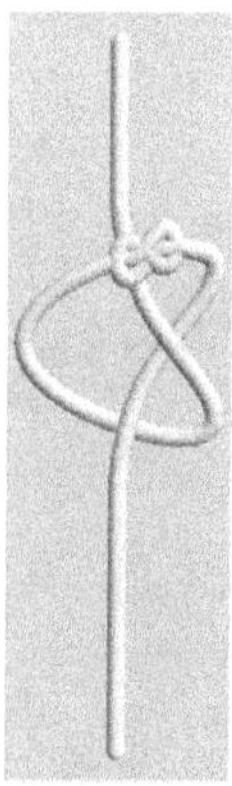

Fig. 10.137

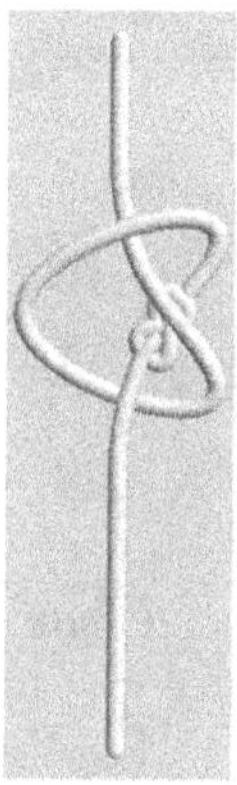

Fig. 10.138

Fig. 10.139

Fig. 10.140

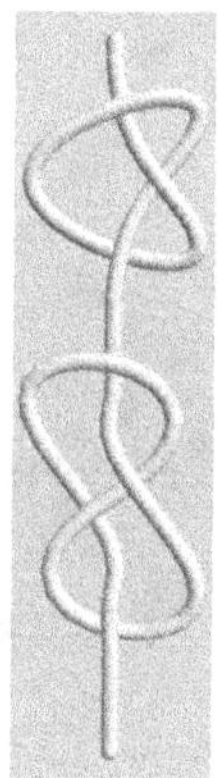

Fig. 10.141

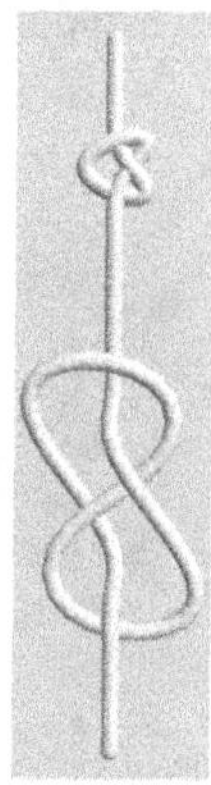

Fig. 10.142

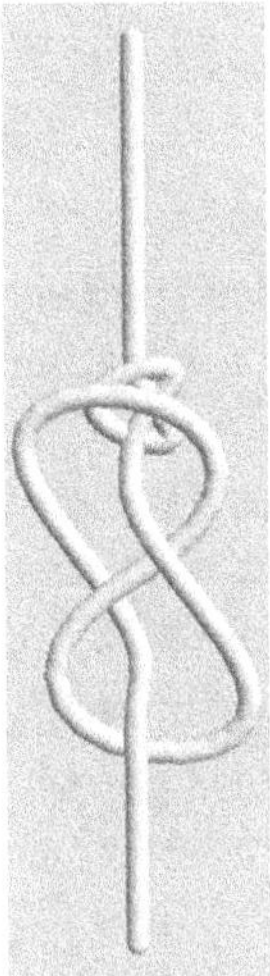
Fig. 10.143

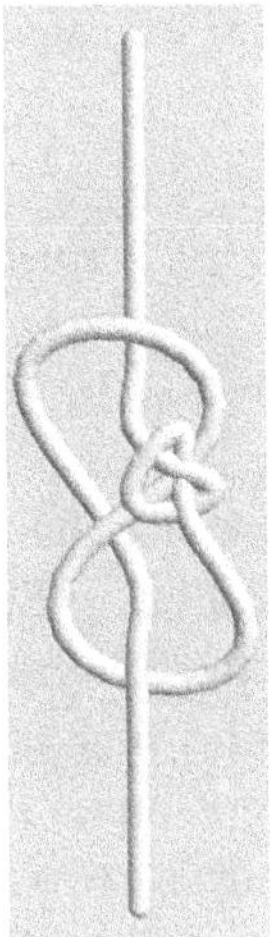
Fig. 10.144

Fig. 10.145

Fig. 10.146

Fig. 10.147

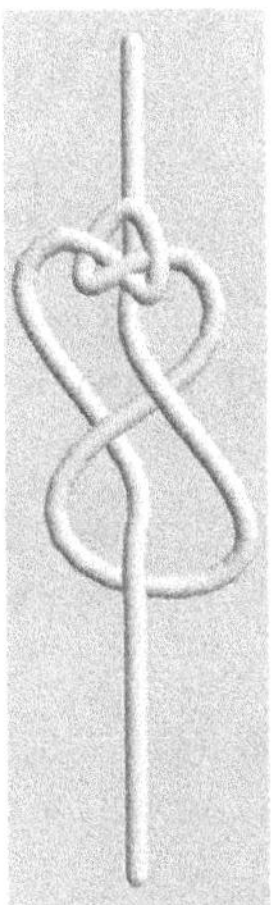
Fig. 10.148

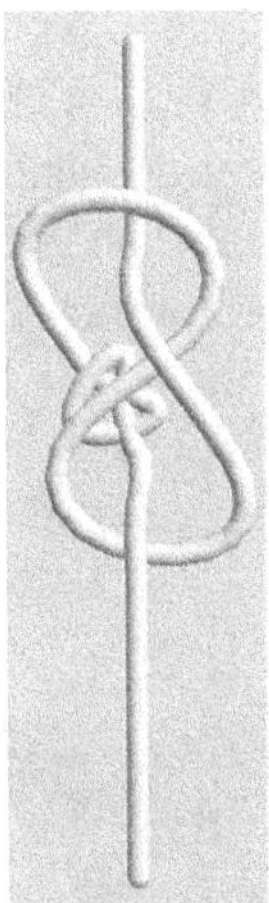
Fig. 10.149

Fig. 10.150

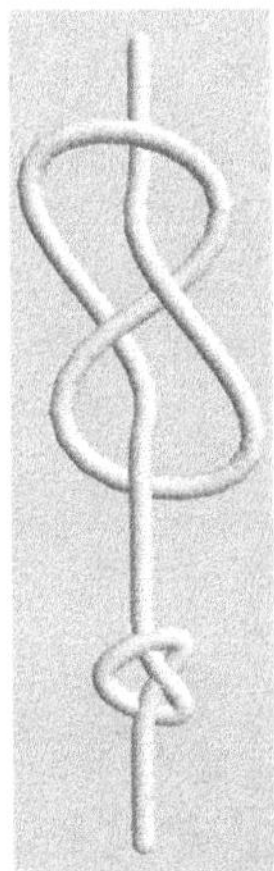
Fig. 10.151

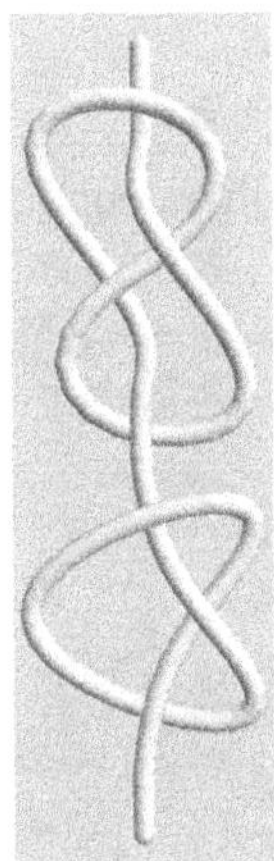
Fig. 10.152

Versions of rolling for higher-dimensions is discussed in Remark 12.17.

10.8 Fractional Turning

In deform spinning of a knotting of a circle $K \subseteq R^3$ we use a deformation $h_t \colon R^3 \to R^3$ where, *as functions*, $h_0 = h_1$. But for an annulus knotting we only really need the images to be the *same subset* $h_0(S^1) = h_1(S^1)$, as happens for knots with symmetry. Symmetry for knottings arises naturally from knots which are branched covers. For example, the trefoil knotting K is a 3-fold branched cyclic cover of a trivial knotting, Example 2.15. A rotation of $\frac{2\pi}{3}$, using the branch set as axis, also called a $\frac{1}{3}$-turn, maps the trefoil knotting K to itself such that no point of K is sent to itself. Furthermore, this self-map of the trefoil knotting is orientation preserving.

Suppose K is produced as a p-fold branched cyclic cover of K' then for any integer q we can define an isotopy K_θ, where $0 \leq \theta \leq 2\pi$ which begins with K_0 corresponding to a standard trefoil in the page H^3_0 and generally $K_\theta \subseteq H^3_\theta$ corresponds to a rotation K_0 of $\theta * (p/q)$ about the branch axis. We call this **a** p/q**-turn**.

Prop 10.4. Suppose K is produced as a p-fold branched cyclic cover of K' The $1/p$-turn spun torus of the p-fold branched cyclic cover of K' is equivalent to the p-turn spinning of the 1-fold branched cyclic cover of K'.

Furthermore, the q/p-turn spun torus of p-fold branched cyclic cover of K is equivalent to the p/q-turn spun torus of the q-fold branched cyclic cover of K'. ∎

Fig. 10.153 Projection of several K_θ.

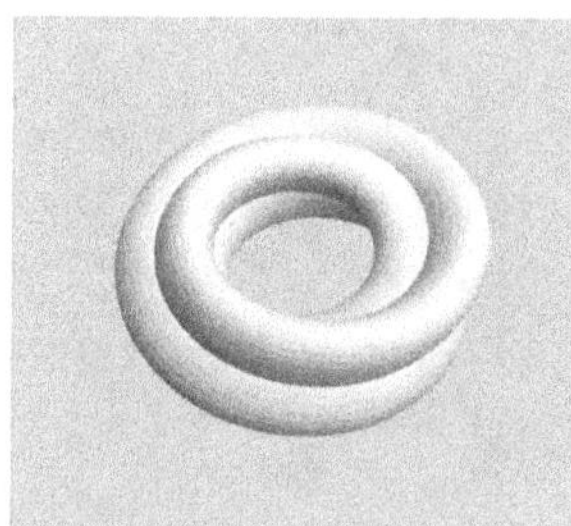

Fig. 10.154 Projection of a $\frac{1}{3}$-turn spun trefoil.

Example 10.20. A proof of Proposition 10.4 can be constructed following the next visual example for the 1/3-turn spun trefoil knotting. In Example

2.15 we saw the trefoil knotting K expressed as a 3-fold branched cyclic cover of a trivial knotting K'. In fact, the standard projection K' has a single crossing. We will show that the third-turned torus T of K is the 1-turn spun knotting of the trivial knotting K'. As a consequence, the third-turned torus of the trefoil is a trivial knotting of the torus.

As with other forms of spinning, we can say that the standard projection T_* of $1/3$-turn spinning of a knotting K is the same as the $1/3$-turn spinning of the standard projection K_* of K. Each K_* is contained in a page-disk in R^3, see Figure 10.153. Figure 10.154 shows that the crossing set of T_* is a single circle C. The familiar projection K_* has three crossings. When we spin, each of these crossings contributes an arc of double points of T_*. But because of the $1/3$-turn, these join together to produce a single circle of crossing points of T_*. In fact, C is a $(3, 1)$-torus curve.

The key is to consider how T_* intersects a cap-disk: C intersects each cap-disk D'_q transversely in a single point and so each $X_q = D'_q \cap T_*$ is a simple closed path with a single self-intersection. If we slice the projection T_* by the $R^2 \subseteq R^3$, we get two disjoint immersed curves, each with a single self-intersection, Figures 10.155 and 10.156. One of these curves is a subset of the cap-disk corresponding to D^2. The other is a subset of the cap disk corresponding to $R^2 - D^2$, Remark 1.10. Similarly, we could view S^2 as a union of two cap-disks and consider $S^2 \cap T_*$, Figures 10.157 and 10.158.

Similarly, the $2/3$-turn spun trefoil is equivalent to the $3/2$ turn of the 2-fold branched cyclic covering of K', a trivial knotting, so the $2/3$-turn spun trefoil is a trivial knotting of a torus.

From this, it is clear that T_* is also a turn-spun knotting for the trivial knotting K'. Looking closer, one can see that this is a 3-turn spinning since the crossing set C appears as a $(1, 3)$ torus curve with respect to the cap-disks. This is also seen in video form in [Roseman (1993b)]. ∎

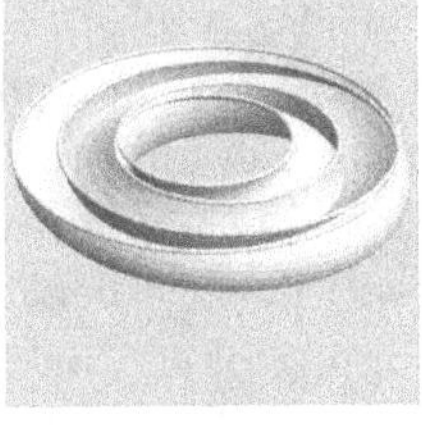

Fig. 10.155 Slice projection of 1/3-turn spun trefoil by R^2.

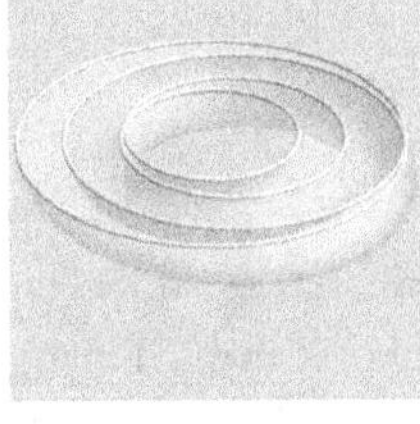

Fig. 10.156 Figure 10.155 with transparency.

Fig. 10.157 Slice projection of 1/3-turn spun trefoil by S^2.

Fig. 10.158 Figure 10.157 with transparency.

Similarly the half-turn spun Hopf link knotting is a trivial torus.

The 1/3-turn Borromean rings is a trivial torus and the 2/3-turn spinning of the Borromean rings is the same as the 3-turn spinning of the figure-eight knot which in turn is equivalent to a 1-turn spinning of the figure-eight knot.

In Example 2.16 we showed the granny knotting is a two-fold cyclic cover of a trefoil knotting. So the half-turn spun granny knotting is equivalent to the turn-spinning of the trefoil knotting. This is entirely to be expected since the half-turn spun granny knotting is the same as spinning a trefoil along a (2,1) torus curve in the standard torus while the turn spun trefoil is the same as spinning along a (1,2)-curve.

Remark 10.7. Representing a 2-knotting as a sequence of slices or slabs can be misleading. The spun trefoil torus and the 1/3-turn spun trefoil are *inequivalent* knottings yet the slabs seem almost the same. We show this in Figures 10.159 to 10.188. For lack of space in the book we show only some of the slices—enough to make this point.

View Figures 10.161 to 10.188 as a single 4×7 array. In all cases we are taking sections of the knotting in the y-direction. The first two columns are from a spinning of a trefoil knotting k (equivalently a 0-turn spinning of k). The second two columns are obtained from a 1/3-turn spinning of k.

The first column is the standard projection (w-projection) of these sections. The second column displays slabs—projections in the y-direction. If we assemble these first column projections they will fit together in R^3 as part of the standard projection of the 0-turn spinning of k. We indicate this in Figure 10.159 alternating grey levels to show sections. The second column shows corresponding slabs for the first column—these do not show the crossing set but do reveal the w-coordinate of our sections. In a similar way we view the 1/3-turn spinning of k in the last two columns—column 3 is the standard projection of sections and column 4 corresponding slabs.

Figure 10.159 shows a standard projection of a 0-turn knotting of a trefoil (i. e. a spun trefoil torus knotting. We will slice this seven times. This figure shows the standard projection of these, with alternating grey levels so one can see the total projection as the union of these slices. Figure 10.160 is produced similarly for the 1/3-turn spun trefoil.

Most would agree that it is very hard to distinguish the graphics of column 2 and the graphics of column 4, while projections, Figures 10.159 to 10.188 seem clearly different. ∎

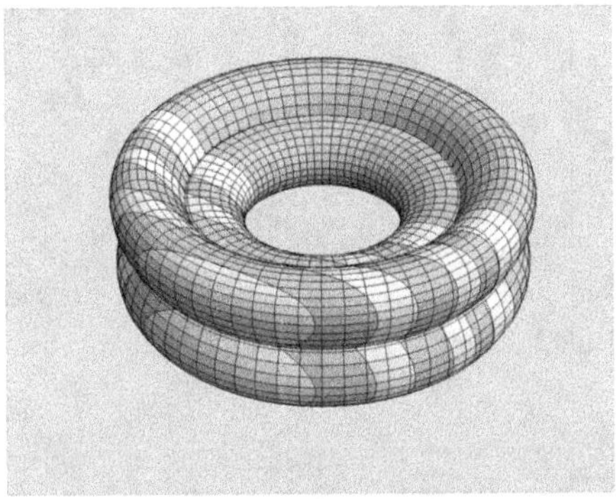

Fig. 10.159 Some slices of 0-turn trefoil.

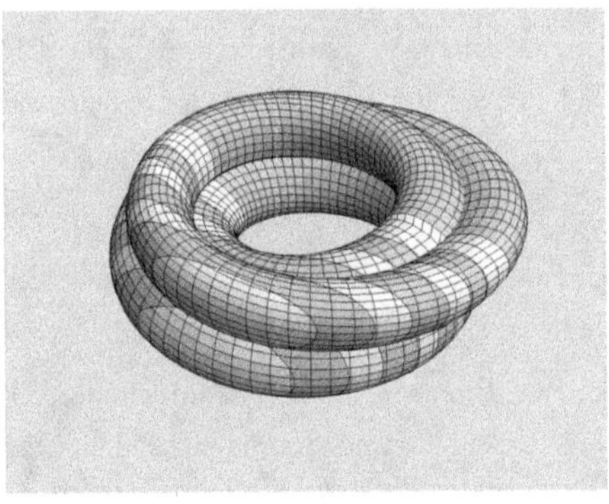

Fig. 10.160 Some slices of third-turn trefoil.

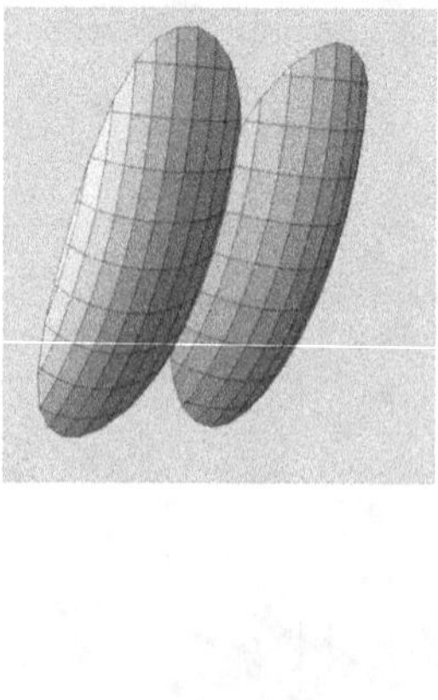

Fig. 10.161 0-turn.

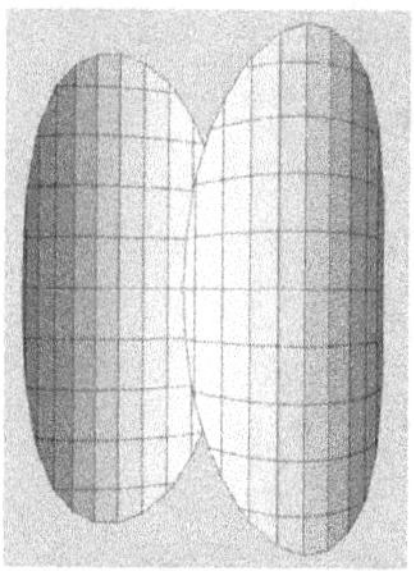

Fig. 10.162 0-turn.

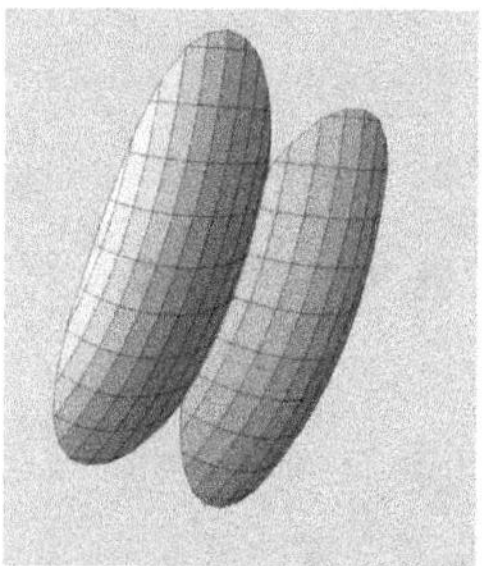

Fig. 10.163 $\frac{1}{3}$-turn.

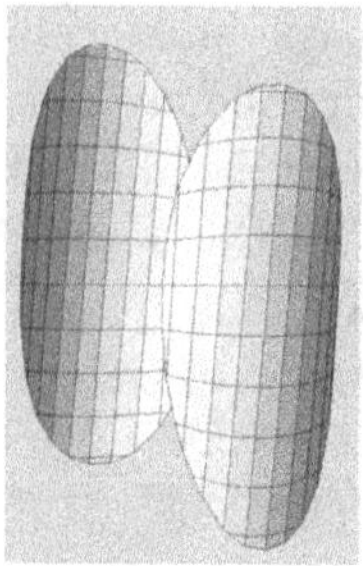

Fig. 10.164 $\frac{1}{3}$-turn.

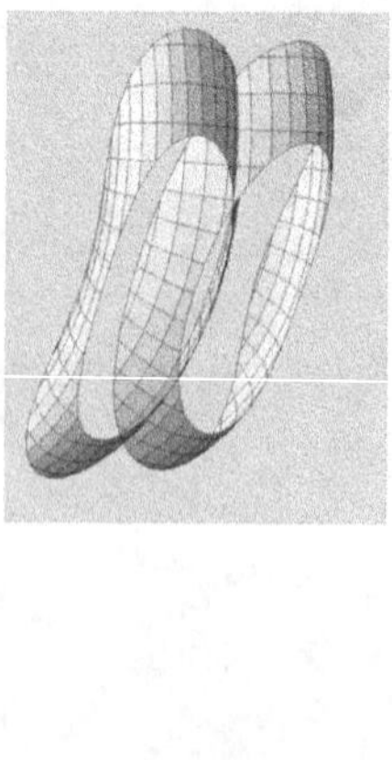

Fig. 10.165 0-turn.

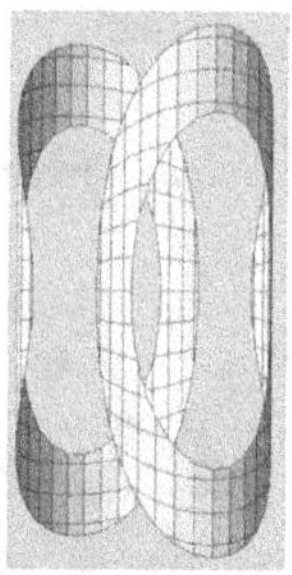

Fig. 10.166 0-turn.

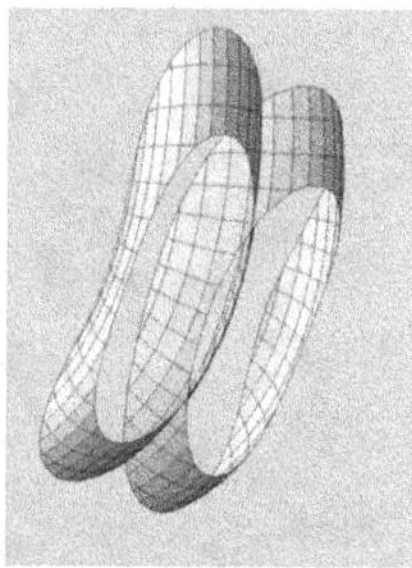

Fig. 10.167 $\frac{1}{3}$-turn.

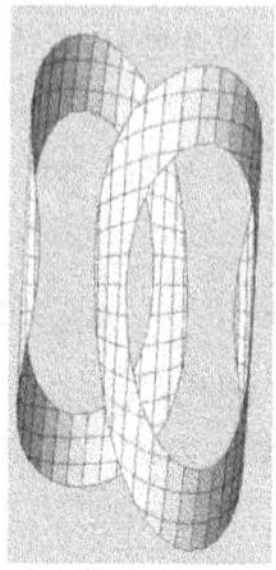

Fig. 10.168 $\frac{1}{3}$-turn.

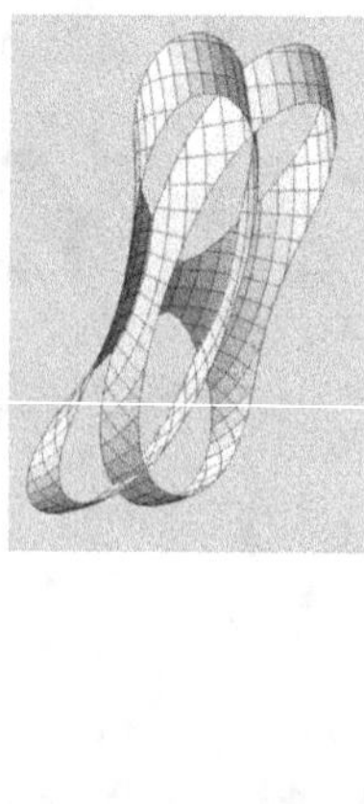

Fig. 10.169 0-turn.

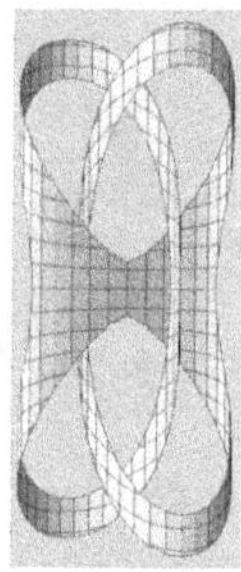

Fig. 10.170 0-turn.

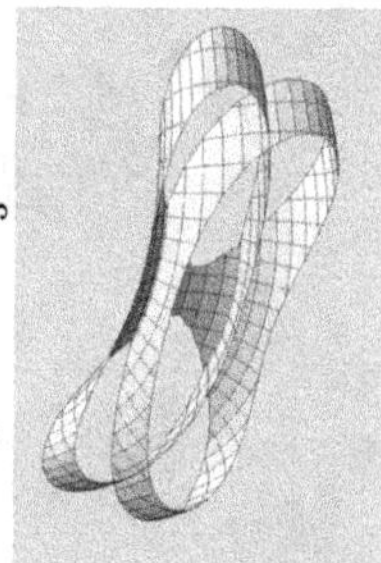

Fig. 10.171 $\frac{1}{3}$-turn.

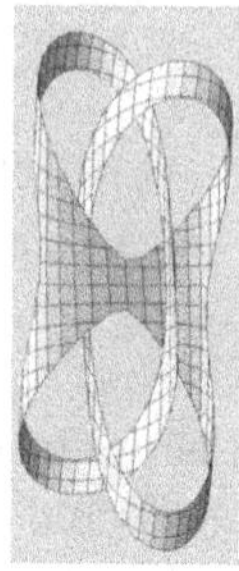

Fig. 10.172 $\frac{1}{3}$-turn.

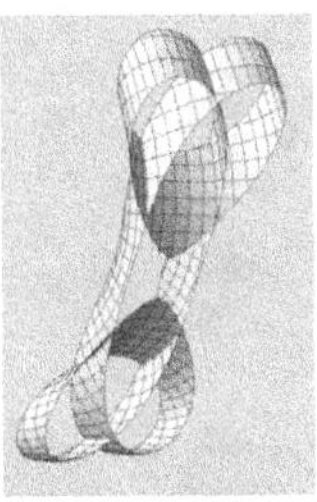

Fig. 10.173 0-turn.

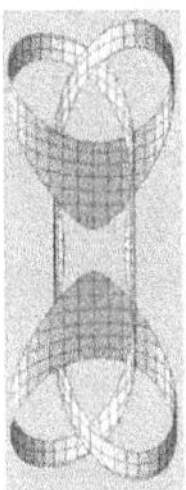

Fig. 10.174 0-turn.

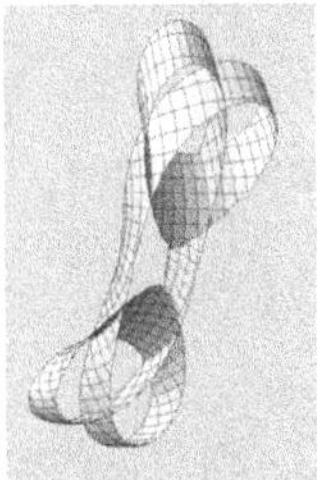

Fig. 10.175 $\frac{1}{3}$-turn.

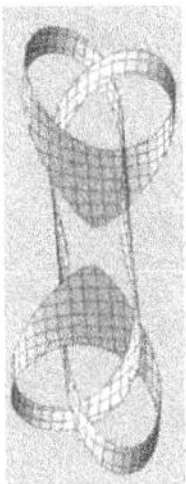

Fig. 10.176 $\frac{1}{3}$-turn.

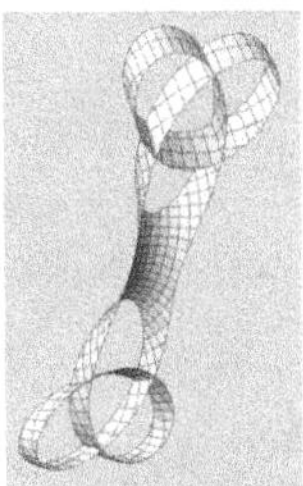

Fig. 10.177 0-turn.

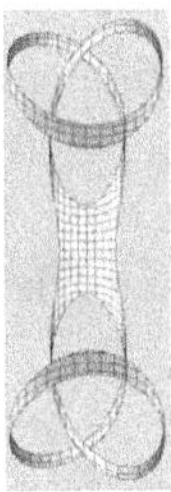

Fig. 10.178 0-turn.

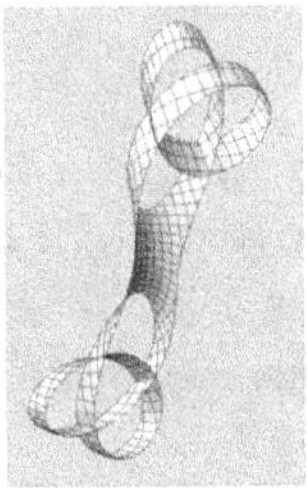

Fig. 10.179 $\frac{1}{3}$-turn.

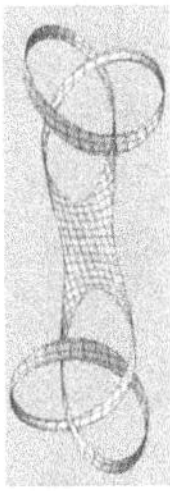

Fig. 10.180 $\frac{1}{3}$-turn.

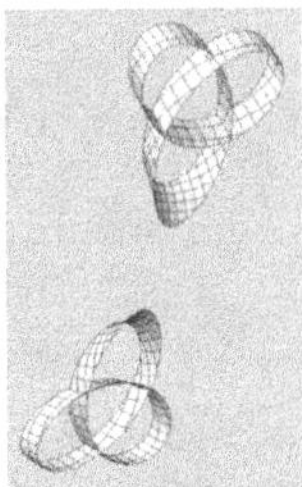

Fig. 10.181 0-turn.

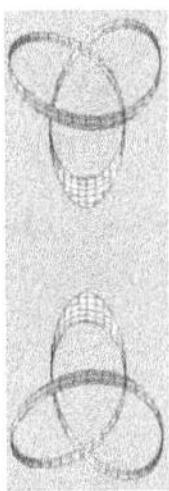

Fig. 10.182 0-turn.

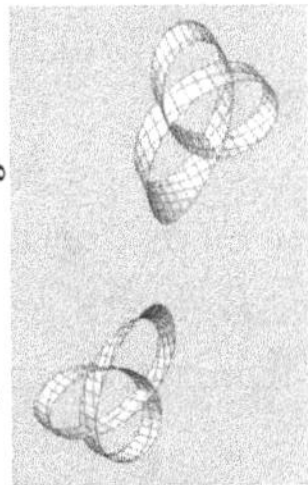

Fig. 10.183 $\frac{1}{3}$-turn.

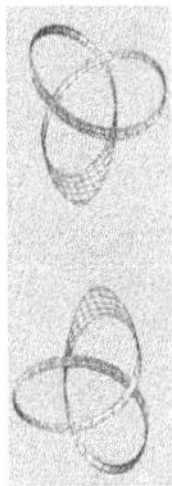

Fig. 10.184 $\frac{1}{3}$-turn.

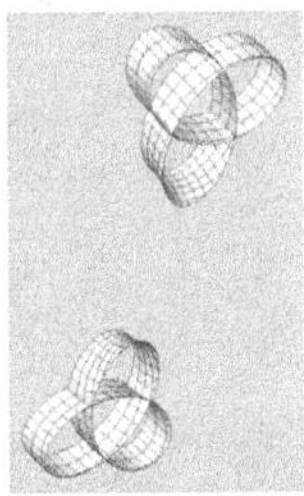

Fig. 10.185 0-turn.

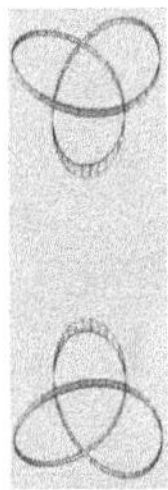

Fig. 10.186 0-turn.

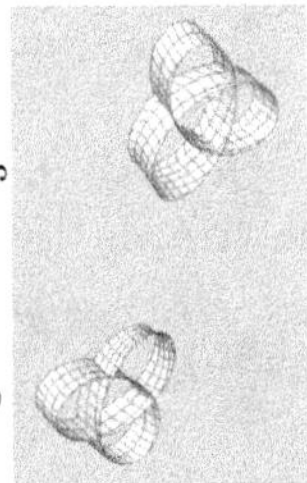

Fig. 10.187 $\frac{1}{3}$-turn.

Fig. 10.188 $\frac{1}{3}$-turn.

Similar examples are explored in [Teragaito (1992)]. As pointed out in [Litherland (1979)] one can consider a fractional turn which he calls a symmetry spin as a deformation. In the literature, sometimes what we call a turned torus is called "twist-spun", for example [Shima (2003)].

10.9 Flip-Spinning, Projective Planes, Klein Bottles

The standard Klein bottle and projective planes, Examples 6.25 and 6.26, were described using a planar circle k and a "half-flip" isotopy H_t of k so that $H_{2\pi}(k) = k$ *and* $H_{2\pi}$ is orientation reversing on k. We obtain non-standard knottings by replacing k with a *non-trivial* knotting K of a circle *if* K has a half-flip orientation reversing symmetry.

Definition 10.9. If $K \subseteq R^3$ is a 1-knotting, K **has a half-flip symmetry**, is **simply invertible** or is **strongly invertible** if there is a half-flip isotopy H_t, $0 \leq t \leq 2\pi$, Definition 10.6, such that for each component $K_0 \subseteq K$, $H_{2\pi}(K_0) = K_0$ in an orientation *reversing* way. ∎

Many 1-knottings are known to be strongly invertible, [Sakuma (1986)]. Our 2-knottings can also be considered as spinning a knotted arc K along a curve C in a 2-manifold $M \subseteq R^4$, except that the regular neighborhood of C in M is a Möbius band and not an annulus.

Example 10.21. A strongly invertible knotting can be used to construct a knotting of a Klein bottle. The trefoil knotting is strongly invertible, Figure 10.189. Figure 10.190 shows the projections of some of the meridians and Figure 10.191 is the projection of the resulting Klein bottle. ∎

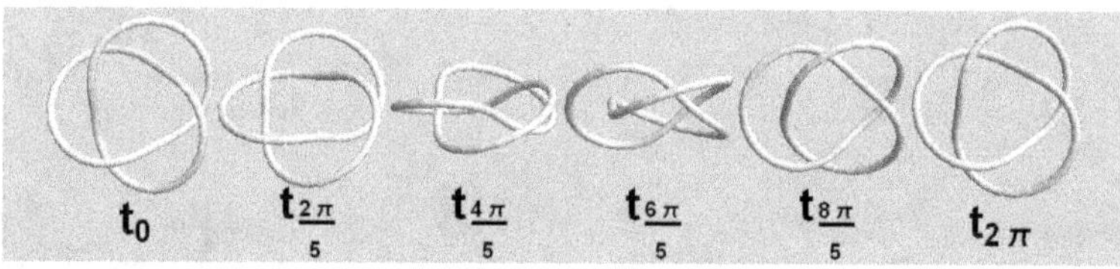

Fig. 10.189 Half-flip isotopy of trefoil knotting.

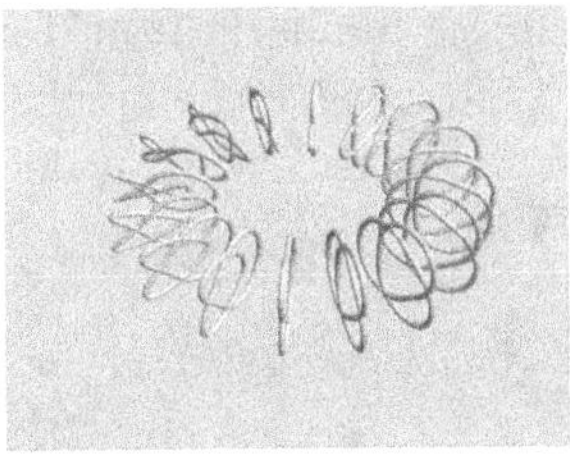

Fig. 10.190 Projection of meridians of half-flip spun trefoil.

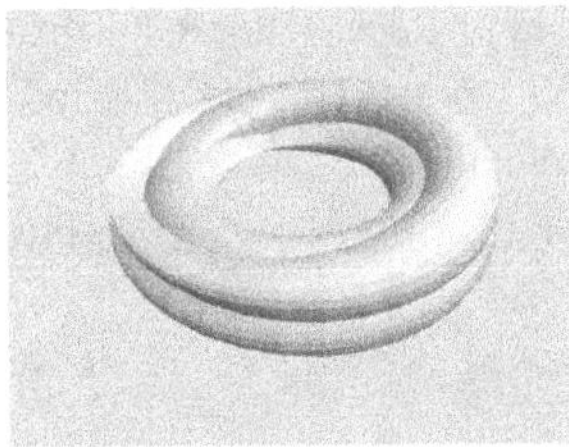

Fig. 10.191 Projection of half-flip spun trefoil.

Example 10.22. A small change in the half-flip isotopy will give a projective plane rather than Klein bottle. We will need our 1-knotting in H^3_+ to contain a single point of ∂H^3_+ (we will take it to be $(0,0,0)$) and also that point must be fixed under the half-flip isotopy. Several steps of this isotopy are shown in Figure 10.192. Projections of the meridians are shown in Figure 10.192. All of these projected circles pass through $(0,0,0)$. The entire projection of this knotting of P^2 is shown in Figure 10.193. ∎

Fig. 10.192 Meridian projections for P^2 made from half-flip of trefoil.

Fig. 10.193 Projection of P^2 knotting for Figure 10.192.

Example 10.23. Using a flipping isotopy of a Hopf link, Figures 10.194 to 10.207, we generate an interesting example of two linked Klein bottles, shown in projection, with wire-frame coding, in Figures 10.208 and 10.209. This projection has two triple points marked "x" in Figure 10.209.

In Figures 10.196 to 10.207, we have colored one component gray, the other white. The triple points of the projection correspond to Reidemeister type $\Omega.3$ moves. Suppose we use this to color our Klein bottles, K_w and K_g, where w and g, are white and gray. One triple point corresponds to Figure 10.200 the gray circle above and below the white one, so that the color type of the corresponding triple point is gwg. Similarly from Figure 10.205 we see the other triple point has color type wgw. In particular, by Proposition 7.3, we can say that our linking $K_w \cup K_g$ is not equivalent to a split linking of two standard Klein bottles. ▮

Fig. 10.194 Projection of meridians for half-flip spun Hopf link.

Fig. 10.195 Projection of half-flip sun Hopf link.

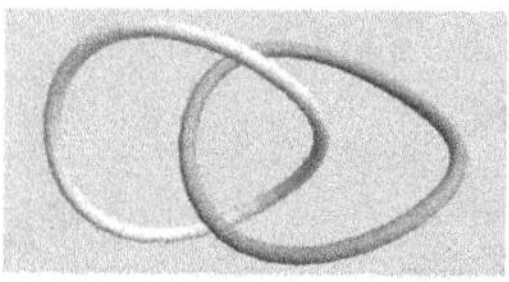

Fig. 10.196

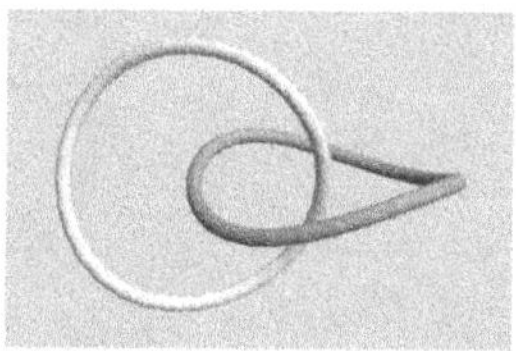

Fig. 10.197

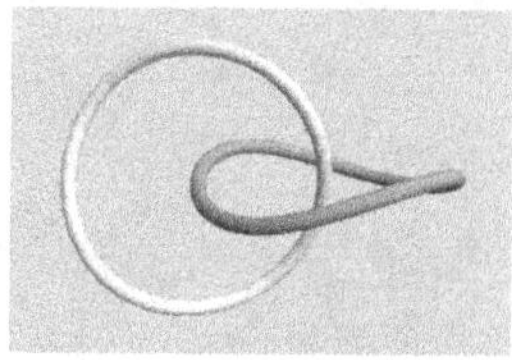

Fig. 10.198

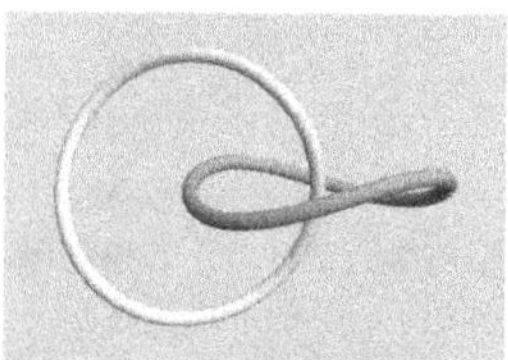

Fig. 10.199

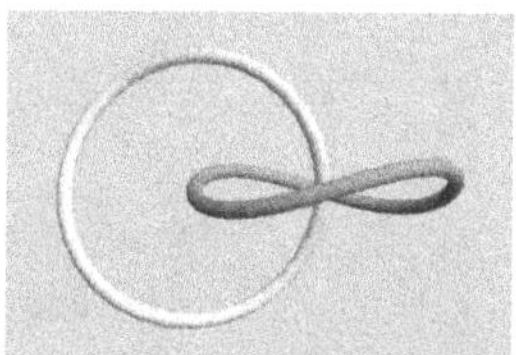

Fig. 10.200

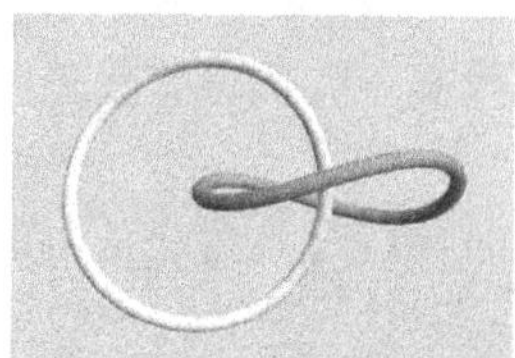

Fig. 10.201

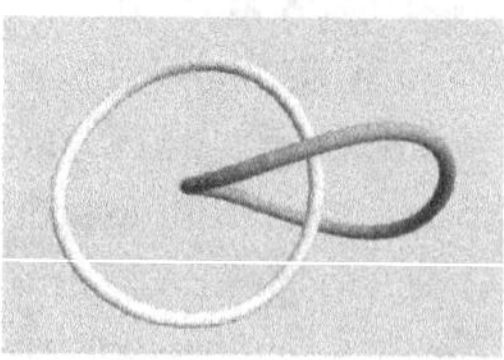

Fig. 10.202

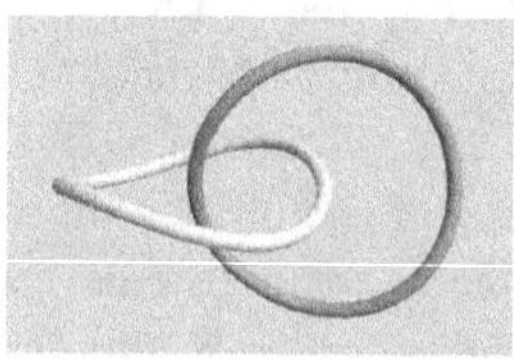

Fig. 10.203

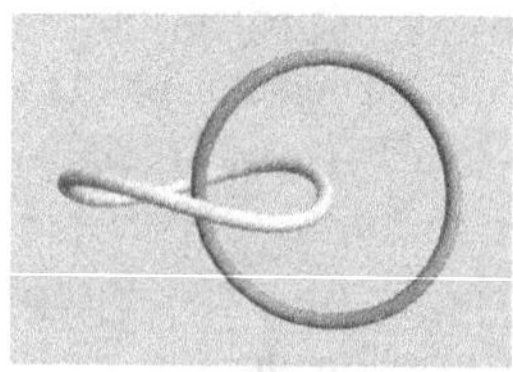

Fig. 10.204

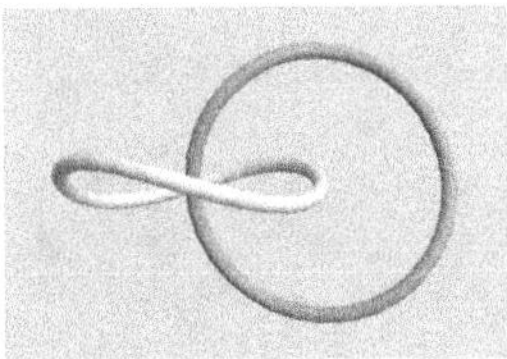

Fig. 10.205

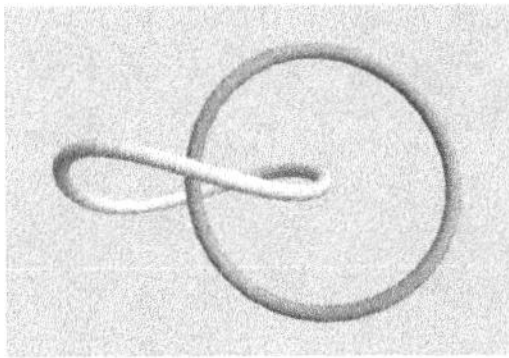

Fig. 10.206

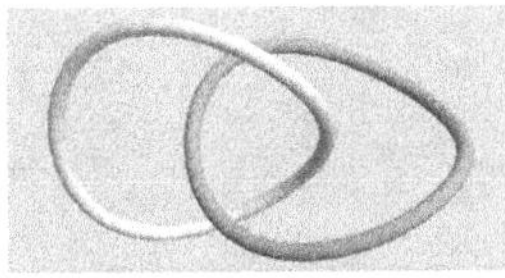

Fig. 10.207

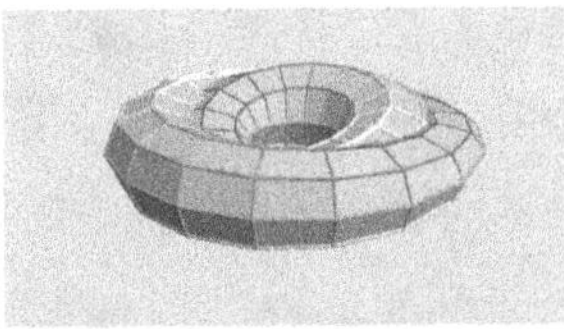

Fig. 10.208 Projection of linked Klein bottles from half-flip spun Hopf link.

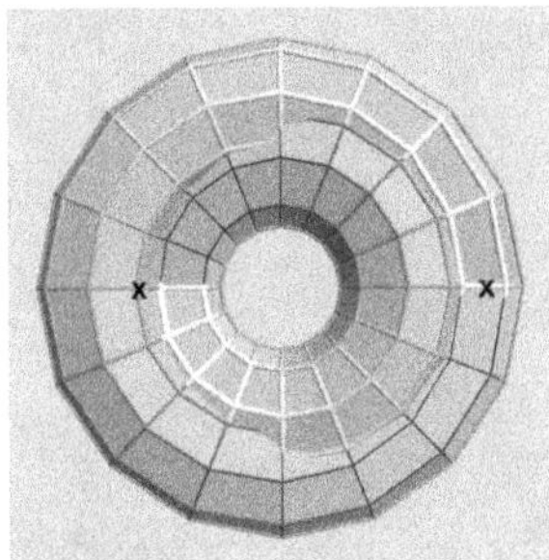

Fig. 10.209 Top view Figure 10.208 with triple points marked x.

Example 10.24. A variation of this construction yields a remarkable pair of linked projective planes that is of interest beyond being a 2-knotting, for example [Massey (1973/74)]. We adapt the knotting of the standard projective plane, Example 6.26.

Figures 10.210 to 10.225 shows an isotopy of a pair of circles C_w and C_g, (white and gray, respectively), in H^3_+. Each circle meets ∂H^3_+ in a single point p_w and p_g fixed during the isotopy. The isotopy for *each circle individually*, can be considered as a half-flip about an axis perpendicular to ∂H^3_+ and through either p_w or p_g.

A closer inspection reveals that the directions of these half-flips are opposite directions. By spinning, we produce two projective planes P_w and P_g. The flip of C_w is counter-clockwise (so a positive rotation) and so *considered by itself*, P_w is equivalent to the standard projective plane $\mathbb{P}$. Similarly P_g *considered by itself*, P_w is equivalent to the other standard projective plane $\mathbb{P}_0$. The projection of $P_w \cup P_c$, has two triple points, corresponding to Reidemeister type $\Omega.3$ moves of our isotopy. The Reidemeister $\Omega.3$ move between Figures 10.212 and 10.214 gives rise to a triple point of type wgw. The Reidemeister $\Omega.3$ move between Figures 10.219 and 10.221 gives rise to a triple point of type gwg. So, as in Example 10.23 we can argue that $P_w \cup P_g$ is not equivalent to a split linking of $\mathbb{P}$ and $\mathbb{P}_0$. ▮

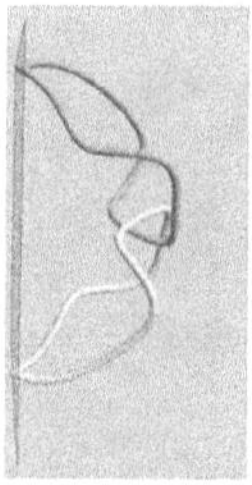

Fig. 10.210

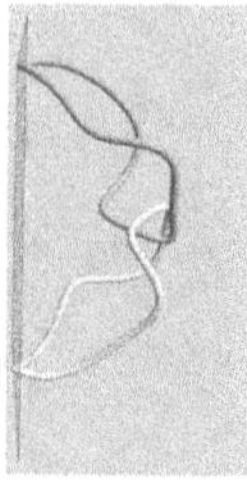

Fig. 10.211

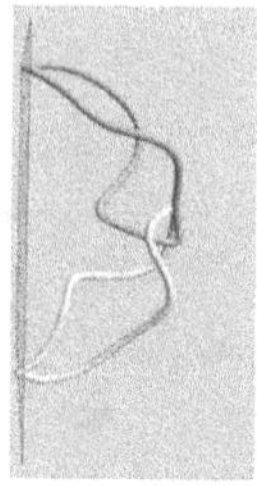

Fig. 10.212

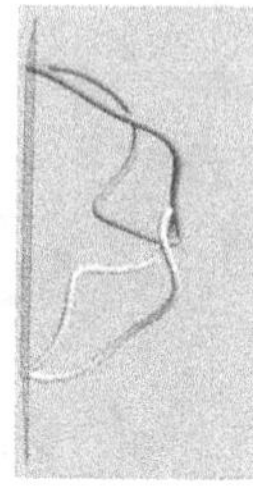

Fig. 10.213

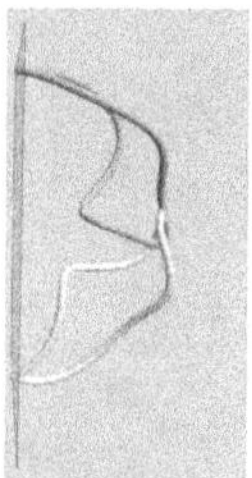

Fig. 10.214

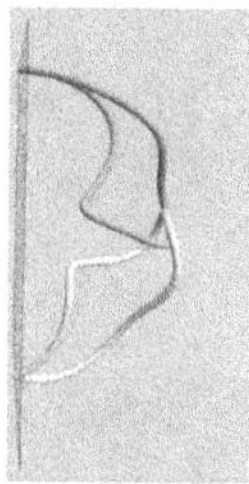

Fig. 10.215

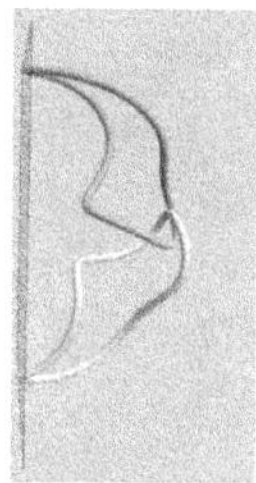

Fig. 10.216

Fig. 10.217

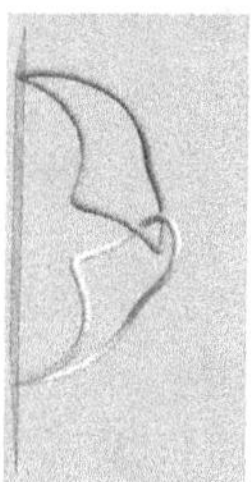

Fig. 10.218

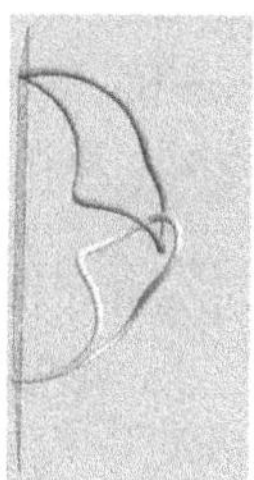

Fig. 10.219

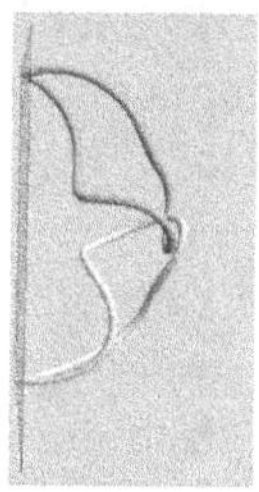

Fig. 10.220

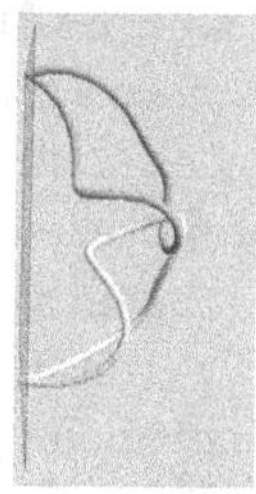

Fig. 10.221

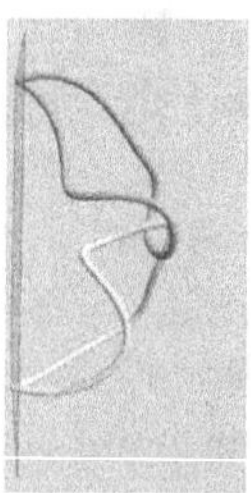

Fig. 10.222

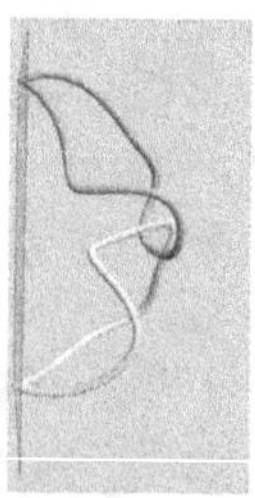

Fig. 10.223

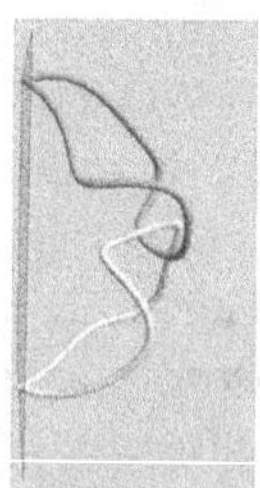

Fig. 10.224

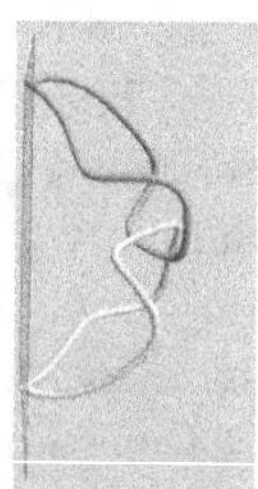

Fig. 10.225

10.10 Spinning Along a Guide

Spinning along a guide combines two ideas that extend spinning a knotting along a submanifold, Definition 12.1, in two distinct ways, possibly at the same time: iteration and localization. Other applications and extensions of this construction are found in Remark 12.17 in the discussion of the toroidal-roll of a knotting.

Iteration in its simplest form is: "spin, then spin again". Take a 2-knotting K, two submanifolds of K, C_0 and C_1 and two 1-knottings α and β. Construct the spun knotting $K' = (K, C_0) \bigotimes \alpha$. As sets $K \approx K'$, and so we can consider $C_1 \subseteq K'$. The **iterated spinning**, or **multi-spinning** is $K'' = (K', C_1) \bigotimes \beta$.

The idea of localization is this. Locally, a 1-manifold C_i looks like a 1-disk, c_i, for $i = 0, 1$. Locally $c_0 \subseteq K \subseteq R^4$ looks like $D^1 \subseteq D^2 \subseteq D^4$, or using cubes $Q^1 \subseteq Q^2 \subseteq Q^4$. We view $Q^2 = Q^1 \times Q^1$ where $Q^1 \times \{0\}$ corresponds to c_0 and each $\{x\} \times Q^1$ is a 1-disk normal to c_0 in K. In spinning, we locally replace $Q^1 \times Q^1$ with $Q^1 \times \alpha$, which in projection is illustrated in Figure 10.9. To get a (global) spinning, we then do this for other sub-disks $c_0 \subseteq C_0$ until we have replaced our original knotting of a tubular neighborhood of $C_0 \subseteq K$ with a new knotting.

When we investigate the localization of an iterated spinning K'', we find that one can locally change the order of the iteration and then assemble such local knottings to get 2-knottings distinct from K'' if we follow certain (global) assembly instructions called a **guide** which consists of C_0, C_1 together with priority information for the order of the local spinning.

In [Roseman (1989)] we defined a construction, **spinning a knotting along a projection of a knotting**. In this Section we discuss this construction, *but use different terminology*, now calling this construction a **spinning along a guide**. Additionally, this new definition drops a self-transversality condition.

Example 10.25. We viewed the spun trefoil torus Figure 10.48 as a union of two annulus knottings: Figures 10.46 and 10.47. Alternatively, we could use three pieces by viewing the knotting of Figure 10.47 as a gluing of two disks, as shown in Figures 10.226 and 10.227. Each of these two smaller pieces can be viewed as a local spinning the trefoil along an arc, see Definition 10.1, which is basically the product knotting of Figure 10.8, "bent into a semi-circle". ∎

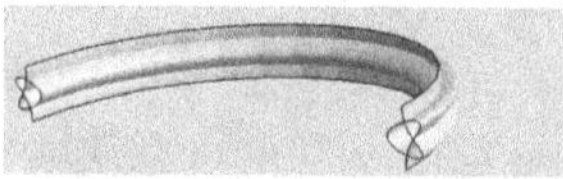

Fig. 10.226 Half the annulus knotting of Figure 10.47.

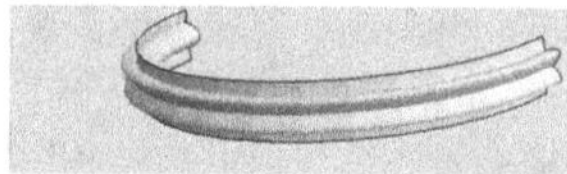

Fig. 10.227 Other half the annulus knotting of Figure 10.47.

Example 10.26. For this example, the ingredients are: an arc knotting α representing a trefoil knotting, β which representing a figure-eight knotting, and smooth disjoint circles C_0 and C_1 which are subsets of $S^2 \subseteq R^4$. Let $N_0 \approx C_0 \times D^1$ be a tubular neighborhood of $C_0 \subseteq S^2$, $N_1 \approx C_1 \times D^1$ a tubular neighborhood of $C_1 \subseteq S^2$, with $N_0 \cap N_1 = \emptyset$, Figure 10.228.

The 2-knotting $K_0 = (S^2, C_0) \otimes \alpha$ is a spun trefoil knotting, Figure 10.229, Since $C_1 \cap N_0 = \emptyset$, we have $C_1 \subseteq K_0 \subseteq R^4$, Figure 10.231. Now spin β using this C_1 resulting in a new 2-knotting of S^2 namely $K_1 = (K_0, C_1) \otimes \beta$, Figure 10.230. (We can alternatively view this as a spinning of a connected sum of a trefoil and a figure-eight as in Figure 10.232.) ∎

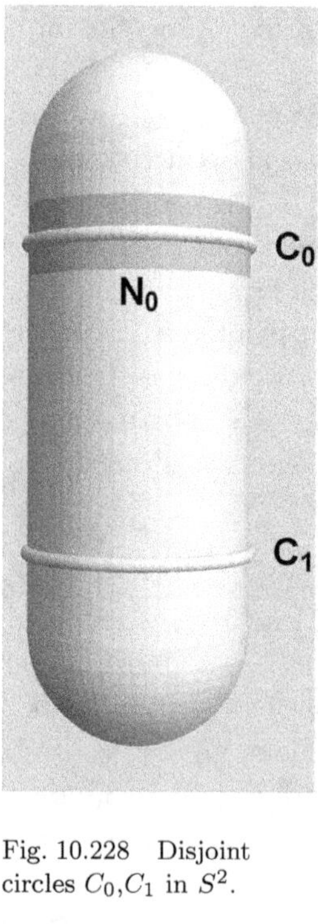

Fig. 10.228 Disjoint circles C_0,C_1 in S^2.

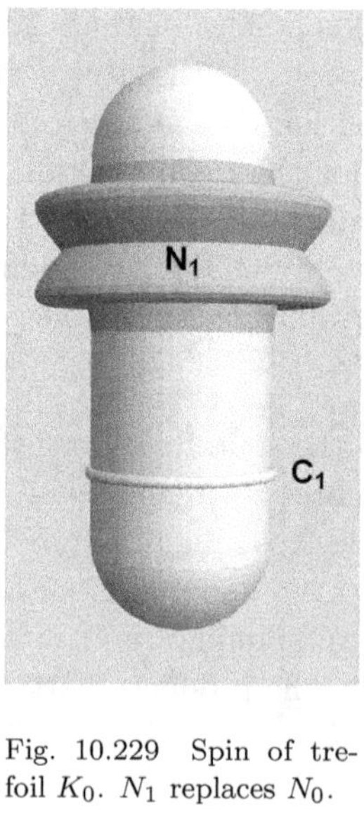

Fig. 10.229 Spin of trefoil K_0. N_1 replaces N_0.

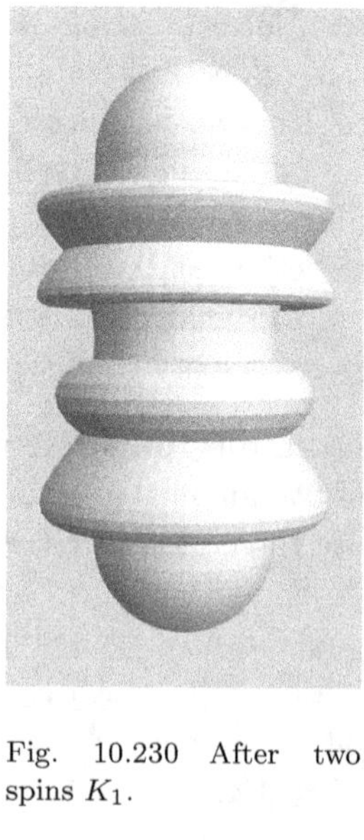

Fig. 10.230 After two spins K_1.

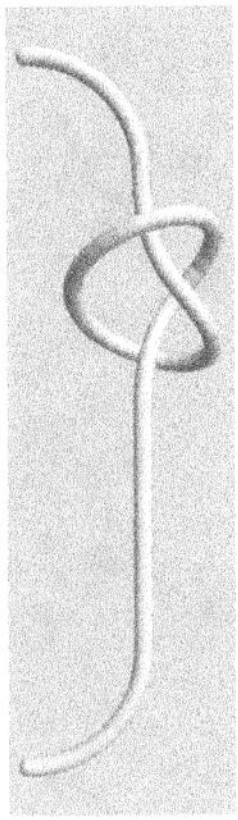

Fig. 10.231 The spin of this arc is K_0.

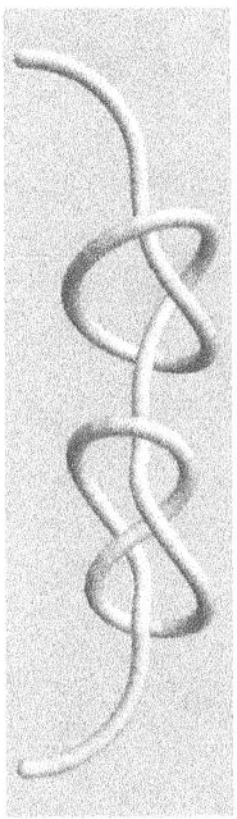

Fig. 10.232 The spin of this arc is K_1.

Example 10.27. A more interesting knotting is a second example where $C_0 \cap C_1 \neq \emptyset$ such as in Figure 10.233. As in Definition 12.1, we can write $K_0 = (S^2, C_0) \otimes \alpha$ as $K_0 = (K - N_0) \cup N_0'$. Since $N_0' \approx N_0 \approx C_0 \times D^1$, there is a natural homomorphism $\phi\colon N_0 \to N_0'$. Now view $C_1 \subseteq K_0$ as $(C_1 - N_0) \cup \phi(C_1 \cap N_0)$, Figure 10.234. To avoid excessive notation we still label it C_1. Figure 10.235 uses transparency for Figure 10.234 to reveal details. We use this C_1 to construct our knotting K_1. In Example 10.26 where $C_0 \cap C_1 = \emptyset$ it is easy to see that the order of our spinning does not matter—that is $\left((S^2, C_0) \otimes \alpha\right), C_1) \otimes \beta = \left((S^2, C_1) \otimes \beta\right), C_0) \otimes \alpha$. In the case $C_0 \cap C_1 \neq \emptyset$ these are not the same knottings, however, we will see that they are *equivalent* knottings, Remark 10.14. ∎

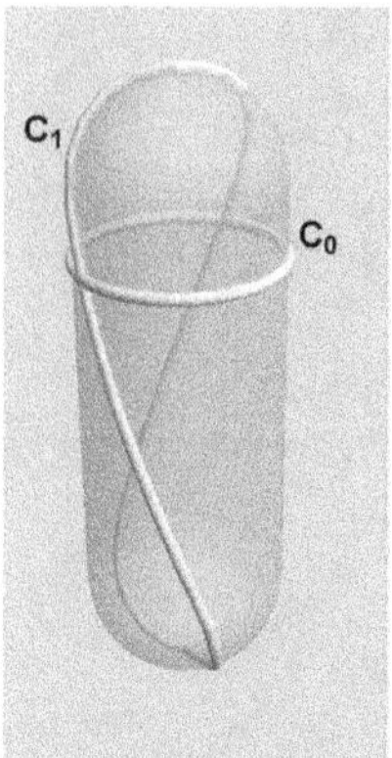

Fig. 10.233 Intersecting circles on S^2.

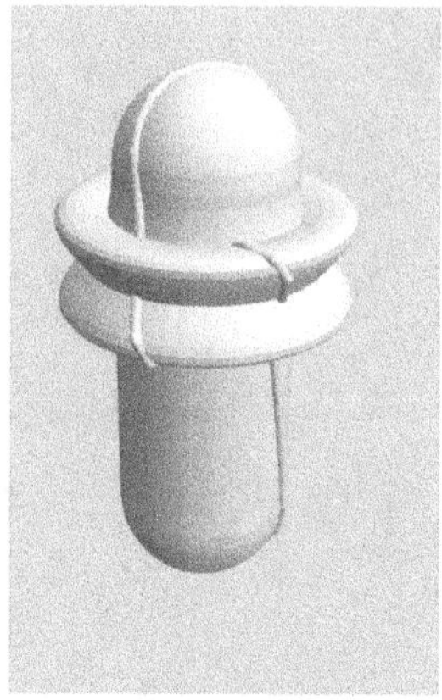

Fig. 10.234 Second circle C_1 after first spin.

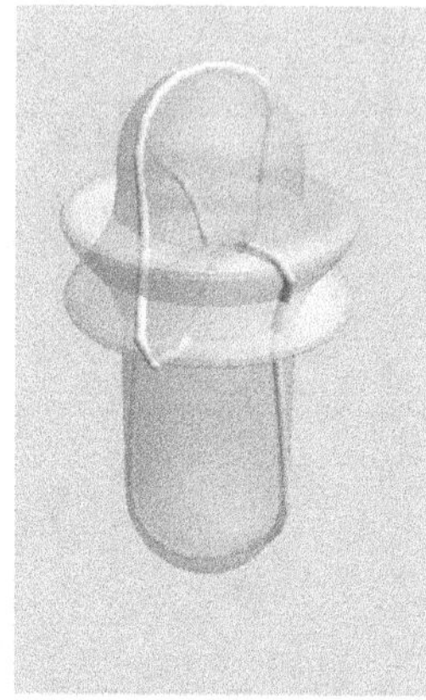

Fig. 10.235 Figure 10.234 transparent.

Example 10.28. We turn our attention to *local* iterated knotting. Focus on a rectangle $R \subseteq K$ containing an intersection point, such as in Figure 10.241. If we do our iterated spinning, we get a new knotting of R, call it $\mathbb{R}$. What does a projection of $\mathbb{R}$ look like?

We first describe a projection of a related knotting of two disjoint 2-disks, Figure 10.239. This figure shows a projection trace of an isotopy of two disjoint proper arcs indicated in Figures 10.236, 10.237 and 10.238 where a trefoil knotting α moves under straight line segment β. In Figure 10.239 the self-intersecting disk A is the projection of the trace of α, and B is the projection of trace of the stationary β. Figure 10.240 is a clipped view revealing that the crossing set resembles the trefoil projection. In particular, the three crossing points of the trefoil projection become three triple points in Figure 10.239. For future reference, we call this **the piercing of an embedded disc by an extruded projection of** α .

For the simplest non-trivial example of a projection of a multi-spinning, we choose $\alpha = \beta$ where α is a (trivial) knotting of an arc whose projection has a single crossing $x = c_0 \cap c_1$, where $c_i \subseteq C_i$ are (small) 1-disks. We spin in two directions—first along c_0, then along c_1. After the first spinning of α the projection of the $(R, c_0) \otimes \alpha$ would simply have a single arc of crossing points as in Figure 10.242. After our second spinning the projection $((R, c_0) \otimes \alpha), c_1) \otimes \alpha$ will look like Figure 10.243. This projection has been put in general position and has two triple points. Locally each of these triple points is modeled after the piercing of an embedded disk by an extruded projection of α. Other views of this projection showing these triple points are Figures 10.244 and 10.245.

Our figures for multi-spinning projection will be clearer if they are in general position, although this will not be mathematically mandated. To this end, we have chosen some convenient parameters—we mention two. Figure 10.243 depends in part on the line segments being transverse, but not exactly orthogonal. There is no mathematical problem for the construction in the orthogonal case, but the natural projection might look like Figure 10.246 which has a six-fold point caused by the two 3-fold points projecting to the same point—so it is not in general position. Similarly using non-transverse segments can give non-typical graphics. Secondly, we have drawn the knotting used for the second spinning relatively small compared with the first. So, the "big"" knotting is done first and the "small" knotting is second. There is no mathematical need to do this, but it makes a clear figure with relatively small and understandable self-intersection set. ∎

Fig. 10.236

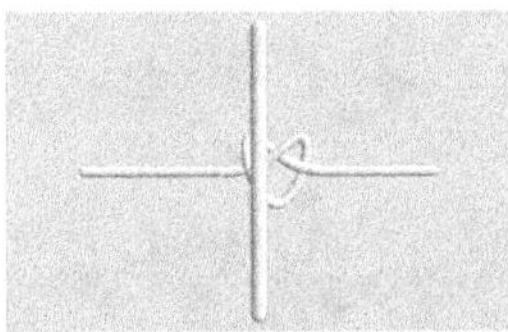

Fig. 10.237

Fig. 10.238

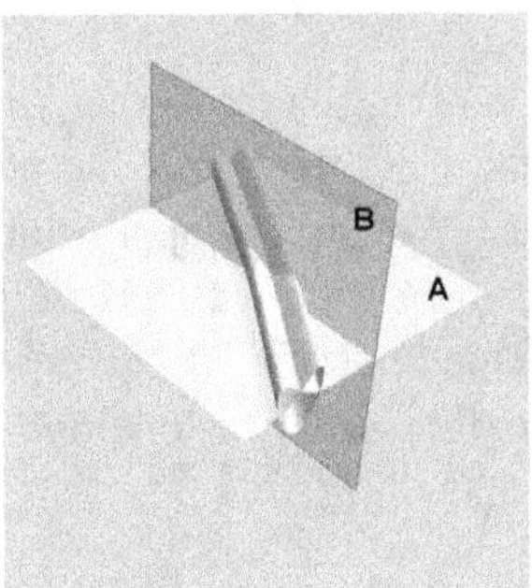

Fig. 10.239 Projection of isotopy trace.

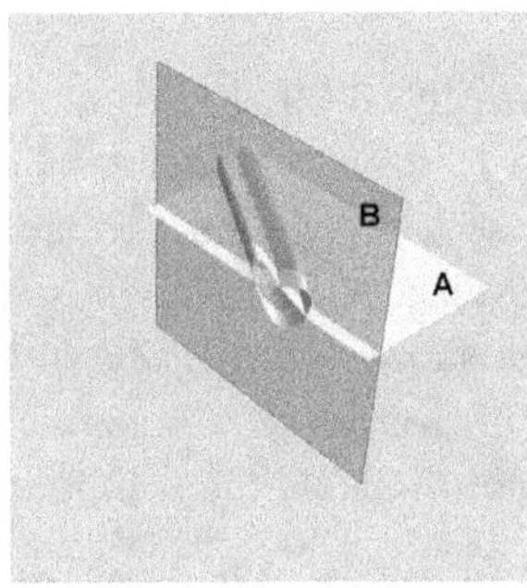

Fig. 10.240 Clipped view of Figure 10.239.

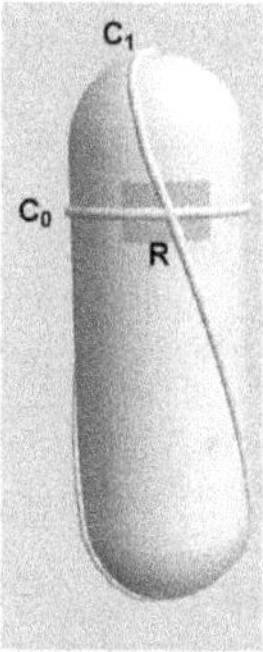

Fig. 10.241 $R \subseteq N_0$ is knotted twice.

Fig. 10.242 After spinning in first direction.

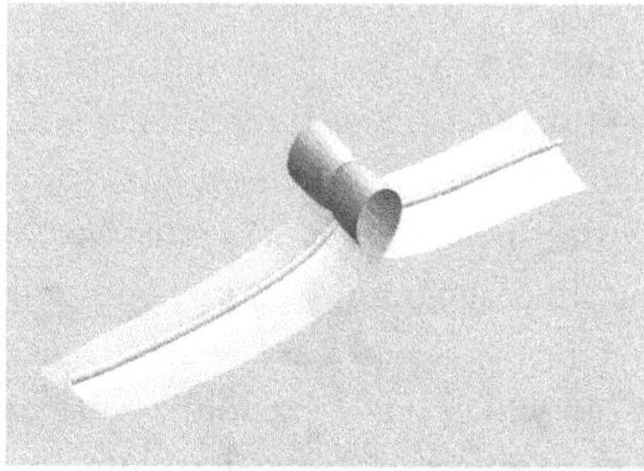

Fig. 10.243 Projection of multi-spinning using a trivial arc with one crossing for both directions.

Fig. 10.244 A second view of Figure 10.243.

Fig. 10.245 A third view of Figure 10.243.

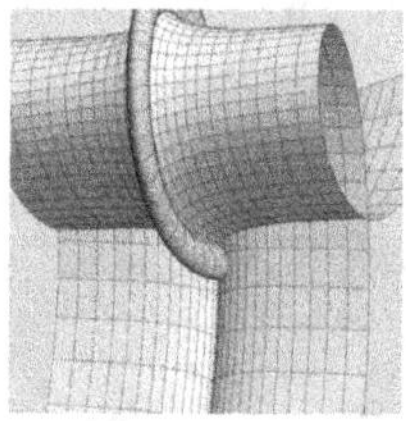

Fig. 10.246 If crossing orthogonal, projection would have a six-fold point.

Example 10.29. For a more substantial local knotting example take the familiar trefoil arc knotting which has a projection with three crossing points and use this twice in two transverse directions. We get a proper knotting of a 2-disk in $R \times D^2$ whose projection is shown in Figure 10.247. Figure 10.248 shows a projection of the boundary of this knotting. It corresponds to a projection of the connected sum of two right-hand trefoils and two left-hand trefoils (as a subset of the 3-sphere $\partial(R \times D^2)$). ∎

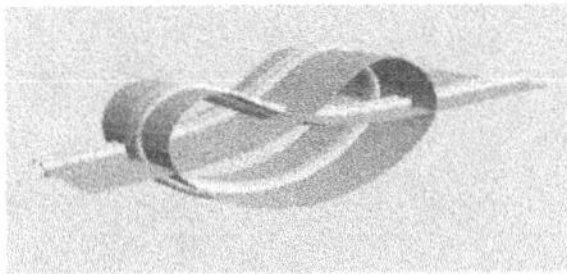

Fig. 10.247 Projection of a multi-spinning of R using a trefoil arc twice.

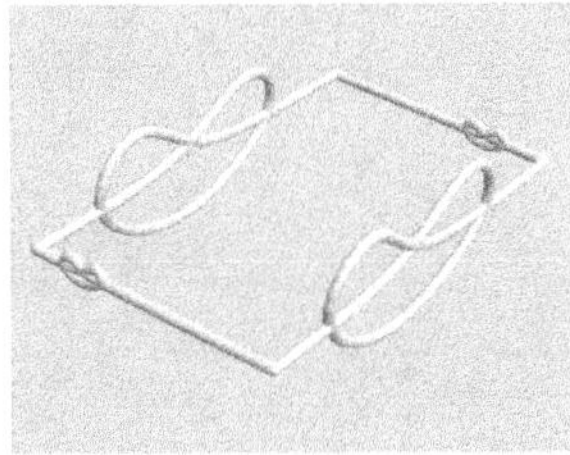

Fig. 10.248 Projection of ∂R of Figure 10.247.

Example 10.30. If first we take the trefoil arc knotting α which has a projection with three crossing points as in the previous example and then use a figure-eight knotting β with projection of four crossing points we will get a knotting of a our 2-disk R we denote as $\mathbb{R} = ((R, c_0) \otimes \alpha), c_1) \otimes \beta \subseteq R \times D^2$ whose projection is shown in Figure 10.251.

Figure 10.252 shows a projection of $\partial\mathbb{R}$. This corresponds to a projection of the connected sum of a right-hand trefoil, a left-hand trefoil and two figure-eight knottings (recall the figure-eight knotting is amphicheiral).

Reversing the order of this multi-spinning gives another knotting of R denoted $\mathbb{R}' = ((R, c_1) \otimes \beta), c_0) \otimes \alpha$, Figure 10.253. Figure 10.254 shows the associated $\partial\mathbb{R}'$. Recall the scaling of the arcs shown is only for graphical convenience and if we use the same scaling on the boundary, we can have $\partial\mathbb{R} = \partial\mathcal{R}'$ as in Figure 10.250. ∎

Fig. 10.249 Rotated view of projection of Figure 10.253.

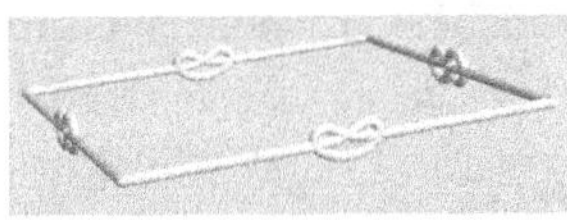

Fig. 10.250 A common model for Figures 10.254 and 10.252.

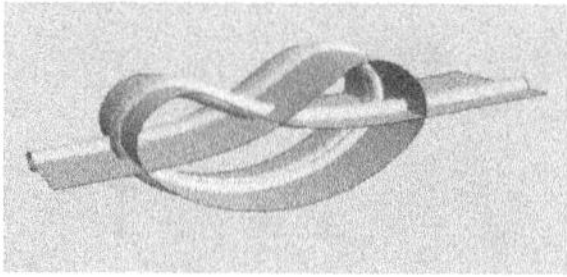

Fig. 10.251 Projection of multi-spinning of $\mathbb{R}$ first using trefoil arc then figure-eight arc.

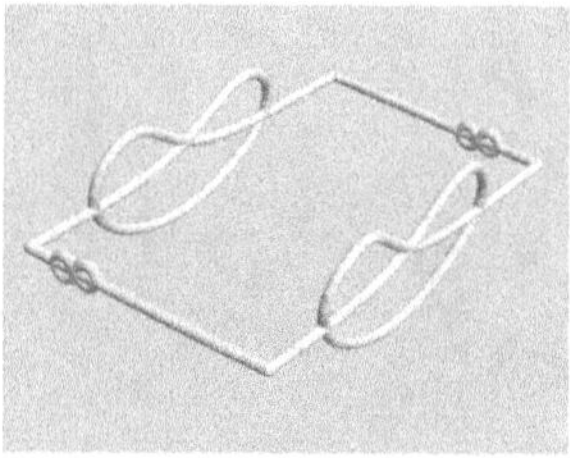

Fig. 10.252 Projection of $\partial\mathbb{R}$ of Figure 10.247.

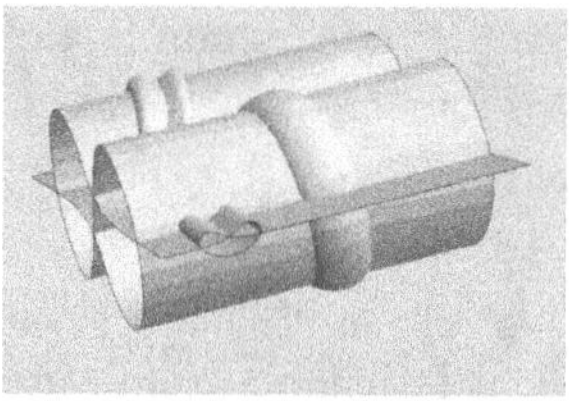

Fig. 10.253 Projection of multi-spinning $\mathbb{R}'$ using figure-eight arc then trefoil arc.

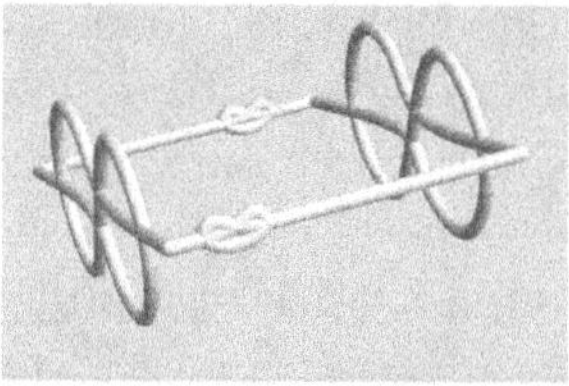

Fig. 10.254 Projection of $\partial\mathbb{R}'$ of Figure 10.253.

Remark 10.8. The order in which we do our knotting matters as is clear comparing the knottings of Figures 10.251 and 10.249—however, the knottings of the boundaries can be taken to be the same. **This observation is critical. It means that if we had a knotting with a rectangle knotted as $\mathbb{R} = ((R, c_0) \otimes \alpha), c_1) \otimes \beta \subseteq R \times D^2$ we could remove it and replace it by the knotting $\mathbb{R}' = ((R, c_1) \otimes \beta), c_0) \otimes \alpha$.** ▮

Remark 10.9. Looking ahead, we will focus on six examples of iterated spinning. The first four are knottings for the 2-torus for Figures 10.255, 10.256, 10.257 and 10.258. The last two are knottings for the 2-sphere, Figures 10.270 and 10.271. (To fit with our discussion, the order in which we consider these will be: Figures 10.256, 10.257, 10.270, 10.271, 10.258 and lastly 10.255.) In each, we are spinning along two intersecting curves. We spin a knotting of a trefoil α_0 along first curve, C_0, drawn as a thicker tube, and spin a knotting of a figure-eight knotting α_1 along the second,

C_1, a thinner tube. (For the first three, we are using tube thickness only to distinguish our two curves from each other. In later Examples we will re-interpret tube-thickness encoding.) In Figures 10.255 and 10.256, the two curves intersect in a single point. In the other Figures, the curves intersect in two points. ∎

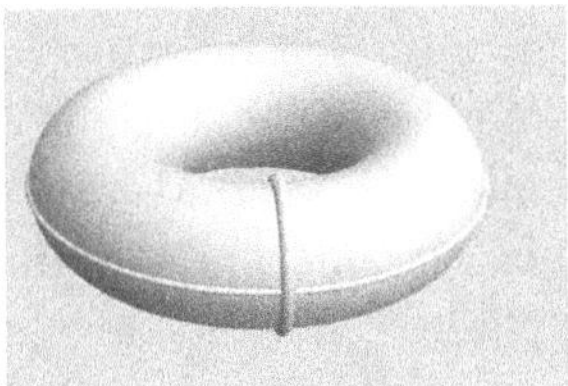

Fig. 10.255 A meridian and longitude.

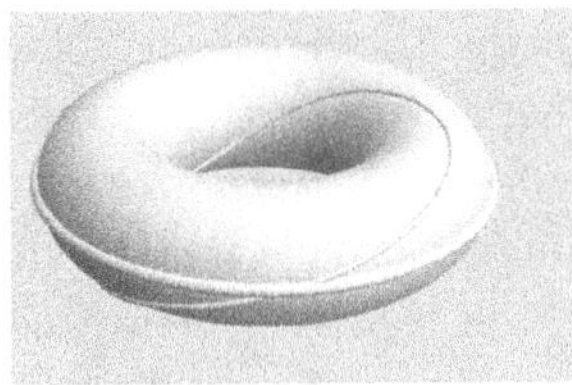

Fig. 10.256 A longitude and diagonal.

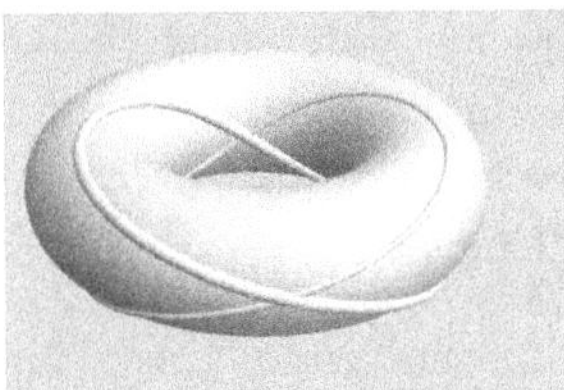

Fig. 10.257 A diagonal and anti-diagonal.

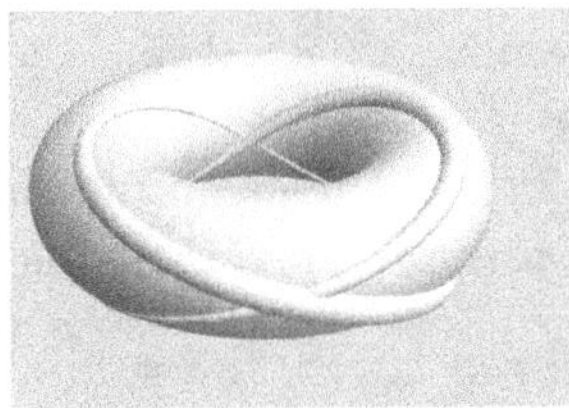

Fig. 10.258 Tube thickness coding at crossings.

Remark 10.10. For the first three examples, we construct iterated spinnings as in Example 10.27. We will see that some of the knottings produced are related to rolling of a composite knotting. In Example 10.19. Figures 10.131 to 10.152 showed steps for a rolling of the connected sum of a trefoil and figure-eight. We take the knotting of Figure 10.131 and complete to get Figure 10.259 which we view as a knotting in H^3_+. Similarly we change Figure 10.132 into Figure 10.265 and Figure 10.133 to Figure 10.264, etc. To conserve space we do not show all 22 such figures. All 22 figures, in sequence describe an isotopy $\mathcal{R}_t$ of a circle in H^3_+ we call a **full-roll** of the composite.

In Remark 10.19 we spoke a roll of composite as a combination of shrinks, swells and slides. For *circle* isotopies, there are two kinds of slides.

For the remainder of this example orient the circle, Figure 10.259, in such a way that the slide of the figure-eight in the following Figures is a slide in the positive direction. We call this isotopy $\mathcal{H}_t$ the **first-half-roll** of our composite knotting. We could also slide the figure-eight knotting in the opposite (i.e. negative) direction. That is a different isotopy and, in terms of our guide-spinning corresponds to altering our knotting by replacing the diagonal of the torus with the anti-diagonal. Informally we view this as: shrink the figure-eight knotting, slide along the trefoil knotting *in the positive direction*, then swell the figure-eight knotting. The circle isotopy corresponding to Figures 10.141 to 10.152 is called the **second-half-roll.** Although we have not provided figures for this, the informal description is: shrink the trefoil knotting, slide along the figure-eight knotting *in the positive direction*, then swell the trefoil knotting. So that a first-half-roll followed by a second-half-roll is the same as a full-roll.

However, the concatenation $\mathcal{H} * \mathcal{H}$ is called the **repeated first-half-roll** of the composite. It is not the same as the full-roll of the composite. Informally this is: shrink the figure-eight knotting, slide along the trefoil knotting *in the positive direction*, then swell the figure-eight knotting, now, shrink the figure-eight knotting, slide along the trefoil knotting *in the positive direction*, then swell the figure-eight knotting. ▌

Example 10.31. In Figure 10.256 C_0 is the longitude and C_1 the diagonal (a $(1,1)$-curve) of the T^2. All of our spun knottings depend on choosing a framing of tubular neighborhoods involved. One knotting produced in this way is the deformation-spinning indicated in Figures 10.259 to 10.266. This isotopy is equivalent to a half-roll of the composite knotting $\alpha_0\#\alpha_1$, in this case using blackboard framing. If other framings were used one would have twisting added to this isotopy. ▌

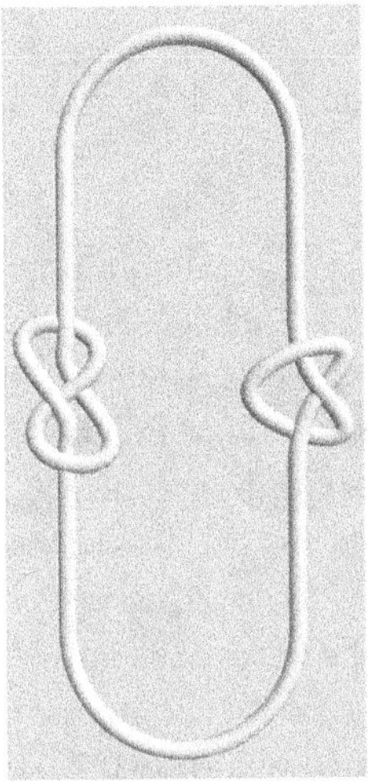

Fig. 10.259

Fig. 10.260

Fig. 10.261

Fig. 10.262

Fig. 10.263

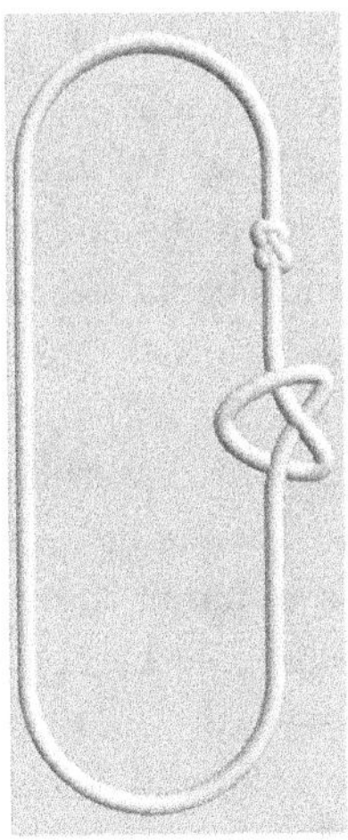

Fig. 10.264

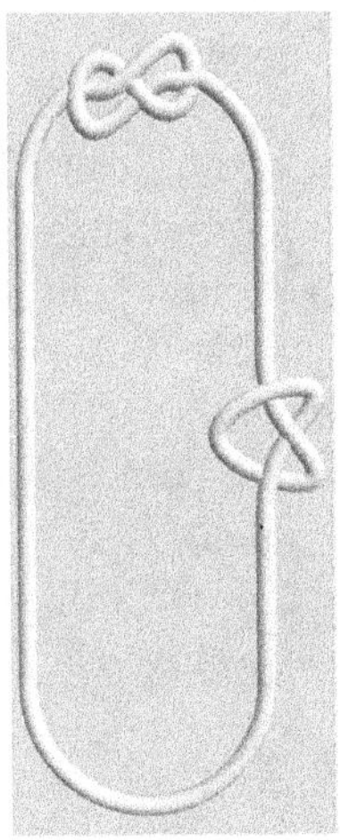

Fig. 10.265

Fig. 10.266

Example 10.32. In Figure 10.257, C_0 is the diagonal (a $(1,1)$-torus curve) and C_1 corresponds to an anti-diagonal (a $(1,-1)$-torus curve). We first spin trefoil α_0 along C_0 and then spin figure-eight α_1 along C_1. Let $\mathbb{K}$ denote this iterated spinning.

The knotting $\mathbb{K}$ can be expressed as the deformation of a circle which is the repeated first-half-roll of the composite. ▮

Remark 10.11. Let us express Example 10.32 in terms of local spinning. For $i = 0, 1$, let N_i be a tubular neighborhood of C_i. Each C_i can be expressed as a union of closed 1-disks. For each If $c \subseteq C_i$ is a closed 1-disk let $N \approx c \times D^1 \subseteq K$ be the corresponding 2-disk in N_i. Points of $K - (N_0 \cup N_1)$ become points of $\mathbb{K}$, so the only changes are points of N_i.

Locally, we obtain $\mathbb{K}$ by replacing each N by a new 2-disk $\mathbb{N}$. If N does not contain one of the two crossing points, then $\mathbb{N} = (N, c_i) \otimes \alpha_i$ is just a local spinning, Definition 10.1. So, for each of our 1-disks c, we take out N and replace it by $\mathbb{N}$. If N *does* contain a crossing point x then (in our case) $x = c_0 \cap c_1$ where c_i are 1-disks in C_i, and $\mathbb{N} = ((N, c_0) \otimes \alpha_0), c_1) \otimes \alpha_1$ is just a local spinning, Definition 10.1. So at each x, we take out our N and replace it by the multi-spun $\mathbb{N}$. When we are done with all our replacements, we will have our new knotting $\mathbb{K}$. ▮

Example 10.33. For Example 10.32, see Remark 10.11, we had two crossing points of $C_0 \cap C_1$, call them $x = c_0^x \cap c_1^x$ and $y = c_0^y \cap c_1^y$, each with their own 2-disk which we now call N_x and N_y. We treated them the same and replaced each of these by the same local multi-spinning, $\mathbb{N}_x$ and $\mathbb{N}_y$, namely $\mathbb{N}_x = ((N_x, c_0^x) \otimes \alpha_0), c_1^x) \otimes \alpha_1$ and $\mathbb{N}_y = ((N_y, c_0^y) \otimes \alpha_0), c_1^y) \otimes \alpha_1$.

However, using Remark 10.8, there are other possible constructions. For example we could use $\mathbb{N}'_y = ((N_y, c_1^y) \otimes \alpha_1), c_0^y) \otimes \alpha_0$ rather than $\mathbb{N}_y = ((N_y, c_0^y) \otimes \alpha_0), c_1^y) \otimes \alpha_1$. That is, we can reverse the order of the local-spinnings at y, spin α_1 along c_1^x first and then spin α_0 along c_0^x (and keep everything else the same). We will get a different knotting and, generally speaking, one that is not equivalent to Example 10.11. ▮

Remark 10.12. We introduce a device that will allow us to graphically express which of the two possible local-spinning iteration order we want at any crossing point. We show a curve on a surface as a tube of continuously *variable* thickness with the condition that at every crossing point, the two thickness will be different. The idea is that the "thick direction" is the first spinning direction and the "thin direction" is the second spinning direction.

In Figure 10.267 we used a tube-thickness, *uniform on each component* curve simply to "label" that curve. So if Figure 10.268 is a crossing point the thick arc indicates that arc is a subset of the first component. So locally we spin R in this direction and then we spin along the thinner arc.

So if we had circles C_0 and C_1 intersecting in p and q as in Figure 10.269. If the tube-thickness coding was as in Figure 10.270, We would spin along C_0 first at both p and q. However if tube-thickness coding was as in Figure 10.271, indicates that at p we spin along C_1 first, then along C_0 but at q the order of local spins is: first C_0, then C_1. (In previous papers such as [Roseman (1998)] we used broken line style to represent local spinning order at a crossing point rather than tube-thickness.) ∎

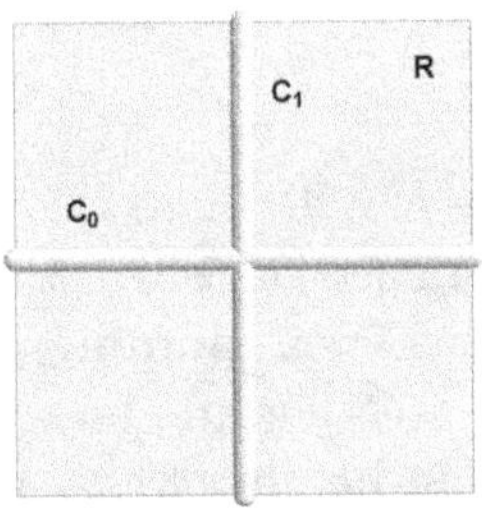

Fig. 10.267 Local picture preparing for multi-crossing.

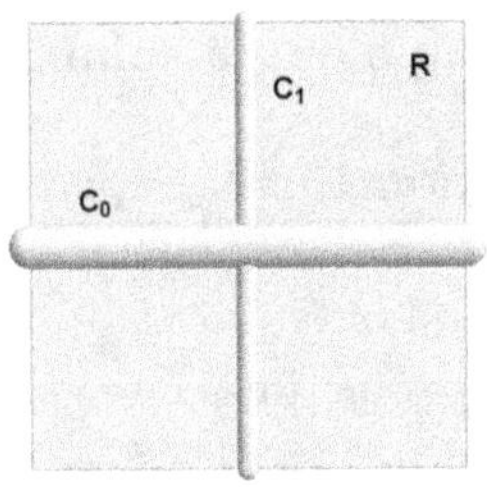

Fig. 10.268 Local picture as a knot-crossing to indicate order of spins.

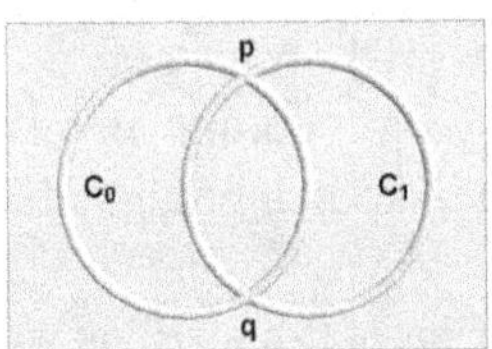

Fig. 10.269 Two intersecting circles.

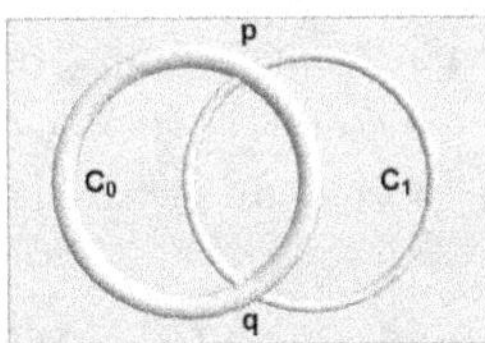

Fig. 10.270 One local iteration choice.

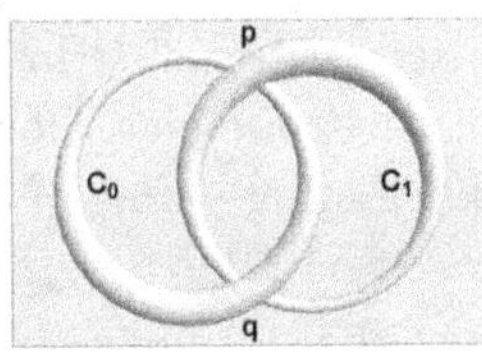

Fig. 10.271 Another local iteration choice.

The next two examples construct similar 2-knottings of a sphere.

Example 10.34. We revisit the iterated spinning associated with Figure 10.233, $\mathbb{S} = \left((S^2, C_0) \otimes \alpha_0\right), C_1) \otimes \alpha_1$. To simplify our graphics, we represent S^2 stereographically by R^2, Figure 10.269. View C_0 and C_1 as two nearly equatorial great circles whose intersection corresponds to p and q. We intend to spin α_0 along C_0 first, and then α_1 along C_1, so our tube-thickness coded graphic is like Figure 10.270. Proposition 10.5 will soon show, that this example is, in fact, equivalent to the connected sum of the spinning of α_0 and the spinning of α_1. ▮

Example 10.35. For this example we will use the same C_0 and C_1 but with tube-thickness coding as in Figure 10.271. As in Example 10.32, we will see the resulting 2-knotting can alternatively be viewed as a roll-spinning of the composite knotting $\alpha_0 \# \alpha_1$. ▮

We are now prepared to formalize our construction. Previously in [Roseman (1989)] we called our process "spinning along the projection of a knot", but in this book we change this definition slightly and, instead, call it a **spinning along a spinning-guide**.

Definition 10.10. Suppose we have a 2-knotting $K \subseteq R^4$, and $G \subseteq K \times R^1$ is a smooth 1-manifold in so that $\pi|_G$ is an immersion where π is the projection of $K \times R$ to the first factor. We refer to G as a **spinning-guide**, for K or sometimes more simply **guide**, and call $\pi(G)$ a **guide image**. For $p \in G$, if $p = (x,t) \in K \times R^1$, t is called the **guide-height** of p. ▮

So Example 10.34 has a guide image Figure 10.270 which is a projection of a knotting equivalent to a split trivial linking of two circles. Example 10.35 has guide image Figure 10.270 which is a projection of a Hopf linking.

In most examples a guide image resembles to a "knot-diagram drawn on some surface". However note that we have not insisted on general position requirements for π—we could have non-transverse crossing points and crossing points of order more than two. Also our 2-knotting might be non-orientable.

Here is terminology and notations for the constriction of 2-knottings we have been discussing:

Definition 10.11. Suppose G is a spinning-guide for a 2-knotting $K \subseteq R^4$. Suppose G has n components, $G_1, \ldots G_n$ and suppose we have n (not necessarily distinct) proper knotted 1-disks, $\alpha_i \subseteq D^3$. The **guide-spinning of K along G of $\mathcal{A}$**, denoted $(K,G)) \otimes \mathcal{A}$, is obtained using spinning along the guide image where we use knotting α_i for component G_i, and the order of multi-spinning at at a n-crossing point $p = \pi(p_1) = \cdots = \pi(p_n)$ is determined by order of the guide-heights of p_i. ∎

Example 10.36. In Figure 10.255 C_1 is a longitude (a $(1,0)$-curve) and C_2 a meridian (a $(0,1)$-curve) of the standard torus with $\widetilde{C}_2$ passing above $\widetilde{C}_1$ at the single crossing point. ∎

Example 10.37. In Figure 10.258 the guide images are the same as Example 10.32 however the local iteration orderings at the two crossing points of C_0 and C_1 are different. So in this example we need to use the local multi-spinning construction and cannot be more simply described as an iterated spinning. ∎

A basic tool for showing two knottings produced by guide-spinnings are isotopic [Roseman (1989)] is the following.

Definition 10.12. Suppose G_0 and G_1 are two spinning-guides for a 2-knotting K. We say G_0 **and** G_1 **are** π**-isotopic guides** if there is an isotopy H_t between knottings G_0 and G_1 in $K \times R^1$ such that $\pi \circ H_t$ is a regular homotopy, where $\pi\colon K \times D^1 \to K$ is projection. ∎

Now, π-isotopic guides give rise to isotopic knottings:

Prop 10.5. Suppose G_0 and G_1 are two π-isotopic spinning-guides for a 2-knotting K and $\mathcal{A} = \{\alpha_i\}$ is a sequence of 1-knottings, one for each component of G_0. Then the associated guide-spinnings $(K, G_0)) \otimes \mathcal{A}$, and $(K, G_1)) \otimes \mathcal{A}$ are isotopic. ∎

It is a standard (non-trivial) exercise for 1-knottings to show that the square knotting $k = 3.1\#3.1$ and the granny knotting $k' = 3.1\# - 3.1$ are inequivalent 1-knottings (nor are they mirror images). However, the two spun knots $\Sigma(k)$ and $\Sigma(k')$ *are* equivalent 2-knottings, [Roseman (1975b)]. A proof is an application involving guide-spinning and in particular in use of Proposition 10.5.

Example 10.38. As a warm up, we revisit an isotopy used in Remark 10.3. Let α denote the left-hand trefoil knotting, $-\alpha$ the right-hand knotting. In that Remark we showed $\Sigma(\alpha) \approx \Sigma - (\alpha)$ The isotopy we used to show this was obtained by a rotation in R^4. We can alternatively obtain an isotopy as of a π-isotopy of a guide. Suppose $G = S^1 \times \{0\} \subseteq S^2 \times R^1$ where $S^1 \subseteq S^2$ is the standard inclusion. Then $\Sigma(\alpha) = (S^2, G) \otimes \alpha$. The isotopy H_t, $0 \le t \le 1$

of G indicated in Figures 10.272 to 10.276 corresponds to a π-isotopy of G. An examination shows that $(S^2, H_1(G)) \otimes \alpha$ (which by Proposition 10.5 is equivalent to $(S^2, G) \otimes -\alpha$) is, in fact, $(S^2, G) \otimes \alpha \approx \Sigma(-\alpha)$. ▌

Fig. 10.272

Fig. 10.273

Fig. 10.274

Fig. 10.275

Fig. 10.276

Example 10.39. Figure 10.277 shows an proper arc representing a granny knotting k'; Figure 10.278 a square knotting. Figures 10.279 and 10.280 show projections of the $\Sigma(k')$ and $\Sigma(k)$ using crossing shadows. These shadows correspond to the shadows drawn on the arcs in Figures 10.277 and 10.278. Figures 10.281 and 10.282 reveal additional details of these projections.

We view $\Sigma(k')$ as a iterated spun knotting $\big((S^2, C_0) \otimes \alpha), C_1\big) \otimes -\alpha$ where these two circles are as shown, in Figure 10.283 and we view $C_0 \cup C_1$ as an image of a guide $G = G_0 \cup G_1$ where perhaps all points of G_0 have guide-height 1 and all points of G_1 have guide height 0. (We spin about the thick-tube circle first.)

Figures 10.283 to 10.287 show tube-thickness coded images of a guide homotopy. In more detail, we get an isotopy shown in projection in Figures 10.288 to 10.299 (from a hand drawing, since this isotopy is difficult to do with computer graphics.) With careful viewing we can follow the crossing shadows during the isotopy. This is an isotopy from a spun granny knotting to a spun square knotting. ▌

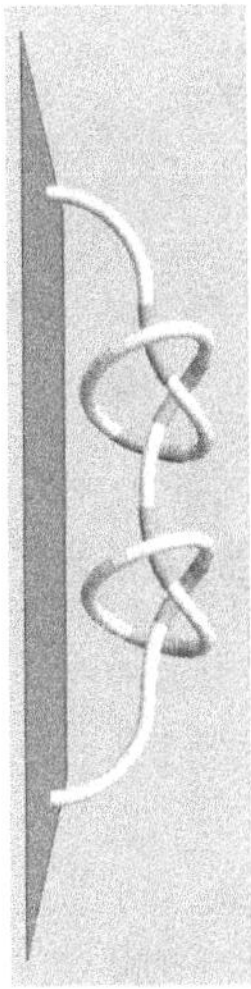

Fig. 10.277 A proper arc representing a left-handed) granny knotting.

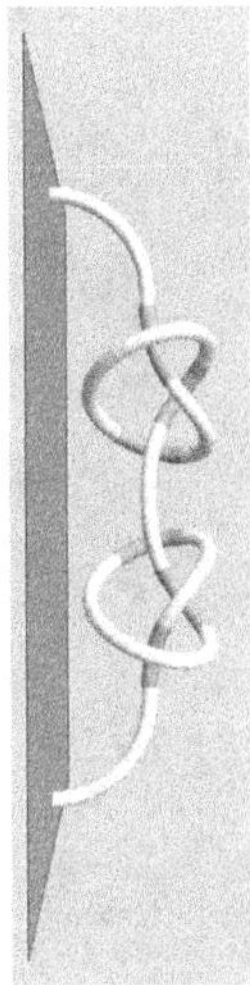

Fig. 10.278 A proper arc representing a square knotting.

Fig. 10.279 Spinning of a granny knotting of Figure 10.277.

Fig. 10.280 Spinning of a square knotting of Figure 10.278.

Fig. 10.281 Half slab of spun granny knotting.

Fig. 10.282 Half slab of spun square knotting.

Fig. 10.283 Can generate Figure 10.288.

Fig. 10.284 Can generate Figure 10.292.

Fig. 10.285 Can generate Figure 10.293.

Fig. 10.286 Can generate Figure 10.295.

Fig. 10.287 Can generate Figure 10.296.

Fig. 10.288

Fig. 10.289

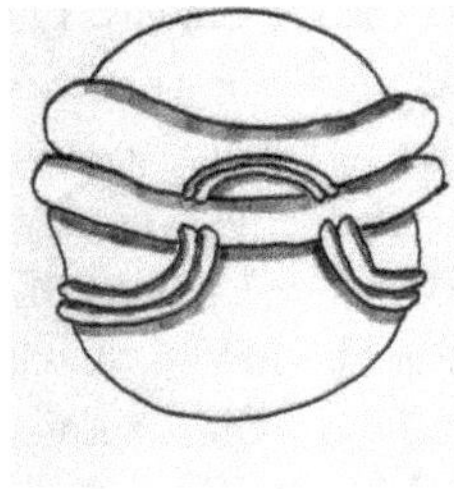

Fig. 10.290

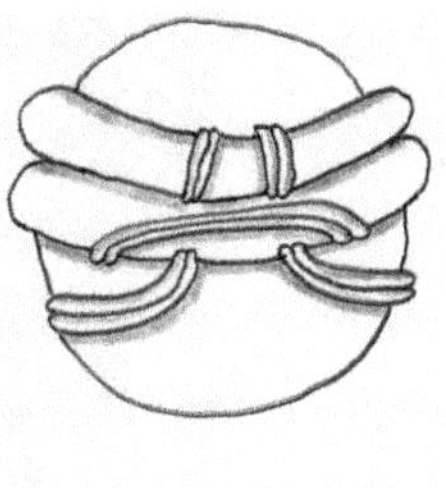

Fig. 10.291

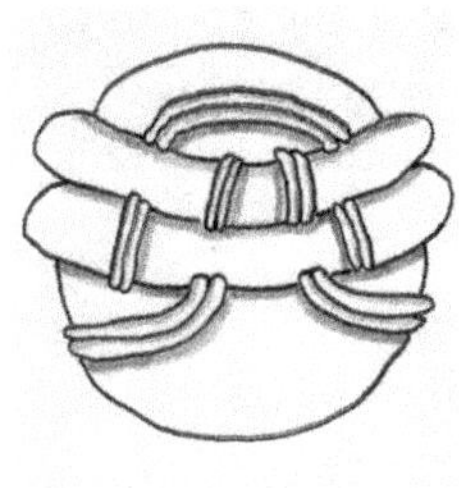

Fig. 10.292

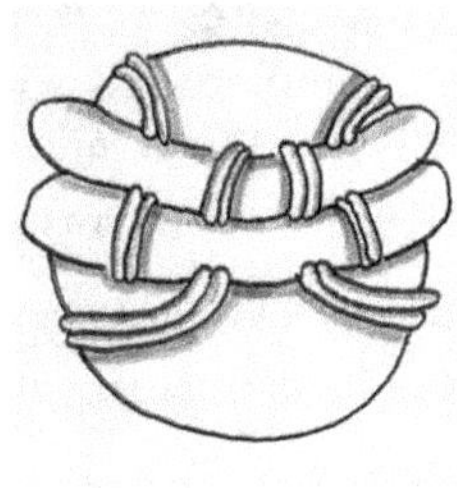

Fig. 10.293

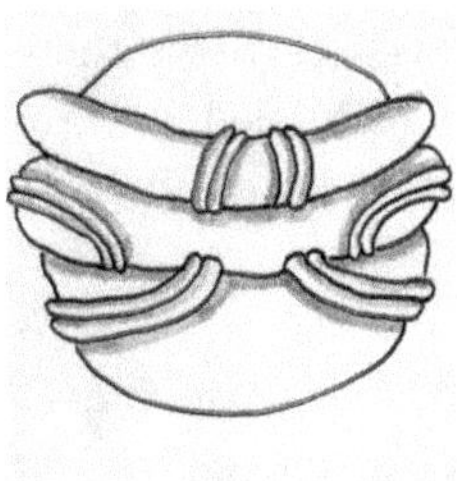

Fig. 10.294

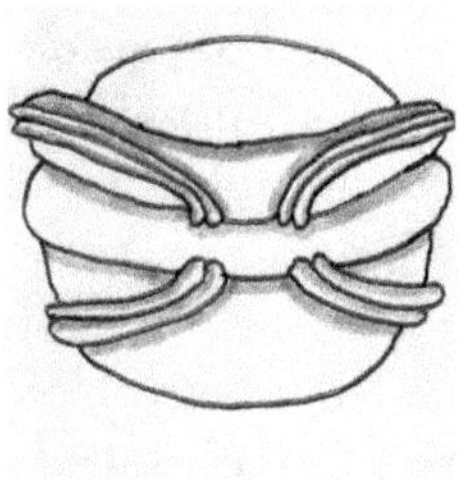

Fig. 10.295

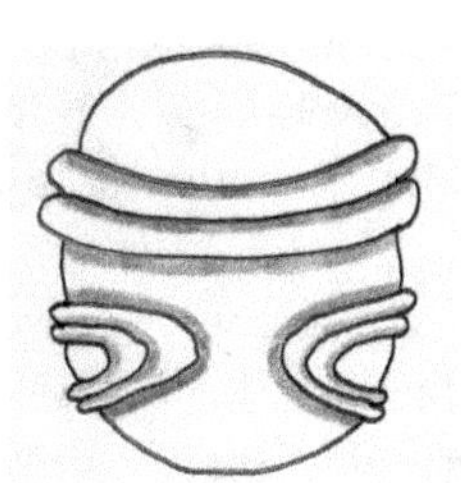

Fig. 10.296

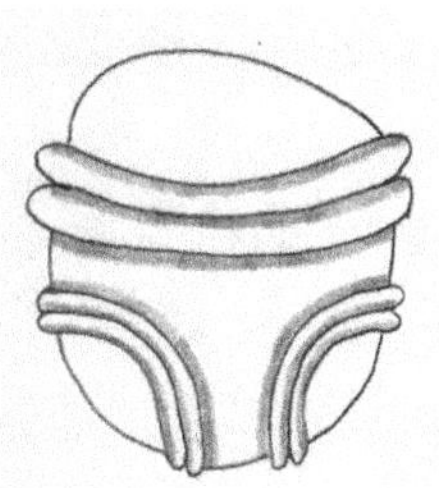

Fig. 10.297

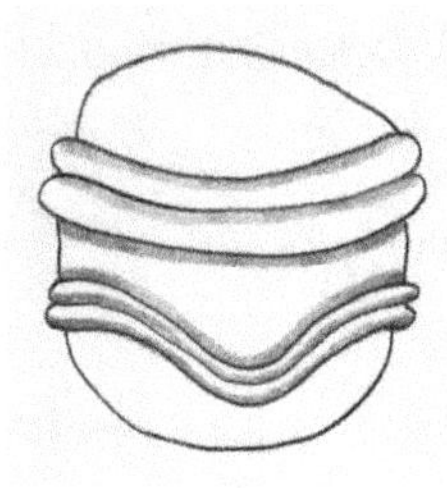

Fig. 10.298

Fig. 10.299

Remark 10.13. The π-isotopy shown in Figures 10.283 to 10.287 can be expressed as two Reidemeister $\Omega.2$ moves, in the sense of Remark 7.4. The first move creates two crossing points as we go from Figure 10.283 to Figure 10.284. The second move deletes two crossing points as we go from Figure 10.286 to Figure 10.287 (we need to use the transparency of the figures to see this on the "back side" of the sphere). It is of visualization interest to note how these moves relate to the projection of the corresponding isotopies in Figures 10.288 to 10.299. ▮

Remark 10.14. We can now see why Example 10.26 and Example 10.27. In both cases we could describe the knottings with a guide corresponding to Example 10.27 would be the union of $C_0 \times \{1\} \subseteq S^2 \times R^1$ and $C_1 \times \{0\} \subseteq S^2 \times R^1$. These guides are π-isotopic by a single Reidemeister $\Omega.2$ move, in the sense of Remark 7.4. ▮

10.11 Spinning and Fibered Knots

For n-knotting K of a sphere in R^{n+2} (or S^{n+2}) an orientable $(n+1)$-manifold Σ is called **Siefert hypersurface for knotting** K if $\partial\Sigma = K$. (Note: Generally a "surface" is a 2-manifold. In higher dimensions. Logically one might expect these to be called "Siefert manifolds". However there is an well-known, unrelated, type of 3-manifold already called a Seifert manifold. Some authors use the term **high-dimensional Seifert surface**.)

In this section, we will only consider n-knottings of spheres in S^{n+2}. As explained for 1-knottings, a Seifert surface for a knotting K corresponds to a proper surface in the exterior $Ext(K)$ called a **Siefert surface for the exterior**. A fiber for a fibered knotting is an especially nice Siefert surface.

Definition 10.13. A knotting of a $K \subseteq S^{n+2}$ is called a **fibered knotting** if the exterior of the knotting is a fibration over a circle where the fiber, F, is a connected orientable $(n+1)$-manifold. ▮

A high-dimensional version for existence of Seifert hypersurfaces is Proposition 4.11 is:

Prop 10.6.

(1) If $K \subseteq S^{n+2}$ is a smooth n-knotting, there exists a smooth map with only non-degenerate critical points $\phi\colon Ext(K) \to S^1$ which sends each

meridian μ_θ of $\partial Ext(K)$ corresponding to $S^1 \times \{\theta\}$ to the point $\theta \in S^1$, [Kervaire (1965a)].

(2) Also, there is such a ϕ that has only non-degenerate critical points.
(3) As a consequence, if $\theta \in S^1$ is a regular point of ϕ, then $\sigma_\theta = \phi^{-1}(\theta)$ is a smooth proper orientable $n+1$-dimensional submanifold—and, if it is connected, it will be a proper Seifert surface for $Ext(K)$.
(4) Additionally if $A \subseteq S^1$ is a closed sub-arc of regular points with $\theta \in A$, then $\phi^{-1}(A) \approx \Sigma_\theta \times D^1$. In other words, $\phi^{-1}(A)$ is a product tubular neighborhood of $\sigma_\theta \subseteq Ext(K)$.
(5) And finally, if ϕ has no critical points. then $Ext(K)$ is a fibration over a circle with fiber σ_θ. ∎

There is an interesting way of using differential geometry to define ϕ, [Dangskul (2015)] [Borodzik *et al.* (2020)].

If a knotting happens to be a fibered then to understand the exterior, we can focus on a single manifold (the fiber F) and a gluing map of F. Although fibered knottings are special, we will see there are plenty of examples we can look at.

The most celebrated theorem on twist-spinning is [Zeeman (1965)]:

Prop 10.7. A twist-spin of knotting k with r-twists with $r \neq 0$ is a fibered knot. For ± 1-twists, the fiber is diffeomorphic to D^3. So, for any k, a ± 1-twist-spun knotting of k is always a trivial knotting.

Distinct proofs of the Fibering Theorem are found in [Friedl and Orson (2015)] and [Goldsmith and Kauffman (1978)] as well as the original [Zeeman (1965)].

Proof (Outline). We outline an alternative proof. To begin with, we view any of our spinnings of a 1-knotting k in S^4 as follows: locally at a point $x \subset S^1 \subset S^2 \subseteq S^4$, replace a local standard (unknotted) ball pair $(D^3, D^1)_x$ with a knotted one $(D^3, \alpha)_x$. In terms of the *exterior* of a knotting, this process replaces a ball with an *unknotted hole* $Ext(D^3, D^1)_x$ with a ball with a ball with a *knotted hole*, $Ext(D^3, \alpha)_x$.

We use subscript notation: X_θ, for the fiber of a fibration over a circle, where $\theta \in S^1$. Also we will be spinning about a submanifold which is a sphere S (in Steps 1 and 2 a 0-sphere, in later steps S^1). In these cases we use an additional superscript to refer to $\psi \in S$, So, for 2-knottings, we have fibers F_θ^ψ where $\psi \in S^1$ and $\theta \in S^1$. At each point $\psi \in S^1 \subset S^2$ we consider a fibration; the fibers of this fibration are each subscripted by θ.

In outline:

Step 1: We view the standard knotting $S^1 \subseteq S^3$ as connected sum of two trivial knottings, develop notations that express $Ext(S^3, S^1)$ and show, in a certain way, that it is fibered.
Step 2: We use structures of Step 1 show a connected sum $k\#(-k)$ of 1-knottings is fibered if k is fibered.
Step 3: We generalize Step 1 to the exterior of the standard knotting $S^2 \subseteq S^4$.
Step 4: We generalize Steps 1 and 2 to show the Artin spinning of a fibered 1-knotting is a fibered 2-knotting.
Step 5: We consider the Artin spinning of a 1-knotting k which is *not* a fibered 1-knotting.
Step 6: We adapt the analysis to examine any twist-spinning of any 1-knotting.

With regards to Step 2, it is more generally true that the connected sum of *any* two fibered 1-knottings is fibered. However to connect to Step 4, the connected $k\#(-k)$ sum can be considered as a low dimensional spinning of k since we can obtain it by "taking $S^1 \subseteq S^3$, removing a tubular neighborhood of $S^0 \subseteq S^1$ (a pair of disjoint sub-arcs) and replacing by two knotted arcs".

Step 1: We consider the standard 1-knotting $S^1 \subseteq S^3$ as a fibered 1-knotting. The considerable notations used in this step will provide a good 3-dimensional model for the 4-dimensional analysis of later steps.

Using $S^3 = S^1 * S^1$, Remark 1.10, associated terminology and notations, the standard $S^1 \subseteq S^3$ is the first factor, and we take $T = (S^1 * S^1)|_0^{\frac{1}{2}}$, the first half of $S^3 = S^1 * S^1$ to be our product tubular neighborhood of $S^1 \subseteq S^3$. Then, the second half of $S^1 * S^1$ will be $X = Ext(S^3, S^1)$. For $\theta \in S^1$, let $X_\theta = (S^1 * \{\theta\})|_{\frac{1}{2}}^1$. Each X_θ is a 2-disk and $X = \bigcup_\theta X_\theta \approx S^1 \times D^2$.

We also have similar subsets of $S^1 = S^0 * S^0$. The standard $S^0 \subseteq S^1$ is the first factor of this. The first half of S^1, call it N, of $S^0 * S^0$ is a product tubular neighborhood $S^0 \subseteq S^1$. Writing $S^0 = \{1, -1\}$, N is just a disjoint pair of arcs: N^1 which contains the point $1 \in S^0 \subseteq S^1 \subset S^3$ and N^{-1} which contains the point $-1 \in S^0 \subseteq S^1 \subset S^3$. So we can write $N = N^1 \cup N^{-1} \approx S^0 \times D^1$. The second half of $S^0 * S^0$, call it M, is a disjoint pair of arcs and we have $S^1 = N \cup M$.

For $\theta \in S^1$, where we are viewing this S^1 as the second factor of $S^1 \times S^1$, we now define subsets of X_θ, Figure 10.300. Let $D_\theta = (S^1 * \{\theta\})|_{\frac{3}{4}}^1$, D_0 will be a 2-disk, shown in light gray in Figure. For $\psi \in S^0$ let $F_\theta^\psi = (N^\psi * \{\theta\})|_{\frac{1}{2}}^{\frac{3}{4}}$, each F_θ^ψ is a 2-disk. Let $G_\theta = \bigcup_\psi F_\theta^\psi \approx S^0 \times D^2$, shown in dark gray. In addition we have $E_\theta = (M * \{\theta\})|_{\frac{1}{2}}^{\frac{3}{4}}$, shown in medium gray. Now write $H_\theta = \overline{X_\theta - F_\theta}$, and $A_\theta = G_\theta \cap H_\theta$ and $A_\theta^\psi = F_\theta^\psi \cap H_\theta$, A_θ, shown with thick dark lines.

Now H_θ is a smooth 2-disk since E_θ is a partial-collar of D_θ, so by Proposition 2.11 part (2) we see $D_\theta \approx D_\theta \cup E_\theta$.

Interpret $X_\theta = G_\theta \cup H_\theta$ as: X_θ is a copy of $D^2 \times S^0$ with a 1-handle $H_0^{(1)}$ attached along A_θ (recall use of superscript notation for handle type). If we want explicit mention of the attaching set we can also write $X_\theta = G_\theta +_{A_0} H_\theta^{(1)}$. In Figure 10.300 we see two components, A_θ^1 and A_θ^{-1}, each of which we call an **attaching staple** since each is roughly shaped like a wire staple.

To visualize these sets, think of the stereographic projection where our X_θ will be spherical caps in R^3. Two, X_0 and X_π will actually be planar. The cap-disk X_π contains the point at infinity and otherwise pictured as a subset of $R^2 \subset R^3$, as in Figure 10.301 where we show the subsets for X_0 and X_π.

We can write

$$\begin{aligned}
Ext(S^3, S^1) &= \bigcup_\theta X_\theta = \bigcup_\theta (G_\theta \cup H_\theta^{(1)}) = \left(\bigcup_\theta F_\theta\right) \cup \left(\bigcup_\theta H_\theta^{(1)}\right) \\
&= \bigcup_\theta (F_\theta^1 \cup F_\theta^{-1}) \cup \bigcup_\theta H_\theta^{(1)} = \left(\bigcup_\theta F_\theta^1\right) \cup \left(\bigcup_\theta F_\theta^{-1}\right) \cup \bigcup_\theta H_\theta^{(1)} \\
&\approx (Ext(S^3, \tau) \times S^0) \cup H.
\end{aligned}$$

At the end of the second line above, we write $Ext(S^3, \tau)^1 = (\bigcup_\theta F_\theta^1)$ and $Ext(S^3, \tau)^{-1} = (\bigcup_\theta F_\theta^{-1})$ where we are viewing each arc of N as representing a (trivial) knotting τ, of $S^1 \subseteq S^3$. We note that $H \approx S^1 \times D^2$ is a solid torus. So in our last line $(Ext(S^3, \tau) \times S^0) \approx Ext(S^3, \tau)^1 \cup Ext(S^3, \tau)^{-1}$.

In Figure 10.303 we see $(Ext(S^3, \tau) \times S^0)$, as subsets of R^3. The thick tube is $\partial Ext(S^3, S^1)$; and $Ext(S^3, S^1)$ is everything on and outside of this tube (plus $S_0 \in S^3$, the "point at infinity" of S^3). We also see two cube-with-holes, each corresponding to $(Ext(S^3, \tau) \times S^0)^\psi = \bigcup_\phi Ext(S^3, \tau)$, where $\psi \in S^0$.

Graphically we view $Ext(S^3, \tau)$ as a 3-ball with an unknotted hole modeled on a solid cylinder with a straight axial hole, Figure 10.302. Shown in dark lines is a subset of $\partial Ext(S^3, \tau)$ that corresponds to some attaching staple. This set consists of a line segment which is a cylinder generator together with two parallel radial line segments, one on the top, one (visually occluded) on the bottom. We may assume, for $\psi \in S^0$, the attaching staple, either A^1_θ or A^{-1}_θ, is a subset of page H^2_θ, see Definition 1.34.

The portions of $Ext(S^3, \tau) \times S^0$ visible in Figure 10.303 are all points of some attaching staple of A^ψ_θ, drawn on the boundaries of two copies of $Ext(S^3, \tau)$.

In summary, the exterior X of the standard knotting $S^1 \subseteq S^3$, (trivially) fibered with fiber $G_\theta +_{A_\theta} H^{(1)}_\theta$ and can also be viewed as a union of two pieces $(Ext(S^3, \tau) \times S^0) \cup H$. Notationally put:

$$Ext(S^3, S^1) = \bigcup_\theta (G_\theta \cup H^{(1)}_\theta) \approx (Ext(S^3, \tau) \times S^0) \cup H.$$

Step 2: Suppose we have a *fibered* knotting k with fiber $\mathbb{F}$. So we know that we can write $Ext(S^3, k) = \bigcup_\phi \mathbb{F}^\phi$, $\phi \in S^1$, a ball with a knotted hole.

We can view $Ext(S^3, k)$ as Figure 10.302. At first this might seem strange since, in Step 1, we viewed this Figure as a ball with an *unknotted* hole and $Ext(S^3, k)$ is a ball with a *knotted* hole. Actually, this is an important point. What we literally can see, in either case, is a portion of the boundary of our knot-exterior. We can see there is some kind of hole, however, we cannot tell, in the Figure, if it is straight or knotted. In either case boundary of a ball with a hole is a union of knot meridians, which are boundaries of our fibers $\mathbb{F}_\phi$. Viewing $Ext(S^3, k)$ as Figure 10.302 we can assume that the dark lines we see are on the boundary of some $\mathbb{F}_\phi$. This means that we can relabel this $\mathbb{F}_\phi$ as $\mathbb{F}_\theta$. So now we can write $Ext(S^3, k) = \bigcup_\theta \mathbb{F}_\theta$. Then $-k$ also fibered with fibers $-\mathbb{F}_\theta$, and $Ext(S^3, -k) = \bigcup_\theta(-\mathbb{F}_\theta)$, also viewed as a ball with a knotted hole.

We get $(k\# - k)$ from $S^1 \subseteq S^3$ by replacing the circular arc $N^1 \subseteq S^1$ with k and, similarly replacing N^{-1} with $-k$. So, we get $Ext(S^3, (k\#-k))$ from $Ext(S^3, S^1) \approx (Ext(S^3, \tau) \times S^0)$ by replacing $(Ext(S^3, \tau) \times \{1\}) = \bigcup_\theta F^1_\theta$ with $Ext(S^3, k) = \bigcup_\theta \mathbb{F}^1_\theta$ and replacing $(Ext(S^3, \tau) \times \{-1\}) = \bigcup_\theta F^{-1}_\theta$ with $Ext(S^3, -k) = \bigcup_\theta \mathbb{F}^{-1}_\theta$. As we do this, for each $\theta \in S^1$ and $\psi \in S^0$, our fibers $\mathbb{F}^\psi_\theta$ have $A^\psi_\theta \subseteq \partial \mathbb{F}^\psi_\theta$ We replace the fibers F^ψ_θ of Step 1 with the corresponding $\mathbb{F}^\psi_\theta$ to give our fibration. So $\mathbb{F}^\psi_\theta$ is a copy of $\mathbb{F}_\theta$ in $Ext(S^3, k)^\psi$.

An appropriate figure is Figure 10.303. Our Figure for $Ext(S^3, (k\#-k)$ is the same as for $Ext(S^3, S^1)$. This is because, in both cases, our hole is visually occluded as in our discussion of Figure 10.302. This means that our 1-handles H_θ, which are uninvolved in our changes, meet our 3-balls (either with unknotted hole or knotted hole) in the same A_θ. But each A_θ lies on the boundary of 3-ball with *knotted* hole, and so A_θ lies on the boundary of the fiber $\mathbb{F}_\theta$ of $Ext(S^3, k)$, whereas in Step 1 it was the boundary of a 2-disk in the solid torus $Ext(S^3, \tau)$.

In summary, the exterior $Ext(S^3, (k\# -k)$ can be written as a union of two pieces $(Ext(S^3, k) \times S^0) \cup H$ in fact, can be obtained by replacing each (if the two) $Ext(S^3, \tau)_\psi$ with $Ext(S^3, k)_\psi$, moreover replacing each 2-disk F_θ^ψ with $\mathbb{F}_\theta^\psi$. So, $Ext(S^3, (k\# -k)$ is fibered with fiber written $\mathbb{G}_\theta +_{A_\theta} H_\theta$, where $\mathbb{G}_\theta \approx \mathbb{F} \times S^0$. Notationally put:

$$Ext(S^3, k\# - k) = \bigcup_\theta (\mathbb{G}_\theta \cup H_\theta^{(1)}) \approx (Ext(S^3, k) \times S^0) \cup H.$$

Step 3: Now we consider 2-knottings of S^2. As a first step we consider the standard 2-knotting $S^2 \subseteq S^4$ as a fibered 2-knotting. To make notations easier to follow we (re)use the notations from Steps 1 and 2.

Using $S^4 = S^2 * S^1$, the first factor corresponds to the standard $S^2 \subseteq S^4$ Then we consider $T = (S^2 * S^1)|_0^{\frac{1}{2}}$, first half of $S^4 = S^2 * S^1$, to be our product tubular neighborhood of the first factor. Then, the second half of $S^2 * S^1$ will be $X = Ext(S^3, S^1)$. For $\theta \in S^1$, let $X_\theta = (S^2 * \{\theta\})|_{\frac{1}{2}}^1$, so each X_θ is a 3-ball and $X = \bigcup_\theta X_\theta$.

Also we consider subsets of $S^2 = S^1 * S^0$. The standard $S^1 \subseteq S^2$ is the first factor of this. The first half of $S^2 = S^1 * S^0$, call it N, is an annulus—a product tubular neighborhood $S^1 \subseteq S^2$. Writing the normal 1-disk to $S^1 \subseteq S^2$ as N^ψ, we have $N = \bigcup_\phi N^\psi \approx S^1 \times D^1$. The second half of $S^2 * S^0$, call it M, is just a disjoint pair of disks—polar caps.

For $\theta \in S^1$, we now define subsets of X_θ and show some in Figure 10.304. Let $D_\theta = (\{S^2 * \theta\})|_{\frac{3}{4}}^1$, D_0 will be a 3-ball, shown as small light gray ball.

For $\psi \in S^0$ let $F_\theta^\psi = (N^\psi * \{\theta\})|_{\frac{1}{2}}^{\frac{3}{4}}$, each F_θ^ψ is a 2-disk. Let $G_\theta = \bigcup_\psi F_\theta^\psi \approx S^1 \times D^2$, shown in dark gray.

In addition we have $E_\theta = (M * \{\theta\})|_{\frac{1}{2}}^{\frac{3}{4}}$, shown in medium gray. Now write $H_\theta = \overline{X_\theta - G_\theta}$, and $A_\theta = G_\theta \cap H_\theta$ and $A_\theta^\psi = F_\theta^\psi \cap H_\theta$, A_θ, shown with thick dark lines in Figure 10.304. In Figure 10.305 we show A_0 separately to avoid occlusion from G_θ. Similar to our Step 1 argument, each H_θ is a 3-ball since it is a 3-ball D_θ with a partial collar, E_θ glued on. So we can consider X_θ to be G_θ with a 2-handle H_θ with attaching set annulus A_θ. Notationally, $X_0 = \Sigma_0 +_{A_0} H_0^{(2)}$.

In summary, $Ext(S^4, S^2)$ can be written as a union of two pieces $(Ext(S^3, \tau) \times S^1) \cup H$ and where $H \approx S^1 \times D^3$ and is (trivially) fibered with fiber $G_\theta +_{A_\theta} H_\theta^{(2)}$. Notationally,

$$\begin{aligned} Ext(S^4, S^2) &= \bigcup_\theta X_\theta = \bigcup_\theta (G_\theta \cup H_\theta^{(2)}) = \left(\bigcup_\theta G_\theta\right) \cup \left(\bigcup_\theta H_\theta^{(2)}\right) \\ &= \bigcup_\theta \left(\bigcup_\psi F_\theta^\psi\right) \cup \bigcup_\theta H_\theta^{(2)} = \bigcup_\psi \left(\bigcup_\theta F_\theta^\psi\right) \cup \bigcup_\theta H_\theta^{(2)} \\ &\approx (Ext(S^3, \tau) \times S^1) \cup H. \end{aligned}$$

Step 4: Consider the Artin spinning $K = (S^2, S^1) \bigotimes k$ where k is a fibered knotting with fiber $\mathbb{F}$. So we can be write $Ext(S^3, k) = \bigcup_\theta \mathbb{F}_\theta$.

We obtain our spinning exterior by replacing each copy of $(Ext(S^3, \tau)$ in $(Ext(S^3, \tau) \times S^1) \cup H$ with $Ext(S^3, k)$. For each θ we replace $F_\theta \times S^1$ in X_θ with $\mathbb{G}_\theta \times S^1$. In fact, each F_θ^ψ is replaced by $\mathbb{F}_\theta^\psi$ so that the attaching staple A_θ^ψ meets $\partial \mathbb{F}_\theta^\psi$. So we have:

$$Ext(S^4, \Sigma(k) = \bigcup_\theta (\mathbb{G}_\theta \cup H_\theta^{(2)}) = \bigcup_\theta \left(\bigcup_\psi \mathbb{F}_\theta^\psi\right) \cup \bigcup_\theta H_\theta^{(2)} \approx (Ext(S^3, k) \times S^1) \cup H.$$

For each $\theta \in S^1$, $\mathbb{G}_\theta = \bigcup_{\psi \in S^1} \mathbb{F}_\theta^\psi \approx \mathbb{F} \times S^1$. So we have a fibration of $Ext(S^4, \Sigma(k)$ with fiber diffeomorphic to $\mathbb{F} \times S^1$ with a 2-handle added.

One cannot show graphics for these fibers other than the trivial case as in Figure 10.304. For example if F_0 is a punctured torus then $S^1 \times \mathbb{F}_0 + \mathbb{H}_0$ cannot embed in R^3, because even the product $S^1 \times \mathbb{F}_0$ cannot embed in R^3 see Remark 2.10.

Step 5: For k-twist-spun of *fibered* knotting, $\mathbb{F}_\theta$ (with $k \neq 0$) will similarly get a fibering of our 2-knotting but with a different fiber. This is because the attaching staple A_θ^ϕ will no longer meet $\mathbb{F}_\theta^\psi$ but instead meets $\mathbb{F}_{\theta+k\psi}^\psi$. Recall our discussion of Figure 10.302, in our hole-in-a-cylinder model. A ball with an unknotted hole looks just like a ball with a knotted hole. A twisting by angle ψ of a ball with an *unknotted* hole corresponds to a rotation of the cylinder about the z-axis of angle ψ. Since we cannot see the "knotted hole", a rotation of a cylinder with a knotted hole will visually still appear, from our view in the Figure, to be rotation of the cylinder about the z-axis of angle ψ.

To simplify notations for the moment, consider a single twist (that is k=1) and focus on X_0. Recall that in Step 4, where we had 0-twists, at A_0^0 we replaced 2-disk F_0^0 with $\mathbb{F}_0^0$, a copy of $\mathbb{F}_0$ placed into $Ext(S^3, k)^0$. More generally, for 0-twists, at A_0^ψ we also replace F_0^ψ by $\mathbb{F}_0$ in $Ext(S^3, k)^\psi$. We note for future reference, that this gives us a smooth map $g\colon \mathbb{G}_0 \to Ext(S^3, k)$ which sends $\mathbb{F}_\theta^\psi$ to $\mathbb{F}_0$. So, as we have seen for 0-twists, Step 4, that $\mathbb{G}_0 = \bigcup_\psi \mathbb{F}_0^\psi \approx \mathbb{F}_0 \times S^1$.

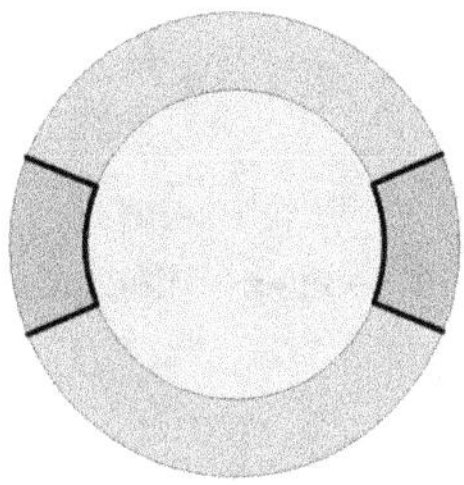

Fig. 10.300 Subsets of X_θ.

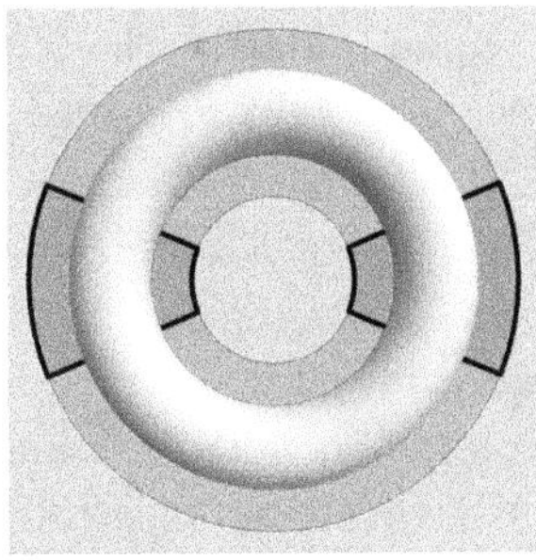

Fig. 10.301 ∂X and subsets of X_0 and X_π.

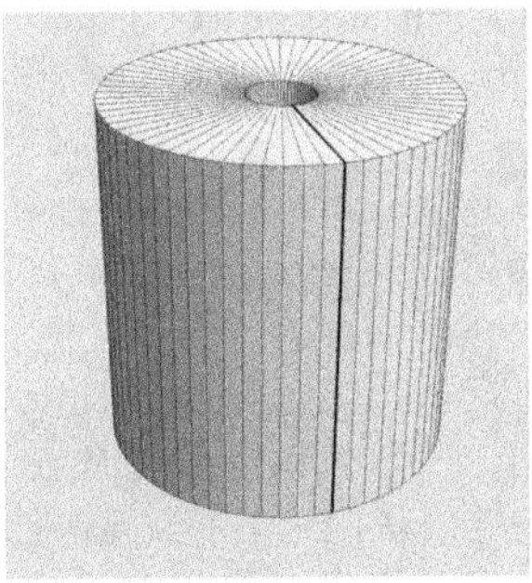

Fig. 10.302 Fibered knot exterior as 3-ball with hole.

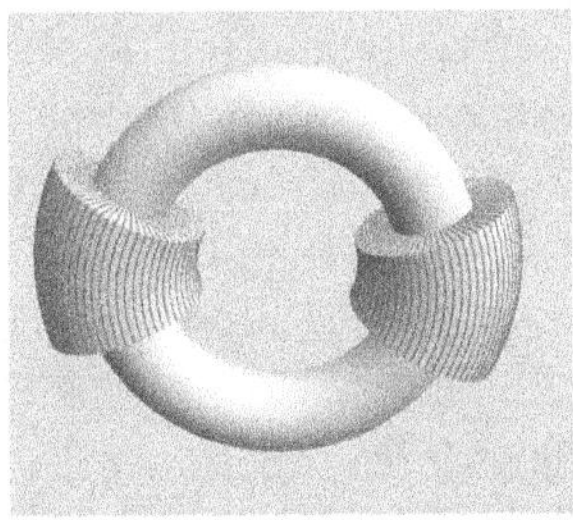

Fig. 10.303 ∂X and two knot exteriors—3-balls with holes.

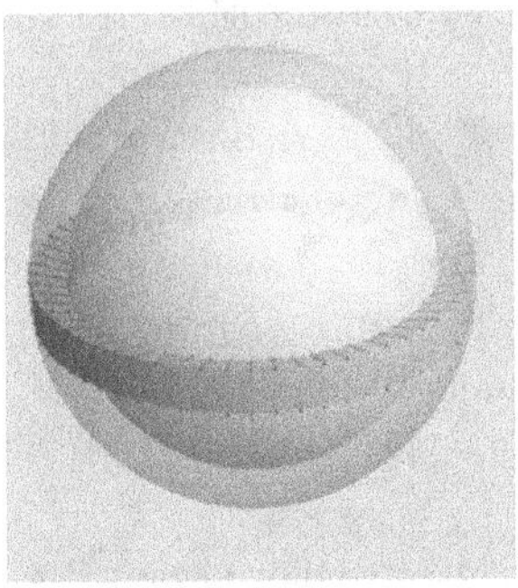

Fig. 10.304 Subsets of X_0, Step 3.

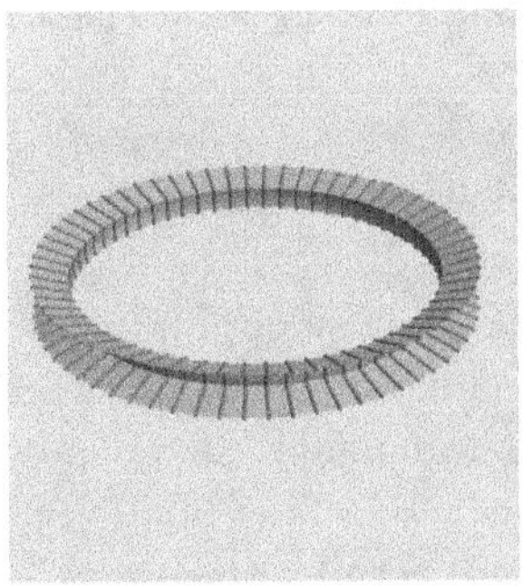

Fig. 10.305 Detail of attaching set $A_0 \subseteq \mathbb{X}_0$.

Now we see the differences with a 1-twist knotting. The very beginning is the same— start with F_0^0 and replace 2-disk F_0^0 with $\mathbb{F}_0$. However, we replace disk F_0^ψ by $\mathbb{F}_\psi$, since we have twisted $Ext(S^3,k)^\psi$ by an angle of ψ so A_0^ψ meets F_ψ^ψ.

This gives us a smooth map $g\colon \mathbb{G}_0 \to Ext(S^3,k)$ which sends $\mathbb{F}_\theta^\psi$ to $\mathbb{F}_\psi$. However, unlike the 0-twist case, this map is 1-1 and so is a diffeomorphism! Similarly, for all $\theta \in S^1$, $\mathbb{G}_\theta \approx Ext(S^3,k)$.

So our fiber for our 2-knotting is diffeomorphic to $Ext(S^3,k)$ with a 2-handle added by an attaching set which meets each $\mathbb{F}_\psi$ once. This a clearly a 2-handle which "fills in the hole" of our ball-with-a-knotted-hole. And so our fiber of our 1-twist spun k is diffeomorphic to D^3, Remark 5.2. And that means K is unknotted!

So a 1-twist spinning of any fibered k is a trivially knotted 2-sphere. For -1-twist we will find that so A_0^ψ meets $F_{-\psi}^\psi$ but still our map $g\colon \mathbb{G}_0 \to Ext(S^3,k)$ will be a diffeomorphism. So, a -1-twist spinning of any fibered k is a trivially knotted 2-sphere.

Now our analysis for the case $k \neq \pm 1$ will be similar. However, g will be a smooth k-to-one map. We *will* get a fibration for K but with a special 3-manifold for the fiber. We discuss this more in Step 6.

Step 6: Finally the general case for twist-spun of any 1-knotting. If k is any 1-knotting of a circle with exterior X. By Proposition 4.11 there is a circular Morse function $f\colon X \to S^1$. Let $\mathbb{F}_\theta = f^{-1}(\theta)$. If f has no critical points then the collection $\mathbb{F}_\theta$ is a fibration of X. If f does have critical points, and suppose $\alpha \in S^1$ is a critical value. We still have $\mathbb{F}_\theta = f^{-1}(\theta)$, and $X = \bigcup_\theta \mathbb{F}_\theta$. This will fail to be a fibering since, F_α is not a 2-manifold. For example if the critical point had index 1, then near the corresponding critical point, F_θ would look like the union of the first and third quadrants of R^2.

Consider a 0-spin K of k. We can still carry out our construction. We can define our G_θ as in Step 4 and still conclude that for any $\theta \in S^1$, $\mathbb{G}_\theta = \bigcup_{\psi \in S^1} \mathbb{F}_\theta^\psi \approx \mathbb{F}_\theta \times S^1$. If θ is a regular value for f, then at least we have $\mathbb{G}_\theta$ is a smooth 3-manifold. But for critical point α, F_α will fail to be a 2-manifold and so G_α will fail to be a 3-manifold. So, $G_\alpha +_{A_\alpha} H_\alpha^{(2)}$ is not a 3-manifold. So our process does not give us a fibering of K if K is a 0-spin.

However our process *will* give a fibration if $k \neq 0$. For simplicity, first consider 1-twist spinning k. Again, we proceed as in Step 5, using $\mathbb{F}_\theta = f^{-1}(\theta)$. As before, we construct a smooth map $g\colon \mathbb{G}_0 \to Ext(S^3,k)$ which sends $\mathbb{F}_\theta^\psi$ to $\mathbb{F}_\psi$. For 1-twist spinning, this will be a diffeomorphism. We can see g is 1-to-1 and onto.

Our argument that for a 1-twist spin our fiber was a 3-ball, was simply that our 2-handle acted as a plug for a cube with a knotted hole and did not use the fact our k was fibered. So the 1-twist spinning of any 1-knotting will give a trivial 2-knotting. ∎

The 3-manifolds obtained as fibers of n-twist-spinning a 1-knotting k are of interest beyond 2-knottings and can be considered as a construction apart from 3-knottings. The common name for such a 3-manifold is the **punctured n-fold branched cyclic cover of S^3 branched about k.** There are punctured 3-manifolds that do not embed in R^4. Any fiber of a fibration of Proposition 10.7 does embed in R^4. Such embeddings are hard to come by. For example, for the 5-twist-spinning of a trefoil knotting, the fiber is the punctured dodecahedral space, and for a (p, q)-torus knotting with p odd, the fiber of the 2-twist-spinning will have as fiber the punctured lens space $L(p, q)$ with p odd.

As might be expected, not all fibered 2-knottings are spun or twist-spun. These use constructions outside the scope of this book. The most interesting is called the Cappell-Shaneson [Cappell and Shaneson (1976)] knotting (also studied by Kirby-Akbulut [Akbulut and Kirby (1985)]) knotting has a fiber punctured $S^1 \times S^1 \times S^1$ and is not ribbon by [Cochran (1983)]. Fibered ribbon knots are discussed in [Larson and Meier (2015)].

Chapter 11

Some Final Topics for 2-Knottings

11.1 Connected Sum of Manifolds and Knottings

The connected sum construction Definition 2.20 expands our collection of previously constructed knottings. A connected sum of two knottings of S^2 is, again, a knotting of S^2—for example the connected sum of a 2-twist-spun trefoil and a 3-roll-spin of a figure-eight. A connected sum of two torus knottings gives a knotting of a genus 2 surface—for example, the connected sum of a 1-turned trefoil torus and a 3-roll-spin of a figure-eight torus.

A knotting is **irreducible** if it cannot be written as a connected sum of non-trivial knottings. Any 1-knotting can be uniquely expressed as a sum of irreducible knottings, except for reordering of summands, for example see [Fox (1962a)]. This is akin to the factoring of integers as a product of primes. This is not true for high-dimensional n-knottings with $n \leq 3$, [Kearton (1979)], but as yet unknown for $n = 2$.

11.2 2-Braids

There is an interesting extension of braids and braid position in higher dimensions first considered in [Rudolph (1983)], later extended in [Viro (1990)]. A highly recommended reference is [Kamada (2002)]. We only present the basic definitions and some examples. These examples are presented from the perspective of projections of knottings in contrast to [Kamada (2002)] and elsewhere, where slicing and 1-dimensional braids are used as shown in Example 11.5.

We extend closed braid position for 1-knottings in the complement of the z-axis, Definition 4.6, to 2-knottings in the complement of the w-axis.

Definition 11.1. Braid projection $\widehat{\pi}\colon (R^4 - axis(w)) \to S^2$ is defined $\widehat{\pi} = (\rho, \theta, \phi, w) = (1, \theta, \phi)$. ▮

It will be useful involve the standard projection and express braid projection in two steps: $\widehat{\pi} = \widehat{\rho} \circ \pi|_{(R^4 - axis(w))}$.

Sometimes 1-braid position is viewed as a knotting in a standard solid torus in R^3. We similarly view 2-braid position as a knotting in a standard $S^2 \times D^2$. For any compact $K \subseteq (R^4 - axis(w)) \approx S^2 \times R^2$ we can find an $0 < M$ such that K is contained in $N = \{(\rho, \theta, \phi, z) \in R^4 \colon 1/M \leq \rho \leq M \text{ and } 1/M \leq z \leq M\}$. We can view N as a product tubular neighborhood of $S^2 \subseteq R^4$ and view $K \subseteq S^2 \times D^2 \approx N$.

Thus we can view $\widehat{\pi}$ as projection onto the first factor either of $S^2 \times R^2$ or $S^2 \times D^2$. The map $\widehat{\rho}\colon (R^3 - \{\mathbb{O}\}) \to S^2$ is called the **radial map**. (If we view $(R^3 - \{\mathbb{O}\}) \approx S^2 \times R^1$ the radial map corresponds to projection onto the first factor.)

Braid projection sends a 2-knotting $M \subseteq R^4$ to S^2. From Definition 4.25 of a 1-braid, it might seem that we should define braid position for a 2-braid $M \subseteq (R^4 - axis(w))$ by requiring braid projection is a covering map from M to S^2. It turns out that the only such covering map (if M is connected) is a diffeomorphism. However there is a related type of mapping, a *branched* covering, that can be used.

Certain critical points of such maps between two 2-manifolds are called **branch points**. This use of the term "branch point" is not identical to Definition 6.1 since that referred to smooth maps from a 2-manifold into R^3. In Example 11.1 we will see how to relate these two notions. As with Definition 6.1 we will define this type of branch point by a local model:

Definition 11.2. The **standard model for a two-fold branch point from R^2 to R^2** is given in polar coordinates by $f(r, \theta) = (r, 2\theta)$. Alternatively, using complex numbers $f(z) = z^2$.

For a smooth map $f : X \to Y$ between 2-manifolds and $x \in X$, if f is locally modeled as a two-fold branch point then both x and $f(x)$ are called **branch points** of f. ▮

So f is a two-to-one map except single valued for $\mathbb{O}$. In the literature such as [Kamada (2002)], one considers n-fold branch points based on $f(r, \theta) = (r, n\theta)$ in which case f is a n-to-one map except for $\mathbb{O}$. If this definition is used, a branched cover as in Definition 11.2 is called a *simple* branched cover.

Definition 11.3. A smooth map $f : X \to Y$ between 2-manifolds is a **branched cover** if, for each $x \in X$, either x is an immersion point of f or x is a branch point of f. ▌

Definition 11.4. Suppose $M \subseteq (R^4 - axis(w))$ is a 2-knotting. We say M is in **closed braid position with respect to** W if $\widehat{\pi}|M\colon M \to S^2$ is a branched cover. ▌

Example 11.1. We describe a subset, $\mathcal{R}_* \subseteq R^3$, which will relate "branch point" of $f\colon X \to Y$ and "branch point of a knotting projection X_*. We examine $f(z) = \sqrt{z}$, the inverse of the square function. This is not a function since every complex number (except $z = 0$) has *two* square roots. (We use complex numbers only to derive two ordinary real-valued smooth function u, found below.)

Using DeMoivre's Formula and polar notation $z = (r\theta)$, for complex numbers so that $z = r(\cos(\theta) + i\sin(\theta))$, the two complex square roots of z are $\pm\sqrt{r}\,(\cos(\theta/2) + i\sin(\theta/2))$. So the real part is $u = \pm\sqrt{r}\cos(\theta/2)$.

The two real two-variable functions $u = +\sqrt{r}\cos(\theta/2)$ and $u = -\sqrt{r}\cos(\theta/2)$ have graphs Figures 11.1 and 11.2, respectively. Our local model $\mathcal{R}_*$ is the union of these graphs, Figure 11.3. A top view of Figure 11.3 provides a projection along the z-axis to R^2, see Figure 11.5.

The set $\mathcal{R}_*$ is an image of a smooth map $g\colon R^2 \to R^3$ which is an immersion at all points except $\mathbb{O}$ since $\mathcal{R}_*$ is the projection of the 2-disk $\mathcal{R} \subseteq R^4$, the graph of the complex function $w = z^2$. Then we have $\pi \circ g = f$ where $\pi : R^3 \to R^2$ is projection and f is as in Definition 11.2. The image under dg of the tangent plane at $\mathbb{O} \in R^2$ is a line—in fact the z-axis, Figures 11.5 and 11.6. The image of this line, when we project to R^2, as in Figure 11.5, is $\mathbb{O} \in R^2$. ▌

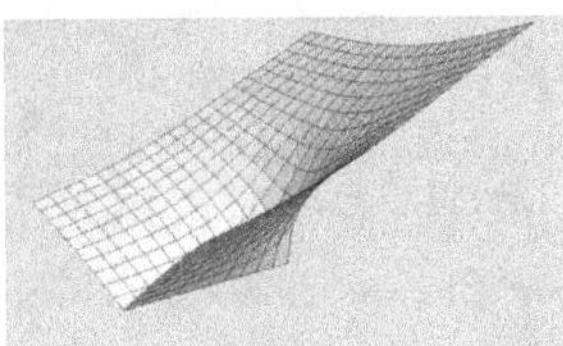

Fig. 11.1 The graph of $u = +\sqrt{(r)}\cos(\theta/2)$.

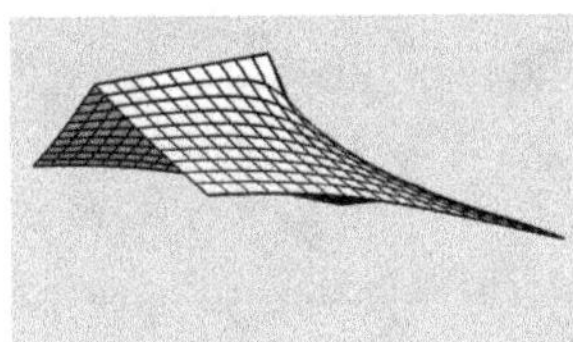

Fig. 11.2 The graph of $u = -\sqrt{(r)}\cos(\theta/2)$.

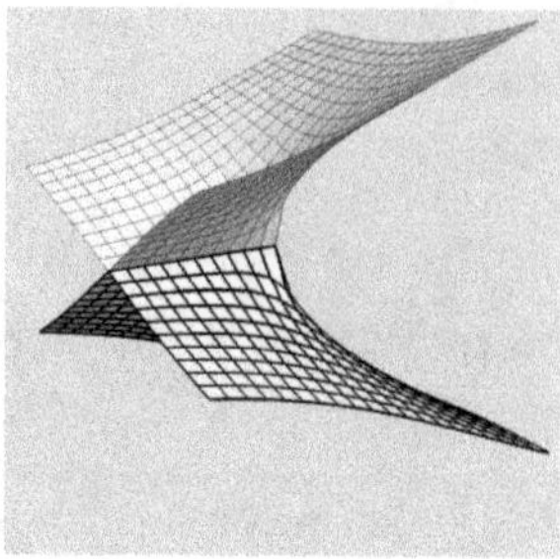

Fig. 11.3 Union of graphs in Figures 11.1 and 11.2 is $\mathcal{R}_*$.

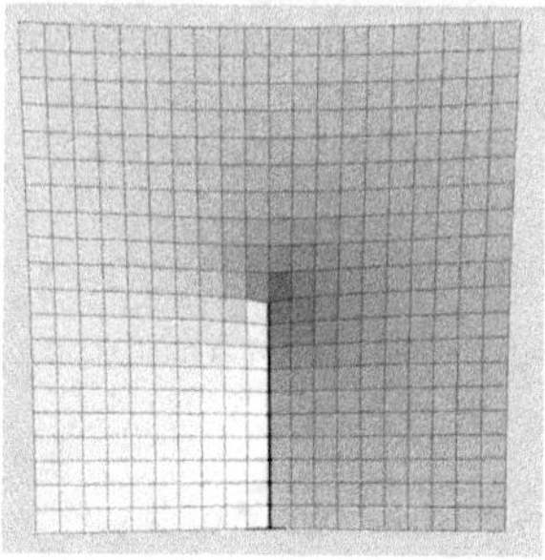

Fig. 11.4 Projection of Figure 11.3 to the plane.

Fig. 11.5 Showing some coordinate curves for Figure 11.3.

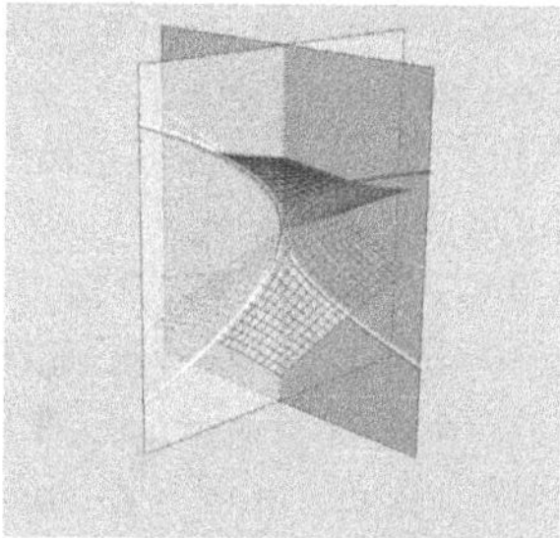

Fig. 11.6 Different view of Figure 11.5.

Remark 11.1. Actually, we have two standard models. If $\mathcal{R}'$ is the graph of $w = -z^2$ we would have $\mathcal{R}'_* = \mathcal{R}_*$. Visually, Figure 11.3 is symmetric via reflection in the plane of the first two coordinates. Equivalently, we could have graphed the real parts of $-\sqrt{z}$. This detail will come up in a later discussion, Example 11.4. ∎

The usual definition of a 2-braid K [Kamada (2002)] (originally defined in [Rudolph (1983)]) is as a certain proper 2-manifold in $D^2 \times D^2$. (A 1-braid is defined as a certain proper 1-manifold in $D^2 \times D^1$, Definition 4.26.) To see the relation to closed braid position, first view $K \subseteq S^2 \times D^2$ with $\widehat{\pi}$ projection onto the first factor. Take a small closed 2-disk $D_0 \subseteq S^2$ which contains no branch points. Then $D_1 = \overline{S^2 - D_0}$ is a 2-disk and our 2-braid is the proper 2-manifold $K_1 = K \cap (D_1 \times D^2) \subseteq S^2 \times D^2$.

Remark 11.2. The torus $T \subseteq R^4$, whose projection $T_* \subseteq R^3$ is as shown in Figure 11.7 is *not* in closed braid position. The set $A = \widehat{\rho}(T_*)$ is the grey annulus. The map $\widehat{\pi}$ clearly not a branched covering. For points in the interior of A, our map is 2-to-1. For points in ∂A, f is 1-to-1 and points not in A correspond to no points of X via f. A small change in T will not give us a branched cover, since the projection will still be an annulus. ∎

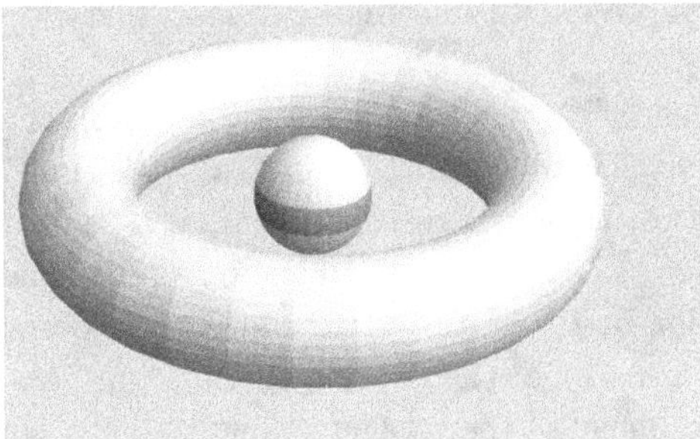

Fig. 11.7 $\widehat{\rho}(T_*)$ an annulus A shown in grey.

Any orientable 2-knotting all can be put in a closed braid position. This is a higher dimension version of Alexander's Theorem, Proposition 4.9—see [Kamada (2002)] for a proof.

Prop 11.1 (Generalized Alexander's Theorem). Any 2-knotting is equivalent to one in closed braid position. ∎

A graphical note: in the examples that follow we scale our projections K_* so that the sphere S^2 can be thought of in our figures as a small sphere with center $\mathbb{O}$.

Example 11.2. We give two knottings of a torus in R^4 as a closed braid. We first consider a (trivial) spun knotting T which is *not* in braid position and then alter this to get our example T' which *is* in braid position.

We begin with a simple closed curve $\gamma \subseteq R^2_+ \subseteq \overset{\circ}{R^3_+}$ with two tangent lines (at $p \in \gamma$ and $q \in \gamma$) intersecting at $\mathbb{O}$, Figure 11.8 and consider the spun torus $T = \Sigma(\gamma) \subseteq R^4$, Definition 10.3. Note: the graphics we use are for spinning with an xw-plane as axis rather than the usual zw-plane, so in projection, our graphics are symmetric with respect to x-axis rather than the z-axis. Each point $u \in \gamma$ gives rise to a circle $C_u \in T$ which projects to C^*_θ as shown in Figure 11.9. Now $T \subseteq (R^4 - axis(w))$, and is equivalent to a standard torus—it simply has γ as a "cross-section" rather than a geometric circle. As in example 11.2, T is not in 2-braid position.

The braid projection of $T = \Sigma(\gamma)$ is an annulus in S^2—in fact between the two circles $\widehat{\rho}(C_p^*)$ and $\widehat{\rho}(C_q^*)$.

We alter knotting T by "putting a half-flip near the circles at p and q" (similar to examples as in Figure 6.11), so that circles C_u will are gradually rotated (in flipping manner) about a diameter in the hyperplane $x = x_0$ and will generally project to an ellipse, except that C_p^* and C_q^* are line segments.

The modification of C_u will be called K_u The result is a new knotting of a torus, $T' = \bigcup_u K_u$. We will see T' is in braid position. The crossing set of T'_* consists of two disjoint line segments $C_p^* \cup C_q^*$ and the branch set B are the four endpoints, B. Figure 11.10 shows some of the K_u^*, and Figure 11.11 is a projection of T'_*. Figure 11.12 shows a clipped version of T'_*; the crossing set of T'_* consists of the two black line segments.

For $b \in B$, the image of the tangent plane to T' at $\widetilde{b}$ is a line L. Since the tangent lines at p and q to γ contain $\mathbb{O}$, see Figure 11.8, L will intersect $\mathbb{O} \in R^3$, see Figure 11.14. Thus these branch points of the standard projection $\pi\colon R^4 \to R^3$ are *also* branch points of the braid projection $\widehat{\pi} = \widehat{\rho} \circ \pi|_{(R^4 - axis(w))}$. So T' is in braid position.

In Figure 11.13 we see that the radial projection of crossing set of T'_* consists of two disjoint circular arcs on S^2. Using stereographic projection, we view these as disjoint arcs in R^2, Figure 11.15. A graphic such as Figure 11.15 corresponds to a **chart**, [Kamada (2002)]. Our terminology is differs slightly from [Kamada (2002)], so we will, later, define a 2-braid chart Definition 11.5. ∎

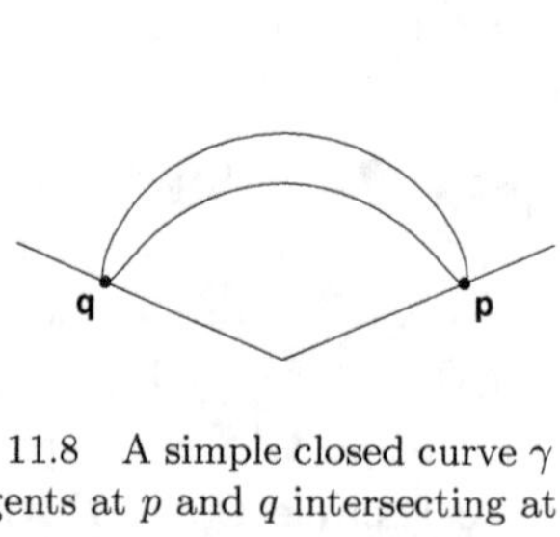

Fig. 11.8 A simple closed curve γ with tangents at p and q intersecting at $\mathbb{O}$.

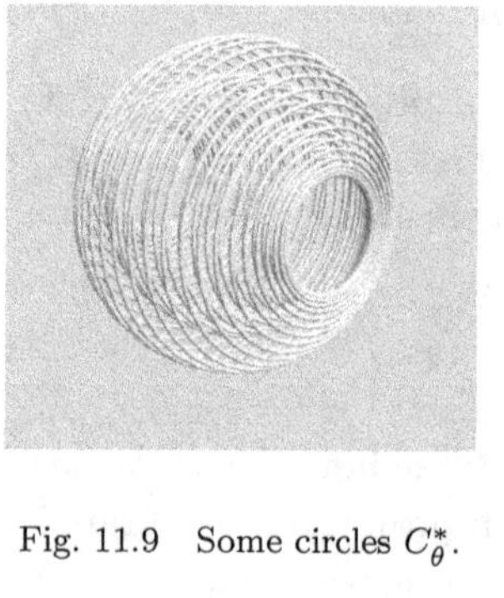

Fig. 11.9 Some circles C_θ^*.

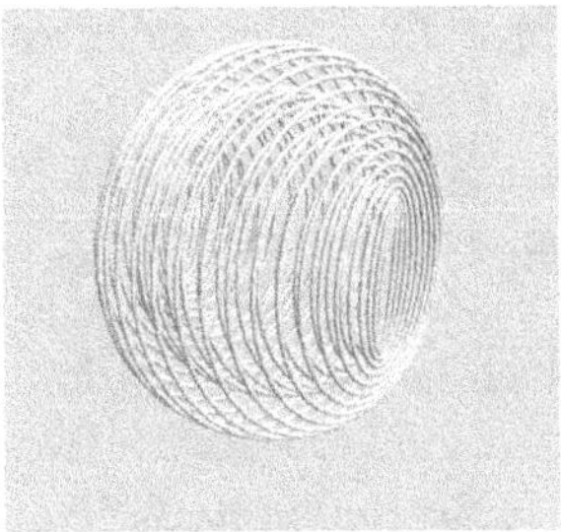

Fig. 11.10 Some subsets K_u^*.

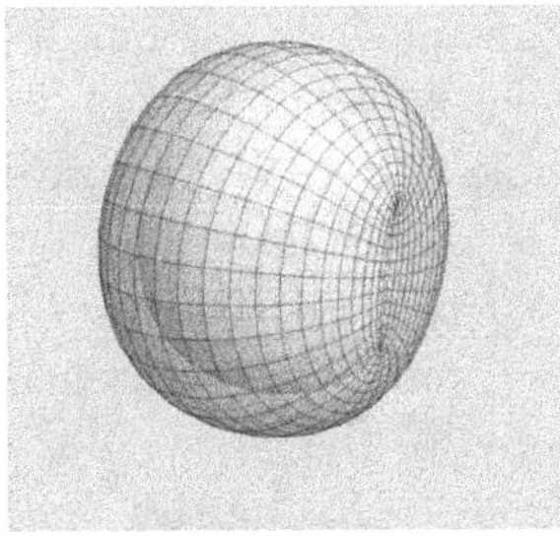

Fig. 11.11 T'_* shown with transparency.

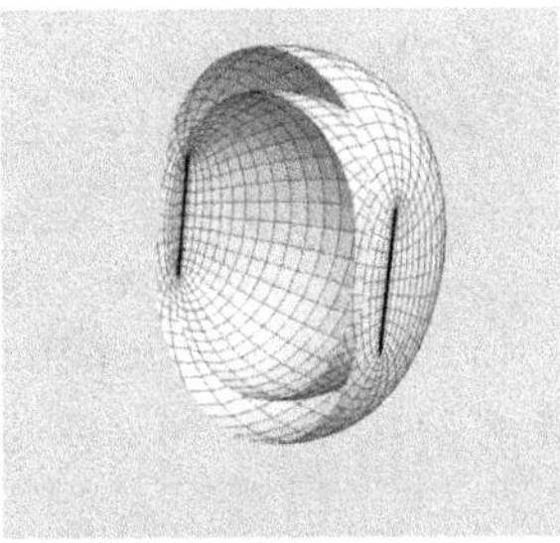

Fig. 11.12 Clipped version of T'_* with crossing set in black.

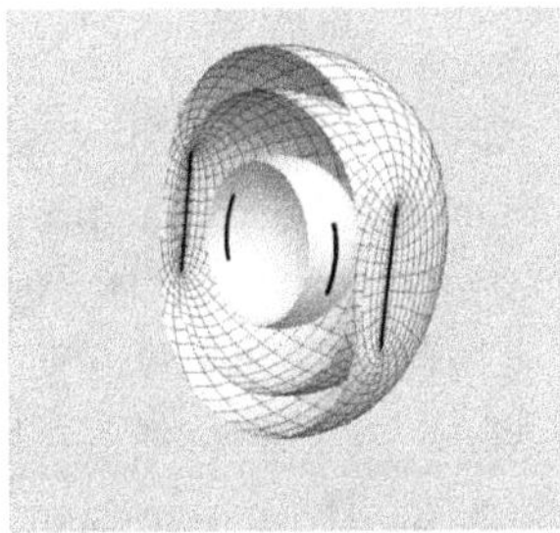

Fig. 11.13 Radial projection of crossing set Projecting crossing set of T'_*.

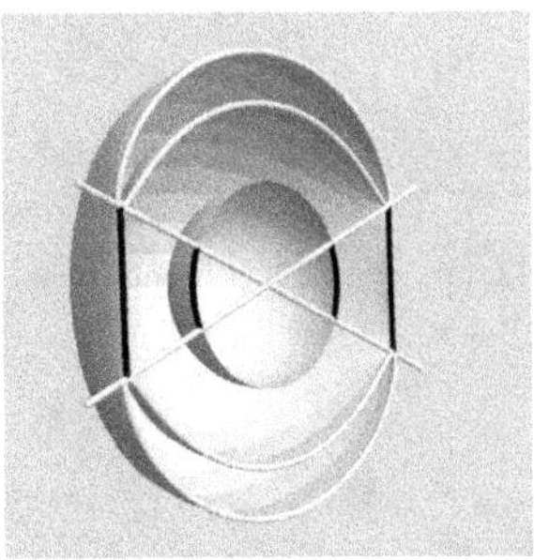

Fig. 11.14 T'_* clipped in half. The two lines are images of tangent planes at branch points.

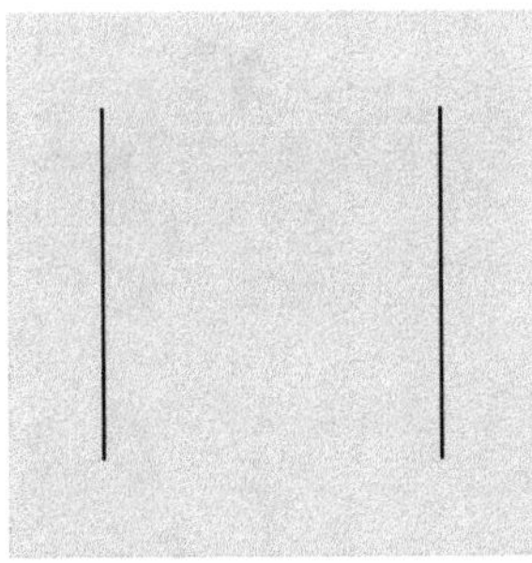

Fig. 11.15 A planar diagram representing radial projection of crossing set.

Example 11.3. More interesting is a 2-braid presentation for a spun trefoil knotting modeled on our methods in Example 11.2. This time we use a proper knotting k of an arc and obtain a knotting of a 2-sphere. As in Example 11.2 we first consider a spun knotting $\Sigma = \Sigma(k) = \bigcup_{u \in k} C_u$ where $C_u \subseteq R^4$ is the geometric circle obtained by spinning $u \in k$. This Σ is *not* in braid position, but then we alter this by "putting a half-flip near two circles C_p and C_q" and get our example Σ' which *is* in braid position.

We alter the circles near C_p and C_q as in Example 11.2 with two half flips and let K_u be the circle obtained from C_u in this process. Now let $\Sigma' = \bigcup_u K_u$. The projections K_p^* and K_q^* will be line segments and nearby K_u^* will be ellipses , Figure 11.18, where we have added in a projection of k to help visually follow K_u^*. Figure 11.19 shows Σ_* with transparency, revealing the crossing set, $C \subseteq \Sigma'_*$, shown in black. This set consists of two line segments K_p^* and K^*q together with three circles corresponding to the three crossing points in the projection of k_* as seen in Figure 11.17.

Figure 11.20 shows a cut-away view of the projection together with a unit sphere. The radial projection $\widehat{\rho}(C)$ consists of the disjoint union of two arcs and three circles, Figures 11.21 and 11.22. The verification that Σ' is in closed 2-braid position is like in Example 11.2. The branch point set of Σ'_* is a four point set B. This time away from $\widehat{\rho}(B)$ we have a covering of degree 3. If we represent $\widehat{\rho}(B)$ in R^2 we get a configuration of three concentric circles and two line segments Figure 11.31. ∎

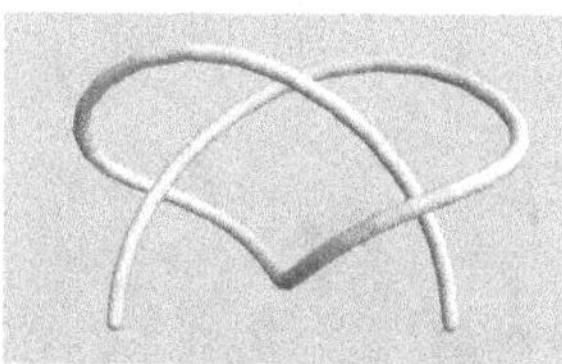

Fig. 11.16 A proper trefoil arc $k \subseteq R^3_+$.

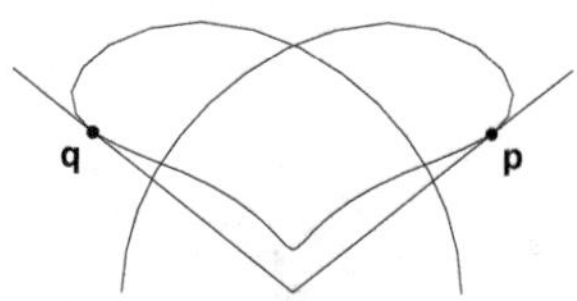

Fig. 11.17 In R^2, projection k_* and two tangent lines which intersect at $\mathbb{O}$.

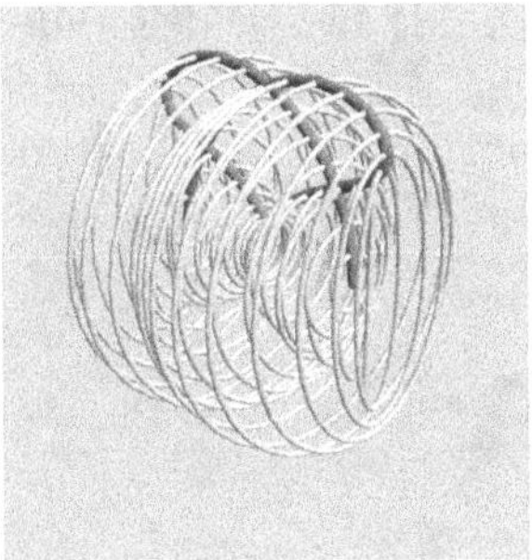

Fig. 11.18 Some K_u^* and copy of k_* (in xz-plane).

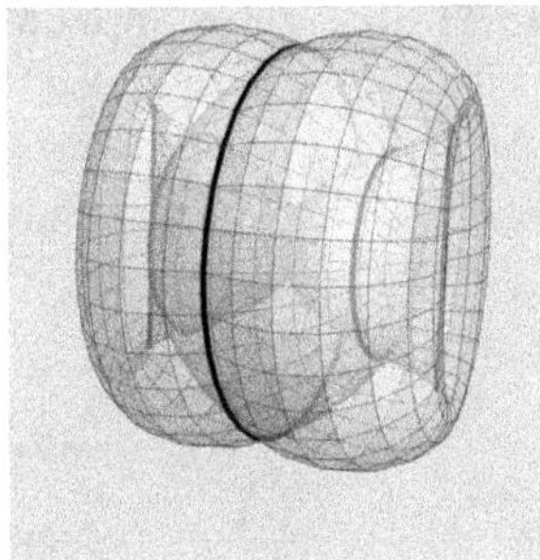

Fig. 11.19 Σ'_* with crossing set shown as black tubes.

Fig. 11.20 A cut-away view of Σ'_* and S^2.

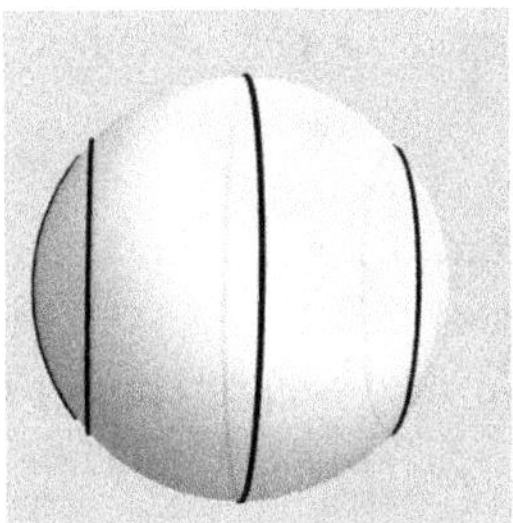

Fig. 11.21 $\widehat{\rho}(C) \subseteq S^2$.

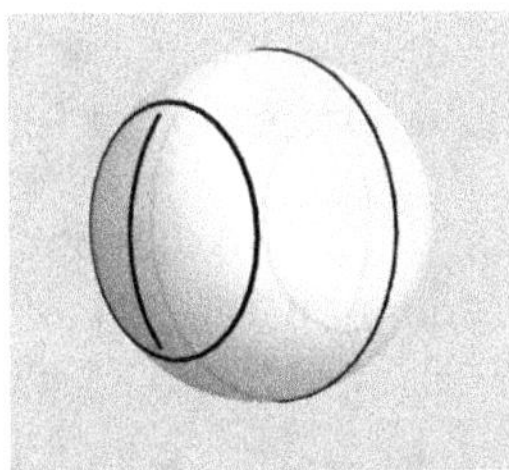

Fig. 11.22 A second view $\widehat{\rho}(C)$.

Remark 11.3. For a knotting in $M2$-braid form it is useful to consider the crossing set $\mathcal{C}$ of M_* and the set $\widehat{\rho}(\mathcal{C})$. This set, considered as a subset of R^2 rather than S^2, with additional notational refinements to distinguish the two branch point types Remark 11.1, gives rise to a diagram for the knotting called a **surface braid chart** or sometimes just **chart**, [Kamada (2002)]. It can be shown that given a 2-knotting M in closed 2-braid position one can isotop M so that, as in our two examples above, the branch point set in S^2 is the radial image of the branch set of the standard projection, $M_* \subseteq R^3$. For other points of $\mathcal{C}$ one can adapt standard general position arguments and assume that $\widehat{\rho}(\mathcal{C})$ consists of a union one of four local models, Figure 11.23: a smooth path, a smooth path ending in a branch point, two transverse paths (images under $\widehat{\rho}$ of disjoint crossing arcs of M_*), three paths intersecting in a single point (the image of a triple point of M_*).

Figure 11.31 is an example which uses only the first two models. Furthermore, if a point is modeled on two crossing line segments one can define "height" relation at each crossing using ρ-value of points of R^3 and derive a series of "moves" for such diagrams, see [Kamada (2002)]. ▮

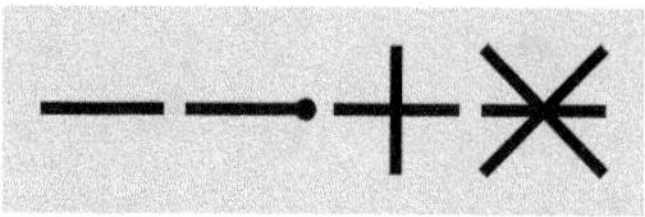

Fig. 11.23 Four local models for $\widehat{\rho}(\mathcal{C})$.

Definition 11.5. Suppose K is a 2-knotting in closed braid position and in general position with respect to standard projection so that the radial projection of the branch points of K_* are the branch points of $\widehat{\rho}$, then the radial projection of the crossing set of K will be called a **2-braid chart.** ▮

The most common graphics for a spun trefoil knotting as a braid (for example [Carter *et al.* (2004)] [Kamada (2002)]) is as in the sequence of Figures 11.33 to 11.43. We next relate these Figures to Figure 11.31. A key to understanding this is to consider "slicing" of a single branch point:

Example 11.4. Figure 11.24 represents a small square containing a branch point in S^2. We visualize this square as a subset of R^2, which is the projection of our standard local branch point model $\mathcal{R}_*$ of Example 11.1 as shown in Figure 11.5. Shown with thick black graphics the second of our local models of Figure 11.23. Also shown are three thin line segments α, β, and γ which meet the image of the crossing set transversely. These three segments are projections of three squares in R^3, a, b and c shown in Figure 11.25 intersecting $\mathcal{R}_*$.

Figure 11.25 shows that plane a intersects $\mathcal{R}$ in two disjoint arcs, and b and c intersect $\mathcal{R}_*$ in a pair of transversely intersecting arcs. In addition, the three squares are projections of cubes in R^4, A, B and C shown in Figures 11.27, 11.28 and 11.29 with intersections with $\mathcal{R}$ as shown. Recall that the projection from $\mathcal{R}$ to $\mathcal{R}_*$ is two-to-one except for the branch point where it is one-to-one.

If our branch point is a projection of $\mathcal{R}'$, see Remark 11.1, then our C we would appear as in Figure 11.30—the distinction being the height relation at the crossing.

So, the two *disjoint* arcs in A project to *disjoint* arcs in a, the two intersecting arcs in B project to two intersecting arcs in b (with the intersection point at the branch point) and the two *disjoint* arcs in C project to two intersecting arcs in c (with intersection point a crossing point of $\mathcal{R}_*$). ∎

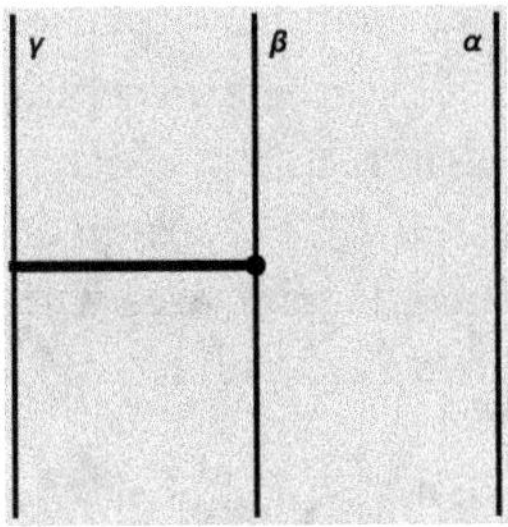

Fig. 11.24 A neighborhood of branch point.

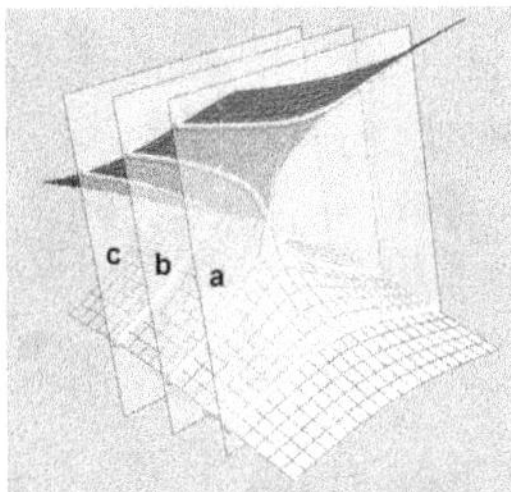

Fig. 11.25 $\mathcal{R}_*$ and three parallel planes.

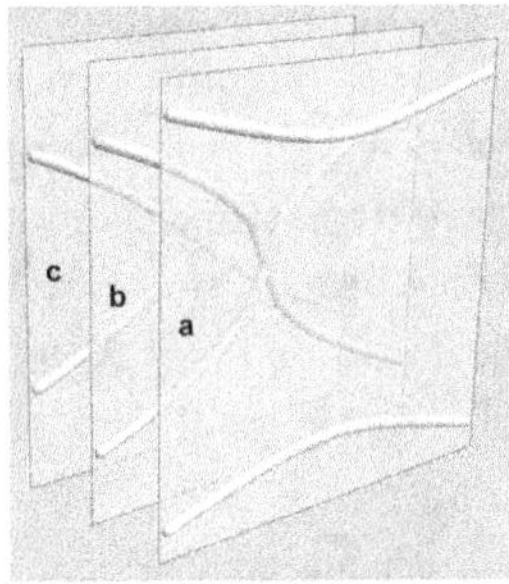

Fig. 11.26 Figure 11.25, without $\mathcal{R}_*$.

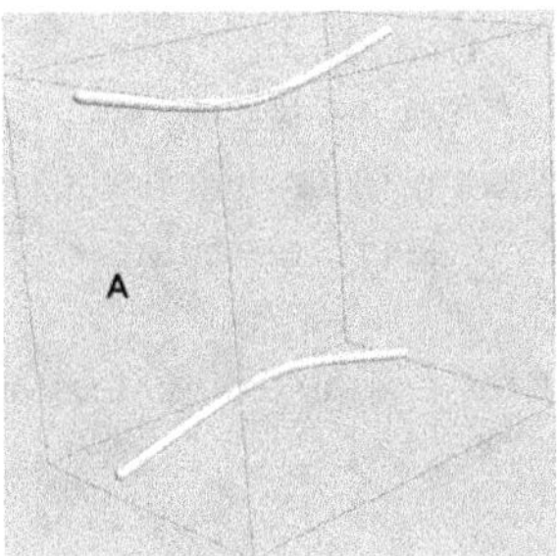

Fig. 11.27 $A \cap \mathcal{R}$ is a *disjoint* pair of arcs.

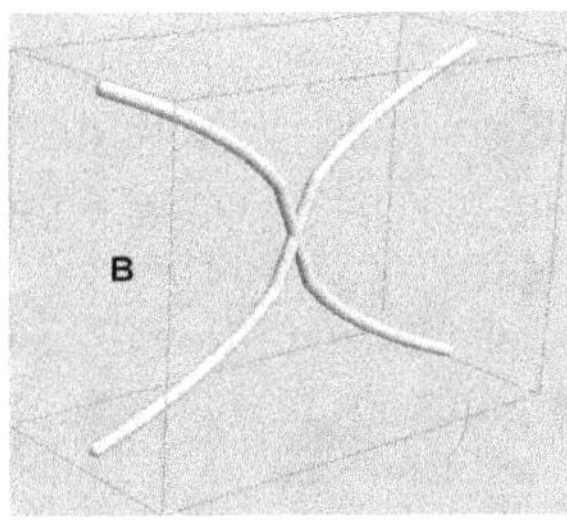

Fig. 11.28 $B \cap \mathcal{R}$ is an *intersecting* pair of arcs.

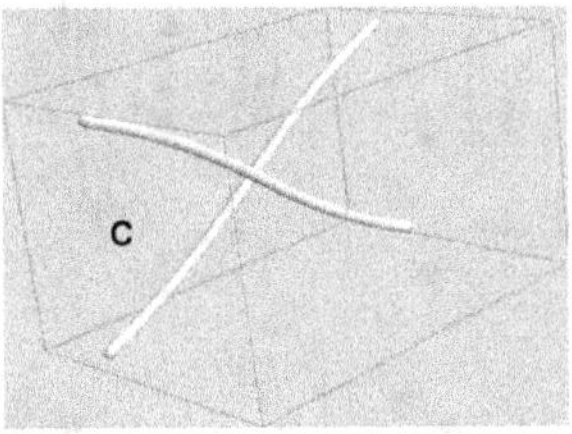

Fig. 11.29 $C \cap \mathcal{R}$ is a *disjoint* pair of arcs.

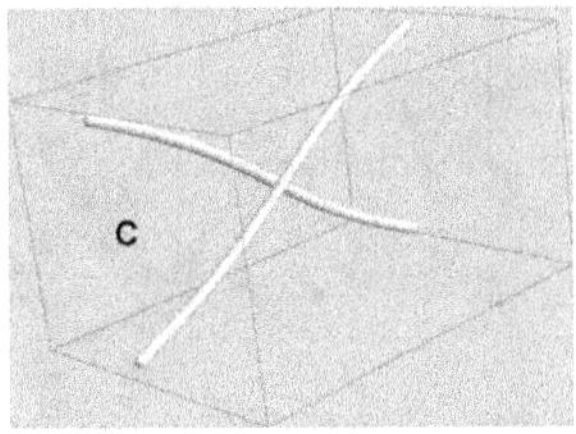

Fig. 11.30 $C \cap \mathcal{R}'$ is a *disjoint* pair of arcs.

Example 11.5. Using Example 11.4 for each branch point, one can derive Figures 11.33 to 11.43 for a spun trefoil knotting from $X = \widehat{\rho}(\mathcal{C})$ using Figure 11.31. We describe our braid as a proper 2-manifold in $D^2 \times D^2$. First, we draw line segments $t_0, \ldots t_{10}$ as shown in Figure 11.32. Each t_i is a projection of a 3-dimensional cube in $t_i \times D^2 \subseteq D^2 \times D^2$. Our branched covering is 3-fold, so that, if t_i contains no branch point, then it is an image of three disjoint arcs in T_i. However, each of t_4 and t_6 contains a pair of branch points so we expect to see, in each of T_4 and T_6, three arcs which have two intersection points. For each point of intersection $X \cap t_i$, which is not a branch point, will correspond to a crossing of arcs in T_i. For example, t_0 does not meet X, and so, in Figure 11.33 we see three disjoint arcs with no crossings. There are two points in $X \cap t_1$ and so we find two crossing points in Figure 11.34.

Basically, one needs to keep track of the branch points where, graphically, there are two possibilities—Figure 11.29 or Figure 11.30. One can decide the correct above/below relations at these crossings by careful examination of the original knotting. This information can also be obtained if one uses Kamada's diagrams notations.

The set $X \cap t_6$ contains two branch points. Note that, in Figure 11.38, we no longer have three disjoint arcs since two of them intersect twice. (Generically, we would expect each t_i to contain only a single branch point—the reason we chose a non-generic situation is only to keep the number of figures small so they can fit on one page.)

Following through in this manner for the rest of the t_i, one can generate Figures 11.33 to 11.43. ∎

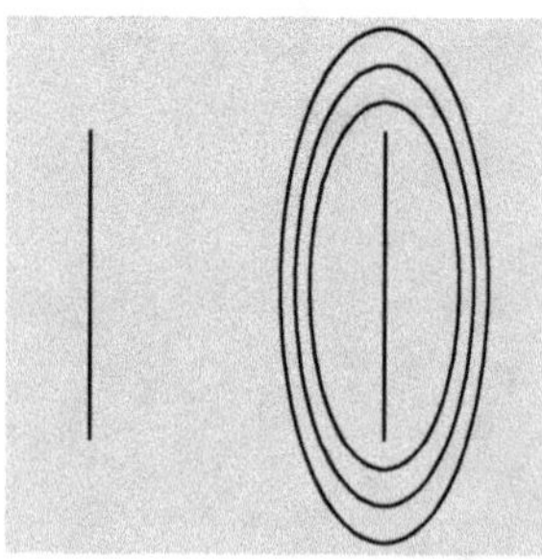

Fig. 11.31 A representation of $\widehat{\rho}(C)$ in R^2.

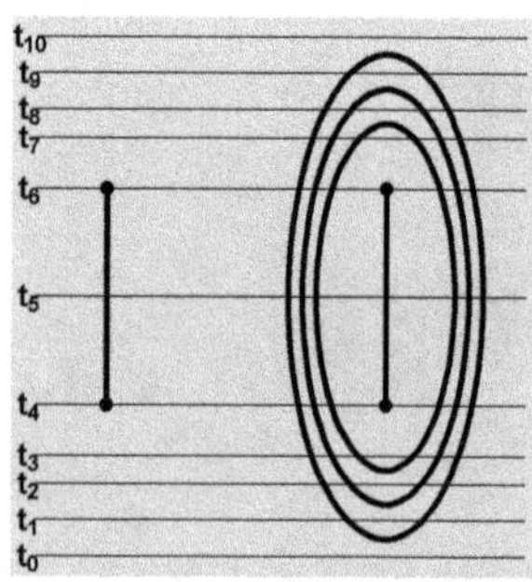

Fig. 11.32 $\widehat{\rho}(C)$ with line segments t_i.

Fig. 11.33

Fig. 11.34

Fig. 11.35

Fig. 11.36

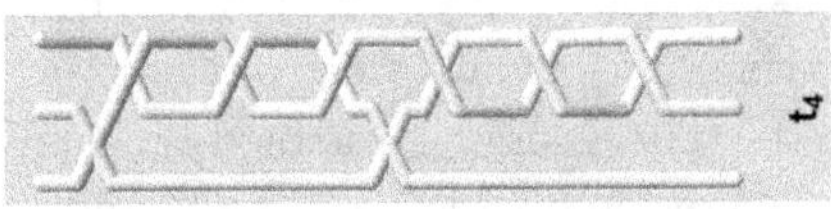

Fig. 11.37

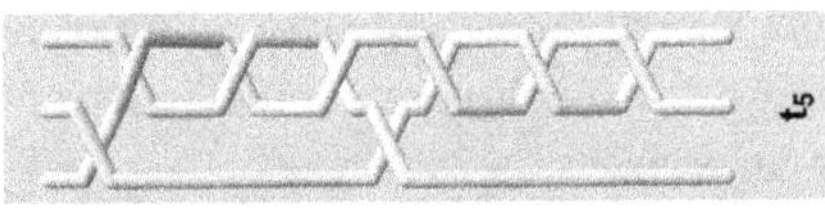

Fig. 11.38

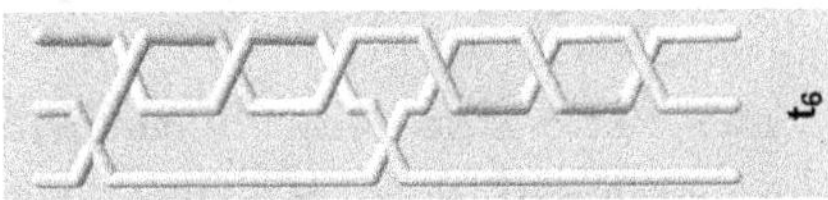

Fig. 11.39

Fig. 11.40

Fig. 11.41

Fig. 11.42

Fig. 11.43

One could also consider projecting a 2-knotting to a 2-manifold other than S^2—for example, to a standard torus, [Nakamura (2012a)].

11.3 Constructing Isotopies of Surfaces

The trace of an isotopy of M^2 in R^4 is a knotting of $M^2 \times I \subseteq R^4 \times I$ and, as such, involves high-dimensional knotting. However, in practice, isotopies are created in other ways. Although any isotopy is some sequence of knot-moves, one cannot often create interesting isotopies of a given knotting by applying knot-moves without considerable guidance.

Many 2-knottings K are created in some way using 1-knottings k. We can generate an isotopy K from an isotopy k. For example, an isotopy of a 1-knotting k_t gives rise to an isotopy of the corresponding spun knottings: $K_t = \Sigma(k_t)$.

Knot-energies for 1-knottings of Section 4.8 have been extended to both smooth and piecewise linear 2-knottings in [Auckly and Sadun (1997)] and [Kusner and Sullivan (1997)]. This is one way of generating isotopies of 2-knottings by allowing the knotting to "flow" in a "natural" energy-reducing fashion. The higher dimensional version of untangling of Remark 4.11 is to ask if a trivial 2-knotting can become a standard knotting in a hyperplane under gradient flow for an energy.

For computation, it is natural to view a knotting as PL. To define energy for a piecewise linear surface $\Sigma \in R^4$, we view vertices $\{v_i\}$ as self-repelling and the area of Σ fixed. The energy is defined as $E(K) = \sum_{i \neq j} 1/d_{ij}^4$ where d_{ij} is the distance between distinct vertices v_i and v_j. Other powers, besides 4, give other energy functionals, but using power 4 results in an energy which is invariant under inversions in a sphere, [Kusner and Sullivan (1997)].

As with 1-knottings few exact energy minima have been found. Unlike 1-knottings, computational approximations are not easy calculations. To date, there are calculations such as [Roseman (2002)]. The most interesting example to date is K, the 1-twist-spinning of the 1-knotting 8.17. (The knotting 8.17 was chosen because it is the simplest non-invertible knotting, and so, might be "typical" in that it lacks this symmetry.) By Proposition 10.7 we know K is equivalent to the standard knotting. There is no known mathematical connection between knot-energy and Proposition 10.7. Nevertheless, a computational PL model of K computationally flowed to a PL model of $S^4 \subseteq R^4$ as shown on the cover of this book. This graphic shows twelve isotopy stages, each a projection presented as two halves; the six stages on the left occur before the six on the right.

11.4 Symmetries of Knotted Surfaces

We briefly explore for 2-knottings the five basic symmetry types knots 2-knottings—fully chiral, reversible, positively amphichiral, negatively amphichiral, fully amphichiral—as in Section 3.5. We focus on the construction question: if we use a 1-knotting k to construct a 2-knotting K, how might symmetry of k relate to symmetry of K?

We can discuss the mirror image and inverse of 2-knottings in terms of their knotting diagrams.

Suppose M is a closed 2-manifold with k components. Given a knotting $K \subset R^4$ of M, a diagram for the mirror image K_R can be obtained by interchanging "above" and "below" at each crossing point of a diagram for K. At a triple point middle points remain middle points.

To graphically indicate orientation for an orientable 2-knotting we ignore branch points of a projection K_*. A point p_* of a 2-shard σ_* is a non-branch point of K_*. It corresponds to a point $p \in K$ which is an immersion point for $\pi|K\colon K \to R^3$ so a tangential 2-frame at p is sent by $d\pi$ to a 2-frame which is tangent to σ_* at p_*. Alternatively, we could use a normal vector to σ at p_*. In most cases, it is simpler to create and to display a single normal vector than a tangential 2-frame. Normal vectors are standardly calculated and used in computer graphics lighting algorithms. A 2-shard in R^3 is two-sided, and we can view our normal vector as indicating a preferred side which might be termed as a **positive side**, the other the **negative side**. We could graphically display such sides of a 2-shard by two distinct colors two colors.

A study of such diagrams is found in [Takase and Tanaka (2016)] where it is shown, for example, that if a projection has no branch points, there is no regular isotopy between this projection and its mirror image.

As with 1-knottings, most 2-knottings (also n-knottings in general) are fully chiral—that is not invertible and not amphicheiral. On the other hand, the most *familiar* knottings are relatively simple constructions and so are often of special types.

Remark 11.4. Ribbon knots are $(-)$-amphicheiral. If K is a ribbon knotting $\overrightarrow{K}$ symmetric with respect to $R^3 \subseteq R^4$ The w-reflection defined $r_w(x, y, z, w) = (x, y, z, -w)$ will map $\overrightarrow{K}$ will map K to K and reverse orientation.

In particular, we can also see this for ribbons which are spun knots, represented by a knotting described in generalized cylindrical coordinates. These knottings are symmetric with respect to the cartesian hyperplane P given by $y = 0$, For generalized cylindrical coordinates, P is the union of pages $P = H^3_0 \cup H^3_\pi$. For this knotting the y-reflection defined $R(x,y,z,w) = (x,-y,z,w)$ will map $\overrightarrow{K}$ to $\overleftarrow{K}$.

For a discussion of (+)-amphicheiral ribbon knottings, see [Yasuda (2020)]. In [Fox (1962b)] (problem 35) it was noted that some ribbon knots, such as the Fox ribbon knot, are not invertible. However, some are invertible, as we will see in Proposition 11.2. Also, in [Shima (1997)] it is shown that torus knottings with only three circles of crossing points are invertible. ∎

In [Farber (1975)] it was shown that two-twist spun trefoil is not invertible. The two-twist spun trefoil is fully chiral. More generally, 2-twist spun knots and r-twist (p,q)-torus knots are fully chiral. For knots that are not invertible nor amphicheiral, see [González-Acuña and Montesinos (1980)].

The 2-twist trefoil is not (-)-amphicheiral, [Ruberman (1983)]. In a continuation of this investigation [Litherland (1985)] the following is shown:

Prop 11.2.

(1) If a k is (-)-amphicheiral, any p-twist-spun $\Sigma^{\circlearrowleft}(k,p)$ is (-)-amphicheiral.
(2) If a k is (+)-amphicheiral, any p-twist-spun $\Sigma^{\circlearrowleft}(k,p)$ is invertible.
(3) If a k is invertible, any p-twist-spun $\Sigma^{\circlearrowleft}(k,p)$ is (+)-amphicheiral. ∎

Remark 11.5. Consider a spun 2-knot $K = (M,C) \otimes k$. An orientation of C and of k determine an orientation of K. At a point of K the tangent space can be oriented by choosing an ordered basis (b_C, b_k) where b_C is tangent to C and b_k is tangent to k. Reversing the orientation of only one of C or k will reverse the orientation of K, but reversing both will result in the same orientation. Symbolically if we write $\overrightarrow{K} = (M,\overrightarrow{C}) \otimes \overrightarrow{k}$ then:

$$\overleftarrow{K} = (M,\overrightarrow{C}) \otimes \overleftarrow{k}, \quad \overleftarrow{K} = (M,\overleftarrow{C}) \otimes \overrightarrow{k} \quad \text{and} \quad \overrightarrow{K} = (M,\overleftarrow{C}) \otimes \overleftarrow{k}. \blacksquare$$

Remark 11.6. Using Remark 11.5 we outline a proof of Proposition 11.2 in the special case of spun knots (that is $k = 0$). We begin with a spun knotting $\overrightarrow{K} = (S^2,\overrightarrow{C}) \otimes \overrightarrow{k}$ for some oriented knotting $\overrightarrow{k}$ and where $\overrightarrow{C}$ corresponds to the standard oriented equator.

We will use two self-diffeomorphisms of R^4. One is the reflection R of Remark 11.4. The other is $\rho(x, y, z, w) = (-x, y, -z, w)$— this is a rotation of π with yw-coordinate plane as axis. The map ρ is seen to be isotopic to the identity using a continuous rotation: write $\rho_t = (x\cos(\pi t) + z\sin(\pi t), y, -x\sin(\pi t) + z\cos(\pi t), w)$. The projection of this isotopy is a rotation of R^3 about the y-axis.

The maps R, ρ, and $\rho \circ R$ all map S^2 to itself, C to itself and all 3-dimensional pages of R^4 to other pages. Thus the images $R(K), \rho(K)$, and $\rho \circ R(K)$ are clearly spun knottings viewed in generalized cylindrical coordinates. The map R is orientation *reversing* on S^2 and on C. The map ρ is orientation *preserving* on S^2 and *reverses* the orientation of C. The knottings K and $\rho(K)$ are equivalent. The knottings $R(K)$ and $\rho \circ R(K)$ are equivalent. We will write $R(K)$ as K_R. If we choose orientation for K and write it $\overrightarrow{K}$ then, since R reverses the orientation of S^2, we write $R(K)$ as the oriented knotting $\overleftarrow{K_R}$.

The hyperplane $P = H^3_y(0)$ of Remark 11.4 will be helpful in understanding our four knottings. The intersection of any of our four 2-knottings with P will represent the square knot. For each of our 2-knottings the arc which we are spinning is the intersection of the knotting H^3_0. In the next set of Figures we use as an example of k an oriented trefoil knotting and depict orientations of 1-knottings with arrowheads along the curves. So we obtain: Figure 11.44 showing $\overrightarrow{K} \cap P$ where $H^3_0 \cap \overrightarrow{K} = \overrightarrow{k}$ is a directed left-hand trefoil, Figure 11.45 showing $R(\overrightarrow{K}) \cap P$ where $H^3_0 \cap R(\overrightarrow{K}) = \overleftarrow{k_R}$ is a directed right-hand trefoil, Figure 11.47 showing $\rho(\overrightarrow{K}) \cap P$ where $H^3_0 \cap \overrightarrow{K} = \overrightarrow{k_R}$ is a right-hand trefoil with the opposite direction from Figures 11.45 and 11.46 showing $\rho \circ R(\overrightarrow{K}) \cap P$ where $H^3_0 \cap \overleftarrow{K} = \overrightarrow{k}$ is a left-hand trefoil with opposite direction from Figure 11.44.

Using the notations of Remark 11.5 we write our oriented knotting as $\overrightarrow{K} = (S^2, \overrightarrow{C}) \otimes \overrightarrow{k}$. The knotting $R(K)$ has induced orientation reversed from $\overrightarrow{K}$ so we may write $R(K)$ as $\overleftarrow{K_R}$.

Using the notations of Remark 11.5 we finish our argument for the three parts of Proposition 11.2:

(1) Figure 11.45 shows that $\overleftarrow{K_R} = (S^2, \overleftarrow{C}) \otimes \overrightarrow{k_R}$. If $\overrightarrow{k_R} = \overleftarrow{k}$ then

$$\overleftarrow{K_R} = (S^2, \overleftarrow{C}) \otimes \overrightarrow{k_R} = (S^2, \overleftarrow{C}) \otimes \overleftarrow{k} = \overrightarrow{K}$$

so if k is $(-)$-amphicheiral then K is $(-)$-amphicheiral.

(2) Figure 11.47 shows that $\overrightarrow{K} = (S^2, \overleftarrow{C}) \otimes \overleftarrow{k_R}$ so if $\overrightarrow{k_R} = \overrightarrow{k}$ then

$$\overleftarrow{K} = (S^2, \overleftarrow{C}) \otimes \overleftarrow{k_R} = (S^2, \overleftarrow{C}) \otimes \overrightarrow{k} = \overleftarrow{K}$$

so if k is $(+)$-amphicheiral then K is invertible.

(3) Figure 11.46 shows that $\overleftarrow{K_R} = (S^2, \overrightarrow{C}) \otimes \overleftarrow{k}$. So if $\overrightarrow{k} = \overleftarrow{k}$ then

$$\overrightarrow{K_R} = (S^2, \overrightarrow{C}) \otimes \overleftarrow{k} = (S^2, \overrightarrow{C}) \otimes \overrightarrow{k} = \overrightarrow{K}.$$

So if k is invertible, then K is $(+)$-amphicheiral ▌

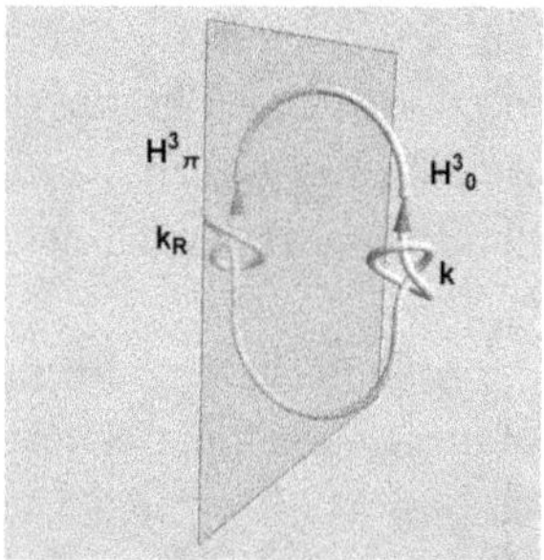

Fig. 11.44 Oriented $K \cap P$ for spun right-hand trefoil K.

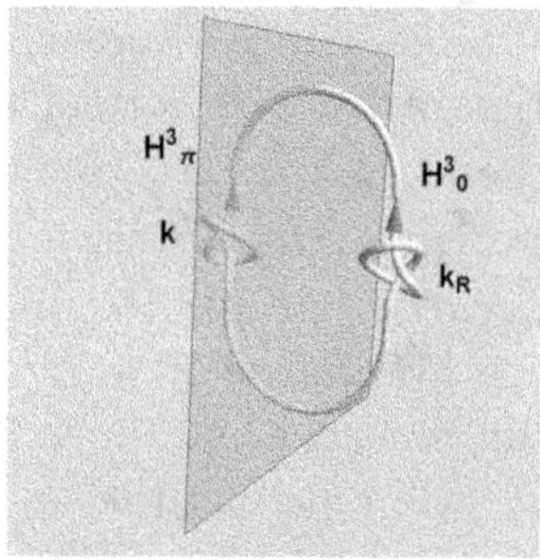

Fig. 11.45 Oriented $R(K) \cap P$ for spun right-hand trefoil K.

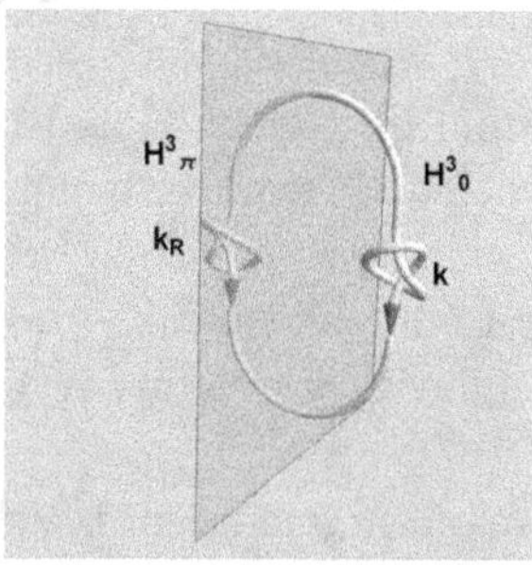

Fig. 11.46 Oriented $\rho(K) \cap P$ for spun right-hand trefoil K.

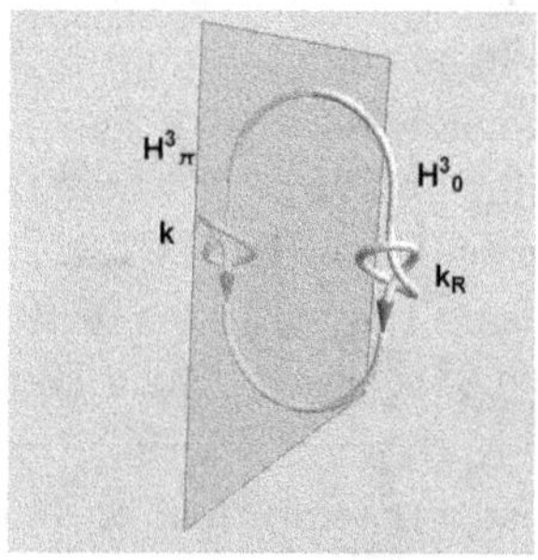

Fig. 11.47 Oriented $\rho \circ R(K) \cap P$ for spun right-hand trefoil K.

One reason knot symmetries have received so much attention is that the difficulty in distinguishing between a knotting and its mirror image or its reverse provides ample challenges for refining the tools of algebraic topology such as [Ruberman (1983)], [Kearton (1973)], [Carter *et al.* (2003)], [Iwakiri (2005)]. Additional valuable information for knottings of spheres is provided in [Gordon (2003)]. One can ask about symmetries of knottings that are not 2-spheres or more complex spinnings. There are a number of results that involve connected sums of twist-spun knots with trivial high-genus surfaces such as [Carter *et al.* (2003)], [Kawauchi (1986, 1990)], [Nakamura (2012a)] [Hillman (2002)], [Hillman and Howie (2013)] and [Asami and Satoh (2005)].

11.5 Lattice 2-Knottings

Lattice knottings are not smooth, there is not much theory, and not much calculation has been done to date. However there is much potential, awaiting more powerful calculational tools. All smooth knottings are equivalent to a lattice knotting. A knotting is a subset of R^n; a lattice knotting is an extremely simple small data set (of integers) that exactly describes such a subset. Also, the few examples we mention give an interesting visualization perspective on smooth constructions.

For 1-knottings, **lattice knottings**, see Section 4.9, are used for statistical studies in biology and physics. These are also called **cubulated** knottings. These studies usually involve lattice knottings with many lattice edges and the length of such knottings. Our focus is on lattice 2-knottings, with small number of lattice squares, that lie in a "small" 4-dimensional lattice box. A lattice 2-knotting can be encoded as finite string of integers: each is a union of a finite number of lattice 2-squares, and each square can be encoded as a finite string. Thus lattice 2-knottings can serve as data for digital computation. Currently however, the number of lattice 2-knottings in even a small lattice box is often prohibitively large with current technology.

In a different direction, there are some very interesting results about the set lattice knottings as the containing box gets large found in [Atapour *et al.* (2015)].

Lattice 2-knottings are the same as smooth or PL knots. A lattice 2-knotting (subdivided into simplexes) will be a PL-knotting [Hinojosa and Valdez (2021)]. A lattice knotting with edges smoothed corresponds to a smooth 2-knotting, [Díaz *et al.* (2018)], and any smooth 2-knotting corresponds to a lattice 2-knotting [Boege *et al.* (2011)].

For smooth 2-knottings, we have moves, Section 7.1. However, given a smooth 2-knotting K, there is no general useful way of finding possible moves to make for K. In contrast for a lattice 2-knotting K, we can generate a list of all possible cubulated moves we can make for K. Similarly, if K is contained in lattice box B, we can, in principle, generate a list of all possible cubulated isotopies of K within B.

In this Section we discuss a few interesting examples of lattice 2-knottings in small lattice boxes and how to visualize them.

Graphically we show a lattice 2-knotting K by displaying slices of K for all integer levels. It is of interest to do this slicing in all four coordinate directions, but in this Section we will slice only in the w-direction. A lattice square of K is a **level-square** if it is contained in an integer level. A lattice edge of K contained in an integer level is a **level-edge**. A lattice square, which is not a level lattice square, will have edges at w-levels q and $q+1$, for some q, and is called a **vertical-square**. The edge of a vertical square at level q is called the **bottom edge**, the other the **top edge**.

Visualization narrative: Viewing integer slices, a vertical-square is displayed as a top edge T and bottom edge B of consecutive slices.

Visualization method: Suppose T is in slice $w = q+1$, B in slice $w = q$. Represent T by a thick transparent tube in the $(q+1)$-slice (thick because T is higher in w) and B by a thin tube in q-slice. Other elements displayed are: level edges, shown as dark lines and lattice squares.

Example 11.6. Slicing a lattice box $B^{[3323]}$ in the direction of the fourth coordinate, we will get four $3 \times 3 \times 2$ lattice boxes as in Figure 11.48. The lattice 2-knotting shown corresponds to the basic sphere-torus knot, Example 6.12. There are two lattice 2-disks at levels 0 and 3. An annulus of vertical-squares joins the boundaries of these disks forming a lattice 2-sphere. In levels 1 and 2 there are lattice annuli. Two vertical annuli join their boundaries to form a lattice torus.

In more detail, in Level 1 we see a lattice annulus and a loop of edges. Each level edge on the boundary of the annulus is a bottom-edge and is thus shown as thin tube and is paired by an un-seen vertical-square with a top-edge in Level 2 shown as a thicker tube. Clearly, each level edge in the loop is in the intersection of two vertical-squares, one going up to Level 2, one going down to level 0. (In addition, close inspection at Level 1 shows, redundantly, a thin tube within a thick tube for these edges.)

It may be of some help to consider the w-projection K_*, see Figure 11.52. This figure shows a lattice version of a ribbon of a disk and annulus with one slot-slice intersection as in Figure 9.2. So, our K can be viewed as a lattice version of a symmetric ribbon knotting, see Definition 9.4. ∎

Projections of lattice knottings. Figures 11.52–11.55, can occasionally be useful but hard to interpret since lattice knottings are not in general position with respect to projection. Even a simple projection can be challenge to understand. For example, vertical-squares project to lattice edges. Also, level-squares at different levels can project to the same lattice square.

Example 11.7. We leave details of the next three examples as exercises for the reader. Figure 11.49 shows a lattice knotting, in a $B^{[3323]}$ lattice-box, equivalent to a spun Hopf link, see Examples 6.20 and 6.21. In projection, Figure 11.53 one can see two intersecting lattice annuli. Their intersection is two disjoint lattice squares. Self-intersections of ribbons must be topological arcs, so technically speaking this projection is not a lattice ribbon. By expanding the notions of slot and slice in an "obvious way", one can adapt the definition of ribbon to a "lattice ribbon" and symmetric ribbon knotting to"lattice symmetric ribbon knotting" so that Figures 11.49–11.55 can be considered to be lattice ribbons which are projections of symmetric lattice ribbon knottings.

Similarly, one can recognize Figure 11.50 as a lattice spun trefoil in a $B^{[3323]}$ lattice-box and Figure 11.51 as a lattice Fox ribbon sphere. ∎

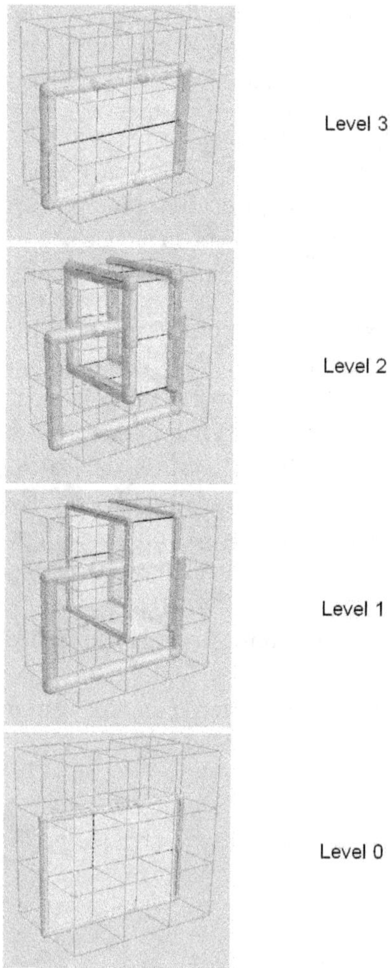

Fig. 11.48 A lattice knotting basic sphere-torus in $B^{[3323]}$.

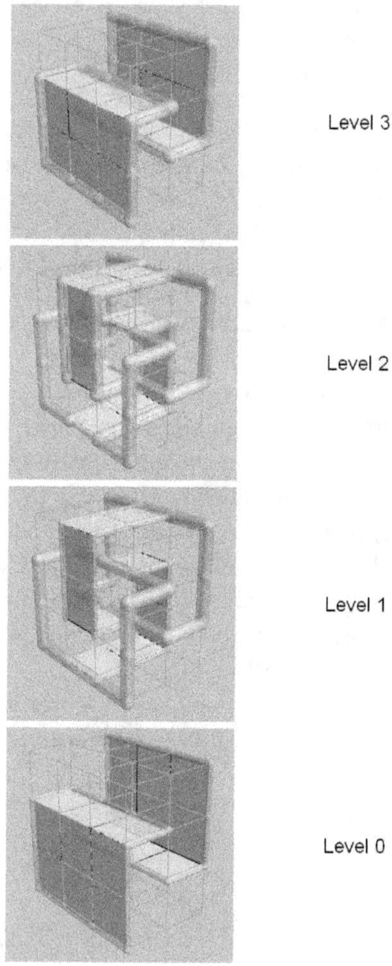

Fig. 11.49 A lattice knotting of the spun Hopf in $B^{[3323]}$.

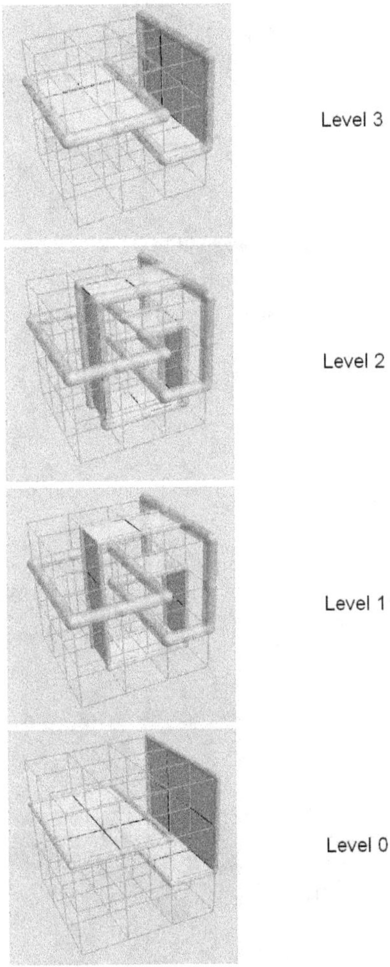

Fig. 11.50 A lattice knotting for spun trefoil in $B^{[3333]}$.

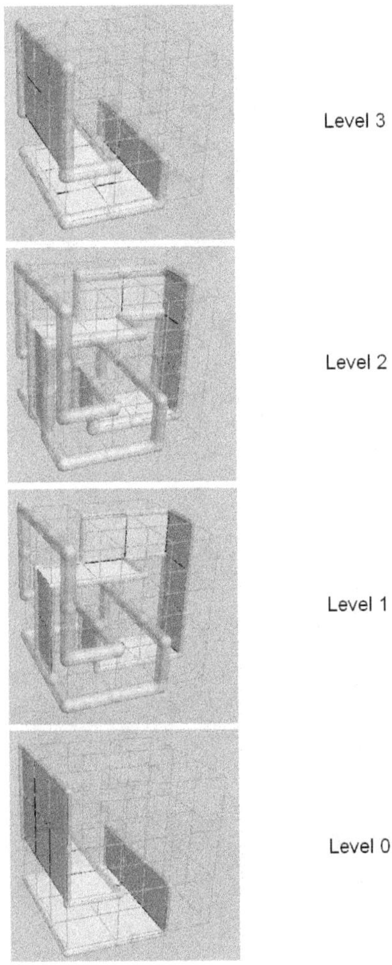

Fig. 11.51 A lattice knotting of the Fox ribbon sphere in $B^{[3333]}$.

Fig. 11.52 Projection for Figure 11.48.

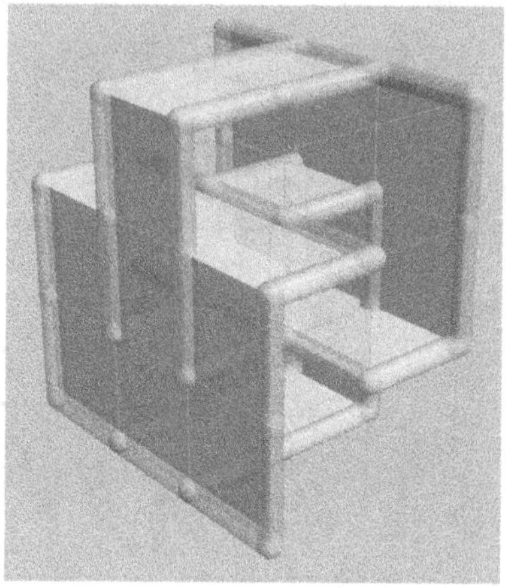

Fig. 11.53 Projection for Figure 11.49.

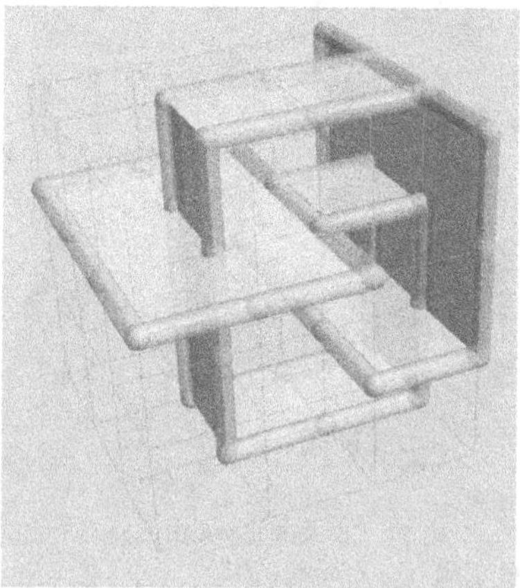

Fig. 11.54 Projection for Figure 11.50.

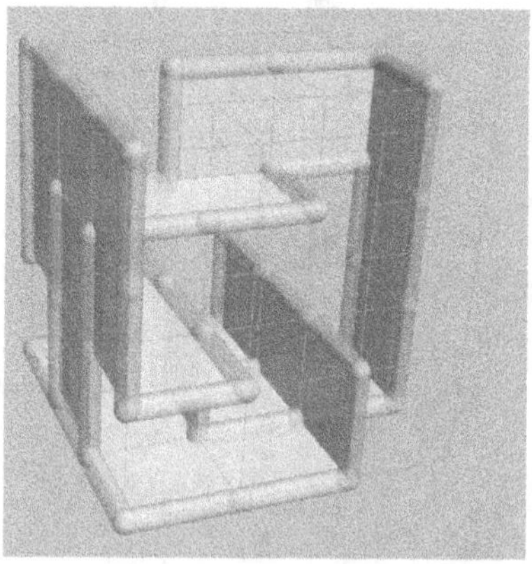

Fig. 11.55 Projection for Figure 11.51.

Remark 11.7. Lattice knottings might provide ways of obtaining answers of some kind basic very difficult questions about smooth knottings. Suppose we had a way of efficiently constructing large set $\mathcal{L}$ of lattice 2-knottings, see Remark 11.8.

Are there 2-knottings not constructible by the methods described in earlier sections? Perhaps a search of $\mathcal{L}$ might reveal one.

Can we make a table of all 2-knottings? (That is knottings with no standard components)? One start on this would be to (somehow) find and list all non-split lattice 2-knottings in some standard lattice box, say, $B^{[3332]}$, then add to this all similar non-split lattice 2-knottings in the next largest box, in this case $B^{[3333]}$, etc.

Given a standard lattice box B can we calculate $N_B(K)$, the number of 2-knottings equivalent to K which are subsets of B? If so, we might define **K is simpler than K' in B** if $N_B(K) < N_B(K')$. Specifically, a challenge question: is the spun trefoil knotting simpler than the Fox ribbon knotting in $B^{[3333]}$? If one could calculate $N_B(K)$ one could use this to help in the construction of a table of knots. ∎

There are many methods of creating lattice 1-knottings, but these do not have analogs in higher dimensions. Also, much literature for lattice 1-knottings is focused on the length of a lattice knotting [Sumners and Whittington (1988)], [van Rensburg and Rechnitzer (2011)]. The next definition is a simple but useful way of combining two lattice 2-knottings to get a third. It easily extends to construction of lattice n-knottings for higher n.

Definition 11.6. For lattice 2-knottings X and Y, the **raw-mod 2-sum, Z, of X and Y**, is the union of all lattice squares that are in either X or Y *but not both.* If Z is a lattice 2-knotting, we write $Z = X \sqcup Y$, and say Z is the **mod-2-lattice-sum of X and Y**. Otherwise, (if it is not a closed 2-manifold, we ignore Y and) we say the mod-2-lattice-sum of X and Y is X. Also we write $(X \sqcup Y) \sqcup Z$ as $X \sqcup Y \sqcup Z$. ∎

It is easy to compute a mod-2-lattice-sum, especially if X or Y has a small number of lattice-squares. A raw-mod-2-sum will be a 2-knotting if and only if, as sets, $X \cap Y$ is a topological 2-manifold (generally with non-empty boundary). A set of lattice squares L is a 2-manifold if and only if for each lattice vertex v of L the union of squares in L containing v forms a lattice 2-disk.

Remark 11.8. Here is a general plan. It is easy to construct small lattice 2-knottings in a lattice box B. The simplest is a boundary of a lattice 3-cube in R^4 we call a **mini-sphere**. It is the union of six lattice squares. It is trivial to list a large number of "random" mini-spheres in B. Generate a thousand mini-spheres in $Y_n \subseteq B$, $n = 1, \cdots, 1000$. We can view $\mathbb{X} = Y_1 \sqcup Y_2 \sqcup \cdots \sqcup Y_{1000}$ as a "random" lattice 2-knotting in B. Such a calculation is quick and easy.

Although not of practical use, in principle every lattice 2-knotting in $X \subseteq B$ is obtainable by some mod-2-lattice-sum simply because $X \sqcup X = \emptyset$, and by Remark 2.1, $\emptyset$ is a 2-manifold.

Mini-sphere results as computed by the author have not been particularly interesting—mostly complicated tube-like split trivial knottings. On the other hand there are many variations on can try such as using more complex 2-knottings for Y or altering the process so that if the raw-mod-2-sum is not a manifold we try to adjust Y so that it will be. ∎

Additionally one can study lattice isotopies of lattice 2-knottings computationally. Analogous to the smooth moves of Section 7.1 there is a set of lattice moves called **cubulated moves**, [Díaz *et al.* (2019)] and [Hinojosa *et al.* (2013)].

Part 3

Higher-Dimensional Knotting

We have learned that the study of 2-knottings sheds light on aspects of 1-knottings such as isotopies and knot-moves. Similarly, some aspects of 3-knottings are significant in understanding 2-knottings, and on it goes. In this book we can only begin to explore higher dimensions and will largely emphasize 3-knottings and their constructions.

Higher-dimensional knottings exhibit high complexity and variety due to the complexity and variety of higher-dimensional manifolds, [Ranicki (1998)]. However, there are additional complexities relating to the basic type of knotting, [Kervaire (1965b)]. For 2-knottings the choice of working with *smooth* knottings rather than PL or locally flat was largely one of technical preference. In higher dimensions, this choice often makes a significant mathematical difference.

Chapter 12

Higher-Dimensional Knotting Basics

12.1 A Beginner's Kit of High-Dimensional Manifolds and Knottings

The first step in exploring higher dimensional knottings $M^n \subseteq R^m$ where for $2 < n < m$, is just to look at possible n-manifolds M^n (we will generally assume smooth closed). Each increase of dimension opens up a new universe of "new" manifolds, each with its own knotting possibilities.

The collection of all 2-manifolds is easy to understand once one becomes acquainted with non-orientable ones such as the Klein bottle and projective plane. For 3-manifolds a good very introductory exposition is [Thurston and Weeks (1984)]. There are important connections between the classification of 3-manifolds and 1-knottings. Good advanced references for 3-manifolds are [Hempel (2004)], [Rolfsen (1976)], [Thurston (1997)] and [Schultens (2014)].

The diversity and complexity of 4 manifolds is difficult on yet another level. Two good (but not easy) books on 4-manifolds are [Freedman and Quinn (1990)] and [Scorpan (2005)]. A rigorous precise mathematical theorem, [Markov (1960)], shows that classification problem for n-manifolds is mathematically not possible if $n \geq 4$. In particular, there is no general algorithm that can be used to answer the question "Given two arbitrary closed 4-manifolds, are they homeomorphic?"

Additionally, for $n \geq 4$, the notions of smooth submanifold, locally flat submanifold and PL-submanifold diverge in a number of ways. There are manifolds that are PL but not smooth [Kervaire (1960)]. Also, examples of smooth manifolds that are homeomorphic but not diffeomorphic—even if the manifolds are spheres, [Milnor (1956)]. For example, there are exactly 28 distinct (non-diffeomorphic) smooth manifolds which are homeomorphic to S^7, [Kervaire and Milnor (1963)] There are no particular books that specifically treat 5-manifolds. For $n \neq 4$ there is only one smooth manifold homeomorphic to R^n, but there are an uncountable number of smoothly distinct manifolds homeomorphic to R^4, [Taubes (1987)].

The open problems for manifold classification do not greatly impact the study of knottings. Knot theorists are not in need of a large supply of manifolds since the knot theory of even the simplest of manifolds is of great interest (and difficulty). In R^5, one often focuses on knottings of S^3, $S^1 \times S^1 \times S^1$ and perhaps $S^1 \times M^2$ where M^2 is any closed 2-manifold—orientable or not.

Some definitions have already been stated for higher dimensions such as spinning Definition 1.34, handle terminology Definition 5.2, handle structure Definition 5.3, orientable handles Definition 5.4, handlebody Definition 5.5, general position with respect to slicing Definition 8.3, cobordism Definition 8.10, topological cobordism Definition 8.11, ribbon disk Definition 9.2, general position with respect to projection Definition 9.6, and fibered knotting Definition 10.13.

Example 12.1. In Definition 1.34 we defined the spinning $\Sigma(M)$ of a subset $M \subseteq H_0^{n-1}$ where, if $X \subseteq \overset{\circ}{H}{}_\alpha^{n-1}$ (that is $M_0 \cap \partial H_0^{n-1} = \emptyset$), then $\Sigma(M)$ is homeomorphic to $M \times S^1$. If, additionally, M is a smooth k-manifold, then $\Sigma(M)$ is a smooth $(k+1)$-manifold. If we take a proper knotting of a smooth $(n-3)$-disk $K \subseteq H_0^{n-1}$, $\Sigma(K)$ will be a smooth knotting of an $(n-2)$-sphere in R^n. We call all these **Artin-spinnings.** ∎

So, any of the 2-knottings of Part 2 give examples of 3-knottings. If we spin a proper knotting of a 2-disk representing any knotting of a 2-sphere K, we get a 3-knotting of S^3. If we spin *all* of K we get a 3-knotting of $S^1 \times S^2$. Spin a 2-knotting of a torus and we get a 3-knotting of $S^1 \times S^1 \times S^1$. Then we could take connected sums of these 3-knottings and get even more 3-knottings. The coloring techniques for projections transfer readily to these Artin-spinnings, Remark 12.10, allowing distinction of many knottings in this family.

Definition 12.1 used generalized cylindrical coordinates to produce n-knottings of S^n from proper knottings of S^{n-1}. Super-spinning is similar but uses generalized spherical coordinates:

Example 12.2. Generalized spherical coordinates for points of R^5 are (r, θ, ψ, z, w). For fixed values of θ and ψ we have a subset $R^3_{(\theta,\psi)}$, homeomorphic to R^3_+. A knotting of a 3-sphere is produced by putting a chosen proper knotted arc in each such subset. This construction was used in [Zeeman (1960a)] and named **super-spinning** [Cappell (1970)] where it was shown Artin-spinning and super-spinning produced distinct knots. Generalized spherical coordinates for points of R^6 are $(r, \theta, \psi, z, w, t)$. For fixed values of θ and ψ we have a subset $R^4_{(\theta,\psi)}$, homeomorphic to R^4_+.

Using an entire 1-knotting $M^1 \subseteq R^3_+$ in each $R^3_{(\theta,\psi)}$, we get a 3-knotting of $M^1 \times S^2 \subseteq R^5$ called a **super-spinning of** $M^1 \times S^2$.

Also using an entire 2-knotting K^2 we get a super-spun 4-knotting of $S^2 \times S^2 \subseteq R^6$. All points of $S^2 \subseteq R^6$ have $r = 1$ so we can write $p \in S^2$ as $p = (\alpha, \beta)$. For $(\alpha, \beta) \in S^2$ we put a copy of K into $R^4(\alpha, \beta)$. This is the **super-spinning of** $K^2 \times S^2$. We will see a variation of this construction used in Example 12.17. ▮

Spinning about a submanifold, Definition 10.5, has a higher-dimensional version, Definition 12.1, [Roseman (1989)]. The ingredients for Definition 10.5 are: a 2-knotting K^2, a 1-manifold $M^1 \subseteq K^2 \subseteq R^4$, and a 1-knotting κ represented by a proper arc knotting α. Additionally, a product tubular neighborhood of $M^1 \subseteq K^2$ was used.

For our generalization, we will have an n-knotting K^n, of an orientable manifold, a k-dimensional submanifold $M^k \subseteq K^n \subseteq R^{n+2}$ and an $(n-k)$-knotting κ represented by an $(n-k)$-disk α in $D^{(n-k)+2}$. Furthermore, we will need a product tubular neighborhood $N \approx M^k \times D^{n-k}$ of $M^k \subseteq K^n$. The role of M^k is to serve as a guide for placing a copy of α; the role of N is to make sure the union of all these copies fits together to form a manifold.

Informally this process can be described as follows. Stand at a point $x \in M$. Sticking out from x into K^n is a normal $(n-k)$-disk D_x. Sticking out from D_x into R^{n+2} we have a normal $(n-k+2)$-disk $\mathbb{D}_x$. As a disk pair, $(\mathbb{D}_x, D_x)$ is standard. Take out D_x and replace it with the knotted disk α. Now, do this for *all* $x \in M$.

Unlike our previous considerations for 2-knottings, M^k is not required to be a codimension 1 submanifold of K; it could be a higher codimension. However we will still want M^k to have a trivial tubular neighborhood $N \approx M^k \times D^{n-k}$. It is not guaranteed that, generally, tubular neighborhoods will be trivial—so in our definition, we add this requirement. However K^n *will* have a trivial tubular neighborhood in R^{n+2} since it is orientable and of co-dimension 2. This gives us a trivial tubular neighborhood T of $M^k \subseteq R^{n+2}$ where $T \approx \left(M^k \times D^{n-k}\right) \times D^2 \approx M^k \times D^{(n-k)+2}$.

At any $x \in M^k$ we have a normal $(n-k)$-disk D^{n-k}_x of N and a normal $(n-k+2)$ disk D^{n-k+2}_x of T. Each $\left(D^{n-k+2}_x, D^{n-k}_x\right)$ is equivalent to the standard codimension 2 disk pair. To do spinning, we replace each D_x^{n-k} by a proper knotted disk.

Definition 12.1. Suppose we have a knotting of a closed orientable n-manifold $K^n \subseteq R^{n+2}$ and a smooth k-dimensional manifold $M^k \subseteq K^n$ with product tubular neighborhood $N \approx M^k \times D^{n-k}$ and let $\alpha^{(n-k)} \subseteq D^{(n-k)+2}$ be a smooth proper knotting of disk corresponding to a knotting of a $(n-k)$-sphere κ. We can write

$$(T, N) \approx \left(M^k \times D^{(n-k)+2}, M^k \times D^{(n-k)}\right) \approx M^k \times \left(D^{(n-k)+2}, D^{(n-k)}\right).$$

Let $N' = M^k \times \alpha \subseteq T \subseteq R^{n+2}$. A new knotting called the **knotting of K^n obtained by spinning κ along M^k, or alternatively, the spinning of α along M^k** is defined $\mathbb{K}^n = (K^n - N) \cup N'$. We use the notation $\mathbb{K}^n = (K^n, M^k) \bigotimes \alpha$ or $(K^n, M^k) \bigotimes \kappa$. ▌

Remark 12.1. Suppose we have an isotopy of f_t of $M^k \subseteq K$, where $0 \leq t \leq 1$ and we let $M_t^k = f_t(M^k)$. This isotopy extends to an isotopy of any tubular neighborhood of $M^k \subseteq K$. Then our construction gives an isotopy of the spun knots $(K, M_t^k) \bigotimes \alpha$. ▌

Remark 12.2. The product tubular neighborhood N of $M^k \subseteq K^n$ is the image of an embedding $\tau\colon N \times D^{n-k} \to K^n$. A close analysis of the definition of $(K, M^k) \bigotimes \alpha$ reveals that the knotting directly uses τ—not simply that such a τ exists. This is detailed in [Suciu (1992)] where τ corresponds to the concept of **framing**. So, technically, the diffeomorphism τ should be incorporated into notations and all discussion of this construction (as is done in detail in [Suciu (1992)]), in which case, the construction is called **frame-spun knotting**. However, we will choose examples where τ is well understood, so this will prove unnecessary.

This consideration was not needed for 2-knottings since circles in orientable 2-manifolds have annular tubular neighborhoods, which give well-determined framings. More generally, any codimension 1 submanifold of an orientable manifold has a product tubular neighborhood. ▌

Example 12.3. Spinning about a circle gives Artin spinning and spinning about higher-dimensional spheres gives super-spinning:

The knotting $(S^3, S^1) \bigotimes \alpha$ for a proper 2-disk (D^4, α) is essentially the Artin spinning of α, Example 12.1. Here (S^3, S^1) is the standard pair of spheres and α is a knotting of a 2-disk. At each $x \in S^1$, we have a normal

2-disk in S^3, and a normal 4-disk in R^5, giving a standard (D^4, D^2) pair at each point. Replace each with a fixed knotting of a 2-disk (D^4, α).

The knotting $(S^3, S^2) \bigotimes \beta$ is essentially the supper spinning of a 1-knotting (D^3, β) for the sphere, Example 12.1. Here, (S^3, S^2) is the standard pair of spheres (S^3, S^2), and $S^1 \subseteq S^3 \subseteq R^5$. At each $x \in S^2$, we have a normal 1-disk in S^3, and a normal 3-disk in R^5, giving a standard (D^3, D^1) pair at each point. Replace each with a knotting of a 1-disk (D^3, β). ▮

Remark 12.3. There are many knottings of circle in S^3, also many knottings of a 2-manifold , Prop 5.3. However, we will show in Example 12.7 this does not matter—that spinning about any 1-knotting is equivalent to spinning about a split trivial knotting and any spinning about a 2-manifold in S^3 is equivalent to spinning about a standard 2-knotting. ▮

Remark 12.4. As with Remark 10.5 we can view a two-point subset of S^n as knotting of S^0. Then for any n-knotting K of S^n we define the **0-spin of** K to be $K\# - K$. ▮

Example 12.4. If we spin a 1-knotting, κ, we get a 2-knotting $\mathcal{K}$; if we spin $\mathcal{K}$ we get a 3-knotting, $\mathbb{K}$. We will see $\mathbb{K}$ is in fact a spinning of κ about a "standard" torus in $T \subseteq S^3$. That is:

$$(S^3, T) \bigotimes \kappa = (S^4, S^1) \bigotimes \left((S^3, S^1) \bigotimes \kappa\right).$$

We consider α, a proper disk in D^3 representing κ, and $\mathcal{A}$, a proper disk in D^4 representing $\mathcal{K}$. To get $\mathcal{A}$, we start with a standard 2-dimensional hemisphere H^2 (H^4_+, H^2). Next take an unknotted circle $M \subseteq H^2$ which has an annulus $A \approx (M \times D^1) \subseteq int(H^2)$ as product tubular neighborhood, projection to H^3_+ is shown in Figure 12.1 For each $\phi \in M$, we have trivial ball pair consisting of $D^1_\phi = \{\phi\} \times D^1 \subseteq A$ and a normal 3-ball D^3_ϕ in the trivial product neighborhood of $H^2 \subseteq H^4_+$.

Our disk knotting $\mathcal{A}$ is obtained by replacing each $D^1_\phi \subseteq D^3_\phi$ by $\alpha \subseteq D^3_\phi$, Figures 12.2 and 12.3.

For each 4-dimensional page H^4_θ of R^5, put a copy H^2, M and A, and labeled with θ subscript, using double labels to the disks, $D^1_{(\phi,\theta)}$ and $D^3_{(\phi,\theta)}$. The spinning of $\mathcal{A}$ is $\mathbb{K} = \bigcup_\theta \mathcal{A}_\theta \subseteq R^5$. In summary we get $\mathbb{K}$ by doing the process " replace each $D^1_\phi \subseteq D^3_\phi$ by $\alpha \subseteq D^3_\phi$" in each page, that is: replace each $D^1_{(\phi,\theta)} \subseteq D^3_{(\phi,\theta)}$ by $\alpha \subseteq D^3_{(\phi,\theta)}$.

But suppose we do not alter H^2, and just spin this hemisphere so that $S^3 = \bigcup_\theta H^2_\theta$. Let $T = \bigcup_\theta M_\theta$, then T is a smooth torus equivalent to the standard torus in S^3 and $N = \bigcup_{(\phi,\theta)\in T} D^3_{(\phi,\theta)}$ is a product tubular neighborhood of $T \subseteq R^5$.

To spin κ about T we replace each $D^1_{(\phi,\theta)} \subseteq D^3_{(\phi,\theta)}$ by $\alpha \subseteq D^3_{(\phi,\theta)}$, for all $(\phi,\theta) \in T$. So this knotting is the same as $\mathbb{K}$. ∎

Fig. 12.1 Projection of hemisphere $H^2 \subseteq H^4_+$, circle M and annulus tubular neighborhood A.

Fig. 12.2 Projection of disk $\Sigma \subseteq H^4_+$.

Fig. 12.3 Half of disk Σ reveals knotting details.

Knottings get more interesting when we consider higher genus surfaces.

Example 12.5. For example, consider the collection of 3-sphere knottings: $(S^3, M^2_g) \bigotimes \alpha$, for $g = 0, 1, 2, \ldots$ where $M^2_g \subseteq S^3$ is a standard genus g 2-submanifold, and α is a proper 1-disk in D^3. For many α (for example, a trefoil knotting or figure-eight knotting) this will be an infinite collection of inequivalent knottings, [Roseman (1989)]. ∎

Deform spinning, Definition 10.7, extends to higher-dimensional knottings in a straightforward way. In Definition 12.1, for each $x \in M^k$, we replaced each normal disk pair $\{x\} \times \left(D^{n-k+2}, D^{n-k}\right)$ by $\{x\} \times \left(D^{n-k+2}, \alpha^{n-k}\right)$. However it is *not* necessary that we replace the disk D^{n-k} by the same knotting α^{n-k} for each x. So we could replace by an *equivalent* proper knotting α^{n-k}_x which varied with x, as long as we take care that $\bigcup_{x\in M^k} \alpha^{n-k}_x$ fit together to form a smooth n-manifold. That said, it is often the case that for there exists only trivial deformations for M^k.

Definition 12.2. Suppose $M^k \subseteq K^n \subseteq R^{(n+2)}$, K^n an n-knotting, M a smooth k-manifold with A a trivial tubular neighborhood of $M \subseteq K$ and N a trivial tubular neighborhood of $M^k \subseteq R^{(n+2)}$. Next suppose we have a spherical knotting $\kappa \subseteq R^{(n-k)+2}$ and a collection of proper $(n-k)$-disk knottings $\mathcal{A} = \{\alpha_x\}_{x \in M^k}$ in $D^{(n-k+2)}$, such that each represents κ. We will say that $\mathcal{A}$ **is a deformation of** κ. Finally suppose that $A'' = \bigcup_{x \in M^k}(\{x\} \times \alpha_x) \subseteq M^k \times D^{(n-k)} \approx N \subseteq R^{(n+2)}$ is a smooth proper submanifold of N diffeomorphic to $M^k \times D^{(n-k)}$.

We say $K' = (K - A) \cup A''$ is a **deform-spun knotting**, or that K' **is the deform-spinning of** $\mathcal{A}$, with notation $(K, M) \bigotimes \mathcal{A}$. ▌

Remark 12.5. In geometric topology literature "deformation" is used in two different ways: as a homotopy of some type or some sort of isotopy—for example [Alexander and Briggs (1926/27)] or [Litherland (1979)]. ▌

Example 12.6. A simple deformation is twist-spinning is a straightforward generalization of the twist-spinning of Section 10.6. Here we are spinning about $S^1 \subseteq S^n \subseteq R^{n+2}$ using a proper knotting α of D^{n-1} in D^{n+1}. The deformation α_θ is the image of α under a rotation by angle θ of D^{n+1} in the last two coordinates of D^{n+1}.

Spinning of k knottings for $k > 1$ are considered in [Hsiang and Sanderson (1965)]. ▌

Example 12.7. We can use spinning and rolling of 1-knottings and get deform-spinnings about the standard torus knotting $S^1 \times S^1 \approx M \subseteq S^3$ Let $\alpha \subseteq D^3$ be a proper knotted arc representing a figure-eight knot. Then the collection of proper knotted arcs $\mathcal{A} = \{\alpha_{(\theta,\phi)}\}$ obtained from α by twisting by angle θ and rolling by angle ϕ is a deformation. One can get 4-knotting examples of deformation spinning can be obtained for spinning about a 3-torus in $S^1 \times S^1 \times S^1 \approx M \subseteq S^4$ in a similar fashion.

Other simple deformations involve composite knottings. Recall rolling a composite, Example 10.19. Consider $M^2 \approx S^1 \times S^1$ be a standard torus in $K^3 = S^3 \subseteq R^5$ and let α be an arc knotting representing sum of figure-8 knotting ϕ and a knotting of trefoil denoted τ knotting as in Figure 10.131. For $0 \leq \theta \leq 2\pi$, let α_θ denote the rolling of this knotting as shown in Figures 10.131–10.152. We next think of an isotopy of Figure 10.131 by giving the figure-8 knotting k full twists while leaving the trefoil untwisted (for example). Finally, for $x = (\theta, \psi) \in S^1 \times S^1$, we define α_x to be arc we obtain from α_θ by twisting the portion corresponding to ϕ by an angle of $k{*}2\pi$. One can check that α_x is a deformation. One gets other deformations if we also allow the trefoil to twist—maybe by a different number of twists, or if we allow the figure-8 to separately roll while we roll the composite. ▮

The generalization of Proposition 10.7 holds:

Prop 12.1. [Zeeman's Fibering Theorem] The τ-twist spin of any knotted n-sphere k in S^{n+2}, where $\tau \neq 0$, produces a knotting of an $(n+1)$-sphere in S^{n+3} which is a fibered knot. whose boundary is the given knotting. The fiber is diffeomorphic to the once punctured τ-fold branched cyclic cover of S^{n+2} branched along k. ▮

Remark 12.6. The ± 1-fold branched cyclic cover of any disk knotting is an $(n+1)$-ball. It follows that the ± 1 twisting of any knotted n-sphere is always a *trivial* knotting.

One can also define symmetry-spinning in higher-dimensions, such as [Kanenobu (1989)]. ▮

Although many knottings are non-fibered, such as [Kearton (1983)], there are many other types if high-dimensional fibered knottings besides spun knottings such as [Saeki (1999)].

One of the most interesting non-trivial knottings is the standard knotting of S^p and S^q in S^{p+q+1}, Definition 1.38. However, it is not a codimension 2 knotting. The next definition is a construction of a codimension 2 knotting based on this. It can be viewed as a generalization of Example 6.22. The basic idea is to replace S^p by a knotting of, A, of $S^p \times S^{q-1}$ and replace S^q by a knotting, B, disjoint from A, of $S^q \times S^{p-1}$, giving a codimension 2 knotting $A \cup B \subseteq S^{p+q+1}$. The knottings A and B are, individually obtained, any of our spinning constructions. For readers familiar with algebraic topology we note that these join-spin knottings inherit algebraic linking of the standard linking. The definition below uses only the most basic spinning. A special variance of this construction is allowed if either $p = 1$ or $q = 1$.

Definition 12.3. Write $S^{p+q+1} = S^p * S^q$ and let $M \approx S^p \times D^{q+1}$ and $N \approx S^q \times D^{p+1}$ be disjoint trivial tubular neighborhoods of the join factors. Let $K \subseteq D^{q+1}$ be a knotting of S^{q-1} and let $K' \subseteq D^{p+1}$ be a knotting of S^{p-1}. If $q = 1$ we allow K to be a (non-empty) discrete set of n points; if $p = 1$ we allow K' to be a (non-empty) discrete set of m points. Let $A = S^p \times K \subseteq M \subseteq S^{p+q+1}$ and let $B = S^q \times K' \subseteq N \subseteq S^{p+q+1}$. The **join-spin of K and K'** is $\mathbb{K} = A \cup B \subseteq S^{p+q+1}$. We say $\mathbb{K}$ has **join-spin type** (p, q). ▮

Example 12.8. The join-spin construction is of interest when even when K and K' are trivial knottings. These knottings are as fundamental in higher-dimensions as the Hopf linking is for 1-knottings. (For $S^1 * S^1$ with K and K' single points, we get exactly the Hopf link of S^3.) For $S^2 * S^1$ we get a basic 2-sphere-torus knotting as in Example 6.12. ∎

Example 12.9. We define a sequence of join-spun q-knottings of type $(1, q)$, $\mathbb{K}_q$, $2 \leq q$. Each $\mathbb{K}_q$ will have q components: one q-sphere and the rest are various sphere products. We start with $\mathbb{K}_2$, we take K to be a Hopf linking and $n = 1$. So, $\mathbb{K}_2$ is equivalent to Example 6.22. For other q, we define K_{q+1} as the join-spun knotting of type $(1, q+1)$, where $K = \mathbb{K}_q$ and $n = 1$. To get the 3-knotting, $\mathbb{K}_3$, we take a spinning $\Sigma(\mathbb{K}_2)$, which is a knotting of two copies of $S^1 \times S^1 \times S^1$ and one of $S^1 \times S^2$, and then add in a 3-sphere which is parallel to the second factor of $S^1 * S^3$. ∎

One final variation of spinning is based on the idea of local iterated spinning (multi-spinning) about a guide, Section 10.10, [Roseman (1989)]. Multi-spinning is a local knotting that occurs naturally in iterated spinning.

Example 12.10. Suppose $\alpha \subseteq D^4$ is a proper knotting of a 2-disk and $\beta \subseteq D^3$ is a proper knotting of a 1-disk. We will construct a knotting of $S^1 \times S^2$ in R^5 which is a higher-dimensional version of Example 10.36. If we chose a point $(x_0, y_0) \in S^1 \times S^2$ we can Artin-spin α about the circle $Y = S^1 \times \{y_0\}$ and obtain $(S^1 \times S^2, Y) \otimes \alpha$. Also, we can super-spin β about the 2-sphere $X = \{x_0\} \times S^2$ and obtain $(S^1 \times S^2, X) \otimes \beta$. If we view in the natural way X as a subset of $(S^1 \times S^2, Y) \otimes \alpha$ (see Example 10.27 for lower-dimensional illustration) we have an iterated knotting $\left(\left((S^1 \times S^2, Y) \otimes \alpha\right), X\right) \otimes \beta$. Similarly, if we view Y as a subset of $(S^1 \times S^2, X) \otimes \beta$, we have an iterated knotting $\left(\left((S^1 \times S^2, X) \otimes \beta\right), X\right) \otimes \alpha$.

An inspection of this construction reveals that if $W \subseteq S^1 \times S^2$ is a small 3-ball containing (x_0, y_0) then, as a subset W_0 of the iterated knotting, it will be a multi-spinning in $W \times D^2$. ∎

One aspect of this higher-dimensional construction is that our spinning-guide Definition 10.10 will need to consider manifold knottings, Definition 3.5 and not just n-knottings 3.4, such as Example 12.10. The notions of general position with respect to projection, height, and shards used for n-knottings clearly extend to manifold-knottings so the only alteration in this next definition is to replace $\widetilde{C}$ as an $(n-1)$-knotting with the more general manifold-knotting.

The higher-dimensional version of guide-spinning about, Definition 10.11 not only extends the class of spinning constructions but also sheds light on some of our previous constructions, see Remark 12.7.

Definition 12.4. Suppose we have an n-knotting $K \subseteq R^{n+2}$, and $G \subseteq K \times R^1$ is a smooth manifold knotting in so that $\pi|_G$ is an immersion where π is the projection of $K \times R$ to the first factor. We refer to G as a **spinning-guide**, for K or sometimes more simply **guide**, and call $\pi(G)$ a **guide image**. For $p \in G$, if $p = (x,t) \in K \times R^1$, t is called the **guide-height** of p. ▮

For spinning about a higher-dimensional guide we need to accommodate the possibility that the components of the guide may have differing dimensions. If a component of a guide is a k-dimensional manifold, then we will want to use a knotting of a $(n-k)$-disk in $D^{(n-k+2)}$ for spinning along that component.

Definition 12.5. Suppose G is a spinning-guide for an n-knotting $K \subseteq R^{n+2}$, and G has p components, $G_1, \ldots G_p$. Suppose also $\mathcal{A} = \{\alpha_i\}$ is a sequence of q knottings of an $(n-n_i)$-dimensional disk in $D^{(n-n_i)+2}$. The **guide-spinning of K along G of $\mathcal{A}$**, denoted $(K,G)) \otimes \mathcal{A}$, is obtained by spinning along the guide image where we use knotting α_i for component G_i, and the order of multi-spinning at at a q-crossing point $p = \pi(p_1) = \cdots = \pi(p_q)$ is determined by order of the guide-heights of p_i. ▮

Example 12.11. Example 12.10 can be viewed as spinning about a guide where we put X at a guide-height above that of Y, for example $X \times \{1\} \cup Y \times \{0\} \subseteq (S^1 \times S^2) \times R^1$. ▮

The definition for **π-isotopy** of higher-dimensional guides is Definition 10.12 but now the lifting of the guide is a manifold-knotting rather than an n-knotting. It can similarly be shown, Proposition 10.5, that π-isotopic guides are isotopic knottings. If components of the guide have codimension greater than one, we need to mention the framing.

Remark 12.7. Suppose $M \subseteq S^3$ is a non-trivial knotting of a circle and α is a proper knotting of a 2-disk. The knotting $K = (S^3, M) \otimes \alpha$ can be considered to be a guide-spun knot by viewing M as a guide: $M = M \times \{0\} \subseteq S^3 \times R^1$. In an isotopy of M we are not allowed to pass M through itself; in contrast a π-isotopy can do this. One can locally change Figure 4.42 to Figure 4.43 by a π-regular homotopy. So M is π-isotopic to $S^1 \subseteq S^3$ and we conclude $K = (S^3, M) \otimes \alpha \approx K = (S^3, S^1) \otimes \alpha$.

Similarly if $M \subseteq S^3$ is a non-standard knotting and α is a proper knotting of a 1-disk and if M' a corresponding standard knotting of M we will have $K = (S^3, M) \otimes \alpha \approx K = (S^3, M') \otimes \alpha$. Each component of M will be a cube with knotted holes and knotted handles. We obtain a π-isotopy between M and M' by unknotting all handles by "passing the handle through itself" much as in the previous example. Similarly we can "unknot" the holes and untangle the components of M. ∎

Now that we have considered a number of higher-dimensional knottings can expand our collection by taking connected sums. The simplest case is $K\#K'$ where K and K' are n-knottings of S^n since $K\#K'$ is also a knotting of S^n. In [Sosinskii (1970)] it is shown that, for $n \geq 3$, any n knotting K of S^n can be written as a connected sum of a finite number of knottings of S^n, $K = K_1\# \cdots \#K_p$ where none of the K_i can be written as a sum of non-trivial S^n knottings. However, unlike the situation for knottings of circles, this expression is not unique, [Kearton (1979)], [Bayer (1980)]. Even the number of factors is not uniquely determined, [Kearton (1982)]. Example 6.28 is an example of non-uniqueness for 2-knottings.

Spun and super-spun knottings are all slice [Levine (1983)] as well as knots spun about a submanifold [Friedman (2004)].

12.2 Projections of n-Knottings

A basic tool for understanding n-knottings is projection. In Section 4.2 we looked at projections for 1-knottings, and Chapter 6 for 2-knottings. A projection of a 3-knotting is a self-intersecting image of a 3-manifold in R^4. This is a difficult object to visualize. Nevertheless, projection is useful in non-visual ways. For example, the trace of an isotopy of a 2-knotting is a 3-knotting, and examination of the projection of the trace reveals the 2-knotting knot-moves of Chapter 7.1. To define knot-moves for 3-knottings we consider the projection for the trace of an isotopy, a projection of a 4-knotting.

(One can, alternatively, look at higher-dimensional knottings via slicing. If one slices a 3-knotting one can obtain a sequence of 2-knottings. One way to proceed is further slice each of these 2-knottings giving a 2-dimensional array of 1-knottings as in [Lomonaco (1983)]. A related slicing is found in [Roseman and Takase (2007)].)

For an n-knotting $M^n \subseteq R^{n+2} \approx R^{n+1} \times R^1$ standard projection corresponds to projection onto the first factor of $R^{n+1} \times R^1$. For future reference, our definitions and theorems apply to other similar knotting/projection situations which we also refer to as **standard projections**:

(1) $M^n \subseteq R_+^{n+2} \approx R_+^{n+1} \times R^1$ with projection onto the first factor
(2) $M^n \subseteq L^{n+1} \times R^1$, L is a compact closed $(n+1)$-manifold with projection on the first factor
(3) proper knotting $(M^n \times I) \subseteq (R^{n+2} \times I) \approx (R^{n+1} \times R^1 \times I)$ with projection to $R^{n+1} \times \{0\} \times I$
(4) proper knotting $(M^n \times I) \subseteq (L^{n+1} \times R^1 \times I)$ with projection to $L^{n+1} \times \{0\} \times I$

In the first two of these cases, projection is on the last factor of the ambient space. For the last two, projection is projection on the middle factor and are involved in analysis of isotopies. The case (3) will be needed to study higher-dimensional knot-moves. The cases (2) and (4) are used in discussing knottings in S^{n+2} (where we consider $S^{n+1} \times R^1 \subseteq S^{n+2}$) and in the construction of spinning a knotting about a submanifold, Section 12.1.

The terminology for projections surface knottings of Section 6 extends to projections of n-knottings $\pi \colon \mathbb{R}^{n+2} \to \mathbb{R}^{n+1}$ in a straightforward way.

Definition 12.6. If $K^n \subseteq R^{n+2}$ and $X \subseteq K^n$, $\pi(X)$ is the **projection of** X and is denoted X_*. For $x \in K_*^n$, $\widetilde{x} = \pi^{-1}(x) \cap K^n$, is the **lifting of** x **to** K^n. If $X \subseteq K_*^n$, $\widetilde{X} = \pi^{-1}(X) \cap K^n$ is **the lifting of** X. If $\phi \colon M^n \to K^n$ is a homeomorphism $\widehat{X} = (\phi \circ \pi)^{-1}(X)$ is the **pullback of** X **to** M^n.
If $\widetilde{x}$ contains exactly n points where $1 < n$ we say that x is a n**-crossing point of** M_*^n and any point of $\widetilde{x}$ is called a **pre-crossing point of** x **in** M^n and a point of $\widehat{x}$ a **crossing pullback.** ∎

As in other dimensions, we want to define general position for projection. The first requirement for a projection to be in general position is that for all $x_* \in X_*$, $\widetilde{x}$ is a finite set, say $\{x_1, \ldots, x_p\}$. We consider the smooth map $\pi|K$ and its differential map $d\pi$. There are p tangent planes T_{x_i}. We require that the dimension of the image via $d\pi$ of this tangent plane to be either n (in which case we are at an **immersion point**), or it is $(n-1)$-dimensional (in which case we are at a **branch point**).

First, consider the case that all $\{x_1, \ldots, x_p\}$ are immersion points—then we have p n-dimensional affine subspaces $T_{x_i}^*$ containing $x_* \in R^{n+1}$. We want these affine spaces to be in general position (roughly "in an unexceptional way").

Example 12.12. Affine intersections are solution sets of (non-homogeneous) linear equations. For example, an $(n-1)$-dimensional affine space is a hyperplane in R^n: a solution space of a single linear equation $a_{11}x_1 + a_{12}x_2 + \cdots a_{1n}x_n = b_1$. An $(n-2)$-dimensional affine space in R^n is an intersection of two hyperplanes; algebraically, the solution space of $a_{11}x_1 + a_{12}x_2 + \cdots + a_{1n}x_n = b_1$ and $a_{21}x_1 + a_{22}x_2 + \cdots + a_{2n}x_n = b_2$.

For example, consider two lines L and L' in R^3. Each line is an intersection of two planes. So, each line is described by a pair of linear equations in three variables. To examine $L \cap L'$, we solve a set of four linear equations in three variables. This is called an **overdetermined system** and generally will be expected to have no solutions and we conclude that, generally, $L \cap L' = \emptyset$. So two lines in R^3 are in general position means they are disjoint.

For an example less familiar, suppose L is a line and L', a 2-dimensional affine plane, both in R^4. Now L is an intersection of three hyperplanes in R^4, and L' the intersection of two. So $L \cap L'$. is the solution space of five linear equations in four variables. Again this system is overdetermined and we conclude that L and L' are in general position if they are disjoint. ∎

The number of equations needed to define an affine subspace is the codimension of that subspace, and so, with the understanding that negative dimension for a set means the set is empty, we define:

Definition 12.7. A set of k-dimensional **affine subspaces** T_i **in** R^n **are in general position** if for any j of them:

$$codim(T_{i_1} \cap \cdots \cap T_{i_j}) = codim(T_{i_1}) + \cdots + codim(T_{i_j}). \blacksquare$$

Recall the Whitney umbrella W for 2-knottings, Definition 6.4, a local model for a branch point in R^3. A local model of a branch set for an n-knotting is $W_* \times D^{n-2} \subseteq R^3 \times D^{n-2} \subseteq D^{n+1}$ an **extruded Whitney umbrella**. Putting this another way:

Definition 12.8. Define a smooth map of a cube $Q = D^1 \times D^1 \times D^n \subseteq R^{n+2}$ into R^{n+2} by $w(x_1, x_2, x_3, \ldots, x_n) = (x_1x_2, x_1, x_2^2, x_2, x_3, \ldots, x_n)$ for $-1 \leq x_1 \leq 1$ and $-1 \leq x \leq 1$. Write $\mathcal{W} = w(D^n)$. The standard projection $\mathcal{W}_* = \pi(\mathcal{W})$ will be called the **extruded Whitney umbrella.** ∎

Suppose K^n is an n-knotting, with branch set in $B \subseteq K$ and projections K_* and B_*, then B will be a smooth $(n-2)$-submanifold K. For 2-knottings we have described B as a union of branch-spheres, Remark 6.3. For our general position, we want the projection of this manifold, $K_* \cap B_*$, to be "as nice as possible".

For a 3-knotting, the branch set B is a 1-manifold, and so, B_* is the image of a 1-manifold in R^3. Our general position will require that B_* is in general position in R^3, *which will mean that* B_* *is embedded in* R^3. However, B_* might intersect other points of K_*

Example 12.13. Let K be the 2-knotting of Example 6.3, translated it so it lies in $\overset{\circ}{R^4_+}$. Then $K_0 = i_0(K)$ is a copy of K in page $H^4_0 \subseteq R^5$. Now $\Sigma(K_0)$ is a 3-knotting of $S^2 \times S^1 \subseteq R^5$. The projection of the spinning is the spinning of the projection. The crossing set of K_* is a line segment with endpoints being branch points. So the crossing set of $\Sigma(K_0)$ is an annulus A, and branch set, B, is the two circles, P and Q, which make ∂A.

The join-spin of type $(1, 3)$ using the knotting K and with $n = 1$ is basically the 3-knotting $\Sigma(K_0)$ with a single linking disjoint geometric 3-sphere S added. We can assume the projection S_* is a geometric ellipsoid. Each circle of branch points, P and Q, of will intersect the S_* in two points. All four points are 2-crossing points; no other points of P and Q are 1-crossing points. We will make a shard structure for B where the four points are four 0-dimensional shards of B and the complements in B are four 1-dimensional shards of B. If p is one of these crossing points and p_1 and p_2 are pre-crossing points at p, one is a branch point and one is not. ∎

Remark 12.8. Here are some things to consider about branch point sets in a full definition of general position for projections in higher-dimensions.

For a projection of a 4-knotting in R^6, the set B might be 2-dimensional. (For example if we take a 2-knotting K which has projection with p branch points, then because "the projection of the spinning is the spinning of the projection", the super-spin $(S^4, S^2) \bigotimes K$ will have a projection whose branch set is a disjoint union of p 2-spheres.) Then B_* is the image of a 2-manifold in R^5 and, generically, we expect a 2-manifold in R^5 to have no self-intersections. So we know that B_* is homeomorphic to a finite disjoint union of closed surfaces in R^5. We will later see that these (as well as higher-dimensional branch manifolds) must be orientable surfaces. *However,* B_* might generically be expected to have points of intersection with the 3-dimensional crossing set, and, if so, in our definition, we will want to require that any such intersection be transverse.

For a projection of a 5-knotting in R^7, the set B might be 3-dimensional. Then B_* is the image of a 3-manifold in R^6 and generically can have a self-intersection set which is a set of isolated points.

Furthermore, we need to consider crossings of the branch set and the crossing set.

For a projection of a 4-knotting in R^6, the crossing set C is 3-dimensional and the set B might be 2-dimensional. Then B_* is the image of a 2-manifold in R^5 and so B and C might intersect transversally. ▮

General position with respect to projection was defined in [Roseman (2000)]. If K is an n-knotting whose projection is an immersion then general position means that there are no crossing points of order more than $n+1$ and that K_* **has self-transverse crossings**. That is, if $x \in K_*$ is a p-crossing point, the tangent that the image of the p n-dimensional tangent planes are in general position. In general, if the branch set B is not empty, then we need to consider possible self-intersections of B and how B_* intersects the rest of K_*:

Definition 12.9. Let $M^n \subseteq \mathbf{R^{n+2}}$ be a knotting, with branch set B and pre-crossing set D. We will say M^n is in general position with respect to the projection if:

(1) D is the union of immersed closed $(n-l)$-dimensional submanifolds of M^n with self-transverse crossings,
(2) B is a closed $(n-2)$-dimensional submanifold of M^n,
(3) B_* is locally modeled on an extruded Whitney umbrella,
(4) B_* meets $M_* - D_*$ transversely,
(5) B_* has self-transverse crossings with itself,
(6) B_* meets the crossing set of D_* transversely,
(7) the crossing set of B^* is transverse to the crossing set of $(D-B)_*$. ▮

An example illustrating Definition 12.9 (4) appears in Example 12.13. Definition 12.9 (6) is only if interest for n-knottings where $3 \leq n$.

The last three conditions only apply for n-knottings where $6 \leq n$. For a 6-knotting, K, the branch set, B, if non-empty, is 4-dimensional, $B_* \subseteq R^7$ so we might possibly have points of self-transverse intersection of B_*. Definition 12.9 (7) is only if interest for n-knottings where $7 \leq n$. For a 7-knotting, the projection of crossing set is 6-dimensional, projection of the branch set is 2-dimensional and both are subsets of R^8, so, generically, we might have some (0-dimensional) transverse intersections.

Prop 12.2 (General position for knottings). Given a knotting $M^n \rightarrow \mathbf{R^{n+2}}$ we may isotop M^n by a smooth isotopy to a knotting which is in general position with respect to projection. In fact this isotopy can be chosen to change points of the knotting by an arbitrarily small amount. ▮

Example 12.14. A simple higher-dimensional version of the example of Remark 4.4 illustrates Proposition 12.2 for 2-knottings. Here our knotting consists of two disjoint spheres $K = \{(x, y, z, w\colon x^2 + z^2 + (w-2)^2 = 1\}$ and $L = \{(x, y, z, w\colon y^2 + z^2 + (w+2)^2 = 1\}$. Each sphere projects to a planar disk and K'_* is the union $\Sigma_* \cup \Sigma'_*$, Figure 12.4. The knotting $K \cup L$ is not in general position.

However, we might rotate K and perhaps shrink L slightly with a small twist and obtain a 2-knotting with projection as shown in Figure 12.5, which *is* a general position projection. ▮

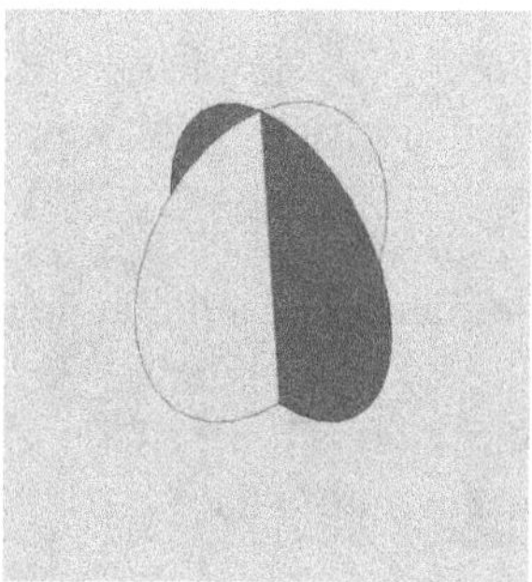

Fig. 12.4 Projection of knotting not in general position $\Sigma_* \cup \Sigma'_*$.

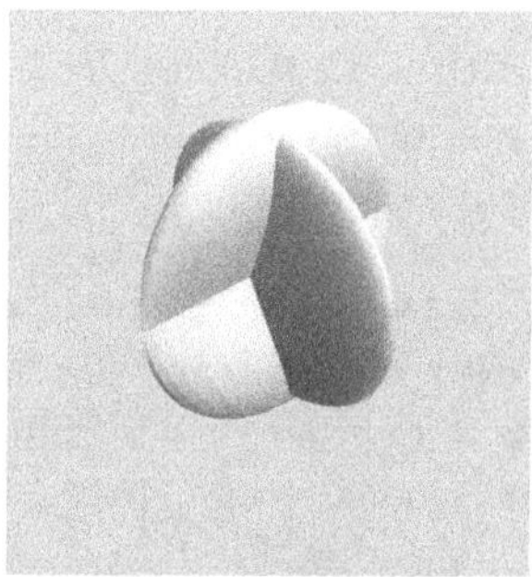

Fig. 12.5 Projection of knotting of Figure 12.4 after small isotopy.

Definition 12.10. Suppose K is an n-knotting of M. with crossing set $\mathcal{C}$ branch set B_*. A 1-**shard** of K_* is a component, X of $K_* - \mathcal{C}$. For $1 \leq q$, a q-**shard** is a component of the set of q-fold points of $\mathcal{C} - B_*$. A bq-**shard** of B_* is a component of the set of q-fold points of B.

For any of these shards, $X \subseteq K_*$, a component of $\widetilde{X}$ is a shard of the same type of K, and a component of $\widehat{S}$ is a shard of the same type of M. ▮

The prefix of a shard is the **type of the shard**. So a $b2$-shard has type $b2$. For a shard S, the type determines the local model of K_* about any point of S.

Branch point sets of a projection are generic—if we change the knotting slightly we will not essentially change the branch point set. However, for *orientable* knottings we can always find an equivalent knotting with an empty branch set, first proven in [Giller (1982)] for surfaces in R^4. In other words:

Prop 12.3. Any orientable n-knotting $K^n \subseteq R^{n+2}$ has an equivalent knotting K' whose K'_* is an immersion in R^{n+1}. ▮

A proof of Proposition 12.3, found in [Roseman (2000)], is a straightforward generalization of the version for 2-knottings sketched in 7.4. For an n-knotting, higher-dimensional branch points look locally like an extruded Whitney umbrella, Definition 12.8. The idea is to replace this with a product of a "pigtail" and an $(n-2)$-disk.

Remark 12.9. We give the idea of a proof of Proposition 2.1 that every orientable n-knotting K has a product tubular neighborhood. We use a trefoil knotting, Figure 4.19. Our construction can be adjusted to construct a trivial tubular neighborhood for any orientable n-knotting.

In general, the image of an immersion of an orientable n-manifold M in R^{n+1} will have a product normal bundle. This is because R^{n+1} is orientable and M is orientable, then this normal bundle must be orientable. But an orientable D^1 bundle must be trivial, [Hirsch (1976)].

We assume that the projection, K_*, of K is an immersion. Using short normal vectors to the immersed smooth planar curve K_*, we can obtain a product tubular neighborhood T with normal 1-disks $D^1_x \subseteq R^2 \subseteq R^3$, for each $x \in K$, see Figure 12.6. A product tubular T' neighborhood of $K_* \subseteq R^3$ can be obtained from T by taking a product with a vertical 1-disk, D_z, in the z-coordinate so that T' has normal 2-disks $D^2_x = D^1_x \times D_z$ as in Figure 12.7. We get a product tubular neighborhood $\mathbb{T}$ of K by lifting each D^2_x with center x_* to a 2-disk D'_x centered at the appropriate $x \in K'$, Figure 12.8. The resulting tubular neighborhood of K is shown in Figure 12.9. We might need shrink disks D'_x, so that if x_* is a crossing point of K_* with above-point x_a and below-point x_b, that D'_{x_a} and D'_{x_b} are disjoint. ▮

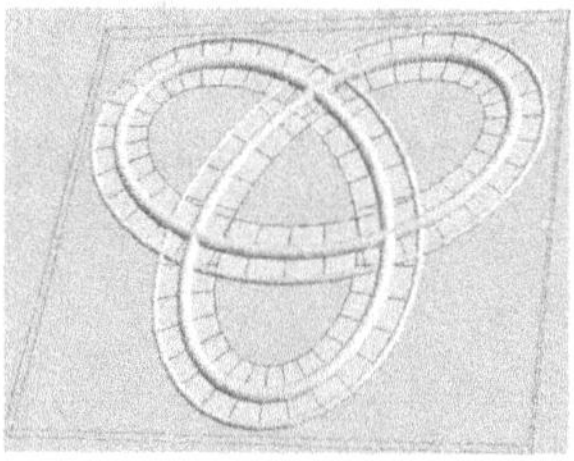

Fig. 12.6 A product tubular neighborhood T of $K_* \subseteq R^2 \subseteq R^3$.

Fig. 12.7 Showing the normal disks D^2_x of T'.

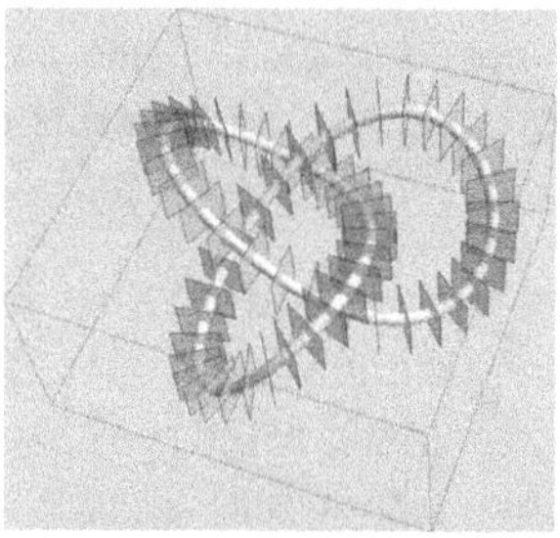

Fig. 12.8 Raising up disks T' to get normal disks, D'_x for K.

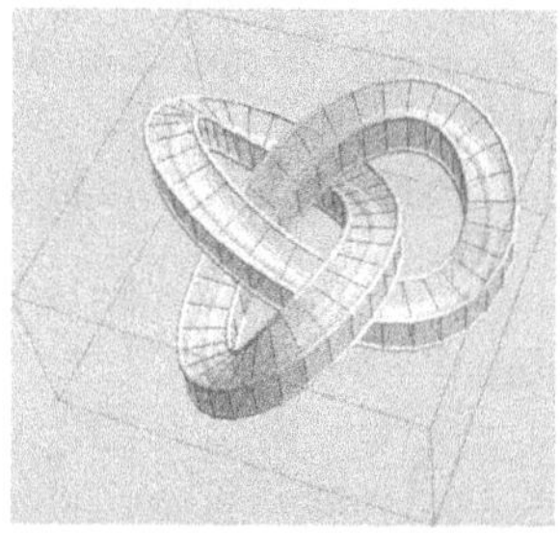

Fig. 12.9 If lifted normal disks D'_x for $\mathbb{T}$ are scaled small enough, they will be disjoint.

Although branch points of a knotting projection (for orientable manifolds) can be eliminated, crossing sets may be persistent.

Definition 12.11. Suppose K is an n-knotting in general position with respect to projection and has K_* some crossings of order $p > 1$ but no crossings of order higher than p. Suppose also that for all equivalent knottings $K' \approx K$ all projections K'_* have some crossings of order p. Then we say the **knot** $[K]$ **is** p**-bulky**. ▌

So for 2-knottings, psuedo-ribbon knots are 2-bulky and knots with non-zero triple-point number are 3-bulky. See Example 12.15 for higher dimensional examples.

Quandles for oriented codimension 2 knottings was first done in [Fenn and Rourke (1992)], Further development, especially in terms of projections is found in [Kamada (2014)] and [Przytycki and Rosicki (2014)]

Remark 12.10. The technique of distinguishing knottings by coloring rules transfers via spinning to distinguishing the corresponding spinnings.

We have seen that a crossing of an arc knotting gives rise to a circle of crossings in the spun knotting. Similarly the shards of the spinning are the spinning of the shards and the coloring of the spinning corresponds to the spinning of a coloring. In terms of shard structures, the shard structure of the spinning of α is the product of the shards with a circle. ▮

12.3 Knot-Moves for High Codimension Smooth Knots

Briefly, for an isotopy of an n-knotting with trace T, in general position with respect to projection, knot-moves correspond to local models of the non-degenerate singularities of the level maps of the various shards of the trace projection T_*. Remark 7.2 is a version of this approach for 1-knottings. For any n there is a finite number of shards of T_* and a finite number of non-degenerate critical point indices, so there are a finite number of move types. Any level map for the trace of a compact manifold has a finite number of critical points, so every isotopy corresponds to a finite set of moves.

Remark 12.11. For any isotopy $F\colon K \times I \to R^n \times I$, with trace T, we outline how to get an equivalent isotopy which is a finite sequence of moves.

(1) Alter T to get an equivalent isotopy, F',whose trace T' is in general position with respect to projection trace projection P.
(2) Further alter F' to get an equivalent isotopy, whose trace, T'', is in general position with respect to projection P *and, in addition,* the level map of T'' is a Morse function on each shard of T''_*.
(3) Finally alter T'' to get a trace of an equivalent isotopy, $\mathbb{T}$, which is in general position with respect to projection P with the level map of $\mathbb{T}$ a Morse function, *and, in addition,* each critical value v is the image of a single critical point m. Label these p values as $\{v_1, \ldots v_p\}$ with critical points $\{m_1, \ldots m_p\}$. We say that an isotopy such as $\mathbb{T}$ is an **isotopy prepared for moves.**
(4) If $[a, b] \subset I$ is a subinterval which contains no critical values, consider the levels $\mathbf{L}_t = \mathbb{T} \cap (R^{n-1} \times \{t\})$ then $\mathbb{T} \cap (R^3 \times J)$ is an isotopy of of the projection $\mathbf{L}^*_a$ in R^{n-1}. We call this an **isotopy with no moves.**
(5) On the other hand, suppose we have a critical point m_j with critical value v_j, where in $[a, b]$. Then there will be a single simple topological change between L_a and L_b. This change is a higher-dimensional knot-move. The **type of the move** will depend on further details. ▮

The alterations of our original isotopy mentioned in the first three steps of the above outline can be done with small changes. Roughly we can say, for two smooth embeddings $f\colon M \to N$, and $f\colon M \to N$, that f **is close to** g, if for all $x \in M$, $f(x)$ is close to $g(x)$ and all derivatives of all orders of f are close to the corresponding derivatives of g. It can be shown that if f is a smooth embedding and g is close to f, then g is a smooth embedding. A trace is an image of an embedding so, if T_1 is a trace and T_2 is close to T_1 then T_2 is a trace. Such small changes to an isotopy result in another isotopy, [Roseman (1998, 2004)].

For Remark 12.11 (1) we can alter our general position arguments as in Proposition 12.2 to apply to the manifolds with non-empty boundary, such as a manifold trace. Since T' is close to T, we have T' is an equivalent isotopy trace.

For Remark 12.11 (2) we next use a version of Proposition 8.2, again adapted to manifolds with non-empty boundary to get our Morse property for P_I. As long as we do not move points of the shard much, we can extend that to an isotopy of the trace. In doing this we are also not changing derivatives at other points much, so we are not unintentionally changing the nature of other critical points.

For Remark 12.11 (3), if we have some value, which is the image of several critical points, we can move them slightly up or down in the I direction without altering the number of and type of any critical points, Since $\mathbb{T}$ is the trace of an isotopy, derivative, at any point, of the level map is 1, in particular, non-zero, so there are no critical points on 1-shards. However, as we will see, the restriction of P_I to a j-shard with $j \neq 1$ certainly can have critical points.

For Remark 12.11 (4) A version of Proposition 8.1 shows the first statement.

For Remark 12.11 (5) it is best to examine first 2-knottings, Remark 12.12, then 3-knottings.

Remark 12.12. For an isotopy of a 2-knotting K, we consider $\mathbb{T}_* \subseteq R^3 \times I$ which is a trace projection of 3-manifold $K \times I$. The crossing set $\mathcal{C}$ has a shard structure with four different types of shard corresponding to 2-crossings, 3-crossings, 4-crossings, and branch $b1$-shards (which are 1-dimensional). For the 2-shards (which are 2-dimensional), there are three possible types of Morse singularities: local max, local min or saddle point. For the 3-shards (which are 1-dimensional), there are two possible types of Morse singularities: local max, local min. For the $b1$-shards (which are

1-dimensional), there are two possible types of Morse singularities: local max, local min. And finally for the 4-shards (which are 0-dimensional), there is only one type. ∎

Prop 12.4. Suppose $K \subseteq R^4$ and $K' \subseteq R^4$ are equivalent 2-knottings via isotopy with trace T. In this case $\mathbb{T}_*$ is projection of a 3-knotting in $R^3 \times I$. Let $\mathcal{C}$ be the crossing set of $\mathbb{T}_*$. In $\mathcal{C}$ there are:

(1) 2-shards
(2) 1-shards of triple points, 1-shards of branch points
(3) 0-shards, quadruple points of $\mathbb{T}_*$ and 0-shards of B where the b-set of meets T_* transversely ∎

Listing possible singularities for each of the above shards we find seven types of elementary knot-moves of for 2-knottings, Section 7.1:

(a)**-move** Local maximum or local minimum of the crossing of a 2-shard of 2-crossing points.
(b)**-move** Saddle point of a 2-shard of 2-crossing points.
(c)**-move** A local maximum or a local minimum in a 1-shard of the branch point set which is a local maximum or minimum (respectively) of the crossing set in the direction transverse to the branch set.
(d)**-move** A local maximum or a local minimum in a 1-shard of the branch point set which is a local minimum or maximum (respectively) of the crossing set in the direction transverse to the branch set.
(e)**-move** A local maximum or a local minimum in a 1-shard of triple points.
(f)**-move** A 0-shard which is a b1-shard.
(g)**-move** A quadruple point 0-shard.

The above list is seven moves is as in [Roseman (1998)]. However this list can be shortened, if we remove either the (c)-move or the (d)-move, Remark 7.7.

Similarly we can consider knot-moves for n-knottings. We briefly consider 3-knottings. The shard structure of the projection trace is a projection of a 4-knotting and so has: 4-dimensional 0-shards, 3-dimensional 1-shards for which we need to consider critical points of index 0 and 1. The branch set B might have dimension 2 and so we need to consider critical points of index 0 and 1. The crossing set C has 3-dimensional 2-shards (so we consider index 0 and 1), 2-dimensional 3-shards (again index 0 and 1), 1-dimensional 4-shards (look at index 0) and 0-dimensional 5-shards. The trace projection is image of a 4-manifold in a 5-manifold. The transverse intersection of B and the 0-shards will be 1-dimensional, X and we will need to consider local min and max on this set. The crossing set of the trace is 3-dimensional, the branch set is 2-dimensional. So we might have isolated points of transverse intersection $B \cap C$, Y. Thus we have the following list of 12 elementary moves for 3-knottings:

(a)**-move** local max/min of 2-shard.
(b)**-move** critical point index 1 or 2 of 2-shard.
(c)**-move** local max/min of 3-shard.
(d)**-move** saddle point of 3-shard.
(e)**-move** local max/min 4-shard.
(f)**-move** 5-shard.
(g)**-move** local max/min of B_* also local max/min respectively transversely.
(h)**-move** local max/min of B_* also local min/max respectively transversely.
(i)**-move** saddle point of B_*.
(j)**-move** a local max/min of B_* on X which is a local min/max respectively transversely.
(k)**-move** a local max/min of B_* on X which is a local max/min respectively transversely.
(l)**-move** a point of Y.

This list is complete but not minimal. Local diagrams for such moves would be 4-dimensional and so are not visually helpful but we can get some idea from the local changes in crossing scheme, see [Roseman (1998)].

Remark 12.13. Next we discuss component coloring in high-dimensions. Suppose we have an n-knotting K with components, K_i. Refer to the subscript i as the **color of** K_i.

Suppose σ is in a p-shard of K_*, with $\widetilde{\sigma}$ consisting of p-shards $\widetilde{\sigma_j}$ of K, labeled so that $\widetilde{\sigma_1} < \cdots < \widetilde{\sigma_p}$, for $j = 1, \ldots p$, (recall use of $<$ for shards, Definition 6.9). Let c_j be the color of $\widetilde{\sigma_j}$.

The sequence $C = (c_1, \ldots, c_p)$, that is, colors *listed in order of increasing height* at a crossing, is called the **color sequence of** σ. If all p colors are distinct, we say that σ **is a colorful shard** with color sequence C and that C **is a colorful sequence.** ▌

The projection of any general position n-knotting K has a finite number of $(n + 1)$-crossing points. In other terms, these are 0-dimensional

$(n+1)$-shards. For any p-shard and any sequence C of p colors, let $N(K, C)$ be the **number p-shards of K_* with color sequence** C. The following is a high-dimensional version of Proposition 7.3. Since we cannot rely on figures for higher-dimensional knot-moves, we prove this by going into the proof for knot-moves:

Proposition 12.1. *Suppose K and K' are two equivalent n-knotting with at least $(n+1)$ components, with C is a color sequence of $(n+1)$ different colors, $N(K, C)$ and $N(K', C)$ are either both even or both odd.*

Proof. Suppose we have general position n-knotting K and $p \in K_*$ is an $(n+1)$-shard (i. e. an $(n+1)$-crossing.) Suppose we have an isotopy with K' and trace $\mathbb{T}$. Consider the $(n+1)$-crossings C of $\mathbb{T}$. This will be the image of a 1-dimensional manifold—a union of topological arcs and circles. If a component, A, of C is an arc, the endpoints must be branch points of some component T_0 of $\mathbb{T}$. If the color of T_0 is x, then the color x must appear twice in A and so A is not colorful a colorful shard. So the union of colorful shards $\mathbb{C}$ is a union of circles. The number $(n+1)$-crossings at level t is the number of points of $\mathbb{C}$ at that level. This number can change only at local max or local min and if so by two and the pairs must be of the same color type. ∎

Example 12.15. We show that $\mathbb{K}_n$ of Example 12.9 is $(n+1)$-bulky.

We outline a proof for n=2, Example 6.22 and show this knotting $K_2 = T \cup T' \cup S$ is 3-bulky. Give torus T color 1, torus T' color 2, and 2-sphere S color 3. We view the Hopf link as having two crossings. When we spin, each generates a circle of 2-fold crossings of T_* and T'_*. Each of these circles intersects S_* in two points. This accounts for the four triple points (3-shards) of K_*. One, p, is seen in Figure 6.111; another, q, shown in Figure 6.112, as well as triple points p' and q', see Figure 6.114. Figures 6.113–6.118 might help. For p, color 3 is highest and p has color sequence $(1, 2, 3)$. For q, color 3 is lowest and q has color sequence $(3, 1, 2)$. Similarly, one can see that p' will have color sequence $(2, 1, 3)$ and q' will have sequence $(3, 2, 1)$.

If this knotting were not 3-bulky, then there would be an isotopy to an embedding with no triple points of any color sequence, and so $N(K, C)$ would be an even number for all color sequences. In fact, $N(K, C)$ is odd (namely 1) for the above four color sequences. So this knotting *is* 3-bulky.

For $n = 3$, we color the 3-sphere S of K_3 with color 4, and chose colors so that the color of the spinning is color of the component being spun, so that, for example, the component of K_3 diffeomorphic to $S^2 \times S^1$ will be 3. Each of p, p', q and q' will correspond to a circle of 3-fold crossings in K_3^*. Each circle will intersect the S_* in a pair of points, so there will be eight (4-shards) of K_*. For the pair of points associated to the circle which is the spinning of p, the two color sequences will be $(1, 2, 3, 4)$ and $(4, 1, 2, 3)$. Similarly, for q we get color sequences $(4, 3, 1, 2)$ and $(3, 1, 2, 4)$. Continuing, we see that each of these eight shards have distinct color sequences. Our previous argument for $n = 2$ will imply that K_3 is 4-bulky.

More generally, for any n, and n-knotting K_n, the argument is similar, beginning with labeling the component diffeomorphic to S^n with color $(n + 1)$, and arguing that the 2^n $(n + 1)$-shards have distinct colors. ▮

12.4 More High-Dimensional Knotting Constructions

In this section we briefly describe constructions for high-dimensional knottings that are generalizations of satellites, ribbons, and roll-spinning.

Higher-dimensional satellites were first defined for codimension-two knottings of spheres in [Shinohara (1971)] by taking a (non-trivial) knotting of $S^n \times D^2$, $N \subseteq R^{n+2}$ and a (non-trivial) knotting of an n-sphere $K' \subseteq N$. If we let $K \subseteq R^{n+2}$ be the subset corresponding to $S^n \times \{\mathbb{O}\}$, then we say K' **is a satellite of** K or that K **is a companion of** K'. See also [Kanenobu (1983)]. One can also use this construction for non-spherical surface knottings. This first appeared in [Litherland (1981)].

Ribbon knottings are a simple class of n-knottings where, as in 1-knottings, one can visualize the knotting largely in terms of "over/under" for crossings.

An n-ribbon, Definition 9.2, is the most common way to generalize a ribbon, Definition 9.1 to higher-dimensions [Yajima (1969)], [Yanagawa (1969)], [Hitt (1979)]. Such a ribbon is an image of an immersion $M^n \looparrowright R^{n+1}$, where $M^n \approx D^n$, such that the self-intersection set is a disjoint union of $(n-1)$-disks which are slots/slices. Then ∂M^n gives a knotting of S^{n-1}. For such ribbons one can show there are slice knots that are not ribbon n-knots, [Hitt (1979)]. Such a ribbon n-disk gives rise to an n-knotting of an n-sphere whose projection whose crossing set is a disjoint union of $(n-1)$-spheres. In papers like [Yasuda (2006)] an n-ribbon is obtained with n-spheres joined with tubes/bands. In [Winter (2014)] and later [Winter (2019)] he talks about a type of equivalence of n-ribbon knots and their quandles.

Example 12.16. An n-knotting with a projection whose crossing set is a disjoint union of $(n-1)$-spheres, is called a **pseudo-ribbon n-knot**, [Tanaka (2004)], [Al Kharusi and Yashiro (2016)]. In general, a 2-bulky n-knotting has a projection whose crossing set is a disjoint union of $(n-1)$-*manifolds*. It is natural to ask if any 2-bulky n-knotting is isotopic to a psuedo-ribbon n-knotting. That is, can we somehow change our $(n-1)$-manifold crossings to be $(n-1)$-spheres? The answer is "no". In [Ogasa (2001)] a very interesting knotting, $\mathcal{O}^3$, of a 3-sphere in R^5 is described. We call $\mathcal{O}^3$ the **Ogasa 3-knotting**, the associated knot the **Ogasa 3-knot** [Ogasa (2001)]. The projection $\mathcal{O}^3_*$ of this knotting is an immersion whose crossing set is a disjoint union of tori. Furthermore, it is shown that $\mathcal{O}^3_*$ is not the projection of a trivial knot. On the other hand [Yoshida (2004)] has shown that a knotting whose projection is an immersion with crossing set a disjoint union of spheres is a projection of a trivial sphere. So, the Ogasa knotting is a 2-bulky knotting that is not a psuedo-ribbon n-knot.

These results extend to higher-dimensional knottings. The n-sphere knotting obtained by iterated spinning, $\mathcal{O}^n$, of the $\mathcal{O}^3$, will have crossing sets homeomorphic to a disjoint union of $(n-1)$-manifolds which are the product of $(n-1)$ circles, [Ogasa (2001)]. Each $\mathcal{O}^n_*$ has the property that the corresponding knot is not a ribbon n-knot. ∎

An immersed 3-manifold in R^4 could have ribbon-like singularities that are 2-manifolds, which are not necessarily all 1-disks. We introduce ribbon-like objects, we call $\mathbb{R}$**-sets**, that have $(n-1)$-manifolds A_i for slots/slices. An $\mathbb{R}$-set gives an n-knotting we call an $\mathbb{R}$**-knotting**. We assume manifolds in our definition are orientable.

Definition 12.12. For a smooth immersion $F\colon M^n \to R^{n+1}$ of orientable manifold M^n, $F(M^n)$ is called an $\mathbb{R}$**-set** if:

(1) $F|_{\partial M^n}$ is a smooth embedding
(2) F is at most 2-to 1
(3) the self-intersection set $\Sigma(F)$ is a finite disjoint union of orientable $(n-1)$-manifolds $\mathbb{A}_i$. Each $\mathbb{A}_i$ is called $\mathbb{R}$**-singularity**
(4) F is self-transverse along each $\mathbb{A}_i$
(5) For each i, $F^{-1}\mathbb{A}_i$ is a disjoint union of two $(n-1)$-manifolds P_i and Q_i
(6) P_i is a proper submanifold of M^n and is called a $\mathbb{R}$**-slice**.
(7) Q_i is entirely contained in the interior of M^n and is called a $\mathbb{R}$**-slot** ∎

Prop 12.5. An $\mathbb{R}$-set for M^n with singularities $\mathbb{A}_i$, gives rise to a 2-bulky n-knotting, $\Sigma(\mathbb{R})$, of the double $D(M^n)$, we call **an $\mathbb{R}$-knotting**. Components of the crossing set of $\Sigma(\mathbb{R})$ are diffeomorphic to the doubles of the singularities $\mathbb{A}_i$.

Proof. This construction follows the construction for symmetric ribbon 2-knottings from a ribbon, R, found in Section 9.1. First construct a simple immersion of $D(M^n)$, Definition 6.16. From this immersion, we can define a number of knottings as we have done for 2-knottings in Section 6.7. In higher-dimensions, we need to account for the fact that our slice/slot manifolds are not disks and rely less on graphics, defining our immersion in terms of product tubular neighborhoods.

Consider one $\mathbb{R}$-singularity $\mathbb{A}$ with slot P and paired slice Q in M^n. Our construction, locally takes a neighborhood of $\mathbb{A}$ as indicated in Figure 12.10 and constructs a local model of our immersion. Figure 12.10 represents $\mathbb{A}$ as a transverse intersection of an two manifolds $\mathbb{N}_P$ and $\mathbb{N}_Q$. Additionally we have, a third manifold $\mathbb{N}_{P'}$ with $\mathbb{N}_{P'} \subseteq \mathbb{N}_Q$. All three are diffeomorphic to $\mathbb{A} \times D^1$.

Let $N_Q \approx Q \times D^1$ be a tubular neighborhood of proper submanifold $Q \subseteq M^n$, then $\mathbb{N}_Q = F(N_Q)$. Next we consider an *improper* product tubular neighborhood N_P of P. Let $P' \subseteq int(M^n)$ correspond to "P with its boundary stretched out, away from P, into M^n". That is $P' = P + C$ where $C \approx \partial P \times I$. We consider an improper tubular neighborhood of P'. and $N_{P'} \approx P' \times D^1$. Define $\mathbb{N}_P = F(N_P)$. and $\mathbb{N}_{P'} = F(N_{P'})$.

Now $\mathbb{N}_{P'} \subset R^{n+1}$ has a product tubular neighborhood $\mathcal{N} \approx \mathbb{N}_{P'} \times D^1$. Our local model of $\Sigma(\mathbb{R})$ is based on subsets of $\mathcal{N}$, see Figure 12.10.

We next define subsets of R^{n+2}, $\mathbb{X}$ and $\mathbb{Y}$, each diffeomorphic to $D(\mathbb{A}) \times D^1$, which will be disjoint subsets of $\Sigma(\mathbb{R})$; $\mathbb{Y} \subseteq \partial(N_P \times D^1)$, and the other is a subset of $\mathbb{X} \subseteq \partial(N_{P'} \times D^1)$. Here $D(\mathbb{A})$ refers to double, Definition 2.21. Additionally we consider $\mathbb{Z} = \mathbb{X} \cap \mathbb{Y}$.

The first, $\mathbb{Y}$, corresponds to $\partial\mathbb{N}_P \times D^1 \subseteq \mathcal{N}$. Consider $\partial P' \times D^1 \subseteq N_{P'}$ then let $\mathbb{B} = F(\partial P' \times D^1)$, bold dark lines in Figure 12.10. Now define:

$$\mathbb{X} = (\mathbb{B} \times D^1) \cup (\mathbb{N}_{P'} \times \{-1\}) \cup (\mathbb{N}_{P'} \times \{1\}) \subseteq \partial\mathcal{N}.$$

Then we can write $\mathbb{X} \cap \mathbb{Y} = \mathbb{Z} \cup \mathbb{Z}'$ where $\mathbb{Z} \subseteq \mathbb{N}_p \times \{1\}$ and $\mathbb{Z}' \subseteq \mathbb{N}_p \times \{-1\}$, Figure 12.11. Both $\mathbb{Z}$ and $\mathbb{Z}'$ are diffeomorphic to $\partial(\mathbb{A} \times D^1) \approx D(\mathbb{A})$.

Suppose $F(M^n)$ is an $\mathbb{R}$-set with slots P_i and slices Q_i and corresponding N_{P_i} and N_{Q_i}, we let $W^n = \overline{M^n - \bigcup_i (N_{P_i} \cup N_{Q_i})}$. Note that $F|_{W^n}$ is an embedding. We also suppose we have an immersed product tubular

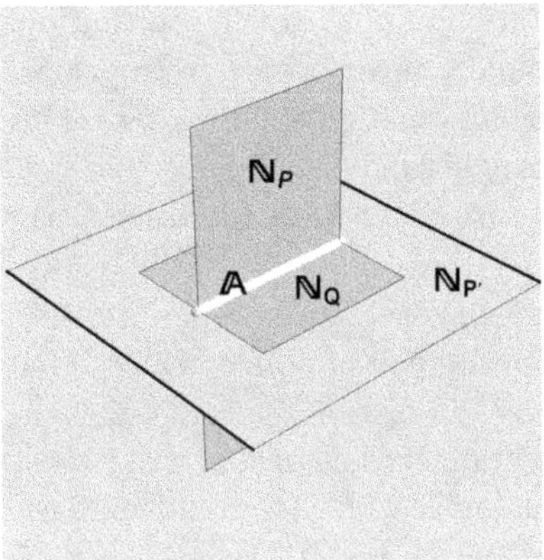

Fig. 12.10 Subsets of $\mathbb{R}$ in $\mathcal{N}$.

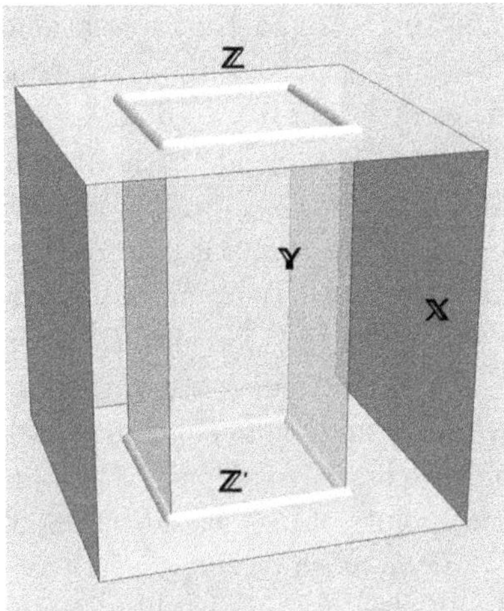

Fig. 12.11 Subsets of $\Sigma(\mathbb{R})$ in $\mathcal{N}$.

neighborhood $\mathbb{F}\colon M^n \times D^1 \to R^{n+1}$ whose restriction to N_{P_i} gives $\mathcal{N}_i \approx \mathbb{N}_{P_i}$. Let

$$\mathbb{W} = \left(\mathbb{F}(\partial M^n) \times D^1\right) \cup \left(\mathbb{F}(M^n \times \{-1\}\right) \cup \left(\mathbb{F}(M^n \times \{1\}\right).$$

Finally, (after rounding corners) $\Sigma(R) = \mathbb{W} \bigcup_i (\mathbb{X}_i \cup \mathbb{Y}_i)$ is an immersion of $D(M^n)$ with self-intersections $\mathbb{Z}_i \approx D(\mathbb{A}_i)$ and $\mathbb{Z}'_i \approx D(\mathbb{A}_i)$. ∎

We have a generalization of Proposition 10.2.

Prop 12.6. If α is a proper 1-knotting and Σ is an n-knotting which is a super-spinning of α, see Example 12.2, then Σ is an n-ribbon knotting.

Suppose α is a proper k-knotting with crossing set a union of disjoint closed $(k-1)$-manifolds $C_1, \ldots, C_q$. The n-knotting, Σ, is an which is an Artin spinning of α is an $\mathbb{R}$-knotting where the slots/slices of the $\mathbb{R}$-set are $C_1 \times D^1, \ldots, C_q \times D^1$. ∎

If we use generalized cylindrical coordinates, $(r, \theta, x_3, \ldots, x_{n+2})$, to define an Artin spun knotting, the image of the projection $\pi_{n+1}(\Sigma(\alpha))$ will be an $\mathbb{R}$-set since each $x \in \alpha$ will give rise to a circle in $\Sigma(\alpha) \subseteq R^{n+2}$ which projects to a line segment via π_{n+1}. As in our proof of Proposition 10.2, at crossing points of α we get a nested pair of line segments.

Example 12.17. We now turn our attention to generalizing the rolling of Section 10.7. This rolling is a deformation spinning that takes a circle knotting, k, and produces a deform-spun 2-knotting of S^2. To roll a 1-knotting k, for each $x \in k$, we translate k so that x is sent to $\mathbb{O}$, use a tangent vector $\overrightarrow{v}_x$ to define a further rotation of our knotting so that our knotting is tangent to the x-axis at $\mathbb{O}$, finally invert to get a proper arc in D^3, α_x so $\partial\alpha_x = S^0$. We produce our deformation spinning by taking $S^1 \subseteq S^2 \subseteq R^4$, with product tubular neighborhoods $N \approx S^1 \times D^1 \subseteq S^2 \subseteq S^2 \times D^2 \subseteq R^4$ and for each $x \in S^1$ replacing the standard local pair $((D^1 \times D^2, D^1)_x) \approx (D^3, D^1)_x$ with (D^3, α_x).

If we had *not* rotated, then inversion would still get a proper arc $\alpha'_x \subseteq D^3$ with endpoints *some* diametrically opposite pair of points of ∂D^3—but not necessarily the pair of points $S^0 \subseteq \partial D^3$. In fact, a line containing the two points of $\partial\alpha'_x$ would be parallel to the tangent line of k at x.

If we attempt this process for a 2-knotting $K \subseteq R^4$ we would similarly need *at least* a smoothly varying unit tangent vector $\overrightarrow{v}_x$ for each $x \in K$. But this is not possible. There is no non-vanishing tangent vector field for S^n if n is even. For S^2, this result frequently goes by the name of "the hairy ball theorem" and elementary proofs can be found in [Eisenberg and Guy (1979)] and[McGrath (2016)]. A close analysis reveals that what we really need to define rolling in this way is that the tangent manifold of S^n is a product—that is S^n is parallelizable.

However, S^n is *only* parallelizable if $n = 1, 3, 7$, Remark 2.4. For $n = 3$ such a roll of 3-knotting K of S^3 produces a 6-knotting of S^6 by taking $S^3 \subseteq S^6$ with product tubular neighborhood $N \approx S^3 \times D^3 \subseteq S^3 \times D^5 \subseteq R^8$ and for each $x \in S^3$ replace each standard $(D^5, D^3)_x$ with (D^5, K_x). For $n = 7$, we similarly roll a 7-sphere in R^9 and produce a 14-sphere in R^{16}.

There *is* a way to generalize rolling S^n for *any* n, the **toroidal-roll of** k written $\mathcal{R}(k)$, which is a certain type of deform-super-spinning, however it is a knotting of $S^n \times S^n$ in R^{2n+2} and not a knotting of a sphere. For 1-knottings, $\mathcal{R}(k)$ is a version of constructions previously encountered, but this is not the case for n-knottings with $n \neq 1, 3, 7$.

We illustrate with a 1-knotting, k, of a circle, where we get 2-knottings of $S^1 \times S^1 \subseteq R^4$. First we translate, if needed, so that $k \subseteq R^3_+$. Write points $x \in k$ as x_θ where $0 \leq \theta \leq 2\pi$. At each $x_\theta \in k$ we invert k in a small sphere with center x_θ we will get an R^1 knotting $\alpha_\theta \subseteq R^3$. We show the result for eight points $x \in k$ where k is a trefoil knotting in Figure 12.12. Each α_θ is asymptotic to the tangent line to k at x_θ. (Note: it would be mathematically better to chose a knotting other than the trefoil. Rolling a trefoil is not typical since it the same as spinning, Remark 10.18. We use the trefoil only because it is a graphically simple knotting.) Now shrink each α_θ, as in Figure 12.13. Next, remove a small sub-arc B_θ of k about x_θ and replace with an slightly bent version of α_θ, A_θ, (bend ends as in Remark 4.2) and we will obtain of knottings, K_θ, Figures 12.14–12.21. (Note: each K_θ is isotopic to $k\# - k$ since inversion reverses orientation.) So K_θ for $0 \leq \theta \leq 2\pi$ represents a deformation of $k\# - k$. The spinning of this deformation, K_θ, of $k\# - k$ gives our 2-knotting of $S^1 \times S^1$, $\mathcal{R}(k)$. So if we let $\mathbb{K}_\theta$ be the natural copy of K_θ in the 3-page H^3_θ, then $\mathcal{R}(k) = \bigcup \mathbb{K}_\theta$.

The 2-knotting $\mathcal{R}(k)$ is not a new construction to us. Recall Example 10.31 where we consider spinning along the two curves of a standard torus $T^2 \subseteq R^4$: C_0, a longitude, and C_1, the diagonal (a $(1,1)$-curve) as in Figure 10.256. We will see that $\mathcal{R}(k)$ is a spinning using C_0 and C_1 as guides where we first do a Artin spin of k along C_0 and then a rolling type deformation along C_1. If $c^\phi_\theta \in \mathbb{K}_\theta$ is the point corresponding to $x_\phi \in k$, we may suppose our meridian is $C_0 = \bigcup_\theta\{c^0_\theta\}$ and we can view the Artin spinning $\Sigma(k)$ as spinning of k along C_0 as in Example 10.10. As explained in Example 10.27 we can then consider $C_1 \subseteq \Sigma(k)$. If we write $\Sigma(k) = \bigcup k_\theta$, then the diagonal is $C_1 = \bigcup_\theta c^\theta_\theta$. We note that $N = \bigcup B^\theta_\theta$ is a product tubular neighborhood of $C_1 \subseteq \Sigma(k)$. If $\mathbb{A}^\phi_\theta$ and $\mathbb{B}^\phi_\theta$ are the subsets of H^3_θ corresponding to A_ϕ and B_ϕ, the we get $\mathbb{K}_\theta$ by removing $\mathbb{B}^\phi_\phi$ and replacing it with $\mathbb{A}^\phi_\phi$.

We can define $\mathcal{R}(k)$ if k is a higher-dimensional knotting. We will just look at the case where k is a 2-knotting. We are going to do a 2-super-spinning so we use generalized spherical coordinates for R^6, $(r, \theta, \psi, w, s, t)$, $0 \leq \theta \leq 2\pi$ and $-\pi/2 \leq \psi \leq \pi/2$. Write points $x \in k$ as $x_{(\theta,\psi)}$ where $0 \leq \theta \leq 2\pi$ and $-\pi/2 \leq \psi \leq \pi/2$. At each $x_{(\theta,\psi)} \in k$ we invert k in a small 3-sphere with center $x_{(\theta,\psi)}$ we will get an R^2 knotting $\alpha_{(\theta,\psi)} \subseteq R^4$, asymptotic to the tangent plane to k at $x_{(\theta,\psi)}$. Next, remove a small sub-disk $B_{(\theta,\psi)}$ of k about $x_{(\theta,\psi)}$ and replace with an slightly bent version of $\alpha_{(\theta,\psi)}$, $A_{(\theta,\psi)}$, and we will obtain knottings, $K_{(\theta,\psi)}$. And now we define: $\mathcal{R}(k) = \bigcup \mathbb{K}_{(\theta,\psi)}$.

This is almost the same as spinning with a guide consisting of C_0 and C_1 as subsets where this time we have $C_0 = S^2 \times \{p\}$ where $p \in S^2$ and C_1 is the diagonal of $S^2 \times S^2$. However, strictly speaking, this is not spinning as we have defined it because the tubular neighborhood of $C_1 \subseteq \Sigma^2(k)$ is not a product. It is equivalent to the tubular neighborhood of the diagonal, Proposition 2.10 which, is not a product, Remark 2.4.

This non-triviality of tangent manifold is not a minor point here— it is exactly why we were unable to roll in a simple way unless $n = 1, 3, 7$. ∎

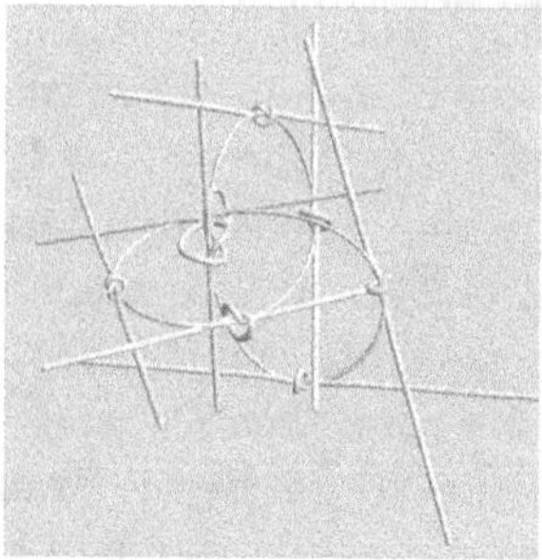

Fig. 12.12 At each $x_\psi \in k$ the local inversion of α_ψ of k is asymptotic to the tangent line L_ψ at x_ψ.

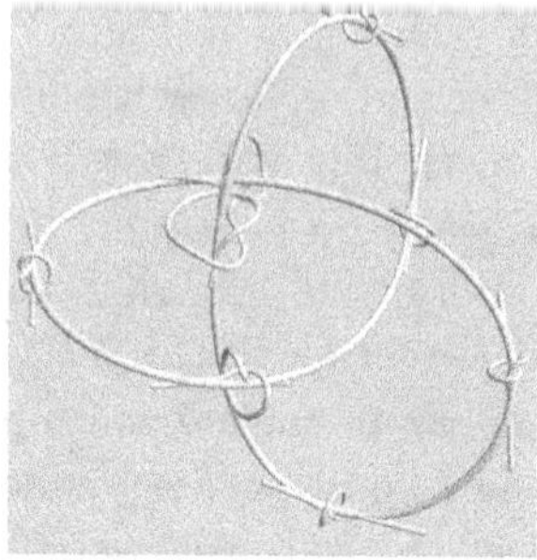

Fig. 12.13 We shorten each α_ψ of Figure 12.12.

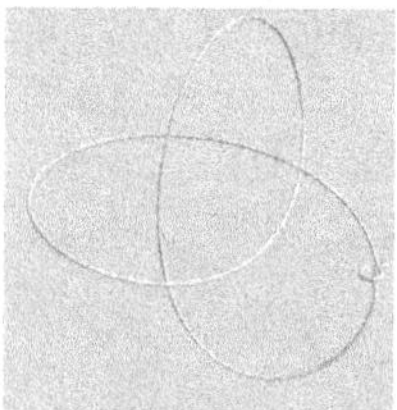

Fig. 12.14 K_0.

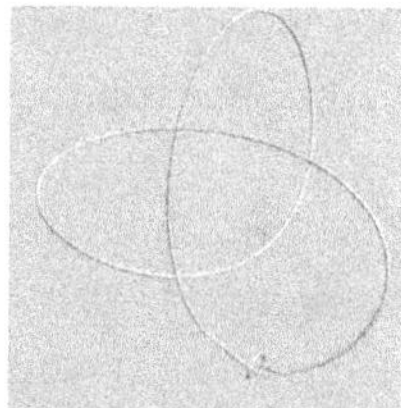

Fig. 12.15 $K_{\pi/4}$.

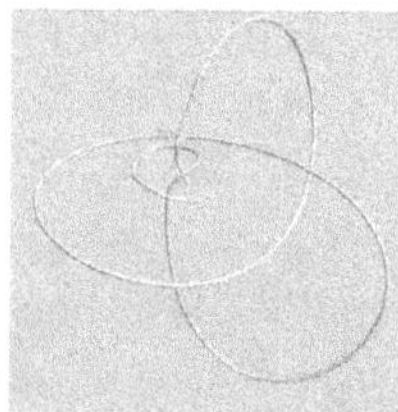

Fig. 12.16 $K_{\pi/2}$.

Fig. 12.17 $K_{3\pi/4}$.

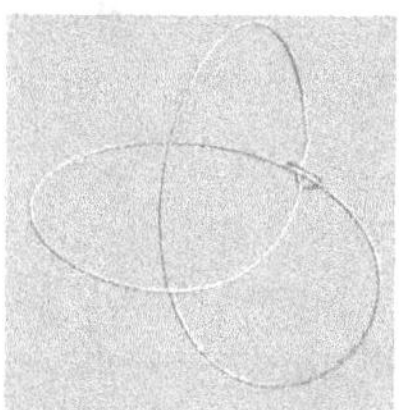

Fig. 12.18 $K_{3\pi/2}$.

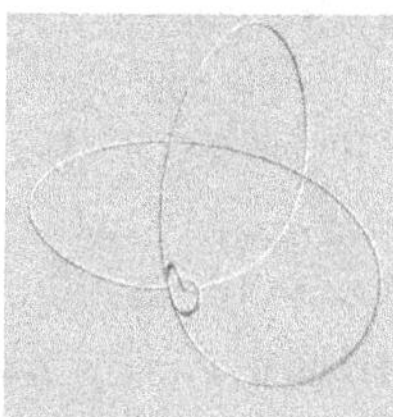

Fig. 12.19 K_6.

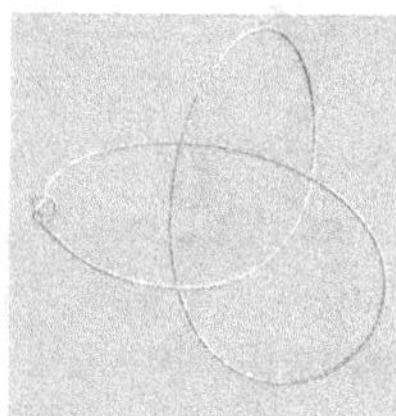

Fig. 12.20 K_7.

Fig. 12.21 K_8.

12.5 Existence Theorems for Knottings and Isotopies

Much of higher dimensional knotting involves mathematical background beyond the level of this book. So we end this book with a brief listing of some key results.

We address two basic questions: (1) For a given k-manifold M^k what is the smallest n such that there exists a knotting M^k into R^n? and (2) what is the smallest n so that any two knottings of M^k will be equivalent?

There are many answers and partial answers to these questions known. In one way or another, they say that "The simpler the manifold, the smaller the minimal n will be" where simpleness is usually quantified using algebraic topology, such as [Penrose *et al.* (1961)]. A good survey is [Skopenkov (2008)].

These powerful theorems, instructive as they are, generally assert existence of knottings and isotopies but lack explicit instructions on how to construct any of them. Four results are:

Prop 12.7. (Easy Embedding Theorem) Any smooth closed k-dimensional manifold M^k can be smoothly embedded in R^{2k+1}.

(Hard Embedding Theorem) Any smooth closed k-dimensional manifold M^k can be smoothly embedded in R^{2k}. [Whitney (1944)] ∎

Prop 12.8. (Easy Knotting Theorem) Any two smooth knottings of a smooth closed k-dimensional manifold M^k in R^{2k+2} are equivalent.

(Hard knotting Theorem) Any two smooth knottings of a smooth closed *connected* k-dimensional manifold M^k in R^{2k+1} are equivalent, [Wu (1958)]a. ∎

See standard texts such as [Hirsch (1976)] for proofs of these theorems. The proofs of the two easy theorems use the general position.

For the Easy Embedding Theorem here is a rough idea of the role of the expression "$2k+1$". By our definition, a smooth manifold $M^k \subseteq R^N$ for some N. We assume $2k+1 < N$, otherwise, we already have our result. Using a general position argument we may suppose the standard projection into R^{2k+1} is an immersion $M^k \looparrowright R^{2k+1}$. We can make this into a knotting by removing possible points of self-intersection. Suppose x is a point of self intersection, then it is the intersection of the images of (at least) two disjoint k-disks of M^k. Since the immersion is smooth we can approximate these two disks intersecting x by two hyperplanes intersecting at x.

In general, two k-dimensional hyperplanes in R^{2k+1}, if they happen to intersect, will have an empty intersection if we move them slightly. This can be explained using linear algebra. For example, two lines in the plane will intersect, with the lone rare exception of parallel lines. Two lines L_1 and L_2 in R^3 rarely intersect and if they do will have empty intersection if we move them slightly. A line in space is an intersection of two planes—algebraically a solution set of a pair of linear equations. Suppose for L_1 these equations are $a_1x + b_1y + c_1z = d_1$ and $a_2x + b_2y + c_2z = d_2$ with equations are

$a_3x + b_3y + c_3z = d_3$ and $a_4x + b_2y + c_4z = d_4$ for L_2. This set of four linear equations in three variables is an over-determined set of equations. Our two lines will fail to intersect if and only if this set of equations is inconsistent. This implies that there is unlikely to be a solution, but if there is, almost any nearby choice of coefficients will result in a set of equations with no solution—geometrically to an empty intersection.

Similarly, in higher-dimensions, a single n-plane L_1 in R^{2n+1} uses $(n+1)$ equations in $(2n+1)$ variables, as does a second one L_2. This gives us an over-determined set of $2(n+1) = 2n+2$ equations in $(2n+1)$ variables.

By the Hard Embedding Theorem we can embed any 3-manifold into R^6 but we can do better than this [Wall (1965a)] [Hirsch (1961)] [Rohlin (1965)]:

Prop 12.9. Every closed 3-manifold can be embedded into R^5. ∎

So, every 3-manifold generates a world of 3-knottings. Some interesting examples of 3-knottings involve manifolds called 3-dimensional knotted homology spheres, see [Saeki (1988b)], [Saeki (1988a)], [Scharlemann (1977a)].

Codimension one embeddings are special in high-dimensions but also rare. In a sense one can say that most 3-manifolds—even simple ones—fail to embed in S^4, [Budney and Burton (2022)].

There are some simple closed orientable manifolds with non-empty boundary that cannot be put into R^4, for example certain punctured lens spaces [Hantzsche (1938)],[Gilmer and Livingston (1983)], [Gilmer and Livingston (1983)],[Edmonds and Livingston (1996)] [Epstein (1965)]. The topic of which punctured orientable 3-manifolds can embed in R^4 is the same as the question of which punctured 3-manifolds can be Seifert surfaces for some knotting of a 2-sphere. This often involves details of the topology of 3-manifolds as in [Ruberman (1990)], [Yanagawa (1969)].

Although many 3-manifolds do not *embed* in R^4, all closed orientable manifolds do *immerse* in R^4 by what is known as the Hirsch-Smale theorem. For a historical overview, see [Spring (2005)].

In the study of manifolds the role of codimension one submanifolds is central—one studies a general manifold by "cutting it into small pieces" where the "cuts" are in fact codimension one submanifolds. We have seen the utility of looking at embedded circles in a surface; for a 3-dimensional manifold one similarly looks at submanifolds which are 2-spheres and 2-tori. This is such a big topic that it is generally considered separate from knot theory. We only address a few remarks to codimension one knottings in R^n that impact our knotting considerations.

Remark 12.14. Any codimension one submanifold of $M^n \subseteq W^{n+1}$ (or submanifold of any orientable $(n+1)$-manifold) must be orientable and have a tubular neighborhood which is a product $M^n \times D^1$. So we can combine this knotting with a knotting of an arc and produce other knottings using the generalized process of spinning a knot about a submanifold. ∎

Remark 12.15. By the Schoenflies Theorem (Propositions 5.1, 5.5)a knotting of an n-sphere in R^{n+1} bounds an $(n+1)$-ball and is therefore a trivial knotting. ∎

Remark 12.16. From any codimensional knotting of a manifold in $K^k \subseteq R^n$ the boundary of a tubular neighborhood is a codimension one knotting in R^n . For1-knottings a tube about a knotting of a circle $K \subseteq R^3$ gives a knotting of a torus in R^3. Similarly for any 2-knottings $K^2 \subseteq S^4$ from Part 2 we obtain a knotting of $K^2 \times S^1 \subseteq S^4$.

When we apply this idea to projective plane in R^4, the tubular neighborhood if a knotting of the projective plane is not a product. The boundary of the tubular neighborhood is homeomorphic to a 3-manifold, **quaternion space**, as studied in [Price (1977)] and [Katanaga and Saeki (1998)]. ∎

Knottings of n-spheres are special and historically were the focus of much attention in high-dimensions. Up until this point we have assumed our knottings are smooth simply as a way of avoiding wild knottings. However even for knottings of S^n, it will make a difference if $3 \leq n$ whether we consider smooth knottings, PL knottings or locally flat topological knottings.

Remark 12.17. We have discussed knotting of circles in R^n knottings of 2-spheres in R^n. The next case is knotting of 2-spheres in R^5. In the PL and smooth cases, there are no non-trivial knottings—in other words 2-spheres always unknot in five dimensions. We present a brief proof of this, adapted from [Zeeman (1960b)]. Our proof is topological. A smooth version obtainable by rounding corners.

As a warmup we consider a proof that a 1-knotting of a circle $k \subseteq R^3$ with one overpass (in the standard projection) A, must be unknotted. Write k as union of closed arcs $k = A \cup B$ with $A \cap B = C$, where C is a two-point subset (C is a 0-sphere), Figures 12.22 and 12.23. Let $v \in R^3$ be a point above all points of k (in last coordinate of R^3). The cone $D = v * A$ is a topological 2-disk which contains no points of k in its interior. Let $k' = (v * C) \cup B$. Using D, we can isotop A to $v * C$ keeping C fixed, so

k is isotopic to k', Figure 12.22. But k' is unknotted since it bounds the embedded 2-disk $v * B$, Figure 12.24.

Now suppose $K \subseteq R^5$ is a 2-knotting of S^2. We suppose it is in general position with respect to standard projection to R^4. The crossing set will be 0-dimensional—a finite set of isolated crossing points x_i with corresponding pre-crossing points $\widetilde{x}_i = \{a_i, b_i\}$ where $a_i \in K$ is above $b_i \in K$. We can find a 2-disk $A \subseteq K$ which contains all a_i. (We can say that K is a knotting with one overpass A.) Let $C = \partial A$ and let $B = \overline{K - A}$. So C is a 1-sphere and B is a 2-disk, $K = A \cup B$ and $A \cap B = C$.

Let $v \in R^5$ be a point above all points of K (in last coordinate of R^5). The cone $D = v * A$ is a topological 3-disk which contains no points of K in its interior. Let $K' = (v*C) \cup B$. Using D, we can isotop A to $v*C$ keeping C fixed, so K is isotopic to K'. But K' is unknotted since it bounds the embedded 3-disk $v * B$. ∎

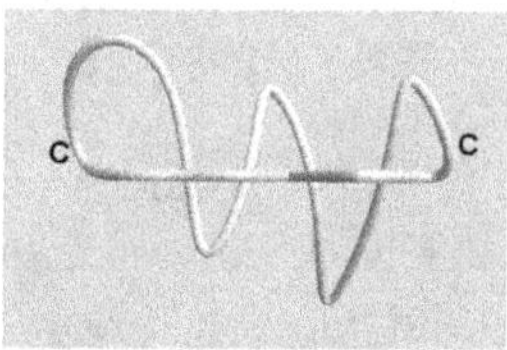

Fig. 12.22 1-knotting projection with one overpass.

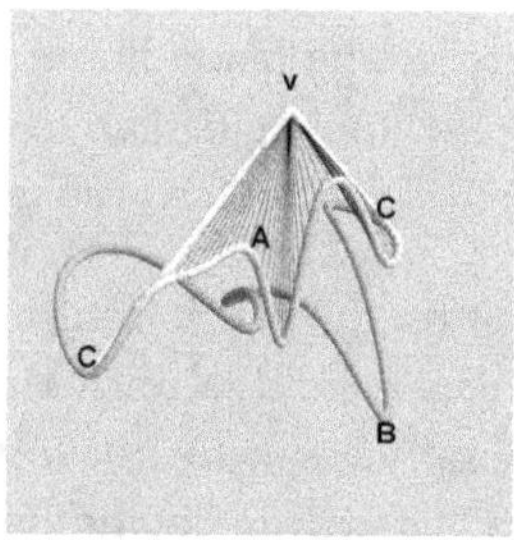

Fig. 12.23 Rotated view with cone on A.

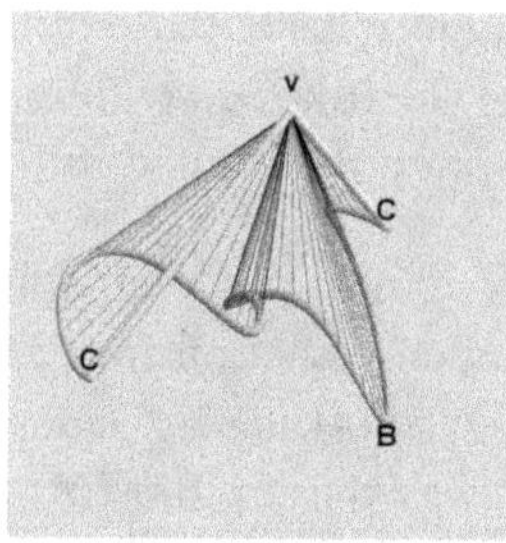

Fig. 12.24 Cone on B.

A fundamental result in higher-dimensional knotting is Zeeman's PL Unknotting Theorem [Zeeman (1960b)]:

Prop 12.10. Any PL knotting of S^k in R^q in codimension three or greater is equivalent to the standard knotting. ∎

Remark 12.18. However the smooth version of Proposition 12.10 is not true. There are smooth knottings of 3-spheres in S^6 which are not equivalent to the standard knotting [Haefliger (1962)]. Another view of such smooth codimension 3 knotted spheres is [Budney (2008)]. Yet another approach uses a version of spinning about a submanifold which can produce knottings that are not codimension 2, [Roseman and Takase (2007)]. In [Stallings (1963)] is a locally flat version of Proposition 12.10. ∎

Here is a small collection of results about non-spherical manifold knottings with codimension more than two.

In [Hudson (1963a)] it is shown that:

Prop 12.11. For piecewise-linear knottings of $S^p \times S^q \subseteq R^{p+2q+1}$ where $p \leq q$ the number of knot-types is:

(1) 2 if $0 < p < q$ and q is odd.
(2) infinite if $0 < p < q$ and q is even.
(3) greater than or equal to 2 if $1 < p = q$ with q odd.
(4) infinite if $1 < p = q$ with q even. ∎

Remark 12.19. In [Zeeman (1962a)] Zeeman shows $S^p \times S^q$ is unknotted in R^n if $p + 2q + 1 \leq n$ ∎

Hudson has theorem that unknotting for torii in high-dimensions is same as bounding solid torus [Hudson (1963b)]. In [Kosiński (1962)] there are several theorems on knottings of torii later improved on in [Wall (1965b)]. Hudson and Zeeman's paper[Hudson and Zeeman (1964)] has good information on isotopies and collar neighborhoods. Homology 3-spheres in R^6 have been explored in [Takase (2006)] and papers leading up to this.

Remark 12.20. In higher-dimensions for knottings of M^n with $3 \leq n$, in R^{n+1} we need to first ask if there is *any* embedding at all? This is a significant problem. Some good summaries are found in [] This is a complex problem with much unknown, so attention is often focused on products of spheres. We know that $S^p \times S^q$ embeds in R^{p+q+1} as the equator of the join $S^p * S^q = S^{p+q+1}$. We see $S^p \times S^q \times S^r$ embeds in $R^{p+q+r+1}$ as the boundary of the tubular neighborhood of $S^p \times S^q \subseteq R^{p+q+1} \subseteq R^{p+q+r+1}$, etc.

In [Kosiński (1962)] it is shown smooth knottings of $S^p \times S^q \subseteq W^{p+q+1}$ is unknotted if $1 < p, 1 < q, 5 < p + q$ and if $q = 2$, then p is odd . Later in [Goldstein (1967)] this was improved to $1 < p \leq q$ and $4 < p + q$.

For the product of three spheres, the situation is much more complicated and there are cases when neither closed complement is simple, see [Lucas and Saeki (2005, 2011)].

We can get an non-trivial knotting of $S^1 \times S^1 \subseteq R^3$ by taking a non-trivial knotting k of S^1. This k will have a tubular neighborhood N which is a knotting of $S^1 \times D^2$ in R^3 and subset of R^3 corresponding to $S^1 \times S^1 \subseteq S^1 \times D^2$ will be a non-trivial knotting. Similarly we can get non-trivial knottings of $S^n \times S^1$ in R^{n+2} by taking a non-trivial knotting k of S^n in

R^{n+2} and looking at the boundary of a product tubular neighborhood of the knotting k.

For higher-dimensional knotted homology spheres [Scharlemann (1977b)]. ▮

A good reference for high-codimensional knottings of products of spheres is [Skopenkov (2020)]. We will mostly focus in codimension two knottings but one important simple non-trivial example that has higher codimension (in fact codimension $n+1$) is:

Definition 12.13. The n**-dimensional Hopf knotting** in S^{2n+1} consisting of to n-spheres consisting of the first and second factors of $S^{2n+1} \approx S^n * S^n$. ▮

For example, this gives us a codimension 3 non-split knotting of two disjoint 2-spheres in R^5.

Existence of wildness of circles in R^3, Figure 3.2 and 2-spheres Section 5.5 and even wild disks [Fox and Artin (1948)] motivate us to restrict our attention to "nicer" knottings such as locally flat, PL or smooth—this continues to be the case for higher-dimensional knottings.

Smooth and PL knottings of 1-manifolds in R^4 are all trivial, Section 5.1. However in the topological case, a circles and arcs can be wildly knotted in R^4 [Wright (1977)] , [McMillan (1978)] and [Brown (1967)].

Definition 12.14. If $M \subseteq R^n$ and $x \in M$ we say M **is locally flat at** x if there is an open subset $U \subseteq R$ with $x \in U$ and a homeomorphism $h\colon U \to R^n$ such that $h(M \cap U) = R^k \subseteq R^n$. If M is not locally flat at x we say M **is wild at** x. ▮

One has a rough measure of how wild a wild knotting is by considering the subset of wild (that is non-locally flat) points The wild examples in R^3 of Fox [Fox and Artin (1948)] are wild a one or several points . The Alexander Horned Sphere and wild sphere based on Antoine's necklace have a wild sets which are Cantor sets. But there are examples more extreme.

Definition 12.15. If $M \subseteq R^n$, M is everywhere wild if it is wild at every point $x \in M$. ▮

There are knottings that are wild knottings and in every codimension. In fact, see [Rushing (1973)]:

Prop 12.12. In R^n and $0 < k < n$ there for any are everywhere wild knottings of D^k and everywhere wild knottings of S^k. ▮.

There is a variety of ways one can construct wild knottings—but all by their nature are infinite constructions of some sort. One can use the higher-dimensional versions of spinning and ribbon constructions found later in this book. For example—even for R^4 one can spin a wild arc and produce a wild 2-sphere. Another fascinating construction is of wild 2-spheres is [Hinojosa (2007)] generalized to higher-dimensions in [Boege *et al.* (2009)].

A generalisation of the Antoine wild sphere is found in [Blankinship (1951)] and [Broussard (1977)].

There are higher dimensional (non-smooth) manifolds which are which have a high-dimensional local knotting similar to a "cone on a non-trivial knotting" such as in the suspension of a knotting, [Cantrell and Edwards (1965)], [Tindell (1966)]. and [Friedman (2003)].

Bibliography

Abrams, A., Cantarella, J., Fu, J. H. G., Ghomi, M., and Howard, R. (2003). Circles minimize most knot energies, *Topology* **42**, 2, pp. 381–394.

Adams, C. (2005). Hyperbolic knots, in *Handbook of knot theory* (Elsevier B. V., Amsterdam), pp. 1–18.

Adams, C., Chu, M., Crawford, T., Jensen, S., Siegel, K., and Zhang, L. (2012). Stick index of knots and links in the cubic lattice, *J. Knot Theory Ramifications* **21**, 5, pp. 1250041, 16.

Adams, C. C. (1994). *The Knot Book* (W. H. Freeman and Company, New York).

Akbulut, S. and Kirby, R. (1985). A potential smooth counterexample in dimension 4, *Topology* **24**, 4, pp. 375–390.

Al Kharusi, A. and Yashiro, T. (2016). On crossing changes for surface-knots. *Kyungpook Math. J.* **56**, 4, pp. 1247–1257.

Alexander, J. W. (1923). A lemma on systems of knotted curves, *Proceedings of the National Academy of Sciences of the United States of America* **9**, 3, pp. 93–95.

Alexander, J. W. (1924a). An example of a simple connected surface bounding a region which is not simply connected. *Proc. Natl. Acad. Sci. USA* **10**, pp. 8–10.

Alexander, J. W. (1924b). On the subdivision of 3-space by a polyhedron. Nat. Acad. Proc. 10, 6-8 (1924).

Alexander, J. W. and Briggs, G. B. (1926/27). On types of knotted curves, *Ann. of Math. (2)* **28**, 1-4, pp. 562–586.

Artin, E. (1925). Zur isotopie zweidimensionalen flachen im $\mathbb{R}^4$, *Abh. Math. Sem. Univ. Hamburg*, pp. 174–177.

Asaeda, M. M., Przytycki, J. H., and Sikora, A. S. (2004). Categorification of the Kauffman bracket skein module of I-bundles over surfaces, *Algebr. Geom. Topol.* **4**, pp. 1177–1210.

Asami, S. and Satoh, S. (2005). An infinite family of non-invertible surfaces in 4-space, *Bull. London Math. Soc.* **37**, 2, pp. 285–296.

Ashihara, S. (2012). Calculating the fundamental biquandles of surface links from their ch-diagrams, *J. Knot Theory Ramifications* **21**, 10, pp. 1250102, 23.

Atapour, M., Soteros, C. E., Sumners, D. W., and Whittington, S. G. (2015). Counting closed 2-manifolds in tubes in hypercubic lattices, *J. Phys. A* **48**, 16, pp. 165002, 19.

Auckly, D. and Sadun, L. (1997). A family of Möbius invariant 2-knot energies, in *Geometric topology (Athens, GA, 1993)*, *AMS/IP Stud. Adv. Math.*, Vol. 2 (Amer. Math. Soc., Providence, RI), pp. 235–258.

Audoux, B., Meilhan, J.-B., and Wagner, E. (2017). On codimension two embeddings up to link-homotopy, *J. Topol.* **10**, 4, pp. 1107–1123.

Baldridge, S. and McCarty, B. (2010). Small examples of cube diagrams of knots, *Topology Proc.* **36**, pp. 213–228.

Banchoff, T. F. (1976). Minimal submanifolds of the bicylinder boundary, *Bol. Soc. Brasil. Mat.* **7**, 1, pp. 37–57.

Bayer, E. (1980). Factorisation is not unique for higher-dimensional knots, *Comment. Math. Helv.* **55**, 4, pp. 583–592.

Bing, R. H. (1964). Some aspects of the topology of 3-manifolds related to the Poincaré conjecture, in *Lectures on modern mathematics, Vol. II* (Wiley, New York), pp. 93–128.

Bing, R. H. (1983). *The geometric topology of 3-manifolds*, *American Mathematical Society Colloquium Publications*, Vol. 40 (American Mathematical Society, Providence, RI).

Birman, J. S. (1974). *Braids, links, and mapping class groups* (Princeton University Press, Princeton, N.J.; University of Tokyo Press, Tokyo), annals of Mathematics Studies, No. 82.

Birman, J. S. and Brendle, T. E. (2005). Braids: a survey, in *Handbook of knot theory* (Elsevier B. V., Amsterdam), pp. 19–103.

Bishop, R. L. (1975). There is more than one way to frame a curve, *Amer. Math. Monthly* **82**, pp. 246–251.

Blankinship, W. A. (1951). Generalization of a construction of Antoine, *Ann. of Math. (2)* **53**, pp. 276–297.

Bleiler, S. and Scharlemann, M. (1988). A projective plane in $\mathbf{R}^4$ with three critical points is standard. Strongly invertible knots have property P, *Topology* **27**, 4, pp. 519–540.

Boege, M., Hinojosa, G., and Verjovsky, A. (2009). Wild knots in higher dimensions as limit sets of Kleinian groups, *Conform. Geom. Dyn.* **13**, pp. 197–216.

Boege, M., Hinojosa, G., and Verjovsky, A. (2011). Any smooth knot $S^n \hookrightarrow R^{n+2}$ is isotopic to a cubic knot contained in the canonical scaffolding of R^{n+2}, *Rev. Mat. Complut.* **24**, 1, pp. 1–13.

Borodzik, M., Dangskul, S., and Ranicki, A. (2020). Solid angles and Seifert hypersurfaces, *Ann. Global Anal. Geom.* **57**, 3, pp. 415–454.

Boyle, J. (1988). Classifying 1-handles attached to knotted surfaces, *Transactions AMS* **306**, 2, pp. 475–487.

Boyle, J. (1993). The turned torus knot in S^4, *J. Knot Theory Ramifications* **2**, 3, pp. 239–249.

Broussard, L. M. (1977). Sums of solid n-spheres, *Trans. Amer. Math. Soc.* **225**, pp. 279–294.

Brown, M. (1967). Wild cells and spheres in higher dimensions, *Michigan Math. J.* **14**, pp. 219–224.

Browne, C. (2006). Wild knots, *Computers & Graphics* **30**, 6, pp. 1027–1032.

Bryant, J. L. (2002). Piecewise linear topology, in *Handbook of geometric topology* (North-Holland, Amsterdam), pp. 219–259.

Budney, R. (2008). A family of embedding spaces, in *Groups, homotopy and configuration spaces, Geom. Topol. Monogr.*, Vol. 13 (Geom. Topol. Publ., Coventry), pp. 41–83.

Budney, R. and Burton, B. A. (2022). Embeddings of 3-manifolds in S^4 from the point of view of the 11-tetrahedron census, *Exp. Math.* **31**, 3, pp. 988–1013.

Budney, R. and Mozgova, A. (2009). An obstruction to a knot being deform-spun via Alexander polynomials, *Proc. Amer. Math. Soc.* **137**, 10, pp. 3547–3552.

Burde, G., Zieschang, H., and Heusener, M. (2014). *Knots, De Gruyter Studies in Mathematics*, Vol. 5, extended edn. (De Gruyter, Berlin).

Burgess, C. E. and Cannon, J. W. (1971). Embeddings of surfaces in E^3, *Rocky Mountain J. Math.* **1**, 2, pp. 259–344.

Cantrell, J. C. and Edwards, C. H., Jr. (1965). Almost locally flat imbeddings of manifolds, *Michigan Math. J.* **12**, pp. 217–223.

Cappell, S. (1970). Superspinning and knot complements, in *Topology of Manifolds (Proc. Inst., Univ. of Georgia, Athens, Ga., 1969)* (Markham, Chicago, Ill.), pp. 358–383.

Cappell, S. E. and Shaneson, J. L. (1976). Some new four-manifolds, *Ann. of Math. (2)* **104**, 1, pp. 61–72.

Carmi, A. Y. and Moskovich, D. (2017). Inca foams, *J. Knot Theory Ramifications* **26**, 1, pp. 1750005, 30.

Carter, J. S. (1995). *How surfaces intersect in space, Series on Knots and Everything*, Vol. 2, 2nd edn. (World Scientific Publishing Co., Inc., River Edge, NJ).

Carter, J. S. (2015). Reidemeister/Roseman-type moves to embedded foams in 4-dimensional space, in *New ideas in low dimensional topology, Ser. Knots Everything*, Vol. 56 (World Sci. Publ., Hackensack, NJ), pp. 1–30.

Carter, J. S., Jelsovsky, D., Kamada, S., Langford, L., and Saito, M. (2003). Quandle cohomology and state-sum invariants of knotted curves and surfaces, *Trans. Amer. Math. Soc.* **355**, 10, pp. 3947–3989.

Carter, J. S., Kamada, S., and Saito, M. (2002). Stable equivalence of knots on surfaces and virtual knot cobordisms, pp. 311–322, knots 2000 Korea, Vol. 1 (Yongpyong).

Carter, J. S., Oshiro, K., and Saito, M. (2010). Symmetric extensions of dihedral quandles and triple points of non-orientable surfaces, *Topology Appl.* **157**, 5, pp. 857–869.

Carter, J. S. and Saito, M. (1992). Canceling branch points on projections of surfaces in 4-space, *Proc. Amer. Math. Soc.* **116**, 1, pp. 229–237.

Carter, J. S. and Saito, M. (1993). Reidemeister moves for surface isotopies and their interpretation as moves to movies, *J. Knot Theory Ramifications* **2**, 3, pp. 251–284.

Carter, S., Kamada, S., and Saito, M. (2004). *Surfaces in 4-space, Encyclopaedia of Mathematical Sciences*, Vol. 142 (Springer-Verlag, Berlin), low-Dimensional Topology, III.

Casson, A. J. and Bleiler, S. A. (1988). *Automorphisms of surfaces after Nielsen and Thurston, London Mathematical Society Student Texts*, Vol. 9 (Cambridge University Press, Cambridge), ISBN 0-521-34203-1.

Cattabriga, A., Manfredi, E., and Mulazzani, M. (2013). On knots and links in lens spaces, *Topology Appl.* **160**, 2, pp. 430–442.

Champanerkar, A., Kumar, A., and Kumaresan, S. (2000). Classification of surfaces via Morse theory, *Expo. Math.* **18**, 1, pp. 37–73.

Clark, W. E., Elhamdadi, M., Saito, M., and Yeatman, T. (2014). Quandle colorings of knots and applications, *J. Knot Theory Ramifications* **23**, 6, pp. 1450035, 29.

Cochran, T. (1983). Ribbon knots in S^4, *J. London Math. Soc. (2)* **28**, 3, pp. 563–576.

Conway, J. H. and Gordon, C. M. (1983). Knots and links in spatial graphs, *J. Graph Theory* **7**, 4, pp. 445–453.

Cotta-Ramusino, P. and Rinaldi, M. (1991). On the algebraic structure of link-diagrams on a 2-dimensional surface, *Comm. Math. Phys.* **138**, 1, pp. 137–173.

Crowell, R. H. and Fox, R. H. (1977). *Introduction to knot theory*, Graduate Texts in Mathematics, No. 57 (Springer-Verlag, New York-Heidelberg).

Dalvit, E., (2016) *Spinning in various dimensions - MathLapse*, (Video) (June 10 2016) YouTube, https://youtu.be/-TuXJswaFAE.

Dalvit, E., (2017) [Knots in 4d], *Knots in 4d Introduction, and Parts 1,2,3* (Video) (Jan. 2, 2017) YouTube, https://youtu.be/Lx85kIGjIoQ?list=PLyxHTRWELFBr9TP8jx400bPGP-ECsBufz.

Dangskul, S. (2015). *A construction of Seifert surfaces via differential geometry*, Ph.D., University of Edinburgh.

Dehn, M. (1914). Die beiden Kleeblattschlingen, *Math. Ann.* **75**, 3, pp. 402–413.

Díaz, J. P., Hinojosa, G., Valdez, R., and Verjovsky, A. (2018). Smoothing closed gridded surfaces embedded in R^4, *J. Knot Theory Ramifications* **27**, 12, pp. 1850065, 14.

Díaz, J. P., Hinojosa, G., and Verjosvky, A. (2019). Cubulated moves for 2-knots, *J. Knot Theory Ramifications* **28**, 1, pp. 1950008, 21.

Dionísio, F. M. and Lopes, P. (2003). Quandles at finite temperatures. II, *J. Knot Theory Ramifications* **12**, 8, pp. 1041–1092.

Dowker, C. H. and Thistlethwaite, M. B. (1983). Classification of knot projections, *Topology Appl.* **16**, 1, pp. 19–31.

Dye, H. A. (2016). *An invitation to knot theory* (CRC Press, Boca Raton, FL).

Dye, H. A., Kaestner, A., and Kauffman, L. H. (2017). Khovanov homology, Lee homology and a Rasmussen invariant for virtual knots, *J. Knot Theory Ramifications* **26**, 3, pp. 1741001, 57.

Edmonds, A. L. and Livingston, C. (1996). Embedding punctured lens spaces in four-manifolds, *Comment. Math. Helv.* **71**, 2, pp. 169–191.

Eisenberg, M. and Guy, R. (1979). A proof of the hairy ball theorem, *Amer. Math. Monthly* **86**, 7, pp. 572–574.

Elhamdadi, M. and Nelson, S. (2015). *Quandles—an introduction to the algebra of knots*, *Student Mathematical Library*, Vol. 74 (American Mathematical Society, Providence, RI).

Epstein, D. B. A. (1960). Linking spheres, *Proc. Cambridge Philos. Soc.* **56**, pp. 215–219.

Epstein, D. B. A. (1965). Embedding punctured manifolds, *Proc. Amer. Math. Soc.* **16**, pp. 175–176.

Farb, B. and Margalit, D. (2012). *A primer on mapping class groups*, *Princeton Mathematical Series*, Vol. 49 (Princeton University Press, Princeton, NJ), ISBN 978-0-691-14794-9.

Farber, M. v. (1975). Linking coefficients and two-dimensional knots, *Dokl. Akad. Nauk SSSR* **222**, 2, pp. 299–301.

Fedoseev, D. A. and Manturov, V. O. (2016). Parities on 2-knots and 2-links, *J. Knot Theory Ramifications* **25**, 14, pp. 1650079, 24.

Feller, P. and McCoy, D. (2016). On 2-bridge knots with differing smooth and topological slice genera, *Proc. Amer. Math. Soc.* **144**, 12, pp. 5435–5442.

Fenn, R. (2012). Tackling the trefoils, *J. Knot Theory Ramifications* **21**, 13, pp. 1240004, 20.

Fenn, R., Ilyutko, D. P., Kauffman, L. H., and Manturov, V. O. (2014). Unsolved problems in virtual knot theory and combinatorial knot theory, in *Knots in Poland III. Part III*, *Banach Center Publ.*, Vol. 103 (Polish Acad. Sci. Inst. Math., Warsaw), pp. 9–61.

Fenn, R. and Rourke, C. (1992). Racks and links in codimension two, *J. Knot Theory Ramifications* **1**, 4, pp. 343–406.

Flapan, E., Mattman, T. W., Mellor, B., Naimi, R., and Nikkuni, R. (2017). Recent developments in spatial graph theory, in *Knots, links, spatial graphs, and algebraic invariants*, *Contemp. Math.*, Vol. 689 (Amer. Math. Soc., Providence, RI), pp. 81–102.

Florens, V. and Gilmer, P. M. (2003). On the slice genus of links, *Algebr. Geom. Topol.* **3**, pp. 905–920.

Foisy, J. S. and Ludwig, L. D. (2009). When graph theory meets knot theory, in *Communicating mathematics*, *Contemp. Math.*, Vol. 479 (Amer. Math. Soc., Providence, RI), pp. 67–85.

Fox, R. H. (1957). Covering spaces with singularities, in *A symposium in honor of S. Lefschetz* (Princeton University Press, Princeton, N.J.), pp. 243–257.

Fox, R. H. (1962a). A quick trip through knot theory, in *Topology of 3-manifolds and related topics (Proc. The Univ. of Georgia Institute, 1961)* (Prentice-Hall, Englewood Cliffs, N.J.), pp. 120–167.

Fox, R. H. (1962b). Some problems in knot theory, in *Topology of 3-manifolds and related topics (Proc. The Univ. of Georgia Institute, 1961)* (Prentice-Hall, Englewood Cliffs, N.J.), pp. 168–176.

Fox, R. H. (1966). Rolling, *Bull. Amer. Math. Soc.* **72**, pp. 162–164.

Fox, R. H. and Artin, E. (1948). Some wild cells and spheres in three-dimensional space, *Ann. of Math. (2)* **49**, pp. 979–990.

Fox, R. H. and Milnor, J. W. (1966). Singularities of 2-spheres in 4-space and cobordism of knots, *Osaka J. Math.* **3**, pp. 257–267.

Francis, G. K. (2007). *A topological picturebook* (Springer, New York).

Francis, G. K. and Weeks, J. R. (1999). Conway's ZIP proof, *Amer. Math. Monthly* **106**, 5, pp. 393–399.

Francis, G. K. *et al.* (1995). Laterna mathemagica, in *Virtual Environments and Distributed Computing at Super Computing '95* (ACM/IEEE), p. 45.

Freedman, M. H. (1982). The topology of four-dimensional manifolds, *J. Differential Geometry* **17**, 3, pp. 357–453.

Freedman, M. H., He, Z.-X., and Wang, Z. (1994). Möbius energy of knots and unknots, *Ann. of Math. (2)* **139**, 1, pp. 1–50.

Freedman, M. H. and Quinn, F. (1990). *Topology of 4-manifolds, Princeton Mathematical Series*, Vol. 39 (Princeton University Press, Princeton, NJ).

Friedl, S. and Orson, P. (2015). Twist spinning of knots and metabolizers of Blanchfield pairings, *Ann. Fac. Sci. Toulouse Math. (6)* **24**, 5, pp. 1203–1218.

Friedman, G. (2003). Alexander polynomials of non-locally-flat knots, *Indiana Univ. Math. J.* **52**, 6, pp. 1479–1578.

Friedman, G. (2004). All frame-spun knots are slice, *Proc. Amer. Math. Soc.* **132**, 10, pp. 3103–3109.

Friedman, G. (2005). Knot spinning, in *Handbook of knot theory* (Elsevier B. V., Amsterdam), pp. 187–208.

Fukuhara, S. (1988). Energy of a knot, in *A fête of topology* (Academic Press, Boston, MA), pp. 443–451.

Gallier, J. and Xu, D. (2013). *A guide to the classification theorem for compact surfaces, Geometry and Computing*, Vol. 9 (Springer, Heidelberg).

Garoufalidis, S. and Teichner, P. (2004). On knots with trivial Alexander polynomial, *J. Differential Geom.* **67**, 1, pp. 167–193.

Gauss, C. (1863–1933). Werke, 6 volumes, *Köigliche Gesellschaft der Wissenshaften, Göttingen* **6**, pp. 106–118.

Giller, C. A. (1982). Towards a classical knot theory for surfaces in $\mathbf{R}^4$, *Illinois J. Math.* **26**, 4, pp. 591–631.

Gilmer, P. M. (1982). On the slice genus of knots, *Invent. Math.* **66**, 2, pp. 191–197.

Gilmer, P. M. and Livingston, C. (1983). On embedding 3-manifolds in 4-space, *Topology* **22**, 3, pp. 241–252.

Gluck, H. (1962). The embedding of two-spheres in the four-sphere, *Trans. Amer. Math. Soc.* **104**, pp. 308–333.

Gluck, H. (1963). Unknotting s^1 in s^4, *Bull. Amer. Math. Soc.* **69**, pp. 91–94.

Goldsmith, D. L. and Kauffman, L. H. (1978). Twist spinning revisited, *Trans. Amer. Math. Soc.* **239**, pp. 229–251.

Goldstein, R. Z. (1967). Piecewise linear unknotting of $S^p \times S^q$ in S^{p+q+1}, *Michigan Math. J.* **14**, pp. 405–415.

Gompf, R. E. and Stipsicz, A. I. (1999). *4-manifolds and Kirby calculus, Graduate Studies in Mathematics*, Vol. 20 (American Mathematical Society, Providence, RI).

González-Acuña, F. and Montesinos, J. M. (1980). Nonamphicheiral codimension 2 knots, *Canadian J. Math.* **32**, 1, pp. 185–194.

Gordon, C. M. (2003). On the reversibility of twist-spun knots, *J. Knot Theory Ramifications* **12**, 7, pp. 893–897.

Gordon, C. M. and Luecke, J. (1989). Knots are determined by their complements, *J. Amer. Math. Soc.* **2**, 2, pp. 371–415.

Gray, A., Abbena, E., and Salamon, S. (2006). *Modern differential geometry of curves and surfaces with Mathematica®*, 3rd edn., Studies in Advanced Mathematics (Chapman & Hall/CRC, Boca Raton, FL).

Greene, J. and Jabuka, S. (2011). The slice-ribbon conjecture for 3-stranded pretzel knots, *Amer. J. Math.* **133**, 3, pp. 555–580.

Guillemin, V. and Pollack, A. (2010). *Differential topology* (AMS Chelsea Publishing, Providence, RI), reprint of the 1974 original.

Haefliger, A. (1962). Knotted $(4k-1)$-spheres in $6k$-space, *Ann. of Math. (2)* **75**, pp. 452–466.

Hantzsche, W. (1938). Einlagerung von Mannigfaltigkeiten in euklidische Räume, *Math. Z.* **43**, 1, pp. 38–58.

Harer, J. (1982). How to construct all fibered knots and links, *Topology* **21**, 3, pp. 263–280.

Hartley, R. (1983). Identifying noninvertible knots, *Topology* **22**, 2, pp. 137–145.

Hatakenaka, E. (2004). An estimate of the triple point numbers of surface-knots by quandle cocycle invariants, *Topology Appl.* **139**, 1-3, pp. 129–144.

Hatcher, A. (2002). *Algebraic topology* (Cambridge University Press, Cambridge).

Hempel, J. (2004). *3-manifolds* (AMS Chelsea Publishing, Providence, RI), reprint of the 1976 original.

Hillman, J. A. (2002). *Four-manifolds, geometries and knots, Geometry & Topology Monographs*, Vol. 5 (Geometry & Topology Publications, Coventry).

Hillman, J. A. and Howie, J. (2013). Seifert fibered knot manifolds, *J. Knot Theory Ramifications* **22**, 14, pp. 1350082, 17.

Hinojosa, G. (2007). A wild knot $\mathbb{S}^2 \hookrightarrow \mathbb{S}^4$ as limit set of a Kleinian group: Indra's pearls in four dimensions, *J. Knot Theory Ramifications* **16**, 8, pp. 1083–1110.

Hinojosa, G. and Valdez, R. (2021). Combinatorial description around any vertex of a cubical n-manifold, *Rev. Colombiana Mat.* **55**, 2, pp. 125–137.

Hinojosa, G., Verjovsky, A., and Marcotte, C. V. (2013). Cubulated moves and discrete knots, *J. Knot Theory Ramifications* **22**, 14, pp. 1350079, 26.

Hirsch, M. W. (1961). The imbedding of bounding manifolds in euclidean space, *Ann. of Math. (2)* **74**, pp. 494–497.

Hirsch, M. W. (1976). *Differential Topology, Graduate Texts in Mathematics*, Vol. 33 (Springer-Verlag, New York).

Hitt, L. R. (1979). Examples of higher-dimensional slice knots which are not ribbon knots, *Proc. Amer. Math. Soc.* **77**, 2, pp. 291–297.

Hocking, J. G. and Young, G. S. (1961). *Topology* (Addison-Wesley Publishing Co., Inc., Reading, Mass.-London).

Hosokawa, F. (1967). A concept of cobordism between links, *Ann. of Math. (2)* **86**, pp. 362–373.

Hosokawa, F. and Kawauchi, A. (1979). Proposals for unknotted surfaces in four-spaces, *Osaka Math. J.* **16**, 1, pp. 233–248.

Hoste, J. (2005). The enumeration and classification of knots and links, in *Handbook of knot theory* (Elsevier B. V., Amsterdam), pp. 209–232.

Hoste, J., Thistlethwaite, M., and Weeks, J. (1998). The first 1,701,936 knots, *Math. Intelligencer* **20**, 4, pp. 33–48.

Howie, J. and Short, H. (1985). The band-sum problem, *J. London Math. Soc. (2)* **31**, 3, pp. 571–576.

Hsiang, W.-c. and Sanderson, B. J. (1965). Twist-spinning spheres in spheres, *Illinois J. Math.* **9**, pp. 651–659.

Hudson, J. and Zeeman, E. (1964). On regular neighbourhoods. *Proc. Lond. Math. Soc. (3)* **14**, pp. 719–745.

Hudson, J. F. P. (1963a). Knotted tori, *Topology* **2**, pp. 11–22.

Hudson, J. F. P. (1963b). A non-embedding theorem, *Topology* **2**, pp. 123–128.

Hudson, J. F. P. (1969). *Piecewise linear topology* (W. A. Benjamin, Inc., New York-Amsterdam).

Hudson, J. F. P. and Zeeman, E. C. (1964). On combinatorial isotopy, *Inst. Hautes Études Sci. Publ. Math.* **19**, pp. 69–94.

Ishii, A. and Kishimoto, K. (2011). The quandle coloring invariant of a reducible handlebody-knot, *Tsukuba J. Math.* **35**, 1, pp. 131–141.

Ishii, A., Kishimoto, K., and Ozawa, M. (2015). Knotted handle decomposing spheres for handlebody-knots, *J. Math. Soc. Japan* **67**, 1, pp. 407–417.

Ishii, A., Nikkuni, R., and Oshiro, K. (2018). On calculations of the twisted Alexander ideals for spatial graphs, handlebody-knots and surface-links, *Osaka J. Math.* **55**, 2, pp. 297–313.

Iwakiri, M. (2005). Calculation of dihedral quandle cocycle invariants of twist spun 2-bridge knots, *J. Knot Theory Ramifications* **14**, 2, pp. 217–229.

Iwakiri, M. (2006). Triple point cancelling numbers of surface links and quandle cocycle invariants, *Topology Appl.* **153**, 15, pp. 2815–2822.

Jablan, S., Radović, L., Sazdanović, R., and Zeković, A. (2012). Knots in art, *Symmetry* **4**, 2, pp. 302–328.

Jablonowski, M. (2012). Knotted surfaces and equivalencies of their diagrams without triple points, *J. Knot Theory Ramifications* **21**, 3, pp. 1250019, 6.

Jackson, A. (2002). The world of blind mathematicians, *Notices Amer. Math. Soc.* **49**, 10, pp. 1246–1251.

Janse van Rensburg, E. J. (2005). A tutorial on knot energies, in *Physical and numerical models in knot theory. Including applications to the life sciences* (Hackensack, NJ: World Scientific), ISBN 981-256-187-0, pp. 19–44.

Kadokami, T. and Kawauchi, A. (2011). Amphicheirality of links and Alexander invariants, *Sci. China Math.* **54**, 10, pp. 2213–2227.

Kamada, S. (1989). Nonorientable surfaces in 4-space, *Osaka J. Math.* **26**, 2, pp. 367–385.

Kamada, S. (2002). *Braid and knot theory in dimension four*, *Mathematical Surveys and Monographs*, Vol. 95 (American Mathematical Society, Providence, RI).

Kamada, S. (2014). Quandles and symmetric quandles for higher dimensional knots, in *Knots in Poland III. Part III*, *Banach Center Publ.*, Vol. 103 (Polish Acad. Sci. Inst. Math., Warsaw), pp. 145–158.

Kamada, S. (2017). *Surface-knots in 4-space*, Springer Monographs in Mathematics (Springer, Singapore), an introduction.

Kanenobu, T. (1983). Groups of higher-dimensional satellite knots, *J. Pure Appl. Algebra* **28**, 2, pp. 179–188.

Kanenobu, T. (1989). Untwisted deform-spun knots: examples of symmetry-spun 2-knots, in *Transformation groups (Osaka, 1987), Lecture Notes in Math.*, Vol. 1375 (Springer, Berlin), pp. 145–167.

Kanenobu, T. and Komatsu, S. (2017). Enumeration of ribbon 2-knots presented by virtual arcs with up to four crossings, *J. Knot Theory Ramifications* **26**, 8, pp. 1750042, 41.

Kanenobu, T. and Shima, A. (2002). Two filtrations of ribbon 2-knots, in *Proceedings of the First Joint Japan-Mexico Meeting in Topology (Morelia, 1999)*, Vol. 121, pp. 143–168.

Kassel, C. and Turaev, V. (2008). *Braid groups, Graduate Texts in Mathematics*, Vol. 247 (Springer, New York).

Katanaga, A. and Saeki, O. (1998). Embeddings of quaternion space in S^4, *J. Austral. Math. Soc. Ser. A* **65**, 3, pp. 313–325.

Kauffman, L. H. (1990). An invariant of regular isotopy, *Trans. Amer. Math. Soc.* **318**, 2, pp. 417–471.

Kauffman, L. H. (2012a). Following knots down their energy gradients, *Symmetry* **4**, 2, pp. 276–284.

Kauffman, L. H. (2012b). Introduction to virtual knot theory, *J. Knot Theory Ramifications* **21**, 13, pp. 1240007, 37.

Kauffman, L. H. (2013). *Knots and physics, Series on Knots and Everything*, Vol. 53, 4th edn. (World Scientific Publishing Co. Pte. Ltd., Hackensack, NJ).

Kauffman, L. H. and Lopes, P. (2018). Colorings beyond Fox: the other linear Alexander quandles, *Linear Algebra Appl.* **548**, pp. 221–258.

Kawamura, K. (2015). On relationship between seven types of Roseman moves, *Topology Appl.* **196**, part B, pp. 551–557.

Kawamura, K., Oshiro, K., and Tanaka, K. (2016). Independence of Roseman moves including triple points, *Algebr. Geom. Topol.* **16**, 4, pp. 2443–2458.

Kawauchi, A. (1986). Three dualities on the integral homology of infinite cyclic coverings of manifolds, *Osaka J. Math.* **23**, 3, pp. 633–651.

Kawauchi, A. (1990). The first Alexander modules of surfaces in 4-sphere, in *Algebra and topology 1990 (Taejon, 1990)* (Korea Adv. Inst. Sci. Tech., Taejŏn), pp. 81–89.

Kawauchi, A. (1996). *A survey of knot theory* (Birkhäuser Verlag, Basel).

Kawauchi, A. (2002). On pseudo-ribbon surface-links, *J. Knot Theory Ramifications* **11**, 7, pp. 1043–1062.

Kawauchi, A., Shibuya, T., and Suzuki, S. (1982). Descriptions on surfaces in four-space. I. Normal forms, *Math. Sem. Notes Kobe Univ.* **10**, 1, pp. 75–125.

Kawauchi, A., Shibuya, T., and Suzuki, S. (1983). Descriptions on surfaces in four-space. II. Singularities and cross-sectional links, *Math. Sem. Notes Kobe Univ.* **11**, 1, pp. 31–69.

Kearton, C. (1973). Noninvertible knots of codimension 2, *Proc. Amer. Math. Soc.* **40**, pp. 274–276.

Kearton, C. (1979). Factorisation is not unique for 3-knots, *Indiana Univ. Math. J.* **28**, 3, pp. 451–452.

Kearton, C. (1982). A remarkable 3-knot, *Bull. London Math. Soc.* **14**, 5, pp. 397–398.

Kearton, C. (1983). Some nonfibred 3-knots, *Bull. London Math. Soc.* **15**, 4, pp. 365–367.

Kervaire, M. (1965a). Les noeds de dimensions superieures, *Bul. Soc. Math. France*, pp. 225–271.

Kervaire, M. A. (1960). A manifold which does not admit any differentiable structure, *Comment. Math. Helv.* **34**, pp. 257–270.

Kervaire, M. A. (1965b). On higher dimensional knots, in *Differential and Combinatorial Topology (A Symposium in Honor of Marston Morse)* (Princeton Univ. Press, Princeton, N.J.), pp. 105–119.

Kervaire, M. A. and Milnor, J. W. (1963). Groups of homotopy spheres. I, *Ann. of Math. (2)* **77**, pp. 504–537.

Kharusi, A. A. and Yashiro, T. (2016). On crossing changes for surface-knots, *Kyungpook Math. J.* **56**, 4, pp. 1247–1257.

Kim, D. and Kusner, R. (1993). Torus knots extremizing the Möbius energy, *Experiment. Math.* **2**, 1, pp. 1–9.

Kinoshita, S. (1972). On elementary ideals of polyhedra in the 3-sphere, *Pacific J. Math.* **42**, pp. 89–98.

Kinoshita, S. and Terasaka, H. (1957). On unions of knots, *Osaka Math. J.* **9**, pp. 131–153.

Knudson, K. P. (2015). *Morse theory* (World Scientific Publishing Co. Pte. Ltd., Hackensack, NJ), smooth and discrete.

Kosiński, A. (1962). On Alexander's theorem and knotted spheres, in *Topology of 3-manifolds and related topics (Proc. The Univ. of Georgia Institute, 1961)* (Prentice-Hall, Englewood Cliffs, N.J.), pp. 55–57.

Kusner, R. B. and Sullivan, J. M. (1997). Möbius energies for knots and links, surfaces and submanifolds, in *Geometric topology (Athens, GA, 1993)*, *AMS/IP Stud. Adv. Math.*, Vol. 2 (Amer. Math. Soc., Providence, RI), pp. 570–604.

Kusner, R. B. and Sullivan, J. M. (1998). Möbius-invariant knot energies, in *Ideal knots*, *Ser. Knots Everything*, Vol. 19 (World Sci. Publ., River Edge, NJ), pp. 315–352.

Larson, K. and Meier, J. (2015). Fibered ribbon disks, *J. Knot Theory Ramifications* **24**, 14, pp. 1550066, 22.

Lee, J. M. (2013). *Introduction to smooth manifolds*, *Graduate Texts in Mathematics*, Vol. 218, 2nd edn. (Springer, New York).

Lefschetz, S. (1942). *Algebraic Topology*, American Mathematical Society Colloquium Publications, v. 27 (American Mathematical Society, New York).

Levine, J. (1965). Unknotting spheres in codimension two, *Topology* **4**, pp. 9–16.

Levine, J. (1970). An algebraic classification of some knots of codimension two, *Comment. Math. Helv.* **45**, pp. 185–198.

Levine, J. (1983). Doubly sliced knots and doubled disk knots, *Michigan Math. J.* **30**, 2, pp. 249–256.

Lewark, L. and McCoy, D. (2019). On calculating the slice genera of 11- and 12-crossing knots, *Exp. Math.* **28**, 1, pp. 81–94.

Lickorish, W. B. R. (1997). *An introduction to knot theory*, *Graduate Texts in Mathematics*, Vol. 175 (Springer-Verlag, New York).

Lin, J. and Zhang, H. (2019). Visualizing mathematical knot equivalence, *Electronic Imaging* **2019**, 1.

Lisca, P. (2007). Lens spaces, rational balls and the ribbon conjecture, *Geom. Topol.* **11**, pp. 429–472.

Litherland, R. A. (1979). Deforming twist-spun knots, *Trans. Amer. Math. Soc.* **250**, pp. 311–331.

Litherland, R. A. (1981). The second homology of the group of a knotted surface, *Quart. J. Math. Oxford Ser. (2)* **32**, 128, pp. 425–434.

Litherland, R. A. (1985). Symmetries of twist-spun knots, in *Knot theory and manifolds (Vancouver, B.C., 1983)*, *Lecture Notes in Math.*, Vol. 1144 (Springer, Berlin), pp. 97–107.

Livingston, C. (1993). *Knot theory*, *Carus Mathematical Monographs*, Vol. 24 (Mathematical Association of America, Washington, DC).

Livingston, C. (2005). A survey of classical knot concordance, in *In Handbook of knot theory* (Elsevier), pp. 319–347.

Lomonaco, S. J. and Kauffman, L. H. (2010). Quantum knots and lattices, or a blueprint for quantum systems that do rope tricks, in *Quantum information science and its contributions to mathematics*, *Proc. Sympos. Appl. Math.*, Vol. 68 (Amer. Math. Soc., Providence, RI), pp. 209–276.

Lomonaco, S. J., Jr. (1981). The homotopy groups of knots. I. How to compute the algebraic 2-type, *Pacific J. Math.* **95**, 2, pp. 349–390.

Lomonaco, S. J., Jr. (1983). Five-dimensional knot theory, in *Low-dimensional topology (San Francisco, Calif., 1981)*, *Contemp. Math.*, Vol. 20 (Amer. Math. Soc., Providence, R.I.), pp. 249–270.

Lucas, L. A. and Saeki, O. (2005). Codimension one embeddings of product of three spheres, *Topology Appl.* **146/147**, pp. 409–419.

Lucas, L. A. and Saeki, O. (2011). The Fox property for codimension one embeddings of products of three spheres into spheres, *Algebr. Geom. Topol.* **11**, 5, pp. 3043–3064.

Maehara, H. (1993). Why is $\mathbf{P}^2$ not embeddable in $\mathbf{R}^3$? *Amer. Math. Monthly* **100**, 9, pp. 862–864.

Markov, A. A. (1960). Insolubility of the problem of homeomorphy, in *Proc. Internat. Congress Math. 1958* (Cambridge Univ. Press, New York), pp. 300–306.

Massey, W. S. (1973/74). Imbeddings of projective planes and related manifolds in spheres, *Indiana Univ. Math. J.* **23**, pp. 791–812.

Matsuda, H. (2022). 2-knot homology and Roseman move, *J. Knot Theory Ramifications* **31**, 6, pp. Paper No. 2250039, 37.

Matsumoto, Y. (2002). *An introduction to Morse theory*, *Translations of Mathematical Monographs*, Vol. 208 (American Mathematical Society, Providence, RI).

McGrath, P. (2016). An extremely short proof of the hairy ball theorem, *Amer. Math. Monthly* **123**, 5, pp. 502–503.

McMillan, D. R., Jr. (1978). An arc in a PL n-manifold with no neighborhood that embeds in S^n, $n \geq 4$, *Michigan Math. J.* **25**, 1, pp. 29–43.

Milnor, J. (1956). On manifolds homeomorphic to the 7-sphere, *Ann. of Math. (2)* **64**, pp. 399–405.

Milnor, J. (1963). *Morse theory,* Based on lecture notes by M. Spivak and R. Wells. Annals of Mathematics Studies, No. 51 (Princeton University Press, Princeton, N.J.).

Milnor, J. (1965). *Lectures on the h-cobordism theorem,* Notes by L. Siebenmann and J. Sondow (Princeton University Press, Princeton, N.J.).

Milnor, J. W. (1997). *Topology from the differentiable viewpoint,* Princeton Landmarks in Mathematics (Princeton University Press, Princeton, NJ).

Mohamad, A. and Yashiro, T. (2009). On triple point numbers of 5-colorable 2-knots, *J. Knot Theory Ramifications* **18**, 11, pp. 1493–1508.

Mohamad, A. and Yashiro, T. (2012). Surface diagrams with at most two triple points, *J. Knot Theory Ramifications* **21**, 1, pp. 1250013, 17.

Moise, E. E. (1977). *Geometric topology in dimensions* 2 *and* 3 (Springer-Verlag, New York-Heidelberg), graduate Texts in Mathematics, Vol. 47.

Müllner, D. (2009). Orientation reversal of manifolds, *Algebr. Geom. Topol.* **9**, 4, pp. 2361–2390.

Murasugi, K. (2008). *Knot theory & its applications,* Modern Birkhäuser Classics (Birkhäuser Boston, Inc., Boston, MA).

Nakamura, I. (2012a). Braiding surface links which are coverings over the standard torus, *J. Knot Theory Ramifications* **21**, 1, pp. 1250011, 25.

Nakamura, I. (2012b). Triple linking numbers and triple point numbers of certain T^2-links, *Topology Appl.* **159**, 5, pp. 1439–1447.

Nakamura, I. (2013). Unknotting numbers and triple point cancelling numbers of torus-covering knots, *J. Knot Theory Ramifications* **22**, 3, pp. 1350010, 28.

Nakamura, I. (2014). Satellites of an oriented surface link and their local moves, *Topology Appl.* **164**, pp. 113–124.

Nanyes, O. (2014). Invariants for everywhere wild knots, *J. Knot Theory Ramifications* **23**, 6, pp. 1450037, 15.

Nicolaescu, L. (2011). *An invitation to Morse theory,* 2nd edn., Universitext (Springer, New York).

Ogasa, E. (2001). The projections of n-knots which are not the projection of any unknotted knot, *J. Knot Theory Ramifications* **10**, 1, pp. 121–132.

O'Hara, J. (1991). Energy of a knot, *Topology* **30**, 2, pp. 241–247.

O'Hara, J. (2008). Energy of knots and the infinitesimal cross ratio, in *Groups, homotopy and configuration spaces, Geom. Topol. Monogr.*, Vol. 13 (Geom. Topol. Publ., Coventry), pp. 421–445.

Oshiro, K. (2010). Triple point numbers of surface-links and symmetric quandle cocycle invariants, *Algebr. Geom. Topol.* **10**, 2, pp. 853–865.

Oshiro, K. (2011). Homology groups of trivial quandles with good involutions and triple linking numbers of surface-links, *J. Knot Theory Ramifications* **20**, 4, pp. 595–608.

Oshiro, K. and Tanaka, K. (2015). On rack colorings for surface-knot diagrams without branch points, *Topology Appl.* **196**, part B, pp. 921–930.

Penrose, R., Whitehead, J. H. C., and Zeeman, E. C. (1961). Imbedding of manifolds in euclidean space, *Ann. of Math. (2)* **73**, pp. 613–623.

Piccirillo, L. (2020). The Conway knot is not slice, *Ann. of Math. (2)* **191**, 2, pp. 581–591.

Pinkall, U. (1985). Regular homotopy classes of immersed surfaces, *Topology* **24**, 4, pp. 421–434.

Price, T. M. (1977). Homeomorphisms of quaternion space and projective planes in four space, *J. Austral. Math. Soc. Ser. A* **23**, 1, pp. 112–128.

Przytycki, J. H. (1998). 3-coloring and other elementary invariants of knots, in *Knot theory (Warsaw, 1995), Banach Center Publ.*, Vol. 42 (Polish Acad. Sci. Inst. Math., Warsaw), pp. 275–295.

Przytycki, J. H. (2012). The Trieste look at knot theory, in *Introductory lectures on knot theory, Ser. Knots Everything*, Vol. 46 (World Sci. Publ., Hackensack, NJ), pp. 407–441.

Przytycki, J. H., Bakshi, R. P., Ibarra, D., Montoya-Vega, G., and Weeks, D. ([2024] ©2024). *Lectures in knot theory—an exploration of contemporary topics*, Universitext (Springer, Cham), ISBN 978-3-031-40043-8; 978-3-031-40044-5, doi:10.1007/978-3-031-40044-5.

Przytycki, J. H. and Rosicki, W. (2014). Cocycle invariants of codimension 2 embeddings of manifolds, in *Knots in Poland III. Part III, Banach Center Publ.*, Vol. 103 (Polish Acad. Sci. Inst. Math., Warsaw), pp. 251–289.

Ramírez Alfonsín, J. L. (2005). Knots and links in spatial graphs: a survey, *Discrete Math.* **302**, 1-3, pp. 225–242.

Ranicki, A. (1998). *High-dimensional knot theory*, Springer Monographs in Mathematics (Springer-Verlag, New York), algebraic surgery in codimension 2, With an appendix by Elmar Winkelnkemper.

Reidemeister, K. (1927). Elementare Begründung der Knotentheorie, *Abh. Math. Sem. Univ. Hamburg* **5**, 1, pp. 24–32.

Rohlin, V. A. (1965). The embedding of non-orientable three-manifolds into five-dimensional Euclidean space, *Dokl. Akad. Nauk SSSR* **160**, pp. 549–551.

Rolfsen, D. (1976). *Knots and links* (Publish or Perish, Inc., Berkeley, Calif.), mathematics Lecture Series, No. 7.

Roseman, D. (1974). Woven knots are spun knots, *Osaka J. Math.* **11**, pp. 307–312.

Roseman, D. (1975a). Projections of knots, *Fund. Math.* **89**, 2, pp. 99–110.

Roseman, D. (1975b). The spun square knot is the spun granny knot, *Bol. Soc. Mat. Mexicana (2)* **20**, 2, pp. 49–55.

Roseman, D. (1989). Spinning knots about submanifolds; spinning knots about projections of knots, *Topology Appl.* **31**, 3, pp. 225–241.

Roseman, D. (1992). Viewing knotted spheres in 4-space, produced at the Geometry Center.

Roseman, D. (1993a). *Discover, The World of Science* **14**, 7.

Roseman, D. (1993b). Twisting and turning in 4 dimensions, (Video) (Jan 8 2018) YouTube. `https://youtu.be/FDTvWgJsErk`.

Roseman, D. (1994). Unraveling in 4 dimensions, (Video) (Jan 8 2018) YouTube `https://youtu.be/FDTvWgJsErk`.

Roseman, D. (1997). Wiener's thought on the computer as an aid in visualizing higher-dimensional forms and its modern ramifications, in *Proceedings of the Norbert Wiener Centenary Congress, 1994 (East Lansing, MI, 1994), Proc. Sympos. Appl. Math.*, Vol. 52 (Amer. Math. Soc., Providence, RI), pp. 441–471.

Roseman, D. (1998). Reidemeister-type moves for surfaces in four-dimensional space, in *Knot theory (Warsaw, 1995), Banach Center Publ.*, Vol. 42 (Polish Acad. Sci., Warsaw), pp. 347–380.

Roseman, D. (2000). Projections of codimension two embeddings, in *Knots in Hellas '98 (Delphi), Ser. Knots Everything*, Vol. 24 (World Sci. Publ., River Edge, NJ), pp. 380–410.

Roseman, D. (2002). Untangling some spheres in $\mathbf{R}^4$ by energy minimizing flow, in *Physical knots: knotting, linking, and folding geometric objects in $\mathbb{R}^3$ (Las Vegas, NV, 2001), Contemp. Math.*, Vol. 304 (Amer. Math. Soc., Providence, RI), pp. 261–268.

Roseman, D. (2004). Elementary moves for higher dimensional knots, *Fund. Math.* **184**, pp. 291–310.

Roseman, D. and Takase, M. (2007). High-codimensional knots spun about manifolds, *Algebr. Geom. Topol.* **7**, pp. 359–377.

Rourke, C. P. and Sanderson, B. J. (1982). *Introduction to piecewise-linear topology*, Springer Study Edition (Springer-Verlag, Berlin-New York).

Ruberman, D. (1983). Doubly slice knots and the Casson-Gordon invariants, *Trans. Amer. Math. Soc.* **279**, 2, pp. 569–588.

Ruberman, D. (1990). Seifert surfaces of knots in S^4, *Pacific J. Math.* **145**, 1, pp. 97–116.

Rudolph, L. (1983). Braided surfaces and Seifert ribbons for closed braids, *Comment. Math. Helv.* **58**, 1, pp. 1–37.

Rushing, T. B. (1973). *Topological embeddings* (Academic Press, New York-London), pure and Applied Mathematics, Vol. 52.

Saeki, O. (1988a). Cobordism classification of knotted homology 3-spheres in S^5, *Osaka J. Math.* **25**, 1, pp. 213–222.

Saeki, O. (1988b). Knotted homology 3-spheres in S^5, *J. Math. Soc. Japan* **40**, 1, pp. 65–75.

Saeki, O. (1999). Theory of fibered 3-knots in S^5 and its applications, *J. Math. Sci. Univ. Tokyo* **6**, 4, pp. 691–756.

Saito, M. and Satoh, S. (2005). The spun trefoil needs four broken sheets, *J. Knot Theory Ramifications* **14**, 7, pp. 853–858.

Sakuma, M. (1986). On strongly invertible knots, in *Algebraic and topological theories (Kinosaki, 1984)* (Kinokuniya, Tokyo), pp. 176–196.

Satoh, S. (2001). Double decker sets of generic surfaces in 3-space as homology classes, *Illinois J. Math.* **45**, 3, pp. 823–832.

Satoh, S. (2002). Triple point invariants of non-orientable surface-links, in *Proceedings of the First Joint Japan-Mexico Meeting in Topology (Morelia, 1999)*, Vol. 121, pp. 207–218.

Satoh, S. (2005a). No 2-knot has triple point number two or three, *Osaka J. Math.* **42**, 3, pp. 543–556.

Satoh, S. (2005b). Non-additivity for triple point numbers on the connected sum of surface-knots, *Proc. Amer. Math. Soc.* **133**, 2, pp. 613–616 (electronic).

Satoh, S. (2009a). Sheet number and quandle-colored 2-knot, *J. Math. Soc. Japan* **61**, 2, pp. 579–606.

Satoh, S. (2009b). Triviality of a 2-knot with one or two sheets, *Kyushu J. Math.* **63**, 2, pp. 239–252.

Satoh, S. (2010). A note on the sheet numbers of twist-spun knots, *Hiroshima Math. J.* **40**, 1, pp. 1–15.

Satoh, S. and Shima, A. (2004). The 2-twist-spun trefoil has the triple point number four, *Trans. Amer. Math. Soc.* **356**, 3, pp. 1007–1024 (electronic).

Satoh, S. and Shima, A. (2005). Triple point numbers and quandle cocycle invariants of knotted surfaces in 4-space, *New Zealand J. Math.* **34**, 1, pp. 71–79.

Scharein, R. G. (1998). *Interactive topological drawing* (ProQuest LLC, Ann Arbor, MI), ISBN 978-0612-27238-5, thesis (Ph.D.)–The University of British Columbia (Canada).

Scharlemann, M. (1977a). Isotopy and cobordism of homology spheres in spheres, *J. London Math. Soc. (2)* **16**, 3, pp. 559–567.

Scharlemann, M. (1977b). Thoroughly knotted homology spheres, *Houston J. Math.* **3**, 2, pp. 271–283.

Scharlemann, M. (1985). Smooth spheres in $\mathbf{R}^4$ with four critical points are standard, *Invent. Math.* **79**, 1, pp. 125–141.

Schultens, J. (2014). *Introduction to 3-manifolds, Graduate Studies in Mathematics*, Vol. 151 (American Mathematical Society, Providence, RI).

Scorpan, A. (2005). *The wild world of 4-manifolds* (American Mathematical Society, Providence, RI).

Seifert, H. and Threlfall, W. (1980). *A textbook of topology. Transl. by Michael A. Goldman. Ed. by Joan S. Birman and Julian Eisner.*

Sequin, C. H. (2013). On the number of Klein bottle types, *J. Math. Arts* **7**, 2, pp. 51–63.

Shima, A. (1997). On simply knotted tori in S^4. *J. Math. Sci., Tokyo* **4**, 2, pp. 279–339.

Shima, A. (2003). Knotted klein bottles with only double points, *Osaka J. Math.* **40**, 3, pp. 779–799.

Shinohara, Y. (1971). Higher dimensional knots in tubes, *Trans. Amer. Math. Soc.* **161**, pp. 35–49.

Skopenkov, A. (2020). Classification of knotted tori, *Proc. R. Soc. Edinb., Sect. A, Math.* **150**, 2, pp. 549–567, doi:10.1017/prm.2018.141.

Skopenkov, A. B. (2008). Embedding and knotting of manifolds in Euclidean spaces, in *Surveys in contemporary mathematics* (Cambridge: Cambridge University Press), ISBN 978-0-521-70564-6, pp. 248–342.

Smale, S. (1958). A classification of immersions of the two-sphere, *Trans. Amer. Math. Soc.* **90**, pp. 281–290.

Smale, S. (1959). The classification of immersions of spheres in Euclidean spaces, *Ann. of Math. (2)* **69**, pp. 327–344.

Sosinskii, A. B. (1970). Decomposition of knots, *Mat. Sb. (N.S.)* **81(123)**, pp. 145–158.

Spanier, E. H. (1966). *Algebraic topology* (Springer-Verlag, New York).

Spring, D. (2005). The golden age of immersion theory in topology: 1959–1973. A mathematical survey from a historical perspective, *Bull. Amer. Math. Soc. (N.S.)* **42**, 2, pp. 163–180.

Stallings, J. (1963). On topologically unknotted spheres, *Ann. of Math. (2)* **77**, pp. 490–503.

Stallings, J. (1971). *Group theory and three-dimensional manifolds* (Yale University Press, New Haven, Conn.-London).

Stasiak, A., Katritch, V., and Kauffman, L. H. (eds.) (1998). *Ideal Knots*, Knots and Everything (World Scientific).

Suciu, A. I. (1992). Inequivalent frame-spun knots with the same complement, *Comment. Math. Helv.* **67**, 1, pp. 47–63.

Sumners, D. W. (1971). Invertible knot cobordisms, *Comment. Math. Helv.* **46**, pp. 240–256.

Sumners, D. W. and Whittington, S. G. (1988). Knots in self-avoiding walks, *J. Phys. A* **21**, 7, pp. 1689–1694.

Suzuki, S. (1969). Local knots of 2-spheres in 4-manifolds. *Proc. Japan Acad.* **45**, pp. 34–38.

Suzuki, S. (1976). Knotting problems of 2-spheres in 4-sphere, *Math. Sem. Notes Kobe Univ.* **4**, 3, pp. 241–371.

Tait, P. G. (1877). On Knots, *Trans. Roy. Soc. Edin.* **28**, pp. 145–190.

Takasaki, M. (1943). Abstraction of symmetric transformations (Japanese), *Tôhoku Math. J.* **49**, pp. 145–207.

Takase, M. (2006). Homology 3-spheres in codimension three, *Internat. J. Math.* **17**, 8, pp. 869–885.

Takase, M. and Tanaka, K. (2016). Regular-equivalence of 2-knot diagrams and sphere eversions, *Math. Proc. Cambridge Philos. Soc.* **161**, 2, pp. 237–246.

Tanaka, K. (2004). Crossing changes for pseudo-ribbon surface-knots, *Osaka J. Math.* **41**, 4, pp. 877–890.

Taubes, C. H. (1987). Gauge theory on asymptotically periodic 4-manifolds, *J. Differential Geom.* **25**, 3, pp. 363–430.

Teragaito, M. (1992). Symmetry-spun tori in the four-sphere, in *Knots 90 (Osaka, 1990)* (de Gruyter, Berlin), pp. 163–171.

Terasaka, H. and Hosokawa, F. (1961). On the unknotted sphere s^2 in e^4, *Osaka Math. J.* **13**, 2, pp. 265–270.

Thompson, A. (1987). Property P for the band-connect sum of two knots, *Topology* **26**, 2, pp. 205–207.

Thurston, W. P. (1997). *Three-dimensional geometry and topology. Vol. 1*, *Princeton Mathematical Series*, Vol. 35 (Princeton University Press, Princeton, NJ).

Thurston, W. P. and Weeks, J. R. (1984). The mathematics of three-dimensional manifolds, *Scientific American* **251**, 1, pp. 108–121.

Tindell, R. (1966). Some wild imbeddings in codimension two, *Proc. Amer. Math. Soc.* **17**, pp. 711–716.

Trotter, H. F. (1963). Non-invertible knots exist, *Topology* **2**, pp. 275–280.

Turaev, V. (2008). Cobordism of knots on surfaces, *J. Topol.* **1**, 2, pp. 285–305.

Van Kampen, E. R. (1928). Zur isotopie zweidimensionaler flächen im R_4, *Abh. Math. Sem. Univ. Hamburg* **6**, 1, p. 216.

van Rensburg, E. J. J. (2009). Thoughts on lattice knot statistics, *Journal of Mathematical Chemistry* **45**, pp. 7–38.

van Rensburg, E. J. J. and Rechnitzer, A. (2011). Minimal knotted polygons in cubic lattices, *Journal of Statistical Mechanics: Theory and Experiment*, 09, p. P09008.

van Rensburg, E. J. J. and Whittington, S. G. (1990). The knot probability in lattice polygons, *Journal of Physics A* **23**, pp. 3573–3590.

Vandermonde, A. (1771). Remarques sur les problems de situation, *Memoires de l'Academie Royale des Sciences, (Paris)*, pp. 566–574.

Viro, O. J. (1973). Local knotting of sub-manifolds, *Mat. Sb. (N.S.)* **90(132)**, pp. 173–183, 325.

Viro, O. Y. (1990). Lecture, Osaka city University.

Vogel, P. (1990). Representation of links by braids: a new algorithm, *Comment. Math. Helv.* **65**, 1, pp. 104–113.

Wall, C. T. C. (1965a). All 3-manifolds imbed in 5-space, *Bull. Amer. Math. Soc.* **71**, pp. 564–567.

Wall, C. T. C. (1965b). Unknotting tori in codimension one and spheres in codimension two, *Proc. Cambridge Philos. Soc.* **61**, pp. 659–664.

Whitney, H. (1937). On regular closed curves in the plane, *Compositio Math.* **4**, pp. 276–284.

Whitney, H. (1944). The self-intersections of a smooth n-manifold in $2n$-space, *Ann. of Math. (2)* **45**, pp. 220–246.

Winter, B. K. (2014). *Virtual, Welded, and Ribbon Links in Arbitrary Dimensions* (ProQuest LLC, Ann Arbor, MI), thesis (Ph.D.)–State University of New York at Buffalo.

Winter, B. K. (2019). Ribbon n-knots with isomorphic quandles, *J. Knot Theory Ramifications* **28**, 7, pp. 1950043, 12.

Wright, D. G. (1977). Sticky arcs in E^n $(n \geq 4)$, *Proc. Amer. Math. Soc.* **66**, 1, pp. 181–182.

Wu, F. Y. (1992). Knot theory and statistical mechanics, *Rev. Mod. Phys.* **64**, pp. 1099–1131.

Wu, W.-t. (1958). On the isotopy of C^r-manifolds of dimension n in euclidean $(2n+1)$-space, *Sci. Record (N.S.)* **2**, pp. 271–275.

Yajima, T. (1964). On simply knotted spheres in R^4, *Osaka J. Math.* **1**, pp. 133–152.

Yajima, T. (1969). On a characterization of knot groups of some spheres in R^4, *Osaka J. Math.* **6**, pp. 435–446.

Yamada, S. (1987). The minimal number of Seifert circles equals the braid index of a link, *Invent. Math.* **89**, 2, pp. 347–356.

Yanagawa, T. (1969). On ribbon 2-knots. The 3-manifold bounded by the 2-knots, *Osaka J. Math.* **6**, pp. 447–464.

Yashiro, T. (2005a). A note on Roseman moves, *Kobe J. Math.* **22**, 1-2, pp. 31–38.

Yashiro, T. (2005b). Triple point numbers of twist spun knots, *J. Knot Theory Ramifications* **14**, 7, pp. 831–840.

Yashiro, T. (2007). Cell-complexes for t-minimal surface diagrams with 4 triple points, in *Intelligence of low dimensional topology 2006*, *Ser. Knots Everything*, Vol. 40 (World Sci. Publ., Hackensack, NJ), pp. 367–374.

Yasuda, T. (2001). Crossing and base numbers of ribbon 2-knots, *J. Knot Theory Ramifications* **10**, 7, pp. 999–1003.

Yasuda, T. (2006). An evaluation of the crossing number on ribbon 2-knots, *J. Knot Theory Ramifications* **15**, 1, pp. 1–9.

Yasuda, T. (2009). Ribbon 2-knots with distinct ribbon types, *J. Knot Theory Ramifications* **18**, 11, pp. 1509–1523.

Yasuda, T. (2018). Ribbon 2-knots of ribbon crossing number four, *J. Knot Theory Ramifications* **27**, 10, pp. 1850058, 20.

Yasuda, T. (2020). Amphicheirality of ribbon 2-knots, *J. Knot Theory Ramifications* **29**, 9, pp. 2050069, 11.

Yoshida, K. (2004). Unknotting of pseudo-ribbon n-knots, *J. Knot Theory Ramifications* **13**, 2, pp. 259–275.

Yoshikawa, K. (1994). An enumeration of surfaces in four-space. *Osaka J. Math.* **31**, 3, pp. 497–522.

Zeeman, E. C. (1960a). Linking spheres, *Abh. Math. Sem. Univ. Hamburg* **24**, pp. 149–153.

Zeeman, E. C. (1960b). Unknotting spheres, *Ann. of Math. (2)* **72**, pp. 350–361.

Zeeman, E. C. (1960c). Unknotting spheres in five dimensions, *Bull. Amer. Math. Soc.* **66**, p. 198.

Zeeman, E. C. (1962a). Isotopies and knots in manifolds, in *Topology of 3-manifolds and related topics (Proc. The Univ. of Georgia Institute, 1961)* (Prentice-Hall, Englewood Cliffs, N.J.), pp. 187–193.

Zeeman, E. C. (1962b). Unknotting 3-spheres in six dimensions, *Proc. Amer. Math. Soc.* **13**, pp. 753–757.

Zeeman, E. C. (1963). Unknotting combinatorial balls, *Ann. of Math. (2)* **78**, pp. 501–526.

Zeeman, E. C. (1964). On the dunce hat, *Topology* **2**, pp. 341–358.

Zeeman, E. C. (1965). Twisting spun knots, *Trans. Amer. Math. Soc.* **115**, pp. 471–495.

Index

Vol. 70: *Geometric Foundations of Design: Old and New*
by J. Kappraff

Vol. 69: *Mathematics of Harmony as a New Interdisciplinary Direction and "Golden" Paradigm of Modern Science*
Volume 3: The "Golden" Paradigm of Modern Science: Prerequisite for the "Golden" Revolution in Mathematics, Computer Science, and Theoretical Natural Sciences
by A. Stakhov

Vol. 68: *Mathematics of Harmony as a New Interdisciplinary Direction and "Golden" Paradigm of Modern Science*
Volume 2: Algorithmic Measurement Theory, Fibonacci and Golden Arithmetic's and Ternary Mirror-Symmetrical Arithmetic
by A. Stakhov

Vol. 67: On Complementarity: A Universal Organizing Principle
by J. S. Avrin

Vol. 66: *Invariants and Pictures: Low-dimensional Topology and Combinatorial Group Theory*
by V. O. Manturov, D. Fedoseev, S. Kim & I. Nikonov

Vol. 65: *Mathematics of Harmony as a New Interdisciplinary Direction and "Golden" Paradigm of Modern Science*
Volume 1: The Golden Section, Fibonacci Numbers, Pascal Triangle, and Platonic Solids
by A. Stakhov

Vol. 64: *Polynomial One-cocycles for Knots and Closed Braids*
by T. Fiedler

Vol. 63: *Board Games: Throughout the History and Multidimensional Spaces*
by J. Kyppö

Vol. 62: *The Mereon Matrix: Everything Connected through (K)nothing*
edited by J. B. McNair, L. Dennis & L. H. Kauffman

Vol. 61: *Numeral Systems with Irrational Bases for Mission-Critical Applications*
by A. Stakhov

Vol. 60: *New Horizons for Second-Order Cybernetics*
edited by A. Riegler, K. H. Müller & S. A. Umpleby

Vol. 59: *Geometry, Language and Strategy: The Dynamics of Decision Processes — Vol. 2*
by G. H. Thomas

Vol. 58: *Representing 3-Manifolds by Filling Dehn Surfaces*
by R. Vigara & A. Lozano-Rojo

Vol. 57: *ADEX Theory: How the ADE Coxeter Graphs Unify Mathematics and Physics*
by S.-P. Sirag

Vol. 56: *New Ideas in Low Dimensional Topology*
edited by L. H. Kauffman & V. O. Manturov

Vol. 55: *Knots, Braids and Möbius Strips*
by J. Avrin

Vol. 54: *Scientific Essays in Honor of H Pierre Noyes on the Occasion of His 90th Birthday*
edited by J. C. Amson & L. H. Kauffman

Vol. 53: *Knots and Physics (Fourth Edition)*
by L. H. Kauffman

Vol. 52: *Algebraic Invariants of Links (Second Edition)*
by J. Hillman

Vol. 51: *Virtual Knots: The State of the Art*
by V. O. Manturov & D. P. Ilyutko

Vol. 50: *Topological Library*
Part 3: Spectral Sequences in Topology
edited by S. P. Novikov & I. A. Taimanov

Vol. 49: *Hopf Algebras*
by D. E. Radford

Vol. 48: *An Excursion in Diagrammatic Algebra: Turning a Sphere from Red to Blue*
by J. S. Carter

Vol. 47: *Introduction to the Anisotropic Geometrodynamics*
by S. Siparov

Vol. 46: *Introductory Lectures on Knot Theory*
edited by L. H. Kauffman, S. Lambropoulou, S. Jablan & J. H. Przytycki

Vol. 45: *Orbiting the Moons of Pluto*
Complex Solutions to the Einstein, Maxwell, Schrödinger and Dirac Equations
by E. A. Rauscher & R. L. Amoroso

Vol. 44: *Topological Library*
Part 2: Characteristic Classes and Smooth Structures on Manifolds
edited by S. P. Novikov & I. A. Taimanov

Vol. 43: *The Holographic Anthropic Multiverse*
by R. L. Amoroso & E. A. Ranscher

Vol. 42: *The Origin of Discrete Particles*
by T. Bastin & C. Kilmister

Vol. 41: *Zero to Infinity: The Fountations of Physics*
by P. Rowlands

Vol. 40: *Intelligence of Low Dimensional Topology 2006*
edited by J. Scott Carter et al.

Vol. 39: *Topological Library*
Part 1: Cobordisms and Their Applications
edited by S. P. Novikov & I. A. Taimanov

Vol. 38: *Current Developments in Mathematical Biology*
edited by K. Mahdavi, R. Culshaw & J. Boucher

Vol. 37: *Geometry, Language, and Strategy*
by G. H. Thomas

Vol. 36: *Physical and Numerical Models in Knot Theory*
edited by J. A. Calvo et al.

Vol. 35: *BIOS — A Study of Creation*
by H. Sabelli

Vol. 34: *Woods Hole Mathematics — Perspectives in Mathematics and Physics*
edited by N. Tongring & R. C. Penner

Vol. 33: *Energy of Knots and Conformal Geometry*
by J. O'Hara

Vol. 32: *Algebraic Invariants of Links*
by J. A. Hillman

Vol. 31: *Mindsteps to the Cosmos*
by G. S. Hawkins

Vol. 30: *Symmetry, Ornament and Modularity*
by S. V. Jablan

Vol. 29: *Quantum Invariants — A Study of Knots, 3-Manifolds, and Their Sets*
by T. Ohtsuki

Vol. 28: *Beyond Measure: A Guided Tour Through Nature, Myth, and Number*
by J. Kappraff

Vol. 27: *Bit-String Physics: A Finite and Discrete Approach to Natural Philosophy*
by H. Pierre Noyes; edited by J. C. van den Berg

Vol. 26: *Functorial Knot Theory — Categories of Tangles, Coherence, Categorical Deformations, and Topological Invariants*
by David N. Yetter

Vol. 25: *Connections — The Geometric Bridge between Art and Science (Second Edition)*
by J. Kappraff

Vol. 24: *Knots in HELLAS '98 — Proceedings of the International Conference on Knot Theory and Its Ramifications*
edited by C. McA Gordon, V. F. R. Jones, L. Kauffman, S. Lambropoulou & J. H. Przytycki

Vol. 23: *Diamond: A Paradox Logic (Second Edition)*
by N. S. Hellerstein

Vol. 22: *The Mathematics of Harmony — From Euclid to Contemporary Mathematics and Computer Science*
by A. Stakhov (assisted by S. Olsen)

Vol. 21: *LINKNOT: Knot Theory by Computer*
by S. Jablan & R. Sazdanovic

Vol. 20: *The Mystery of Knots — Computer Programming for Knot Tabulation*
by C. N. Aneziris

Vol. 19: *Ideal Knots*
by A. Stasiak, V. Katritch & L. H. Kauffman

Vol. 18: *The Self-Evolving Cosmos: A Phenomenological Approach to Nature's Unity-in-Diversity*
by S. M. Rosen

Vol. 17: *Hypercomplex Iterations — Distance Estimation and Higher Dimensional Fractals*
by Y. Dang, L. H. Kauffman & D. Sandin

Vol. 16: *Delta — A Paradox Logic*
by N. S. Hellerstein

Vol. 15: *Lectures at KNOTS '96*
by S. Suzuki

Vol. 14: *Diamond — A Paradox Logic*
by N. S. Hellerstein

Vol. 13: *Entropic Spacetime Theory*
by J. Armel

Vol. 12: *Relativistic Reality: A Modern View*
edited by J. D. Edmonds, Jr.

Vol. 11: *History and Science of Knots*
edited by J. C. Turner & P. van de Griend

Vol. 10: *Nonstandard Logics and Nonstandard Metrics in Physics*
by W. M. Honig

Vol. 9: *Combinatorial Physics*
by T. Bastin & C. W. Kilmister

Vol. 8: *Symmetric Bends: How to Join Two Lengths of Cord*
by R. E. Miles

Vol. 7: *Random Knotting and Linking*
edited by K. C. Millett & D. W. Sumners

Vol. 6: *Knots and Applications*
edited by L. H. Kauffman

Vol. 5: *Gems, Computers and Attractors for 3-Manifolds*
by S. Lins

Vol. 4: *Gauge Fields, Knots and Gravity*
by J. Baez & J. P. Muniain

Vol. 3: *Quantum Topology*
edited by L. H. Kauffman & R. A. Baadhio

Vol. 2: *How Surfaces Intersect in Space — An Introduction to Topology (Second Edition)*
by J. S. Carter

Vol. 1: *Knots and Physics (Third Edition)*
by L. H. Kauffman

www.ingramcontent.com/pod-product-compliance
Lightning Source LLC
LaVergne TN
LVHW011655200926
848484LV00004B/950